Silicon Nanomaterials Sourcebook

VOLUME ONE

Series in Materials Science and Engineering

Recent books in the series:

SILICON NANOMATERIALS
SOURCEBOOK VOLUME I
Low-Dimensional Structures, Quantum Dots, and Nanowires

Editor: Klaus D. Sattler

CRC Press
Taylor & Francis Group
Boca Raton London New York

CRC Press is an imprint of the
Taylor & Francis Group, an **informa** business

CRC Press
Taylor & Francis Group
6000 Broken Sound Parkway NW, Suite 300
Boca Raton, FL 33487-2742

First issued in paperback 2019

ISBN-13: 978-1-4987-6377-6 (hbk)
ISBN-13: 978-0-367-87759-0 (pbk)

Library of Congress Cataloging-in-Publication Data

Names: Sattler, Klaus D., editor.
Title: Silicon nanomaterials sourcebook / edited by Klaus D. Sattler.
Other titles: Series in materials science and engineering.
Description: Boca Raton, FL: CRC Press, Taylor & Francis Group, [2017] |
Series: Series in materials science and engineering | Includes
bibliographical references and index. Contents: volume 1. Low-dimensional
structures, quantum dots, and nanowires
Identifiers: LCCN 2016059471| ISBN 9781498763776 (v. 1; hardback; alk.
paper) | ISBN 1498763774 (v. 1; hardback; alk. paper)
Subjects: LCSH: Nanosilicon. | Nanostructured materials.
Classification: LCC TA418.9.N35 S5556 2017 | DDC 620.1/15--dc23
LC record available at https://lccn.loc.gov/2016059471

Visit the Taylor & Francis Web site at
http://www.taylorandfrancis.com

and the CRC Press Web site at
http://www.crcpress.com

Contents

Series Preface

This international series covers all aspects of theoretical and applied optics and optoelectronics. Active since 1986, eminent authors have long been choosing to publish with this series, and it is now established as a premier forum for high-impact monographs and textbooks. The editors are proud of the breadth and depth showcased by published works, with levels ranging from advanced undergraduate and graduate student texts to professional references. Topics addressed are both cutting edge and fundamental, basic science and applications-oriented, on subject matter that includes: lasers, photonic devices, nonlinear optics, interferometry, waves, crystals, optical materials, biomedical optics, optical tweezers, optical metrology, solid-state lighting, nanophotonics, and silicon photonics. Readers of the series are students, scientists, and engineers working in optics, optoelectronics, and related fields in the industry.

Proposals for new volumes in the series may be directed to Lu Han, senior publishing editor at CRC Press, Taylor & Francis Group (lu.han@taylorandfrancis.com).

Preface

Silicon is one of the most technologically important materials today owing to its omnipresent significance in microelectronics. Its nanoscale forms, such as nanocrystals, porous silicon, quantum wells, or nanowires, have stimulated great interest among scientists because of their special physical properties, such as light emission, field emission, and quantum confinement effects. The progress made in the synthesis of silicon nanostructures in recent years has attracted considerable attention. Today, large quantities of silicon nanomaterials can be produced, and they are investigated with the most advanced analytical instruments available.

While silicon is the essential semiconductor material for modern microelectronic devices, this sourcebook shows a much wider range of applications, which are possible for silicon on the nanometer scale. Mostly inspired by the discovery of new carbon allotropes, methods have been developed in the last two decades for silicon to obtain similar low-dimensional and nanoscale morphologies and structures. When the size of silicon is reduced to the 1–100 nm range, quantum confinement can significantly affect the properties and performance of the material. Another inspiration came from the discovery of visible light emission from porous silicon, which inspired many scientists to start research on nanoscale silicon. Electronic and photonic studies of these materials have revealed peculiar effects and have subsequently been extended toward biomedicine with applications in tissue engineering, drug delivery, biosensing, radiotherapy, and sonodynamic therapy. This is possible because of the good biocompatibility and biodegradability of nanoscale silicon and tunable surface derivatization. Fabrication of subwavelength nanostructures has allowed for development of antireflection materials as well as other photon management structures such as materials with light-trapping properties, with applications in optoelectronic devices, photodetectors, and phototransistors.

Silicon Nanomaterials Sourcebook provides an introduction to synthetic methods used for the production of various silicon nanoscale morphologies and structures. Among these methods are solution synthesis and microwave-assisted synthesis, pulsed laser ablation, electrodeposition and plasma synthesis, metal-assisted chemical edging, interface functionalization, and nanoscale interface manipulations.

Volume One of the sourcebook covers low-dimensional silicon nanostructures such as nanosheets, clusters, nanoparticles, nanocrystals, nanowires, and nanotubes. Structural, electronic, and photonic properties of these materials may differ significantly from the silicon bulk properties.

Volume Two focuses on functional and industrial nanosilicon, describing materials such as nanowire, nanopencil and nanopore arrays, core–shell nanostructures, or porous silicon templates. These nanostructures have interesting antireflection, photonic, and thermoelectric properties. They have a wide range of applications as sonosensors, solar cells, for Li-ion batteries, for energy storage, biomedicine, solar energy conversion, chemical and biological sensing techniques, DNA sequencing, or quantum information.

The Sourcebook comprehensively covers the many aspects of silicon nanomaterials. It reflects the interdisciplinary nature of this field bringing together physics, chemistry, materials science, molecular biology, engineering, and medicine. Its contents include growth mechanisms and fundamental properties as well as electronic device, energy storage, biomedical, and environmental applications. It is a unique reference for industrial professionals and university students, offering deep insight into a wide range of areas from science to engineering. While addressing the current knowledge and the latest advances, it also includes basic mathematical equations, tables, and graphs. This provides the reader with the tools necessary to understand the current status of the field as well as future technology development of nanoscale silicon materials and structures.

Editor

Klaus D. Sattler pursued his undergraduate and master's courses at the University of Karlsruhe in Germany. He received his PhD under the guidance of Professors G. Busch and H.C. Siegmann at the Swiss Federal Institute of Technology (ETH) in Zurich, where he was among the first to study spin-polarized photoelectron emission. In 1976, he began a group for atomic cluster research at the University of Konstanz in Germany, where he built the first source for atomic clusters and led his team to pioneering discoveries such as "magic numbers" and "Coulomb explosion." He was at the University of California, Berkeley, for 3 years as a Heisenberg fellow, where he initiated the first studies of atomic clusters on surfaces with a scanning tunneling microscope.

Dr. Sattler accepted a position as professor of physics at the University of Hawaii, Honolulu, in 1988. There, he initiated a research group for nanophysics, which, using scanning probe microscopy, obtained the first atomic-scale images of carbon nanotubes directly confirming the graphene network. In 1994, his group produced the first carbon nanocones. He has also studied the formation of polycyclic aromatic hydrocarbons (PAH) and nanoparticles in hydrocarbon flames in collaboration with ETH Zurich. Other research has involved the nanopatterning of nanoparticle films, charge density waves on rotated graphene sheets, band gap studies of quantum dots, and graphene folds. His current work focuses on novel nanomaterials and solar photocatalysis with nanoparticles for the purification of water.

He is the editor of the sister reference, *Carbon Nanomaterials Sourcebook* (CRC Press, 2016), *Fundamentals of Picoscience* (CRC Press, 2014), and the seven-volume *Handbook of Nanophysics* (CRC Press, 2011). Among his many other accomplishments, Dr. Sattler was awarded the prestigious Walter Schottky Prize from the German Physical Society in 1983. At the University of Hawaii, he teaches courses in general physics, solid state physics, and quantum mechanics.

Contributors

Neda Ahmadi
Department of Basic Sciences
Garmsar Branch
Islamic Azad University
Garmsar, Iran

Atif Mossad Ali
Research Center for Advanced Materials Science
King Khalid University
Abha, Saudi Arabia

and

Department of Physics
Faculty of Science
Assiut University
Assiut, Egypt

Hua Bao
University of Michigan–Shanghai Jiao Tong
    University Joint Institute
Shanghai Jiao Tong University
Shanghai, China

Mirko Battaglia
Laboratorio di Chimica Fisica Applicata
Dipartimento dell'Innovazione Industriale e
    Digitale (DIID)
Università di Palermo
Palermo, Italy

Ashkan Momeni Bidzard
Department of Physics
Iran University of Science and Technology
Tehran, Iran

Fernando Brandi
Intense Laser Irradiation Laboratory (ILIL)
Istituto Nazionale di Ottica (INO)
Consiglio Nazionale delle Ricerche (CNR)
Pisa, Italy

and

Istituto Italiano di Technologia (IIT)
Genova, Italy

Jie Cao
National Laboratory of Solid State
    Microstructures
and
Department of Physics
Nanjing University
Nanjing, China

and

College of Science
Hohai University
Nanjing, China

Zhongfang Chen
Department of Chemistry
Institute for Functional Nanomaterials
University of Puerto Rico
San Juan, Puerto Rico

Jeffery L. Coffer
Department of Chemistry
Texas Christian University
Fort Worth, Texas

Yaping Dan
Department of Electrical and Computer
    Engineering
University of Michigan–Shanghai Jiao Tong
    University Joint Institute
Shanghai Jiao Tong University
Shanghai, China

Yi Ding
Institute of Photoelectronic Thin Film Devices and
    Technology
Tianjin Key Laboratory of Photoelectronic Thin
    Film Devices and Technology
Nankai University
Tianjin, China

Kateřina Dohnalová
Van der Waals Zeeman Institute
University of Amsterdam
Amsterdam, The Netherlands

Samantha K. Ehrenberg
Department of Mechanical Engineering
University of Minnesota
Minneapolis, Minnesota

Minoru Fujii
Department of Electrical and Electronic Engineering
Graduate School of Engineering
Kobe University
Kobe, Japan

Gian G. Guzmán-Verri
Materials Research Science and Engineering Center
University of Costa Rica
San José, Costa Rica

and

Materials Science Division
Argonne National Laboratory
Argonne, Illinois

Klaus von Haeften
Department of Physics and Astronomy
University of Leicester
Leicester, UK

and

K-nano
Leicester, UK

Ming Hu
Institute of Mineral Engineering
Division of Materials Science and Engineering
Faculty of Georesources and Materials Engineering
RWTH Aachen University
Aachen, Germany

and

Aachen Institute for Advanced Study in
    Computational Engineering Science (AICES)
RWTH Aachen University
Aachen, Germany

Danilo Roque Huanca
Laboratório de Sensores e Dispositivos
Instituto de Física e Química
Universidade Federal de Itajubá
Minas Gerais, Brazil

Katharine I. Hunter
Department of Mechanical Engineering
University of Minnesota
Minneapolis, Minnesota

Rosalinda Inguanta
Laboratorio di Chimica Fisica Applicata
Dipartimento dell'Innovazione Industriale e
    Digitale (DIID)
Università di Palermo
Palermo, Italy

Takao Inokuma
Graduate School of Natural Science and Technology
Kanazawa University
Kanazawa, Japan

Firman Bagja Juangsa
Department of Mechanical Science and
    Engineering
Tokyo Institute of Technology
Tokyo, Japan

Uwe R. Kortshagen
Department of Mechanical Engineering
University of Minnesota
Minneapolis, Minnesota

Canan Kurşungöz
UNAM-National Nanotechnology
    Research Center
and
Institute of Materials Science and Nanotechnology
Bilkent University
Ankara, Turkey

Kateřina Kůsová
Institute of Physics
Academy of Sciences of the Czech Republic
Prague, Czech Republic

Nguyen T. Le
Department of Chemistry
Texas Christian University
Fort Worth, Texas

Lok C. Lew Yan Voon
College of Science and Mathematics
University of West Georgia
Carrollton, Georgia

Pu Liu
State Key Laboratory of Optoelectronic Materials
and Technologies
Nanotechnology Research Center
School of Materials Science and Engineering
Sun Yat-sen University
Guangdong, China

Nastaran Mansour
Department of Physics
Shahid Beheshti University
Tehran, Iran

Davide Mariotti
Nanotechnology and Integrated Bio-Engineering
Centre (NIBEC)
Ulster University
Newtownabbey, UK

Calum McDonald
Nanotechnology and Integrated Bio-Engineering
Centre (NIBEC)
Ulster University
Newtownabbey, UK

Gerardo Morell
Institute for Functional Nanomaterials
University of Puerto Rico
San Juan, Puerto Rico

and

Department of Physics
University of Puerto Rico at Río Piedras
San Juan, Puerto Rico

Hideyuki Nakano
Toyota Central R&D Labs., Inc.
Aichi, Japan

Tomohiro Nozaki
Department of Mechanical Science and
Engineering
Tokyo Institute of Technology
Tokyo, Japan

Masataka Ohashi
Toyota Central R&D Labs., Inc.
Aichi, Japan

Bülend Ortaç
UNAM-National Nanotechnology
Research Center
and
Institute of Materials Science and Nanotechnology
Bilkent University
Ankara, Turkey

Javier Palomino
Institute for Functional Nanomaterials
University of Puerto Rico
San Juan, Puerto Rico

and

Department of Physics
University of Puerto Rico at Río Piedras
San Juan, Puerto Rico

and

Department of Physics
University of Puerto Rico at Mayagüez
Mayagüez, Puerto Rico

Peter J. Pauzauskie
Department of Materials Science and Engineering
University of Washington
Seattle, Washington

Xiaodong Pi
State Key Laboratory of Silicon Materials
School of Materials Science and Engineering
Zhejiang University
Hangzhou, China

Salvatore Piazza
Laboratorio di Chimica Fisica Applicata
Dipartimento dell'Innovazione Industriale e
Digitale (DIID)
Università di Palermo
Palermo, Italy

Didier Pribat
Department of Energy Science
Sungkyunkwan University
Suwon, Korea

and

Laboratoire de Physique des Interfaces et des
Couches Minces (LPICM)
Ecole polytechnique
Université Paris-Saclay
Palaiseau, France

Guangzhao Qin
Institute of Mineral Engineering
Division of Materials Science and Engineering
Faculty of Georesources and Materials Engineering
RWTH Aachen University
Aachen, Germany

Yanoar Pribadi Sarwono
Department of Physics and Materials Science
City University of Hong Kong
Hong Kong SAR, China

Bennett E. Smith
Department of Materials Science and Engineering
University of Washington
Seattle, Washington

Sean C. Smith
Integrated Materials Design Centre (IMDC)
School of Chemical Engineering
University of New South Wales
Sydney, New South Wales, Australia

Hiroshi Sugimoto
Department of Electrical and Electronic Engineering
Graduate School of Engineering
Kobe University
Kobe, Japan

Carmelo Sunseri
Laboratorio di Chimica Fisica Applicata
Dipartimento dell'Innovazione Industriale e
Digitale (DIID)
Università di Palermo
Palermo, Italy

Vladimir Svrcek
Research Center for Photovoltaics
National Institute of Advanced Industrial Science
and Technology
Tsukuba, Japan

Elif Uzcengiz Şimşek
UNAM- National Nanotechnology Research Center
and
Institute of Materials Science and Nanotechnology
Bilkent University
Ankara, Turkey

Xin Tan
Integrated Materials Design Centre (IMDC)
School of Chemical Engineering
UNSW Australia
Sydney, New South Wales, Australia

Chi-Pui Tang
Lunar and Planetary Science Laboratory
Macau University of Science and Technology
Macau, China

Deepak Varshney
Institute for Functional Nanomaterials
University of Puerto Rico
San Juan, Puerto Rico

and

Department of Physics
University of Puerto Rico at Río Piedras
San Juan, Puerto Rico

and

Advanced Green Innovations
Chandler, Arizona

Tamilselvan Velusamy
Nanotechnology and Integrated Bio-Engineering
Centre (NIBEC)
Ulster University
Newtownabbey, UK

Alexandru Vlad
Division of Molecules, Solids and Reactivity
(MOST)
Institute of condensed Matter and Nanoscience
(IMCN) Université catholique de Louvain
Louvain-la-Neuve, Belgium

Brad R. Weiner
Institute for Functional Nanomaterials
University of Puerto Rico
San Juan, Puerto Rico

and

Department of Chemistry
University of Puerto Rico at Río Piedras
San Juan, Puerto Rico

Morten Willatzen
Department of Photonics Engineering,
Technical University of Denmark
Lyngby, Denmark

Jun Xiao
State Key Laboratory of Optoelectronic Materials
    and Technologies
Nanotechnology Research Center
School of Materials Science and Engineering
Sun Yat-sen University
Guangdong, China

Han Xie
University of Michigan–Shanghai Jiao Tong
    University Joint Institute
Shanghai Jiao Tong University
Shanghai, China

Guowei Yang
State Key Laboratory of Optoelectronic Materials
    and Technologies
Nanotechnology Research Center
School of Materials Science and Engineering
Sun Yat-sen University
Guangdong, China

Ritsuko Yaokawa
Toyota Central R&D Labs., Inc.
Aichi, Japan

Krzysztof Zberecki
Faculty of Physics
Warsaw University of Technology
Warsaw, Poland

Rui-Qin Zhang
Department of Physics and Materials Science
City University of Hong Kong
Hong Kong SAR, China

Shu Zhou
Department of Mechanical Science and
    Engineering
Tokyo Institute of Technology
Tokyo, Japan

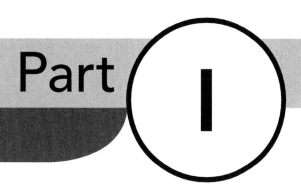

Part I

Low-dimensional structures

1
One-dimensional porous silicon photonic crystals

Danilo Roque Huanca

Contents

1.1 INTRODUCTION

In 1865, James Clark Maxwell presented his work entitled *A Dynamical Theory of the Electromagnetic Field* which showed the connection between electricity, magnetism, and light (Maxwell 1865). Maxwell's equations for handling periodic systems, with the help of Bloch's theorem, led Yablonovitch to predict the existence of periodic materials in which photonic states are not allowed in the photonic structure (Yablonovitch 1987). The region in which the photonic state is forbidden is the photonic band gap (PBG) where the structures are photonic crystals (PCs). However, because of technological difficulties, the first PC fabrication was only reported 4 years after the first theoretical work and was named Yablonovite (Yablonovitch 1991). To fabricate these structures, different materials and techniques were employed (Lopez 2003; Rodriguez et al. 2005); however, from the universe of materials, crystalline silicon (c-Si) is the most important material because the complementary metal–oxide–semiconductor (CMOS) technology opens up the possibility of engineering optical circuits to manipulate photons in an analogous form to electrons in semiconductors (Yablonovitch 2001). For introducing photonic states within the PBG, a defect structure, microcavity (material with different optical and structural properties), is placed within the periodic structure (Joannopoulos 2008; John 1987; Lopez 2003;

Yablonovitch 1987). Its presence produces the confinement of photons with a wavelength proportional to the microcavity optical thickness (John 1987), which are subsequently emitted by spontaneous emission. This can be used to fabricate excellent light-emitting devices (Birner et al. 2001; Pavesi et al. 1996). In particular, in silicon-based PCs, this aim is achieved by inclusion of rare earths (Lopez and Fauchet 2001; Zhou et al. 2000b) or another photoluminescent material, as well as by employing the photoluminescent properties of porous silicon (PS) itself as active material (Kim et al. 2003; Xu et al. 2002). Thus, silicon-based PCs can enable the fabrication of photonic circuits that could be integrated with light-emitting devices using CMOS technology (Gaburro et al. 2000) and thereby enable the fabrication of high-performance computers with high information transfer speed and lossless by heat dissipation.

1.1.1 GENERAL ASPECT OF PHOTONIC CRYSTALS

By definition, a PC is a medium in which the refractive index or dielectric constant varies periodically along the whole medium. This structure can be classified in one-dimensional (1D) PC structures (Figure 1.1a) because the refractive index varies in one direction—a typical example is a Bragg mirror. Figure 1.1b shows the two-dimensional (2D) PC structures which are composed of rods embedded in air. In this structure, the dielectric constants vary along the x–y plane, whereas for the three-dimensional (3D) PC, the variation is along the three directions; one example is the woodpile-like structure (Figure 1.1c).

Some marked differences between the photonic and electronic lattices are as follows: (1) The PC unit cell size has dimensions ranging from the nanometric to micrometric scale. In contrast, for an electronic structure the unit cell is of the order of Angstroms. (2) In a crystal lattice, the electrons' behavior is well described by the Schrödinger equation. All information about this can be extracted from the scalar wave function, while in the photonics case, the photons' behavior in the material medium is well described by

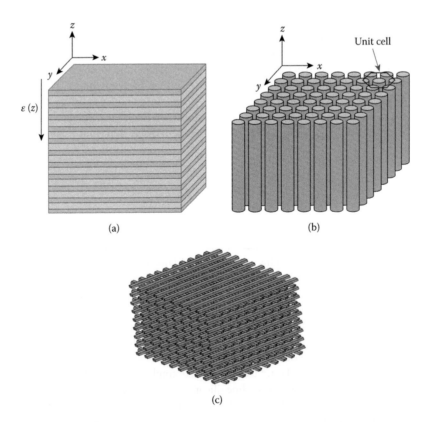

Figure 1.1 Schematic representation of some examples of (a) one-dimensional photonic crystal (1D-PC); (b) two-dimensional photonic crystal (2D-PC), showing its hexagonal unit cell; and (c) three-dimensional (3D-PC) woodpile-like photonic crystal structures.

the master equation whose wave function has a vector nature. (3) For photons, the electromagnetic wave scattering coefficient is always positive (Joannopoulos et al. 2008; Lopez 2003), while for electrons it can also assume negative values. (4) Whereas the master equation was deduced by mathematical handling of Maxwell's equations, Schrödinger's equation cannot be deduced by this means because this is the result of experimental observations. More details about the mathematical treatment of PCs can be found in the literature (Joannopoulos et al. 2008; Lopez 2003).

1.1.2 FABRICATION METHODS AND LIMITATIONS

From the technological aspect, PCs can be fabricated using diverse methods, such as those based on self-organizing silica particles, colloidal crystals, X-ray photolithography, and holographic laser. Some materials employed for its fabrication are metals, dielectrics, semiconductors, polymers, organic materials, and so on (Edrington et al. 2001; Lopez 2003; Meseguer 2005). However, depending on the geometrical features of the unit cell, their fabrication can be a complicated task, especially in 3D structures with the unit cell having nanometric dimensions (Lopez 2003). Despite that only a 3D-PC shows a complete PBG (Lopez and Fauchet 2001; Yablonovitch 1991), the search for alternative structures with less fabrication complexity is needed. In this sense, for some applications, 1D-PC structures are useful, and for this reason we will deal here with 1D-PCs focusing on those made by electrochemical corrosion of c-Si.

1.1.3 ONE-DIMENSIONAL PHOTONIC CRYSTAL

Among the range of PCs, the 1D-PC structure is perhaps the most known and studied because it is nothing more than a multilayer structure composed of two materials with different thickness and different refractive indices that are stacked periodically over the entire structure. They were studied alongside the thin film theory (Macleod 2010; Stenzel 2005) and can be easily attained through different means (Edrington et al. 2001; Lopez 2003; Meseguer 2005). Among the different methods (Canham 2014), the electrochemical corrosion method is particularly useful for the fabrication of PCs (Birner et al. 2001; Lopez 2003). In this sense, the PS technology appears as the most suitable for fabrication of these devices (porous silicon photonic crystals [PSPCs]) because of its compatibility with the CMOS processes that allow the integration of c-Si devices into a single chip (Barillaro et al. 2007; Gaburro et al. 2000; Lopez 2003). PS is usually obtained by electrochemical corrosion at room temperature and does not require vacuum or special atmospheres, but requires an exhaust fan to remove the toxic gases from the working environment. However, one of the main disadvantages is related to the use of hydrofluoric acid (HF) as a key element. HF is highly dangerous to health, so it is necessary to take further precautions (Lehmann 2002; Sailor 2012).

1.2 GENERAL ASPECTS ABOUT POROUS SILICON

For PS fabrication by electrochemical means, the c-Si is employed as a working electrode in an HF-based solution in either a galvanostat or a potentiostat setup (Föll et al. 2002; Lehmann 2002; Sailor 2012). Regardless of the c-Si type used (n- or p-type), it is widely accepted that for pore formation, the participation of positive charge carriers (holes) is needed. In the absence of them, the pore formation does not occur (Lehmann 2002; Sailor 2012). This can be seen clearly through the potential versus current density curves (Figure 1.2). In this figure, the solid line corresponds to the anodization of a p$^+$-type c-Si in a nonfluoride electrolyte and shows the passivation of its surface by a silicon oxide (SiO$_2$) layer. When HF is added, the voltage–current density shows an increasing region until reaching the maximum J_{PS} where pores are formed by c-Si divalent dissolution. For values beyond this, ($J_{PS} < J < J_{ox}$), the tetravalent c-Si promotes the electropolishing phenomenon; the region between J_{ps} and J_{ox} is known as critical region and corresponds to the formation of silicon oxide in the structure and reaches its maximum value in Jox (Lehmann 2002; Sailor 2012). For an n-type substrate, in the absence of light it remains without significant modification (open circles), but under illumination the voltage–current density profile changes following similar behavior of that observed in the p-type substrate, showing that for pore formation the participation of holes is necessary. Experimentally it was found that the J_{PS} and J_{ox} positions can be modified by changing the electrolyte composition (presence of oxidizing agents, etc.) and substrate

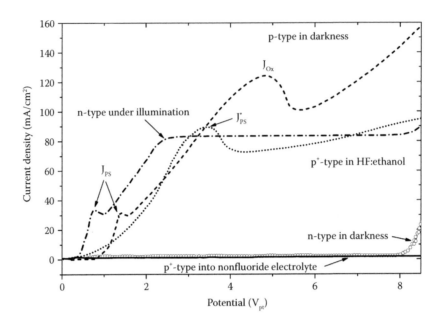

Figure 1.2 Potential versus current density curves for a p⁺-type c-Si anodized into a nonfluoride electrolyte (solid line) and in HF:ethanol (1:1) in darkness (dotted line). In the case of the n-type substrate, they were anodized into HF:H₂O in darkness (open circles) and under illumination (dash dotted line). The effect of the electrolyte type upon both the Js and Jox peaks of the characteristic curve is shown in the dashed line.

features (type, doping, and crystallographic orientation), as well as by external factors which promote the electron–hole pairs formation such as light or temperature (dashed line). Although the effect of these parameters on their physicochemical properties is relatively known, there is no unanimity about how PS is formed. A review of the different proposed mechanisms can still be found in recently published literature (Föll et al. 2002; Lehmann 2002; Sailor 2012; Zhang 2001).

An interesting feature of PS that allows the formation of multilayer structures composed by layers with different porosities is associated to its self-limiting property. This property comes from the fact that c-Si divalent dissolution reaction occurs only at the pore tips by consuming holes (Lehmann 2002; Sailor 2012). Once the first porous layer is formed, it cannot be dissolved by electrochemical means during the formation of the second one with different porosity because the first porous structure is lacking in holes (Lehmann 2002). However, it can still be dissolved by chemical routes because when c-Si is immersed in a fluoride electrolyte its dissolution occurs by competition of chemical and electrochemical processes. The predominance of one depends on the characteristics of the electrolyte solution (Zhang 2001). After anodization, c-Si remains as an island (crystallites) with shape and size determined by the anodization conditions linked to the electrolyte and substrate features (Lehmann 2002; Zhang 2001). The different parameters that influence the c-Si crystallites' shape, as well as pore morphology and porosity, are schematically summarized in Figure 1.3.

1.2.1 OPTICAL AND STRUCTURAL CHARACTERIZATION OF POROUS SILICON MONOLAYERS

Prior to the fabrication of the 1D-PSPCs, it is necessary to know the dependency of both the thickness and porosity on the anodization time and current density for a given electrolyte composition. This is because these parameters along with the presence of other phases (SiO₂, amorphous Si) determine the optical properties of the porous layers (Pickering et al. 1985; Sailor 2012; Strashnikova et al. 2001; Theiss 1997), which in turn will determine the optical properties of the multilayer photonic devices (Pavesi 1997). For measuring these parameters, there are various techniques, such as the gravimetric and the spectroscopy liquid infiltration methods (SLIMs) (Sailor 2012), for example.

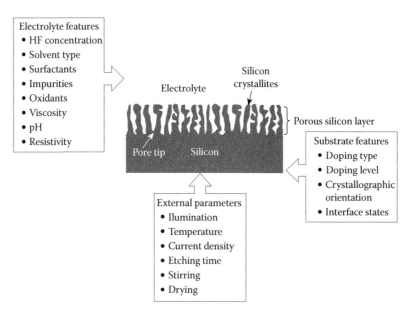

Figure 1.3 Schematic diagram about the parameter that determines the morphological features of the PS films.

1.2.2 POROSITY AND THICKNESS MEASUREMENTS

In the gravimetric method, the porosity (P) and thickness (d) of the porous layers are determined by Equations 1.1 and 1.2, respectively:

$$P = \frac{m_1 - m_2}{m_1 - m_3} \tag{1.1}$$

$$d = \frac{m_1 - m_3}{A\rho_{Si}} \tag{1.2}$$

where m_1 and m_2 are the substrate masses before and after anodization, respectively, whereas m_3 is the mass after removing the porous layer from the substrate. A is the area where the pores are formed, and ρ_{Si} is the mass density of c-Si (≈ 2.33 g/cm³). This method is characterized by being straightforward to use and practical to compute the required parameters because it does not need sophisticated mathematics, but its main disadvantage is linked to its destructive nature. In addition, the accuracy of this method depends strongly on the precise porous layer removal (Sailor 2012). However, this procedure cannot be a trivial task, principally in macroporous structures in which the interpore distance ranges from about 0.2–3.5 μm (Huanca et al. 2010; Trifonov et al. 2007).

The porosity, as well as the etching rate, is shown as a function of the current density in Figure 1.4a for a p-type c-Si(100) with resistivity equal to 0.010 Ω · cm, anodized in an electrolyte solution based on a mixture of ethanol and 20% HF. According to this, both the porosity and etching rate follow a nonlinear dependence with the current density. However, for some electrolyte types the porosity–current density curve shows an almost linear dependence, independently of the substrate doping level, as shown in Figure 1.4b for monolayers yielded on c-Si(100) with three different doping levels (Table 1.1). According to this (Figure 1.4b), the porosity range is wider in the porous layers formed on the p⁺-type (open triangles), whereas in the case of the p⁻-type (open squares), it becomes narrower varying only between 60 and 73% for a narrow current density range. Similar results were achieved by using p⁺-type substrate (Charrier et al. 2007), but with different slope (open stars), showing the electrolyte composition effect on the porosity. The use of p⁺⁺-type substrate (0.6–1.0 mΩ · cm) shifts the porosity range and extends the applied current density holding its almost linear behavior

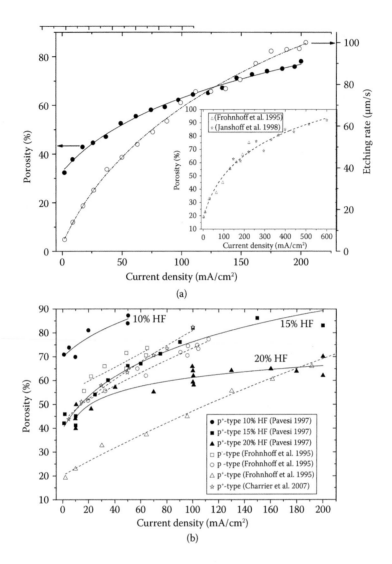

Figure 1.4 (a) Nonlinear dependence of the porosity and etching rate as a function of the current density. Inset shows the nonlinear behavior of the junction of data from two sources. (b) Comparison of the porosity for different anodization process versus current density.

Table 1.1 PS structures fabricated in different solutions showing the electrolyte and substrate dependence of porosity

SUBSTRATE TYPE	RESISTIVITY ($\Omega \cdot$ cm)	ELECTROLYTE COMPOSITION	POROSITY RANGE (%)	CURRENT DENSITY (mA/cm²)	REFERENCE
p⁺-type	0.01	HF:ethanol (1:1)	30–75	10–240	Frohnhoff et al. 1995
p-type	0.2	HF:ethanol (1:1)	55–78	10–120	Frohnhoff et al. 1995
p⁻-type	8.0	HF:ethanol (1:1)	60–73	15–74	Frohnhoff et al. 1995
p⁺-type	0.005	HF:H₂O:ethanol:glycerol	44–82	5–100	Charrier et al. 2007
p⁺⁺-type	0.001	HF:ethanol (1:3)	60–92	150–600	Janshoff et al. 1998

(Janshoff et al. 1998), as shown in the inset in Figure 1.4a (open stars). The superposition of this result with that marked by the black-filled circle shows that the porosity obtained on p^{++}-type can be seen as the continuation of that achieved in p^{+}-type, confirming the logarithmic trend of the porosity with the applied current density. The importance of the HF concentration in porosity can be seen in Figure 1.4b and shows that the HF concentration imposes the applied current density range. The lower the HF concentration, the narrower the applied current density range. The porosity is not only determined by the current density, etching time, or electrolyte composition, but also by the etching temperature (Figure 1.3). PS layers produced at low temperatures have larger porosity than those obtained at room temperature, which is the reason that the applied current density is also narrow (Huanca and Salcedo 2015; Setzu et al. 1998, 2000).

Another useful method for porosity and thickness measurement is the SLIM. Unlike the gravimetric method, it is nondestructive, and the porosity and thickness are determined by comparing the reflectance spectrum measured in an air environment with that of the porous film immersed into a liquid substance with a known refractive index. The thickness and porosity are determined by simultaneously solving the following equations (Sailor 2012):

$$P = 1 - \frac{\left[\left(OT_{air}/d\right)^2 - n_{air}^2\right]\left[2\left(OT_{air}/d\right)^2 - n_{Si}^2\right]}{\left[3\left(OT_{air}/d\right)^2\right]\left[n_{Si}^2 - n_{air}^2\right]} \tag{1.3}$$

$$P = 1 - \frac{\left[\left(OT_{liq}/d\right)^2 - n_{liq}^2\right]\left[2\left(OT_{liq}/d\right)^2 - n_{Si}^2\right]}{\left[3\left(OT_{liq}/d\right)^2\right]\left[n_{Si}^2 - n_{liq}^2\right]} \tag{1.4}$$

where $OT_{air} = n_{PS}d$ and $OT_{lig} = n_{PS}d$ are the optical thicknesses measured from the reflectance spectrum of the PS film in air and immersed in an organic or inorganic liquid substance, respectively.

The fundamentals, advantages, limitations, and additional information about the SLIM can be found in Sailor (2012) in which additional methods for measuring the porosity based on gas absorption isotherms of nitrogen at cryogenic temperature—such as Brunauer–Emmett–Teller (BET) and Barrett–Joyner–Halenda (BJT)—are also described.

Although the gravimetric and SLIM methods are useful for thickness measurement, the most suitable method for this task is the scanning electron microscope (SEM) analysis, in spite of its destructive nature. Regardless of the method used for measuring the thickness, it has been observed that the PS thickness has an almost linear dependence on the anodization time, for any value of the applied current density during the anodization, electrolyte composition, and crystallographic orientation of c-Si (Charrier et al. 2007; Pavesi 1997; Riley and Gerhardt 2000; Svyakhoskiy et al. 2012). Figure 1.5a corresponds to thickness computed from PS layers obtained in p-type c-Si(001) with resistivity ranging between 0.002 and 0.005 $\Omega \cdot$ cm anodized in a solution composed of HF (21%):H_2O:ethanol (2:4:3). The thickness was estimated from the reflectance spectrum (Svyakhoskiy et al. 2012), while in Figure 1.5b the linear correlation of the thickness for different values of current density ($J = 1, 2, 4, 8$, and 10 mA/cm^2) is observed for anodization time ranging from 0 to 160 s. In this case, the porous structure was formed by anodization of p-type c-Si(100) with c.a. $\rho = 0.6$ $\Omega \cdot$ cm and the thickness was measured by SEM.

Experimentally, it was observed that the PS thickness depends on etching time and porosity (Equation 1.5), which in turn depends on current density, J, and HF concentration, c, and etching time, t (Thönissen et al. 1997). In Equation 1.5, porosity is labeled $P(J, c, t)$ and $n_v(J)$ is the current-dependent dissolution valence and equal to 2 during pore formation (Lehmann 2002; Sailor 2012); m_{si} and ρ_{Si} are the relative mass and mass density of c-Si. N is the molar density; e is the electric charge.

$$d = \frac{1}{n_v(J)} \frac{m_{Si}Jt}{Ne\rho_{Si}P(J,c,t)} \tag{1.5}$$

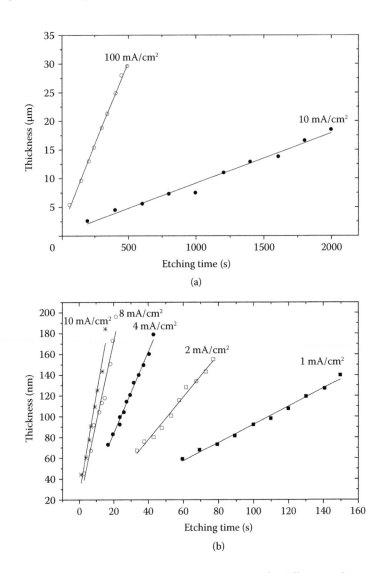

Figure 1.5 Linear dependency of the thickness on the anodization time for different etching conditions on (a) p-type c-Si(001) (after Svyakhoskiy, S.E., et al., *J. Appl. Phys.* 112, 013106, 2012) and (b) p-type c-Si (100) (after Riley, D.W., and Gerhardt, R.A., *J. Appl. Phys.* 87, 2169–2177, 2000).

1.2.3 EFFECTIVE REFRACTIVE INDEX MEASUREMENTS

For systems in which only the porosity is taken into account, the effective refractive index (ERI) is computed using models based on the effective medium approach (EMA), which regards the pore structure as composed of a host medium with dielectric constant ε_h, where particles (pores) with different geometrical features (L) and dielectric constants ε_i are embedded. The general form for determining the effective dielectric constant of the medium ε is deduced by the general relation (Stenzel 2005):

$$\frac{\varepsilon - \varepsilon_h}{\varepsilon_h + (\varepsilon - \varepsilon_h)L} = \sum_{i=1}^{n} \frac{(\varepsilon_i - \varepsilon_h)f_i}{\varepsilon_h + (\varepsilon_i - \varepsilon_h)L} \tag{1.6}$$

Under certain conditions, Equation 1.6 can be rewritten to obtain the well-known relations of the Maxwell–Garnett approach (MG), Lorentz–Lorenz (LL), and Bruggeman effective medium approach (BEMA). Among them, for calculating the ERI of PS, the most widely used model is the BEMA, along with the also well-known Looyenga model (LM) (Equation 1.7) (Looyenga 1965) because they take account of the percolation effect (Theiss 1997), while the MG model considers only few isolated pores embedded into a matrix, c-Si, for example (Stenzel 2005; Theiss 1997).

$$\varepsilon^{1/3} = \frac{1}{V} \sum_{i=1}^{n} V_i \varepsilon_i^{1/3} \tag{1.7}$$

In fact, the comparative study of these models reported in Theiss (1997) shows that for high-porosity structures the LM and LL models provide more accurate results, and for moderate porosity (up to about 40%) the BEMA is the most suitable, whereas for structures with low or high porosities the more useful is the MG model. For a system where the host medium is the system itself ($\varepsilon = \varepsilon_h$), the general form of the BEMA can be written as (Stenzel 2005):

$$\sum_{i=1}^{n} \frac{(\varepsilon_i - \varepsilon_h) f_i}{\varepsilon_h + (\varepsilon_i - \varepsilon_h) L} = 0 \tag{1.8}$$

In general, for the ERI computation ($n = \sqrt{\varepsilon}$), the pore structure is assumed to be composed of pores with spherical shape ($L = 1/3$) although the nonporous structure obtained by anodization shows pores with this shape (Zhang 2001). For instance, during the anodization of c-Si(100), pores with a cylindrical shape are formed, so L must be equal to zero. Despite this, the profile of the ERI versus wavelength diminishes its level as a function of the porosity, as seen in Figure 1.6a (Huanca and Salcedo 2015). The diminution is marked within the visible region, so that for larger porosities the ERI is almost constant along the electromagnetic spectrum. The curves were computed using Equation 1.8 for a mesoporous structure composed of spherical voids ($L = 1/3$), but these results must be viewed carefully since in the case of a microporous structure in which the c-Si crystallites have an average diameter of less than 2 nm, quantum confinement effects become important and can modify the ERI. The pore-shape effect on the ERI can be observed in Figure 1.6b in which it is shown that for low porosity systems the effect of the geometrical shape of the pores is negligible, but becomes important for high-porosity structures.

On the other hand, it is important to emphasize that the models based on the EMA are only valid for porous systems where the average pore diameter, ϕ, is much smaller than the wavelength of the incident light ($\phi \ll \lambda$), so that it can "see" the system as a continuous medium, as in the case of the micro- and mesoporous structures. However, for a macroporous silicon structure, in which the average pore diameter can be equal or larger than the incident light wavelength ($\phi \geq \lambda$), the part that focuses at the pores edge is scattered by it, while the another part which focuses within the pores are absorbed by them in such a way that the structure behaves as an excellent absorbing material (Ernst et al. 2012); therefore, these models are no longer valid. A detailed review about the light/PS interaction can be found in Kocherging and Föell (2006), whereas an improved model that takes into account not only the percolation effect but also the percolation shape and size effect can be found in Theiss (1996).

A faster estimate of the ERI can be made using the models based on the thin films physics principles, but they are valid only for certain restricted cases. For example, for a high-absorbing porous layer, the reflectance at perpendicular incidence can be regarded as coming only from the PS/air interface, whereas the contribution of the PS/c-Si interface has vanished because of the absorption losses in the PS structure. This can be observed in the visible region because of the high absorption of both c-Si and PS in this region (Pavesi 1997; Torres-Costa et al. 2003).

$$n_{PS} = \frac{1 + \sqrt{R}}{1 + \sqrt{R}} \tag{1.9}$$

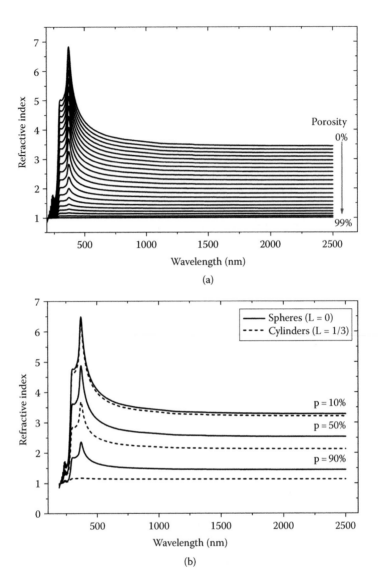

Figure 1.6 (a) Real part of the effective refractive index computed by the Bruggeman effective medium approach for a porous system composed of spherical pores (after Huanca, D.R., and Salcedo, W.J., *Phys. Status Solidi A*, 212, 1975–1983, 2015). (b) Comparison between the result for pores with spherical and cylindrical shapes.

For porous layers with low absorbance, which is frequently observed in the infrared region, the reflectance or transmittance spectra display typical interference fringes coming from the superposition of both the reflected light beam at air/PS and PS/c-Si interfaces (Pavesi 1997; Stenzel 2005). Hence, the ERI can be computed by the relationship between the optical thickness, $n(\lambda)d$, and the position of the interference fringes maxima (or minima), which for perpendicular light incidence is written as

$$1 = \frac{1}{2}\left(\frac{n(\lambda_{i+1})d}{\lambda_{i+1}} - \frac{n(\lambda_i)d}{\lambda_i} \right)^{-1} \tag{1.10}$$

where λ_i and $n(\lambda_i)$ are the wavelength position and ERI at the i-th maximum peak of the interference fringes. Equation 1.10 is useful for ERI calculation within the infrared region because there the system behaves as a nondispersive medium (Stenzel 2005), so that $n(\lambda_{i+1}) \approx n(\lambda_i)$ as shown in Figure 1.6a for $\lambda \geq 900$ nm. For this aim, the thickness d is an input parameter. However, the accuracy of this method depends on the uniformity of both the thickness and porosity along the PS layer.

Equations 1.9 and 1.10 are useful for determining the real part of the ERI, but does not allow the computation of its imaginary part. For computing the complex ERI ($\tilde{n}_{PS} = n_{PS} - ik_{PS}$) of a porous layer with homogeneous porosity and thickness having smooth interfaces along the whole structure, the more suitable means is the fitting procedure of the experimental reflectance or transmittance spectrum. For the case of the reflectance spectrum, the fitting procedure for a single smooth layer can be made by

$$R = \left| \frac{r_{01} + r_{12}e^{-2i\delta}}{1 + r_{01}r_{12}e^{-2i\delta}} \right| \tag{1.11}$$

where r_{01} and r_{12} are the Fresnel reflectance coefficients at the air/PS and PS/c-Si interfaces that are defined in terms of their complex ERI ($\tilde{n}_1$, $\tilde{n}_2$ and $\tilde{n}_3$) by $r_{ij} = (\tilde{n}_i - \tilde{n}_j) / (\tilde{n}_i + \tilde{n}_j)$, whereas $\delta = 4\pi\tilde{n} / \lambda$. To fit the reflectance spectrum, the layer thickness is an input parameter into Equation 1.12. The advantage of this method is the possibility to take into account the interface roughness effect into Equation 1.11 by including the Davies–Bennett factor in each Fresnel's coefficient (Dariani and Ebrahimnasab 2014; Lérondel et al. 1997; Lérondel and Romestain 1997), which represents the roughness by a Gaussian distribution. The effect of the interface roughness on the reflectance spectrum is to diminish the reflectance quality and is marked in layers yielded at room temperature (Mulloni et al. 1999; Setzu et al. 1998), as is observed in Figure 1.7. The causes for the interface roughness will be described in the next section.

1.2.4 ROUGHNESS AT THE POROUS SILICON INTERFACES

During the pore formation, the air/PS and PS/c-Si interfaces are formed, each of them with a determined level of roughness. The roughness formation is an undesirable effect in the 1D-PSPCs because it decreases the optical quality of the devices of both single and multilayer structures (Lérondel et al. 1997; Setzu et al. 1998).

Experimentally was observed that the roughness level depends on the electrolyte chemical composition (Charrier et al. 2007; Servidori et al. 2001), electrolyte viscosity (Kan et al. 2005; Servidori et al. 2001), anodization temperature (Setzu et al. 1998), ohmic contact quality (Huanca et al. 2009), porosity (Charrier et al. 2007; Setzu et al. 1998), and crystallographic orientation (Zhang 2001). In this sense,

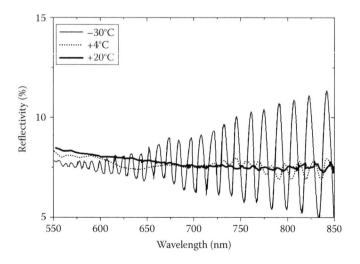

Figure 1.7 Reflectance spectra from porous silicon monolayers made at different temperatures. (After Setzu, S., et al., *J. Appl. Phys.* 84, 3129–3133, 1998.)

several authors have shown that air/PS interface roughness is greater in samples with larger porosity (Charrier et al. 2007; Setzu et al. 1998). The higher the porosity, the higher the roughness. Further works have shown that the roughness level is linked to the silicon consumption speed during the pore formation. This in turn is dependent on temperature and electrolyte viscosity, so that in structures formed at low temperatures it is lower than in samples obtained at room temperature (Figure 1.8a), for instance, even when both were formed under the same anodization conditions and the porosity level of the first structure was lower than the second one (Servidori et al. 2001; Setzu et al. 1998). According to the work undertaken

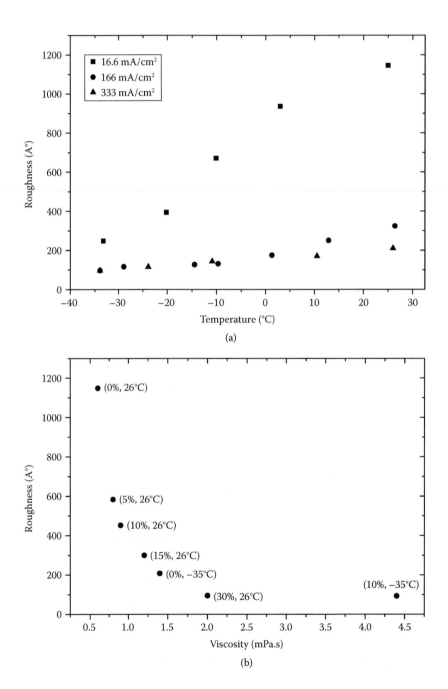

(a)

(b)

Figure 1.8 Roughness dependence with the (a) temperature and (b) electrolyte viscosity for a p-type c-Si. (After Setzu, S., et al., *Mater. Sci. Eng. B*, 69–70, 34–42, 2000.)

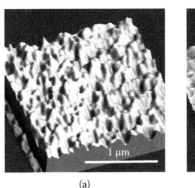

(a)   (b)

Figure 1.9 Three-dimensional AFM profile of the PS/c-Si interface (a) before and (b) after oxidation procedure. (After Charrier, J., et al., *Appl. Surf. Sci.* 253, 8632–8636, 2007.)

by Mulloni et al. (1999), Servidori et al. (2001), and Setzu et al. (2000), the reaction rate for silicon consumption can be further reduced by increasing the electrolyte viscosity. In the case of the p-type c-Si (ρ = 1–10 $\Omega \cdot$ cm), adding 10% of glycerol yielded a marked reduction in roughness, but for different values the effect is contrary (Servidori et al. 2001). However, the results shown by Charrier et al. (2007) suggest that, for p+-type c-Si, the interface smoothness is proportional to the glycerol concentration in the electrolyte. Noticeable results are obtained combining both the temperature diminution and viscosity increment (Figure 1.8b) (Setzu et al. 1998).

On the other side, the roughness can be reduced after posterior annealing treatment in an oxygen atmosphere giving rise to the roughness consumption by SiO_2 formation at 600–900°C, approximately (Charrier et al. 2007; Pap et al. 2005), as shown in Figure 1.9. In this is shown the superficial morphology obtained before and after the oxidation process. For annealing temperatures lower than 400°C or even for high one, but in an inert atmosphere, the roughness becomes greater (Ott and Nerding 2004; Pap et al. 2005). The oxidation method presents the disadvantage of reducing the ERI of the porous structure (Charrier et al. 2007; Zangooie et al. 1998), which could be undesirable for some optical purposes.

1.3 POROUS SILICON MULTILAYER STACK

1.3.1 NONPERIODIC STRUCTURES

As mentioned, the multilayer structures formation is possible only because of the self-limiting property of PS by applying current density, which varies discretely. High current density (J_H) is applied to yield porous layers with high porosity (H), whereas low porosity (L) is achieved when delivering low current density (J_L). Their thicknesses (d_H and d_L) are controlled by the etching time t_H and t_L. In the literature, the fabrication of structures with quasiperiodic sequence has recently been reported in which the H and L layers are distributed following mathematical patterns such as the Fibonacci, Cantor, and Thue–Morse structures, for example, Argawal and Mora-Ramos (2007), Escorcia-Garcia et al. (2012), and Perez et al. (2012). The study of these structures is outside the aims of this chapter.

1.3.2 PERIODIC STRUCTURES: ONE-DIMENSIONAL POROUS SILICON PHOTONIC CRYSTAL OR BRAGG MIRRORS

The formation of 1D-PSPCs occurs by stacking H and L layers following the HLHLHL...HL or LHLHLH...LH sequences along the entire thickness (Huanca and Salcedo 2015; Pavesi 1997). This structure is usually achieved by anodizing a p-type c-Si into a given HF-based electrolyte, applying J_H and J_L periodically to form periodic H and L layers with ERI equal to n_H and n_L, respectively, and

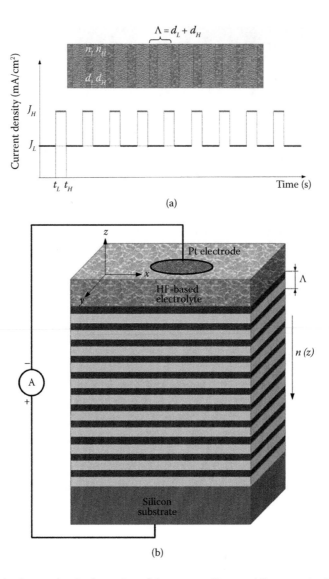

Figure 1.10 (a) Schematic diagram for the formation of the porous silicon multilayer stack with periodic thickness of the H and L layers and the effective refractive index by applying periodic current density. (b) Schematic picture of the 1D-PSPC.

the thicknesses d_H and d_L which are defined by the etching times t_H and t_L, as shown schematically in Figure 1.10.

The unit cell of the 1D-PSPCs is composed of the junction of two near H and L layers such that the thickness of the unit cell, labeled Λ, is given by (Figure 1.10b):

$$\Lambda = d_H + d_L \tag{1.12}$$

For this structure, Λ is projected by regarding the values of d_H and d_L computed for single layers for a specific anodization condition by controlling the etching time t_H and t_L.

The optical features of the 1D-PSPC devices are shown in Figure 1.11a, in which the PBG is clearly observed where the reflectance is maximum, as well as the interference fringes and the secondary maximum peaks. The PBG center is correlated with the unit cell optical thickness by Equation 1.13,

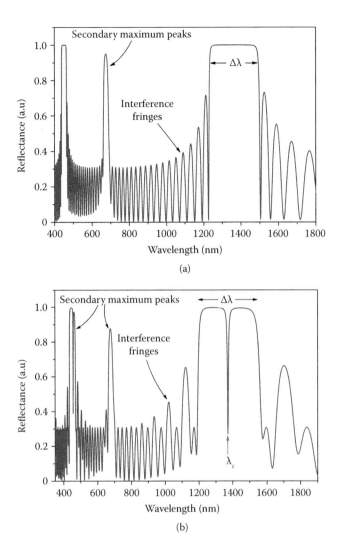

Figure 1.11 Theoretical reflectance spectrum of (a) one-dimensional photonic crystal with flat interfaces, and (b) the reflectance spectrum of one-dimensional photonic crystal in which an aperiodic layer was included.

while its width is given by the difference between the superior (λ_2) and inferior (λ_1) PBG edges (Equation 1.14):

$$m\lambda_m = 2\left(d_H \sqrt{n_H^2 - n_0^2 \sin^2 \theta} + d_L \sqrt{n_L^2 - n_0^2 \sin^2 \theta}\right) \qquad (1.13)$$

$$\Delta\lambda = \lambda_2 - \lambda_1 \qquad (1.14)$$

where m is the order of the maximum peak, n_0 corresponds to the incident medium refractive index, and θ is the light incident angle.

The values of λ_2 and λ_1 can be obtained by solving Equation 1.15 numerically and placing it in Equation 1.14:

$$\cos^2\left[\frac{\pi(n_H d_H + n_L d_L)}{\lambda}\right] = \rho^2 \cos^2\left[\frac{\pi(n_H d_H - n_L d_L)}{\lambda}\right] \qquad (1.15)$$

Low-dimensional structures

However, for devices in which the optical thickness of the H and L layers is equal or nearly equal to the L layers ($d_H n_H \approx d_L n_L$), the PBG width depends solely on the effective refractive index contrast (ERIC) of the H and L layers, defined as $\Delta n = n_L - n_H$, and can be written as:

$$\Delta\lambda = \frac{4\lambda_0}{\pi} \arcsin\left(\frac{n_L - n_H}{n_L + n_H}\right) \tag{1.16}$$

For an ideal 1D-PSPC composed of H and L layers with flat interfaces, the reflectance depends only on the number of layers, N, and the n_L/n_H ratio, as well as on the c-Si refractive index (n_{si}) of which the 1D-PSPC was produced as is given by (Macleod 2010; Pavesi 1997; Stenzel 2005):

$$R = \left[\frac{(n_H / n_{Si})^2 (n_L / n_H)^2 - 1}{(n_H / n_{Si})^2 (n_L / n_H)^2 + 1}\right]^2 \tag{1.17}$$

According to this equation, the reflectance is higher when the number of layers is greater (Figure 1.12a), but this condition can also be achieved in a structure with a small number of layers with a larger n_L/n_H ratio ($n_H < n_L$), as shown in Figure 1.12b. In the case of Figure 1.12a, the curves were simulated for n_L ranging between 2.95 and 2.15, and n_H is equal to 1.95 and $n_s = 3.5$.

The theoretical reflectance spectrum of a 1D-PSPC for different numbers of layers (N = 6, 15, 31, and 41) (Figure 1.13a) shows that as N increases, R becomes equal to 1 and the PBG shape and interference fringes are well defined. In Figure 1.13b, the PBG width ($\Delta\lambda$) of devices having $d_H n_H \approx d_L n_L$ varies in a nonlinear way as a function of λ_0 and n_L/n_H.

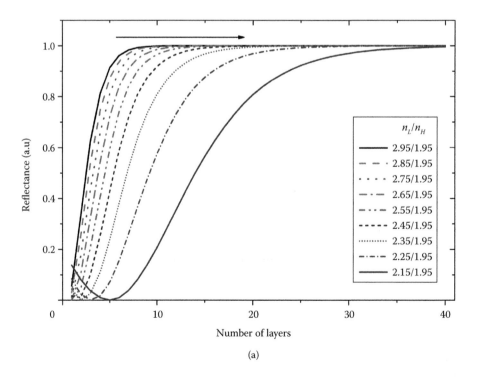

(a)

Figure 1.12 Reflectance dependence with (a) the number of layers of the structure. *(Continued)*

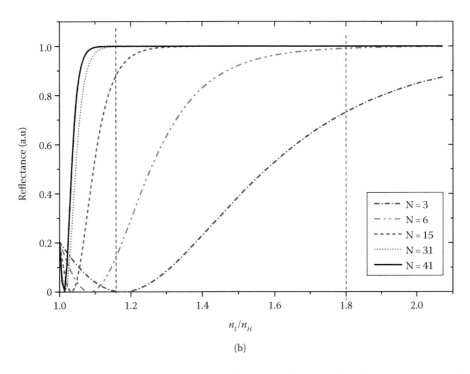

Figure 1.12 (Continued)   Reflectance dependence with (b) the ratio between the refractive indexes, n_H/n_L.

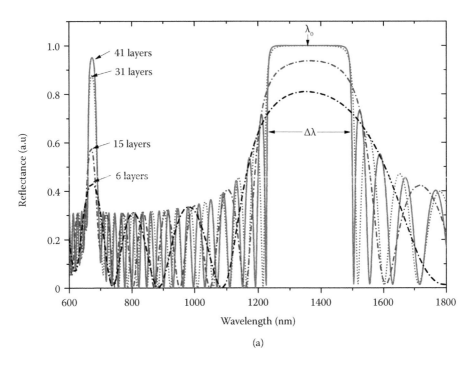

Figure 1.13 (a) Reflectance with respect to the number of layers.   *(Continued)*

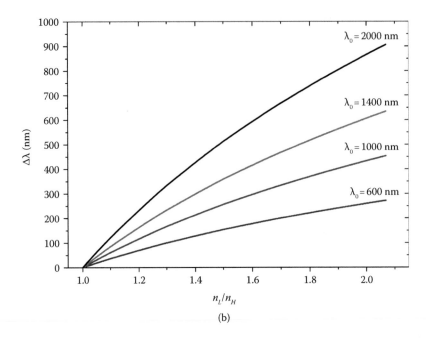

Figure 1.13 (Continued) (b) PBG width as a function of the effective refractive index ratio of the H and L layers.

1.3.3 POROUS SILICON MICROCAVITIES: FABRY–PEROT DEVICES

A 1D-PSPC where a defect layer is placed in the periodic structure is known as a microcavity or Fabry–Perot device (Huanca et al. 2009; Pavesi 1997). In the 1D-PSPC field, the defect layer has similar geometrical features to the unit cell but different dielectric properties. The introduction of this defect layer gives rise to resonance effects that allows the spontaneous emission of photons (transmittance) (Joannopoulos et al. 2008; John 1987). In practice, it is made by applying a different current density or different etching time to those employed to form the periodic structure, as shown schematically in Figure 1.14. The theoretical reflectance spectrum (Figure 1.11b) shows clearly the resonance effect as a narrow transmission slit, λ_c, located in the middle of the PBG. In general, the spectral position of the resonance peak can be tuned by controlling the microcavity optical thickness as shown by Equation 1.18, whereas the finesse coefficient, which measures the emission width, is given by Equation 1.19 (Born and Wolf 2005; Pavesi 1997):

$$\lambda_c = 2d_c \sqrt{n_c^2 - n_0^2 \sin^2 \theta} \tag{1.18}$$

$$F = \frac{\Delta\lambda}{\lambda_c} \tag{1.19}$$

Equations 1.13 through 1.19 show that the optical quality of both 1D-PSPCs and microcavities depends uniquely on the geometrical features and optical properties of the H and L individual layers, which in turn can be easily controlled by proper control of the anodization parameters (Huanca et al. 2009; Huanca and Salcedo 2015; Pavesi 1997). Figure 1.15a shows different Fabry–Perot devices showing different colors which are associated to different H and L thicknesses. The cross-section SEM image of a 1D-PSPC is shown in Figure 1.15b, and in Figure 1.15c the Fabry–Perot profile is shown. Both devices have 31 layers and they were yielded on p⁺-type c-Si substrate having resistivity equal to 0.005 $\Omega \cdot$ cm and electrolyte HF (48%):ethanol (3:7). The 1D-PSPC was formed following the HLHL… HLH sequence, whereas for the Fabry–Perot it was L(HL)⁷HH(LH)⁷L. The electrochemical parameters are summarized in Table 1.2, where also are placed the PBG width and center. Their reflectance spectrum is shown in Figure 1.16, where it is clearly observed that the PBG center of the 1D-PSPC is placed

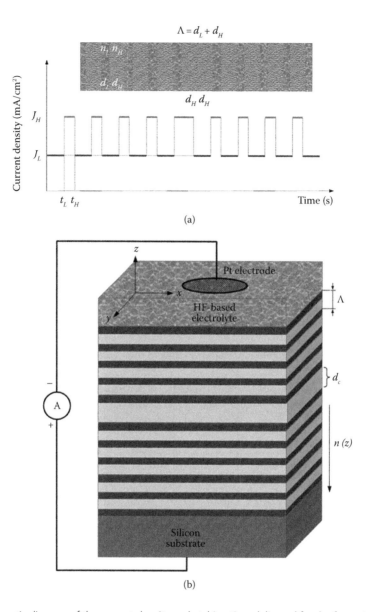

(a)

(b)

Figure 1.14 (a) Schematic diagram of the current density and etching time delivered for the formation of the periodic structure in which a microcavity (defect) is included, and (b) schematic representation of the microcavity structure.

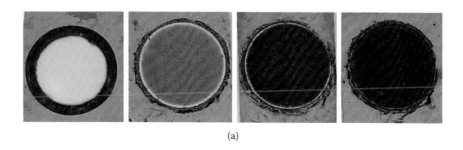

(a)

Figure 1.15 (a) Set of 1D-PSPCs made with different parameters.   (Continued)

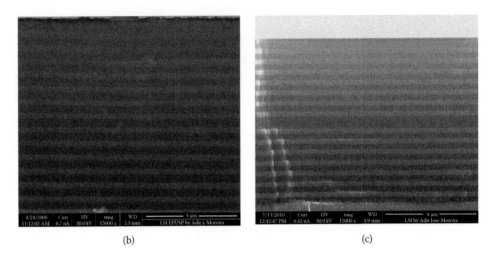

(b)

(c)

Figure 1.15 (Continued) (b) SEM image from the transversal section of (a) 1D-PSPC and (c) microcavity device achieved by anodizing p$^+$-type c-Si(100) having $\rho = 0.005\ \Omega \cdot$ cm in HF:ethanol (3:7). The electrochemical parameters are summarized in Table 1.2.

Table 1.2 Electrochemical parameter, thickness extracted from SEM images and PBG features from 1D-PSPC and microcavity devices

DEVICE	CURRENT DENSITY (mA/cm²)		ETCHING TIME (s)		THICKNESS (nm)		PBG FEATURES (nm)			λ_c	d_c(nm)
	J_H	J_L	t_H	t_L	d_H	d_L	$\Delta\lambda$	λ_0	λ_t		
1D-PSPC	100	10	9	39	310	260	352	1544	1600	–	–
Fabry–Perot	30	3	9	39	247	180	250	933	950	917	495

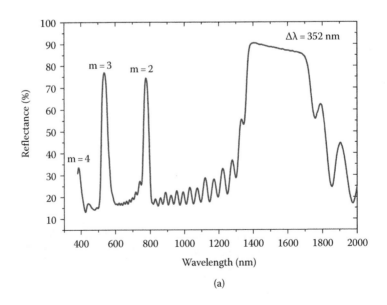

(a)

Figure 1.16 Reflectance spectra of (a) the 1D-PSPC showing the tilted PBG and the secondary maximum peaks for m = 2, 3, and 4.

(Continued)

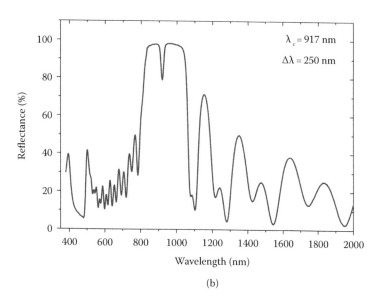

(b)

Figure 1.16 (Continued) Reflectance spectra of (b) the Fabry–Perot device in which the resonance effect of the microcavity is clear.

at $\lambda = 1544$ nm and the width is $\Delta\lambda = 352$ nm. It is also possible to see the different maximum secondary peaks for $m = 2$, 3, and 4 are positioned at $\lambda = 779$, 539, and 386 nm, respectively. The tilted aspect of the PBG upper edge is produced by the greater roughness of the first H layer (Huanca and Salcedo 2015). In the case of the microcavity device, the PBG center is positioned at $\lambda = 933$ nm and its width is $\Delta\lambda = 250$ nm. In addition, the resonance peak—transmission peak—appears slightly shifted from the center ($\lambda_c = 917$ nm). For these values the finesse is $F = 0.273$. Despite the resonance peak presence, the marked difference between the reflectance of these two devices is linked to the PBG width. It is greater for the 1D-PSPC due to its larger unit cell in relation to the microcavity device, because the PBG of the 1D-PSPC is larger. The PBG center of both devices differs from that projected (λ_r) (Table 2) and is associated to the etching rate variation in depth. It has been observed that the etching rate of the individual H and L layers which belongs to the 1D-PSPCs becomes smaller than that observed in single H and L layers, even though both, single and multilayer stacks, are anodized under the same electrochemical parameters (Maehama et al. 2005).

1.4 MICRO- AND MESOPOROUS STRUCTURES FOR PHOTONIC CRYSTALS: p-TYPE VERSUS p⁺-TYPE SUBSTRATE

Although the 1D-PSPC can be fabricated on any p-type substrate, in practice the most used is the heavily doped substrate ($\rho = 1$–5 m$\Omega \cdot$ cm), because in slightly doped one was observed: (1) narrow porosity range (Figure 1.4b) that allows to obtain poor values of n_L/n_H and n_L-n_H which yields devices with narrow PBG and weak mechanical stability (Frohnhoff et al. 1995); (2) interfaces with larger roughness (Setzu et al. 2000), which increases the extinction coefficient of the system (Lérondel et al. 1997; Setzu et al. 1998); (3) high-porosity gradient in depth owing to the difficulty in spreading the HF through the pores with average diameter smaller than 2 nm (Föll et al. 2002; Herino et al. 1987; Lehmann 2002); (4) larger chemical dissolution that becomes critical in devices with larger number of layers (Herino et al. 1987; Huanca and Salcedo 2007); (5) larger aging effect associated to its higher porosity and very reactive inner surface (Mulloni et al. 1999); and (6) weak mechanical stability due to the smaller c-Si interconnections (Lehmann 2002; Sailor 2012).

The importance of the substrate doping level is highlighted by comparing the optical response of two different 1D-PSPCs made using two different substrates which are anodized in the same electrolyte HF:H$_2$O:C$_2$H$_5$OH (1:1:2) to have 20 periods. The reflectance spectra of them are observed in Figure 1.17a

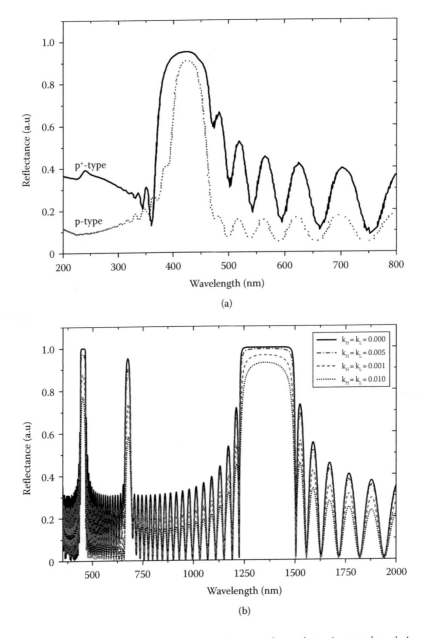

Figure 1.17 (a) Reflectance spectra from a one-dimensional porous silicon photonic crystal made in p-type (blue line) and p⁺-type (black line) c-Si (after Berger, M.G., et al., *Thin Solid Films*, 297, 237–240, 1997). (b) Theoretical simulation of a 1D-PSPCS with different extinction coefficients k_H and k_L.

(Berger et al. 1997). The maximum value of n_L/n_H was equal to 1.18 in the device yielded on p-type (p-1D-PSPC), whereas for that achieved on p⁺-type (p⁺-1D-PSPC) it was about 1.8. For these values, the PBG width of the p⁺-1D-PSPC is about 3.5 times larger than the p-1D-PSPC, for the reflectance is also lower. For both cases, the reflectance differs from that computed by Equation 1.17 for devices with 20 periods (see the perpendicular dashed lines in Figure 1.12b). It is more marked in p-1D-PSPC due to the greater extinction coefficient of their H and L layers (k_H and k_L), which are enhanced by the interface roughness of the H and L layers because and larger intrinsic extinction coefficient within the visible region (Lérondel et al. 1997; Setzu et al. 1998, 2000). This fact was confirmed by theoretical simulation for different values of k_H and k_L (Figure 1.17b).

1.4.1 ELECTROLYTE EFFECT ON THE OPTICAL RESPONSE OF THE ONE-DIMENSIONAL PHOTONIC CRYSTAL BASED ON POROUS SILICON

Equation 1.17 predicts that for a given value of $n_L/n_H = 2.95/1.95$, for instance, the maxima reflectance is achieved for about 20 periods (Figure 1.12). However, in practice it is not always possible due to factors linked to the electrolyte features that define the morphology of the layers and also could promote the chemical dissolution of it. In Figure 1.18a is shown the reflectance spectra of devices produced by anodizing p$^+$-type c-Si(100) with $\rho = 0.005\ \Omega \cdot$ cm into HF:ethanol (1:3). The H and L layers were formed by $J_H = 28.3$ mA/cm^2 and $J_L = 2.83$ mA/cm^2 during $t_H = 6.3$ s and $t_L = 32.0$ s, respectively, following the HLHL...HL configuration. Table 1.3 summarizes the PBG features for them extracted from Figure 1.18a. According to this, the PBG center position of the device 24N ($\lambda_0 = 610$ nm) appears shifted in relation to the 21N ($\lambda_0 = 630$ nm) in about 20 nm, whereas its PBG width is shrunk. This means that the optical thickness of the unit cell has changed by electrolyte action. Since the etching time for the H and L layers is equal for all devices, the role of the total etching time is important because the larger the number of layers, the larger the total etching time. The PBG shrink reveals reduction of the ERIC ($n_L - n_H$) associated to the change on the porosity by chemical dissolution of the upper layers, as shown in Figure 1.19 (Herino et al. 1987; Huanca and Salcedo 2007; Unno et al. 1987), as well as to the porosity gradient effect. This phenomenon is mainly attributed to difficulties on HF diffusion through the porous structure (Billat et al. 1997; Thönissen et al. 1997), and also modifies the etching rate in depth, as predicted by Equation 1.5. This fact was confirmed by measuring the thickness and the n_H and n_L in depth of the multilayer structures (Maehama et al. 2005; Zangooie et al. 1998). The dissolution of the upper layers increases the value of k_H in depth, rounding the PBG left edge, as shown by the simulation (Figure 1.18b).

In the case of the 1D-PSPCs with microcavity, known as Fabry–Perot devices, the theoretical simulation shows that the number of layers improves both the PBG shape and resonance peak of the finesse ($F = \Delta\lambda/\lambda_c$) (Figure 1.20a). The effect of it on the finesse, as well as on the reflectance can be viewed in Figure 1.20b. The spectra were simulated using $n_L = 2.55$, $n_H = 1.89$, $d_H = 109$, and $d_L = 103$ nm, respectively, whereas for the microcavity optical thickness the values are $n_c = 1.99$ and $d_c = 220$ nm, which

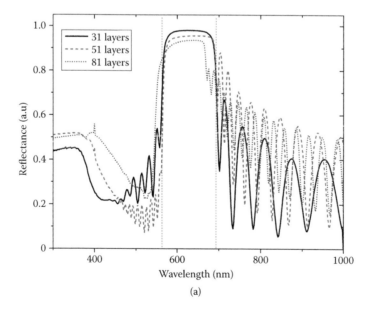

(a)

Figure 1.18 (a) Experimental reflectance spectra from 1D-PSPCs having 31, 51, and 81 layers produced by anodizing p$^+$-type c-Si(100) into HF:ethanol (1:3) following the LHLH...LHL sequence applying (etching time) $J_H = 28.3$ mA/cm^2 ($t_H = 6.3$ s) and $J_L = 2.83$ mA/cm^2 ($t_L = 32.0$ s) (after Huanca, D.R., et al., *Microelectron. J.* 39, 499–506, 2008).

(Continued)

Low-dimensional structures

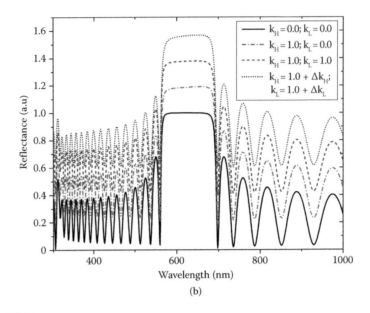

Figure 1.18 (Continued) (b) Theoretical spectra for photonic crystal having 31 layers including the effect of the extinction coefficient on the PBG shape.

Table 1.3 Central position and width of the PBG extracted from the reflectance spectra of the 1D-PSPCs made in HF:ethanol (1:3) by applying $J_H = 28.3$ mA/cm^2 and $J_L = 2.83$ mA/cm^2 during $t_H = 6.3$ s and $t_L = 32.0$ s

SAMPLE	LAYERS	λ (nm)	λ_2 (nm)	λ_1 (nm)	$\Delta\lambda$ (nm)
21N	31	630	696	564	132
22N	51	628	691	565	128
24N	81	610	670	549	121

(a)

Figure 1.19 SEM image from the cross-section profile of the devices with (a) 51 layers (22N). *(Continued)*

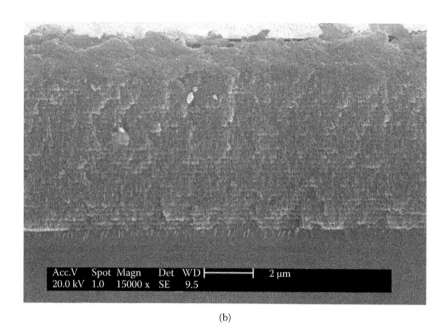

(b)

Figure 1.19 (Continued) SEM image from the cross-section profile of the devices with (b) 81 layers (24N), showing clearly the partial dissolution of the upper layers. The devices were obtained on p⁺-type cSi(100) anodized into HF:ethanol (1:3). (After Huanca, D.R., et al., *Microelectron. J.* 39, 499–506, 2008.)

are extracted from the reflectance spectrum of experimental Fabry–Perot devices with 31 layers shown in Figure 1.21a, where also is observed that for larger number of layers the resonance peak disappears (68N). According to the theoretical simulation for 31 layers (Figure 1.21b), the resonance peak disappearance is promoted by the increment of the microcavity extinction coefficient (k_c). The larger the k_c, the smaller the finesse depth (Figure 1.21b). The spectra in Figure 1.21b corresponds to devices fabricated on c-Si(100) having $\rho = 0.005\ \Omega \cdot$ cm anodized in HF:ethanol (3:7) applying $J_H = 50$ mA/cm² and $J_L = 5$ mA/cm² during $t_H = 9$ s and $t_L = 39$ s, following the L(HL)ᵖHH(LH)ᵖL, for $p = 4, 7$, and 15, as shown in Table 1.4.

On the other side, for larger numbers of layers, another difficulty is to avoid the chemical dissolution effect (Herino et al. 1987; Huanca and Salcedo 2007). Recently it has been shown that the key factor for solving this drawback is to add deionized water into the electrolyte (Bruyant et al. 2003; Qian et al. 2006; Svyakhoskiy et al. 2012). In this sense, aqueous electrolyte permits the formation of 1D-PSPC with more than 120 layers (Figure 1.22; Qian et al. 2006) or even structures having thousands of layers, as reported in Svyakhoskiy et al. (2012). In that work, the 1D-PSPC structure has 2500 periods (Figure 1.22b) than that was achieved by anodizing heavily doped c-Si(001) into electrolyte composed by HF (21% w/w):H₂O:ethanol (2:4:3), but to diminish the increment of the interface roughness as function of the number of layers and also to avoid the porosity gradient effect, the anodization was carried out under permanent electrolyte mixing to keep the HF concentration in the electrolyte bulk equal to the PS/c-Si interface. The method success was confirmed by measuring the reflectance spectra of the devices formed with and without stirring the electrolyte (Figure 1.23). The reflectance was measured in free-stand structure impinging the light at the facing side air/PS interface (solid line) and at the backside PS/air interface (dotted line) that was detached from the c-Si substrate. The results show strong diminution not only of the porosity gradient but also the roughness interface, as is seeing through the position and shape of its PBG.

1.4.2 OPTICAL REFLECTANCE DEPENDENCE ON THE SUBSTRATE BACKSIDE CONTACT

For pore formation, the c-Si substrate backside is usually metalized by deposition of Al and then annealed in inert atmosphere to promote a good ohmic contact (Wolf 1990). However, this procedure modifies the

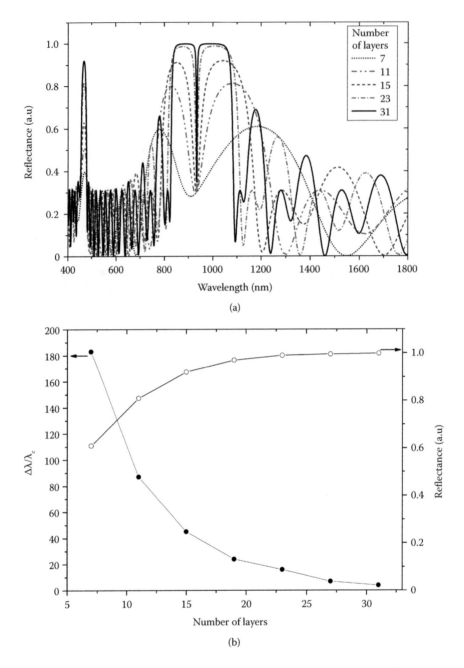

Figure 1.20 (a) Theoretical simulation of reflectance spectra of a Fabry–Perot device for different number of layers. (b) Finesse and reflectance dependence of the number of layers of the Fabry–Perot device.

morphological and optical properties of the 1D-PSPCs yielded on it (Huanca et al. 2009). The backside contact importance was studied by comparing a set of Fabry–Perot devices obtained on p⁺-type c-Si(100) with ρ = 0.005 Ω · cm in the (48%) into HF:ethanol (3:7) electrolyte using a single electrochemical cell (SEC) and a double electrochemical cell (DEC). In the first cell, the c-Si is placed over a cupper plate, whereas in the second one the backside contact was made by some liquid acids (Table 1.5). The H and L layers were formed by delivering J_H = 50 mA/cm² and J_L = 5 mA/cm² for 9 s. and 39 s, respectively, following the L(HL)⁷HH(LH)⁷L sequence (Huanca et al. 2009).

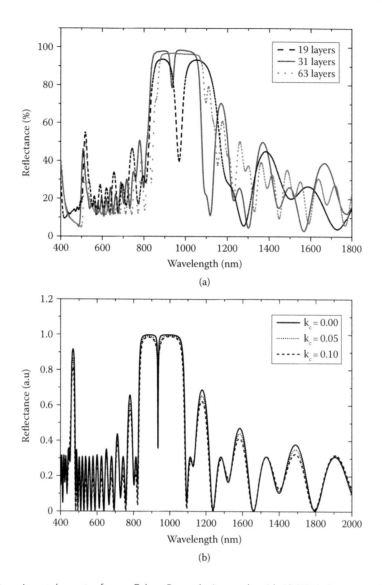

Figure 1.21 (a) Experimental spectra from a Fabry–Perot device made with 19 (67N), 31 (T2), and 63 layers (68N). (b) Theoretical reflectance of a Fabry–Perot device having 31 layers for different values of k_c.

Table 1.4 Optical parameters extracted from the Fabry–Perot reflectance spectra

SAMPLE	LAYERS	λ_0 (nm)	λ_2 (nm)	λ_1 (nm)	$\Delta\lambda$ (nm)	λ_c (nm)
67N	19	950	696	564	232	967
T2	31	977	691	565	222	935
68N	63	988	670	549	315	–

The SEM images of devices D2, D30, and P32 obtained in SEC are shown in Figure 1.24a through c. The thicknesses d_H and d_L are thinner in the sample D30 and thicker in Al-backside contact device (D2), showing the strong effect of the backside contact on the morphology of the devices, as well as on the etching rate of c-Si. For the devices achieved in the DEC (Figure 1.24d through f), the d_H and d_L are thinner than those achieved in solid backside (Figure 1.24d through f), but there is no linear correlation between the pKa and the thickness of the H and L layers. For instance, the thickness for the device made

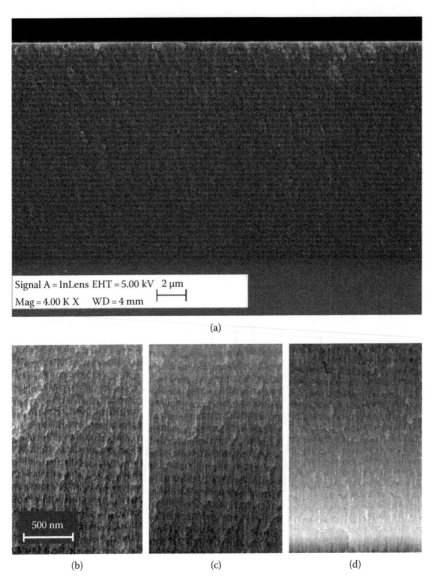

Signal A = InLens EHT = 5.00 kV   2 μm
Mag = 4.00 K X   WD = 4 mm

(a)

500 nm

(b)   (c)   (d)

Figure 1.22 SEM image of the 1D-PSPC obtained by (a) Qian et al. (2006) and (b, c, and d) Svyakhovskiy et al. (2012) in stirred solution. (b) corresponding to the first upper layers placed; (c) in the middle region; and (d) layers in the deeper region closer to the backside. (After Svyakhoskiy, S.E., et al., *J. Appl. Phys.* 112, 013106, 2012.)

using the H_2SO_4 (pKa = −3) is thinner than that obtained using HCl as backside contact (pKa = −4) or even than that achieved using weak acids as contact (Huanca et al. 2009).

The optical reflectance is shown in Figure 1.25. For the devices obtained on solid backside substrate (Figure 1.25a), the higher reflectance and well-defined PBG are achieved in nonmetalized substrate (P32), whereas the device made on Al-metallized substrate depicts lowest reflectance with resonance peak position placed at about λ_c = 1060 nm, which is 63 nm less than that observed from Fabry–Perot P32 (λ_c = 1123 nm), showing that the Al diffusion not only introduce largest roughness, but also modifies the etching rate and the porosity, as is clearly viewed comparing Figure 1.24a and c. Since Pt as ohmic contact promotes intermediate values, we can conclude that the main reason for the lowest reflectance is linked to the nonhomogeneous diffusion of Al into the c-Si due to the Kinkerdall effect (Wolf 1990), which is not observed in Pt-deposited substrate. The backside contact role on the optical response quality can be observed in the reflectance spectra measured from Fabry–Perot devices made

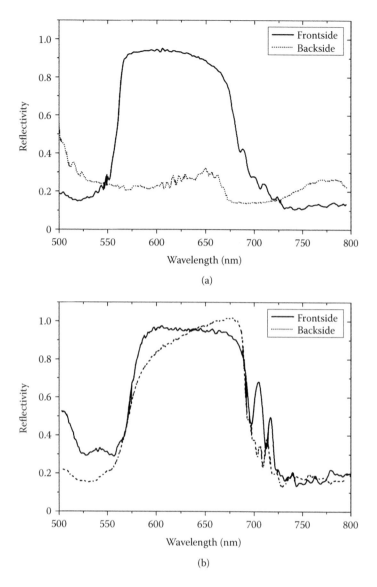

Figure 1.23 Reflectivity spectra of a device yielded in (a) not stirred and (b) stirred electrolyte. In these spectra, the solid line corresponds to the optical response measured from the front side and the dashed one for the backside of the free-standing 1DPSPC. (After Svyakhoskiy, S.E., et al., *J. Appl. Phys.* 112, 013106, 2012.)

Table 1.5 Electrochemical parameters used to obtain the Fabry–Perot devices in the simple and double electrochemical cells

SAMPLE	BACKSIDE CONTACT	pKa OF ACIDS	ELECTROCHEMICAL CELL
D2	Al	–	Simple
P32	Si	–	Simple
D30	Pt	–	Simple
P16	H_2SO_4	–3.00	Double
D41	HCl	–4.00	Double
D42	H_3PO_4	2.15	Double
T2	HF	3.17	Double

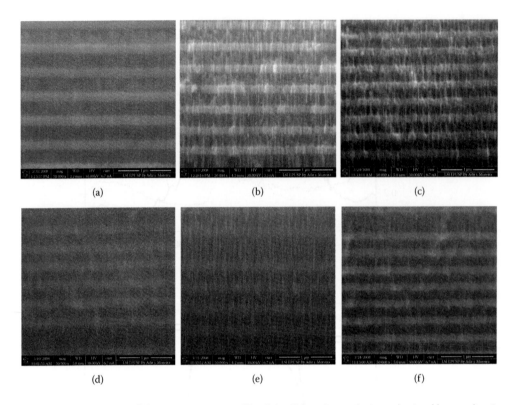

Figure 1.24 The SEM images of the cross-section profile of the Fabry–Perot devices obtained by anodization of c-Si with solid backside contact made by (a) Al, (b) Pt, (c) nonmetalized, and with liquid contact by (d) H_2SO_4, (e) HCl, and (f) HF. (After Huanca, D.R., et al., *Microelectron. J.* 40, 744–748, 2009.)

on liquid contact substrate (Figure 1.25b). In all the cases, the reflectance coefficient at PBG regions has high values (~100%), as well as the PBG appears well defined, except in device D41, which have H and L layers with rougher interfaces.

To validate the useful DEC for improving the optical quality of the 1D-PSPC, hence, the interface roughness reduction, two 1D-PSPCs with microcavity were made on nonmetallized p$^+$-type c-Si for the same current density and etching time, but different electrolyte conditions, and using the HF as liquid ohmic contact. The first one (T2) was anodized into HF:ethanol (3:7), while for the second (57N) 10% of glycerol was added into the solution, HF:ethanol:glycerol (3:6:1). It is known that the addition of this level of glycerol yields smoother interfaces promoting the improvement the optical quality of the 1D-PSPC devices (Huanca et al. 2008, 2009). The optical response of them is shown in Figure 1.25c. In this picture, the PBG shape and reflectance level within this region are similar for both T2 and 57N, as seen in Table 1.6, validating the DEC method as an alternative way to reduce the interface roughness and, hence, to improve the optical response. The slight shift of the resonance peak of device 57N in relation to that observed in the T2 sample indicates that glycerol reduces the etching rate slightly, whereas the inclusion of 5 s as an etch-stop reduces the optical thickness of the H and L layers, so shifts the PBG position to 849 nm and shrinks its width (Table 1.6).

1.4.3 EXPERIMENTAL VERSUS MODEL: OPTICAL PARAMETERS EXTRACTION FROM EXPERIMENTAL SPECTRA

The reflectance spectrum from 1D-PSPC contains all information about the structural and optical features and about how light interacts with it. To extract this information, the experimental curves are fitted with theoretical models based on different methods. However, the most useful method for 1D-PSPCs and any multilayers structure is the transfer matrix method (TMM) (Huanca and Salcedo 2015; Pavesi 1997).

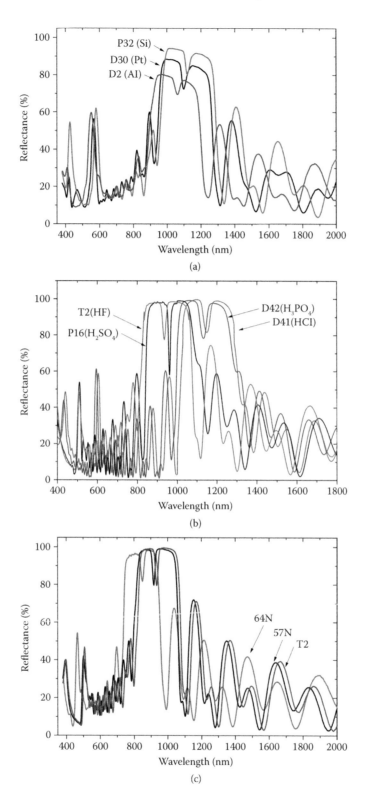

Figure 1.25 Reflectance spectra from the Fabry–Perot devices formed on (a) solid backside contact with Al (D2), Pt (D30), and nonmetalized (P32), and (b) liquid backside contact made by H_2SO_4 (P16), HF (T2), HCl (D41), and H_3PO_4 (D42) (from Huanca, D.R., et al., *Microelectron. J.* 40, 744–748, 2009). (c) Reflectance spectra measured from the devices obtained by anodization of c-Si using the DEC.

Low-dimensional structures

Table 1.6 Electrolyte composition and optical parameter extracted from the reflectance spectra of the devices fabricated into DEC

SAMPLE	ELECTROLYTE	PAUSE	$\Delta\lambda$ (nm)	λ_C (nm)
64N	HF:ethanol	Yes	219	849
57N	HF:ethanol:glycerol	No	250	917
T2	HF:ethanol	No	252	936

For this procedure, it is necessary to know the input parameters (n_H, n_L, n_c, d_H, d_L, and d_c). In some cases, only either the ERI (n_H and n_L) or their physical thicknesses (d_H and d_L) are known, so the fitting procedure allows to extract the remaining parameters, as well as to study the gradient of them in depth. Within the TMM frame, the amplitude of the incoming electrical field (E_+) and the beam leaving the photonic structure (E_-) in the s-polarization are linked to the of the beam traveling through the staked structure by Equation 1.20 (Macleod 2010; Stenzel 2005), which is longer for a system with the physical and optical thickness of the unit cell constant at incidence angle φ.

$$\begin{pmatrix} E_+ \\ E_- \end{pmatrix} = \begin{pmatrix} \cos\left(\dfrac{2\pi\tilde{n}_j d_j}{\lambda}\cos\varphi\right) & -\dfrac{i}{\tilde{n}_j\cos\varphi}\sin\left(\dfrac{2\pi\tilde{n}_j d_j}{\lambda}\cos\varphi\right) \\ -i\tilde{n}_j\sin\left(\dfrac{2\pi\tilde{n}_j d_j}{\lambda}\cos\varphi\right) & \cos\left(\dfrac{2\pi\tilde{n}_j d_j}{\lambda}\cos\varphi\right) \end{pmatrix}\begin{pmatrix} 1 \\ \tilde{n}_s \end{pmatrix} \tag{1.20}$$

From which the reflectance can be computed by $R = |r|^2$ and the transmittance $T = n_s|tr|^2$, where n_s is the c-Si substrate refractive index, and r is computed by

$$r = \frac{(m_{11} + \acute{n}_s m_{12}\cos\varphi_s)\acute{n}_0\cos\varphi - (m_{21} + \acute{n}_s m_{22}\cos\varphi_s)}{(m_{11} + \acute{n}_s m_{12}\cos\varphi_s)\acute{n}_0\cos\varphi + (m_{21} + \acute{n}_s m_{22}\cos\varphi_s)} \tag{1.21}$$

In which $\tilde{n}_j = \sqrt{N_j^2 - N_0^2\sin^2\theta_0}$ and N_j is the complex refractive index and is defined as $N_j = n_j - ik_j$.

For the p-polarization, the equation is similar (Stenzel 2005), and both p- and s-polarizations become equal for light incoming at 90°.

For a 1D-PSPC structure in which the optical thickness of the H layers is equal or very close to the L layers, the remaining input parameter can be estimated by simultaneously solving Equations 1.13 and 1.16. In other cases, the numerical solution of Equations 1.13 and 1.15 will be needed. The precision of these estimated values depends on the PBG sharpness and porosity gradient level (Huanca and Salcedo 2015). In any case, these values can be used as initial input parameter and will be improved by the fitting procedure, in which also can be included their variation in depth. The electrochemical parameters for the fabrication of them are summarized in Table 1.7 and the values of n_H, n_L, d_H, and d_L as function of the porosity in Table 1.8 (Huanca and Salcedo 2015).

The fitting procedure was made regarding the variation of the complex ERI for the H and L layers in depth assuming a linear relation expressed by $n_i = n_{i0} - \alpha_i N_i d_i^2 / T$ and $k_i = k_{i0} - \beta_i N_i d_i^2 / T$, where the term $n_{o,i}$ is the ERI obtained by solving the Equations 1.15 and 1.18, and $k_{o,i}$ is the extinction coefficient. α_i and β_i are fitting parameters ($i = H, L$); N_i is the number of the i-th layer (i.e., the deep position of i-th layer); and $T = N(d_H + d_L)$ is the total thickness of photonic device, where N is the total number of the H and L layers. These parameters are summarized in Table 1.9.

The results shown in Figure 1.26 indicates that the reduction of the PBG width in relation to the projected one (dotted line) is caused by the presence of porosity gradient (refractive index gradient) within the photonic structure. However, according to the work made by Ariza-Flores et al. (2011), it can also be caused by the formation of interface gradients between the H and L layers. The interface gradient not only shrank the PBG width, but also decreases the reflectance level of the whole spectrum. The larger the interface gradient, the narrower the PBG width.

Table 1.7 Electrochemical parameters used to fabricate the 1D-PSPC with different layers

DEVICE	ELECTROLYTE	CURRENT DENSITY (mA/cm²)		ETCHING TIME (s)		LAYERS
	HF:ETHANOL	J_H	J_L	t_H	t_L	
PSPC1	1:1	100	20	2.6	6.0	31
PSPC2[a]	1:1	100	20	2.6	6.0	18
PSPC3[b]	1:3	28.3	2.8	6.3	59.1	20
PSPC4	1:3	28.3	2.8	8.5	43.7	41
PSPC5	1:3	28.3	2.8	6.3	32.1	31
PSPC6	1:3	28.3	2.8	6.3	32.1	51
PSPC7[b]	3:7	100	10	9.0	39.0	31
PSPC8	3:7	100	10	9.0	39.0	20

[a] Device anodized at −15°C.
[b] Device yielded following the HLHL… sequence.

Table 1.8 Porosity and thicknesses estimated by the etch rate along with the effective refractive index estimated by the BEMA

SAMPLES	POROSITY (%)		REFRACTIVE INDEX		THICKNESS (nm)		λ_0 (nm)	$\Delta\lambda$ (nm)
	P_H	P_L	n_H	n_L	d_H	d_L		
PSPC1	51	34	2.106	2.631	218	127	1600	227
PSPC2[a]	67	43	1.603	2.316	130	100	880	207
PSPC3[b]	82	60	1.283	1.849	205	93	870	201
PSPC4	72	49	1.505	2.138	172	66	800	178
PSPC5	82	64	1.285	1.736	134	88	650	124
PSPC6	82	65	1.285	1.736	134	88	650	124
PSPC7[b]	88	43	1.166	1.687	320	253	1600	374
PSPC8	88	43	1.166	1.687	320	253	1600	374

[a] Device anodized at −15°C
[b] Device yielded the HLHL… sequence

Table 1.9 Optical parameters extracted by simultaneously solving Equations 1.13 and 1.16 and by fitting the experimental curves, along with the geometrical parameters extracted by SEM analysis

SAMPLE	d_H (nm)	d_L (nm)	n_H	n_L	λ_0 (nm)	$\Delta\lambda$ (nm)	α_H (cm⁻¹)	α_L (cm⁻¹)	k_H (×10⁻³)	k_L (×10⁻³)	β_H (cm⁻¹)	β_L (cm⁻¹)
PSPC1	213	123	2.107	2.651	1550	226	1.8	9.5	5.0	5.0	0.1	0.5
PSPC2	120	96	1.666	2.423	870	205	0.0	8.0	3.0	1.0	0.0	0.1
PSPC3	200	90	1.313	1.894	864	203	0.1	7.5	0.8	0.1	0.0	0.1
PSPC4	168	64	1.536	2.187	796	178	1.3	2.0	0.1	0.1	0.03	0.1
PSPC5	130	86	1.274	1.737	630	124	0.4	3.5	3.0	1.0	0.1	0.1
PSPC6	130	86	1.274	1.737	630	124	0.5	3.7	1.0	1.0	0.1	0.1
PSPC7	310	250	1.146	1.679	1550	374	1.18	0.01	0.01	0.3	0.1	0.0
PSPC8	310	250	1.146	1.679	1550	370	0.05	1.0	0.01	0.8	0.0	0.2

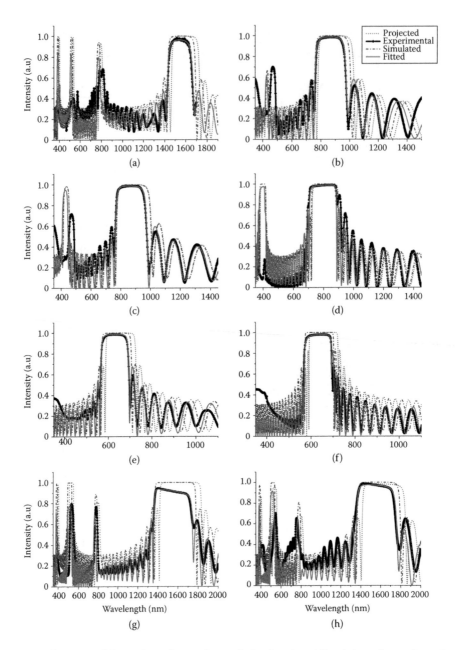

Figure 1.26 Optical spectra of the projected, experimental, simulated, and fitted data of one-dimensional porous silicon photonic crystals made by anodizing in HF:ethanol (1:1) labeled (a) PSPC1 and (b) PSPC2 and of those yielded in (c) HF:ethanol (1:3): PSPC3, (d) PSPC4, (e) PSPC5, (f) and PSPC6, along with those obtained in (g) HF:ethanol (3:7): PSPC7 and (h) PSPC8 using the electrochemical parameters summarized in Table 1.1. (After Huanca, D.R., and Salcedo, W.J., *Phys. Status Solidi A*, 212, 1975–1983, 2015.)

1.5 APPLICATIONS

The application of PCs comprises different fields from optoelectronic, sensor, to solar cells (De la Mora et al. 2009; Gupta et al. 2013; Huanca et al. 2008; Lopez and Fauchet 2001; Mulloni and Pavesi 2000; Pacholski 2013; Yablonovitch 2001). Nevertheless, the larger number of applications of these structures are in the field of sensors, because their porous nature makes the 1D-PSPC a perfect transducer for sensing chemical and biological species within the porous matrix at a liquid or gas state (Pacholski 2013; Snow et al. 1999). Since the as-etched structure is highly reactive, the interaction with its surrounding environment makes the 1D-PSPCs suitable for very sensitive optical sensors. The sensor parameters can be based on changes of its effective dielectric, photoluminescent properties, or other related properties (Mulloni and Pavesi 2000; Pacholski 2013; Snow et al. 1999). Unfortunately, one of its drawbacks is associated to the thermodynamic stability of the hydrogenated inner surface which rapidly reacts with air forming the SiO_2 phase in time (Huanca et al. 2008; Krüger et al. 1998; Mulloni et al. 1999). This aging phenomenon also produces changes on its electrical and optical properties in time, disabling this material for sensor application. For sensors with optimal performance, the following features are required: high selectivity, larger sensitivity, reproducibility, repeatability, stability, and so on (Gründler 2007). To avoid the aging effect, the porous structure must be passivated, and for this task different strategies have been suggested such as nitration, carbonization, thermal and electrochemical oxidation, polymers, and so on (De Stefano et al. 2009; Mawhinney et al. 1997; Morazzani et al. 1995; Torres-Costa et al. 2009; Vasin et al. 2011). The most easier, rapid, and cheap method to passivate the multilayer stack is the thermal oxidation which is usually made at 900°C (Huanca et al. 2008; Krüger et al. 1998). However, the presence of SiO_2 within the structure decreases the value of n_H and n_L layers and also promotes the contraction and expansion of the L and H layers, respectively (Huanca et al. 2011; Zhou et al. 2000a).

When the 1D-PSPC (or Fabry–Perot) is placed into environment filled by organic solvent, for instance, the presence of this species increases the ERI of both H and L layers, its refractive index is larger than air one, so that its reflectance spectrum is shifted toward regions with larger wavelength (Figure 1.27a; Huanca et al. 2011; Pacholski 2013). It was observed that not only the PBG center (or resonance peak)

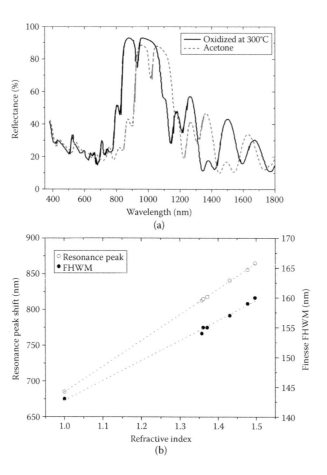

Figure 1.27 (a) Reflectance spectra from the Fabry–Perot devices oxidized at 300°C in an acetone environment, showing the redshift phenomenon. (b) Linear dependence of both resonance peak shift and finesse with the refractive index of organic solvents. *(Continued)*

Low-dimensional structures

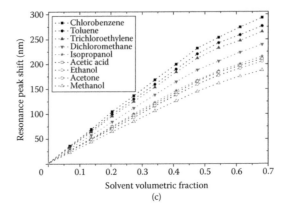

(c)

Figure 1.27 (Continued) (c) Resonance peak shift as a function of the volumetric fraction of organic solvents.

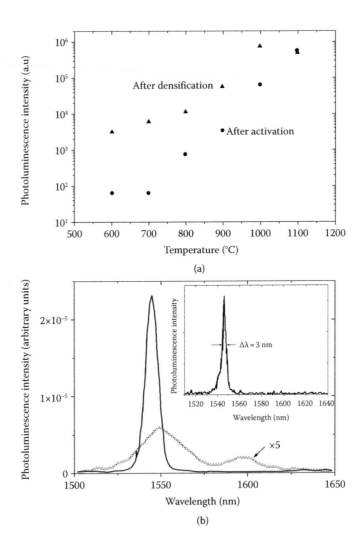

Figure 1.28 (a) Photoluminescence intensity as a function of the activation (circles) and densification (triangles) temperatures (from Lopez and Fauchet 2001). (b) Photoluminescence spectra measured at 10 K from Er-doped single-layer (triangles) and multilayer stacks (solid line) with microcavity. The narrow peak emission is shown in the inset. (After Reece, P.J., et al., *Appl. Phys. Lett.* 85, 3363–3365, 2004.)

position shift follows a linear dependence with the solvent refractive index (Huanca et al. 2008; Volk et al. 2005), as is shown in Figure 1.27b, but also the finesse width. Furthermore, according to the results shown by several search groups (De Stefano et al. 2003; Snow et al. 1999), for a given target species, the PBG shift is species molar fraction dependent and this dependence is not linear, as seen in Figure 1.27c, for several organic solvents. The results are similar for biological species and can be viewed in the recent review in Gupta et al. (2013) and Pacholski (2013).

In the field of the optoelectronics, 1D-PSPC has been proposed to fabricate light-emitting devices (Lopez et al. 2001; Pavesi et al. 1996; Zhou et al. 2000b). For this aim, the photoluminescent properties of PS or another chemical or biological species which is placed into the microcavity was employed. It is well known that PS photoluminescent of the microcavity is enhanced in Fabry–Perot structure (Kim et al. 2003; Xu et al. 2002). Similar phenomenon was observed when organic or inorganic species are placed into the microcavity. One species of particular interest within the optical communication field is Erbium (Er), which has maximum photoluminescent peak at $\lambda = 1.54$ μm. Nonetheless, when Er is introduced into the active layer (microcavity), which can be made by implantation (Reece et al. 2004) or electrochemical method (Zhou et al. 2000b), it is necessary to activate the Er ions by thermal annealing in an O_2 environment, then it is annealed at 1100°C for densification in an N_2 atmosphere (Lopez and Fauchet 2001.). It was found that the photoluminescence intensity depends strongly on the activation temperature (Figure 1.28a), which is also further enhanced by the densification procedure. However, its electroluminescent properties achieve their maximum peak at about 800°C of annealing temperature (Lopez et al. 2001). The Er photoluminescence is enhanced about 10 times and has a narrow emission width c.a. 3 nm at 1.54 mm in comparison to that observed on single PS layers (Figure 1.28b) (Reece et al. 2004).

In the photovoltaic field, coating has recently been proposed by Selj et al. (2011) and Osorio et al. (2011) the use of 1D-PSPCs as antireflecting for solar cell applications. According to this, the structure increases the light absorption and thereby increases the photocurrent output. So in this case, the most important parameter is associated to its absorbance because to optimize the solar cell effectiveness it is necessary to collect larger amounts of incoming solar radiation.

1.6 SUMMARY

PS is a suitable material useful in the PCs field, because it permits the fabrication of 1D-, 2D-, and 3D-PCs. However, it is more useful for the fabrication of 1D-PSPCs because this device is nothing more than a stack of H and L layers arranged in a periodic way so that the unit cell optical thickness follows the Bragg relation. Under this condition, the optical response of the structure depicts a PBG in which the incoming light is completely reflected. The shape and width of its PBG depends on the structural and morphological features of the H and L layers, as well as on the interface roughness. In this sense, the main drawbacks are associated with the difficulty in spread of HF toward deeper layers giving rise to the formation of ERI gradient and thinner layers in depth and is more critical in structure with pores smaller than 2 nm. The addition of surfactants within the electrolyte or etch-stop times promotes the formation layers with almost homogeneous thickness and porosity in depth. Another problem of these structures is linked to its chemical dissolution, which was overcome by the addition of water into the electrolyte. This strategy allows the formation of 1D-PSPCs with thousands of layers with about constant thickness. For solving the interface roughness problem, the most common strategy is that based on electrolyte viscosity diminution by adding glycerol or lowering the etching temperature. However, the use of a DEC during the anodization of nonmetalized substrates has shown to be useful for this task. Finally, the more marked applications of these devices are in the optoelectronic and sensor fields. Several types of optical sensor are reported probing the suitability of this material for this purpose. The inclusion of elements with luminescent properties into the active layer enhances it, so it is possible to fabricate light-emitting devices.

REFERENCES

Argawal V, Mora-Ramos ME (2007). Optical Characterization of Polytype Fibonacci and Thue-Morse Quasiregular Dielectrics Structures Made of Porous Silicon Multilayers. *Journal of Physics D: Applied Physics* 40: 3203–3211.

Ariza-Flores AD, Gaggero-Sager LM, Agarwal V (2011). Effect of interface Gradient on the Optical Properties of Multilayered Porous Silicon Photonic Structures. *Journal of Physics D: Applied Physics* 44: 155102.

Barillaro G, Bruschi P, Pieri F, Strambini LM (2007). CMOS-Compatible Fabrication of Porous Silicon Gas Sensors and Their Redout Electronics on the Same Chip. *Physica Status Solidi A* 204: 1423–1428.

Berger MG, Arens-Fischer R, Thönissen M, Krüger M, Billat S, Lüth H, Hilbrich S, et al. (1997). Dielectric Filters Made of PS: Advanced Performance by Oxidation and New Structures. *Thin Solid Films* 297: 237–240.

Billat S, Thönissen M, Arens-Fischer R, Berger MG, Krüger M, Lüth H (1997). Influence of Etch Stop on the Microstructure of Porous Silicon Layers. *Thin Solid Films* 297: 22–25.

Birner A, Wehrspohn RB, Gösele UM, Busch K (2001). Silicon-Based Photonic Crystals. *Advanced Materials* 13: 377–388.

Born M, Wolf E (2005). *Principles of Optics: Electromagnetic Theory of Propagation Interference and Diffraction of Light.* London: Cambridge University Press.

Bruyant A, Lérondel G, Reece PJ, Gal M (2003). All-Silicon Omnidirectional Mirrors Based on One-Dimensional Photonic Crystals. *Applied Physics Letters* 82: 3227–3229.

Canham L (2014). Routes of Formation for Porous Silicon. In *Handbook of Porous Silicon* (Canham L, eds.), pp 3–9. London: Springer.

Charrier J, Alaiwan V, Piratesh P, Najar A, Gadonna M (2007). Influence of Experimental Parameters on Physical Properties of Porous Silicon and Oxidized Porous Silicon Layers. *Applied Surface Science* 253: 8632–8636.

Dariani RS, Ebrahimnasab S (2014). Root Mean Square Roughness of Nano Porous Silicon by Scattering Spectra. *The European Physical Journal Plus* 129: 129.

De la Mora MB, Jaramillo OA, Nava R, Tagüeña-Martinez J, Del Rio JA (2009). Viability Study of Porous Silicon Photonic Mirrors as Secondary Reflectors for Solar Concentration Systems. *Solar Energy Material & Solar Cells* 93: 1218–1224.

De Stefano L, Rendina I, Moretti L, Mario A, Rossi M (2003). Optical Sensing of Flammable Substances Using Porous Silicon Microcavities. *Materials Science and Engineering B* 100: 271–274.

De Stefano L, Rotiroti L, De Tommasi E, Rea I, Rendina I, Canciello M, Maglio G, et al. (2009). Hybrid Polymer-Porous Silico Photonic Crystals for Optical Sensing. *Journal of Applied Physics* 106: 023109.

Edrington AC, Urbas AM, DeRege P, Chen CX, Swager TM, Hadjichristidis N, Xenidou M, et al. (2001). Polymer-based photonic crystals. *Advanced Materials* 13: 422–425.

Ernst M, Brendel R, Ferre R, Harder N-P (2012). Thin Macroporous Silicon Heterojunction Solar Cells. *Physica Status Solid RRL* 6: 187–189.

Escorcia-Garcia J, Gaggero-Sager LM, Palestino-Escobedo AG, Argawal V (2012). Optical Properties of Cantor Nanostructures Made from Porous Silicon: A Sensing Application. *Photonics and Nanostructures-Fundamentals and Applications* 10: 452–458.

Föll H, Christophersen M, Cartensen J, Hasse G (2002). Formation and Application of Porous Silicon. *Materials Science and Engineering R* 39: 93–141.

Frohnhoff St, Berger MG, Thönissen M, Dieker C, Vescan L, Münder H, Lüth H (1995). Formation Techniques for Porous Silicon Superlattices. *Thin Solid Films* 255: 59–62.

Gaburro Z, Bellutti P, Pavesi L (2000). CMOS Fabrication of a Light Emitting Diode Based on Porous/Porous Silicon Heterojunction. *Physica Status Solidi A* 182: 407–412.

Gründler P (2007). *Chemical Sensors—An introduction for Scientists and Engineers.* Dresden: Springer.

Gupta B, Zhu Y, Guan B, Reece PJ, Gooding JJ (2013). Functionalised Porous Silicon as a Biosensor: Emphasis on Monitoring Cells in *Vivo* and in *Vitro*. *Analyst* 138: 3593–3615.

Herino R, Bomchil G, Barla K, Bertrand C, Ginoux JL (1987). Porosity and Pore Size Distributions of Porous Silicon Layers. *Journal of Electrochemical Society* 134: 1994–2000.

Huanca DR, Elias VF, Salcedo WJ (2011). Study of the Thermal Oxidation of Photonic Crystals for Sensor Applications. *ECS Transactions* 39: 321–328.

Huanca DR, Raimundo DS, Salcedo WJ (2009). Backside Contact Effect on the Morphological and Optical Features of Porous Silicon Photonic Crystals. *Microelectronics Journal* 40: 744–748.

Huanca DR, Ramirez-Fernandez FJ, Salcedo WJ (2008). Porous Silicon Optical Cavity Structure Applied to High Sensitivity Organic Solvent Sensor. *Microelectronics Journal* 39: 499–506.

Huanca DR, Ramirez-Fernanadez J, Salcedo WJ (2010). Morphological and Structural Effect of Aluminum on Macroporous Silicon Layers. *Journal of Materials Science and Engineering* 4: 55–59.

Huanca DR, Salcedo WJ (2007). Effect of Number of Layers on the Optical Response of Porous Silicon Bragg's Mirrors. *ECS Transactions* 9: 525–530.

Huanca DR, Salcedo WJ (2015). Optical Characterization of One-dimensional Porous Silicon Photonic Crystals with Effective Refractive Index Gradient in Depth. *Physica Status Solidi A* 212: 1975–1983.

Janshoff A, Dancil K-PS, Steinem C, Greiner DP, Lin VS-Y, Gurtner C, Motesharei K, et al. (1998). Macroporous p-Type Silicon Fabry-Perot Layers. Fabrication, Characterization, and Applications in Biosensing. *Journal of American Chemistry Society* 120: 12108–12116.

Joannopoulos JD, Johnson SG, Winn JN, Meade RD (2008). *Photonic Crystals-Molding the Flow of Light.* Singapore: Princeton University Press.

John, S (1987). Strong Localization of Photons in Certain Disordered Dielectric Superlattices. *Physics Review Letters* 58: 2486–2489.

Kan PYY, Foss SE, Fistand TG (2005). The Effect of Etching with Glycerol, and Interferometric Measurements on the Interface Roughness of Porous Silicon. *Physica Status Solidi A* 202: 1533–1538.

Kim Y-Y, Lee K-W, Lee C-W, Hong S, Ryu J-W, Jeon J-H (2003). Photoluminescence Resonance Properties of Porous Silicon Microcavity. *Journal of the Korean Physical Society* 42: S329–S332.

Kochergin V, Foell H (2006). Novel Optical Elements Made from Porous Si. *Materials Science and Engineering R* 52: 93–140.

Krüger M, Hilbrich S, Thönissen M, Scheyen D, Theiss W, Lüth H (1998). Suppression of Ageing Effect in Porous Silicon Interference Filters. *Optics Communication* 146: 309–315.

Lehmann V (2002). *Electrochemistry of Silicon.* Weinheim: Wiley-VCH.

Lerondel G, Romestain R (1997). Roughness of the Porous Silicon Dissolution Interface. *Journal of Applied Physics* 81: 6171–6178.

Lérondel G, Romestain R, Barret S (1997). Quantitative Analysis of the Light Scattering Effect on Porous Silicon Optical Measurements. *Thin Solid Films* 297: 114–117.

Looyenga H (1965). Dielectric Constants of Heterogeneous Mixtures. *Physica* 31: 401–406.

Lopez C (2003). Materials Aspects of Photonic Crystals. *Advanced Materials* 15: 1679–1704.

Lopez HA, Fauchet PM (2001). Infrared LEDs and Microcavities Based on Erbium-Doped Silicon Nanocomposites. *Materials Science and Engineering B* 81: 91–96.

Macleod HA (2010). *Thin-Film Optical Filters.* New York: Taylor & Francis CRC Press.

Maehama T, Teruya T, Moriyama Y, Sonegawa T, Higa A, Toguchi M (2005). Analysis of Layer Structure Variation of Periodic Porous Silicon Multilayer. *Japanese Journal of Applied Physics* 44: L391–L393.

Mawhinney DB, Glass JA Jr, Yates JT (1997). FTIR Study of the Oxidation of Porous Silicon. *Journal of Physical Chemistry B* 101: 1202–1206.

Maxwell JC (1865). A dynamical Theory of the Electromagnetic Field. *Philosophical Transactions* 155: 459–512.

Meseguer F (2005). Colloidal Crystals as Photonics Crystals. *Colloids and Surfaces A: Physicochemistry and Engineering Aspects* 270–271: 1–7.

Morazzani V, Cantin JL, Ortega C, Pajot B, Rahbi R, Rosenbauer M, Von Bardeleben KI, et al. (1995). Thermal Nitridation of p-Type Porous Silicon in Ammonia. *Thin Solid Films* 276: 32–35.

Mulloni V, Mazzoleni C, Pavesi L (1999). Elaboration, Characterization and Aging Effects of Porous Silicon Microcavities Formed on Lightly p-Type Doped Substrate. *Semiconductor Science and Technology* 14: 1052–1059.

Mulloni V, Pavesi L (2000). Porous Silicon Microcavities as Optical Chemical Sensors. *Applied Physics Letters* 76: 2523–2525.

Osorio E, Urteaga R, Acquaroli LN, García-Salgado G, Juaréz H, Koropecki RR (2011). Optimization of Porous Silicon Multilayers as Antireflection Coatings for Solar Cells. *Solar Energy Materials & Solar Cells* 95: 3069–3073.

Ott N, Nerding M, Müller G, Brendel R, Strunk HP (2004). Evolution of the Microstructure During Annealing of Porous Silicon Multilayers. *Journal Applied Physics* 95: 497–503.

Pacholski C (2013). Photonic Crystal Sensors Based on Porous Silicon. *Sensors* 13: 4694–4713.

Pap AE, Kordas K, Toth G, Levoska J, Uusimaki A (2005). Thermal Oxidation of Porous Silicon: Study on Structure. *Applied Physics Letters* 86: 041501.

Pavesi L (1997). Porous Silicon Dielectric Multilayers and Microcavities. *Rivista del Nuevo Cimento* 20: 1–76.

Pavesi L, Guardini R, Mazzoleni C (1996). Porous Silicon Resonant Cavity Light Emitting Diodes. *Solid State Communications* 97: 1051–1053.

Perez KS, Estevez JO, Mendez-Blas A, Arriaga J, Palestino G, Mora-Ramos ME (2012). Tunable Resonance Transmission Modes in Hybrid Heterostructures Based on Porous Silicon. *Nanoscale Research Letters* 7: 392.

Pickering C, Beale MIJ, Robbins DJ (1985). Optical Properties of Porous Silicon. *Thin Solid Films* 125: 157–163.

Qian M, Bao XQ, Wang LW, Lu X, Shao J, Chen XS (2006). Structural Tailoring of Multilayer Porous Silicon for Photonic Crystals Application. *Journal of Crystal Growth* 292: 347–350.

Reece PJ, Gal M, Tan HH, Jagadish C (2004). Optical Properties of Erbium-Implanted Porous Silicon Microcavities. *Applied Physics Letters* 85: 3363–3365.

Riley DW, Gerhardt RA (2000). Microstructure and Optical Properties of Submicron Porous Silicon Thin Films Grown at Low Current Densities. *Journal of Applied Physics* 87: 2169–2177.

Rodriguez I, Atienzar P, Ramiro-Manzano F, Meseguer F, Corma A, Garcia H (2005). Photonic Crystals for Applications in Photoelectrochemical Process: Photoelectrochemical Solar Cells with Inverse Opal Topology. *Photonics and Nanostructures—Fundamentals and Applications* 3: 148–154

Sailor MJ (2012). *Porous Silicon in Practice: Preparation, Characterization and Applications.* Weinheim: Wiley-VCH.

Selj JH, Marstein ES, Thegersen A, Foss SE (2011). Porous Silicon Multilayer Antireflection Coating for Solar Cells; Process Considerations. *Physica Status Solidi C* 8: 1860–1864.

Servidori M, Ferrero C, Lequien S, Milita S, Parisini A, Romestain R, Sama S, et al. (2001). Influence of the Electrolyte Viscosity on the Structural Features of Porous Silicon. *Solid State Communication* 118: 85–90.

Setzu S, Ferrand P, Romestain R (2000). Optical Properties of Multilayered Porous Silicon. *Materials Science and Engineering B* 69–70: 34–42.

Setzu S, Lerondel G, Romestain R (1998). Temperature Effect on the Roughness of the Formation Interface of p-Type Porous Silicon. *Journal of Applied Physics* 84: 3129–3133.

Snow PA, Squire EK, Russell PStJ, Canham LT (1999). Vapor Sensing Using the Optical Properties of Porous Silicon Bragg Mirrors. *Journal of Applied Physics* 86: 1781–1784.

Stenzel O (2005). *The Physics of Thin Films Optical Spectra.* Berlin: Springer.

Strashnikova MI, Voznyi VL, Reznichenko VYa, Gaivoronskii VYa (2001). Optical Properties of Porous Silicon. *Journal of Experimental and Theoretical Physics* 93: 363–371.

Svyakhoskiy SE, Maydykosky AI, Murzina TV (2012). Mesoporous Silicon Photonic Structures with Thousands of Periods. *Journal of Applied Physics* 112: 013106.

Theiss W (1996). The Dielectric Function of Porous Silicon-How to Obtain it and How to Use it. *Thin Solid Films* 276: 7–12.

Theiss W (1997). Optical Properties of Porous Silicon. *Surface Science Reports* 29: 91–192.

Thönissen M, Bergera MG, Billata S, Arens-Fischera R, Krügera M, Lütha H, Theißb W, et al. (1997). Analysis of the Depth Homogeneity of p-PS by Reflectance Measurements. *Thin Solid Films* 297: 92–96.

Torres-Costa V, Gago R, Martin-Palma RJ, Vinnichenko M, Grötzschel R, Martínez-Duart JM (2003). Development of Interference Filters Based on Multilayer Porous Silicon Structures. *Materials Science and Engineering C* 23: 1043–1046.

Torres-Costa V, Salonen J, Jalkanen T, Lehto V–P, Martín-Palma R J, Martínez-Duart JM (2009). Carbonization of Porous Silicon optical Gas Sensors for Enhanced Stability and Sensitivity. *Physica Status Solidi A* 206: 1306–1308.

Trifonov T, Garin M, Rodriguez A, Marsal L F, Alcubilla R (2007). Tuning the Shape of Macroporous Silicon. *Physica Status Solidi A* 204: 3237–3242.

Unno H, Imai K, Muramoto S (1987). Dissolution Reaction Effect on Porous-Silicon Density. *Journal of Electrochemical Society* 134: 645–648.

Vasin AV, OkholinI PN, Verovsky IN, Nazarov AN, Lysenko VS, Kholostov KI, Bondarenko VP, et al. (2011). Study of the Process of Carbonization and Oxidation of Porous Silicon by Raman and IR Spectroscopy. *Semiconductors* 45: 350–354.

Volk J, Grand TL, Bársony I, Gomköto J, Ramsden JJ (2005). Porous Silicon Multilayer Stack for Sensitive Refractive Index Determination of Pure Solvents. *Journal of Physics D: Applied Physics* 38: 1313–1317.

Wolf S (1990). *Silicon Processing for the VLSI Era. Vol II.* Sunset Beach, CA: Lattice Press.

Xu SH, Xiong ZH, Gu LL, Liu Y, Ding XM, Zi J, Hou XY (2002). Narrow-Line Light Emission from Porous Silicon Multilayers and Microcavities. *Semiconductor Science and Technology* 17: 1004–1007.

Yablonovitch E (1987). Inhibited Spontaneous Emission in Solid-state Physics and Electronics. *Physics Review Letter* 58: 2059–2062.

Yablonovitch E (1991). Photonic Band Structure: The Face-centered-cubic Case Employing Non-spherical Atoms. *Physical Review Letters* 67: 2295–2298.

Yablonovitch E (2001). Photonic Crystals: Semiconductors of Light. *Scientific American* 285(6): 47–55.

Zangooie S, Jansson R, Arwin H (1998). Reversible and Irreversible Control of Optical Properties of Porous Silicon Superlattices by Thermal Oxidation, Vapor Adsorption, and Liquid Penetration. *Journal of Vacuum Science and Technology A* 16: 2901–2912.

Zhang XG (2001). *Electrochemistry of Silicon and Its Oxide.* New York: Kluwer Academic.

Zhou Y, Snow PS, Russell PStJ (2000a). The Effect of Thermal Processing on Multilayer Porous Silicon Microcavity. *Physica Status Solidi A* 182: 319–324.

Zhou Y, Snow PA, Russell PStJ (2000b). Strong Modification of Photoluminescence in Erbium-Doped Porous Silicon Microcavities. *Applied Physics Letters* 77: 2440–2442.

Low-dimensional structures

2 Two-dimensional silicon

Guangzhao Qin, Han Xie, Ming Hu, and Hua Bao

Contents

2.1 INTRODUCTION

Graphene, a two-dimensional (2D) material with a honeycomb structure, exhibits some extraordinary physical properties and can, in principle, be considered as an elementary building block for all carbon allotropes (Lin et al. 2008). Developments in the science of graphene prompted an unprecedented surge of activity and demonstration of new physical phenomena. Despite its success, graphene still faces some severe problems in its nature of a semimetal or zero band gap semiconductor and its incompatibility with the current Si-based technology (Han et al. 2007). Given that the honeycomb geometry is related to some of the exceptional properties of graphene, there is strong motivation to investigate its analog, 2D Si, or silicene (Vogt et al. 2012). Silicene, exhibiting many superior properties such as a modulable gap due to the asymmetric structure, can solve the above problems smoothly and thus has received intense interest (Ni et al. 2011; Drummond et al. 2012). Given the fact that thermal transport plays a critical role in many applications such as heat dissipation in nanoelectronics and heat prohibition in thermoelectric (TE) energy conversion, there has been an emerging demand in characterizing the thermal (mainly phonon) transport property of silicene. Moreover, research results have shown that silicene exhibits a few novel thermal transport properties, which are fundamentally different from that of graphene, despite the similarity of their honeycomb lattice structure (Pei et al. 2013; Gu and Yang 2015; Xie et al. 2016). Therefore, the anomalous physical properties, primarily stemming from its unique low-buckled structure, may enable silicene to open entirely new possibilities for revolutionary electronic devices and energy conversion materials.

This chapter aims to present the state-of-the-art advances in silicene, in particular its formation (fabrication and synthesis), physical and chemical properties, and relevant emerging applications. The chapter is organized as follows. The recent advances in experimental investigations are presented in Section 2.2, focusing on the fabrication, structural characterization, and electronic properties. The theoretical investigations, including the structural, electronic, optical, thermal, and mechanical properties, are reviewed in Section 2.3. In Section 2.4, the potential applications of silicene in various fields are discussed.

2.2 EXPERIMENTAL FABRICATION AND CHARACTERIZATION OF SILICENE

Despite the fact that silicene was theoretically predicted back in 1994, the breakthrough in experimental synthesis of silicene started in 2010. Several groups reported successful fabrication of monolayer/few-layer silicene sheet on different substrates simultaneously (Lalmi et al. 2010; Song et al. 2010). In contrast to graphene usually displaying a honeycomb, nonreconstructed surface, silicene on various substrates often displays complicated surface superstructures. Following the successful fabrication, silicene attracted lots of research interest in the following few years (Song et al. 2010; Liu et al. 2011a; Ni et al. 2011). In this section, we will review the research on the experimental synthesis and characterization of monolayer/few-layer/nanoribbons of silicene on different substrates, including Ag(111), ZrB$_2$(0001), and so on. We will discuss in detail the surface structures, growth mechanism, and characterization of electronic properties, which are currently the focus of the experimental studies of silicene.

2.2.1 SUPPORTED SILICENE

Supported silicene is, in fact, a common form in real applications. For example, silicon atoms can be transferred onto an insulating substrate and gated electrically. Considering the sensitivity of Si to oxidation, it is necessary to fabricate silicene sheets in an ultrahigh vacuum (UHV) environment (Lalmi et al. 2010; Sone et al. 2014). With the technological advances in keeping environments clean and the easy control of the deposition coverage with sub-monolayer precision, molecular beam epitaxy (MBE) is becoming one of the most preferable approaches for fabricating silicene sheets via epitaxial growth of silicon on solid substrates (Prévôt et al. 2014; Salomon et al. 2014). Currently, with the simplest Si source involving a silicon bar cut from a silicon wafer and clamped between two electrodes, silicon can be deposited on a substrate by thermal evaporation. During the operation, stable silicon flux can be obtained with the silicon bar heated by passing a direct current, which is due to the sufficiently high vapor pressure below the melting temperature of silicene. The substrate can be a clean Ag(111) substrate that is achieved by a repeated argon ion sputtering and annealing process. Besides the single crystal Ag(111), Ag(111) films prepared on a mica substrate have also been used as the substrate for silicene fabrication (Tao et al. 2015). The advantage is that the Ag films can be more easily peeled off from the mica substrate and etched away. Moreover, Ag(111) film on mica is commercially available at a larger size and lower price. Such advantages make it a promising substrate for mass production of potential silicene devices.

In 2012, several works (Chen et al. 2012; Enriquez et al. 2012; Feng et al. 2012; Jamgotchian et al. 2012; Lin et al. 2012; Vogt et al. 2012) simultaneously reported the successful preparation of monolayer

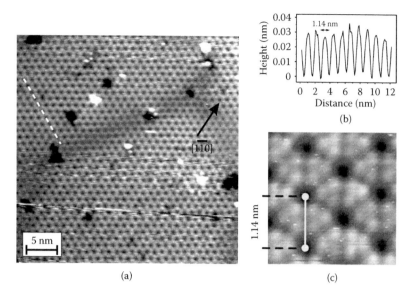

Figure 2.1 (a) Filled-states STM image of the 2D Si layer on Ag(111)-(1 × 1) (U_{bias} = −1.3 V, I = 0.35 nA). The honeycomb-like structure is clearly visible. (b) Line profile along the dashed white line indicated in (a). The dark centers in the STM micrograph are separated by 1.14 nm, corresponding to four times the Ag(111) lattice constant, in agreement with the (4 × 4) symmetry. (c) High-resolution STM topograph (3 × 3 nm, U_{bias} = −1.3 V, I = 0.35 nA) of the Si adlayer. (Reprinted with permission from Vogt, P., et al., *Phys. Rev. Lett.*, 108, 155501, 2012. Copyright 2016 by the American Physical Society.)

silicene sheets on Ag (111), as shown in Figure 2.1. Silicene was also successfully synthesized on another substrate, ZrB_2(0001), independently by Fleurence et al. (2012) around the same time. Soon afterwards, silicene monolayer sheets were also fabricated on various substrates, including ZrB_2(0001) (Fleurence et al. 2012, 2014; Friedlein et al. 2014; Aizawa et al. 2015), ZrC(111) (Aizawa et al. 2014), Ir(111) (Meng et al. 2013), and MoS_2 (Chiappe et al. 2014). Strong Si–Si covalent bonding throughout the silicene layer was demonstrated based on the analysis of electron localization functions. It is highly desirable to fabricate silicene on a nonmetallic substrate for the purpose of device applications. Using the MBE technique, 2D-Si nanosheets have been successfully fabricated on a semiconducting substrate of MoS_2 (Chiappe et al. 2014), which is important for its future applications in nanoelectronic, although silicene sheets on an MoS_2 surface are severely distorted from the freestanding form. Fabrications of silicene on substrates other than Ag(111) suggest that there are possible opportunities of finding suitable substrates for silicene growth, and even for germanene and stanene.

Besides these successful fabrications of silicene on various substrates in experiments, the interactions between silicene and many other metal substrates, such as Al, Mg, Au, Cu, Pt, Pb, Ca, and Ca-functionalized Si(111) surface, have been extensively explored from first principles (Morishita et al. 2013; Quhe et al. 2014; Pflugradt et al. 2014a; Podsiadły-Paszkowska and Krawiec 2015). The goals are to find other possible substrates for fabrications of silicene and to elucidate the influence of metal substrates on the electronic structure of silicene. Besides the metal substrates, there have also been some efforts from theoretical calculations devoted to searching for suitable nonmetal substrates for growth of silicene, such as Cl-passivated Si(111) and clean CaF_2(111) (Bhattacharya et al. 2013; Neek-Amal et al. 2013; Kokott et al. 2014). Das et al. explored the bonding, stability, and electronic structure of silicene epitaxially grown on various semiconductor substrates with bare surfaces of AlAs(111), AlP(111), GaAs(111), GaP(111), ZnS(111), and ZnSe(111) (Bhattacharya et al. 2013). It was found that silicene undergoes n-type (p-type) doping on the metal/nonmetal-terminated surface of the semiconductor substrates depending on the surface work function.

Due to the fact that the buckling of silicene lattice will change due to the interactions with the substrate, silicene on these substrates usually displays complicated surface superstructures, which is in contrast to the situation of graphene that usually displays a honeycomb, nonreconstructed surface (Lin et al. 2008). Depending on the Si coverage and substrate temperature, several silicene films superstructures can be formed on an Ag(111) surface (Chen et al. 2012; Chiappe et al. 2012; Enriquez et al. 2012; Feng et al. 2012; Jamgotchian et al. 2012; Lin et al. 2012; Vogt et al. 2012; Cinquanta et al. 2013; Resta et al. 2013; Sone et al. 2014; Xu et al. 2014b; Grazianetti et al. 2015; Zhuang et al. 2015). For example, silicon atoms deposited on Ag(111) tend to form clusters or disordered structures if the substrate temperature is below 400 K during the deposition (Feng et al. 2012). If the substrate temperature is above 400 K, the silicene sheets will typically have five ordered phases with increasing substrate temperature, from the so-called T phase to 4 × 4, $(\sqrt{13} \times \sqrt{13})$ R ± 13.9° $(\sqrt{13} \times \sqrt{13})$ for short, $2\sqrt{3} \times 2\sqrt{3}$, and finally to a $\sqrt{3} \times \sqrt{3}$ phase (Figure 2.2). Above all, other ordered structures of silicene have also been reported (Lalmi et al. 2010; Acun et al. 2013; Kawahara et al. 2014; Liu et al. 2014b). The underlying mechanism for silicene exhibiting so many reconstructions compared with the nonreconstructed graphene structure lies in its buckled lattice structure. The buckling patterns are rearranged in these phases and thus show apparently distinct scanning tunneling microscopy (STM) images although the honeycomb lattices are all preserved on Ag substrate in these phases (Gao and Zhao 2012).

To date, however, the reports of different phases of the silicene grown on Ag(111) and discussions of their structural models in literature remain controversial. For instance, the STM image of silicene sheet on Ag(111) surface with (4 × 4) superstructure reported by Vogt et al. (2012) shows that the silicene sheet sits on the top of the Ag(111) surface in such a way that a (3 × 3) silicene supercell coincides with

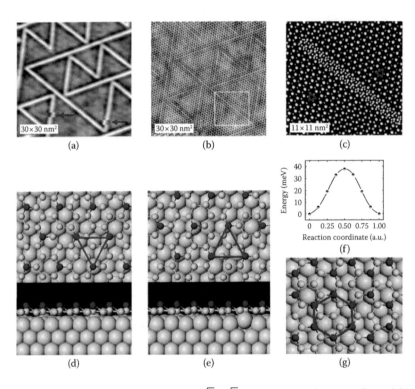

Figure 2.2 (a, b) The STM images of the same area on $(\sqrt{3} \times \sqrt{3})$R30° silicene taken at tip bias −2.0 V and 0.1 V, respectively, at 5 K. (c) The filtered high-resolution STM image with high contrast taken at 0.1 V. (d, e) Models of two energy-degenerated $\sqrt{3}$ reconstructed structures of silicene sheet on Ag(111) surface with an orientation angle of θ = 30°, which are obtained from DFT. Color code: Blue, yellow, and red spheres denote Ag atoms, Si atoms in lower layer, and Si atoms in higher layer, respectively. The red triangles denote the units of $\sqrt{3}$ silicene structures. (f) The interpolated potential energy curve for structural transition between the two mirror-symmetric $\sqrt{3}$ geometries on Ag(111). (g) The intermediate structure between the two rhombic $\sqrt{3}$ structures of silicene shown in (d, e). (Reprinted with permission from Chen, L., et al., *Phys. Rev. Lett.*, 110, 85504, 2013b. Copyright 2016 by the American Physical Society.)

a (4 × 4) supercell of the Ag(111) substrate, forming a so-called (4 × 4) superstructure (Figure 2.1). The Si(3 × 3)/Ag(4 × 4) is the first perfectly ordered phase observed in experiments, which exhibits triangular half-unit cells with each consisting of three bright spots quite similar to the well-known Si(111)-5 × 5 surface structure.

The interactions between silicene sheets and Ag(111) surface have been theoretically studied to better understand the growth mechanism of silicene sheets on the substrate (Gao and Zhao 2012; Kaltsas et al. 2012, 2014; Cahangirov et al. 2013, 2014; Guo et al. 2013a; Houssa et al. 2013, 2014; Lin et al. 2013; Wang and Cheng 2013; Chen and Weinert 2014; Guo and Oshiyama 2014, 2015; Johnson et al. 2014; Mahatha et al. 2014; Quhe et al. 2014; Scalise et al. 2014; Stephan et al. 2014; Yuan et al. 2014; Pflugradt et al. 2014b; Ishida et al. 2015). These studies focused on the issues of hybridization between Si and Ag atoms, the reconstructed silicene phases, the electronic band structures, the existence or not of a Dirac cone, and the structural evolution of silicene film versus coverage. For example, the continuous growth of high-quality silicene sheets was found to have benefited from the homogeneous interactions between silicon atoms and the Ag(111) surface (Gao and Zhao 2012; Tao et al. 2015). There are three typical stages for the epitaxial growth of silicene on metal surfaces: (1) Si atoms are deposited on the metal surface; (2) Si monomers or dimers aggregate into small Si clusters, most of which dissociate into monomers, dimers, or small pieces again; (3) Si clusters grow beyond nucleation size until they are covering the whole metal surface. The second stage (nucleation of Si) among these three stages is crucial for the understanding and controlling of silicene growth on a metal surface. The nucleation rate of Si atoms was found to be sensitive to both chemical potential and growth temperature (Shu et al. 2014). Moreover, various silicene superstructures can also be formed on the Ag(111) surface during the third stage (epitaxial growth of silicene), depending on the dose of deposited Si atoms and the substrate temperature (Kaltsas et al. 2012; Acun et al. 2013; Xu et al. 2014b). Above all, the diffusion barrier of the Si atom on the Ag(111) surface is another important factor affecting the growth of silicene sheets.

2.2.2 FEW-LAYER SILICENE

To date, there have been many theoretical studies from first principles on bilayer (Bai et al. 2010; Morishita et al. 2010; Pan et al. 2011; Fu et al. 2014; Naji et al. 2014; Liu et al. 2014a; Padilha and Pontes 2015; Sakai and Oshiyama 2015) and multilayer (Spencer et al. 2012; Kamal et al. 2013; Fu et al. 2015; Guo et al. 2015) silicene sheets, mainly focusing on the stacking patterns, interlayer interactions, and electronic structure properties.

In experiments, few-layer silicene can only be fabricated on an Ag(111) substrate so far (Feng et al. 2012; De Padova et al. 2013a, 2013b; Resta et al. 2013; Chen et al. 2014; De Padova et al. 2014; Salomon et al. 2014; Sone et al. 2014; Vogt et al. 2014; Fu et al. 2015; Johnson et al. 2015; Zhuang et al. 2015). The work by Feng et al. (2012) was the first that reported the existence of bilayer and multilayer silicene films (Figure 2.3). The surfaces were found to exhibit a thickness-independent $\sqrt{3} \times \sqrt{3}$ honeycomb superstructure with respect to the Si(111) lattice. With the monolayer silicene serving as a buffer layer, few-layer silicene film can be successively grown on the top of the initial silicene surface, displaying a $\sqrt{3} \times \sqrt{3}$ superstructure (Salomon et al. 2014; Vogt et al. 2014), of which the mechanism has also been theoretically discussed (Guo and Oshiyama 2014; Pflugradt et al. 2014c).

For few-layer silicene, debate still exists about whether the $\sqrt{3} \times \sqrt{3}$ superstructure on the top surface is pristine silicon structure, or just a well-known $\sqrt{3} \times \sqrt{3}$ phase formed by Ag atoms on a bulk Si(111) substrate. Recently, the structure of the surface was found identical to the well-known inequivalent triangle model of Si(111)-$\sqrt{3} \times \sqrt{3}$-Ag from the low-energy electron diffraction (LEED) intensity versus voltage (IV) measurement by Shirai et al. (2014). In fact, the $\sqrt{3} \times \sqrt{3}$ phase and Si(111)-$\sqrt{3} \times \sqrt{3}$-Ag surface actually shows very similar features from STM experiments. It was also argued that Si grown on Ag(111) substrate at high temperatures are not few-layer silicene, but cubic diamond-like silicon instead (Borensztein et al. 2015). Considering the fact that the stacking structure of few-layer silicene is quite close to bulk Si(111), it was suggested that the $\sqrt{3} \times \sqrt{3}$ phase is due to the segregation of the Ag atom to the Si(111) film surface, which means that the so-called "few-layer silicene" would be just a Si(111) film with an Ag-terminated surface. However, this argument contradicts many experimentally observed facts. For example, it cannot explain the Dirac cone in few-layer silicene on Ag(111) observed by APRES (De Padova 2013a, 2013b; Zhuang et al. 2015) which

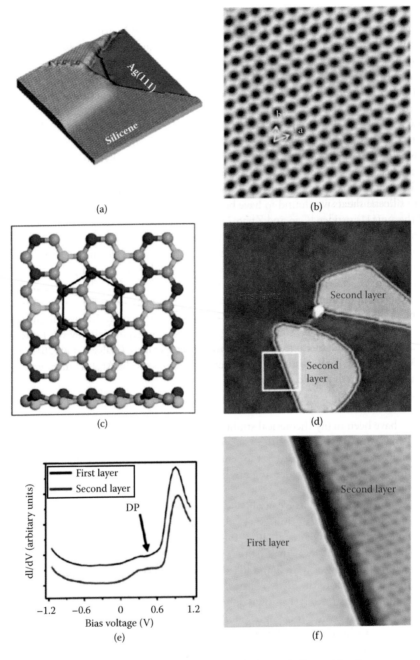

Figure 2.3 (a) 3D STM image (30 × 30 nm², V_{tip} = 1.0 V) of a single layer of silicene island across a step edge of Ag(111). (b) High-resolution STM image (8 × 8 nm², V_{tip} = 1.2 V) of one monolayer silicene terrace showing the $\sqrt{3} \times \sqrt{3}$ honeycomb superstructure with the period of 0.64 nm. (c) Top and side views of schematic model of $\sqrt{3} \times \sqrt{3}$ superstructure of silicene. The red, gray, and green balls represent the upper buckled, in plain, and lower buckled Si atoms, respectively. The $\sqrt{3} \times \sqrt{3}$ honeycomb superstructure is indicated by the black hexagon. (d) STM image (54 × 54 nm², V_{tip} = 1.5 V) of 1.2 ML silicon atoms deposited on Ag(111) surface at substrate temperature of 500 K showing the second layer of silicene formed on the first layer of silicene. (e) dI/dV spectra taken on the first (black) and second (red) layers of silicene, respectively. (f) High-resolution STM image (10.5 × 10.5 nm², V_{tip} = 1.5 V) of area as marked by the white rectangle in (d) showing atomic structure of the first and second layers of silicene simultaneously. (Reprinted with permission from Feng, B., et al., 2012, 3507–3511, Copyright 2016 American Chemical Society.)

is definitely different from the well-known angle-resolved photoemission spectroscopy (ARPES) spectrum of the Si(111)- $\sqrt{3} \times \sqrt{3}$-Ag surface. These contradicting experimental results might arise from the possible existence of a narrow temperature window for few-layer silicene growth, where Ag atoms do not segregate to the surface.

For monolayer silicene, recent studies revealed a strong influence of the substrate on its electronic structure properties, so the Dirac cone could no longer exist if it is fabricated on metal substrates, which is challenging for further researches and applications of silicene. Nevertheless, few-layer silicene is more favorable because the Dirac-like electronic structure properties stemming from the $\sqrt{3} \times \sqrt{3}$-reconstructed surface are more robust, and the metal substrates are isolated by the underlying Si layers. Apart from that, the $\sqrt{3} \times \sqrt{3}$ surface of few-layer silicene was found to be surprisingly stable when exposed to air (24h) (De Padova et al. 2014). Therefore, few-layer silicene is expected to play a significant role in future applications.

Moreover, the fabrication of a novel form of bilayer silicene (denoted as w-BLSi) from calcium-intercalated monolayer silicene ($CaSi_2$) was recently reported by Yaokawa et al. (2016). Compared with monolayer silicene, the number of unsaturated silicon bonds in the w-BLSi is reduced, and an indirect band gap of 1.08 eV is opened. This progress provides a new way to synthesize few-layer silicene on Ag(111) substrate in addition to the conventional MBE method.

2.2.3 SILICENE NANORIBBONS

Due to the quantum confinement of the finite width, tailoring graphene into nanoribbons makes it possible to open an energy band gap (Han et al. 2007; Jia et al. 2009). Similarly, one may also expect such an effect in silicene nanoribbons (SiNRs). Besides, the buckled structure of silicene (different from the planar structure of graphene) makes it possible to tune the electronic property by an external field. Combining the effects due to finite size and an external field, SiNRs have been theoretically predicted to possess numerous extraordinary properties, such as size-dependent energy gap (Cahangirov et al. 2009; An et al. 2013a), magnetic ordering (Cahangirov et al. 2009), and unusual transport properties (Kang et al. 2012; Ezawa 2013). SiNRs have been found to be formed spontaneously and perfectly ordered on Ag(001) (Léandri et al. 2007), Ag(110) (Léandri et al. 2005; Kara et al. 2009; Aufray et al. 2010; De Padova et al. 2010, 2011; De Padova et al. 2012b; Bernard et al. 2013; Colonna et al. 2013; Tchalala et al. 2014; Feng et al. 2016), and Au(110) surfaces (Tchalala et al. 2013), which is in contrast to the difficulty in obtaining graphene nanoribbons. In fact, the experimental reports of SiNRs on Ag(001) and Ag(110) substrates were even prior to silicene sheets on Ag(111) substrate, although whether these SiNRs are pure and pristine remains an open question. Below we will mainly focus on the experimental results of SiNRs on Ag(110) substrate, including structure models, electronic properties, and the existing debates in these systems (Figure 2.4).

Two types of SiNRs can be formed on Ag(110) substrate for a wide range of substrate temperature from 300 K to about 500 K. At room temperature, SiNRs with two rows of protrusions are found in STM images, exhibiting identical width of ~1.0 nm, and length from a few nanometers to tens of nanometers (Bernard et al. 2013; Colonna et al. 2013; Feng et al. 2016). When the substrate temperature is raised, the second type of SiNRs with four rows of protrusions are found in the STM image, which will dominate the surface with the substrate temperature greater than 480 K. The width of the second SiNRs is 2.0 nm and the length can extend up to hundreds of nanometers, which is mainly limited by the size of the Ag(110) substrates terraces. The ARPES measurements of surface fully covered by 2.0 nm SiNRs gratings were reported by De Padova et al. (2010). Both types of SiNRs are oriented along the [110] direction of the Ag(110) surface, which can also pack tightly in the [100] direction and fully cover the surface if properly annealed, forming monolayer SiNRs gratings. If the Si coverage is further increased, few-layer structures will be formed preserving the highly anisotropic shapes (De Padova et al. 2012a; Colonna et al. 2013).

Commensurate structures of SiNRs are formed on Ag(110) substrate, as evidenced by both the STM images and the 5 × 2 LEED patterns (in the case of 2 nm SiNRs), which is unlike the possibly incommensurate structures of monolayer silicene on Ag(111) substrate. Due to the smaller lattice constant (3.86 Å) of freestanding silicene sheets compared to the column–column distance (4.08 Å) on the Ag (110) surface along the [001] direction, tensile stress is caused by the mismatch between the lattices of silicene and Ag substrate, which increases with the increasing width of the SiNRs along the Ag[001] direction. The tensile stress might be responsible for the limited width, narrower than 2.0 nm, of SiNRs on the Ag(110) surface observed in experiments.

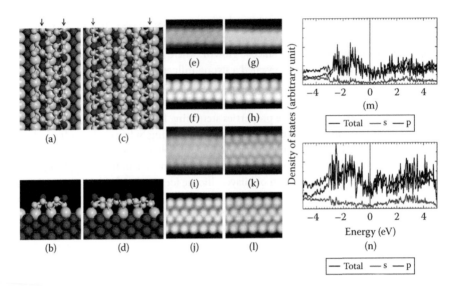

Figure 2.4 (a, b) Top and side view of the relaxed structural model of 1 nm wide SiNRs on top of Ag(110). (c, d) Top and side view of the relaxed structural model of 2 nm wide SiNRs on top of Ag(110). Light blue balls: topmost Ag atoms; dark blue balls: underlying Ag atoms; red balls: upper buckled silicon atoms that can be probed by STM; yellow balls: other silicon atoms. (e–l) Simulated STM images compared with the experimental STM images at different bias voltages. Bias voltage: (e) 1.1 V; (g) –1.0 V; (i) -1.5 V; (k) 1.0 V; (f, j) 0–1.5 V; (h and l) –1.5 to 0 V; (m, n). Calculated partial density of states of the relaxed SiNRs without Ag(110) for the 1 nm and 2 nm SiNRs, respectively. (Reprinted from *Surface Science*, 645, Feng, B., et al., Structure and quantum well states in slicene nanoribbons on Ag (110), 74–79, Copyright 2016, with permission from Elsevier B.V.)

There are some debates on the morphology during the growth of Si atoms on Ag substrates. For example, the morphology of the Ag(110) substrate was reported to be substantially modified during the growth of SiNRs (Bernard et al. 2013; Ronci et al. 2014), suggesting possible alloying of Si atoms with Ag substrates. The situation is similar to the case of silicene sheets on Ag(111) surface, where the Ag substrate was also modified during the growth (Prévôt et al. 2014). However, some theoretical calculations and extensive experiments proved the validity of the pristine silicene model consisting of 3 × 3 buckling Si structures on Ag(111) without alloying with substrates. Note that Si atoms desorb completely when annealing to ~700 K on both the Ag(110) and Ag(111) substrates. Further studies are demanded to make clear the composition of the SiNRs on Ag(110) substrate.

Another issue is the atomic structure of the SiNRs, because the existing structural models based on pure and unreconstructed SiNRs interpreted as a cut of silicene sheets cannot explain the observed STM images. In addition, a recent Raman study also indicated that the excitation spectrum is not consistent with the honeycomb silicene structure (Speiser et al. 2014). SiNRs with reconstructed edges could be a possible explanation to this puzzling issue. Based on the evaluation of a range of new and existing reconstruction models for SiNRs formed on Ag(110) using a combination of STM imaging and total energy calculations within density functional theory (DFT), two models were proposed consistent with experimental estimates of the Si coverage and the previously reported formation of Ag missing rows (Hogan et al. 2015). The models are thermodynamically stable and yield STM images in excellent agreement with experiment, which provides clear evidence for a strong bond Si–Ag reconstruction on Ag(110) substrate.

2.2.4 CHARACTERIZATION OF ELECTRONIC PROPERTIES

Following the successful preparation of monolayer and few-layer silicene sheets on various substrates, the question of interest is whether silicene indeed exhibits the theoretically predicted exotic electronic structure, especially the Dirac electronic state (readers may refer to Section 2.3.2 to find more details about the theoretical investigations on the electronic structure). To date, ARPES and scanning tunneling microscopy (STM)/scanning tunneling spectroscopy (STS) are two major spectroscopic techniques, which were used to characterize the electronic structure of silicene.

2.2.4.1 Monolayer silicene

Together with the obtained silicene structure, possible existence of a Dirac cone and linear dispersion in the 4 × 4 phase of silicene was reported by Lay et al. from ARPES measurements (Vogt et al. 2012). By using a compiled set of LEED, high-energy resolution core levels, and ARPES data, Avila et al. (2013) studied the electronic structures of the 3 × 3 silicene (4 × 4 using the present notion) on Ag(111) substrate. The existence of a silicene-derived band at several Γ points in the 2D Brillouin zones was demonstrated by the ARPES spectra, which highlighted two bands with linear dispersions crossing over each other at 0.3 eV below the Fermi level, and have a Fermi velocity of ~1.3 × 10^6 m/s. The linear dispersion was further confirmed to arise from silicene rather than the Ag substrate based on the analysis of the momentum distribution curves taken at different photon energies of the bands around the $\Gamma_{3\times3}$ points (Avila et al. 2013) (Figure 2.5).

Soon after the first ARPES experimental result for 4 × 4 silicene on Ag(111) substrate, the bone of contention is whether the Dirac cone in freestanding silicene can be preserved when fabricated on Ag(111) substrate (Vogt et al. 2012; Avila et al. 2013; Cahangirov et al. 2013; Guo et al. 2013a; Lin et al. 2013; Wang and Cheng 2013; Chen and Weinert 2014; Johnson et al. 2014; Mahatha et al. 2014; Quhe et al. 2014; Yuan et al. 2014; Pflugradt et al. 2014b; Ishida et al. 2015). It is argued that once silicene sheet is placed on the Ag(111) substrate, the strong interactions between the silicene and the Ag substrate may significantly influence the electronic structure. The atomic and electronic structures of five distinct epitaxial silicene morphologies on Ag(111) substrate were examined using the complementary techniques of DFT and soft X-ray spectroscopy (Johnson et al. 2014). It was found that the *p*-states of Si atoms span the Fermi level due to the strong *sp*^{3}-like hybridization between Si atoms and the *d*-states of Ag as well as the presence of the Ag *sp*-states on the substrate. As a consequence, epitaxial silicene on the Ag(111) substrate is metallic and does not possess the Dirac cone-like electronic structure. The argument that the Dirac cone of silicene is destroyed while fabricated on an Ag(111) substrate due to strong interactions and mixing between Si and Ag orbitals is also supported by DFT calculations from different groups. The linear dispersions observed in

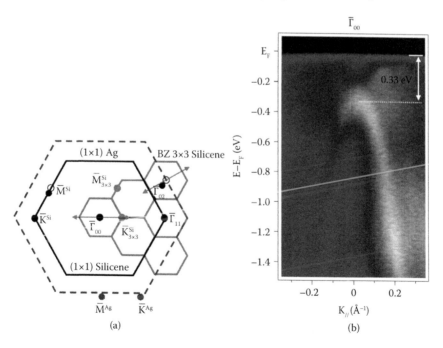

Figure 2.5 Experimental valence band electronic structure of a (3 × 3) silicene sub-monolayer film on the Ag(111) surface. (a) BZs of the (1 × 1) unreconstructed silicene and surface (111) silver lattices, together with the SBZ of the (3 × 3) supercell. (b) ARPES spectrum measured at the $\bar{\Gamma}_{00}$ point, along the *K*–Γ–*K* direction of the (3 × 3) lattice. The arrows indicate the directions in which the ARPES measurements have been carried out at each symmetry point of the reciprocal space. (The source of the material Avila, J., et al., Presence of gapped silicene-derived band in the prototypical (3 × 3) silicene phase on silver (111) surfaces, *J. Phys. Condens. Matter*, 25, 262001, 2013, IOP Publishing, is acknowledged.)

Low-dimensional structures

experiments are suggested to originate either from the Ag(111) substrate or from the hybridization states between Si and Ag atoms (Cahangirov et al. 2013; Gori et al. 2013; Lin et al. 2013; Wang and Cheng 2013; Guo et al. 2013a, 2013b; Chen and Weinert 2014; Mahatha et al. 2014; Pflugradt et al. 2014b; Quhe et al. 2014; Yuan et al. 2014; Ishida et al. 2015). However, 12 Dirac cones presented at the edges of the Brillouin zone were directly observed in 4×4 silicene sheet on Ag(111) by ARPES recently (Feng et al. 2016b), which is in contrary to the theoretical expectation for freestanding silicene that there are 6 Dirac cones at the K (K') points of the BZ. The observed unusual Dirac cone electronic structure is considered to come from the interaction at the silicene–Ag(111) interface, which may open up new possibilities for investigating quantum phenomena in low-dimensional systems.

In contrast to the debates on the electronic structure of 4×4 silicene on Ag(111) substrate, the $\sqrt{3} \times \sqrt{3}$ phase was revealed exhibiting a metallic surface state generating quasiparticle interference (QPI) patterns from the low-temperature electronic structure STS measurements (Chen et al. 2012). The decaying behavior of the QPI patterns was also measured. A large decaying factor was found, different from conventional 2D gas systems, which is typical for backscattering-suppressed systems like a topological insulator or graphene (Feng et al. 2013). The large QPI-decaying factor provides further evidences for the Dirac fermions in the $\sqrt{3} \times \sqrt{3}$ phase. While for other monolayer silicene phases on substrates that are also expected to be metallic, the QPI patterns have never been observed. Moreover, plenty of other intriguing phenomena have also been observed in the $\sqrt{3} \times \sqrt{3}$ phase of monolayer silicene, such as the low-temperature phase transition (Chen et al. 2013b) and a superconducting-like gap at zero bias (Chen et al. 2013a). All these results demonstrated that silicene is a promising system possessing exotic electronic properties.

In addition to the ARPES and STS results as presented above, the atomic structures, chemical bonding, vibrational properties, and electronic structures of epitaxial silicene sheets on Ag(111) substrate have also been characterized by other spectroscopic methods, such as X-ray photoemission spectroscopy (XPS) (Chiappe et al. 2012; Molle et al. 2013; Grazianetti et al. 2015), soft X-ray spectroscopy (Johnson et al. 2014), reflection high-energy positron diffraction (Fukaya et al. 2013), Raman (Cinquanta et al. 2013, 2015; Molle et al. 2013; Tao et al. 2015; Zhuang et al. 2015), and extended X-ray absorption fine structure (EXAFS) (Lagarde et al. 2016). For example, the Raman spectra of various silicene phases on Ag(111) substrate were obtained by different groups (Cinquanta et al. 2013; Molle et al. 2013; Tao et al. 2015; Zhuang et al. 2015), and the Raman spectra of the encapsulated silicene sheets were interpreted by comparison with the DFT-simulated spectra. The Raman spectra of the $\sqrt{13} \times \sqrt{13} / 4 \times 4$ phase of epitaxial silicene on Ag(111) substrate with different structures and coverages measured from *in situ* is consistent with that from *ex-situ* results (Cinquanta et al. 2013; Molle et al. 2013; Zhuang et al. 2015). Unlike the case of freestanding low-buckled silicene possessing only one peak (Scalise 2014), more Raman active modes are found. The E_{2g} mode around 530 cm^{-1} is the major peak while five other Raman peaks in the range of 200–500 cm^{-1} (marked as D_1–D_5) exist in the low-coverage samples. Based on the analysis for the shift of Raman peaks, it was concluded that the electron–phonon coupling (EPC) in silicene can be significantly affected by the tensile strain due to the lattice mismatch between silicene and substrate as well as the charge doping from the substrate.

2.2.4.2 Few-layer silicene

As reviewed in Section 2.2.2 few-layer silicene can only be fabricated on Ag(111) substrate, which is different from the monolayer silicene that has been succesfully grown on many substrates. The electronic properties of bilayer silicene with $\sqrt{3} \times \sqrt{3}$ superstructure were first examined by STS (Chen et al. 2012; Feng et al. 2013). The existence of Dirac fermions is supported by the linear energy–momentum dispersion relation, which is extracted from the pronounced QPI based on the dI/dV maps at different tunneling bias voltage. Similar QPI patterns from all films with different thicknesses were observed on the surface of few-layer silicene films with thickness up to 40 layers, indicating that the metallic surface state on the Si(111) surface is delocalized and originates from the $\sqrt{3} \times \sqrt{3}$ superstructure on the surface rather than from the Ag(111) substrate (Chen et al. 2014). Using ARPES technique, De Padova et al. investigated the electronic structures of few-layer silicene sheets with different thicknesses (De Padova et al. 2013a, 2013b). The Dirac cone structure at the Γ point of the surface Brillouin zone in silicene films with different thicknesses was observed, which is different from monolayer silicene on substrate where its electronic structure may be significantly affected by the substrate. The presence of a gapless Dirac cone is clearly shown in Figure 2.6 where the quasilinear π^* and π bands

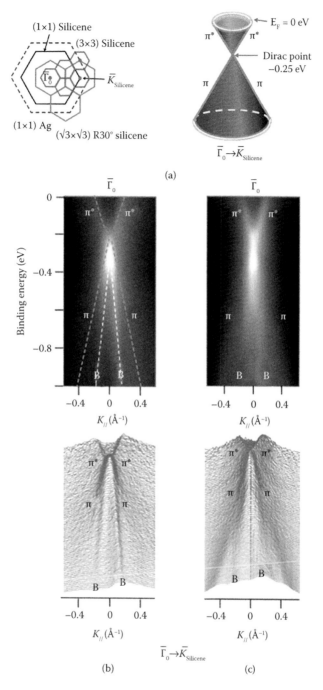

Figure 2.6 Dirac cones for multilayer ($\sqrt{3} \times \sqrt{3}$) R30° silicene islands on top the first (3 × 3) wetting layer. (a) Left: scheme of the Brillouin zones of (1 × 1) Ag(111) (blue hexagon), (1 × 1) standalone silicene (black hexagons), ($\sqrt{3} \times \sqrt{3}$) R30° silicene (green hexagons) and ($\sqrt{3} \times \sqrt{3}$) silicene (red hexagons). Right: scheme of a Dirac cone, with C- and V-shaped linear $\pi^\star$ and π silicene bands, where the green lines represent the dispersion along the $\bar{\Gamma}_0 \rightarrow \bar{K}_{silicene}$ direction. (b ,c) Top: band dispersions around, measured along the $\bar{\Gamma}_0 \rightarrow \bar{K}_{silicene}$ direction, collected at 1.5 ML, (3 × 3) plus ($\sqrt{3} \times \sqrt{3}$) symmetries in LEED patterns, and 3 MLs (dominant $\sqrt{3} \times \sqrt{3}$ symmetry). Bottom: their waterfall line profiles. For clarity, we traced the dashed darker lines in (b) (not in c), which are the C- and V-shaped linear $\pi^\star$ and π silicene bands, and the dashed white lines B, which are the structures related to the (3 × 3) symmetry. (The source of the material De Padova, P., et al., The quasiparticle band dispersion in epitaxial multilayer silicene, *J. Phys. Condens. Matter*, 25, 382202, 2013a, IOP Publishing, is acknowledged.)

cross at 0.25 eV below the Fermi level. Similar results were obtained (0.33eV below the Fermi level) from a recent ARPES study of $\sqrt{3} \times \sqrt{3}$ silicene layer on top of $\sqrt{13} \times \sqrt{13}$ / 4 × 4 buffer layers grown on Ag(111) substrate (Zhuang et al. 2015). Such bands should be ascribed to the silicene due to their not coinciding with any of the bands of Ag(111) substrate. The thickness-independent character further indicates that the Dirac cone stems from the top surface layer of silicene. However, there are still debates on the issue. Johnson et al. (2015) demonstrated that the epitaxial silicene bilayer on the Ag(111) substrate is metallic and strongly interacts with the underlying substrate, showing no Dirac cone around the Fermi level.

For silicene, the partially sp^2/sp^3 hybridized Si bonds result from the buckled structure, which is different from the sp^2-hybridized C bonds in graphene with planar structure. It is well known that sp^3-hybridization favors a diamond-like structure, and it has been indicated from STM images that the interlayer spacing in few-layer silicene is about 3.14 Å, which is very close to the interlayer spacing of bulk Si(111) planes with diamond-like structure. Besides, most silicene films are of ABC stacking as revealed by high-resolution STM images, which is different from the AB stacking in graphite (Chen et al. 2014). These results suggest strong interactions between silicene layers, although whether it has a diamond-like structure is still under debate. The sp^3 nature of Si bonds is largely relaxed in few-layer silicene as indicated by Raman studies (Zhuang et al. 2015). Moreover, as opposed to weak van der Walls interactions between graphene layers, strong covalent bonding between the silicene layers is found in few-layer silicene, in which there is no Dirac cone (Bai et al. 2010; Morishita et al. 2010; Pan et al. 2011; Spencer et al. 2012; Kamal et al. 2013; Fu et al. 2014; Liu et al. 2014a; Naji et al. 2014; Padilha and Pontes 2015; Sakai and Oshiyama 2015).

2.3 THEORETICAL INVESTIGATIONS ON THE PHYSICAL PROPERTIES OF SILICENE

In spite of the experimental efforts that have been devoted to the growth and characterization of supported silicene, systematic measurements on its physical properties are still difficult. Most of the investigations on the physical properties are still based on theoretical approaches. Moreover, theoretical methods allow us to study the properties of freestanding silicene, which has not yet been synthesized in experiments. These works investigate various aspects of physical properties of silicene, including its structure, electronic, optical, thermal, and mechanical properties, and are reviewed below.

2.3.1 STRUCTURE

From both theoretical investigation and experimental observations, silicene has a honeycomb structure similar to graphene (Takeda and Shiraishi 1994; Cahangirov et al. 2009; Şahin et al. 2009; Aufray et al. 2010; Kara et al. 2010; Enriquez et al. 2012; Fleurence et al. 2012; Meng et al. 2013). In contrast to the planar structure of graphene, theoretical calculations based on DFT predict that a buckled structure with the neighboring atoms in different planes is more stable for silicene (Takeda and Shiraishi 1994; Cahangirov et al. 2009; Şahin et al. 2009).

Freestanding silicene has only been investigated with theoretical methods because its experimental synthesis is not available yet. In 1994, Takeda and Shiraishi (1994) employed first-principles calculations and predicted that silicene would prefer the low-buckled structure rather than the planar one. They found that the total energy reached a local minimum with a lattice constant of 3.855 Å and a deformation angle of 9.9°. After more than one decade, Cahangirov et al. (2009) and Şahin et al. (2009) applied first-principles calculations to planar, low-buckled, and high-buckled honeycomb structures of silicene. The planar and buckled honeycomb structures are shown in Figure 2.7a, with Δ indicating the buckling distance between the two planes. Planar and high-buckled silicene were found to have imaginary frequencies in their dispersion, therefore were not stable. Low-buckled silicene was found to be stable with an equilibrium lattice constant of 3.83 Å and an equilibrium buckling distance of 0.44 Å. As the silicon counterpart of graphene, silicene has a weaker π bonding because of the longer bond length. The sp^2 bonding will be dehybridized into sp^3-like bonding (Cahangirov et al. 2009; Şahin et al. 2009). Therefore, silicene cannot have a complete planar structure as graphene.

In addition to freestanding silicene, supported silicene has also been investigated by DFT calculations (Aufray et al. 2010; Kara et al. 2010; Enriquez et al. 2012; Fleurence et al. 2012; Meng et al. 2013). Guo-Min He (2006) was the first to study SiNRs on Ag(110) with first-principles DFT calculations. Silicon atoms were

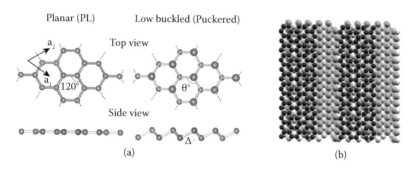

Figure 2.7 (a) Planar and low-buckled honeycomb structures. (Reprinted with permission from Şahin, H., et al., 2009, Monolayer honeycomb structures of group-IV elements and III-V binary compounds: First-principles calculations, *Phys. Rev. B*, 80, 155453. Copyright 2016 by the American Physical Society.) (b) Top view of silicene nanoribbon on Ag(110). The silicon atoms are indicated as red and the surface silver atoms as green. (The source of the material Kara, A., et al., Silicon nano-ribbons on Ag (110): A computational investigation, *J. Phys. Condens. Matter*, 22, 45004, 2010, IOP Publishing, is acknowledged.)

proposed to form a rectangular arrangement. However, this was not compatible with experimental observation. Later, Kara et al. (2010) found that the starting configuration for SiNR on Ag(110) can be either hexagonal or rectangular. On a 4 × 6 Ag(110) configuration, Kara et al. (2010) found that rectangular configuration would be relaxed to a hexagonal structure, which was in agreement with experimental results. The top view of the SiNR honeycomb structure on Ag(110) is shown in Figure 2.7b. Aufray et al. (2010) obtained SiNR in a honeycomb arrangement on Ag(110) in experiment and DFT calculations with generalized gradient approximation (GGA) justified the experimental observation. In addition to the Ag(110) substrate, DFT calculations have also been used to confirm the honeycomb structure of silicene synthesized on other substrates, as described in Section 2.2.1.

Furthermore, the strain effect on the variation of freestanding silicene structure parameters has also been investigated with DFT calculations (Liu et al. 2012; Zhao 2012; Peng et al. 2013; Wang et al. 2014a; Xie et al. 2016). The bond length was found to increase with increasing tensile strain, regardless of uniaxial or biaxial strain (Liu et al. 2012; Zhao 2012; Peng et al. 2013; Xie et al. 2016). Zhao (2012) found that the buckling distance would decrease with increasing uniaxial strain in the range of 0 to 0.2, either for armchair or zigzag uniaxial strain. Peng et al. (2013) also observed decreasing buckling distance in the range of –0.09 to 0.08 uniaxial strain, without any chirality effects. The minus sign here denotes compressive strain. The buckling distance would stay unchanged from 0.08 to 0.15 and then decrease with even greater uniaxial strain. Similarly, Wang et al. (2014a) found that under zigzag strain, the buckling distance kept decreasing in the range of 0 to 0.2. Planar structure with zero buckling was attained under 0.2 zigzag strain and was predicted to be stable. Under armchair strain, the buckling distance showed an anomalous behavior, first decreased, then increased, then decreased again to a minimum value, and finally increased to a maximum value at 0.4 zigzag uniaxial strain.

Under biaxial tensile strain, Liu et al. (2012) found that the buckling was reduced from 0.46 Å to 0.32 Å when the strain was increased from 0 to 0.125. Peng et al. (2013) showed that the buckling would first decrease and then increase with the critical transition point near 0.08. Xie et al. (2016) observed that in the range of 0 to 0.1 biaxial strain, the buckling would first decrease and then stay almost unchanged, as show in Figure 2.8. Generally, silicene would become more planar (Xie et al. 2016) with increasing strain in the range of 0 to 0.1, regardless of uniaxial or biaxial strain. However, it should be noted that under very large strain, the structure may fracture.

2.3.2 ELECTRONIC AND OPTICAL PROPERTIES

As the silicon counterpart of graphene, silicene exhibits some similar electronic properties to those of graphene, such as massless Dirac fermion, quantum spin Hall effect (QSHE), and high Fermi velocity. However, unlike graphene, silicene has a tunable band gap. Optical properties can be derived from electronic band structure and they are discussed together with the aforementioned electronic properties in this part.

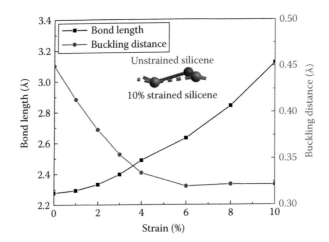

Figure 2.8 Bond length and buckling distance versus strain. The inset shows the primitive unit cell structures for unstrained and 10% strained silicene. (Reprinted with permission from Xie, H., et al., 2016, Large tunability of lattice thermal conductivity of monolayer silicene via mechanical strain, *Phys. Rev. B*, 93, 075404. Copyright 2016 by the American Physical Society.)

2.3.2.1 Massless Dirac fermion

Similar to graphene, the charge carrier in freestanding silicene is characterized as massless Dirac fermion (Guzmán-Verri and Voon 2007; Cahangirov et al. 2009). Guzmán-Verri and Lew Yan Voon (2007) developed a tight-binding approach to investigate the electronic properties of planar and buckled freestanding silicene (indicated as silicene and Si(111), respectively, in the reference paper). The band gaps for both structures are predicted to be zero with a Dirac cone near K point and electrons in both structures resemble massless Dirac fermions. Later, Cahangirov et al. (2009) adopted first-principles calculations to show the stability of low-buckled freestanding silicene and calculated its band structure. In the band structure, π and π^* bands cross at the Fermi level at K and K' points in the Brillouin zone. The linear behavior in the vicinity of crossing points, as shown in Figure 2.9, implies the massless Dirac fermion characteristic of the charge carrier. It should be noted that although the lattice parameters might be different for Guzmán-Verri's and Cahangirov's simulations, both reach the same conclusion of massless Dirac fermion.

For supported silicene on substrates, whether Dirac cones exist or not is a matter of debate (Liu et al. 2011a; Liu et al. 2011b; Chen et al. 2012; Arafune et al. 2013; Chen and Weinert 2014; Mahatha et al. 2014; Ishida et al. 2015). For example, Chen et al. (2012) previously synthesized $\left(\sqrt{3} \times \sqrt{3}\right)$ SiNR on

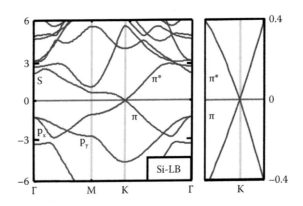

Figure 2.9 Band structure of low-buckled silicene. The crossing at K or K' points are amplified to show their linear behavior. (Reprinted with permission from Cahangirov, S., et al., 2009, Two- and one-dimensional honeycomb structures of silicon and germanium, *Phys. Rev. Lett.*, 102, 236804. Copyright 2016 by the American Physical Society.)

Ag(111) substrate and proved the existence of massless Dirac fermions. Later, Arafune et al. (2013) commented on this work and argued that the dispersion relation was not linear. Some other recent works also show that the Dirac cone does not exist for (3 × 3) silicene (Chen and Weinert 2014) and (4 × 4) silicene (Mahatha et al. 2014; Ishida et al. 2015) on Ag(111). Similar to the case of silicene fabricated on Ag(111), strong interaction between silicene and $ZrB_2(0001)$ or ZrC(111) substrates is also revealed by DFT calculations, so that silicene will lose its Dirac cone when it is synthesized on these two substrates (Lee et al. 2013, 2014; Aizawa et al. 2014).

2.3.2.2 Quantum spin Hall effect

In the discussions on massless Dirac fermion reviewed above, a spin-orbit coupling (SOC) effect is not included in numerical calculations. Inclusion of SOC effect shows that a small band gap exists in free-standing silicene (Liu et al. 2011a; Liu et al. 2011b; Drummond et al. 2012), which induces QSHE. Liu et al. (2011a) carried out first-principles calculations based on DFT with Perdew–Burke–Ernzerhof (PBE) exchange correlation for planar and low-buckled silicene. The band gap opened by SOC increases when the structure evolves from planar to low-buckled geometry, reaching 1.55 meV with a buckling distance of 0.48 Å. The band structure for low-buckled silicene with 0.48 Å buckling distance is shown in Figure 2.10 and the SOC band gap can be observed. The SOC band gap and direct calculation of Z_2 topological invariant confirm the existence of QSHE. Drummond et al. (2012) later predicted SOC band gaps of 1.4 meV and 1.5 meV with local-density approximation (LDA) and PBE, respectively, which was in agreement with the result of Liu et al. (2011a) and confirmed it. Liu et al. (2011b) also used a tight-binding model to demonstrate QSHE in silicene with a low-energy effective Hamiltonian they derived. Silicene was predicted to have much larger gaps than graphene, explained by the low-buckled geometry and larger atomic intrinsic SOC strength. This QSHE in silicene should be observable at an experimentally accessible temperature.

For zigzag SiNRs, the transport properties were predicted by An et al. (2013b) with the tight-binding model; it was found that QSHE could be induced by applying an electric field. Actually, two types of QSHE were found in their work—one had a wide bulk gap while the other had a narrow bulk gap.

2.3.2.3 Fermi velocity

Similar to graphene, silicene has a high Fermi velocity. Cahangirov et al. (2009) estimated the Fermi velocity to be ~10^6 m/s by DFT calculations with LDA. Drummond et al. (2012) used DFT calculations with different functionals to extract the Fermi velocity. The result for LDA or PBE exchange correlation functional was approximately $5.3 × 10^5$ m/s. With the screened Heyd–Scuseria–Ernzerhof 06 (HSE06) hybrid functional, the result −$6.75 × 10^5$ m/s—was slightly higher. In addition, Huang et al. (2013) predicted the Fermi velocity with DFT calculations. Both LDA and GGA have been tested to make sure the results will not be affected by the choice of functional. The corresponding results for planar and buckled

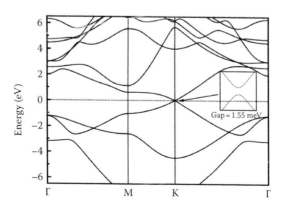

Figure 2.10 Band structure of low-buckled silicene with SOC effect included. (Reprinted with permission from Liu, C.-C., et al., 2011a, Quantum spin Hall effect in silicene and two-dimensional germanium, *Phys. Rev. Lett.*, 107, 76802. Copyright 2016 by the American Physical Society.)

silicene are 5.6×10^5 m/s and 5.4×10^5 m/s, respectively. Self-energy corrections by the GW approximation can enhance these values by about 37%, which is attributed to the depressed screening effect (Huang et al. 2013). To compare the Fermi velocities of graphene, silicene, and germanene, first-principles calculations were carried out without considering self-energy correction (Bechstedt et al. 2012; Matthes et al. 2013a, 2013b). The value for silicene is consistently smaller than that for graphene and about 4% larger than for germanene. The smaller Fermi velocity of silicene compared with graphene was explained by the larger Si–Si atomic distance (Liu et al. 2011a).

For supported silicene, since it has been successfully synthesized on substrates, the experimental measurements on Fermi velocity can be conducted, as mentioned in Section 2.2.4.1. For example, Chen et al. (2012) measured the Fermi velocity of silicene on Ag(111) substrate to be $\sim 1.2 \times 10^6$ m/s. De Padova et al. (2012a) measured the Fermi velocity of multilayer silicene on Ag(110) substrate and obtained a similar result of $\sim 1.3 \times 10^6$ m/s.

2.3.2.4 Tunable band gap

It has been predicted theoretically that applying an external electric field can modulate the electronic band gap of freestanding silicene (Ni et al. 2011; Drummond et al. 2012; Gürel et al. 2013). Ni et al. (2011) and Drummond et al. (2012) adopted first-principles calculations to predict the electronic properties of freestanding silicene and found that a tunable band gap can be opened under a perpendicular electric field. Both results showed that the band gap would increase linearly with the increasing electric field strength. The slope of the linear curve for band gap versus electric field strength was predicted to be 0.157 eÅ by Ni et al. (2011) and 0.0742 eÅ by Drummond et al. (2012). This discrepancy was partially explained by the extrapolation to infinite box length by Drummond et al. (2012) and the different plane wave basis sets used in these calculations. The band gap opening under a perpendicular electric field was explained by the symmetry breaking (Gürel et al. 2013).

For SiNR, the electronic properties under an electric field have also been investigated by numerical simulations (Ding and Ni 2009; Liang et al. 2012; Song et al. 2013). Ding and Ni (2009) adopted first-principles calculation to investigate the electronic properties of SiNRs under a transverse electric field and found that the band gap of up spin and down spin for 6-zigzag silicene nanoribbon (6-ZSiNR) changes differently. Liang et al. (2012) used first-principles calculations and the generalized nearest neighboring approximation method to study the electronic properties of ZSiNRs under a perpendicular electric field. The band gap for 12-ZSiNR would increase almost linearly with increasing electric field strength. The rate of change for band gap versus electric field strength was predicted to be 0.1046 eÅ. Song et al. (2013) adopted first-principles calculations to investigate the effect of a transverse electric field on the electronic properties of armchair silicene nanoribbon (ASiNR). The results showed that the band gap of ASiNR could be tuned under different magnitudes of transverse electric field but the trend was quite anomalous.

2.3.2.5 Strain effect on the electronic properties

The electronic properties of silicene can be modified under strain (Liu et al. 2012; Qin et al. 2012; Zhao 2012). It was predicted with first-principles calculations that silicene would undergo a semimetal–metal transition with increasing tensile strain (Liu et al. 2012; Qin et al. 2012). This was explained by the lowering of the lowest conduction band at Γ point (Liu et al. 2012). Liu et al. (2012) showed that the critical transition point was 0.075 strain but Qin et al. (2012) showed it to be 0.07.

Qin et al. (2012) found that the work function would first increase with increasing uniform biaxial strain and then nearly saturates around 0.15 strain, which can be seen in Figure 2.11a. However, the same group found that the work function increases monotonically under uniaxial strain (Qin et al. 2014), as shown in Figure 2.11b. Saturation point was not observed at 0.15 strain. The work function would increase more rapidly under zigzag strain than under armchair strain.

Regarding the strain effect on the band gap, there was controversy about whether uniaxial strain could induce a band gap (Zhao 2012) or not (Qin et al. 2014). Zhao (2012) observed the band gap opening under uniaxial strain while Qin et al. (2014) only found band energy shift. Recently, Voon et al. (2015) obtained the analytical forms of the Hamiltonians and band structure of silicene with the method of invariants. They concluded that band gap will not be opened by applying an in-plane uniaxial strain and provided a resolution for the controversy.

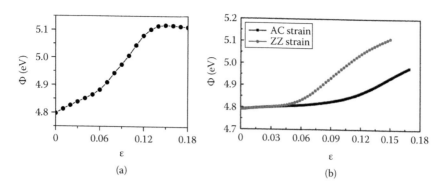

Figure 2.11 Strain-dependent work function of silicene under (a) biaxial strain and (b) uniaxial strain. (Reprinted from Qin, R., et al., *Aip Adv.*, 2, 22159, 2012; Qin, R., et al., *Nanoscale Res. Lett.*, 9, 1, 2014, under Creative Commons Attribution License.)

2.3.2.6 Optical properties

The optical properties of silicene can be derived from the electronic band structure. Matthes and co-workers (Bechstedt et al. 2012; Matthes et al. 2013a, 2013b; Matthes et al. 2014) adopted first-principles calculations to investigate the optical properties of silicene. Firstly, optical absorbance was calculated under independent-particle approximation in the absence of self-energy and excitonic corrections (Bechstedt et al. 2012; Matthes et al. 2013a). Their calculations were based on DFT with GGA exchange correlation. It was shown that in the low-frequency region, silicene has the same absorbance as graphene, despite its buckled structure compared to graphene's planar construction. In the limit of zero frequency, the absorbance of silicene was predicted to be $A(0) = \pi\alpha$ with $\alpha = e^2/\hbar c = 1/137.036$, which is the same as graphene. This value was also confirmed by analytical derivation using tight-binding model (Matthes et al. 2013a). Later, Matthes et al. (2013b) took quasiparticle and spin-orbit effects into account to investigate the optical properties of silicene. Quasiparticle many-body effects were incorporated in the HSE06 exchange correlation in DFT calculations, and it would induce the blueshift of the optical absorption peaks. Spin-orbit effects would lead to a band gap near the Dirac point. The optical absorbance did not change much overall but increased significantly near the fundamental absorption edge to $A\left(\omega \approx \dfrac{E_g}{\hbar}\right) = 2\pi\alpha$. Wei et al. (Wei et al. 2013; Wei and Jacob 2013) also investigated many-body effects in silicene using many-body Green's function theory. Strong excitonic effects were observed in the optical absorption. In silicene, the $\pi \rightarrow \pi^*$ resonant excitation arose at 1.23 eV and $\sigma \rightarrow \pi^*$ resonant excitation appeared at 3.75 eV.

2.3.3 THERMAL AND MECHANICAL PROPERTIES

Thermal properties of a material include thermal conductivity, heat capacity, thermal expansion coefficient, and so forth. In the following discussion, we mainly focus on the thermal conductivity of silicene, including the predicted thermal conductivity of freestanding silicene and SiNR, as well as the substrate and strain effect on thermal conductivity. Mechanical properties of silicene are also summarized.

2.3.3.1 Thermal conductivity of freestanding silicene and silicene nanoribbon

From the theoretical side, the thermal conductivity of freestanding silicene has been predicted with several numerical simulations (Li and Zhang 2012; Hu et al. 2013; Pei et al. 2013; Xie et al. 2014, 2016; Zhang et al. 2014; Gu and Yang 2015; Wang et al. 2015b; Kuang et al. 2016) and the results range from 5 to 43 W/mK at room temperature. These simulations have been conducted with classical molecular dynamics (MD) simulations or the first-principles-based Boltzmann transport equation (BTE) method.

Classical MD simulations solve Newton's second law of motion to extract all the atomic trajectories and then thermal conductivity can be obtained. Li and Zhang (2012) predicted the thermal conductivity of silicene to be around 20 W/mK with equilibrium MD (EMD)—the Green–Kubo method. Hu et al. (2013) and Wang et al. (2015b) adopted nonequilibrium MD (NEMD) and predicted the thermal conductivity of freestanding silicene to be around 40 W/mK and 43 W/mK, respectively. Pei et al. (2013) adopted NEMD with modified embedded-atom method (MEAM) potential and obtained a value of 41 W/mK for thermal conductivity, but the buckling distance predicted by the MEAM potential is 0.85 Å, and different from the more accurate, first-principles-based result of 0.44 Å. Because all the above simulations employed the empirical potential for bulk silicon, the low-buckled structure of silicene cannot be reproduced well. More recently, the SW potential was optimized to fit the first-principles results and this modified potential was used to predict the thermal conductivity (Zhang et al. 2014). Two sets of SW potential parameters are obtained by fitting to first-principles results, which can reproduce the lattice structure well and provide a reasonable phonon dispersion curve. The thermal conductivity values predicted are ranging from 5 to 12 W/mK, depending on the potential sets and the detailed numerical method for thermal conductivity prediction.

A more accurate approach to predict the thermal conductivity of silicene is the first-principles-based BTE method, which extracts interatomic force constants with first-principles calculations and solves phonon BTE to obtain the thermal conductivity. Xie et al. (2014) predicted the thermal conductivity of silicene with a relaxation time approximation (RTA) of BTE for the first time, but later refined their result (Xie et al. 2016) to about 25 W/mK. Gu and Yang (2015) used an iterative solution of BTE and obtained the thermal conductivity to be about 28 W/mK with sample width L = 30 μm. Kuang et al. (2016) also predicted the thermal conductivity with an iterative solution and the result was 19.34 W/mK.

At the present time, whether the thermal conductivity is converged or diverged with respect to the sample size is still a matter of debate. For example, it was claimed by Gu and Yang (2015) that the thermal conductivity of unstrained silicene would diverge with sample size, while Kuang et al. (2016) argued that it was converged. The major reason for the diverged thermal conductivity is that the dispersion curve for the long-wavelength flexural acoustic (ZA) phonons (i.e., phonons near the Γ point in the Brillouin zone) has a linear component. However, Kuang et al. (2016) argued that this linear component arose from the finite sample size, for infinitely large silicene, so the ZA dispersion should be quadratic around the Γ point. Therefore, the group velocity should be zero for long-wavelength phonons and the thermal conductivity must be converged. Xie et al. (2016) showed a possibly diverged thermal conductivity for infinitely large silicene but did not arrive at a conclusion. As Gu and Yang (2015) showed, the diverged result is obtained within a three-phonon scattering framework. Inclusion of even higher order phonon scattering terms may give a converged result. It should also be noted that all the MD simulation results are claimed to be converged. Future work is expected to be done to reach a conclusion about this issue, by considering higher order phonon scattering terms for the BTE method or more accurate potential developed for MD simulation.

Although consensus has not been reached about the converged/diverged issue, all the reported values for the thermal conductivity of silicene are much smaller than that of bulk silicon (~150 W/mK). This result is contrary to its 2D carbon counterpart, graphene, whose thermal conductivity is much greater than bulk diamond. This difference is further explored by considering the contribution to thermal conductivity from the flexural acoustic (ZA) branch. For graphene, the ZA branch is generally believed to give the dominant contribution to the total thermal conductivity (Lindsay et al. 2010, 2011; Seol et al. 2010). However, the result for silicene is quite different; the contribution from the ZA branch is significantly less than in graphene (Gu and Yang 2015; Wang et al. 2015b; Kuang et al. 2016; Xie et al. 2016). The reason is that the buckled structure breaks the symmetry selection rules (Gu and Yang 2015).

The thermal conductivity of SiNRs has been predicted by Ng et al. (2013), and the effect of uniaxial strain on its thermal conductivity has been investigated by Hu et al. (2013). Ng et al. (2013) adopted original Tersoff potential and NEMD to predict the thermal conductivity of SiNR at 300 K with width of approximately 5 nm. The thermal conductivity of either armchair or zigzag nanoribbon increases with the sample length. The result was about 65 W/mK with length greater than 40 nm. This value is lower than that of bulk silicon and explained by the limited number of vibrational frequencies. Hu et al. (2013) used NEMD to investigate the uniaxial strain effect on the thermal conductivity of SiNR with the original Tersoff potential. They found that the thermal conductivity first increases and then fluctuates with increasing uniaxial strain, for nanoribbon width of either 3.44 nm or 6.88 nm. This anomalous effect is explained by two competing mechanisms: the phonon-softening effect of the in-plane phonon modes, and the phonon-stiffening effect of the flexural phonon modes. Both Ng et al. (2013) and Hu et al. (2013) used NEMD with the original Tersoff potential to investigate the thermal conductivity of SiNR, but as mentioned above, the accuracy of the Tersoff potential need to be verified.

2.3.3.2 Substrate effect on thermal conductivity

For the substrate effect on the thermal conductivity of silicene, Wang et al. (2015b) conducted NEMD simulations to investigate the thermal conductivity of silicene on an amorphous silicon dioxide (SiO_2) substrate. They employed the Tersoff potential that was originally developed for bulk silicon and the equilibrium silicene structure was planar. Compared with freestanding silicene, significant reduction in the thermal conductivity was observed with the presence of a substrate. By further spectral energy density (SED) analysis, the reduction is explained by the decrease of phonon lifetimes. Zhang et al. (2015) also conducted NEMD to explore the effect of silicon carbide (SiC) substrate on the thermal conductivity of silicene. Different structures of SiC have different effects on the thermal conductivity of silicene. On a 3C-SiC substrate, supported silicene has lower thermal conductivity than freestanding silicene. However, on a 6H-SiC substrate, the thermal conductivity of supported silicene is higher. SED analysis is adopted for deeper understanding about this phenomenon and the increase in thermal conductivity is explained by the augmented lifetime of the majority of acoustic phonons, while the reduction is explained by the suppression in the lifetime of almost all the phonons. Given the limitation of available methods for thermal conductivity prediction, both works use empirical MD simulation to study the thermal transport of supported silicene. Both works also assume van der Waals bonding for the silicene and substrate interaction. We should note that for the current fabrication technology as mentioned above, the interaction between silicene and the substrate is fairly strong, so the assumption of the van der Waals nature of this interaction needs to be further verified.

2.3.3.3 Strain effect on thermal conductivity

In real applications, residue strain usually exists, and applying strain is a simple way to tune electronic and thermal properties. Therefore, the strain dependence of silicene thermal conductivity needs to be investigated. Pei et al. (2013) and Hu et al. (2013) adopted MD simulations to investigate the effect of uniaxial strain on the thermal conductivity. Pei et al. found that the thermal conductivity first increases and then decreases with the increment of strain. Hu et al. observed monotonic increase of the thermal conductivity of silicene sheet and SiNR by a factor of 2. However, both results are obtained with empirical potential developed for bulk silicon. With the first-principles-based BTE method, Xie et al. (2016) predicted that the thermal conductivity can increase dramatically within 10% tensile strain. The dispersion curves of unstrained silicene and strained silicene under 4 and 10% tensile strain are shown in Figure 2.12a. With a finite q mesh for infinite silicene, the thermal conductivity first increases greatly and then decreases slightly. Figure 2.12b shows the strain-dependent thermal conductivity of silicene for different sizes. For infinite silicene, both RTA and the iterative solution of BTE show that the highest thermal conductivity appears at about 4% tensile strain and the value is about 7 times that of the unstrained case. For all the finite sizes considered, the tunability (defined as the ratio of the highest thermal conductivity of strained silicene to the thermal conductivity of the unstrained silicene) is smaller than infinite size. With smaller size, the tunability also decreases. Even for 0.3 μm, the tenability is still around 4.3, which is much greater than for other materials (Li et al. 2010; Parrish et al. 2014).

Kuang et al. (2016) also investigated the strain effect on the thermal conductivity of silicene with the first-principles-based BTE method. They found that the thermal conductivity of unstrained silicene was converged but for strained silicene it would diverge. As we mentioned before, the converged/diverged issue is under debate and needs further investigation. From their result, the thermal conductivity for infinitely large (with $N = 301 \times 301 \times 1$) also increases significantly first and then decreases slightly. This trend is similar to Xie et al.'s result.

2.3.3.4 Mechanical properties

The in-plane stiffness, Poisson's ratio, ultimate strength, and yield point of freestanding silicene have been investigated intensively. However, there are limited researches on the supported silicene, and therefore it is not discussed in the following part.

Mechanical properties can be calculated from the strain energy $E_s = E(\varepsilon) - E(0)$, that is, the difference of total energy between strained and unstrained silicene. Strain energy was calculated with first-principles

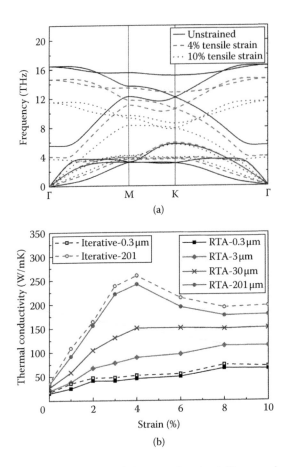

(a)

(b)

Figure 2.12 (a) Phonon dispersion curves of unstrained silicene and strained silicene under 4 and 10% tensile strain. (b) Thermal conductivity of infinite (201 × 201 × 1 q mesh) and finite-size (0.3, 3, and 30 μm) silicene as a function of strain, computed with RTA and iterative methods. Note that since RTA and iterative methods give similar thermal conductivity values for different cases, we only show the iterative results for the 201 × 201 × 1 infinite and the 0.3 μm size cases. (Reprinted with permission from Xie, H., et al., 2016, Large tunability of lattice thermal conductivity of monolayer silicene via mechanical strain, *Phys. Rev. B*, 93, 075404 Copyright 2016 by the American Physical Society.)

Table 2.1 **Mechanical properties of silicene and graphene**

	IN-PLANE STIFFNESS (n/m)		POISSON'S RATIO	
	SILICENE	GRAPHENE	SILICENE	GRAPHENE
Şahin et al. 2009	62	335	0.3	0.16
Zhang and Wang 2011	61	—	0.33	—
Qin et al. 2012	63	—	0.31	—
Peng et al. 2013	63.8	—	0.325	—
Xu et al. 2014a	63.3	334.77	0.32	0.18

calculations (Şahin et al. 2009; Zhang and Wang 2011; Qin et al. 2012; Peng et al. 2013; Xu et al. 2014a). The results are listed in Table 2.1.

In 2009, Şahin et al. (2009) calculated E_s in the harmonic range of the elastic deformation and fitted E_s to $E_s(\varepsilon_x, \varepsilon_y) = a_1\varepsilon_x^2 + a_2\varepsilon_y^2 + a_3\varepsilon_x\varepsilon_y$. Then the in-plane stiffness C and Poisson's ratio ν can be obtained from $C = \left[2a_1 - (a_3)^2/2a_1\right]/A_0$ and $\nu = a_3/2a_1$, respectively. In-plane stiffness and Poisson's ratio for silicene were predicted to be 62 N/m and 0.3. Compared with the results of graphene in the same work,

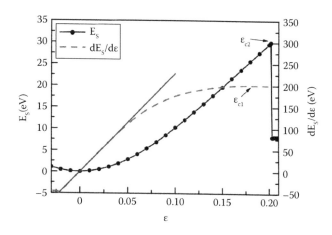

Figure 2.13 Strain energy and its dervative versus uniform strain ε. ε_{c1} and ε_{c2} denote the two critical strain points. (Reprinted from Qin, R., et al., *Aip Adv.*, 2, 22159, 2012, under Creative Commons Attribution License.)

in-plane stiffness of silicene was smaller and Poisson's ratio was larger. Later, Zhang and Wang (2011) considered both second-order harmonic and third-order anharmonic elastic constants in their calculation and predicted $C = 61$ N/m, $\nu = 0.33$. Anharmonic effects were demonstrated to play a role for strain larger than 0.035. The ultimate strength, which is the maximum stress that silicene can withstand, occurred at a strain of 0.231. Then in 2012, Qin et al. (2012) showed that the harmonic range of the strain is −0.02 to 0.04 and obtained $C = 63$ N/m, $\nu = 0.31$. One critical strain point is the ultimate strain ε_{c1} where the stress reaches the ultimate strength. It was the point where $dE_s/d\varepsilon$ achieves its maximum value, $\varepsilon_{c1} = 18\%$. Another critical point, called the yield point ε_{c2}, was at 20% strain. The strain energy reaches its maximum value at ε_{c2} and beyond this point decreases sharply. The two critical points are shown in Figure 2.13.

Peng et al. (2013) investigated the nonlinear elasticity of silicene and up to fifth-order elastic constants were considered in their calculation. They found that in the strain range of -0.03 to 0.03, harmonic elastic constants were enough to characterize the mechanical properties. Their results for in-plane stiffness and Poisson's ratio were 63.8 N/m and 0.325, respectively. The ultimate strength was predicted to occur at 0.17 for uniaxial armchair strain and biaxial strain, and at 0.21 for uniaxial zigzag strain. Xu et al. (2014a) comparatively studied the mechanical properties of silicene and graphene. Similar to Şahin, they also found that the in-plane stiffness of silicene was smaller than graphene while the Poisson's ratio of silicene would be larger.

2.4 APPLICATIONS

The experimental success stimulated many efforts to explore the intrinsic properties as well as potential device applications of silicene, including QSHE, quantum anomalous Hall effect, quantum valley Hall effect, superconductivity, band engineering, magnetism, TE effect, gas sensor, tunneling field-effect transistor (TFET), spin filter, and spin FET, which in turn promote the applications of silicene in many aspects. In this section, we will mainly focus on the applications of silicene in FET, energy storage, and TEs.

2.4.1 FIELD-EFFECT TRANSISTORS

Recently, silicene FETs have been fabricated, which shows the expected ambipolar Dirac charge transport and paves the way toward silicene-based nanoelectronics. Opening a sizable band gap (>0.4 eV) in silicene without degrading the electronic properties is critical for its applications in nanoelectronics. To date, two mechanisms are available for opening band gap in silicene while preserving its electronic properties. One is applying a vertical electric field to break the inversion symmetry in the two sublattices in silicene (Ni et al. 2011) as discussed above, and another is creating a nanomesh for producing periodic holes for intervalley interaction (Pan et al. 2015). Besides, adsorption of a foreign atom to silicene is a way to open a band gap based on the mixture of the two mechanisms (Quhe et al. 2012; Ni et al. 2014). With the opened band gap, various kinds of silicene FETs can be theoretically designed and their performances will be discussed below in detail.

Because of the buckled structures of silicene, applying a vertical external electric field $E_\perp$ will give rise to a difference in the electrostatic potentials of the two sublattices in monolayer silicene, which will open a direct band gap E_g at the K and K' point in the Brillouin zone by breaking the inversion symmetry between the two sublattices. As shown in Figure 2.14, the opened band gap increases linearly with the strength of vertical external electric field ($E_\perp$) (Ni et al. 2011; Drummond et al. 2012). A dual-gated silicene-based FET with SiO_2 dielectric and h-BN buffer layer was theoretically investigated, in which the doping level as well as the vertical electric field can be controlled by the dual gate (Ni et al. 2011). Besides the h-BN layer, other substrates such as hydrogenated Si-terminated SiC(0001), graphene, and MoS_2 also interact with silicene via weak van der Waals intereactions (Li et al. 2013; Liu et al. 2013; Gao et al. 2014a, 2014b). Therefore, the geometry structure and excellent electronic properties of silicene can be retained when fabricated on these substrates (Kaloni et al. 2013; Kamal et al. 2014).

We will review several types of silicene-based FETs in detail in the following part. The Dirac cone of silicene can be restored due to the insertion of alkali atoms between silicene and a metal substrate with n-type doping character (Quhe et al. 2014). Moreover, an ultrahigh on/off current ratio can be obtained in the alkali metal-adsorbed silicene FET with a large supply voltage V_{dd} (Quhe et al. 2012). Power dissipation is a fundamental issue in nanoelectronics. Due to the drawback of this type of FET that a large supply voltage (V_{dd}) of ~30 V is required for FET applications, designing new types of silicene-based FET with a high on/off ratio under low supply voltage is highly desirable. Besides silicene nanomesh FET (Pan et al. 2015), TFETs, based on band-to-band tunneling instead of thermal fluctuation, can also possess less power dissipation due to the smaller subthreshold swing (SS) below the limit of 60 mV/dec in MOSFET and supply voltage V_{dd}.

Zero band gap leads to low on/off current ratio, which will limit the applications of pure Dirac materials in nanoelectronics at room temperature. Besides searching for effective approaches to open band gap in silicene, all-metallic FET is another solution for the applications of silicene in nanoelectronics. For example, the Dirac cone of silicene can be preserved when embedded between two graphene layers (Neek-Amal et al. 2013). Because both silicene and graphene are semimetallic and the minima of their projected density of

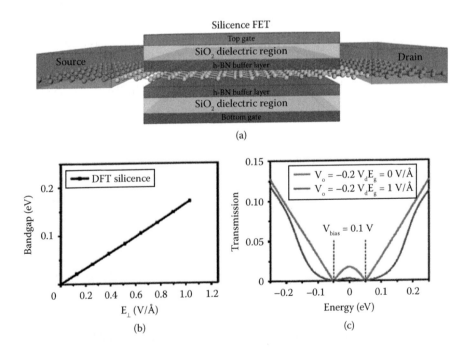

Figure 2.14 (a) Schematic model of dual-gated silicene FET. (b) Band gap as a function of vertical electric field. (c) Transmission spectra. (Reprinted with permission from Ni, Z., et al., 2011, 113–118. Copyright 2016 American Chemical Society.)

states do not coincide with each other, the hybrid structure is metallic. Without opening the band gap, the vertically stacked silicene/graphene as a Dirac material can be directly used in high-performance devices (Wang et al. 2015a), where the on/off switch effect comes from the electron transport between graphene and silicene near E_F being forbidden without the assistance of a phonon, due to momentum mismatch.

Based on electron transport simulations, the ASiNR FETs show tunable properties with respect to the change of width and length of the nanoribbon channels (Li et al. 2012). Due to the dependence of band gap of SiNRs on the nanoribbon width, the on/off current ratio (with the highest value being 10^6) of SiNR FET can be controlled by modulating the nanoribbon width, which also positively relates to the length of the channel. While the subthreshold decreases swing monotonously with increasing L, the lowest value is 90 mV/dec. Compared with carbon-based (e.g., graphene [Lin et al. 2008, 2010], graphene nanoribbons [Yan et al. 2007], and carbon nanotubes [Franklin and Chen 2010]) FETs which exhibit output characteristics of linear shape without any saturation or only weak saturation, the output characteristic of the SiNR FETs is saturated, although along with some unsaturated characteristics (Sacconi et al. 2007; Colinge et al. 2010; Yi et al. 2011).

The key parameters for the performance of silicene-based FETs are channel length, band gap, I_{on}/I_{off}, and SS (mV/dec). In spite of excellent and promising performances predicted in silicene FETs, it remains highly challenging to realize these devices in experiments due to the difficulty of isolating silicene from its substrate and the unstability of silicene exposed to air. Considering that the widely used wet transfer technique cannot be used for transferring silicene, the key role of Si–Ag interactions in stabilizing silicene has been clearly pointed out recently. The first conceptual silicene FET was successfully fabricated using a synthesis–transfer–fabrication process by epitaxially growing silicene on an Ag(111) substrate (Figure 2.15; Tao et al. 2015). A band gap in silicene can be opened without

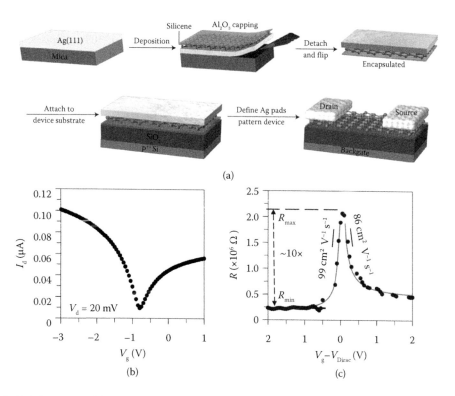

Figure 2.15 (a) Schematics of the synthesis–transfer–fabrication process of silicene. (b) Drain current I_d versus gate voltage V_g curve displays ambipolar electron–hole symmetry expected from silicene. (c) R versus gate overdrive voltage ($V_g - V_{dirac}$) of the silicene device. Measured transfer characteristics (dots) are in good agreement with a widely used ambipolar diffusive transport model (line). (Reprinted by permission from Macmillan Publishers Ltd., *Nat. Nanotechnol.*, Tao, L., et al., 2015, copyright 2016, Nature Publishing Group.)

degrading its high carrier mobility, and high performance is expected in silicene-based FETs. The experimentally achieved silicene FET possesses a band gap of 0.21 eV and shows an I_{on}/I_{off} of 10. There is still much room for improving the performance of silicene FETs by taking advantage of the proposed approaches for band gap opening.

2.4.2 ENERGY STORAGE

Hydrogen can be generally used for energy storage applications in the form of either chemical hydrides (chemical storage) or physisorbed gaseous H_2 molecules (physical storage). Silicene is considered as a promising hydrogen storage material similar to its carbon analog, graphene. For example, the chemical hydrogen storage properties of a number of silicene nanoflakes have been explored by Jose and Datta based on first-principles calculations (Jose and Datta 2011). It was found that the weight percentage of hydrogen ranges from 4.5 ($Si_{70}H_{92}$) to 6.6% (Si_6H_{12}) when fully hydrogenated on both upper/lower sides and at the edge-dangling Si atoms. Furthermore, the fully hydrogenated silicene sheet on Ag(111) substrate was found able to restore to the pristine silicene in experiment when annealing to a moderate temperature of ~450 K (Qiu et al. 2015). Such easily reversible hydrogenation of silicene suggests its potential applications for controllable chemical storage of hydrogen.

Due to the weak interactions of silicene itself with the H_2 molecule, silicene cannot be used for physical storage of hydrogen gas. However, considerable computational efforts have been devoted to designing physical hydrogen storage material using metal-decorated silicene, where a (2 × 2) supercell with eight Si atoms is constructed. For example, Na-decorated silicene ($NaSi_8$) adsorbed with different amount of hydrogen is a prototype model system (Figure 2.16; Wang et al. 2014b). When Na atoms sit at two opposite sides of silicene, the maximum gravimetric density of H_2 can be achieved of 9.40 wt% since each Na atom can store seven H_2 molecules.

By applying external strain or electric field, the hydrogen storage performance of metal-decorated silicene can be further enhanced. Under the biaxial symmetric strain of 10%, a drastic increase (~80%) of the binding energy of Mg adatoms on the silicene monolayer sheet can be achieved, ensuring the uniform distribution of Mg atoms over the substrate (Hussain et al. 2014). Under a perpendicular electric field (E), the average binding energy per H_2 (0.19 eV) on a Ca atom supported on monolayer or bilayer silicene can be enhanced to 0.37 eV ($E = 0.004$ a.u.) or reduced

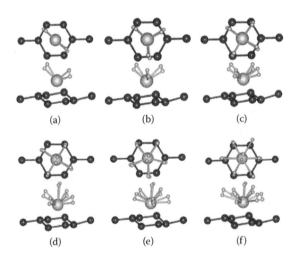

(a)   (b)   (c)

(d)   (e)   (f)

Figure 2.16 Schematics of adsorption of H_2 molecules on single Na atom decorated (2 × 2) unit cell of silicene ($NaSi_8$). Panels (a–f) correspond to the number of H_2 varying from 2 to 7. For each panel, top view, and side view are displayed. (Reprinted from *Int. J. Hydrogen Energy*, 39, Wang, Y., et al., Metal adatoms-decorated silicene as hydrogen storage media, 14027–14032, Copyright 2016, with permission from Elsevier B.V.)

to 0.02 eV ($E = -0.004$ a.u.). Thus with external electric field utilized as a switch, the adsorption and desorption of the hydrogen molecules on Ca-decorated silicene systems can be effectively controlled.

2.4.3 THERMOELECTRIC

Owing to the excellent scalability and compatibility with current silicon-based technology, integration of silicene into nanoscale TE devices is tempting and currently of great interest. TE materials can perform a direct solid-state conversion from thermal to electrical energy or vice versa, and have a number of valuable applications such as TE generators, waste heat recovery, and TE cooling and heating devices, which may make crucial contributions to the crises of energy and environment. In general, the TE performance and efficiency are characterized by the dimensionless figure of merit $ZT = S^2\sigma T/\kappa$, where S, σ, T, and κ are the Seebeck coefficient (thermopower), electrical conductivity, absolute temperature, and thermal conductivity, respectively. The thermal conductivity ($\kappa = \kappa_e + \kappa_{ph}$) consists of those from electrons (κ_e) and phonons (κ_{ph}). Accordingly, a high ZT value at a given temperature requires a high thermopower, a suitable combination of electrical conductivity and electronic thermal conductivity (both related to carrier mobility), and a low lattice thermal conductivity. Typically, materials with $ZT \sim 1$ are regarded as good TE materials, while a device with $ZT > 3$ is competitive compared to the conventional energy conversion techniques (Yang et al. 2014). Improvement of ZT can be achieved by either increasing the power factor ($S^2\sigma$) or decreasing the thermal conductivity.

Silicene and its nanostructures show great potential for TE applications in two aspects: (1) The electronic structure of silicene is similar to that of graphene (i.e., the characteristic Dirac cone and high carrier mobility); (2) The lattice thermal conductivity of silicene is much lower than that of graphene due to its buckled structure (Xie et al. 2014; Gu and Yang 2015). The TE performance of silicene sheets and nanoribbons have been intensively studied (Pan et al. 2012; Zberecki et al. 2013; Yang et al. 2014; Sadeghi et al. 2015; Wierzbicki et al. 2015; Zhao et al. 2016). A small thermopower of ~87 μV/K at 300 K was found in silicene (Yan et al. 2013), close to that of graphene (~100 μV/K) (Kaloni and Schwingenschlögl 2014), which is due to their similar gapless feature. When neglecting the contribution to thermal conductivity from phonon transport, the maximum of the electronic ZT ($ZT_e = 0.36$) can be achieved at chemical potentials of about ±0.08 eV, where positive and negative chemical potentials correspond to electron and hole doping, respectively (Yang et al. 2014).

The ZT of silicene can be enhanced by placing it on metallic substrates due to the induced distortions to the atomic structure and the consequent modification of the electronic structures. For example, silicene on Ag(111) substrate exhibits a buckling distance of 0.79 Å, which is larger than the 0.43 Å in freestanding silicene. Due to the opened band gap of ~0.3 eV in the supported silicene, the ZT_e is enhanced to 0.81 at 300 K (Yang et al. 2014). The total TE figure of merit of silicene (ZT) is much smaller than the upper bound given by ZT_e, which is due to the fact that the lattice thermal conductivity contributed from phonon transport of freestanding and supported silicene (3–60 W/mK) is at least one order of magnitude greater than the contribution from electron transport (Ng et al. 2013; Xie et al. 2014; Zhang et al. 2014).

Nanostructuring of silicene is also an effective way to enhance the ZT by simultaneously enhancing the thermopower and reducing the lattice thermal conductivity (Pan et al. 2012; Zberecki et al. 2013; Sadeghi et al. 2015). In SiNRs with the edges passivated by hydrogen, the power factor can be greatly enhanced compared to that of infinite silicene sheets due to the band gap opening (Song et al. 2010; Pan et al. 2012; Zberecki et al. 2013; Sadeghi et al. 2015). On the other hand, the phonon thermal conductance of SiNRs with a smaller width has a lower value due to stronger phonon-boundary scattering, and it would increase with temperature (Figure 2.17) (Yang et al. 2014). At room temperature, the phonon contribution to the thermal conductance is of the same order of magnitude or slightly greater than the electron contribution. The ZT of armchair SiNRs shows an oscillatory variation with ribbon width and reaches maximum value of 1.04 for a width of 3, while for zigzag SiNRs, ZT decreases with increasing ribbon width and achieves the largest value of 0.6 with same width for hole doping.

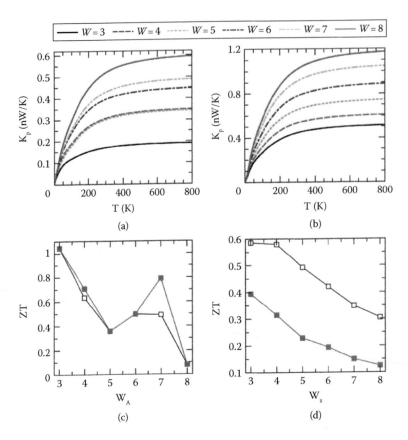

Figure 2.17 The phonon thermal conductance as a function of temperature for (a) ASiNR and (b) ZSiNR, respectively. The thermoelectric figure of (c) ASiNR and (d) ZSiNR as a function of ribbon width, respectively. (Reprinted with permission from Yang, K., et al., 2014, Thermoelectric properties of atomically thin silicene and germanene nanostructures, *Phys. Rev. B*, 89, 125403. Copyright 2016 by the American Physical Society.)

2.5 CONCLUDING REMARKS AND OUTLOOK

This chapter has surveyed the recent advances of silicene, which is one of the fastest-growing nanostructured semiconductors, and nanocrystals. The experimental investigations of silicene are focused on the fabrication, structural characterization, and electronic properties. Following the successful fabrication, silicene has attracted a lot of research interest in the past few years. The studies are mainly about the experimental synthesis and characterization of monolayer/few-layer/nanoribbons of silicene on different substrates. The surface structures, growth mechanism, and characterization of electronic properties are discussed in detail, which are currently the focus of the experimental studies of silicene. Due to the fact that systematic measurements on the physical properties of silicene are still difficult, most of the investigations on the physical properties are still based on theoretical approaches, mainly on the structural, electronic, optical, thermal, and mechanical properties of silicene. The outstanding properties of silicene as revealed in experimental and theoretical studies promise potential device applications for silicene, including QSHE, quantum anomalous Hall effect, quantum valley Hall effect, superconductivity, band engineering, magnetism, TE effect, gas sensor, TFET, spin filter, and spin FET, which in turn promote the applications of silicene in many aspects where the FET, energy storage, and TE applications are discussed in detail.

Besides the currently achieved advances of silicene in the past few years, further important physical fundamentals still remain to be clarified. For example, how the phonon transport would be in few-layer silicene is still unclear. Whether there are other methods besides mechanical strain to efficiently modulate the electrical and thermal transport properties in silicene remains an open question. In addition, silicene-based heterostructures could possess even more exceptional electronic properties and phonon transport behaviors, which could open up new building blocks for the next generation of advanced functional devices.

REFERENCES

Acun, A., Poelsema, B., Zandvliet, H.J.W., and Van Gastel, R. (2013). The instability of silicene on Ag (111). *Appl. Phys. Lett.*, **103**, 263119.

Aizawa, T., Suehara, S., and Otani, S. (2014). Silicene on zirconium carbide (111). *J. Phys. Chem. C*, **118**, 23049–23057.

Aizawa, T., Suehara, S., and Otani, S. (2015). Phonon dispersion of silicene on ZrB2 (0 0 0 1). *J. Phys. Condens. Matter*, **27**, 305002.

An, X.-T., Zhang, Y.-Y., Liu, J.-J., and Li, S.-S. (2013a). Interplay between edge and bulk states in silicene nanoribbon. *Appl. Phys. Lett.*, **102**, 213115.

An, X.-T., Zhang, Y.-Y., Liu, J.-J., and Li, S.-S. (2013b). Quantum spin Hall effect induced by electric field in silicene. *Appl. Phys. Lett.*, **102**, 043113.

Arafune, R., Lin, C.-L., Nagao, R., Kawai, M., and Takagi, N. (2013). Comment on "Evidence for dirac fermions in a honeycomb lattice based on silicon." *Phys. Rev. Lett.*, **110**, 229701.

Aufray, B., Kara, A., Vizzini, S., Oughaddou, H., Leandri, C., Ealet, B., and Le Lay, G. (2010). Graphene-like silicon nanoribbons on Ag (110): A possible formation of silicene. *Appl. Phys. Lett.*, **96**, 183102.

Avila, J., De Padova, P., Cho, S., Colombo, I., Lorcy, S., Quaresima, C., Vogt, P., Resta, A., Le Lay, G., and Asensio, M.C. (2013). Presence of gapped silicene-derived band in the prototypical (3 × 3) silicene phase on silver (111) surfaces. *J. Phys. Condens. Matter*, **25**, 262001.

Bai, J., Tanaka, H., and Zeng, X.C. (2010). Graphene-like bilayer hexagonal silicon polymorph. *Nano Res.*, **3**, 694–700.

Bechstedt, F., Matthes, L., Gori, P., and Pulci, O. (2012). Infrared absorbance of silicene and germanene. *Appl. Phys. Lett.*, **100**, 261906.

Bernard, R., Leoni, T., Wilson, A., Lelaidier, T., Sahaf, H., Moyen, E., Assaud, L., et al. (2013). Growth of Si ultrathin films on silver surfaces: Evidence of an Ag (110) reconstruction induced by Si. *Phys. Rev. B*, **88**, 121411.

Bhattacharya, A., Bhattacharya, S., and Das, G.P. (2013). Exploring semiconductor substrates for silicene epitaxy. *Appl. Phys. Lett.*, **103**, 123113.

Borensztein, Y., Curcella, A., Royer, S., and Prévot, G. (2015). Silicene multilayers on Ag (111) display a cubic diamondlike structure and a $\sqrt{3} \times \sqrt{3}$ reconstruction induced by surfactant Ag atoms. *Phys. Rev. B*, **92**, 155407.

Cahangirov, S., Audiffred, M., Tang, P., Iacomino, A., Duan, W., Merino, G., and Rubio, A. (2013). Electronic structure of silicene on Ag (111): Strong hybridization effects. *Phys. Rev. B*, **88**, 035432.

Cahangirov, S., Özçelik, V.O., Xian, L., Avila, J., Cho, S., Asensio, M.C., Ciraci, S., and Rubio, A. (2014). Atomic structure of the 3×3 phase of silicene on Ag (111). *Phys. Rev. B*, **90**, 035448.

Cahangirov, S., Topsakal, M., Aktürk, E., Şahin, H., and Ciraci, S. (2009). Two-and one-dimensional honeycomb structures of silicon and germanium. *Phys. Rev. Lett.*, **102**, 236804.

Chen, J., Li, W., Feng, B., Cheng, P., Qiu, J., Chen, L., and Wu, K. (2014). Persistent Dirac fermion state on bulk-like Si (111) surface. *ArXiv Prepr. ArXiv14057534*.

Chen, L., Feng, B., and Wu, K. (2013a). Observation of a possible superconducting gap in silicene on Ag (111) surface. *Appl. Phys. Lett.*, **102**, 081602.

Chen, L., Li, H., Feng, B., Ding, Z., Qiu, J., Cheng, P., Wu, K., and Meng, S. (2013b). Spontaneous symmetry breaking and dynamic phase transition in monolayer silicene. *Phys. Rev. Lett.*, **110**, 85504.

Chen, L., Liu, C.-C., Feng, B., He, X., Cheng, P., Ding, Z., Meng, S., Yao, Y., and Wu, K. (2012). Evidence for Dirac fermions in a honeycomb lattice based on silicon. *Phys. Rev. Lett.*, **109**, 056804.

Chen, M.X., and Weinert, M. (2014). Revealing the substrate origin of the linear dispersion of silicene/Ag (111). *Nano Lett.*, **14**, 5189–5193.

Chiappe, D., Grazianetti, C., Tallarida, G., Fanciulli, M., and Molle, A. (2012). Local electronic properties of corrugated silicene phases. *Adv. Mater.*, **24**, 5088–5093.

Chiappe, D., Scalise, E., Cinquanta, E., Grazianetti, C., van den Broek, B., Fanciulli, M., Houssa, M., and Molle, A. (2014). Two-dimensional Si nanosheets with local hexagonal structure on a MoS(2) surface. *Adv. Mater.*, **26**, 2096–2101.

Cinquanta, E., Fratesi, G., dal Conte, S., Grazianetti, C., Scotognella, F., Stagira, S., Vozzi, C., Onida, G., and Molle, A. (2015). Optical response and ultrafast carrier dynamics of the silicene-silver interface. *Phys. Rev. B*, **92**, 165427.

Cinquanta, E., Scalise, E., Chiappe, D., Grazianetti, C., van den Broek, B., Houssa, M., Fanciulli, M., and Molle, A. (2013). Getting through the nature of silicene: An sp2–sp3 two-dimensional silicon nanosheet. *J. Phys. Chem. C*, **117**, 16719–16724.

Colinge, J.-P., Lee, C.-W., Afzalian, A., Akhavan, N.D., Yan, R., Ferain, I., Razavi, P., et al. (2010). Nanowire transistors without junctions. *Nat. Nanotechnol.*, **5**, 225–229.

Colonna, S., Serrano, G., Gori, P., Cricenti, A., and Ronci, F. (2013). Systematic STM and LEED investigation of the Si/Ag (110) surface. *J. Phys. Condens. Matter*, **25**, 315301.

De Padova, P., Avila, J., Resta, A., Razado-Colombo, I., Quaresima, C., Ottaviani, C., Olivieri, B., et al. (2013a). The quasiparticle band dispersion in epitaxial multilayer silicene. *J. Phys. Condens. Matter*, **25**, 382202.

De Padova, P., Kubo, O., Olivieri, B., Quaresima, C., Nakayama, T., Aono, M., and Le Lay, G. (2012a). Multilayer silicene nanoribbons. *Nano Lett.*, **12**, 5500–5503.

De Padova, P., Ottaviani, C., Quaresima, C., Olivieri, B., Imperatori, P., Salomon, E., Angot, T., et al. (2014). 24 h stability of thick multilayer silicene in air. *2D Mater.*, 1, 21003.

De Padova, P., Perfetti, P., Olivieri, B., Quaresima, C., Ottaviani, C., and Le Lay, G. (2012b). 1D graphene-like silicon systems: Silicene nano-ribbons. *J. Phys. Condens. Matter*, **24**, 223001.

De Padova, P., Quaresima, C., Olivieri, B., Perfetti, P., and Le Lay, G. (2011). Sp2-like hybridization of silicon valence orbitals in silicene nanoribbons. *Appl. Phys. Lett.*, **98**, 081909.

De Padova, P., Quaresima, C., Ottaviani, C., Sheverdyaeva, P.M., Moras, P., Carbone, C., Topwal, D., et al. (2010). Evidence of graphene-like electronic signature in silicene nanoribbons. *Appl. Phys. Lett.*, **96**, 261905.

De Padova, P., Vogt, P., Resta, A., Avila, J., Razado-Colombo, I., Quaresima, C., Ottaviani, C., et al. (2013b). Evidence of Dirac fermions in multilayer silicene. *Appl. Phys. Lett.*, **102**, 163106.

Ding, Y., and Ni, J. (2009). Electronic structures of silicon nanoribbons. *Appl. Phys. Lett.*, **95**, 083115.

Drummond, N.D., Zolyomi, V., and Fal'Ko, V.I. (2012). Electrically tunable band gap in silicene. *Phys. Rev. B*, **85**, 075423.

Enriquez, H., Kara, A., Lalmi, B., Oughaddou, H., and Vizzini, S. (2012). Silicene structures on silver surfaces. *J. Phys. Condens. Matter*, **24**, 314211.

Ezawa, M. (2013). Quantized conductance and field-effect topological quantum transistor in silicene nanoribbons. *Appl. Phys. Lett.*, **102**, 172103.

Feng, B., Ding, Z., Meng, S., Yao, Y., He, X., Cheng, P., Chen, L., and Wu, K. (2012). Evidence of silicene in honeycomb structures of silicon on Ag (111). *Nano Lett.*, **12**, 3507–3511.

Feng, B., Li, H., Liu, C.-C., Shao, T.-N., Cheng, P., Yao, Y., Meng, S., Chen, L., and Wu, K. (2013). Observation of Dirac cone warping and chirality effects in silicene. *ACS Nano*, **7**, 9049–9054.

Feng, B., Li, H., Meng, S., Chen, L., and Wu, K. (2016a). Structure and quantum well states in silicene nanoribbons on Ag (110). *Surf. Sci.*, **645**, 74–79.

Feng, Y., Liu, D., Feng, B., Liu, X., Zhao, L., Xie, Z., Liu, Y., et al. (2016b). Direct evidence of interaction-induced Dirac cones in monolayer silicene/Ag (111) system. *Proc. Natl. Acad. Sci. U. S. A.*, **13**(51), 14656–14661.

Fleurence, A., Friedlein, R., Ozaki, T., Kawai, H., Wang, Y., and Yamada-Takamura, Y. (2012). Experimental evidence for epitaxial silicene on diboride thin films. *Phys. Rev. Lett.*, **108**, 245501.

Fleurence, A., Yoshida, Y., Lee, C.-C., Ozaki, T., Yamada-Takamura, Y., and Hasegawa, Y. (2014). Microscopic origin of the π states in epitaxial silicene. *Appl. Phys. Lett.*, **104**, 021605.

Franklin, A.D., and Chen, Z. (2010). Length scaling of carbon nanotube transistors. *Nat. Nanotechnol.*, **5**, 858–862.

Friedlein, R., Fleurence, A., Aoyagi, K., de Jong, M.P., Van Bui, H., Wiggers, F.B., Yoshimoto, S., et al. (2014). Core level excitations-a fingerprint of structural and electronic properties of epitaxial silicene. *J. Chem. Phys.*, **140**, 184704.

Fu, H., Chen, L., Chen, J., Qiu, J., Ding, Z., Zhang, J., Wu, K., Li, H., and Meng, S. (2015). Multilayered silicene: The bottom-up approach for a weakly relaxed Si (111) with Dirac surface states. *Nanoscale*, **7**, 15880–15885.

Fu, H., Zhang, J., Ding, Z., Li, H., and Meng, S. (2014). Stacking-dependent electronic structure of bilayer silicene. *Appl. Phys. Lett.*, **104**, 131904.

Fukaya, Y., Mochizuki, I., Maekawa, M., Wada, K., Hyodo, T., Matsuda, I., and Kawasuso, A. (2013). Structure of silicene on a Ag (111) surface studied by reflection high-energy positron diffraction. *Phys. Rev. B*, **88**, 205413.

Gao, J., and Zhao, J. (2012). Initial geometries, interaction mechanism and high stability of silicene on Ag (111) surface. *Sci. Rep.*, **2**, 861.

Gao, N., Li, J.C., and Jiang, Q. (2014a). Bandgap opening in silicene: Effect of substrates. *Chem. Phys. Lett.*, **592**, 222–226.

Gao, N., Li, J.C., and Jiang, Q. (2014b). Tunable band gaps in silicene–MoS2 heterobilayers. *Phys. Chem. Chem. Phys.*, **16**, 11673–11678.

Gori, P., Pulci, O., Ronci, F., Colonna, S., and Bechstedt, F. (2013). Origin of Dirac-cone-like features in silicon structures on Ag (111) and Ag (110). *J. Appl. Phys.*, **114**, 113710.

Grazianetti, C., Chiappe, D., Cinquanta, E., Fanciulli, M., and Molle, A. (2015). Nucleation and temperature-driven phase transitions of silicene superstructures on Ag (1 1 1). *J. Phys. Condens. Matter*, **27**, 255005.

Gu, X., and Yang, R. (2015). First-principles prediction of phononic thermal conductivity of silicene: A comparison with graphene. *J. Appl. Phys.*, **117**, 025102.

Guo, Z. -X., Furuya, S., Iwata, J., and Oshiyama, A. (2013a). Absence of Dirac electrons in silicene on Ag (111) surfaces. *J. Phys. Soc. Jpn.*, **82**, 63714.

Guo, Z.-X., Furuya, S., Iwata, J., and Oshiyama, A. (2013b). Absence and presence of Dirac electrons in silicene on substrates. *Phys. Rev. B*, **87**, 235435.

Guo, Z.-X., and Oshiyama, A. (2014). Structural tristability and deep Dirac states in bilayer silicene on Ag (111) surfaces. *Phys. Rev. B*, **89**, 155418.

Guo, Z.-X., and Oshiyama, A. (2015). Crossover between silicene and ultra-thin Si atomic layers on Ag (111) surfaces. *New J. Phys.*, **17**, 45028.

Guo, Z.-X., Zhang, Y.-Y., Xiang, H., Gong, X.-G., and Oshiyama, A. (2015). Structural evolution and optoelectronic applications of multilayer silicene. *Phys. Rev. B*, **92**, 201413.

Gürel, H.H., Özçelik, V.O., and Ciraci, S. (2013). Effects of charging and perpendicular electric field on the properties of silicene and germanene. *J. Phys. Condens. Matter*, **25**, 305007.

Guzmán-Verri, G.G., and Voon, L.L.Y. (2007). Electronic structure of silicon-based nanostructures. *Phys. Rev. B*, **76**, 075131.

Han, M.Y., Özyilmaz, B., Zhang, Y., and Kim, P. (2007). Energy band-gap engineering of graphene nanoribbons. *Phys. Rev. Lett.*, **98**, 206805.

He, G. (2006). Atomic structure of Si nanowires on Ag (110): A density-functional theory study. *Phys. Rev. B*, **73**, 035311.

Hogan, C., Colonna, S., Flammini, R., Cricenti, A., and Ronci, F. (2015). Structure and stability of Si/Ag (110) nanoribbons. *Phys. Rev. B*, **92**, 115439.

Houssa, M., van den Broek, B., Scalise, E., Ealet, B., Pourtois, G., Chiappe, D., Cinquanta, E., et al. (2014). Theoretical aspects of graphene-like group IV semiconductors. *Appl. Surf. Sci.*, **291**, 98–103.

Houssa, M., van den Broek, B., Scalise, E., Pourtois, G., Afanas'ev, V., and Stesmans, A. (2013). Theoretical study of silicene and germanene. *ECS Trans.*, **53**, 51–62.

Hu, M., Zhang, X., and Poulikakos, D. (2013). Anomalous thermal response of silicene to uniaxial stretching. *Phys. Rev. B*, **87**, 195417.

Huang, S., Kang, W., and Yang, L. (2013). Electronic structure and quasiparticle bandgap of silicene structures. *Appl. Phys. Lett.*, **102**, 133106.

Hussain, T., Chakraborty, S., De Sarkar, A., Johansson, B., and Ahuja, R. (2014). Enhancement of energy storage capacity of Mg functionalized silicene and silicane under external strain. *Appl. Phys. Lett.*, **105**, 123903.

Ishida, H., Hamamoto, Y., Morikawa, Y., Minamitani, E., Arafune, R., and Takagi, N. (2015). Electronic structure of the 4$\times$ 4 silicene monolayer on semi-infinite Ag (111). *New J. Phys.*, **17**, 15013.

Jamgotchian, H., Colignon, Y., Hamzaoui, N., Ealet, B., Hoarau, J.Y., Aufray, B., and Bibérian, J.P. (2012). Growth of silicene layers on Ag (111): Unexpected effect of the substrate temperature. *J. Phys. Condens. Matter*, **24**, 172001.

Jia, X., Hofmann, M., Meunier, V., Sumpter, B.G., Campos-Delgado, J., Romo-Herrera, J.M., Son, H., et al. (2009). Controlled formation of sharp zigzag and armchair edges in graphitic nanoribbons. *Science*, **323**, 1701–1705.

Johnson, N.W., Muir, D., Kurmaev, E.Z., and Moewes, A. (2015). Stability and electronic characteristics of epitaxial silicene multilayers on Ag (111). *Adv. Funct. Mater.*, **25**, 4083–4090.

Johnson, N.W., Vogt, P., Resta, A., De Padova, P., Perez, I., Muir, D., Kurmaev, E.Z., Le Lay, G., and Moewes, A. (2014). The metallic nature of epitaxial silicene monolayers on Ag (111). *Adv. Funct. Mater.*, **24**, 5253–5259.

Jose, D., and Datta, A. (2011). Structures and electronic properties of silicene clusters: A promising material for FET and hydrogen storage. *Phys. Chem. Chem. Phys.*, **13**, 7304–7311.

Kaloni, T.P., and Schwingenschlögl, U. (2014). Effects of heavy metal adsorption on silicene. *Phys. Status Solidi RRL-Rapid Res. Lett.*, **8**, 685–687.

Kaloni, T.P., Tahir, M., and Schwingenschlögl, U. (2013). Quasi free-standing silicene in a superlattice with hexagonal boron nitride. *Sci. Rep.*, **3**, 3192.

Kaltsas, D., Tsetseris, L., and Dimoulas, A. (2012). Structural evolution of single-layer films during deposition of silicon on silver: A first-principles study. *J. Phys. Condens. Matter*, **24**, 442001.

Kaltsas, D., Tsetseris, L., and Dimoulas, A. (2014). Silicene on metal substrates: A first-principles study on the emergence of a hierarchy of honeycomb structures. *Appl. Surf. Sci.*, **291**, 93–97.

Kamal, C., Chakrabarti, A., and Banerjee, A. (2014). Ab initio investigation on hybrid graphite-like structure made up of silicene and boron nitride. *Phys. Lett. A*, **378**, 1162–1169.

Kamal, C., Chakrabarti, A., Banerjee, A., and Deb, S.K. (2013). Silicene beyond mono-layers—Different stacking configurations and their properties. *J. Phys. Condens. Matter*, **25**, 85508.

Kang, J., Wu, F., and Li, J. (2012). Symmetry-dependent transport properties and magnetoresistance in zigzag silicene nanoribbons. *Appl. Phys. Lett.*, **100**, 233122.

Kara, A., Léandri, C., Dávila, M.E., De Padova, P., Ealet, B., Oughaddou, H., Aufray, B., and Le Lay, G. (2009). Physics of silicene stripes. *J. Supercond. Nov. Magn.*, **22**, 259–263.

Kara, A., Vizzini, S., Leandri, C., Ealet, B., Oughaddou, H., Aufray, B., and LeLay, G. (2010). Silicon nano-ribbons on Ag (110): A computational investigation. *J. Phys. Condens. Matter*, **22**, 45004.

Kawahara, K., Shirasawa, T., Arafune, R., Lin, C.-L., Takahashi, T., Kawai, M., and Takagi, N. (2014). Determination of atomic positions in silicene on Ag (111) by low-energy electron diffraction. *Surf. Sci.*, **623**, 25–28.

Kokott, S., Pflugradt, P., Matthes, L., and Bechstedt, F. (2014). Nonmetallic substrates for growth of silicene: An ab initio prediction. *J. Phys. Condens. Matter*, **26**, 185002.

Kuang, Y.D., Lindsay, L., Shi, S.Q., and Zheng, G.P. (2016). Tensile strains give rise to strong size effects for thermal conductivities of silicene, germanene and stanene. *Nanoscale*, **8**, 3760–3767.

Lagarde, P., Chorro, M., Roy, D., and Trcera, N. (2016). Study by EXAFS of the local structure around Si on silicene deposited on Ag (1 1 0) and Ag (1 1 1) surfaces. *J. Phys. Condens. Matter*, **28**, 75002.

Lalmi, B., Oughaddou, H., Enriquez, H., Kara, A., Vizzini, S., Ealet, B., and Aufray, B. (2010). Epitaxial growth of a silicene sheet. *Appl. Phys. Lett.*, **97**, 223109.

Leandri, C., Le Lay, G., Aufray, B., Girardeaux, C., Avila, J., Davila, M.E., Asensio, M.C., Ottaviani, C., and Cricenti, A. (2005). Self-aligned silicon quantum wires on Ag (110). *Surf. Sci.*, **574**, L9–L15.

Léandri, C., Oughaddou, H., Aufray, B., Gay, J.M., Le Lay, G., Ranguis, A., and Garreau, Y. (2007). Growth of Si nanostructures on Ag (001). *Surf. Sci.*, **601**, 262–267.

Lee, C.-C., Fleurence, A., Friedlein, R., Yamada-Takamura, Y., and Ozaki, T. (2013). First-principles study on competing phases of silicene: Effect of substrate and strain. *Phys. Rev. B*, **88**, 165404.

Lee, C.-C., Fleurence, A., Yamada-Takamura, Y., Ozaki, T., and Friedlein, R. (2014). Band structure of silicene on zirconium diboride (0001) thin-film surface: Convergence of experiment and calculations in the one-Si-atom Brillouin zone. *Phys. Rev. B*, **90**, 075422.

Li, H., Wang, L., Liu, Q., Zheng, J., Mei, W.-N., Gao, Z., Shi, J., and Lu, J. (2012). High performance silicene nanoribbon field effect transistors with current saturation. *Eur. Phys. J. B*, **85**, 1–6.

Li, H., and Zhang, R. (2012). Vacancy-defect–induced diminution of thermal conductivity in silicene. *EPL Europhys. Lett.*, **99**, 36001.

Li, L., Wang, X., Zhao, X., and Zhao, M. (2013). Moiré superstructures of silicene on hexagonal boron nitride: A first-principles study. *Phys. Lett. A*, **377**, 2628–2632.

Li, X., Maute, K., Dunn, M.L., and Yang, R. (2010). Strain effects on the thermal conductivity of nanostructures. *Phys. Rev. B*, **81**, 245318.

Liang, Y., Wang, V., Mizuseki, H., and Kawazoe, Y. (2012). Band gap engineering of silicene zigzag nanoribbons with perpendicular electric fields: A theoretical study. *J. Phys. Condens. Matter*, **24**, 455302.

Lin, C.-L., Arafune, R., Kawahara, K., Kanno, M., Tsukahara, N., Minamitani, E., Kim, Y., Kawai, M., and Takagi, N. (2013). Substrate-induced symmetry breaking in silicene. *Phys. Rev. Lett.*, **110**, 076801.

Lin, C.-L., Arafune, R., Kawahara, K., Tsukahara, N., Minamitani, E., Kim, Y., Takagi, N., and Kawai, M. (2012). Structure of silicene grown on Ag (111). *Appl. Phys. Express*, **5**, 45802.

Lin, Y.-M., Dimitrakopoulos, C., Jenkins, K.A., Farmer, D.B., Chiu, H.-Y., Grill, A., and Avouris, P. (2010). 100-GHz transistors from wafer-scale epitaxial graphene. *Science*, **327**, 662–662.

Lin, Y.-M., Jenkins, K.A., Valdes-Garcia, A., Small, J.P., Farmer, D.B., and Avouris, P. (2008). Operation of graphene transistors at gigahertz frequencies. *Nano Lett.*, **9**, 422–426.

Lindsay, L., Broido, D.A., and Mingo, N. (2010). Flexural phonons and thermal transport in graphene. *Phys. Rev. B*, **82**, 115427.

Lindsay, L., Broido, D.A., and Mingo, N. (2011). Flexural phonons and thermal transport in multilayer graphene and graphite. *Phys. Rev. B*, **83**, 235428.

Liu, C.-C., Feng, W., and Yao, Y. (2011a). Quantum spin Hall effect in silicene and two-dimensional germanium. *Phys. Rev. Lett.*, **107**, 76802.

Liu, C.-C., Jiang, H., and Yao, Y. (2011b). Low-energy effective Hamiltonian involving spin-orbit coupling in silicene and two-dimensional germanium and tin. *Phys. Rev. B*, **84**, 195430.

Liu, G., Wu, M.S., Ouyang, C.Y., and Xu, B. (2012). Strain-induced semimetal-metal transition in silicene. *EPL Europhys. Lett.*, **99**, 17010.

Liu, H., Gao, J., and Zhao, J. (2013). Silicene on substrates: A way to preserve or tune its electronic properties. *J. Phys. Chem. C*, **117**, 10353–10359.

Liu, H., Han, N., and Zhao, J. (2014a). Band gap opening in bilayer silicene by alkali metal intercalation. *J. Phys. Condens. Matter*, **26**, 475303.

Liu, Z.-L., Wang, M.-X., Liu, C., Jia, J.-F., Vogt, P., Quaresima, C., Ottaviani, C., Olivieri, B., De Padova, P., and Le Lay, G. (2014b). The fate of the $2\sqrt{3}\times 2\sqrt{3}R$ (30°) silicene phase on Ag (111). *APL Mater.*, **2**, 92513.

Mahatha, S.K., Moras, P., Bellini, V., Sheverdyaeva, P.M., Struzzi, C., Petaccia, L., and Carbone, C. (2014). Silicene on Ag (111): A honeycomb lattice without Dirac bands. *Phys. Rev. B*, **89**, 201416.

Matthes, L., Gori, P., Pulci, O., and Bechstedt, F. (2013a). Universal infrared absorbance of two-dimensional honeycomb group-IV crystals. *Phys. Rev. B*, **87**, 035438.

Matthes, L., Pulci, O., and Bechstedt, F. (2013b). Massive Dirac quasiparticles in the optical absorbance of graphene, silicene, germanene, and tinene. *J. Phys. Condens. Matter*, **25**, 395305.

Matthes, L., Pulci, O., and Bechstedt, F. (2014). Optical properties of two-dimensional honeycomb crystals graphene, silicene, germanene, and tinene from first principles. *New J. Phys.*, **16**, 105007.

Meng, L., Wang, Y., Zhang, L., Du, S., Wu, R., Li, L., Zhang, Y., et al. (2013). Buckled silicene formation on Ir (111). *Nano Lett.*, **13**, 685–690.

Molle, A., Chiappe, D., Cinquanta, E., Grazianetti, C., Fanciulli, M., Scalise, E., van den Broek, B., and Houssa, M. (2013). Structural and chemical stabilization of the epitaxial silicene. *ECS Trans*, **58**(7), 217–227.

Morishita, T., Russo, S.P., Snook, I.K., Spencer, M.J., Nishio, K., and Mikami, M. (2010). First-principles study of structural and electronic properties of ultrathin silicon nanosheets. *Phys. Rev. B*, **82**, 04519.

Morishita, T., Spencer, M.J., Kawamoto, S., and Snook, I.K. (2013). A new surface and structure for silicene: Polygonal silicene formation on the Al (111) surface. *J. Phys. Chem. C*, **117**, 22142–22148.

Naji, S., Khalil, B., Labrim, H., Bhihi, M., Belhaj, A., Benyoussef, A., Lakhal, M., and El Kenz, A. (2014). Interdistance effects on flat and buckled silicene like-bilayers. *J. Phys. Conf. Ser.*, **491**, 12006.

Neek-Amal, M., Sadeghi, A., Berdiyorov, G.R., and Peeters, F.M. (2013). Realization of free-standing silicene using bilayer graphene. *Appl. Phys. Lett.*, **103**, 261904.

Ng, T.Y., Yeo, J., and Liu, Z. (2013). Molecular dynamics simulation of the thermal conductivity of shorts strips of graphene and silicene: A comparative study. *Int. J. Mech. Mater. Des.*, **9**, 105–114.

Ni, Z., Liu, Q., Tang, K., Zheng, J., Zhou, J., Qin, R., Gao, Z., Yu, D., and Lu, J. (2011). Tunable bandgap in silicene and germanene. *Nano Lett.*, **12**, 113–118.

Ni, Z., Zhong, H., Jiang, X., Quhe, R., Luo, G., Wang, Y., Ye, M., Yang, J., Shi, J., and Lu, J. (2014). Tunable band gap and doping type in silicene by surface adsorption: Towards tunneling transistors. *Nanoscale*, **6**, 7609–7618.

Padilha, J.E., and Pontes, R.B. (2015). Free-standing bilayer silicene: The effect of stacking order on the structural, electronic, and transport properties. *J. Phys. Chem. C*, **119**, 3818–3825.

Pan, F., Wang, Y., Jiang, K., Ni, Z., Ma, J., Zheng, J., Quhe, R., et al. (2015). Silicene nanomesh. *Sci. Rep.*, **5**, 9075.

Pan, L., Liu, H.J., Tan, X.J., Lv, H.Y., Shi, J., Tang, X.F., and Zheng, G. (2012). Thermoelectric properties of armchair and zigzag silicene nanoribbons. *Phys. Chem. Chem. Phys.*, **14**, 13588–13593.

Pan, L., Liu, H.J., Wen, Y.W., Tan, X.J., Lv, H.Y., Shi, J., and Tang, X.F. (2011). First-principles study of mono-layer and bilayer honeycomb structures of group-IV elements and their binary compounds. *Phys. Lett. A*, **375**, 614–619.

Parrish, K.D., Jain, A., Larkin, J.M., Saidi, W.A., and McGaughey, A.J. (2014). Origins of thermal conductivity changes in strained crystals. *Phys. Rev. B*, **90**, 235201.

Pei, Q.-X., Zhang, Y.-W., Sha, Z.-D., and Shenoy, V.B. (2013). Tuning the thermal conductivity of silicene with tensile strain and isotopic doping: A molecular dynamics study. *J. Appl. Phys.*, **114**, 33526.

Peng, Q., Wen, X., and De, S. (2013). Mechanical stabilities of silicene. *RSC Adv.*, **3**, 13772–13781.

Pflugradt, P., Matthes, L., and Bechstedt, F. (2014a). Silicene on metal and metallized surfaces: Ab initio studies. *New J. Phys.*, **16**, 75004.

Pflugradt, P., Matthes, L., and Bechstedt, F. (2014b). Silicene-derived phases on Ag (111) substrate versus coverage: Ab initio studies. *Phys. Rev. B*, **89**, 035403.

Pflugradt, P., Matthes, L., and Bechstedt, F. (2014c). Unexpected symmetry and AA stacking of bilayer silicene on Ag (111). *Phys. Rev. B*, **89**, 205428.

Podsiadły-Paszkowska, A., and Krawiec, M. (2015). Dirac fermions in silicene on Pb (111) surface. *Phys. Chem. Chem. Phys.*, **17**, 2246–2251.

Prévôt, G., Bernard, R., Cruguel, H., and Borensztein, Y. (2014). Monitoring Si growth on Ag (111) with scanning tunneling microscopy reveals that silicene structure involves silver atoms. *Appl. Phys. Lett.*, **105**, 213106.

Qin, R., Wang, C.-H., Zhu, W., and Zhang, Y. (2012). First-principles calculations of mechanical and electronic properties of silicene under strain. *AIP Adv.*, **2**, 22159.

Qin, R., Zhu, W., Zhang, Y., and Deng, X. (2014). Uniaxial strain-induced mechanical and electronic property modulation of silicene. *Nanoscale Res. Lett.*, **9**, 1.

Qiu, J., Fu, H., Xu, Y., Oreshkin, A.I., Shao, T., Li, H., Meng, S., Chen, L., and Wu, K. (2015). Ordered and reversible hydrogenation of silicene. *Phys. Rev. Lett.*, **114**, 126101.

Quhe, R., Fei, R., Liu, Q., Zheng, J., Li, H., Xu, C., Ni, Z., et al. (2012). Tunable and sizable band gap in silicene by surface adsorption. *Sci. Rep.*, **2**, 853.

Quhe, R., Yuan, Y., Zheng, J., Wang, Y., Ni, Z., Shi, J., Yu, D., Yang, J., and Lu, J. (2014). Does the Dirac cone exist in silicene on metal substrates? *Sci. Rep.*, **4**, 5476.

Resta, A., Leoni, T., Barth, C., Ranguis, A., Becker, C., Bruhn, T., Vogt, P., and Le Lay, G. (2013). Atomic structures of silicene layers grown on Ag (111): Scanning tunneling microscopy and noncontact atomic force microscopy observations. *Sci. Rep.*, **3**, 2399.

Ronci, F., Serrano, G., Gori, P., Cricenti, A., and Colonna, S. (2014). Silicon-induced faceting at the Ag (110) surface. *Phys. Rev. B*, **89**, 115437.

Sacconi, F., Persson, M.P., Povolotskyi, M., Latessa, L., Pecchia, A., Gagliardi, A., Balint, A., Fraunheim, T., and Di Carlo, A. (2007). Electronic and transport properties of silicon nanowires. *J. Comput. Electron.*, **6**, 329–333.

Sadeghi, H., Sangtarash, S., and Lambert, C.J. (2015). Enhanced thermoelectric efficiency of porous silicene nanoribbons. *Sci. Rep.*, **5**, 9514.

Şahin, H., Cahangirov, S., Topsakal, M., Bekaroglu, E., Akturk, E., Senger, R.T., and Ciraci, S. (2009). Monolayer honeycomb structures of group-IV elements and III-V binary compounds: First-principles calculations. *Phys. Rev. B*, **80**, 155453.

Sakai, Y., and Oshiyama, A. (2015). Structural stability and energy-gap modulation through atomic protrusion in free-standing bilayer silicene. *Phys. Rev. B*, **91**, 201405.

Salomon, E., El Ajjouri, R., Le Lay, G., and Angot, T. (2014). Growth and structural properties of silicene at multilayer coverage. *J. Phys. Condens. Matter*, **26**, 185003.

Scalise, E., Houssa, M., Pourtois, G., van den Broek, B., Afanas'ev, V., Stesmans, A. (2013). Vibrational properties of silicene and germanene. Nano Research 6, 19–28.

Scalise, E., Cinquanta, E., Houssa, M., van den Broek, B., Chiappe, D., Grazianetti, C., Pourtois, G., et al. (2014). Vibrational properties of epitaxial silicene layers on (111) Ag. *Appl. Surf. Sci.*, **291**, 113–117.

Seol, J.H., Jo, I., Moore, A.L., Lindsay, L., Aitken, Z.H., Pettes, M.T., Li, X., et al. (2010). Two-dimensional phonon transport in supported graphene. *Science*, **328**, 213–216.

Shirai, T., Shirasawa, T., Hirahara, T., Fukui, N., Takahashi, T., and Hasegawa, S. (2014). Structure determination of multilayer silicene grown on Ag (111) films by electron diffraction: Evidence for Ag segregation at the surface. *Phys. Rev. B*, **89**, 241403.

Shu, H., Cao, D., Liang, P., Wang, X., Chen, X., and Lu, W. (2014). Two-dimensional silicene nucleation on a Ag (111) surface: Structural evolution and the role of surface diffusion. *Phys. Chem. Chem. Phys.*, **16**, 304–310.

Sone, J., Yamagami, T., Aoki, Y., Nakatsuji, K., and Hirayama, H. (2014). Epitaxial growth of silicene on ultra-thin Ag (111) films. *New J. Phys.*, **16**, 95004.

Song, Y.-L., Zhang, S., Lu, D.-B., Xu, H., Wang, Z., Zhang, Y., and Lu, Z.-W. (2013). Band-gap modulations of armchair silicene nanoribbons by transverse electric fields. *Eur. Phys. J. B*, **86**, 1–6.

Song, Y.-L., Zhang, Y., Zhang, J.-M., and Lu, D.-B. (2010). Effects of the edge shape and the width on the structural and electronic properties of silicene nanoribbons. *Appl. Surf. Sci.*, **256**, 6313–6317.

Speiser, E., Buick, B., Esser, N., Richter, W., Colonna, S., Cricenti, A., and Ronci, F. (2014). Raman spectroscopy study of silicon nanoribbons on Ag (110). *Appl. Phys. Lett.*, **104**, 161612.

Spencer, M.J., Morishita, T., and Snook, I.K. (2012). Reconstruction and electronic properties of silicon nanosheets as a function of thickness. *Nanoscale*, **4**, 2906–2913.

Stephan, R., Hanf, M.-C., and Sonnet, P. (2014). Spatial analysis of interactions at the silicene/Ag interface: First principles study. *J. Phys. Condens. Matter*, **27**, 15002.

Takeda, K., and Shiraishi, K. (1994). Theoretical possibility of stage corrugation in Si and Ge analogs of graphite. *Phys. Rev. B*, **50**, 014916.

Tao, L., Cinquanta, E., Chiappe, D., Grazianetti, C., Fanciulli, M., Dubey, M., Molle, A., and Akinwande, D. (2015). Silicene field-effect transistors operating at room temperature. *Nat. Nanotechnol.*, **10**, 227–231.

Tchalala, M.R., Enriquez, H., Mayne, A.J., Kara, A., Dujardin, G., Ali, M.A., and Oughaddou, H. (2014). Atomic structure of silicene nanoribbons on Ag (110). *J. Phys.: Conf. Ser.*, **491**, 012002.

Tchalala, M.R., Enriquez, H., Mayne, A.J., Kara, A., Roth, S., Silly, M.G., Bendounan, A., et al. (2013). Formation of one-dimensional self-assembled silicon nanoribbons on Au (110)-(2$\times$ 1). *Appl. Phys. Lett.*, **102**, 083107.

Vogt, P., Capiod, P., Berthe, M., Resta, A., De Padova, P., Bruhn, T., Le Lay, G., and Grandidier, B. (2014). Synthesis and electrical conductivity of multilayer silicene. *Appl. Phys. Lett.*, **104**, 21602.

Vogt, P., De Padova, P., Quaresima, C., Avila, J., Frantzeskakis, E., Asensio, M.C., Resta, A., Ealet, B., and Le Lay, G. (2012). Silicene: Compelling experimental evidence for graphenelike two-dimensional silicon. *Phys. Rev. Lett.*, **108**, 155501.

Voon, L.L.Y., Lopez-Bezanilla, A., Wang, J., Zhang, Y., and Willatzen, M. (2015). Effective Hamiltonians for phosphorene and silicene. *New J. Phys.*, **17**, 25004.

Wang, B., Wu, J., Gu, X., Yin, H., Wei, Y., Yang, R., and Dresselhaus, M. (2014a). Stable planar single-layer hexagonal silicene under tensile strain and its anomalous Poisson's ratio. *Appl. Phys. Lett.*, **104**, 081902.

Wang, Y., Ni, Z., Liu, Q., Quhe, R., Zheng, J., Ye, M., Yu, D., et al. (2015a). All-metallic vertical transistors based on stacked Dirac materials. *Adv. Funct. Mater.*, **25**, 68–77.

Wang, Y., Zheng, R., Gao, H., Zhang, J., Xu, B., Sun, Q., and Jia, Y. (2014b). Metal adatoms-decorated silicene as hydrogen storage media. *Int. J. Hydrog. Energy.*, **39**, 14027–14032.

Wang, Y.-P. and Cheng, H.-P. (2013). Absence of a Dirac cone in silicene on Ag (111): First-principles density functional calculations with a modified effective band structure technique. *Phys. Rev. B*, **87**, 245430.

Wang, Z., Feng, T., and Ruan, X. (2015b). Thermal conductivity and spectral phonon properties of freestanding and supported silicene. *J. Appl. Phys.*, **117**, 084317.

Wei, W., Dai, Y., Huang, B., and Jacob, T. (2013). Many-body effects in silicene, silicane, germanene and germanane. *Phys. Chem. Chem. Phys.*, **15**, 8789–8794.

Wei, W., and Jacob, T. (2013). Strong many-body effects in silicene-based structures. *Phys. Rev. B*, **88**, 045203.

Wierzbicki, M., Barnaś, J., and Swirkowicz, R. (2015). Zigzag nanoribbons of two-dimensional silicene-like crystals: Magnetic, topological and thermoelectric properties. *J. Phys. Condens. Matter*, **27**, 485301.

Xie, H., Hu, M., and Bao, H. (2014). Thermal conductivity of silicene from first-principles. *Appl. Phys. Lett.*, **104**, 131906.

Xie, H., Ouyang, T., Germaneau, É., Qin, G., Hu, M., and Bao, H. (2016). Large tunability of lattice thermal conductivity of monolayer silicene via mechanical strain. *Phys. Rev. B*, **93**, 075404.

Xu, P., Yu, Z., Yang, C., Lu, P., Liu, Y., Ye, H., and Gao, T. (2014a). Comparative study on the nonlinear properties of bilayer graphene and silicene under tension. *Superlattices Microstruct.*, **75**, 647–656.

Xu, X., Zhuang, J., Du, Y., Eilers, S., Peleckis, G., Yeoh, W., Wang, X., Dou, S.X., and Wu, K. (2014b). Epitaxial growth mechanism of silicene on Ag (111). In *2014 International Conference on Nanoscience and Nanotechnology*. Adelaide, South Australia, 2014, pp. 28–30.

Yan, Q., Huang, B., Yu, J., Zheng, F., Zang, J., Wu, J., Gu, B.-L., Liu, F., and Duan, W. (2007). Intrinsic current-voltage characteristics of graphene nanoribbon transistors and effect of edge doping. *Nano Lett.*, **7**, 1469–1473.

Yan, Y., Wu, H., Jiang, F., and Zhao, H. (2013). Enhanced thermopower of gated silicene. *Eur. Phys. J. B*, **86**, 1–5.

Yang, K., Cahangirov, S., Cantarero, A., Rubio, A., and D'Agosta, R. (2014). Thermoelectric properties of atomically thin silicene and germanene nanostructures. *Phys. Rev. B*, **89**, 125403.

Yaokawa, R., Ohsuna, T., Morishita, T., Hayasaka, Y., Spencer, M.J., and Nakano, H. (2016). Monolayer-to-bilayer transformation of silicenes and their structural analysis. *Nat. Commun.*, **7**, 10657.

Yi, K.S., Trivedi, K., Floresca, H.C., Yuk, H., Hu, W., and Kim, M.J. (2011). Room-temperature quantum confinement effects in transport properties of ultrathin Si nanowire field-effect transistors. *Nano Lett.*, **11**, 5465–5470.

Yuan, Y., Quhe, R., Zheng, J., Wang, Y., Ni, Z., Shi, J., and Lu, J. (2014). Strong band hybridization between silicene and Ag (111) substrate. *Phys. E Low-Dimens. Syst. Nanostructures*, **58**, 38–42.

Zberecki, K., Wierzbicki, M., Barnaś, J., and Swirkowicz, R. (2013). Thermoelectric effects in silicene nanoribbons. *Phys. Rev. B*, **88**, 115404.

Zhang, H., and Wang, R. (2011). The stability and the nonlinear elasticity of 2D hexagonal structures of Si and Ge from first-principles calculations. *Phys. B Condens. Matter*, **406**, 4080–4084.

Zhang, X., Bao, H., and Hu, M. (2015). Bilateral substrate effect on the thermal conductivity of two-dimensional silicon. *Nanoscale*, **7**, 6014–6022.

Zhang, X., Xie, H., Hu, M., Bao, H., Yue, S., Qin, G., and Su, G. (2014). Thermal conductivity of silicene calculated using an optimized Stillinger-Weber potential. *Phys. Rev. B*, **89**, 054310.

Zhao, H. (2012). Strain and chirality effects on the mechanical and electronic properties of silicene and silicane under uniaxial tension. *Phys. Lett. A*, **376**, 3546–3550.

Zhao, W., Guo, Z.X., Zhang, Y., Ding, J.W., and Zheng, X.J. (2016). Enhanced thermoelectric performance of defected silicene nanoribbons. *Solid State Commun.*, **227**, 1–8.

Zhuang, J., Xu, X., Du, Y., Wu, K., Chen, L., Hao, W., Wang, J., Yeoh, W.K., Wang, X., and Dou, S.X. (2015). Investigation of electron-phonon coupling in epitaxial silicene by in situ Raman spectroscopy. *Phys. Rev. B*, **91**, 161409.

3
Two-dimensional silicon nanosheets

Hideyuki Nakano, Ritsuko Yaokawa, and Masataka Ohashi

Contents

3.1 INTRODUCTION

Since the discovery of graphene, a single-atom-thick honeycomb carbon structure, a great deal of effort has been both theoretically and experimentally dedicated to the search for similar two-dimensional (2D) materials comprising group-IV elements, especially silicon. The silicon equivalent of graphene was first theoretically proposed and later reinvestigated, which renamed it silicene (Takeda and Shiraishi 1994; Guzman-Verri and Voon 2007). In those reports, silicene was described as a single-atom-thick silicon crystalline material with a honeycomb lattice comprising sp^2 silicon bonds. Therefore, silicene is not a naturally occurring material and a solid phase of silicon similar to graphite is lacking in nature. Thus, pure silicene layers cannot be obtained by exfoliation methods as initially performed in the case of graphene, and more sophisticated methods have to be considered for the growth or synthesis of silicene.

Silicene was first synthesized by following a bottom-up approach in which silicon was deposited on metal surfaces that do not strongly interact with Si atoms or the resultant compounds. Ag surfaces have been recently used to grow Si superstructures and to grow massively parallel Si nanoribbons on Ag(001) and Ag(110) surfaces, respectively (Sahaf et al. 2007; De Padova et al. 2008; De Padova et al. 2010). These nanoribbons were shown to comprise hexagonal honeycomb-like Si-based unit cells, in agreement with density functional theory (DFT) calculations (Kara et al. 2009; De Padova et al. 2011).

These structures were also shown to have a linear electronic dispersion similar to massless relativistic Dirac fermions, which can be seen as the first evidence of a local silicene-like arrangement with sp^2-hybridized Si atoms. Subsequently, in 2012, actual 2D silicene sheets were formed on Ag(111) surfaces with a sixfold top-layer symmetry (Vogt et al. 2012). In 2015, the first silicene transistor was synthesized by a tricky process with modest performance and lifetime (i.e., measured in minutes) characteristics (Tao et al. 2015). This work was surprising to many researchers, and the silicene transistor field is rapidly progressing.

Although graphene is the most conductive substance known, it lacks a crucial characteristic. Unlike the semiconductors used in computer chips, graphene lacks the band gaps required to switch on and off and to perform "logical" operations on bits. On the other hand, silicene has a band gap since some of its atoms buckle upwards to form corrugated ridges, thereby slightly modifying the energy state of some electrons. However, handling silicene in the laboratory has been extremely challenging. Thus, silicene deposited under high-vacuum conditions almost comprises sp^2 silicon, which is easily oxidized under ambient conditions. Therefore, if silicene is to be used under ambient conditions, 2D silicon nanosheets (SiNSs) with sp^3 silicon bonds should be selected. The SiNSs described here are defined as compounds having silicon bonding with sp^3 hybridization.

In contrast to conventional physical methods (i.e., deposition of silicene), soft chemical synthesis is one candidate for methods and they are anticipated to receive significant attention for a wide range of applications. The chemical reaction of organosilicon compounds is a simple, gentle method to build up silicon nanomaterials. For example, silicon nanoparticles can be synthesized by reacting tetrachlorosilane with the Zintl compound (Teo and Sun 2007). However, the synthesis of 2D nanocrystalline silicon structures is complex compared with cluster structures. Thus, the reaction of trihalosilane with alkali metals was expected to synthesize 2D SiNS, but amorphous bridged polysilane compounds and/or small silicon clusters were obtained instead.

According to previous reports on metal-oxide nanosheets (Wang and Sasaki 2014), exfoliation is a reliable approach for synthesizing SiNS from isotropic 2D silicon compounds. The complete list of layered silicon compounds includes Zintl $CaSi_2$, layered polysilane (Si_6H_6), and siloxene ($Si_6H_3(OH)_3$), which are crystalline layered structures composed of buckled 2D silicon sublattices similar to the (111) plane of diamond-type Si. Single silicon layers of Si_6H_6 and $Si_6H_3(OH)_3$ were stabilized through termination with hydrogen or hydroxyl groups placed out of the layer plane. These layered silicon compounds are composed of sp^3 silicon bonding. Although the single layers in these materials weakly interact with each other, they are not easily exfoliated, therefore, the solubility into organic or inorganic solvents is very low. Thus, some modifications of the silicon surface are required to obtain 2D SiNS. Here we summarize the main characteristics of 2D silicon compounds, along with the synthetic methods of SiNS and some properties of these materials.

3.2 MONOLAYER SILICENE COMPOUNDS

3.2.1 GROWTH OF CALCIUM-INTERCALATED SILICENE, $CaSi_2$

$CaSi_2$, one of the Zintl phases (Zintl 1939; Schäfer et al. 1973), is an intermetallic compound formed between a strongly electropositive metal (e.g., alkali metals and alkaline earth metals) and a somewhat less electropositive atom such as Si. $CaSi_2$ has a 2D silicon subnetwork resembling buckled Si(111) planes in which the Si_6 rings are interconnected with sp^3 bonds while the puckered $(Si^-)_n$ polyanion layers are separated by planar monolayers of Ca^{2+}. $CaSi_2$ has a trigonal structure (space group R-3m) at ambient pressure, which can only vary in its stacking sequence to give rise to tr6 ($a = 0.3855$ nm, $c = 3.062$ nm) (Evers 1979) and tr3 ($a = 0.3829$ nm, $c = 1.590$ nm) (Dick and Ohlinger 1998) polymorphs. In tr6 $CaSi_2$, the stacking of the trigonal Ca layers follows an AABBCC sequence with a six-layer repeat distance (Figure 3.1a and b). Tr3 $CaSi_2$ has a three-layer repeat distance, with the Ca layers stacked following an ABC sequence (Figure 3.1c). $CaSi_2$ ingots typically crystallize in the tr6 polymorph. Thus, it has been reported that only 3 out of more than 200 samples crystallized in the tr3 polymorph (Dick and Ohlinger 1998). Tr6 $CaSi_2$ is more stable than its tr3 counterpart. Additionally, the calculated Gibbs energy relation suggests that tr6 $CaSi_2$ stabilizes over tr3 with higher temperature (Nedumkandathil et al. 2015). It has been speculated that impurities and/or defects stabilize the tr3 $CaSi_2$ phase (Evers 1979; Vogg et al. 1999; Nedumkandathil et al. 2015).

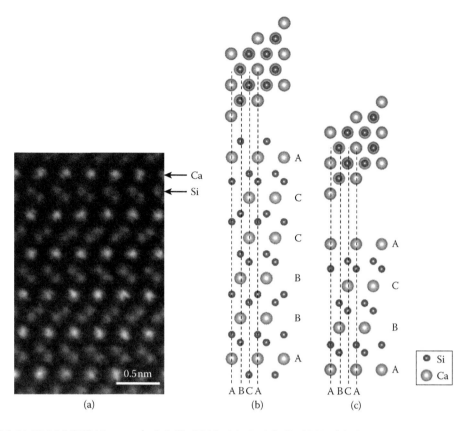

Figure 3.1 (a) HAADF-STEM image of tr6 CaSi$_2$. (b) Model of tr6 CaSi$_2$. (c) Model of tr3 CaSi$_2$.

Several CaSi$_2$ single-crystal growth experiments have been reported. Single-crystal specimens with lateral sizes of 8 × 8 mm and 5 × 5 mm were obtained by slow cooling of ingots (Evers and Weiss 1974; Yamanaka et al. 1981). Using the floating zone method, a CaSi$_2$ single crystal with a diameter of 10 mm and length of 100 mm was also fabricated (Hirano 1991). The generally accepted Ca–Si binary phase diagram indicates that the process CaSi + liquid → CaSi$_2$ occurs as a metastable peritectic reaction in the high-purity melt, on the other hand, the stable peritectic reaction Ca$_{14}$Si$_{19}$ + liquid → CaSi$_2$ takes place at CaSi$_2$ stoichiometric compositions when the melt contains some impurities (Figure 3.2a) (Yaokawa et al. 2014). A CaSi$_2$ single phase is formed from the high-purity stoichiometric CaSi$_2$ melt by using 5N Ca and 5N Si as starting materials while slightly undercooling (by only 5°C). Thus, a tr6 CaSi$_2$ single-phase ingot can be obtained by casting at 1–100°C/min cooling rates (Figure 3.2b through d).

3.2.2 ELECTRONIC PROPERTIES OF MONOLAYER SILICENE IN CaSi$_2$

CaSi$_2$, considered a calcium-intercalated silicene, provides a good opportunity to investigate the intrinsic electronic structure of genuine silicene. While the Si atoms in CaSi$_2$ form a hexagonal 2D layer, silicene synthesized on a silver substrate has been reported to have a different structure (i.e., dumbbell structure) (Cahangirov et al. 2014). In CaSi$_2$, the Ca atoms are considered to play a role in stabilizing the buckled honeycomb network structure of Si atoms by donating electrons to the silicene layers (Cahangirov et al. 2009; Rui et al. 2013).

The electronic properties (i.e., band structure) of silicene in a tr3 CaSi$_2$ single crystal can be analyzed by high-resolution angle-resolved photoemission spectroscopy (ARPES). From the X-ray diffraction (XRD) patterns of the powder and the single-crystal samples (Figure 3.3a and b), the grown tr3 CaSi$_2$ crystal reveals the high-quality crystal. Low-energy electron diffraction (LEED) measurements of the cleaved surface also show a clear 1 × 1 pattern, thereby indicating that no reconstruction of the crystal structure took

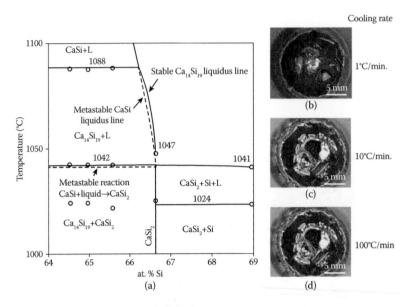

Figure 3.2 (a) Phase diagram of the Ca–Si binary system around $CaSi_2$ obtained on the basis of differential thermal analysis (DTA) results. Top-surface pictures of ingots fabricated from $Ca_{1.00}Si_2$ melt at varying cooling rates: (b) 1, (c) 10, and (d) 100°C/min, respectively.

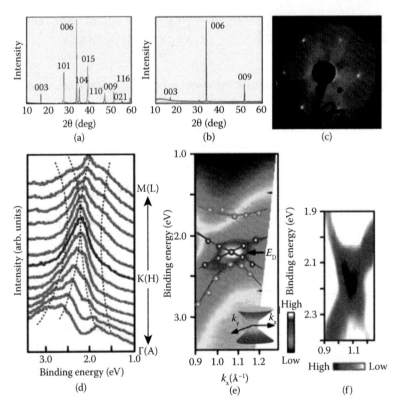

Figure 3.3 (a) XRD pattern from the tr3 $CaSi_2$ crystal powder sample. (b) XRD pattern from the (111) cleavage plane of a $CaSi_2$ single crystal. (c) LEED pattern from the (111) cleavage plane of a $CaSi_2$ crystal measured with a primary electron energy of 80 eV. (d) High-resolution ARPES spectra near the Dirac point at *ca.* the K(H) point in tr3 $CaSi_2$. (e) ARPES intensity plot around the K(H) point as a function of the wave vector and the binding energy. The peak position in the ARPES spectra, obtained by fitting with Lorentzians, is shown by blue and red circles. (f) Second-derivative ARPES intensity plot around E_D.

place at the surface (Figure 3.3c). In the ARPES-derived band structure, a massless Dirac cone of dispersed π electrons at the K(H) point in the Brillouin zone is clearly observed, together with σ-band dispersions at the Γ point (Noguchi et al. 2015). Furthermore, the Dirac point is located at *ca.* 2 eV from the Fermi level (E_F), thereby revealing a substantial charge transfer from the Ca atoms to the silicene layers (Figure 3.3d through f). The ARPES results indicate that the sp^2 bonding framework essentially holds the $CaSi_2$ structure, thereby producing the massless Dirac-cone state at the K point despite the strongly buckled structure of the silicene layers (i.e., the graphene-like electronic structure is stably formed in this metal-intercalated multilayer silicene).

3.3 BILAYER SILICENE COMPOUNDS

Although it is required for the preparation of practical and adaptable silicene transistors (Tao et al. 2015), the development of significantly more facile and practical processing methods has remained a challenging issue. The most relevant issue is that silicene grows on specific substrates and is stable only under vacuum conditions (Fleurence et al. 2012; Gao and Zhao 2012; Vogt et al. 2012; Morishita et al. 2013). Another issue lies in avoiding the influence of the substrate. Thus, the strong hybridization between Si and the substrate may stabilize silicene grown on specific substrates (Gao and Zhao 2012; Vogt et al. 2012; Cahangirov et al. 2013). Provided these limitations are overcome, silicene could be seriously considered as a next-generation high-efficiency platform for a wide variety of electronic applications. The experimental challenge with forming 2D silicon compounds is that the silicon tetrahedral bonding makes it difficult to exfoliate into silicene without using capping or functional-ization with organic molecules (Okamoto et al. 2010; Sugiyama et al. 2010). However, the existence of an unfunctionalized 2D bilayer Si structure has been predicted by molecular dynamics (MD) calculations. Thus, several types of bilayer silicene were predicted to form in a slit-like pore or vacuum (Morishita et al. 2008; Bai et al. 2010; Johnston et al. 2011; Morishita et al. 2011; Cahangirov et al. 2014; Guo and Oshiyama 2014; Pflugradt et al. 2014; Sakai and Oshiyama 2015). One type of bilayer Si structure that is predicted to form under vacuum conditions (i.e., re-BLSi or reconstructed bilayer Si) has an unusual atomic arrangement consisting of two Si monolayers connected via four-, five-, and six-membered rings while retaining the tetrahedral coordination. However, such a bilayer Si structure has been predicted exclusively in computer simulations, and no one has succeeded in experimentally preparing this material to date.

3.3.1 SYNTHETIC METHODS

Experimentally, three types of bilayer silicene can be synthesized from $CaSi_2$. After annealing in an ionic liquid (i.e., [BMIM][BF_4]) at 250–300°C, a $CaSi_2$ single crystal is transformed into a $CaSi_2F_x$ ($0 \leq x \leq 2.3$) compound through the diffusion of F^-. The concentration gradually decreases from the crystal edge to the interior (Figure 3.4a and b). STEM-EDX elemental mapping identifies the dark and bright crystal domains as CaF_2 and Si phases, respectively (Figure 3.4c through g). These planar domains are identified as trilayer CaF_2, trilayer Si, bilayer CaF_2, and a novel bilayer silicene (denoted as w-BLSi in Figure 3.4g and h) in the $CaSi_2F_{1.8}$ and $CaSi_2F_{2.0}$ compounds shown in Figure 3.4b. Furthermore, two types of bilayer silicenes (i.e., inversion symmetry [i-BLSi] and mirror symmetry [m-BLSi] silicenes) are recognized in the $CaSi_2F_{0.6-1.0}$ composition area (Figure 3.4i). The formation of m-BLSi is in accordance with previous MD predictions (Morishita et al. 2008). The i- and m-BLSi structures must be adjacent to a pair of CaF_2 and $CaSi_2$ crystal layers. The abundance ratio of i-BLSi to m-BLSi is 124:3, as observed in the HAADF-STEM images. Since the calculated energy of i-BLSi is 0.03 eV/atom lower than that of m-BLSi under vacuum conditions, the calculated abundance ratio is qualitatively reasonable.

3.3.2 STRUCTURAL DETERMINATION OF BILAYER SILICENE

Bilayer silicene generally exists as a multiphase compound in $CaSi_2F_x$. Thus, the lattice constants and atomic positions of w-BLSi cannot be directly characterized by XRD. HAADF-STEM images at high magnification with atomic resolution often suffer distortion owing to specimen drift during scanning. Thus, the atomic structure of the bilayer silicene can be determined from HAADF-STEM images taken at

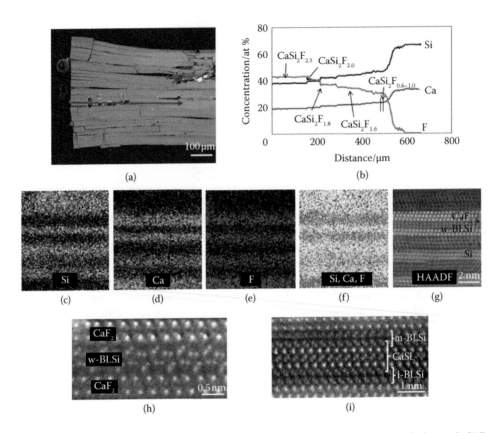

Figure 3.4 (a) Cross-sectional back-scatter detector (BSE) image of the crystal grain including a $CaSi_2F_x$ compound. (b) Electron probe micro-analysis (EPMA) quantitative line analysis result along the red arrow in (a). (c–f) STEM-EDX elemental mapping results of the $CaSi_2F_2$ composition region. One-element mapping (c: Si, d: Ca, and e: F). (f) Overlapped mapping of Si, Ca, and F. (g) HAADF-STEM image of the STEM-EDX elemental mapping area. (h) An enlarged HAADF-STEM image taken from a $CaSi_2F_2$ region in (b); red arrows indicate an F-vacancy site. (i) HAADF-STEM image taken from a $CaSi_2F_{0.6-1.0}$ region in (b); a bright spot contrast, corresponding to the projected atomic positions of m- and i-BLSi, can be observed in the image.

different incident electron beam directions (Figure 3.5a through c). As shown in Figure 3.5d, the w-BLSi structure has 2D translational symmetry and a wavy morphology. This structure consists of two silicenes in alternating chair and boat conformations, which are vertically connected via four-, five-, and six-membered rings. Figure 3.5e through h indicates the corresponding structural models projected in each direction. Since w-BLSi exclusively consists of Si atoms with tetrahedral coordination, the top atom of the five-membered silicon ring possess unsaturated silicon bonds (i.e., dangling bonds). Therefore, compared with monolayer silicene and i- (or m-) BLSi, the density of unsaturated silicon bonds in w-BLSi decreases by 25 and 50%, respectively.

The 2D translational periods of w-BLSi are calculated to be $a = 0.661(2)$ nm and $b = 0.382(3)$ nm, with the two translational axes being normal to each other. The a period of w-BLSi is similar to the triple lattice spacing of d_{11-2} in CaF_2 (0.223 nm), while the b period is similar to that of d-110 in CaF_2 (0.386 nm). Thus, the difference between w-BLSi and CaF_2(111) is lower than the observation error. Since the atomic arrangement of the (111) plane in a CaF_2 crystal exhibited threefold symmetry, three equivalent relative rotation angles are observed between w-BLSi and the CaF_2(111) plane (Figure 3.6a). In addition, the angle between the [01]w-BLSi and the [11]w-BLSi directions is nearly 60° (Figure 3.6b). Nearly all the HAADF-STEM images reveals w-BLSi facing the (111) plane of CaF_2, with the F vacancies (red arrows in Figure 3.4h) on the CaF_2(111) surface being recognized at special positions associated with the wavy structure of w-BLSi.

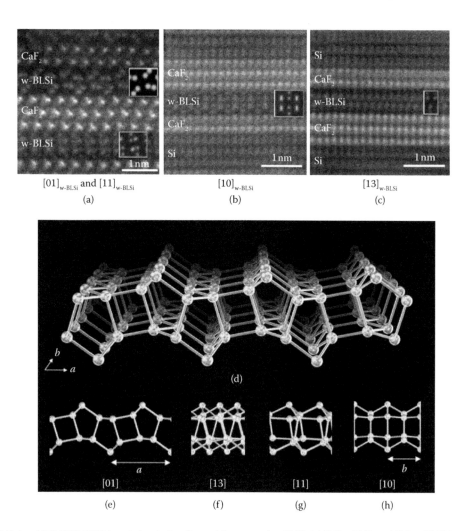

Figure 3.5 (a–c) HAADF-STEM and simulation (insets) images of w-BLSi. (a) [01]w-BLSi and [11] w-BLSi incident directions ([1-10]CaF$_2$). (b) [10]w-BLSi, [11-2]Si, and CaF$_2$ directions. (c) [13]w-BLSi, [11-2]Si, and CaF$_2$ directions. (d) Schematic illustration of the w-BLSi atomic structure. (e–h) Schematic structures projected in each direction in e [01], f [13], g [11], and h [10] directions.

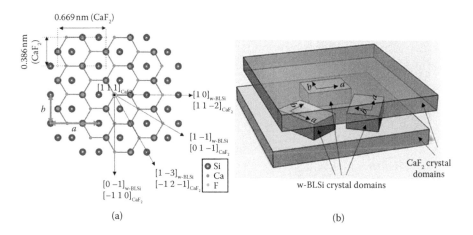

Figure 3.6 (a) Atomic positions in the interface between the w-BLSi (001) and the CaF$_2$ (111) plane. (b) Three equivalent relative rotation angles between w-BLSi and the CaF$_2$(111) plane.

Low-dimensional structures

3.3.3 BAND STRUCTURE

The w-BLSi structure apparently resembles that of re-BLSi (Morishita et al. 2011), although their atomic arrangements are clearly different. *Ab initio* MD calculations were performed for BLSi under the conditions corresponding to the experimentally observed structure (i.e., BLSi sandwiched between two CaF_2 layers with an F-site surface vacancy of 0.5 at the interface). If the MD calculations were started with i-BLSi, this structure would have been immediately transformed to another BLSi structure. The system is subsequently equilibrated, and the resultant BLSi structure is found to perfectly agree with the experimentally observed w-BLSi structure shown in Figure 3.7a. The electronic density of states (EDOSs) for w-BLSi was calculated by the structure in Figure 3.7a, and the decomposed DOS for Si, Ca, and F are shown in Figure 3.7b. The Ca and F bands are located far below the Fermi level, with the valence bands exclusively consisting of Si bands. Unlike a zero-gap semiconductor such as a monolayer (Huang et al. 2013), the band gap of these materials opens to ca. 0.65 eV. The presence of the F vacancies allows the electrons on Ca to be transferred to Si, thereby enhancing the stability of the w-BLSi structure by saturating the dangling bonds. The CaF_{2-x} domains (i.e., ionic crystalline domains) surrounding the Si layers are key to the formation of the w-BLSi structure.

The optical band gap of w-BLSi can be measured from the absorption spectrum of the powder $CaSi_2F_{1.8-2.3}$ sample, although this sample is a mixture of w-BLSi, two types of trilayer silicene, and a CaF_2 layer. The DOS for the 2D crystal is constant with energy ($D(E)$ = const.) (in a three-dimensional (3D) crystal: $D(E) \cong E^{1/2}$). Therefore, the relationship between the absorption coefficient and the band gap energy can be described by the equations $\alpha h\nu$ = const. (direct gap) and $\alpha h\nu = A (h\nu - Eg)$ (indirect gap), where α, h, ν, A, and Eg are the absorption coefficient, Planck's constant, the light frequency, the proportional constant, and the band gap, respectively (Lee et al. 1969; Gaiser et al. 2004; Mak et al. 2010; Bianco et al. 2013). In the case of the 3D crystal, this equation becomes $(\alpha h\nu)^n = A (h\nu - Eg)$, where n = 1/2 and 2 indicate direct and indirect transitions, respectively. Thus, the absorption coefficient of an indirect gap is proportional to the energy, whereas that of a direct gap is constant. The diffuse reflectance spectrum is converted to the Kubelka–Munk function (K-M), which is proportional to the absorption coefficient, as shown in Figure 3.7c. Assuming indirect transitions, the absorption edges of the $CaSi_2F_{1.8-2.3}$ compound are 1.08 and 1.78 eV. From previous DFT results on monolayer and multilayer silicenes terminated with atoms (Morishita et al. 2010; Gao et al. 2012), it is conjectured that the band gap of F-terminated trilayer silicenes should be *ca.* 1 eV within the framework of DFT and the Perdew, Burke, and Ernzerhof technique. It should be noted that DFT calculations using a standard generalized gradient approximation functional tend to underestimate the band gap (by ca. 2/3 in crystal Si). This indicates that the band gap experimentally measured for the trilayer silicene should be ca. 1.5 eV. The band gap for w-BLSi, which was estimated to be ca. 0.65 eV via DFT–PBE calculations, is expected to reach ca. 1 eV in experimental measurements. Therefore, the measured gaps are estimated to be 1.08 and 1.78 eV for w-BLSi and F-terminated trilayer silicenes, respectively.

Low-dimensional structures

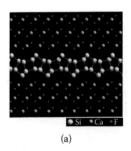
(a)

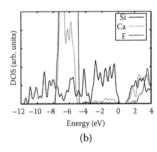
(b)

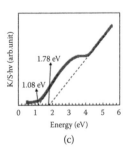
(c)

Figure 3.7 (a) Structure of w-BLSi sandwiched between two CaF_2 crystals, with vacancies at half of the F sites at the interface; this structure was used to calculate the DOS and was obtained from the transformation of i-BLSi in the *ab initio* MD simulation and the subsequent quenching process. (b) Decomposed DOS for Si, Ca, and F in the w-BLSi displayed in (a). (c) Plot of multiplication of the Kubelka–Munk function and energy as a function of energy for $CaSi_2F_{1.8-2.3}$ consisting of w-BLSi, trilayer silicene with dangling bonds, and F-terminated trilayer silicenes.

3.4 EXFOLIATION OF LAYERED SILICON COMPOUNDS

3.4.1 EXFOLIATION OF $CaSi_2$

The formula of $CaSi_2$ including the formal charges is $Ca^{2+}(Si^-)_2$. Thus, the electrostatic interactions between the Ca^{2+} and Si^- layers are strong. However, it is very important to decrease the amount of negative charge on the silicon layers to facilitate exfoliation. With this aim, Mg-doped $CaSi_2$ is used as a starting material to decrease the interaction strength between the Ca and Si layers. Thus, the immersion of bulk $CaSi_{1.85}Mg_{0.15}$ in a solution of propylamine hydrochloride (PA·HCl) leads to deintercalation of the calcium ions, which is accompanied by the evolution of hydrogen (Figure 3.8a). $CaSi_{1.85}Mg_{0.15}$ is subsequently converted into a mixture of Mg-doped SiNS and an insoluble black metallic solid. A light brown suspension containing SiNS can be separated after the sediment is removed from the bottom (Figure 3.8b). Using X-ray photoelectron spectroscopy (XPS), the composition of the as-obtained SiNS is determined to be Si:Mg:O in a ratio of 7.0:1.3:7.5. The Si:Mg ratio is appreciably lower in the starting material $CaSi_{1.85}Mg_{0.15}$, thereby indicating that exfoliation into individual SiNS occurs preferentially in the section of the silicon layer where magnesium atoms are present.

The dimensions of the SiNS can be directly determined by atomic force microscopy (AFM), and the sheets are found to be 0.37 nm thick, with lateral dimensions ranging from 200 to 500 nm (Figure 3.8c and d). The crystallographic thickness of SiNS (0.16 nm) is calculated from its atomic architecture, and the

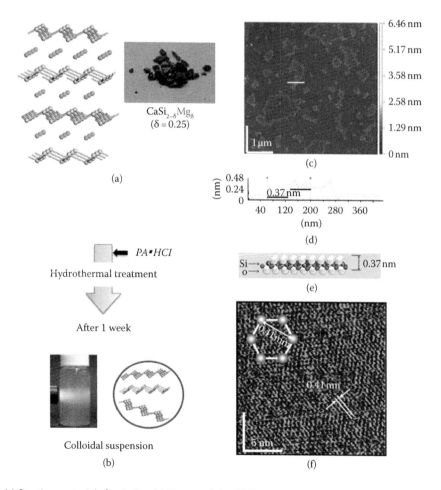

Figure 3.8 (a) Starting materials for $CaSi_2$. (b) Picture of the SiNS suspension. (c) AFM image of Mg-doped SiNS. (d) Line profile taken along the white line in (c). (e) Schematic illustration of Mg-doped SiNS. (f) Atomically resolved AFM image of SiNS.

difference between this value and that obtained by AFM indicates that the surface of the SiNS is stabilized via capping with oxygen atoms as shown in Figure 3.8e. As revealed by the high-resolution AFM images, the closest distance between atoms (i.e., dot-like marks in the AFM image) is 0.41 ± 0.02 nm (Figure 3.8f), which is slightly larger than the distance between Si atoms in the Si(111) plane of bulk crystalline silicon (0.38 nm). These sheets are considered to be very similar to silicene. The overall exfoliation reaction comprised the following steps: (1) The oxidation of $CaSi_{1.85}Mg_{0.15}$ is initiated by the oxidation of the Ca atoms with PA·HCl, which is accompanied by the release of PA; (2) The resulting Mg-doped Si_6H_6 is likely very reactive and thus easily oxidized with water to form gaseous hydrogen; (3) Mg-doped layered silicon with capping oxygen atoms is exfoliated by reaction with the aqueous PA solution, thereby resulting in a stable colloidal suspension of SiNS.

The absorption spectrum of the silicon sheets shows a peak at 268 nm (4.8 eV), corresponding to the L→L critical point in the silicon band structure, which is strongly blue shifted with respect to the bulk indirect band gap of 1.1 eV (Figure 3.9). The absorbance at 268 nm is observed to be linearly dependent on the silicon content and excludes a possible association of nanosheets in this concentration range (Figure 3.9, inset). The photoluminescence (PL) spectrum obtained using an excitation wavelength of 350 nm shows an emission peak at 434 nm (2.9 eV). These results directly indicate that the 2D silicon backbone is maintained, since the PL spectra of the silicon quantum dots with diameters lower than 2 nm show a peak at 3 eV that red shift by as much as 1 eV upon exposure to oxygen. Moreover, the absence of PL from defect or trap-state recombination, which typically occurs near 600 nm, supports the hypothesis that the observed PL is produced by direct electron–hole recombination in the silicon sheets, as occurs in silicon nanocrystals. As shown above, the exfoliation of Mg-doped $CaSi_2$ with PA·HCl results in SiNS with capping oxygen atoms.

3.4.2 LAYERED SILICON COMPOUNDS

Deintercalation of calcium from the interlayer of $CaSi_2$ can be achieved using HCl solutions at low temperatures (Figure 3.10). Thus, $CaSi_2$ changes into amorphous silica with the evolution of significant amounts of hydrogen upon reaction with a 1N HCl solution at room temperature. Treatment with ice-cold concentrated aqueous HCl resulted in $CaSi_2$ being topotactically transformed into a green-yellow solid (i.e., layered siloxene $[Si_6H_3(OH)_3]$) with the release of hydrogen gas, as described in Reaction (3.1):

$$3CaSi_2 + 6HCl + 3H_2O \rightarrow Si_6H_3(OH)_3 + 3CaCl_2 + 3H_2 \qquad (3.1)$$

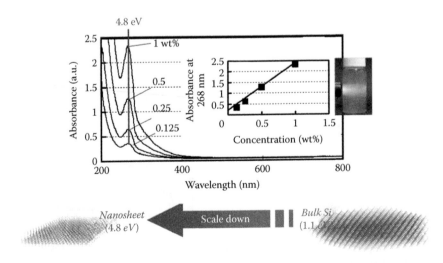

Figure 3.9 Room-temperature optical properties of Mg-doped SiNS. UV/Vis spectra of SiNS suspensions at various concentrations. Inset: the absorbance at 268 nm was plotted against the concentration of SiNS.

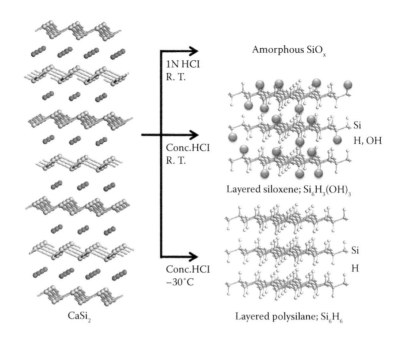

Figure 3.10 Schematic of the reaction of CaSi₂ with HCl aqueous solutions.

The crystalline sheets contain puckered 2D Si layers similar to crystalline Si(111) layers, with Si atoms being stabilized by terminal hydrogen or hydroxide groups located out of the layer plane. At temperatures below –30°C, CaSi₂ is transformed into layered polysilane (Si₆H₆) without the release of hydrogen, as shown in Reaction (3.2). This means that additional reactions between the Si layer and water do not occur.

$$3CaSi_2 + 6HCl \rightarrow Si_6H_6 + 3CaCl_2 \tag{3.2}$$

When compared with CaSi₂, the interlayer bonding between adjacent 2D Si layers is weaker in both $Si_6H_3(OH)_3$ and Si_6H_6 since the bonding takes place by weak hydrogen bonds and van der Waals forces, respectively. Using a solution of HCl in methanol, ethanol, butanol, $C_{12}H_{25}$, benzyl alcohol, or CH_2COOMe, the corresponding alkoxide-terminated organosiloxenes (Si_6H_5OR, where R = methanol, ethanol, butanol, $C_{12}H_{25}$, benzyl alcohol, or CH_2COOMe) can also be obtained. However, the organosiloxenes are insoluble in any organic solvent and hence do not exfoliate.

3.5 FUNCTIONALIZED SILICON NANOSHEETS

The chemical modification–exfoliation process provides a generic approach for preparing organo-chapped silicenes. In this method, layered polysilane (Si_6H_6) was used as a starting layered silicon compound with sp^3 silicon bonding. The hydrosilyl (Si-H) groups on the surface of Si_6H_6 are polarized as $Si^{\delta+}$-$H^{\delta-}$ owing to the lower electronegativity of silicon compared with hydrogen, and can thus be utilized as a reaction point for the chemical modification of silicenes. The reactivity of the Si-H groups has been widely studied in the field of silicon chemistry and various methods for chemically modifying these groups have been developed. Thus, several chemically modified SiNS were successfully synthesized by reacting Si_6H_6 with various chemicals such as Grignard reagent (Sugiyama et al. 2010), hydrosilanes (Nakano et al. 2012), alkyl- or aryl-amines (Okamoto et al. 2015; Ohshita et al. 2016), metallic lithium (Ohashi et al. 2014), and ionic liquids (Nakano et al. 2014), as summarized in Figure 3.11.

3.5.1 SYNTHESIS AND CHARACTERIZATIONS OF Ph-SiNS

The Grignard reaction is one of the important organometallic chemical reactions for C–C and C–Si bond formation (Ashish et al. 1996). In this reaction, aryl-magnesium halides (i.e., Grignard reagents) serve as

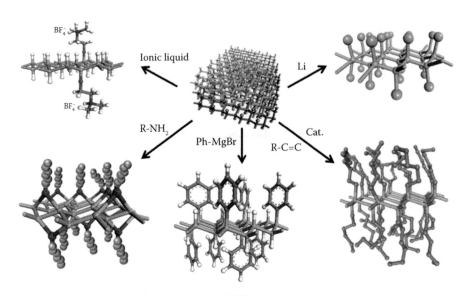

Figure 3.11 Chemical modifications of Si-H groups on layered polysilane (Si_6H_6) surfaces with various chemical moieties.

nucleophiles and react with the electrophilic atoms. Thus, phenyl-modified SiNS (Ph-SiNS) was successfully synthesized by the reaction of Si_6H_6 with phenyl magnesium bromide (Ph-MgBr) in tetrahydrofuran (Sugiyama et al. 2010). Halogenation of the silicon substrates is generally required for Si–C bond formation using Grignard reagents owing to the low reactivity of the Si-H group (Dahn et al. 1993). However, in the case of Si_6H_6, the Si-H groups react with Grignard reagents without halogenation. The as-obtained Ph-SiNS is a colorless paste that dissolves in typical organic solvents. FTIR and ^{1}H-NMR analyses indicate that the composition of Ph-SiNS is $Si_6H_4Ph_2$, thereby revealing the substitution of approximately 30% of the Si-H groups with phenyl groups. The XANES spectrum of this material exhibits two peaks at 1841 and 1844 eV, attributed to the Si–Si and Si–C bonds, respectively, with no peak derived from Si–O bonds being observed.

The AFM images of $Si_6H_4Ph_2$ show a flexible and flat-plane monolayer sheet with 1.11 nm thickness, which is in good agreement with the thickness value of the structural model of Ph-SiNS (Figure 3.12a through c). In addition, atomically resolved AFM images show a periodic arrangement of phenyl groups on the sheet surface as atom-like dots on the $Si_6H_4Ph_2$ surface (Figure 3.12d). The closest distance between the dots is 0.96 nm, which is in good agreement with the distance between phenyl groups in the $Si_6H_4Ph_2$ structural model (Figure 3.12e and f).

3.5.2 OPTICAL PROPERTIES OF AMINO-MODIFIED SiNS

An aryl-methyl moiety, having photoelectric characteristics, can also be attached to the surface of Si_6H_6. Amino-modified SiNS surfaces treated with benzene- (Ph-), naphthalene- (Np-), and carbazole- (Cz-) containing amino substituents are prepared by dehydrogenative coupling of Si_6H_6 with the corresponding amines (Ohshita et al. 2016). XRD analysis of the obtained samples reveals a clear diffraction peak at 1.32 nm for Np-SiNS, which is likely ascribed to the stacked structure. In contrast, no clear XRD peaks are observed for Ph-SiNS and NS-SiNS. This is indicative of the strong tendency of Np-SiNS to form stacking structures, likely originating from the extended conjugation of naphthalene.

The obtained samples exhibit different optical characteristics depending on the nature of the aromatic unit. Thus, UV/V is absorption and PL bands produced by both the aromatic units and the SiNS consisting of 2D silicon frameworks are observed in each sample. The strong stacking nature of Np-SiNS is also confirmed from the PL spectrum, which shows a broad band likely ascribed to π stacking. Furthermore, photoinduced currents are observed for the Np-SiNS and Cz-SiNS films, which are prepared on the electrode by a drop-drying method (Figure 3.13). The photocurrent is generated by UV light irradiation, thereby suggesting that the aromatic substituents and/or silicon frameworks absorb the light for photocurrent generation.

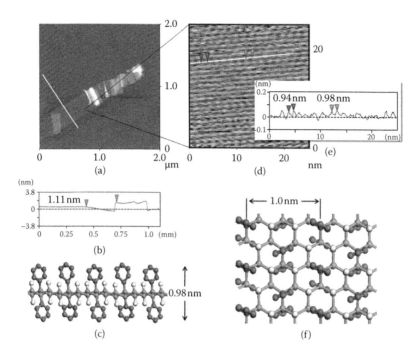

Figure 3.12 (a) AFM image of Ph-SiNS. (b) Line profile along the black line in (a). (c) Side view of the model structure of Ph-SiNS. (d) Atomically resolved AFM image of the surface of Ph-SiNS. (e) Line profile along the black line in (d). (f) Top view of the model structure of Ph-SiNS.

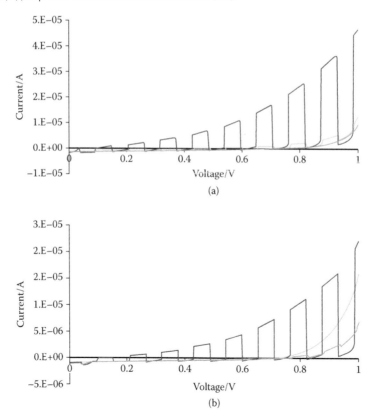

Figure 3.13 Photoinduced current generation behaviors of (a) naphthalene- (Np-) and (b) carbazole- (Cz-) modified SiNS films.

3.5.3 SELF-STACKING PROPERTIES OF AMINO-MODIFIED SiNS

The dehydrogenative coupling reaction can be widely applied to other organoamines for synthesizing other alkyl- or aryl-modified SiNS. Thus, other primary *n*-alkylamines, such as *n*-butyl- (C4-), *n*-hexyl- (C6-), *n*-decyl- (C10-), *n*-dodecyl- (C12-), and *n*-hexadecyl- (C16-) amines, also react with Si_6H_6 to produce the corresponding alkylamine-modified SiNS (Okamoto et al. 2015). The XRD peak attributed to the stacking structure, namely (001) planes, shifts to lower angles depending on the length of the alkyl chain and the interlayer distance [calculated from the (001) reflection] and expands to 1.3–3.35 nm (Figure 3.14a). This result indicates that the interlayer distance can be controlled by the length of the attached alkyl chains. The linear proportional relation between the interlayer distance (*d* spacing) of the obtained samples and the carbon number of the *n*-alkylamine suggests that the attached *n*-alkylamines are regularly arrayed, taking the same conformation. The slope of 0.172 nm suggests a bilayered alkyl chain structure at a tilt angle of *ca.* 47° with respect to the stacking layers (Figure 3.14b and c).

The reaction of Si_6H_6 with α, ω-alkyldiamines, except for diaminopropane, readily progressed to produce alkyldiamine-modified SiNS (Okamoto et al. 2015). Even in this case, the interlayer distance expands depending on the length of the alkyl chain, but the degree of expansion of the interlayer distance is lower than that of *n*-alkylamine-derived SiNS. Since both ends of amino groups in the alkyldiamines react with the SiNS, a single-layer structure of alkyldiamine is formed in the interlayer space. The relationship between the interlayer distance and the carbon number in the alkyldiamine suggests a single-layered alkyldiamine structure at a tilt angle of *ca.* 66° with respect to the stacking layers (Figure 3.15a and b). In addition, the alkyldiamine-modified SiNS shows a low dispersibility against organic solvents, since the SiNS layers should be covalently bonded by the alkyldiamines.

By using alkylamine derivatives such as ω-aminocarboxylic acids, novel functional groups can be introduced on the SiNS surface via the alkyl chains. Thus, 12-aminododecanoic acid ($C_{12}COOH$) reacted with Si_6H_6 in pyridine to form a carboxylic group-containing alkyl chain fixed on the SiNS surface by Si–N linkages (Okamoto et al. 2015). The XRD pattern indicates that the interlayer distance of 12-aminododecanoic acid-modified SiNS ($C_{12}COOH$-SiNSs) is approximately half that of *n*-dodecylamine-modified SiNS. This result implies that $C_{12}COOH$ moieties attached to adjacent silicon layers mutually overlap each other (Figure 3.15c). The introduction of carboxylic groups on SiNS via alkylamine-linkage creates new possibilities for the development of applications such as combinations with other types of layered materials.

3.5.4 THEORETICAL PROPERTIES

Theoretical studies of SiNS, such as DFT or tight-binding calculations, can be used to explore possible stable 2D structures of Si, which are often modeled by considering graphene analogs of Si. DFT and *ab initio* MD simulations were used to investigate the properties of single-layer organo-modified SiNS [$Si_6H_4Ph_2$]. The calculations can be performed within the framework of DFT using the projector-augmented wave method and the generalized gradient approximation using the exchange–correlation functional PBE as implemented in the Vienna *ab initio* simulation package.

The optimized structure of the organo-modified SiNS obtained from DFT calculations is presented in Figure 3.16a. The structure is stable, and the model structure for the $Si_6H_4Ph_2$ nanosheet is highly plausible, because each of the phenyl groups can easily rotate and tilt at 300 K. The band structure (Figure 3.16b) shows that the organo-modified nanosheet has a direct band gap of 1.92 eV, which represents a widened band gap compared with that of bulk Si (0.7 eV) calculated using the same computational parameters. It is interesting to note that the hydrogenated Si single-layer nanosheet shows a comparable indirect band gap of *ca.* 2 eV, while the $Si_6H_4Ph_2$ nanosheet shows a direct band gap of *ca.* 2 eV, thereby demonstrating the significant effect of molecular functionalization (or doubling the thickness of the hydrogenated nanosheet) on its electronic structure.

The total EDOSs is presented in Figure 3.16c through f along with the partial orbital-decomposed EDOS for Si, C, and H atoms. The major contribution to the bands at and just below the highest occupied level is from the Si $2p_x$ and $2p_y$ states (Figure 3.16d), which would involve bonding

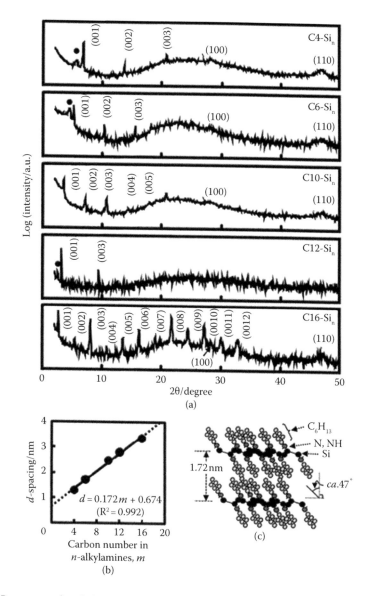

Figure 3.14 (a) XRD patterns of n-alkylamine-modified SiNS having different lengths of alkyl chain. (b) Relationship between the d space and m for Cm-SiNS stacked structures. Peaks marked with circles are derived from the (001) plane of Cm-HCl salts. (c) Structure model of regularly stacked C6-SiNS.

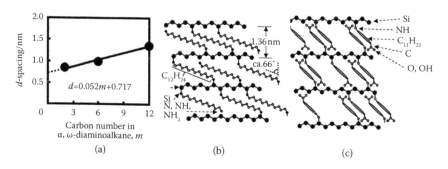

Figure 3.15 (a) Relationship between the d space and m for DiCm-SiNS stacked structures. (b) Structure model of DiC$_{12}$-SiNS. (c) Approximate structure model of C$_{12}$COOH-SiNS.

Low-dimensional structures

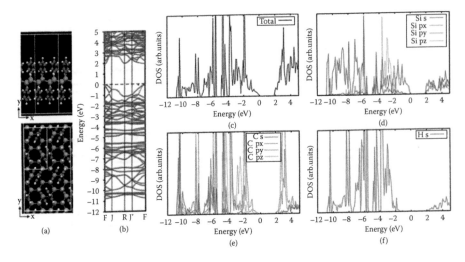

Figure 3.16 (a) Structure of the relaxed $Si_6H_4Ph_2$ showing both side views. (Si, C, and H atoms shown in yellow, blue, and white, respectively) (b) Band structure of $Si_6H_4Ph_2$. The band energy was measured from the highest occupied level at $T = 0$ K. (c) Total EDOS of the organosilicon nanosheet. (d–f) Orbital-decomposed density of states of Si, C, and H atoms, respectively. The zero of energy was aligned to the highest occupied level.

between Si atoms of the nanosheet. In contrast, the main contribution to the conduction band is from the C $2p_x$ and $2p_y$ orbitals (Figure 3.16e). The deeper valence states, between *ca.* 4 and 10 eV below the highest occupied level, are primarily comprised of Si s states, with minor contributions from the C orbitals. The H s bands (Figure 3.16f) are generally coincident with the C p_x, p_y, and p_z bands, as expected from their bonding in the phenyl groups. In addition, the overlap of the H s and Si p_z states, primarily between 2 and 4 eV, indicates the formation of strong Si–H bonds, as indicated by the electron localization function (ELF) plot. Overall, the spiky profile of the CDOS indicates a weak electronic interaction between these atoms, which is also suggested by the band structure and the ELF profile.

3.5.5 ELECTRICAL PROPERTIES OF LITHIATED SiNS

Apart from the abovementioned methods, the Si-H groups present on the Si_6H_6 surface can be modified by mechanochemical solid-phase reaction (Ohashi et al. 2014). Thus, lithiated SiNS (Li-SiNS) were synthesized by mechanical kneading of Si_6H_6 with metallic lithium (Li) using an agate mortar (Figure 3.17a). This mechanochemical lithiation of Si_6H_6 proceeded at room temperature under an argon atmosphere with the release of hydrogen gas, with Li-SiNS being obtained as a homogeneous powder after kneading for 30 min (Figure 3.17b). The substitution rate of the Si-H groups with Li is controlled by the additive amount of metallic lithium. Thus, Li-SiNS having different lithium content ($Si_6H_{6-n}Li_n$) can be prepared (Figure 3.17c).

FTIR measurements indicate that the intensity of the Si–H stretching vibration at 2100 cm^{-1} is reduced in accordance with the rate of lithiation, while the Si–Li stretching vibration is confirmed at 450 cm^{-1} (Figure 3.18a). XANES measurements also support the formation of Si–Li bonds. XRD measurements reveal typical diffraction peaks of Si_6H_6, with layered ((001) and (002) planes) and hexagonal crystal structures of the Si framework ((100) and (110) planes) changing, depending on the extent of the lithiation process. In addition, no diffraction peak attributed to Li metal is observed in the lithiated samples. These results reveal the distortion of the Si framework and the collapse of the layered structure upon lithiation. The electron band structure of Si_6H_6 is also changed as a result of the lithiation process. As shown in Figure 3.18b, the diffuse reflection absorption spectra of $Si_6H_{6-n}Li_n$ reveal a shifting of the absorption band edge of $Si_6H_{6-n}Li_n$ to longer wavelengths while showing lower band gap energies (0.85 eV) compared with Si_6H_6 (2.2 eV). The completely lithiated sample (Si_6Li_6) shows semiconductor-like conductivity values.

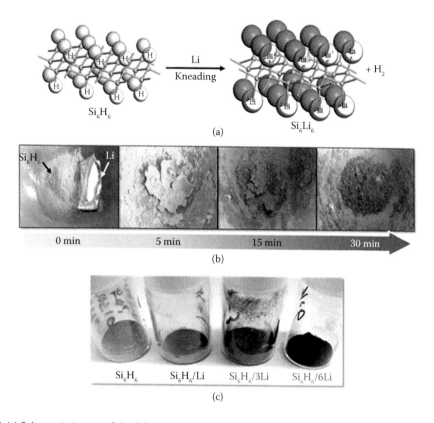

Figure 3.17 (a) Schematic image of the lithiation reaction. (b) Pictures of the lithiation and mechanochemical reaction processes of Si_6H_6 with Li. (c) Pictures showing the color of the obtained composites.

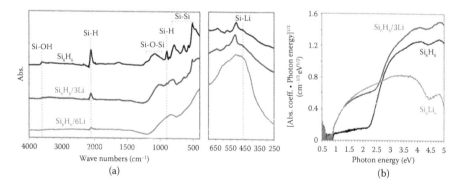

Figure 3.18 (a) FTIR and (b) diffuse reflection absorption spectra of layered polysilane (black) and Li-SiNS (blue: Si_6H_6/3Li, red: Si_6H_6/6Li).

3.6 CONCLUSIONS

Silicene and SiNS are additional examples of 2D novel materials beyond graphene. When compared with graphene, they have some differences (since the sheets are puckered), such as the point-group symmetry and pseudospin. Thus, these materials are predicted to have Dirac cones similar to graphene but with a larger spin-orbital gap and with a band gap opening under a perpendicular electric field. Their physical properties are generally less extreme compared with graphene (e.g., slightly lower mechanical strengths and electron Fermi velocities), although they can present better physical properties, such as thermoelectric figure of merit and a richer topological phase diagram.

Currently, freestanding silicene (i.e., without functional groups) has not been achieved experimentally. Nevertheless, there is a strong resemblance between the organo-modified SiNS reviewed in this text and the predicted freestanding silicene. Therefore, knowledge of the properties of SiNS is a good starting point for understanding the properties of silicene on a substrate or on a functionalized material. Finally, on-demand molecular design and control of the surface modification of 2D silicon nanomaterials are key processes for developing electric devices and energy storage materials in the near future. 2D silicon materials could trigger a revolution in the semiconductor industry to achieve the ultimate in miniaturization if they could be used to build electronic devices.

ACKNOWLEDGMENTS

We thank H. Okamoto and Y. Sugiyama for their helpful discussions. This work was supported in part by PRESTO, the Japan Science and Technology Agency and JSPS KAKENHI Grant Number 60466331.

REFERENCES

Ashish B, Xiuling L, Iver L, Nathan SL, Sang IY, Weinberg WH. (1996). Alkylation of Si surfaces using a two-step halogenation/Grignard route. *Journal of the American Chemical Society* 118: 7225–7226.

Bai J, Tanaka H, Zeng XC. (2010). Graphene-like bilayer hexagonal silicon polymorph. *Nano Research* 3: 694–700.

Bianco E, Butler S, Jiang S, Restrepo OD, Windl W, Goldberger JE. (2013). Stability and exfoliation of germanane: A germanium graphane analogue. *ACS Nano* 7: 4414–4421.

Cahangirov S, Audiffred M, Tang P, Iacomino A, Duan W, Merino G, Rubio A. (2013). Electronic structure of silicene on Ag(111): Strong hybridization effects. *Physical Review B* 88: 035432.

Cahangirov S, Ozcelik VO, Xian L, Avila J, Cho S, Asensio MC, Ciraci S, Rubio A. (2014). Atomic structure of the root 3 x root 3 phase of silicene on Ag(111). *Physical Review B* 90: 035448.

Cahangirov S, Topsakal M, Akturk E, Sahin H, Ciraci S. (2009). Two- and one-dimensional honeycomb structures of silicon and germanium. *Physical Review Letters* 102: 236804.

Dahn JR, Way BM, Fuller E, Tse JS. (1993). Structure of siloxene and layered polysilane (Si6H6). *Physical Review B* 48: 17872–17877.

De Padova P, Quaresima C, Olivieri B, Perfetti P, Le Lay G. (2011). sp(2)-like hybridization of silicon valence orbitals in silicene nanoribbons. *Applied Physics Letters* 98: 081909.

De Padova P, Quaresima C, Ottaviani C, Sheverdyaeva PM, Moras P, Carbone C, Topwal D, et al. (2010). Evidence of graphene-like electronic signature in silicene nanoribbons. *Applied Physics Letters* 96: 261905.

De Padova P, Quaresima C, Perfetti P, Olivieri B, Leandri C, Aufray B, Vizzini S, Le Lay G. (2008). Growth of straight, atomically perfect, highly metallic silicon nanowires with chiral asymmetry. *Nano Letters* 8: 271–275.

Dick S, Ohlinger G. (1998). Crystal structure of calciumdisilicide, 3R-CaSi$_2$. *Zeitschrift Fur Kristallographie-New Crystal Structures* 213: 232–232.

Evers J. (1979). Transformation of 3-connected silicon nets in Casi$_2$. *Journal of Solid State Chemistry* 28: 369–377.

Evers J, Weiss A. (1974). Electrical properties of alkaline earth disilicides and digermanides. *Materials Research Bulletin* 9: 549–553.

Fleurence A, Friedlein R, Ozaki T, Kawai H, Wang Y, Yamada-Takamura Y. (2012). Experimental evidence for epitaxial silicene on diboride thin films. *Physical Review Letters* 108: 245501.

Gaiser C, Zandt T, Krapf A, Serverin R, Janowitz C, Manzke R. (2004). Band-gap engineering with HfSxSe2-x. *Physical Review B* 69: 075205.

Gao J, Zhao J. (2012). Initial geometries, interaction mechanism and high stability of silicene on Ag(111) surface. *Scientific Reports* 2: 861.

Gao N, Zheng WT, Jiang Q. (2012). Density functional theory calculations for two-dimensional silicene with halogen functionalization. *Physical Chemistry Chemical Physics* 14: 257–261.

Guo Z-X, Oshiyama A. (2014). Structural tristability and deep dirac states in bilayer silicene on Ag(111) surfaces. *Physical Review B* 89: 155418.

Guzman-Verri GG, Voon LCLY. (2007). Electronic structure of silicon-based nanostructures. *Physical Review B* 76: 075131.

Hirano T. (1991). Single-crystal growth and electrical-properties of CaSi$_2$. *Journal of the Less-Common Metals* 167: 329–337.

Huang ST, Kang W, Yang L. (2013). Electronic structure and quasiparticle bandgap of silicene structures. *Applied Physics Letters* 102: 133106.

Johnston JC, Phippen S, Molinero V. (2011). A single-component silicon quasicrystal. *Journal of Physical Chemistry Letters* 2: 384–388.

Kara A, Leandri C, Davila M, Padova P, Ealet B, Oughaddou H, Aufray B, Lay G. (2009). Physics of silicene stripes. *Journal of Superconductivity and Novel Magnetism* 22: 259–263.

Lee PA, Said G, Davis R, Lim TH. (1969). On the optical properties of some layer compounds. *Journal of Physics and Chemistry of Solids* 30: 2719–2729.

Mak KF, Lee C, Hone J, Shan J, Heinz TF. (2010). Atomically thin MoS2: A new direct-gap semiconductor. *Physical Review Letters* 105: 136805.

Morishita T, Nishio K, Mikami M. (2008). Formation of single- and double-layer silicon in slit pores. *Physical Review B* 77: 081401.

Morishita T, Russo SP, Snook IK, Spencer MJS, Nishio K, Mikami M. (2010). First-principles study of structural and electronic properties of ultrathin silicon nanosheets. *Physical Review B* 82: 045419.

Morishita T, Spencer MJS, Kawamoto S, Snook IK. (2013). A new surface and structure for silicene: Polygonal silicene formation on the Al(111) aurface. *Journal of Physical Chemistry C* 117: 22142–22148.

Morishita T, Spencer MJS, Russo SP, Snook IK, Mikami M. (2011). Surface reconstruction of ultrathin silicon nanosheets. *Chemical Physics Letters* 506: 221–225.

Nakano H, Nakano M, Nakanishi K, Tanaka D, Sugiyama Y, Ikuno T, Okamoto H, Ohta T. (2012). Preparation of alkyl-modified silicon nanosheets by hydrosilylation of layered polysilane (Si6H6). *Journal of the American Chemical Society* 134: 5452–5455.

Nakano H, Sugiyama Y, Morishita T, Spencer MJS, Snook IK, Kumai Y, Okamoto H. (2014). Anion secondary batteries utilizing a reversible BF4 insertion/extraction two-dimensional Si material. *Journal of Materials Chemistry A* 2: 7588–7592.

Nedumkandathil R, Benson DE, Grins J, Spektor K, Haussermann U. (2015). The 3R polymorph of CaSi$_2$. *Journal of Solid State Chemistry* 222: 18–24.

Noguchi E, Sugawara K, Yaokawa R, Hitosugi T, Nakano H, Takahashi T. (2015). Direct observation of dirac cone in multilayer silicene intercalation compound CaSi$_2$. *Advanced Materials* 27: 856–860.

Ohashi M, Nakano H, Morishita T, Spencer MJS, Ikemoto Y, Yogi C, Ohta T. (2014). Mechanochemical lithiation of layered polysilane. *Chemical Communications* 50: 9761–9764.

Ohshita J, Yamamoto K, Tanaka D, Nakashima M, Kunugi Y, Ohashi M, Nakano H. (2016). Preparation and photocurrent generation of silicon nanosheets with aromatic substituents on the surface. *The Journal of Physical Chemistry C* 120: 10991–10996.

Okamoto H, Kumai Y, Sugiyama Y, Mitsuoka T, Nakanishi K, Ohta T, Nozaki H, Yamaguchi S, Shirai S, Nakano H. (2010). Silicon nanosheets and their self-assembled regular stacking structure. *Journal of the American Chemical Society* 132: 2710–2718.

Okamoto H, Sugiyama Y, Nakanishi K, Ohta T, Mitsuoka T, Nakano H. (2015). Surface modification of layered polysilane with n-alkylamines, alpha,omega-diaminoalkanes, and omega-aminocarboxylic acids. *Chemistry of Materials* 27: 1292–1298.

Pflugradt P, Matthes L, Bechstedt F. (2014). Unexpected symmetry and AA stacking of bilayer silicene on Ag (111). *Physical Review B* 89: 205428.

Rui W, Shaofeng W, Xiaozhi W. (2014). The formation and electronic properties of hydrogenated bilayer silicene from first principles. *Journal of Applied Physics* 116: 024303.

Sahaf H, Masson L, Leandri C, Aufray B, Le Lay G, Ronci F. (2007). Formation of a one-dimensional grating at the molecular scale by self-assembly of straight silicon nanowires. *Applied Physics Letters* 90: 263110.

Sakai Y, Oshiyama A. (2015). Structural stability and energy-gap modulation through atomic protrusion in freestanding bilayer silicene. *Physical Review B* 91: 201405.

Schäfer H, Eisenmann B, Müller W. (1973). Zintl phases: Transitions between metallic and ionic bonding. *Angewandte Chemie International Edition* 12: 694–712.

Sugiyama Y, Okamoto H, Mitsuoka T, Morikawa T, Nakanishi K, Ohta T, Nakano H. (2010). Synthesis and optical properties of monolayer organosilicon nanosheets. *Journal of the American Chemical Society* 132: 5946–5947.

Takeda K, Shiraishi K. (1994). Theoretical possibility of stage corrugation in Si and Ge analogs of graphite. *Physical Review B* 50: 14916–14922.

Tao L, Cinquanta E, Chiappe D, Grazianetti C, Fanciulli M, Dubey M, Molle A, Akinwande D. (2015). Silicene field-effect transistors operating at room temperature. *Nature Nanotechnology* 10: 227–231.

Teo BK, Sun XH. (2007). Silicon-based low-dimensional nanomaterials and nanodevices. *Chemical Reviews* 107: 1454–1532.

Vogg G, Brandt MS, Stutzmann M, Albrecht M. (1999). From CaSi$_2$ to siloxene: Epitaxial silicide and sheet polymer films on silicon. *Journal of Crystal Growth* 203: 570–581.

Vogt P, De Padova P, Quaresima C, Avila J, Frantzeskakis E, Asensio MC, Resta A, Ealet B, Le Lay G. (2012). Silicene: Compelling experimental evidence for graphenelike two-dimensional silicon. *Physical Review Letters* 108: 155501.

Wang L, Sasaki T. (2014). Titanium oxide nanosheets: Graphene analogues with versatile functionalities. *Chemical Reviews* 114: 9455–9486.

Yamanaka S, Suehiro F, Sasaki K, Hattori M. (1981). Electrochemical deintercalation of calcium from $CaSi_2$ layer structure. *Physica B & C* 105: 230–233.

Yaokawa R, Nakano H, Ohashi M. (2014). Growth of $CaSi_2$ single-phase polycrystalline ingots using the phase relationship between $CaSi_2$ and associated phases. *Acta Materialia* 81: 41–49.

Zintl E. (1939). Intermetallische Verbindungen. *Angewandte Chemie* 52: 1–6.

4 Nanocrystalline silicon thin films

Atif Mossad Ali and Takao Inokuma

Contents

4.1 INTRODUCTION

Silicon is at the heart of the microelectronics. Its dominance over other semiconductors is intimately tied to its superior materials and process, and to the tremendous base of technology that has developed around it. A recent trend toward development of nanocrystalline silicon (nc-Si) lies in preparing nc-Si thin films exhibiting strong photoluminescence, based on a quantum size effect (Ali et al. 2002). This technique is expected to have a potential for application to optoelectronics. In addition, as crystallites are decreased in size to the nanoscale level, their electronic and vibrational properties will be modified, then the surface and quantum size effects play an important role. Formation of nc-Si structures has been tried utilizing various techniques: anodic oxidation of crystalline Si, that is, formation of porous Si formation of nc-Si thin films using plasma-enhanced chemical vapor deposition (PECVD), sputtering, evaporation, and ion beam synthesis. Furthermore, nc-Si thin films can be obtained by the thermal crystallization of amorphous Si films or Si-rich oxide films.

Study of the effects of the different and various deposition parameters on the growth of the material, and also the growth mechanisms, are therefore important both for newer device applications and also for understanding the basic physics of the growth process of Si thin films (Ali et al. 2002). Several deposition parameters, such as substrate temperature, gas flow rate, radio frequency (rf) power, dilution of the source gas (silane) with other gases (argon, hydrogen, or helium), plasma energy and density, and deposition pressure will strongly influence the structure and properties of the grown nc-Si thin films.

In the case of PECVD nc-Si thin films deposited using $SiH_4/SiF_4/H_2$ gas mixtures, the etching effects of H- and F-radicals in plasma will play important roles. Such etching effects may result in different mechanisms that affect crystallinity. However, the growth mechanisms of the nc-Si thin films are still unclear. In this chapter, we focus on the mechanisms of crystallization, through the understanding of the roles of H and F atoms in plasma under different plasma conditions.

4.2 EXPERIMENT

Nc-Si thin films were deposited using a $SiH_4/SiF_4/H_2$ gas mixture by PECVD. The substrates were cleaned for 40 min using nitrogen plasma and hydrogen plasma, respectively, just before deposition of films. The samples were deposited on corning glass substrates for measurements of X-ray diffraction (XRD) and Raman scattering, and on n-type (100) Si substrates for measurements of Fourier transform infrared (FT-IR) absorption. The deposition temperature (T_d) values were varied from ~100 to 500°C. Furthermore, we adopted two different series for H_2 values: for series A, a large H_2 condition was selected, as SiF_4 = 0.38 sccm and H_2 = 30 sccm, and we also selected a condition without H_2 addition for series B, as SiF_4 = 0.5 sccm.

The structural properties were investigated using an XRD instrument (SHIMADZU XD-D1). The average grain size, $<\delta>$, was estimated using Scherrer's formula (Cullity 1978) from the width of the XRD spectra.

$$<\delta> = \frac{0.9\lambda}{B\cos\Theta_B}, \tag{4.1}$$

where λ is wavelength of the X-ray (1.54 Å), B the half-value width of the XRD spectral peak, and Θ_B the Bragg angle.

The Raman spectra were measured by a Raman spectrometer having a double monochrometer (Jobin Yvon RAMANOR HG 2S) coupled with a cooled photomultiplier tube (Hamamatsu R649S) and excited with an Ar-ion laser light at 488 nm. The crystallinity of the films, ρ, was estimated from the intensity of Raman spectra by the procedure proposed by Tsu et al. (1982) that is, a Raman spectrum was decomposed into two components of crystalline Si (c-Si) phase occurring at around 520 cm^{-1} and the amorphous Si (a-Si) phase at around 480 cm^{-1}, and then the ρ values were estimated from the intensity ratio of the above two components using the ratio of the integrated Raman cross section for crystalline and amorphous phases as follows:

$$\rho = \frac{I_c}{(I_c + I_a)} \tag{4.2}$$

where I_c is the Raman integrated intensity for the crystalline component (sharp peak at 520 cm^{-1}), and I_a is for amorphous phase (smooth peak at 480 cm^{-1}).

The vibrational spectra were measured by an FT-IR spectrometer (JASCO FT/IR-610). The density of given bonds can be estimated by the following method (Milovzorov et al. 1998):

$$N_{SiM} = \sum_v A_\Omega^v I_\Omega^v (M = H \text{ or } O) \tag{4.3}$$

where A_Ω^v is proportionality coefficient and I_Ω^v is the intensity of the absorption IR spectrum from SiM bonds with the frequency v, Ω configuration of Si atom's bonds.

4.3 CRYSTALLINE PROPERTIES DEPENDING ON DEPOSITION CONDITIONS FOUND IN PREVIOUS WORKS

As mechanisms causing influence on the crystalline properties, the following might be proposed: (1) an effect of chemical etching on the growing surface of the films (Okada et al. 1989; Tsai et al. 1989; Hasegawa et al. 1990; Kim et al. 1995; Ali et al. 2002), (2) an effect of chemical cleaning for removing impurities (Meyerson et al. 1990; Nagahara et al. 1992; Syed et al. 1997), (3) an effect of different surface morphology of the substrate (Arai et al. 1996; Hu et al. 1996; Kondo et al. 1996; Syed et al. 1997; Hasegawa et al. 1998a; Syed et al. 1999), and (4) an effect of hydrogen coverage, being related to the surface migration of adsorbates (Matsuda 1983; Nagamine et al. 1987; Kim et al. 1995). However, it has been suggested that the etching effects (mechanism 1) are more important in forming a crystalline structure rather than the effect of the hydrogen coverage (mechanism 4) (Okada et al. 1989; Tsai et al. 1989; Baert et al. 1992). Then hydrogen and fluorine, included in the feed gases used in the present work, are known to act as etchants for Si. So, in the abovementioned mechanisms, mechanism

(1) should accompany a change in the deposition rate. Indeed, the increase in the H_2 or SiF_4 flow rate, H_2 (Kim et al. 1995) or SiF_4 (Ali et al. 1999), decreased the deposition rates. Furthermore, as mechanism (1) acts to roughen the substrate surface at the initial stage of the film growth, mechanism (1) should then be closely related to mechanism (3). This roughened surface would also increase the nucleation rate, and the roughness depends on the range of T_d used. By contrast, the smooth surface of substrates would result in an increase in δ (Syed et al. 1997). This is because the etching efficiency due to H- and F-related radicals for Si films depends on T_d (Kim et al. 1991; Lim et al. 1996), as mentioned at a later stage.

On the other hand, crystalline Si films can be prepared by repeating, in cycles, film deposition and H-plasma (Srinivasan and Parsons 1998) or He-plasma exposure (Lee et al. 1996). Based on the former results, it has been suggested that the removal of hydrogen from the growing surface of films is essential for the improvement in the crystalline properties (Srinivasan and Parsons 1998). Based on the latter results, the crystallization of the surface layer in films after He-plasma exposure was important to the crystalline properties of the resultant Si films (Lee et al. 1996). These results also support the above mentioned model that effects of the etching along with those of ion-bombardment, due to H- or F-radicals, are more important. In addition, it has been reported that the etch rate by H-radicals decreases with increasing T_d (Kim et al. 1991), but the etch rate by F-radicals increases with T_d (Lim et al. 1996). Therefore, the addition of H- or F-related molecules to the feed gases under different T_d conditions would play an important role in the crystallization of Si films (Ali et al. 2002).

The H_2 addition under $T_d \geq 300°C$ resulted in smaller δ and greater ρ values with increasing H_2 values (Kim et al. 1995; Milovzorov et al. 1998). In addition, both the (111)- and (110)-textured δ values were found to decrease with H_2 (Milovzorov et al. 1998). By contrast, under $T_d < 300°C$, only the (111) texture was observed (Hasegawa et al. 1998b; Milovzorov et al. 1998; Ali et al. 1999; Ali et al. 2001). However, both δ and ρ values have the respective maximum values at a given H_2 value (Milovzorov et al. 1998), or increase with increasing H_2 (Ali et al. 2001). For these films, the following conditions were used: H_2 = 50–300 sccm, SiF_4 = 3 sccm, SiH_4 = 0.09 sccm, and T_d = 330°C (Kim et al. 1995), H_2 = 10–46 sccm, SiF_4 = 0.13 sccm, SiH_4 = 0.6 sccm, and T_d = 100 and 300°C (Milovzorov et al. 1998), and H_2 = 5–30 sccm, SiF_4 = 0.1 sccm, SiH_4 = 0.5 sccm, and T_d = 100 and 220°C (Ali et al. 2001). Thus, rather small fixed SiF_4 values, compared with the range of H_2 used, were applied for these films so that the effect of F atoms may be eliminated, due to the formation of more stable HF bonds in the gas phase. The increase in δ and ρ with H_2 became slow as T_d increased from 100 to 220°C (Ali et al. 2001). Therefore, it is suggested that both δ and ρ values are likely to increase with increasing H_2 under lower T_d conditions. However, as a result of the present work, the increase in H_2 under low T_d was found to cause only the increase in the XRD intensities or ρ. Furthermore, the crystalline properties should also be affected by F-radicals as etchants as well as the effect of H_2 addition, causing mechanism (1) or (3).

The δ and ρ values as a function of SiF_4 exhibited different behaviors, depending on SiH_4 and/or T_d (Syed et al. 1997; Ali et al. 1999). For conditions of T_d = 400°C and H_2 = 0 sccm, Syed et al. (1997) found that both the δ and ρ values for films with SiH_4 = 1 sccm increased as SiF_4 increased from 0 to 0.5 sccm, while those for SiH_4 = 0.15 sccm monotonically decreased with SiF_4 (Syed et al. 1997). Furthermore, they found that the (110)-textured grains for SiH_4 = 1 sccm were preferentially grown as SiF_4 increased, while that for SiH_4 = 0.15 sccm weakened. Based on these results, they proposed that the changes in δ and ρ values were controlled by a change in the surface morphology of substrates, due to F-radicals (mechanism 3). Furthermore, in a previous work (Hasegawa et al. 1998a), in which the surface of substrates were pretreated by exposing them in H_2, N_2, and/or CF_4-He plasma, all Si films were deposited under the same conditions. As a result, the increases in δ and ρ were found as the Si films were deposited on substrates with a proper degree of surface roughness. In addition, it has also been reported that an excess supply of F-radicals will in turn deteriorate the crystalline properties (Kakinuma et al. 1995). On the other hand, when T_d decreased to 100°C under the conditions of SiH_4 = 0.6 sccm and H_2 = 40 sccm, the δ and ρ values as a function of SiF_4 (= 0 – 0.5 sccm) had the respective minimum values at SiF_4 = 0.1 sccm (Ali et al. 1999). The changes in δ and ρ were interpreted in terms of a difference in the etch rate due to H- and F-radicals (Ali et al. 1999), depending on T_d (Kim et al. 1991; Lim et al. 1996), which is related to mechanism (1) or (3).

On the other hand, Toyoshima et al. (1989) examined the changes in the crystallization behavior of microwave PECVD Si films by varying SiH_4 (= 5–50 sccm) and T_d (= 100–300°C) under fixed H_2 (= 450 sccm). They found that formation of (110) or (111) grains, respectively, will be enhanced as T_d increases or SiH_4 decreases, in which they proposed that the decrease in SiH_4 is tantamount to the excess supply of H-radicals. Furthermore, under the conditions without SiF_4 addition (SiH_4 = 1 sccm, SiF_4 = 0, H_2 = 3 sccm, and T_d = 150–750°C) (Hasegawa et al. 1998b), it has been shown that the dominant texture changed from the (111) orientation under low T_d conditions below 550°C and to the (110) orientation under high T_d above 550°C, in agreement with the results found in a previous work (Toyoshima et al. 1989). In addition, the addition of H- and F-related molecules to the feed gases has been reported to lower T_d for obtaining crystalline Si films (Kakinuma et al. 1995; Lim et al. 1996; Hasegawa et al. 1998a; Milovzorov et al. 1998). Furthermore, the crystalline properties of PECVD Si films appear to vary as a direct function of T_d. In previous articles, both δ and ρ values have been shown to increase as T_d increases from 100 to 300°C under SiH_4 = 0.6 sccm, SiF_4 = 0.13 sccm, and H_2 = 10 to 46 sccm (Milovzorov et al. 1998), and from 250 to 400°C under SiH_4 = 10 sccm, SiF_4 = 400 sccm, and H_2 = 500 sccm (Kakinuma et al. 1995). By contrast, Lim et al. (1996) have found a decrease in δ and a fixed ρ value as T_d increases from 280 to 450°C under SiH_4 = 0.1 sccm, SiF_4 = 3.5 sccm, and H_2 = 1 sccm. Furthermore, Syed et al. (1999) have found rather complex behaviors of δ and ρ values as a function of T_d (= 150–400°C) under SiH_4 = 1 sccm, SiF_4 = 0.5 sccm, and H_2 = 0 or 5 sccm, that is, both the (110) δ and the ρ values for H_2 = 5 sccm had the maximum values at around T_d = 300°C and monotonically increased with T_d for H_2 = 0 sccm. As a consequence, the (110)-textured δ value appears to be enhanced with an increase in T_d (Toyoshima et al. 1989; Kakinuma et al. 1995; Hasegawa et al. 1998a; Syed et al. 1999) or SiF_4 (Syed et al. 1997), or a decrease in H_2 (Milovzorov et al. 1998), or an increase in SiH_4 (Toyoshima et al. 1989). In addition, the (110)-XRD intensities have also been reported to increase with increasing T_d (Hasegawa et al. 1998b; Syed et al. 1999), or SiF_4 (Syed et al. 1997). Thus, the change in the texture (or occurrence of different textures in crystal grains) of the films may be closely related to the change in the crystallization process. So, the changes in the crystalline properties such as δ, ρ, and the texture, depending on the deposition conditions, are rather complex, but the mechanisms are not clear.

4.4 RESULTS AND DISCUSSION

4.4.1 STRUCTURAL CHARACTERIZATION

Figure 4.1 shows the average grain size, (a) $<\delta(111)>$ and (b) $<\delta(110)>$, obtained from the (111) and (110) XRD spectra, respectively, as a function of T_d, for series A films (closed triangles) and series B films (closed circles) (Ali et al. 2002). As shown in Figure 4.1, both the $<\delta(111)>$ and the $<\delta(110)>$ values for series A films and the $<\delta(110)>$ values for series B films monotonically increase with increasing T_d, while the $<\delta(111)>$ values for series B films have a maximum value at around T_d = 200–250°C. Figure 4.2 shows (a) the (111) and (b) the (110) XRD intensities, as a function of T_d, for series A films (closed triangles) and series B films (closed circles) shown in Figure 4.1. As shown in Figure 4.2a, the (111) XRD intensities for series A films monotonically decrease with increasing T_d, while those for series B films appear to have a minimum value at around T_d = 200–250°C, in good correspondence to the change in $<\delta(111)>$ (Figure 4.1a). By contrast, the (110) XRD intensities for series A films show the increase with T_d up to 300°C, followed by the saturation, but those for series B films monotonically increase with T_d. Furthermore, it is found that both the (111) and (110) XRD intensities for series A films are larger than those for series B films, especially under low T_d conditions as seen in Figure 4.2, that is, the volume of both textured crystalline phases in the films is expected to be enhanced as H_2 increases.

Figure 4.3 shows (a) the peak frequency, E_R, of the Raman signal arising from the c-Si phase and (b) ρ, as a function of T_d, for series A films (closed triangles) and series B films (closed circles) shown in Figure 4.1 (Ali et al. 2002). As shown in Figure 4.3a, the values of E_R for series A and B films increase with an increase in T_d up to T_d = 150°C. After the saturation of E_R at around E_R = 519 cm^{-1} for both series films, the E_R values for series A films increase with T_d up to E_R = 522 cm^{-1} at T_d = 300°C and then slightly decrease again.

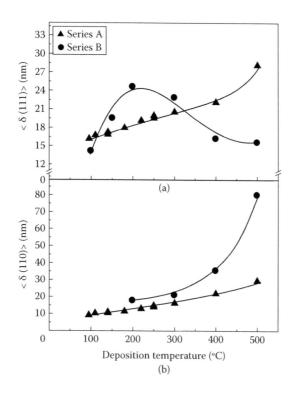

Figure 4.1 (a) Average grain size, <δ(111)> and (b) <δ(110)>, obtained from the <111> and <110> XRD spectra, respectively, as a function of Td, for series A films (closed triangles) and series B films (closed circles). (From Ali, A.M., et al., *Jpn. J. Appl. Phys.*, 41, 169–175, 2002.)

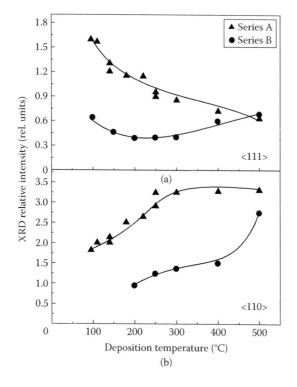

Figure 4.2 Integrated intensities for the <111> and <110> XRD spectra, respectively, (a) <111> XRD intensities and (b) <110> XRD intensities, as a function of Td, for series A films (closed triangles) and series B films (closed circles).

Low-dimensional structures

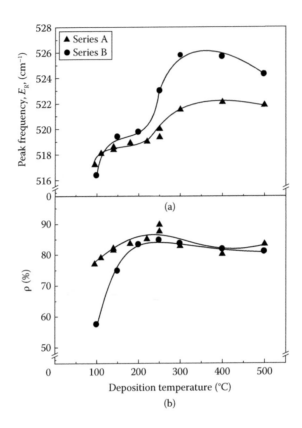

Figure 4.3 (a) Peak frequency, E_R, of the Raman signal arising from the crystalline Si phases and (b) crystalline volume fraction, r, as a function of T_d, for series A films (closed triangles) and series B films (closed circles). (From Ali, A.M., et al., *Jpn. J. Appl. Phys.*, 41, 169–175, 2002.)

On the other hand, the E_R values for series B films increase with T_d up to $E_R = 526$ cm^{-1} at $T_d = 300°$C and then decrease. As shown in Figure 4.3b, the ρ values for both series A and B films increase as T_d increases in $T_d < 250°$C and then slightly decrease again, in correspondence with the change in the XRD intensities (Figure 4.2). However, a relationship between the changes in XRD intensities and that in ρ for $T_d > 250°$C are rather complex. Figure 4.4 shows the full width at half maximum (FWHM) of the 520 cm^{-1} component in a Raman signal, as a function of T_d, for series A films (closed triangles) and series B films (closed circles) shown in Figure 4.1. As seen in Figure 4.4, the FWHM values for both series films monotonically decrease with increasing T_d.

Figure 4.5 illustrates the IR absorption spectra over the range 400–4000 cm^{-1}, which were measured under vacuum for (a) series A films and (b) series B films, with different T_d. In these spectra, the film thickness values for series A films and those for series B films were almost the same. These spectra were measured within 2 days after deposition. As seen in Figure 4.5a, the absorption bands for both series A and B films were observed at around 650, 800–900, and 2000–2100 cm^{-1}, which are assigned to the wagging, bending, and stretching motions of Si–H bonds, respectively. In addition, for series B films, the stretching motion due to Si–F bonds can be found at around 930 cm^{-1}. The absorption band around 2100 cm^{-1}, observed for series A films, is known to arise from dihydrides (Si–H$_2$) (Tsu et al. 1989). On the other hand, series B films exhibit an absorption line at around 2100 cm^{-1} having a shoulder at 2000 cm^{-1} that may be due to isolated monohydride (Si–H) (Tsu et al. 1989). The Si–H$_2$ bonds should exist in the grain boundary regions. Based on these results, the grain boundary in series B films may be more heavily damaged, including a large number of Si–H$_2$ and Si–H bonds, compared with that in series A films. This result is consistent with the wider Raman spectra for series B films than those for series A films, as seen in Figure 4.4.

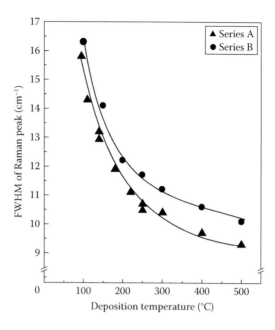

Figure 4.4 Full width at half maximum (FWHM) values, which were obtained from the 520 cm⁻¹ Raman signal arising from the crystalline Si phases, as a function of Td, for series A films (closed triangles) and series B films (closed circles).

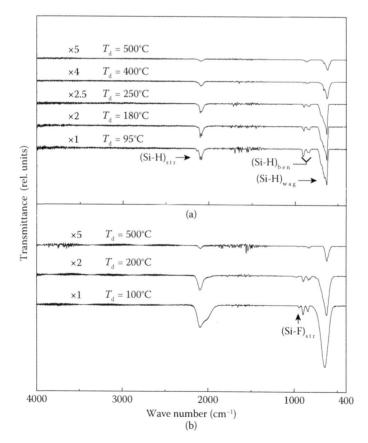

Figure 4.5 IR absorption spectra over the range 400–4000 cm⁻¹, which were measured under vacuum for (a) series A films and (b) series B films, with different Td values.

Low-dimensional structures

4.4.2 ADDITIONAL DISCUSSION OF THE GROWTH MECHANISMS OF NC-SI

As revealed in Figures 4.1 and 4.2, for the (110) texture, the change in δ with T_d for series A and B films is the same direction with that in the XRD intensity. However, the δ values and the XRD intensity with T_d for the (111) texture of both series films are found to change in the opposite direction. Thus, the (111) nucleation is suggested to be enhanced as the growth of the (111) nuclei is suppressed (Ali et al. 2002). Furthermore, the $<\delta(110)>$ values for series B films without H_2 addition are always larger than those for series A films, while the (110) XRD intensities for series B films are smaller than those for series A films. When we examine the results shown in Figures 4.1 through 4.3 in Section 4.1, along with the results shown in Section 4.3 (Toyoshima et al. 1989; Kim et al. 1995; Hasegawa et al. 1998b; Ali et al. 1999; Syed et al. 1999), it is suggested that the $<\delta(110)>$ values increase with decreasing H_2 under high T_d (Figure 4.1 and Kim et al. 1995; Milovzorov et al. 1998), but the increase in H_2 under low T_d acts to increase both the (111) and (110) XRD intensities and ρ (Figures 4.2 and 4.3, and Toyoshima et al. 1989).

On the other hand, the (110) XRD intensities are found to increase with increasing T_d (Figure 4.2 and Toyoshima et al. 1989; Syed et al. 1999). In addition, an increase in SiF_4, under high SiH_4 (or low H_2) conditions and high T_d conditions, is likely to result in an increases in the $<\delta(110)>$ values and the (110) XRD intensities (Figure 4.1 and Syed et al. 1997), but in a decrease in the $<\delta(111)>$ values (Figure 4.1a). However, under low T_d a dominant (111) texture is likely to occur (Figure 4.2 and Hasegawa et al. 1998b; Milovzorov et al. 1998; Ali et al. 1999), though the strength depends on the deposition conditions, such as H_2 (Milovzorov et al. 1998), SiF_4 (Ali et al. 1999), and T_d (Hasegawa et al. 1998b). Supposing that the change in the dominant textures is due to a chemical etching effect (Section 4.3; mechanism 1 or 3), the dominant (110) texture with increasing T_d, corresponding to the conditions for obtaining the results shown in Figure 4.2 and Syed et al. (1997), may be caused by the enhancement in the etching due to the SiF_4 addition to the feed gases (Lim et al. 1996). However, the dominant (111) texture is due to the conditions of low T_d.

Based on these results, the occurrence and the growth of <110>-textured grains and the $<\delta(111)>$ values appear to be enhanced and reduced, respectively, as T_d increases under low H_2 and high SiF_4 conditions. On the other hand, the occurrence and the growth of (111) grains appear to be enhanced as T_d decreases under high H_2 and low SiF_4 conditions. Furthermore, a large number of H atoms were incorporated in films with high SiF_4 values under low T_d conditions. However, as H_2 gas is added to the feed gases along with SiF_4, similar to those for series A films, the effect of the F-radicals will be eliminated as stated in Section 4.3. Such a difference in the crystalline properties due to the SiF_4 or H_2 addition under different T_d conditions may be caused by a difference in the etching rate depending on T_d, that is, under low T_d conditions the etching due to H-radicals is enhanced (Kim et al. 1991), and under high T_d the etching due to F-radicals is enhanced (Lim et al. 1996). Since F-radicals are more reactive than H-radicals, F-radical may cause more roughened surface than H-radicals. This model (mechanism 3 in Section 4.3) may be why the role of H-radicals is different from that of F-radicals, depending on T_d. However, the roles of H- and F-radicals in plasma on the crystalline properties appear to be different from each other. So, in order to make clear the roles of H- and F-radicals in plasma, more detailed examinations for films with different H_2 or SiF_4 values will be required.

As shown in Figure 4.3a, the Raman spectra for both series films exhibit similar T_d dependence, and E_R for series A films are found to approach $E_R = 522$ cm^{-1} at $T_d = 300°C$, whose E_R value is close to that found for single c-Si. However, the maximum E_R value for series B films is 526 cm^{-1} at $T_d = 300°C$. If the Raman shifts are controlled by a fluctuation of the electronic polarization for constituents in the films, depending on the bonding structure such as atomic distance, the Raman peak shifts would be related to a change in the stress of the films: an increase in the compressive stress or a decrease in the tensile stress should result in a positive Raman shift. Furthermore, if the Raman shift is due only to the confinement of optical phonons in spherical small grains, the peak shift can be expressed as a function of the effective grain size, D_R (Edelberg et al. 1997). Thus, the changes of E_R may reflect a change in stress or in δ. Then, the change in E_R smaller than 522 cm^{-1} found for series A films may be interpreted in terms of different δ values (Edelberg et al. 1997). However, the values of $E_R = 526$ cm^{-1} found for series B films may be interpreted in terms of a change in the stress rather than that due to different δ value. The result found for the series B films may be related to the high etching rate by F-radicals under high T_d (Lim et al. 1996), causing mechanism 3 in Section 4.3. This effect will also result in occurrence of the dominant (110) texture as stated above.

By contrast, as seen in Figure 4.4, the FWHM of the Raman spectrum for both series films monotonically decrease with increasing T_d. Based on the mechanisms causing the Raman shift, the FWHM values will, in general, broaden as random stress exists in a film or as grains with different δ values are widely distributed. As stated above, the Raman shift for series A films may be mainly caused by different δ, and the shift under high T_d conditions for series B films may be due to the different stress. In addition, the FWHM values for series B films in a high T_d range are found to be significantly larger than those for series A films, as shown in Figure 4.4. Therefore, the excess roughening of the substrate surface (mechanism 3 in Section 4.3) due to F-radicals for series B films may also cause the increase in the random stress, in addition to the effect of the change in stress causing a large positive Raman shift under high T_d conditions (Figure 4.3a), as stated above.

4.5 CONCLUSION

We deposited nc-Si films by a PECVD method using $SiF_4/SiH_4/H_2$ gas mixtures. The structural properties of the nc-Si films were examined by increasing T_d from ~100 to 500°C for two different series with $H_2 = 30$ sccm and $SiF_4 = 0.38$ sccm as series A films, and with $H_2 = 0$ sccm and $SiF_4 = 0.5$ sccm as series B films. The interesting feature in the present work lies in making clear the crystallization process, through the understanding of the roles of H- and F-radicals in plasma under different T_d conditions. So, the structural properties were examined for PECVD nc-Si films with different H_2, SiF_4, and T_d values, comparing with those in the previously published works. As described in Sections 4.3 and 4.4.2, occurrence of the different textures may have a close relationship with a change in the structural properties, such as the shifts of the Raman spectra due to the c-Si phase, the changes in the FWHM, and the different stress and bonding properties.

The occurrence and the growth of (110)-textured grains and the $<\delta(111)>$ values appear to be enhanced and reduced, respectively, as T_d increases under low H_2 and high SiF_4 conditions. On the other hand, the occurrence and the growth of (111)-textured grains are enhanced as T_d decreases under high H_2 and low SiF_4 conditions. Such a difference in the crystalline properties due to the SiF_4 or H_2 addition under different T_d conditions may be caused by a difference in the etching rate depending on T_d. The changes in other structural properties, such as the above-stated physical parameters, with varying H_2 and SiF_4 values under different T_d conditions were also examined, and the results were interpreted in terms of the model for interpreting the changes in the texture as stated above. This model would be closely related to mechanism 3 shown in Section 4.3.

REFERENCES

Ali AM, Inokuma T, Kurata Y, Hasegawa S. (1999). Effects of addition of SiF_4 during growth of nanocrystalline silicon films deposited at 100°C by plasma-enhanced chemical vapor deposition. *Jpn. J. Appl. Phys.* 38: 6047–6053.

Ali AM, Inokuma T, Kurata Y, Hasegawa S. (2001). Luminescence properties of nanocrystalline silicon films. *Mate. Sci. Eng. C.* 15: 125–128.

Ali AM, Inokuma T, Kurata Y, Hasegawa S. (2002). Structural and optical properties of nanocrystalline silicon films deposited by plasma-enhanced chemical vapor deposition. *Jpn. J. Appl. Phys.* 41: 169–175.

Arai T, Nakamura T, Shirai H. (1996). Initial stage of microcrystalline silicon growth by plasma-enhanced chemical vapor deposition. *Jpn. J. Appl. Phys.* 35: L1161–L1164.

Baert K, Deschepper P, Poortmans J, Nijs J, Mertens R. (1992). Selective Si epitaxial growth by plasma-enhanced chemical vapor deposition at very low temperature. *Appl. Phys. Lett.* 60: 442–444.

Cullity BD. (1978). *Elements of X-Ray Difffraction*, 2nd edn., Addison-Wesley, Reading, p. 102.

Edelberg E, Bergh S, Naone R, Hall M, Aydil ES. (1997). Luminescence from plasma deposited silicon films. *J. Appl. Phys.* 81: 2410–2417.

Hasegawa S, Sakata M, Inokuma T, Kurata Y. (1998b). Effects of deposition temperature on polycrystalline silicon films using plasma-enhanced chemical vapor deposition. *J. Appl. Phys.* 84: 584–588.

Hasegawa S, Uchida N, Takenaka S, Inokuma T, Kurata Y. (1998a). Initial growth of polycrystalline silicon films on substrates subjected to different plasma treatments. *Jpn. J. Appl. Phys.* 37: 4711–4717.

Hasegawa S, Yamamoto S, Kurata Y. (1990). Control of preferential orientation in polycrystalline silicon films prepared by plasma-enhanced chemical vapor deposition. *J. Electrochem. Soc.* 137: 3666–3674.

Hu YZ, Zhao CY, Basa C, Gao WX, Irene EA. (1996). Effects of hydrogen surface pretreatment of silicon dioxide on the nucleation and surface roughness of polycrystalline silicon films prepared by rapid thermal chemical vapor deposition. *Appl. Phys. Lett.* 69: 485–487.

Kakinuma H, Mohri M, Tsuruoka T. (1995). Mechanism of low-temperature polycrystalline silicon growth from a SiF$_4$/SiH$_4$/H$_2$ plasma. *J. Appl. Phys.* 77: 646–652.

Kim SC, Jung MH, Jang J. (1991). Growth of microcrystal silicon by remote plasma chemical vapor deposition. *Appl. Phys. Lett.* 58: 281–283.

Kim SK, Park KC, Jang J. (1995). Effect of H$_2$ dilution on the growth of low temperature as-deposited poly-Si films using SiF$_4$/SiH$_4$/H$_2$ plasma. *J. Appl. Phys.* 77: 5115–5118.

Kondo M, Toyoshima Y, Matsuda A. (1996). Substrate dependence of initial growth of microcrystalline silicon in plasma-enhanced chemical vapor deposition. *J. Appl. Phys.* 80: 6061–6063.

Lee KE, Lee WH, Shin SC, Lee C. (1996). Microcrystalline silicon films deposited by electron cyclotron resonance plasma chemical vapor deposition using helium gas. *Jpn. J. Appl. Phys.* 35: L1241–L1244.

Lim HJ, Ryu BY, Ryu JI, Jang J. (1996). Structural and electrical properties of low temperature polycrystalline silicon deposited using SiF$_4$-SiH$_4$-H$_2$. *Thin Solid Films* 289: 227–233.

Matsuda A. (1983). Formation kinetics and control of microcrystallite in μc-Si:H from glow discharge plasma. *J. Non-Cryst. Solids* 59&60: 767–774.

Meyerson BS, Himpsel F, Uram K. (1990). Bistable conditions for low-temperature silicon epitaxy. *Appl. Phys. Lett.* 57: 1034–1036.

Milovzorov D, Inokuma T, Kurata Y, Hasegawa S. (1998). Relationship between structural and optical properties in polycrystalline silicon films prepared at Low temperature by plasma-enhanced chemical vapor deposition. *J. Electrochem. Soc.* 145: 3615–3620.

Nagahara T, Fujimoto K, Kano N, Kashiwagi Y, Kakinori H. (1992). In-situ chemically cleaning poly-Si growth at low temperature. *Jpn. J. Appl. Phys.* 31: 4555–4558.

Nagamine K, Yamada A, Konagai M, Takahashi K. (1987). Epitaxial growth of silicon by plasma chemical vapor deposition at a very low temperature of 250°C. *Jpn. J. Appl. Phys.* 26: L951–L953.

Okada Y, Chen J, Campbell IH, Fauchet PM, Wagbner S. (1989). Mechanism of microcrystalline silicon growth from silicon tetrafluoride and hydrogen. *J. Non-Cryst. Solids* 114: 816–818.

Srinivasan E, Parsons GN. (1998). Hydrogen abstraction kinetics and crystallization in low temperature plasma deposition of silicon. *Appl. Phys. Lett.* 72: 456–458.

Syed M, Inokuma T, Kurata Y, Hasegawa S. (1997). Effects of the addition of SiF$_4$ to the SiH$_4$ feed gas for depositing polycrystalline silicon films at low temperature. *Jpn. J. Appl. Phys.* 36: 6625–6632.

Syed M, Inokuma T, Kurata Y, Hasegawa S. (1999). Temperature effects on the structure of polycrystalline silicon films by glow-discharge decomposition using SiH$_4$/SiF$_4$. *Jpn. J. Appl. Phys.* 38: 1303–1309.

Toyoshima Y, Arai K, Matsuda A. (1989). Lattice orientation of microcrystallites in μc-Si:H. *J. Non-Cryst. Solids* 114: 819–821.

Tsai CC, Anderson GB, Thompson R, Wacker B. (1989). Control of silicon network structure in plasma deposition. *J. Non-Cryst. Solids* 114: 151–153.

Tsu DV, Lucovsky G, Davidson BN. (1989). Effects of the nearest neighbors and the alloy matrix on SiH stretching vibrations in the amorphous SiOr:H (0<r<2) alloy system. *Phys. Rev. B* 40: 1795–1805.

Tsu R, Gonzalez-Hernandez J, Chao SS, Lee SC, Tanaka K. (1982). Critical volume fraction of crystallinity for conductivity percolation in phosphorus-doped Si: F: H alloys. *Appl. Phys. Lett.* 40: 534–535.

Low-dimensional structures

5 Fundamentals of silicene

Gian G. Guzmán-Verri, Lok C. Lew Yan Voon, and Morten Willatzen

Contents

5.1 INTRODUCTION

Silicene[1] is a single atomic layer of silicon (Si) much like graphene, the first example of an elemental two-dimensional (2D) nanomaterial whose study led to the 2010 Nobel Prize in Physics.[2] Until 2010 or so, the only known crystalline form of elemental silicon was the one with the so-called diamond structure (a three-dimensional [3D] cubic structure with sp^3-bonded Si atoms). That silicon could potentially form a 2D structure was first postulated by Takeda and Shiraishi.[3,4] This early work and others, both theoretical[5–7] and experimental,[8,9] went mostly unnoticed until the prediction that silicene could have similar exotic properties as graphene by Guzmán-Verri and Lew Yan Voon in 2007,[1] and silicene nanoribbons were reported to have been fabricated on a silver substrate by Kara et al. in 2009.[10] Since then, the study of silicene has exploded, mainly theoretically[11–527] but also experimentally.[528–612]

The interest in silicene is exactly the same as that for graphene[2]: in being 2D and possessing a linear band structure, the so-called Dirac cone.[1] One advantage relies on its possible application in electronics, whereby its natural compatibility with the current Si technology might make fabrication much more of an industrial reality.

We will concentrate this tutorial on the properties of a single freestanding silicene sheet. Freestanding means that the silicene sheet is not chemically or physically bonded to any other material. The feat with graphene was the ability to peel off a single layer of graphene from a piece of graphite, a process known as mechanical exfoliation. Such a layered precursor is missing for silicene. A close analogue, though, is calcium disilicide, indeed a layered material and the related process of chemical exfoliation has been tried,[8] with only partial success as the resulting product was mostly multilayers and functionalized.

Not surprisingly, a number of review articles have already appeared that includes extensive discussions of silicene.[309,603,613–646] The current review is more tutorial in nature.

5.2 BACKGROUND

In many ways, silicene share much more with graphene than with silicon. Thus, we introduce graphene briefly in this section.

Ever since the discovery of carbon nanotubes,[647] graphene has attracted attention, even though the band structure was already obtained back in 1947 using tight-binding theory.[648] The prediction of the structure using density functional theory (DFT) goes back to 1982.[649] What was transformational was its fabrication in 2004.[2] Its importance lies mostly in its linear electronic dispersion relation near the Fermi energy, so different from conventional semiconductors, and its 2D nature—thus allowing one to realize quantum electrodynamics in the laboratory. Those two properties can be summarized by

$$E(\mathbf{k}) = \pm v_F |\mathbf{k}|, \tag{5.1}$$

$$\rho(\omega) = \frac{2|\omega|}{D^2}, \tag{5.2}$$

for the energy and density of states (DOS) per spin, respectively, where v_F is the Fermi velocity, $\mathbf{k}$ is the wave vector, ω is the angular frequency, k_c is the cutoff wave number, and $D = v_F k_c$; they lead to electronic properties much different from Si inversion layers and quasi-2D heterostructures. These are sketched in Figure 5.1.[650] The band structure is seen to be linear and gapless. The DOS goes to zero at the Fermi energy but is gapless.

We refer to review articles by Neto et al.[651] and by Sarma et al.[650] on the basic properties of graphene.

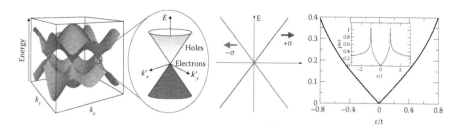

Figure 5.1 (a and b) Graphene dispersion relation and (b) density of states. (Reprinted with permission from Sarma, S.D., et al., 2011, *Rev. Mod. Phys.*, 83, 407. Copyright 2011 by the American Physical Society.)

5.3 STRUCTURAL PROPERTIES

The most fundamental question about a material is its structure. A crystal structure can be defined by giving the Bravais lattice, which describes the geometrical periodicity in space, and the associated basis of atoms in a unit cell. We start with silicon and then describe silicene.

5.3.1 SILICON

Bulk silicon has what is known as the diamond structure, with a face-centered cubic Bravais lattice and a basis of two atoms per primitive unit cell (Figure 5.2). The cell pictured is a nonprimitive one.

5.3.2 SILICENE

Silicene is also made purely of Si atoms arranged in a 2D structure. According to Kittel,[652] there are five possible 2D Bravais lattices and the one assumed by silicene is the hexagonal one. A hexagonal unit cell has equal sides with an angular separation of 60°. The silicene structure is completed by placing at each of the corners of the unit cell a basis of two Si atoms. The Si atoms thus assume a honeycomb arrangement and each set of atoms forms a hexagonal sublattice (Figure 5.3). In fact, the silicene structure could be viewed as a (111) plane of bulk silicon.

The above structure was arrived at from theoretical predictions using first-principles calculations. First-principles calculations are invariably based upon DFT. For crystalline solids, the theory implements the spatial periodicity by introducing a wave vector and only the atoms inside a unit cell need to be studied.

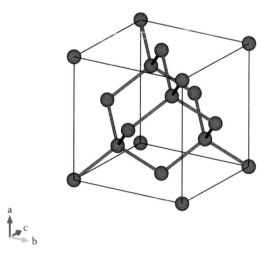

Figure 5.2 Crystal structure of silicon. Lattice is cubic, a unit cell is the dotted box. Each sphere represents an Si atom.

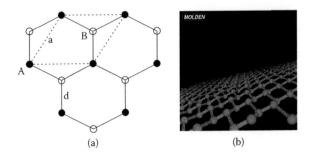

Figure 5.3 Proposed crystal structure of silicene. (a) Lattice is hexagonal, a unit cell is the dotted parallelogram, and the basis consists of two Si atoms labeled A and B, separated by a distance Δz perpendicular to the plane. (b) Perspective view of silicene.

A trial initial geometry for the unit cell is used and the atoms are moved so as to minimize the total energy of the crystal while preserving the lattice symmetry. Takeda and Shiraishi[4] first carried out this process for a single layer of Si by assuming, as is the case for graphene, a hexagonal lattice for Si as well (with a superiodicity perpendicular to the plane with a large vacuum layer, typically at least ~10 Å) and allowed the in-plane lattice constant a to vary, as well as the position of the basis atom (B in Figure 5.3) within the unit cell while preserving the imposed D_{3d} symmetry.

They found a corrugated or buckled structure, with one of the sublattices out of the plane, to have a lower total energy than the flat one. The equilibrium lattice constant is predicted to be $a = 3.82$ Å.[26] The out-of-plane height Δz of the Si atom is 0.44 −0.45 Å.[15,26] In bulk Si, the out-of-plane Si atom is 0.78 Å from the (111) plane. Thus, the bonding in silicene can be viewed as in between sp^2 and sp^3. The lowering in energy for the buckled structure compared to the flat one is 30 meV/atom and the binding energy is 4.9 eV/atom, which is lower than for bulk silicon (diamond structure) by 0.6 eV/atom.[7]

As an excellent example of how a trial geometry could lead to a local minimum but not necessarily a global one, work starting with other trial configurations have actually led to lower energy geometries. Thus, Kaltsas and Tsetseris,[137] by taking the surface layer of various Si surface reconstructions and optimizing the structure, found that structures based on the $\sqrt{3} \times \sqrt{3}$, 5×5, and 7×7 reconstructions are actually all more stable than the perfect silicene structure, by 48, 17, and 6 meV per atom, respectively. A large honeycomb dumbbell structure has been found to be the most stable to date.[427] Nonetheless, due to the paucity of work on these alternate structures and the simplicity of the ideal silicene structure, we focus primarily on the properties of the latter in the rest of the chapter. External growth conditions would also influence which structure is eventually realized.

5.4 MECHANICAL PROPERTIES

We next consider how the structure might change under the influence of external stresses. Mechanical properties refer to the change in the structure of a solid in response to external forces. The mechanical properties of 2D materials can be expected to be significantly different from 3D solids, particularly in the direction perpendicular to the sheet. Indeed, it can be expected that the sheets would be quite flexible in that direction leading to enhanced bending, whereas a covalently bonded flat sheet might be expected to be quite rigid if compressed in the plane since the horizontally aligned bonds can only change in length and not in orientation as well. For the latter reason, the 2D materials are predicted to have much higher mechanical strengths than bulk materials.[174] Additionally, the ability to support large strain means the non-linear elastic regime can easily be attained. The above qualitative predictions have been borne by quantitative calculations. Mechanical properties of semiconductor materials can be determined theoretically using *ab initio* DFT, empirical atomistic methods based on a dynamical matrix formalism or continuum elasticity methods. All methods make use of the full symmetry properties of the 3D or 2D crystal. Experimental methods rely on transmission–reflection studies of elastic waves[653–655] and optical methods such as Raman scattering[656–658] to determine the elastic properties.

5.4.1 SILICON

The most fundamental physical quantities characterizing the mechanical properties are the linear elastic constants or stiffness tensor C_{ijkl} appearing in Hooke's law:

$$T_{ij} = C_{ijkl} S_{kl},$$ (5.3)

where T_{ij} and S_{ij} denote stress and strain, respectively, and the Einstein summation convention is used. It is common to use the so-called Voigt notation, whereby two indices are combined into one as follows: '11'→'1', '22'→'2', '33'→'3', '23'→'4', '13'→'5', '21'→'6'. For bulk silicon, due to its cubic crystal structure, the stiffness tensor then has the form (indices in the Voigt notation are henceforth denoted by capital letters)

$$c_{IJ} = \begin{pmatrix} c_{11} & c_{12} & c_{12} & 0 & 0 & 0 \\ c_{12} & c_{11} & c_{12} & 0 & 0 & 0 \\ c_{12} & c_{12} & c_{11} & 0 & 0 & 0 \\ 0 & 0 & 0 & c_{44} & 0 & 0 \\ 0 & 0 & 0 & 0 & c_{44} & 0 \\ 0 & 0 & 0 & 0 & 0 & c_{44} \end{pmatrix}.$$ (5.4)

The stiffness coefficients for bulk silicon are[659]

$$c_{11} = 1.66 \times 10^{11}\,\text{Pa}, c_{12} = 0.64 \times 10^{11}\,\text{Pa}, c_{44} = 0.80 \times 10^{11}\,\text{Pa}.$$ (5.5)

5.4.2 SILICENE

The elastic properties of a hexagonal 2D sheet can be characterized by elastic constants C_{11} and C_{12} instead of the three for bulk Si. Within this elastic regime, one can compute the elastic constants by first obtaining the strain energy as a quadratic function of the applied strain ϵ_{ij}, where the strain energy is defined as the difference in the total energy with and without strain:

$$E_s(\epsilon_{xx}, \epsilon_{yy}) = a_1 \epsilon_{xx}^2 + a_2 \epsilon_{yy}^2 + a_3 \epsilon_{xx} \epsilon_{yy}.$$ (5.6)

This is known as the harmonic approximation. Then the elastic constants are given by

$$C_{11} = \frac{2}{hA_0} a_1 = \frac{2}{hA_0} a_2 = \frac{1}{hA_0} \frac{\partial^2 E_s}{\partial \epsilon_{xx}^2}\bigg|_{\epsilon=0},$$ (5.7)

$$C_{12} = \frac{1}{hA_0} a_3 = \frac{1}{hA_0} \frac{\partial^2 E_s}{\partial \epsilon_{xx} \partial \epsilon_{yy}}\bigg|_{\epsilon=0},$$ (5.8)

where h and A_0 are the effective thickness and equilibrium area of the supercell, respectively. Some authors introduce the Poisson ratio, and an in-plane stiffness C:

$$\nu = -\frac{\epsilon_{\text{trans}}}{\epsilon_{\text{axial}}} = \frac{C_{12}}{C_{11}} = \frac{a_3}{2a_1},$$ (5.9)

$$C = hC_{11}\left[1 - \left(\frac{C_{11}}{C_{12}}\right)^2\right] = \frac{[2a_1 - (a_3)^2 / 2a_1]}{A_0}.$$ (5.10)

Low-dimensional structures

Sahin et al.[19] computed the mechanical properties by stretching the sheets in the plane (biaxial strain). They obtained $v = 0.3$ and $C = 62\text{J/m}^2$ for silicene.

In theory, the nonlinear elastic regime can be reached. Thus, Qin et al.[83] showed that the anharmonic regime is reached for a biaxial strain larger than −2 and 4% (Figure 5.4), where the negative sign corresponds to compressive strain. Beyond the harmonic regime, they identified two critical points. At the first one ($\epsilon_{c1} \approx 18\%$), which they also refer to as the "ultimate strain,"[297] $dE_s/d\epsilon$ reaches a maximum value—it takes less tension to stretch the structure and it is unstable under certain acoustic waves (imaginary phonon frequencies), a phenomenon known as phonon instability. Beyond the second critical point ($\epsilon_{c2} \approx 20\%$), the strain energy decreases sharply and this corresponds to the yield point.

The Poisson ratio characterizes the fact that the layer thickness would be expected to decrease as the sheet is stretched. While an initial decrease with expansion is found, the buckling was found to increase again for a strain larger than 10%.[134,174]

A uniaxial strain provides additional information on the mechanical properties since the hexagonal structure has clear anisotropies in the plane due to the difference in bond orientations. Generally, strain along perpendicular directions (conventionally taken to be x and y and chosen to be either the zigzag [ZZ] or armchair [AC] directions) could lead to different responses. For example, the Poisson ratio is isotropic and constant for low strain (below 2%) but then decreases (increases) for AC (ZZ) strain (Figure 5.5a).[297,348]

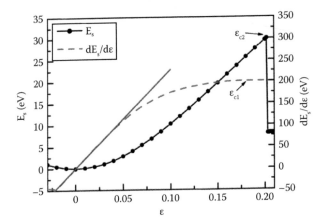

Figure 5.4 Strain energy and its first differential as a function of biaxial tensile strain. (Reprinted from Qin, R., et al., *AIP Adv.*, 2, 022159, 2012. Used in accordance with the Creative Commons Attribution license.)

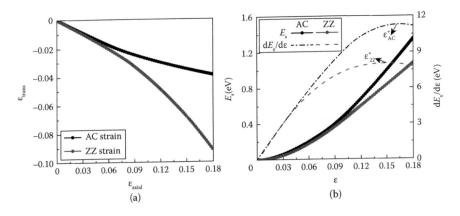

Figure 5.5 (a) Transverse strain as a function of axial strain. (b) Strain energy and its derivative as a function of axial strain. (Reprinted from Qin, R., et al., *Nano. Res. Lett.*, 9, 521, 2014. Used in accordance with the Creative Commons Attribution license.)

The mechanical response also becomes anisotropic beyond the harmonic regime (Figure 5.5b). The ultimate strain was computed to be 0.17 (0.15) for AC (ZZ) strain.

The bending of the silicene sheet has also been studied, though using classical molecular dynamics (MD) studies of the mechanical properties of large but finite nanosheets.[211,291,660] For example, Roman and Crawford[660] used a simulation region of 10×10 nm, the ReaxFF potential to describe Si–Si interactions, a microcanonical ensemble, and a nominal temperature of 10 K to limit temperature fluctuations but also to observe failure events. They computed the out-of-plane bending stiffness by deforming the sheets into partial cylindrical tubes and minimizing the strain energy with respect to the curvature:

$$E_s = \frac{1}{2} D \kappa^2, \tag{5.11}$$

where κ, the curvature, was in the range of $0.05–0.3 \text{nm}^{-1}$, and D is the bending modulus per unit width. D was obtained to be 38.63 ± 0.501 eV. These classical simulations show that silicene is more flexible but harder to break than silicon.[291]

5.5 PHONONS

A knowledge of phonons is important for a variety of reasons including characterization, and for studying transport processes. A plot of the phonon modes versus the wave vector is known as the phonon band structure or dispersion relation.

5.5.1 SILICON

A typical plot of the phonon modes of Si is given in Figure 5.6.[679]

The three acoustic modes (longitudinal LA and two transverse TA) are all characterized by linear dispersion relations away from the $k = 0$ (Γ) point. The transverse optical (TO) and longitudinal optical (LO) modes are degenerate at the Γ point but disperse away from it. This result can be found using the dynamical matrix formalism accounting for individual atomic displacements in the unit cell or, for the acoustic modes, by solving the classical elastic equations

$$\rho \frac{\partial^2 u_x}{\partial t^2} = \frac{\partial T_1}{\partial x} + \frac{\partial T_6}{\partial y} + \frac{\partial T_5}{\partial z}, \tag{5.12}$$

$$\rho \frac{\partial^2 u_y}{\partial t^2} = \frac{\partial T_6}{\partial x} + \frac{\partial T_2}{\partial y} + \frac{\partial T_4}{\partial z}, \tag{5.13}$$

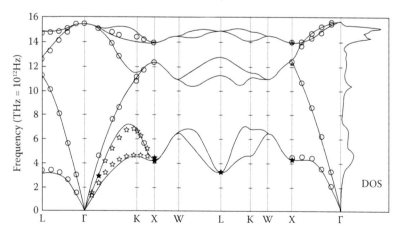

Figure 5.6 Phonon band structure for Si. (Reprinted with permission from Wei, S., and Chou, M.Y., 1994, *Phys. Rev. B*, 50, 2221. Copyright 1994 by the American Physical Society.)

$$\rho \frac{\partial^2 u_z}{\partial t^2} = \frac{\partial T_5}{\partial x} + \frac{\partial T_4}{\partial y} + \frac{\partial T_3}{\partial z}, \tag{5.14}$$

where $u = (u_x, u_y, u_z)$ is the displacement and ρ is the mass density.

The dynamical matrix formalism is an atomistic, empirical method to determine both optical and acoustic phonon modes. Following Ashcroft and Mermin,[659] the harmonic potential can be written as

$$U^{harm} = \frac{1}{2} \sum_{RR'} \mathbf{u}_\mu(\mathbf{R}) D_{\mu\nu}(\mathbf{R} - \mathbf{R}') \mathbf{u}_\nu(\mathbf{R}'), \tag{5.15}$$

and the equation-of-motion becomes

$$M \frac{\partial^2 u_\mu}{\partial t^2} = -\frac{\partial U^{harm}}{\partial u_\mu(\mathbf{R})} = -\sum_{R'\nu} D_{\mu\nu}(\mathbf{R} - \mathbf{R}') u_\nu(\mathbf{R}'), \tag{5.16}$$

with $\mathbf{R}$ the atomic position, M the total cell mass and D the dynamical matrix. Normal modes have plane-wave solutions of the form

$$\mathbf{u}(\mathbf{R}, t) = \boldsymbol{\epsilon} e^{i(\mathbf{k} \cdot \mathbf{R} - \omega t)}, \tag{5.17}$$

where $\boldsymbol{\epsilon}$ is the so-called polarization vector and gives the direction in which the ions move. Substitution of Equation 5.17 into Equation 5.16 gives the dynamical matrix equation

$$M\omega^2 \boldsymbol{\epsilon} = \mathbf{D}(\mathbf{k})\boldsymbol{\epsilon}, \tag{5.18}$$

where

$$\mathbf{D}(\mathbf{k}) = \sum_R \mathbf{D}(\mathbf{R}) e^{-i\mathbf{k} \cdot \mathbf{R}}. \tag{5.19}$$

The dynamical matrix obeys the symmetry properties of the crystal. By solving the dynamical matrix equation, Equation 5.18, the phonon dispersion relations are obtained together with the polarization vectors $\boldsymbol{\epsilon}$. For each of the N allowed values of the wavevector $\mathbf{k}$, there are three solutions giving $3N$ normal modes.

5.5.2 SILICENE

The phonon band structures can again be determined using the standard dynamical matrix formalism. The computed phonon dispersion is shown in Figure 5.7 for graphene, silicene, and germanene.[19] It is

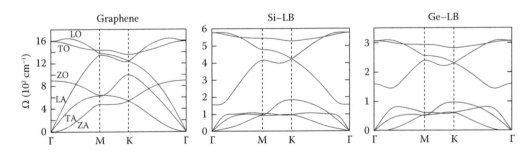

Figure 5.7 Phonon dispersion relations of graphene, silicene, and germanene. (Reprinted with permission from Şahin, H., et al., 2009, *Phys. Rev. B*, 80, 155453. Copyright 2009 by the American Physical Society.)

well known that as k → 0, phonon dispersions of LA and TA branches are linear and the out-of-plane ZA branch is quadratic since forces are zero at the surfaces of the honeycomb structure. We note, as a consequence of the symmetry of honeycomb group-IV elements nanosheets, that ZO and TO branches cross at the K point. Silicene has a flexural mode, as does graphene. However, because of the symmetry reduction due to the buckling, this mode for silicene has both a z and an xy component,[344] whereas it is pure z-like for graphene.

5.6 ELECTRONIC PROPERTIES

A central quantity that significantly impacts a wide range of physical properties is the electronic energies. The most direct property would be whether the material is an electrical conductor, semiconductor, or insulator. Hence, much effort is often expended in determining the electronic energies. It has already been alluded to that, for a periodic solid, a good quantum number is the wave vector and a lot of properties can then be parametrized with respect to the latter. A plot of the electronic energy with respect to the wave vector is known as the dispersion relation or band structure. The wave vector has two (three) components in 2D (3D), and the reciprocal space formed in wave vector space is also known as a Brillouin zone.[652]

5.6.1 SILICON

A plot of the band structure of silicon is given in Figure 5.8.[680]

Silicon is known as a semiconductor due to the presence of an energy gap separating the filled (valence) bands from the unfilled (conduction) bands, but a gap that is only about 1.1 eV which allows fairly easy excitation of electrons from the valence to conduction band (e.g., at room temperature), giving the material a small conducting property.

The shape of the bands near the band extrema is important. For most standard semiconductors such as Si and GaAs, the bands near the extrema have a parabolic shape

$$E(\mathbf{k}) = \frac{\hbar^2 \mathbf{k}^2}{2m^*}, \tag{5.20}$$

where m^* is known as the effective mass. The carriers located at those extrema thus behave semiclassically as nonrelativistic particles of mass m^*. For Si, the top of the valence band is at k = 0 (so-called Γ point) and the bottom of the conduction band is near k = $(1,0,0)2\pi/a_0$, where a_0 is the lattice constant, the so-called X point.

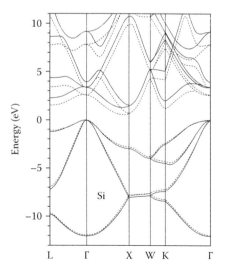

Figure 5.8 Band structure of silicon. (Reprinted with permission from Rohlfing, M., et al., 1993, *Phys. Rev. B*, 48, 17791. Copyright 1993 by the American Physical Society.)

5.6.2 SILICENE

We now describe the band structure of silicene (Figure 5.9).[19] The silicene lattice is hexagonal leading to a hexagonal Brillouin zone. The unique band structure feature of silicene and graphene (in the absence of spin-orbit (SO) interaction) is the presence of valence and conduction bands with linear dispersions, the so-called Dirac cones (since they form circular cones in the wave vector space), crossing at the Fermi energy and at the so-called K and K' points in the Brillouin zone. This means that silicene is a semimetal. This is reflected in the DOS going to zero at the Fermi energy. In contrast, standard semiconductors such as Si have a small but finite energy gap and the bands near the gap have a dispersion relation that is quadratic in the wave vector. The difference in the shape of the dispersion relation for silicene and silicon is the primary reason for the qualitatively different physical properties of the two materials. While the Dirac cones have been observed for graphene, it remains a prediction for freestanding silicene[1] and whether it is present for silicene on silver remains a controversy[121,225,307,409,564,569,581,584,661] in spite of early claims of observation.[539,549,562]

All the above calculations were done by ignoring the SO coupling. Inclusion of the latter effect has shown that silicene would open a small gap of 1.55 meV and therefore, might be better than graphene at displaying the quantum spin Hall effect.[41]

The Fermi velocity of the linear bands is an important parameter. Using DFT, Cahangirov et al.[15] estimated it to be ~10^6 m/s for silicene, basically the same value as for graphene. However, Guzmán-Verri and Lew Yan Voon,[1] using tight-binding models, evaluated it to be ~10^5 m/s for silicene. Dzade et al.[23] also obtained a smaller Fermi velocity for silicene. More recent calculations also find a slightly smaller value for silicene than for graphene. This can be easily understood from the reduced hopping in silicene since the Si atoms are more distant from each other.

One approach to studying the properties of a new material is by applying external fields and determining the response. We now discuss the electronic properties of silicene in strain, electric, and magnetic fields.

5.6.2.1 Strain

Calculations of a freestanding silicene sheet under strain have been extensively carried out due to the attempt at creating a band gap.[74,83,92,134,145,194,229,283,297,331,347,348,424] The standard approach is to first obtain the unstrained relaxed structure and then distort the unit cell in the appropriate direction such that the strain in that direction is given by

$$\epsilon = \frac{a - a_0}{a_0}, \tag{5,21}$$

where a_0 (a) is the equilibrium (strained) lattice parameter.

A relatively large strain can be applied to the nanosheets and a biaxial strain is a natural expectation if the sheets are deposited on a substrate. A biaxial tensile strain was found to lead to a semimetal–metal transition when the strain is larger than 7%.[74,83,134,194] This is due to the lowering of the conduction band at the Γ-point; the Dirac point was also found to increase in energy (but remaining

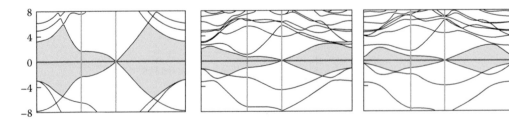

Figure 5.9 DFT band structures of graphene, silicene, and germanene. (Reprinted with permission from Şahin, H., et al., 2009, *Phys. Rev. B*, 80, 155453. Copyright 2009 by the American Physical Society.)

degenerate),[83] leading to the possibility of *p*-type self-doping.[134] Similarly, biaxial compressive strain leads to a lowering of the Dirac point below the Fermi level, leading to *n*-type doping.[194] These changes have been correlated with the changing character of the bonding between sp^3 and sp^2. The above change in character does not occur for graphene since the bonding is pure sp^2 and the atoms remain in a plane. The Fermi velocity is found to decrease slowly with strain, decreasing to 94% of the unstrained value for strain up to 7%.[83] Beyond changing the band structure, biaxial strain is also predicted to lead to such exotic phenomena as superconductivity.[229] In particular, for an electron doping of 3.5×10^{14} cm^{-2} and a tension of 5%, the critical temperature was calculated to be 18.7 K using the Eliashberg theory.

On the other hand, a uniaxial strain is expected to lead to a gap opening due to the symmetry lowering. Indeed, a gap was reported to open up for uniaxial tensile strain, up to 0.08 eV for strain along the ZZ direction and up to 0.04 eV for strain along the AC direction,[92] at about 8 and 5%, respectively (Figure 5.10).[92]

Mohan et al.[283] similarly obtained a small band gap for tensile strain but found that a direct band gap of 389 meV is formed for 6% uniaxial compression. However, Qin et al.[297] and Yang et al.[348] did not obtain a gap and only obtained the Dirac point to shift. The latter interpreted the disagreement of Zhao to the latter not using sufficient k points near the crossing. The lack of band gap opening has been confirmed using $k \cdot p$ theory.[418]

The work function is another important property determining, for example, band alignment with contacting materials. The work function is defined as the minimum energy to remove an electron and is given by

$$\Phi = E_{\mathrm{vac}} - E_F. \tag{5.22}$$

In a DFT calculation, the vacuum energy E_{vac} was determined as the average electrostatic potential energy in a plane parallel to the silicene layer and asymptotically away. For a biaxial strain, the work function initially increased from the unstrained value of 4.8 eV, then saturates to around 5.1 eV for a strain above 15%.[83] For a uniaxial strain,[297] the change is isotropic up to 3% beyond which the ZZ strain leads to a larger work function but no saturation is observed. The strain dependence has been interpreted in terms of the change in the Fermi level.

5.6.2.2 Electric field

Silicene exhibits interesting and unusual behavior in the presence of external electric fields. Due to its buckled lattice structure, the two Si atoms in the unit cell experience different electric potentials (they are at different heights). Ni et al.[76] found that a vertical electric field opened a band gap in single-layer silicene, contrary to graphene, and the gap increased linearly with the field up to about 1 V/Å. The reason for the

Low-dimensional structures

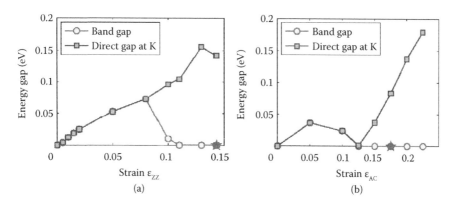

Figure 5.10 Band gap with uniaxial tensile strain for silicene. (Reprinted from *Phys. Lett. A*, 376, Zhao, H., Strain and chirality effects on the mechanical and electronic properties of silicene and silicane under uniaxial tension, 3546, Copyright 2012 with permission from Elsevier.)

gap opening is because the two atoms in the unit cell experience different electric potentials due to the different heights. Ni et al. obtained a rate of 0.157 eÅ while Drummond et al.[55] achieved 0.0742 eÅ. The latter also indicated that the gap actually starts closing for $E_\perp \approx 0.5 \text{VÅ}^{-1}$ due to the overlap of the conduction band at Γ and the valence band at K. The electric field also leads to an almost linear increase in the effective mass; for example, for a field of 0.4 V/Å, the whole mass was found to be $0.015m_0$ ($0.033m_0$) along the $K\Gamma$ (MK) direction and about 2% different for the electron mass.

More complicated studies combining external electric fields with mechanical strain fields have been carried out.[477,662] Using first-principles methods, Yan et al. have shown that the gap opened by an electric field increases with compressive biaxial strain and decreases with tensile strain. Combining electric and strain fields can tune the topological phases.[662] Yan et al.[662] have calculated the dependence of the gap on both the strain and electric fields by means of DFT and have constructed a strain–electric field phase diagram. In particular, they have shown that for moderate biaxial strains ($\lesssim \pm 7\%$), so-called topological-to-band insulator phase transition can occur for electric field strengths below E_c where the SO gap closes. For strong biaxial strains ($\gtrsim \pm 7\%$), the gap closes and silicene becomes a metal for any electric field.

5.6.2.3 Magnetic properties

Here, we only describe the effect of an external magnetic field on the electronic properties. For standard semiconductors, this is typically observed by performing a magneto-optical experiment, whereby the Landau levels formed are probed by optical transitions. To first order, the transition energies are linear in the external magnetic field. This is a very sensitive approach to characterizing the band structure. For freestanding silicene, magneto-optical properties have been studied theoretically revealing interesting spin-valley effects.[187,188,313,324] For example, the low-energy dispersion is obtained with the neglect of the so-called Rashba spin-orbit term as

$$E_n = sgn(n)\sqrt{\frac{1}{4}\Delta_{SO}^2 + 2|n|v_F^2\hbar eB},$$ (5.23)

for $n \neq 0$ and $E_n = \sigma\eta\Delta_{so}/2$ for $n = 0$, where n is the subband index and B is the magnetic field (Figure 5.11).[188] The selection rules for interband transitions are found to be $\Delta n = \pm 1$.

For graphene, Chang,[663] using a simple effective mass approach and the Luttinger–Kohn formalism,[664] showed that the Landau levels are given by

$$E_n = sgn(n)\sqrt{2}\hbar v_F \sqrt{|n|B}.$$ (5.24)

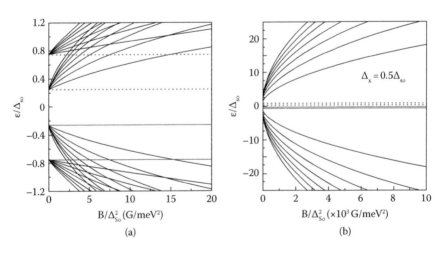

(a)

(b)

Figure 5.11 Landau levels as a function of magnetic field for small (a) and large (b) field. (Reprinted with permission from Tabert, C.J., and Nicol, E.J., 2013, *Phys. Rev. B*, 88, 085434. Copyright 2013 by the American Physical Society.)

The basic result is that the positions and intensities of the absorption lines scale with $\sqrt{B}$ and that the line corresponding to the transition from the lowest $n = 0$ to the $n = 1$ Landau level is anomalous due to the Dirac nature.[665] This behavior has been observed in infrared (IR) spectroscopy.[666–668]

The effect on the band structure in the presence of an external electric field and a perpendicular magnetic field has also been studied.[310] It was found that, for finite E_z, the spin and valley degeneracies of the Landau levels are lifted and this leads to additional plateaus in the Hall conductivity, at half-integer values of $4e^2/h$, due to spin intra-Landau-level transitions that are absent in graphene. The Hall and longitudinal conductivities in the presence of off-resonant light has also been studied.[460] The behavior is found to be different from graphene with only a single Dirac cone state.

5.7 OPTICAL PROPERTIES

Silicon is not typically known as an optical material even though it is widely used as a photodetector material. We first present the theory of linear optics for Si and then show how those are modified for silicene.

5.7.1 GENERAL THEORY AND SILICON

The polarization vector of a dielectric medium is given by[669]

$$P_i(\mathbf{r}',t') = \int \epsilon_0 \chi_{ij}(\mathbf{r},\mathbf{r}',t,t') E_j(\mathbf{r},t) d\mathbf{r} dt, \tag{5.25}$$

and assuming homogeneity in space and time allows us to write

$$P_i(\mathbf{r}',t') = \int \epsilon_0 \chi_{ij}(|\mathbf{r}-\mathbf{r}'|,|t-t'|) E_j(\mathbf{r},t) d\mathbf{r} dt. \tag{5.26}$$

In wave-vector-frequency space, the dielectric tensor ϵ_{ij} relates to the susceptibility tensor χ_{ij} via

$$\epsilon_{ij}(\mathbf{q},\omega) = 1 + \chi_{ij}(\mathbf{q},\omega). \tag{5.27}$$

An important relation exists for a complex, in the upper half-space, analytical function known as the Kramers–Kronig relations. These are for the dielectric constant,

$$\epsilon_r(\omega) = \frac{2}{\pi} P \int_0^\infty \frac{\omega' \epsilon_i(\omega') d\omega'}{\omega'^2 - \omega^2}, \tag{5.28}$$

$$\epsilon_i(\omega) = -\frac{2\omega}{\pi} P \int_0^\infty \frac{\epsilon_r(\omega') d\omega'}{\omega'^2 - \omega^2}. \tag{5.29}$$

Using the Fermi golden rule, the electric dipole transition rate becomes (for direct-gap transitions)

$$R = \frac{2\pi}{\hbar} \left(\frac{e}{m\omega}\right)^2 \left|\frac{E(\omega)}{2}\right|^2 \sum_k |P_{cv}|^2 \, \delta(E_c(\mathbf{k}) - E_v(\mathbf{k}) - \hbar\omega), \tag{5.30}$$

based on the one-electron Hamiltonian

$$H = \frac{1}{2m_0}\left[\mathbf{p} + \frac{e\mathbf{A}}{c}\right]^2 + V(\mathbf{r}), \tag{5.31}$$

and the electric dipole approximation

$$\frac{e}{m_0 c} \mathbf{A} \cdot \mathbf{p} = -e\mathbf{r} \cdot \mathbf{E},$$ (5.32)

where

$$A = -\frac{E}{2q}\left[\exp(i\mathbf{q} \cdot \mathbf{r} - i\omega t) + \exp(-i\mathbf{q} \cdot \mathbf{r} + i\omega t)\right].$$ (5.33)

In the preceding equations, the momentum matrix element P_{cv} is defined by

$$\langle c | \hat{\mathbf{e}} \cdot \mathbf{p} | v \rangle = \int_{\text{unit cell}} u_{c\mathbf{k}}^*(\hat{\mathbf{e}} \cdot \mathbf{p}) u_{v\mathbf{k}} d\mathbf{r},$$ (5.34)

and the u's are the cell-periodic part of the electron wave function.

The imaginary and real parts of the dielectric constant are then given by (using the Kramers–Kronig relations)

$$\epsilon_i(\omega) = \frac{1}{4\pi\epsilon_0}\left(\frac{2\pi e}{m_0\omega}\right)^2 \sum_{\mathbf{k}} |P_{cv}|^2 \, \delta(E_c(\mathbf{k}) - E_v(\mathbf{k}) - \hbar\omega),$$ (5.35)

$$\epsilon_r(\omega) = 1 + \frac{4\pi e^2}{4\pi\epsilon_0 m_0}\left[\sum_{\mathbf{k}}\left(\frac{2}{m_0\hbar\omega_{cv}}\right)\frac{|P_{cv}|^2}{\omega_{cv}^2 - \omega^2}\right],$$ (5.36)

with $\omega_{cv} = \omega_c - \omega_v$.

In Figure 5.12, we show the real and imaginary parts of the dielectric constant of silicon,[670] measured at room temperature.

5.7.2 SILICENE

A fascinating property of graphene is the universal low-frequency optical absorbance predicted by a noninteracting Dirac fermion theory in 2D and equal to $\pi\alpha$, where $\alpha = 1/137.076$ is the Sommerfeld fine structure constant; this has been observed experimentally.[671] Since this result is independent of the atomic species, buckling, and orbital hybridization, it should apply to silicene as well. A proof based upon the independent-particle approximation is now provided.[672]

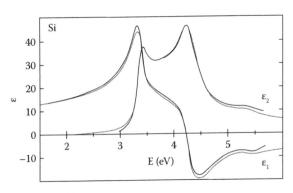

Figure 5.12 Experimental real (ε_1) and imaginary (ε_2) parts of the dielectric constant of silicon. (Reprinted with permission from Lautenschlager, M., et al., 1987, *Phys. Rev. B, 36*, 4821. Copyright 1987 by the American Physical Society.)

The direct, yet small, band gap and the crystal structure D_{3d} of silicene makes its optical properties fundamentally different from cubic 3D silicon with crystal structure O_h having an indirect band gap. Following the independent-particle approximation.[673] the optical properties of silicene are derived from the in-plane dielectric function $\epsilon(\omega)$ at normal incidence using the Ehrenreich–Cohen formula for empty conduction bands $E_c(\mathbf{k})$ and filled valence bands $E_v(\mathbf{k})$ (assuming the undoped case)[162]

$$\epsilon(\omega) = 1 + \frac{8\pi}{LA} \sum_{c,v} \sum_{\mathbf{k}} |M_{cv}(\mathbf{k})|^2 \frac{1}{E_c(\mathbf{k}) - E_v(\mathbf{k}) + \beta(\hbar\omega + i\gamma)}, \tag{5.37}$$

where A is the 2D area and L is the distance between the single-layer silicene sheets. The latter distance is needed for a periodic computational treatment. The dipole matrix element is obtained in the longitudinal approach as

$$M_{cv}(\mathbf{k}) = \lim_{\mathbf{q}\to 0} \frac{e}{|\mathbf{q}|} \langle c, \mathbf{k} | e^{i\mathbf{q}\cdot\mathbf{r}} | v, \mathbf{k}+\mathbf{q} \rangle. \tag{5.38}$$

Using the average value of the squared matrix element of the dipole operator between pure π and π^* at the K or K' points in the Brillouin zone, $|M_{cv}(\mathbf{k})|^2 = \frac{1}{2} m_0^2 v_F^2$ where m_0 is the free-electron mass, one finds the approximate result for the absorbance

$$A(\omega) = \frac{\omega}{c} L \, Im(\epsilon(\omega)) = \pi\alpha. \tag{5.39}$$

Optical properties have been computed using DFT.[162,163,282,373,672] Figure 5.13 compares the results for graphene, silicene, and germanene obtained in the absence of self-energy and excitonic corrections, and at normal incidence.[162] The optical properties at higher energies do differ due to differences in the band structure. Inclusion of quasiparticle many-body effects was found to have no influence on the low-frequency absorbance but led to a blueshift of the peaks at higher energies, mainly because of the increase in the interband energies.[163] However, inclusion of SO coupling leads to important changes at low frequency.[163] First, an SO gap is opened leading to transparency. Second, the near-gap absorbance is enhanced by a factor of 2; this is due to the fact that we now have the equivalent of a massive Dirac particle in 2D.

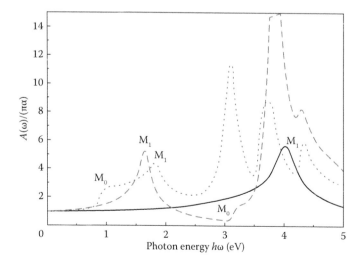

Figure 5.13 Spectral absorbance of graphene (black solid line), silicene (red dashed line), and germanene (red dotted line). (Reprinted with permission from Matthes, L., et al., 2013, *Phys. Rev. B*, 87, 035438. Copyright 2013 by the American Physical Society.)

Low-dimensional structures

Third, the absorbance decays back down to the universal value between the gap and the higher-energy peaks. Deviations from the Dirac dispersion lead to the absorbance increasing as $\sim\omega^2$ above $\omega = 0$ in the IR frequency region but the slope is material dependent (silicene and germanene slopes are close but quite different from that of graphene). Besides the noticeable peak in the absorbance for silicene at an energy around 2 eV, which is absent in silicon, the major peak is located around 4 eV for both materials.

The dielectric functions also reveal the nature of the plasmons (collective excitations of the electron gas) since the latter are probed using electron energy loss spectroscopy (EELS) and peaks in the latter correspond to dips in the imaginary part of the dielectric function, $\varepsilon_2(\omega)$. Two features in EELS of silicene for in-plane polarization are below 5 eV and above 5 eV, corresponding to the π and $\pi + \sigma$ peaks, respectively. The π plasmon results from the collective $\pi - \pi^*$ transitions while $\pi + \sigma$ plasmon results from the $\pi - \pi^*$ and $\sigma - \sigma^*$ transitions.[166] The energies are lower than for graphene.[282] Calculations of the dielectric properties[283] show that the π plasmon in silicene disappears with tensile and asymmetric strains; the $\pi + \sigma$ plasmons are red shifted for tensile strains; and the $\pi + \sigma$ plasmons are blue shifted for compressive strains. Plasmons have been proposed as probes of topological phase transitions.[222,318]

Temperature effects on plasmons have been studied.[338] At $T = 0$ and in the absence of the SO gap, only interband transitions are possible. At finite temperature, intraband transitions become possible as well. The low DOS at the Dirac point reduces the likelihood of low-frequency plasmons. However, the SO gap for silicene increases the DOS and, at finite temperatures, allows intraband plasmons as well as interband ones. When the zero of ε_1 is in the gap of ε_2, the plasmon is undamped. At a critical temperature, the interband transition is negligible and the intraband one dominates, leading to damped plasmons.

5.8 THERMAL PROPERTIES

The two most important thermal properties studied to date are the thermal conductivity and thermoelectric properties.

5.8.1 LATTICE THERMAL CONDUCTIVITY

Much work has been done on the thermal conductivity of silicene. As a semimetal, the thermal conductivity of silicene has been assumed to be mainly due to phonon transport. Initial calculations have used classical MD and gave values in the range 5–65 W/mK at 300 K, compared to a value of ~3000–5000 W/mK for suspended graphene[674] and 150–200 W/mK for silicon. Kamatagi and Sankeshwar[414] similarly found an order of magnitude smaller compared to bulk Si. It should be noted that they observed a small anisotropy in the value of the conductivity of the order of 1–4 W/mK and a general trend of increasing κ with increasing supercell size.

Nonequilibrium MD simulations have also been carried out.[88,125,170,173,269,354] Hu et al.[125] used a simulation region with 1000 unit cells in one direction and the atomic positions were relaxed. Thermal conductivity was then calculated using nonequilibrium MD and extracted from the Fourier law by computing the heat flux.[125] They found that κ (~40 W/mK) is significantly influenced by the out-of-plane flexural modes. Xie et al.[344] reported a value of 9.4 W/mK at 300 K using first-principles calculations; out-of-plane vibrations were found to only contribute less than 10% of the overall thermal conductivity. The difference compared to graphene was related to the buckling which breaks the reflection symmetry and leads to scattering of the flexural modes with other modes as the former are no longer pure out-of-plane vibrations. Graphene was found to possess significantly higher thermal conductivities than silicene at every length scale and chirality, and this was attributed to the higher phonon group velocities of the dominant acoustic modes in graphene.

The temperature dependence of the thermal conductivity of freestanding and SiO_2-supported silicene has been investigated using both equilibrium and nonequilibrium MD from 300 to 900 K,[467] inclusive of quantum corrections. The substrate leads to a 78% reduction at 300 K. However, Zhang et al.[488] found that the thermal conductivity could either increase or decrease depending on the substrate. Thus, they found an increase in the thermal conductivity of silicene supported on the 6H-SiC substrate and attributed this increase to the augmented lifetime of the majority of the acoustic phonons. They found a significant decrease in the thermal conductivity of silicene supported on the 3C-SiC substrate results, due to the reduction in the lifetime of almost the entire phonon spectrum. This is in contrast to graphene where substrates have been found to always lead to a decrease in the phonon transport.[488] For suspended silicene, it has been computed over $0 < T < 400$ K.[675] They obtained a maximum value of about 100 W/mK near K, compared to around 1000 W/mK for bulk Si[675] and 3000–5000 W/mK for graphene.

The effect of defects,[70,414] strain,[125,173] and isotopic doping[173,269] on κ has been studied. Not surprisingly, defects were found to reduce κ due to the increased defect-phonon scattering. In fact, Li and Zhang[70] found that removing a single atom out of 448 atoms from the domain led to a 78% reduction in κ. Under tensile strain, κ was found to initially increase at small strains (below 4%) and then decrease with larger strains[173]; this was done on a 33nm × 33nm silicene sheet. The initial increase could be attributed to the reduced buckling under strain and, therefore, an increase in in-plane stiffness. Higher strains led to stretching of the Si–Si bonds and, therefore, a decrease in in-plane stiffness. Isotopic doping is found to decrease κ as well; for example, 50% of ^{30}Si in a ^{28}Si lattice led to a 20% reduction. Germanium doping of 6% reduces the thermal conductivity by 62%[397] due to phonon softening and enhanced phonon scattering.

5.8.2 THERMOELECTRICITY

The lower thermal conductivity of silicene compared to graphene could be advantageous for thermoelectric properties since the latter are improved by lower thermal conductivity and higher thermopower. Thermopower is defined by the voltage drop that appears across a material due to an imposed temperature gradient, and a measure is via the Seebeck coefficient,

$$S = \frac{1}{eT} \frac{L_1}{L_0}, \tag{5.40}$$

where T is the temperature and

$$L_m(\mu) = \frac{2}{h} \int_{-\infty}^{\infty} dE \, T(E)(E - \mu)^m \left(-\frac{\partial f(E, \mu)}{\partial E} \right), \tag{5.41}$$

where $T(E)$ is the transmission at energy E, $F(E, \mu)$ is the Fermi–Dirac distribution, and μ is the chemical potential. The thermopower was calculated with and without an external field.[203] A four-band tight-binding model was used for the electronic band structure, transport properties were calculated using the nonequilibrium Green's function (NEGF) method, and the simulation length was 1000 unit cells. They found the thermopower to be somewhat insensitive to temperature changes between 100 and 500 K and to have a peak value of 87μ V/K. In the presence of an electric field, and with the chemical potential in the band gap, there is an increase in κ up to 300μ V/K at 300 K.

Thermoelectricity is measured by the figure of merit,

$$ZT = \frac{\sigma S^2 T}{\kappa}, \tag{5,42}$$

where σ is the electrical conductance. The thermoelectric coefficients of silicene at room temperature have been computed[349] using the BoltzTrap and VASP codes. They were found to have a ZT_e less than 0.5. This is much better than for many common bulk materials but still smaller than the goal of $ZT > 1$.

5.9 APPLICATIONS

The original motivation for silicene is its relation and compatibility to silicon electronics and the potential of pushing Moore's Law to the ultimate limit of a single monolayer of silicon. Nonetheless, silicene is qualitatively very different from silicon, particularly with respect to the dimensionality of the material. Thus, the monolayer thickness motivates its potential use as a DNA sequencer when such an application would not make sense for bulk Si. Indeed, what is being envisioned for silicene is similar to what has been considered for graphene. To date, only one possible device has been realized in the lab, a silicene-based field-effect transistor (FET).[607] In addition to discussing the latter application, we provide a brief survey of a few other potential applications.

5.9.1 TRANSISTORS

The use of silicene as an FET material is actually counterintuitive. The channel material is a semiconductor, such as Si, whereas silicene is a semimetal with a zero energy gap. A band gap is needed in order to switch the device on and off. However, a band gap could be opened using a variety of approaches, such as by applying a perpendicular electric field, by depositing on a substrate or by inducing quantum confinement of the electronic energies by reducing the lateral width into a nanoribbon. Thus, a study of FET characteristics[461] found a switching effect with high on/off current ratio exceeding 10^5. They also found that the device output characteristic displays a very good saturation due to improved pinch-off of the channel, stemming from the electrically induced band gap.

A schematic of how the silicene FET device was fabricated is given in Figure 5.14.[607] To date, silicene has only been grown on a substrate. In this case, the decision was made to grow on silver, which can then be patterned into metallic contacts. The silicene was first capped to improve its stability, after which it was flipped to expose the silver which is then etched into contacts. Room temperature operation was achieved with a mobility of 100 cm²/Vs; while low, it is expected that performance will improve.

5.9.2 SPIN AND VALLEY FILTERS

The advent of spintronics signaled a new era whereby the spins of electrons are used to carry information rather than the charge. This requires that one is able to produce, transmit, and measure spins. In the discussion of band structures, it was mentioned that bands have extrema in energy which allow charge carriers to populate them under appropriate conditions. For n-doped Si, these extrema, or valleys, are near the so-called X points in the Brillouin zone. Thermal excitations of carriers would populate the six X valleys equivalently (corresponding to the six equivalent <100> directions of the cubic lattice). The ability of populating those equivalent valleys differently gives rise to the additional degree of freedom of encoding information into the valley populations, thus giving rise to valleytronics. An enabler of these novel phenomena is the ability to produce carriers in specific spin and/or valley states.

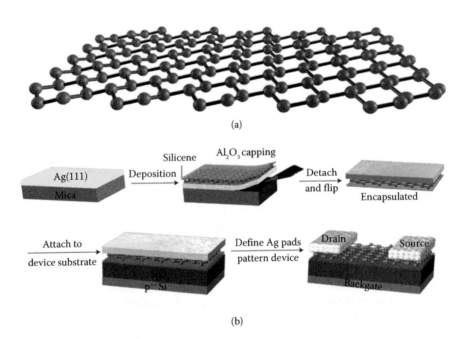

Figure 5.14 (a, b) Fabrication of a silicene FET. (Reprinted by permission from Macmillan Publishers Ltd., *Nature Nanotechnology*, L. Tao, L., et al., 10, 227, 2015, copyright 2015.)

Silicene has been predicted to be a perfect spin filter,[317] as well as a perfect spin-valley filter.[372,433,444,474] For example, the direction of the valley polarization can be switched while the direction of spin polarization is unchanged[474]; this was achieved using a magnetic barrier. Perfect spin and valley-polarized conductance can be achieved by adjusting the barrier potential.[444] Valley and spin-resolved transport through a ferromagnetic/ferromagnetic/ferromagnetic junction was shown to be 100% polarized with the appropriate photoirradiation.

5.9.3 GAS SENSORS

The large surface-to-volume ratio of a 2D material provides a relatively large surface for the adsorption of molecules per weight and, hence, makes them attractive as gas sensors.

Silicene has been proposed as a sensitive sensor for NH_3, NO, and NO_2,[247,439] CO,[439] and for a variety of other gases.[235] The gas detection performance of silicene nanosensors for four different gases (NO, NO_2, NH_3, and CO) have been studied in terms of sensitivity and selectivity, using DFT and NEGF methods.[439] They found that pristine silicene exhibited strong sensitivity for NO and NO_2, while it was incapable of sensing CO and NH_3.

5.9.4 HYDROGEN STORAGE

Hydrogen as an energy source promises to be cheap, abundant, and environmentally friendly. However, the current storage technique of using a high-pressure tank is less attractive due to the weight and large footprint. Much work has been ongoing on storing hydrogen in porous materials with large surface-to-volume ratios. The US Department of Energy has set the goal of reaching 9wt%.

The adsorption of hydrogen on the surface motivates the concept of a hydrogen storage material. Silicene would have a much better storage capacity per weight than bulk Si. On the other hand, the formation of strong covalent bonds between the Si atoms and hydrogen[26] would reduce the practically of silicene as a hydrogen storage material since it would be difficult to recover the hydrogen. Thus, alternate schemes have been considered, such as having the hydrogen bond to adatoms on the surface of silicene.

The hydrogen storage properties of 10 different adatom-decorated silicene were carried out using DFT calculations with van der Waals dispersion correction[337]. The authors found that the binding energy between metal adatoms and the silicene is greater than the cohesive energy of the bulk metal so that clustering of adatom will not occur once it is bonded with silicene. The adsorption of H_2 on Li-, Na-, K-, Mg-, Ca-, and Au-decorated silicene is a weak physisorption process. A weaker chemisorption is responsible for the adsorption of H_2 on Be-, Sc-, Ti-, and V-decorated silicene. In particular, silicene with Na-, K-, Mg-, and Ca-decorating on both sides is predicted to have 7.31–9.40wt% hydrogen storage capacity with desirable adsorption energy. Further work on Ca-decorated silicene[419] has been promising as well. For adsorption on both sides of silicene, the hydrogen storage capacity was found to reach 6.17wt% (gravimetric density) with an average adsorption energy of $0.265\ eVH_2^{-1}$, optimal for reversible hydrogen adsorption and desorption at ambient conditions. The storage capacity was further improved to 8.43wt% with the average adsorption energy in the range of $0.182 - 0.269\ eVH_2^{-1}$ in Ca-decorated ZZ silicene nanoribbons (ZSiNRs). Calcium atoms were found to prefer to disperse on the silicene or at the edges of ZSiNRs without clustering, due to the strong binding between Ca and Si atoms. Thus, metal-decorated silicene could serve as a high-capacity hydrogen storage medium.

Hussain et al.[250] additionally considered Mg functionalized silane and also with strain present. They found that a significant positive charge on each Mg accumulates a maximum of six H_2 molecules with H_2 storage capacity of 8.10 and 7.95% in the case of SiMg and SiHMg, respectively.

5.9.5 DNA SEQUENCING

The idea of using silicene is similar to other nanofilms, by passing the nucleobases through nanopores and measuring the electrical properties, with the goal that individual nucleobases could be resolved through different electrical responses. DFT calculations show that bases could be discriminated.[304] Amorim and Scheicher[676] have used a combination of DFT with van der Waals correction and NEGF in order to study the adsorption of individual nucleobases on silicene. In spite of the weak interaction, they found significant changes in the transmittance at zero bias for cytosine and guanine; the latter are predicted to chemisorp to silicene via the single oxygen atom present. Adenine and Thymine have even weaker interactions with silicene leading to physisorption and a weaker differential transmission response. On the other hand, others found that the dips due to different nucleobases are somehow less clear.[321]

5.9.6 BATTERIES

Boron-doped silicene is predicted to be metallic and a potential for a high-capacity anode for Li-ion batteries.[320] Li metal and ions adsorption onto freestanding silicene has been found to be better than for graphene.[403,677] Thus, Guo et al.[403] found that the diffusion energy barrier of Li adatoms on silicene sheet is 0.25 eV, much lower than on graphene and Si bulk. A carbon-free cathode based on silicene has been proposed for Li–O_2 batteries.[251]

5.10 FABRICATION AND GROWTH

5.10.1 SILICON

The traditional method of growing single crystals of Si for the electronics industry is via the Czochralski method,[669] whereby a seed crystal is introduced into melted Si and a single crystal forms on it as it is pulled from the melt. Thus, large single crystals of very high purity is made.

5.10.2 SILICENE

Unlike graphene, there does not appear to exist a starting material for mechanical exfoliation. Thus, other techniques must be used in order to make silicene. Three methods have been used to date: wet chemistry, chemical vapor deposition (CVD), and ultrahigh vacuum (UHV). In this section, we provide a brief review of them.

5.10.2.1 Wet chemistry

In chemical exfoliation, a foreign ion or molecule is used to separate layers. Layered silicon structures have been made in the past, such as siloxene $Si_6 H_3 (OH)_3$ and polysilane $Si_6 H_6$,[614] but single sheets could not be exfoliated even though the bonding is via van der Waals and hydrogen bonds. Interlayer spacing had been increased by intercalating with organic compounds such as amides and amines. In 2005, individual sheets of siloxene were identified from a shaken colloidal suspension though the sheets restacked upon drying.[8] Zintl phase calcium disilicide ($CaSi_2$) was used to prepare Weiss siloxene which was then exfoliated to produce siloxene nanosheets, a form of functionalized silicene.

The first report of the fabrication of a crystalline Si monolayer sheet was possibly in 2006.[9] In that paper, the authors stated that "Although many researchers have attempted to prepare two-dimensional Si sheets, there have been no reports of a successful fabrication of Si monolayer sheets." They reported the fabrication of the nanosheets by chemical exfoliation of magnesium-doped calcium disilicide. Given that the bonding between Ca and Si is ionic and charged, Mg was used to reduce the charge on the Si. The doped material was then immersed in a solution of propylamine hydrochloride which deintercalated the Ca ions. They concluded that the sheets were oxidized and magnesium-doped. From high-resolution atomic-force microscopy (AFM) and transmission electron microscopy (TEM) and from electron diffraction, they deduced that the sheets are single crystalline and (111) oriented with a lattice constant of 0.82 nm and thickness of a slightly squashed Si(111) plane. They observed the sheets to be almost transparent and having lateral dimensions in the range 200–500 nm. They did find the Si nanosheets to oxidize easily and, therefore, aggregate, making it difficult to disperse in a solvent.

Organic solvents are expected to allow for high dispersion and stability against oxidation due to the formation of a hydrophobic phase on the Si surface. Oxygen-free (111)-oriented nanosheets were, therefore, obtained by exfoliating polysilane using a mixture of n-decylamine and chloroform for 1 hour at 60°C in a nitrogen atmosphere.[532] They have lateral dimensions of 1–2μ m and thicknesses of the order of nm. Subsequent work[536] has revealed the possibility of depositing a functionalized monolayer (total thickness of 2.98 nm) onto a graphene substrate by dissolving the nanosheets in chloroform and then evaporating away the latter. Perpendicular electron transport was investigated using an AFM in vacuum at room temperature. The transport was inferred to be nonresonant tunneling below approximately 1 eV, with high-field breakdown above that.

The same group succeeded in synthesizing oxygen-free, phenyl-modified organosilicon nanosheets with atomic thickness.[533] This was achieved by reacting layered polysilane with phenyl magnesium bromide and exfoliating in an organic solvent. Using[1] H nuclear-magnetic resonance (NMR) spectroscopy, they determined the nanosheets to have the composition $Si_6 H_4 Ph_2$ and, together with IR spectroscopy, for the Si–Ph and Si–H bonds to be covalent. The thickness was determined to be 0.98 nm using AFM. Photoluminescence data in an organic solvent again indicated a monolayer structure with an optical band gap of 3 eV (415 nm) but not so in solid phase.

They also obtained alkyl-modified Si nanosheets by the hydrosilylation of layered polysilane.[548] The polysilane was formed starting from $CaSi_2$ in HCl, rinsed in HF solution, and dried in vacuum. It was then functionalized using a Pt-catalyzed hydrosilylation reaction with 1-hexene, producing a stable colloidal suspension. AFM measurements of the nanosheet dispersed in hexane and then deposited on graphite gave a thickness of 3.1 nm. X-ray diffraction revealed a crystalline structure with a hexagonal unit cell with $a = 0.383$ nm and $c = 0.63$ nm. Optical measurements revealed a band gap of 4.2 eV. From the results presented in this section, it is clear that the band gap is dependent upon the functional groups. Nevertheless, much work remains to be done in this area.

5.10.2.2 Chemical vapor deposition

One group has reported the CVD growth of thin silicon nanosheets in a high-flux H_2 environment on various substrates including Si, sapphire, GaN, and quartz.[537] The growth mechanism was inferred from scanning electron microscope (SEM) images and explained to first involve the formation of a bridge-like framework and then the filling of the space in between. The films were less than 2 nm in thickness but likely multilayers. Indeed, the optical emission observed at 435 nm reflects a smaller optical band gap.

5.10.2.3 Ultrahigh vacuum

The study of silicene started taking off in 2009, with the reported deposition of silicon nanoribbons (SiNRs) on (110) and (100) Ag.[10,528,529] This was achieved at room temperature under UHV conditions from a direct-current heated piece of silicon wafer. The nanoribbons had widths of 1.6 nm.

This same group had subsequently reported whole sheets on (111) Ag[531] though this has been disputed,[549] primarily because of a discrepancy between the measured Si–Si distance (0.19 ± 0.01 nm) and the theoretically predicted one, as this was the only characterization performed. In both cases, the substrate temperature was 220–260°C and a slow deposition rate was crucial—less than 0.1 monolayer per minute. The Lalmi work proposed a $(2\sqrt{3} \times 2\sqrt{3})R30^0)$ superstructure.

In 2012, five distinct groups reported the fabrication of silicene on Ag(111).[540,543,545,547,549] In 2013, two additional groups did so.[550,569] Atomic structure has been measured using low-energy electron diffraction (LEED) for the (4×4) superstructure on Ag(111),[582] with the Si–Si bond length ranging from 2.29 to 2.31 Å. More reports of growth by other groups have now appeared.[586,588,593] Post-deposition annealing temperature has been shown to influence the silicene phases.[579]

Further fabrication has been reported on ZrB_2,[544] Ir,[566] ZrC,[570] and MoS_2.[573] The (0001) ZrB_2 was grown epitaxially on Si(111) and silicene formed by surface segregation with a $(\sqrt{3} \times \sqrt{3})$ superstructure. For Ir, silicon was first deposited on Ir(111) at room temperature under UHV conditions and then annealed at 670 K for 30 min. A $(\sqrt{3} \times \sqrt{3})$ structure was established. For ZrC(111), a 2×2 structure was formed. They measured the phonon dispersion relations using high-resolution electron energy loss spectroscopy (HREELS). The growth on MoS_2 is significant in that it was the first time silicene had been grown on a nonmetallic substrate. The structure obtained was different from the ideal silicene structure even though it was a high-buckled monolayer silicene.

In 2016, Du et al.[678] reported the epitaxial fabrication of quasi-freestanding silicene from a Ag(111) substrate by oxygen intercalation. Bilayer silicene was first epitaxially grown on Ag(111) with a $\sqrt{3} \times \sqrt{3}$ top layer with respect to 1×1 silicene and a $\sqrt{13} \times \sqrt{13} / 4 \times 4$ bottom layer with respect to Ag(111). The latter is regarded as a buffer layer. Then, oxygen molecules were intercalated into the buffer layer, which, according to Du et al., effectively reduces the interaction between the top $\sqrt{3} \times \sqrt{3}$ layer and the substrate to the point where some of the intrinsic properties of freestanding silicene begin to be observable. By means of scanning tunneling microscopy (STM), X-ray photoemission spectroscopy (XPS), angle-resolved photoemission spectroscopy (ARPES), and *ab initio* calculations, they found evidence that this may be indeed the case as the top layer exhibits the 1 × 1 lattice structure of freestanding silicene and a Dirac cone.

5.11 SUMMARY

Allotropes of silicon beyond the diamond structure have always fascinated materials scientists. A 2D form of silicon has been anticipated for a long time. Yet, interest in the subject did not fully materialize until the discovery of graphene in 2004.

Active silicene research directions include achieving growth on insulating substrates, obtaining freestanding samples, and making silicene-based devices. Properties of functionalized silicene and of multilayer silicene are also being vigorously investigated. Interest in silicene is expected to remain constant for a long time.

ACKNOWLEDGMENTS

Work at the University of Costa Rica is supported by Vicerrectoría de Investigación under the project no. 816-B5-220. MW and LLYV acknowledge partial financial support from the Office of International Affairs at the University of Costa Rica.

REFERENCES

1. G. G. Guzmán-Verri and L. C. Lew Yan Voon, *Phys. Rev. B Condens. Matter Mater. Phys.* **76**, 075131 (Pages 10) (2007), http://link.aps.org/abstract/PRB/v76/e075131.
2. K. S. Novoselov, A. K. Geim, S. V. Morozov, D. Jiang, Y. Zhang, S. V. Dubonos, I. V. Grigorieva, and A. A. Firsov, *Science* **306**, 666 (2004).
3. K. Takeda and K. Shiraishi, *Phys. Rev. B* **39**, 11028 (1989).
4. K. Takeda and K. Shiraishi, *Phys. Rev. B* **50**, 14916 (1994).
5. Y. Wang, K. Scheerschmidt, and U. Gösele, *Phys. Rev. B* **61**, 12864 (2000), http://link.aps.org/doi/10.1103/PhysRevB.61.12864.
6. X. Yang and J. Ni, *Phys. Rev. B* **72**, 195426 (2005).
7. E. Durgun, S. Tongay, and S. Ciraci, *Phys. Rev. B* **72**, 075420 (2005).
8. H. Nakano, M. Ishii, and H. Nakamura, *Chem. Commun.* 2945–2947 (2005), http://dx.doi.org/10.1039/B500758E.
9. H. Nakano, T. Mitsuoka, M. Harada, K. Horibuchi, H. Nozaki, N. Takahashi, T. Nonaka, Y. Seno, and H. Nakamura, *Angew. Chem.* **118**, 6451 (2006).
10. A. Kara, C. Léandri, M. E. Dávila, P. de Padova, B. Ealet, H. Oughaddou, B. Aufray, and G. L. Lay, *J. Supercond. Nov. Magn.* **22**, 259 (2009).
11. M. T. Yin and M. L. Cohen, *Phys. Rev. B* **29**, 6996 (1984), http://link.aps.org/doi/10.1103/PhysRevB.29.6996.
12. M. M. Roberts, L. J. Klein, D. E. Savage, K. A. Slinker, M. Friesen, G. Celler, M. A. Eriksson, and M. G. Lagally, *Nat. Mater.* **5**, 388 (2006).
13. T. Miyazaki and T. Kanayama, *Appl. Phys. Lett.* **91**, 082107 (Pages 3) (2007), http://link.aip.org/link/?APL/91/082107/1.
14. T. Morishita, K. Nishio, and M. Mikami, *Phys. Rev. B* **77**, 081401 (2008).
15. S. Cahangirov, M. Topsakal, E. Aktürk, H. Şahin, and S. Ciraci, *Phys. Rev. Lett.* **102**, 236804 (Pages 4) (2009).
16. Y. Ding and J. Ni, *Appl. Phys. Lett.* 95, 083115 (2009).
17. S. Lebègue and O. Eriksson, *Phys. Rev. B Condens. Matter Mater. Phys.* **79**, 115409 (Pages 4) (2009), http://link.aps.org/abstract/PRB/v79/e115409.
18. A. Lu, X. Yang, and R. Zhang, *Solid State Commun.* **149**, 153 (2009).
19. H. Şahin, S. Cahangirov, M. Topsakal, E. Bekaroglu, E. Akturk, R. T. Senger, and S. Ciraci, *Phys. Rev. B* **80**, 155453 (2009).
20. H. Behera and G. Mukhopadhyay, *AIP Conf. Proc.* **1313**, 152 (2010).
21. S. Cahangirov, M. Topsakal, and S. Ciraci, *Phys. Rev. B* **81**, 195120 (2010).

22. S. Y. Davydov, *Phys. Solid State* **52**, 184 (2010).
23. N. Y. Dzade, K. O. Obodo, S. K. Adjokatse, A. C. Ashu, E. Amankwah, C. D. Atiso, A. A. Bello, et al., *J. Phys. Condens. Matter* **22**, 375502 (2010).
24. M. Houssa, G. Pourtois, V. V. Afanas'ev, and A. Stesmans, *Appl. Phys. Lett.* **97**, 112106 (2010).
25. A. Kara, S. Vizzini, C. Leandri, B. Ealet, H. Oughaddou, B. Aufray, and G. LeLay, *J. Phys. Cond. Matter* **22**, 045004 (2010).
26. L. C. Lew Yan Voon, E. Sandberg, R. S. Aga, and A. A. Farajian, *Appl. Phys. Lett.* **97**, 163114 (2010).
27. T. Morishita, S. P. Russo, I. K. Snook, M. J. S. Spencer, K. Nishio, and M. Mikami, *Phys. Rev. B* **82**, 045419 (2010).
28. O. Pulci, P. Gori, M. Marsili, V. Garbuio, A. P. Seitsonen, F. Bechstedt, A. Cricenti, and R. D. Sole, *Phys. Status Solidi A* **207**, 291 (2010).
29. Y.-L. Song, Y. Zhang, J.-M. Zhang, and D.-B. Lu, *Appl. Surf. Sci.* **256**, 6313 (2010), http://www.sciencedirect.com/science/article/B6THY-4YXK071-1/2/8e4afeb566c5ebd2577d348667e8152e.
30. T. Suzuki and Y. Yokomizo, *Physica E Low Dimens. Syst. Nanostruct.* **42**, 2820 (2010).
31. M. Topsakal and S. Ciraci, *Phys. Rev. B* **81**, 024107 (2010).
32. S. Wang, *J. Phys. Soc. Jpn.* **79**, 064602 (2010).
33. X.-D. Wen, T. J. Cahill, and R. Hoffmann, *Chem. A Eur. J.* **16**, 6555 (2010), http://dx.doi.org/10.1002/chem.200903128.
34. R. Brazhe, A. Karenin, A. Kochaev, and R. Meftakhutdinov, *Phys. Solid State* **53**, 1481 (2011).
35. Y. C. Cheng, Z. Y. Zhu, and U. Schwingenschlögl, *Europhys. Lett.* **95**, 17005 (2011).
36. G. G. Guzmán-Verri and L. C. Lew Yan Voon, *J. Phys. Condens. Matter* **23**, 145502 (2011).
37. M. Houssa, G. Pourtois, M. M. Heyns, V. V. Afanas'ev, and A. Stesmans, *J. Electrochem. Soc.* **158**, H107 (2011).
38. M. Houssa, E. Scalise, K. Sankaran, G. Pourtois, V. V. Afanas'ev, and A. Stesmans, *Appl. Phys. Lett.* **98**, 223107 (2011).
39. A. Ince and S. Erkoc, *Comput. Mater. Sci.* **50**, 865 (2011).
40. D. Jose and A. Datta, *Phys. Chem. Chem. Phys.* **13**, 7304 (2011).
41. C.-C. Liu, W. Feng, and Y. Yao, *Phys. Rev. Lett.* **107**, 076802 (2011), http://link.aps.org/doi/10.1103/PhysRevLett.107.076802.
42. C.-C. Liu, H. Jiang, and Y. Yao, *Phys. Rev. B* **84**, 195430 (2011), http://link.aps.org/doi/10.1103/PhysRevB.84.195430.
43. M. Miller and F. J. Owens, *Chem. Phys.* **381**, 1 (2011), http://www.sciencedirect.com/science/article/B6TFM-51GHWYX-2/2/83825f54385319fafb56a40a5bc60139.
44. T. Morishita, M. J. Spencer, S. P. Russo, I. K. Snook, and M. Mikami, *Chem. Phys. Lett.* **506**, 221 (2011), http://www.sciencedirect.com/science/article/B6TFN-52B118G-2/2/5d28b41be976542c42fd8277ad97d12f.
45. T. H. Osborn, O. V. Pupysheva, R. S. Aga, and L. C. Lew Yan Voon, *Chem. Phys. Letts.* **511**, 101 (2011).
46. Y.-L. Song, Y. Zhang, J.-M. Zhang, D.-B. Lu, and K.-W. Xu, *Eur. Phys. J. B Condens. Matter Complex Syst.* **79**, 197 (2011), http://dx.doi.org/10.1140/epjb/e2010-10627-5.
47. Y.-L. Song, Y. Zhang, J.-M. Zhang, D.-B. Lu, and K.-W. Xu, *J. Mol. Struct.* **990**, 75 (2011), http://www.sciencedirect.com/science/article/B6TGS-520TJYM-T/2/ef395b85b94e6a8d3a80c2dca67e17f6.
48. Y.-L. Song, Y. Zhang, J.-M. Zhang, D.-B. Lu, and K.-W. Xu, *Physica B Condens. Matter* **406**, 699 (2011), http://www.sciencedirect.com/science/article/B6TVH-51MDSFN-4/2/9dfb577692af2b1c9b5f96754d807cf5.
49. M. J. S. Spencer, T. Morishita, M. Mikami, I. K. Snook, Y. Sugiyama, and H. Nakano, *Phys. Chem. Chem. Phys.* **13**, 15418 (2011), http://dx.doi.org/10.1039/C1CP21544B.
50. S. Wang, L. Zhu, Q. Chen, J. Wang, and F. Ding, *J. Appl. Phys.* **109**, 053516 (2011).
51. S. Wang, *Phys. Chem. Chem. Phys.* **13**, 11929 (2011).
52. X. An, Y. Zhang, J. Liu, and S. Li, *New J. Phys.* **14**, 083039 (2012).
53. Y. Cui, J. Gao, L. Jin, J. Zhao, D. Tan, Q. Fu, and X. Bao, *Nano Res.* **5**, 352 (2012), http://dx.doi.org/10.1007/s12274-012-0215-4.
54. L. B. Drissi, E. H. Saidi, M. Bousmina, and O. Fassi-Fehri, *J. Phys. Condens. Matter* **24**, 485502 (2012), http://stacks.iop.org/0953-8984/24/i=48/a=485502.
55. N. D. Drummond, V. Zólyomi, and V. I. Fal'ko, *Phys. Rev. B* **85**, 075423 (2012), http://link.aps.org/doi/10.1103/PhysRevB.85.075423.
56. A. Dyrdal and J. Barnaś, *Phys. Status Solidi Rapid Res. Lett.* **6**, 340 (2012), http://dx.doi.org/10.1002/pssr.201206202.
57. H. Enriquez, S. Vizzini, A. Kara, B. Lalmi, and H. Oughaddou, *J. Phys. Condens. Matter* **24**, 314211 (2012), http://stacks.iop.org/0953-8984/24/i=31/a=314211.
58. M. Ezawa, *Eur. J. Phys. B* **85**, 1 (2012).
59. M. Ezawa, *New J. Phys.* **14**, 033003 (2012), http://stacks.iop.org/1367-2630/14/i=3/a=033003.
60. M. Ezawa, *Phys. Rev. Lett.* **109**, 055502 (2012), http://link.aps.org/doi/10.1103/PhysRevLett.109.055502.

61. M. Ezawa, *J. Phys. Soc. Jpn.* **81**, 064705 (2012), http://dx.doi.org/10.1143/JPSJ.81.064705.
62. J. Gao and J. Zhao, *Sci. Rep.* **2**, 861 (2012).
63. N. Gao, W. Zheng, and Q. Jiang, *Phys. Chem. Chem. Phys.* **14**, 257 (2012).
64. A. Ince and S. Erkoc, *Phys. Status Solidi B* **249**, 74 (2012).
65. D. Jose and A. Datta, *J. Phys. Chem. C* **116**, 24639 (2012), http://pubs.acs.org/doi/pdf/10.1021/jp3084716; http://pubs.acs.org/doi/abs/10.1021/jp3084716.
66. D. Kaltsas, L. Tsetseris, and A. Dimoulas, *J. Phys. Condens. Matter* **24**, 442001 (2012), http://stacks.iop.org/0953-8984/24/i=44/a=442001.
67. J. Kang, F. Wu, and J. Li, *Appl. Phys. Lett.* **100**, 233122 (2012).
68. J. Kim, M. V. Fischetti, and S. Aboud, *Phys. Rev. B* **86**, 205323 (2012), http://link.aps.org/doi/10.1103/PhysRevB.86.205323.
69. H. Li, L. Wang, Q. Liu, J. Zheng, W.-N. Mei, Z. Gao, J. Shi, and J. Lu, *Eur. Phys. J. B* **85**, 1 (2012), http://dx.doi.org/10.1140/epjb/e2012-30220-2.
70. H.-P. Li and R.-Q. Zhang, *Europhys. Lett.* **99**, 36001 (2012), http://stacks.iop.org/0295-5075/99/i=3/a=36001.
71. C. Lian and J. Ni, *Physica B Condens. Matter* **407**, 4695 (2012), http://www.sciencedirect.com/science/article/pii/S0921452612008228.
72. Y. Liang, V. Wang, H. Mizuseki, and Y. Kawazoe, *J. Phys. Condens. Matter* **24**, 455302 (2012), http://stacks.iop.org/0953-8984/24/i=45/a=455302.
73. X. Lin and J. Ni, *Phys. Rev. B* **86**, 075440 (2012), http://link.aps.org/doi/10.1103/PhysRevB.86.075440.
74. G. Liu, M. S. Wu, C. Y. Ouyang, and B. Xu, *Europhys. Lett.* **99**, 17010 (2012), http://stacks.iop.org/0295-5075/99/i=1/a=17010.
75. S. Naji, A. Belhaj, H. Labrim, H. Labrim, A. Benyoussef, and A. Kenz, *Eur. J. Phys. B* **85**, 1 (2012).
76. Z. Ni, Q. Liu, K. Tang, J. Zheng, J. Zhou, R. Qin, Z. Gao, D. Yu, and J. Lu, *Nano Lett.* **12**, 113 (2012), http://pubs.acs.org/doi/pdf/10.1021/nl203065e; http://pubs.acs.org/doi/abs/10.1021/nl203065e.
77. A. O'Hare, F. V. Kusmartsev, and K. I. Kugel, *Nano Lett.* **12**, 1045 (2012), http://pubs.acs.org/doi/pdf/10.1021/nl204283q; http://pubs.acs.org/doi/abs/10.1021/nl204283q.
78. A. O'Hare, F. Kusmartsev, and K. Kugel, *Physica B Condens. Matter* **407**, 1964 (2012), http://www.sciencedirect.com/science/article/pii/S0921452612000804.
79. T. H. Osborn and A. A. Farajian, *J. Phys. Chem. C* **116**, 22916 (2012), http://pubs.acs.org/doi/pdf/10.1021/jp306889x; http://pubs.acs.org/doi/abs/10.1021/jp306889x.
80. L. Pan, H. Liu, Y. Wen, X. Tan, H. Lv, J. Shi, and X. Tang, *Appl. Surf. Sci.* **258**, 10135 (2012), http://www.sciencedirect.com/science/article/pii/S0169433212011555.
81. L. Pan, H. J. Liu, X. J. Tan, H. Y. Lv, J. Shi, X. F. Tang, and G. Zheng, *Phys. Chem. Chem. Phys.* **14**, 13588 (2012), http://dx.doi.org/10.1039/C2CP42645E.
82. O. Pulci, P. Gori, M. Marsili, V. Garbuio, R. D. Sole, and F. Bechstedt, *Europhys. Lett.* **98**, 37004 (2012), http://stacks.iop.org/0295-5075/98/i=3/a=37004.
83. R. Qin, C.-H. Wang, W. Zhu, and Y. Zhang, *AIP Adv.* **2**, 022159 (2012).
84. R. Quhe, R. Fei, Q. Liu, J. Zheng, H. Li, C. Xu, Z. Ni, et al., *Sci. Rep.* **2**, 853 (2012).
85. F. Schwierz and J. Pezoldt, in *2012 IEEE 11th International Conference on Solid-State and Integrated Circuit Technology (ICSICT)*, (2012), pp. 1–4.
86. M. J. Spencer, T. Morishita, and I. K. Snook, *Nanoscale* **4**, 2906 (2012), http://dx.doi.org/10.1039/C2NR30100H.
87. M. Tahir and U. Schwingenschlogl, *Appl. Phys. Lett.* **101**, 132412 (2012).
88. L. Wang and H. Sun, *J. Mol. Model.* **18**, 4811 (2012), http://dx.doi.org/10.1007/s00894-012-1482-4.
89. Y. Wang, J. Zheng, Z. Ni, R. Fei, Q. Liu, R. Quhe, C. Xu, J. Zhou, Z. Gao, and J. Lu, *Nano* **7**, 1250037 (2012).
90. C. Xu, G. Luo, Q. Liu, J. Zheng, Z. Zhang, S. Nagase, Z. Gao, and J. Lu, *Nanoscale* **4**, 3111 (2012), http://dx.doi.org/10.1039/C2NR00037G.
91. C.-W. Zhang and S.-S. Yan, *J. Phys. Chem. C* **116**, 4163 (2012), http://pubs.acs.org/doi/pdf/10.1021/jp2104177; http://pubs.acs.org/doi/abs/10.1021/jp2104177.
92. H. Zhao, *Phys. Lett. A* **376**, 3546 (2012), http://www.sciencedirect.com/science/article/pii/S0375960112010523.
93. F.-B. Zheng and C.-W. Zhang, *Nanoscale Res. Lett.* **7**, 422 (2012).
94. E. C. Anota, A. B. Hernández, M. Castro, and G. H. Cocoletzi, *J. Comput. Theor. Nanosci.* **10**, 2264 (2013).
95. A. Belhaj, *Afr. Rev. Phys.* **8**, 11 (2013).
96. R. Bernard, T. Leoni, A. Wilson, T. Lelaidier, H. Sahaf, E. Moyen, L. C. Assaud, et al., *Phys. Rev. B* **88**, 121411 (2013), http://link.aps.org/doi/10.1103/PhysRevB.88.121411.
97. B. Bishnoi and B. Ghosh, *RSC Adv.* **3**, 26153 (2013), http://dx.doi.org/10.1039/C3RA43491E.
98. V. Q. Bui, T.-T. Pham, H.-V. S. Nguyen, and H. M. Le, *J. Phys. Chem. C* **117**, 23364 (2013), http://pubs.acs.org/doi/pdf/10.1021/jp407601d; http://pubs.acs.org/doi/abs/10.1021/jp407601d.

99. Y. Cai, C.-P. Chuu, C. M. Wei, and M. Y. Chou, *Phys. Rev. B* **88**, 245408 (2013), http://link.aps.org/doi/10.1103/PhysRevB.88.245408.

100. S. Cahangirov, M. Audiffred, P. Tang, A. Iacomino, W. Duan, G. Merino, and A. Rubio, *Phys. Rev. B* **88**, 035432 (2013), http://link.aps.org/doi/10.1103/PhysRevB.88.035432.

101. K. Chen, X. Wan, and J. Xu, *J. Mater. Chem. C* **1**, 4869 (2013), http://dx.doi.org/10.1039/C3TC30567H.

102. G. Cheng, P.-F. Liu, and Z.-T. Li, *Chin. Phys. B* **22**, 046201 (2013).

103. J. Dai, Y. Zhao, X. Wu, J. Yang, and X. C. Zeng, *J. Phys. Chem. Lett.* **4**, 561 (2013), http://pubs.acs.org/doi/pdf/10.1021/jz302000q; http://pubs.acs.org/doi/abs/10.1021/jz302000q.

104. A. Datta, A. Nijamudheen, and D. Jose, *Phys. Chem. Chem. Phys.* **15**, 8700 (2013), http://dx.doi.org/10.1039/C3CP51028J.

105. J. Deng, J. Z. Liu, and N. V. Medhekar, *RSC Adv.* **3**, 20338 (2013), http://dx.doi.org/10.1039/C3RA43326A.

106. Y. Ding and Y. Wang, *Appl. Phys. Lett.* **102**, 143115 (2013).

107. Y. Ding and Y. Wang, *J. Phys. Chem. C* **117**, 18266 (2013), http://pubs.acs.org/doi/pdf/10.1021/jp407666m; http://pubs.acs.org/doi/abs/10.1021/jp407666m.

108. Y. Ding and Y. Wang, *Appl. Phys. Lett.* **103**, 043114 (2013).

109. M. Ezawa, *Phys. Rev. Lett.* **110**, 026603 (2013), http://link.aps.org/doi/10.1103/PhysRevLett.110.026603.

110. M. Ezawa, *Phys. Rev. B* **87**, 155415 (2013), http://link.aps.org/doi/10.1103/PhysRevB.87.155415.

111. M. Ezawa, *Eur. Phys. J. B* **86**, 1 (2013), http://dx.doi.org/10.1140/epjb/e2013-31029-1.

112. M. Ezawa, Y. Tanaka, and N. Nagaosa, *Sci. Rep.* **3**, 2790 (2013).

113. M. Ezawa, *Sci. Rep.* **3**, 3435 (2013).

114. M. Ezawa, *AIP Conf. Proc.* **1566**, 199 (2013), http://scitation.aip.org/content/aip/proceeding/aipcp/10.1063/1.4848354.

115. D.-Q. Fang, S.-L. Zhang, and H. Xu, *RSC Adv.* **3**, 24075 (2013), http://dx.doi.org/10.1039/C3RA42720J.

116. C. Gang, L. Peng-Fei, and L. Zi-Tao, *Chin. Phys. B* **22**, 046201 (2013).

117. J. Gao, J. Zhang, H. Liu, Q. Zhang, and J. Zhao, *Nanoscale* **5**, 9785 (2013), http://dx.doi.org/10.1039/C3NR02826G.

118. F. Geissler, J. C. Budich, and B. Trauzettel, *New J. Phys.* **15**, 085030 (2013), http://stacks.iop.org/1367-2630/15/i=8/a=085030.

119. F. Geissler, J. C. Budich, and B. Trauzettel, *New J. Phys.* **17**, 119401 (2015), http://stacks.iop.org/1367-2630/17/i=11/a=119401.

120. E. Golias, E. Xenogiannopoulou, D. Tsoutsou, P. Tsipas, S. A. Giamini, and A. Dimoulas, *Phys. Rev. B* **88**, 075403 (2013), http://link.aps.org/doi/10.1103/PhysRevB.88.075403.

121. P. Gori, O. Pulci, F. Ronci, S. Colonna, and F. Bechsted, *J. Appl. Phys.* **114**, 113710 (2013).

122. Z.-X. Guo, S. Furuya, J.-I. Iwata, and A. Oshiyama, *Phys. Rev. B* **87**, 235435 (2013), http://link.aps.org/doi/10.1103/PhysRevB.87.235435.

123. H. H. Gürel, V. O. Özçelik, and S. Ciraci, *J. Phys. Condens. Matter* **25**, 305007 (2013), http://stacks.iop.org/0953-8984/25/i=30/a=305007.

124. M. Houssa, E. Scalise, B. van den Broek, G. Pourtois, V. V. Afanas'ev, and A. Stesmans, *ECS Trans.* **58**, 209 (2013), http://ecst.ecsdl.org/content/58/7/209.full.pdf+html, http://ecst.ecsdl.org/content/58/7/209.abstract.

125. M. Hu, X. Zhang, and D. Poulikakos, *Phys. Rev. B* **87**, 195417 (2013), http://link.aps.org/doi/10.1103/PhysRevB.87.195417.

126. W. Hu, X. Wu, Z. Li, and J. Yang, *Phys. Chem. Chem. Phys.* **15**, 5753 (2013). http://dx.doi.org/10.1039/C3CP00066D.

127. W. Hu, Z. Li, and J. Yang, *J. Chem. Phys.* **139**, 154704 (2013), http://scitation.aip.org/content/aip/journal/jcp/139/15/10.1063/1.4824887.

128. W. Hu, X. Wu, Z. Li, and J. Yang, *Nanoscale* **5**, 9062 (2013). http://dx.doi.org/10.1039/C3NR02326E.

129. J. Huang, H.-J. Chen, M.-S. Wu, G. Liu, C.-Y. Ouyang, and B. Xu, *Chin. Phys. Lett.* **30**, 017103 (2013).

130. S. Huang, W. Kang, and L. Yang, *Appl. Phys. Lett.* **102**, 133106 (2013).

131. B. Huang, H. J. Xiang, and S.-H. Wei, *Phys. Rev. Lett.* **111**, 145502 (2013), http://link.aps.org/doi/10.1103/PhysRevLett.111.145502.

132. T. Hussain, T. Kaewmaraya, S. Chakraborty, and R. Ahuja, *Phys. Chem. Chem. Phys.* **15**, 18900 (2013), http://dx.doi.org/10.1039/C3CP52830H.

133. Y. Jing, Y. Sun, H. Niu, and J. Shen, *Phys. Status Solidi B* **250**, 1505 (2013). http://dx.doi.org/10.1002/pssb.201349023.

134. T. P. Kaloni, Y. C. Cheng, and U. Schwingenschlögl, *J. Appl. Phys.* **113**, 104305 (2013).

135. T. P. Kaloni, M. Tahir, and U. Schwingenschlögl, *Sci. Rep.* **3**, 3192 (2013).

136. T. P. Kaloni, S. Gangopadhyay, N. Singh, B. Jones, and U. Schwingenschlögl, *Phys. Rev. B* **88**, 235418 (2013), http://link.aps.org/doi/10.1103/PhysRevB.88.235418.

137. D. Kaltsas and L. Tsetseris, *Phys. Chem. Chem. Phys.* **15**, 9710 (2013), http://dx.doi.org/10.1039/C3CP50944C.

138. D. Kaltsas, T. Tsatsoulis, O. G. Ziogos, and L. Tsetseris, *J. Chem. Phys.* **139**, 124709 (2013), http://scitation.aip.org/content/aip/journal/jcp/139/12/10.1063/1.4822263.

139. C. Kamal, A. Chakrabarti, A. Banerjee, and S. K. Deb, *J. Phys. Condens. Mat.* **25**, 085508 (2013).

140. E. Kogan, *Graphene* **2**, 74 (2013).

141. S. Kokott, L. Matthes, and F. Bechstedt, *Phys. Status Solidi Rapid Res. Lett.* **7**, 538 (2013), http://dx.doi.org/10.1002/pssr.201307215.

142. N. Konobeeva and M. Belonenko, *Tech. Phys. Lett.* **39**, 579 (2013), http://dx.doi.org/10.1134/S1063785013060205.

143. V. V. Kulish, M. Ng, O. I. Malyi, P. Wu, and Z. Chen, *Chemphyschem.* **14**, 1161 (2013).

144. M. Lan, G. Xiang, C. Zhang, and X. Zhang, *J. Appl. Phys.* **114**, 163711 (2013), http://scitation.aip.org/content/aip/journal/jap/114/16/10.1063/1.4828482.

145. C.-C. Lee, A. Fleurence, R. Friedlein, Y. Yamada-Takamura, and T. Ozaki, *Phys. Rev. B* **88**, 165404 (2013), http://link.aps.org/doi/10.1103/PhysRevB.88.165404.

146. C. Li, S. Yang, S.-S. Li, J.-B. Xia, and J. Li, *J. Phys. Chem. C* **117**, 483 (2013), http://pubs.acs.org/doi/pdf/10.1021/jp310746m; http://pubs.acs.org/doi/abs/10.1021/jp310746m.

147. F. Li, R. Lu, Q. Yao, E. Kan, Y. Liu, H. Wu, Y. Yuan, C. Xiao, and K. Deng, *J. Phys. Chem. C* **117**, 13283 (2013), http://dx.doi.org/10.1021/jp402875t.

148. F. Li, C.-W. Zhang, H.-X. Luan, and P.-J. Wang, *J. Nanopart. Res.* **15**, 1 (2013), http://dx.doi.org/10.1007/s11051-013-1972-z.

149. L. Li, X. Wang, X. Zhao, and M. Zhao, *Phys. Lett. A* **377**, 2628 (2013), http://www.sciencedirect.com/science/article/pii/S0375960113006877.

150. R. Li, J. Zhou, Y. Han, J. Dong, and Y. Kawazoe, *J. Chem. Phys.* **139**, 104703 (2013).

151. X. Li, J. T. Mullen, Z. Jin, K. M. Borysenko, M. Buongiorno Nardelli, and K. W. Kim, *Phys. Rev. B* **87**, 115418 (2013), http://link.aps.org/doi/10.1103/PhysRevB.87.115418.

152. Y. Li and Z. Chen, *J. Phys. Chem. Lett.* **4**, 269 (2013), http://pubs.acs.org/doi/pdf/10.1021/jz301821n; http://pubs.acs.org/doi/abs/10.1021/jz301821n.

153. C. Lian and J. Ni, *AIP Adv.* **3**, 052102 (2013).

154. M. P. Lima, A. Fazzio, and A. J. R. da Silva, *Phys. Rev. B* **88**, 235413 (2013), http://link.aps.org/doi/10.1103/PhysRevB.88.235413.

155. H. Liu, J. Gao, and J. Zhao, *J. Phys. Chem. C* **117**, 10353 (2013), http://pubs.acs.org/doi/pdf/10.1021/jp311836m; http://pubs.acs.org/doi/abs/10.1021/jp311836m.

156. F. Liu, C.-C. Liu, K. Wu, F. Yang, and Y. Yao, *Phys. Rev. Lett.* **111**, 066804 (2013), http://link.aps.org/doi/10.1103/PhysRevLett.111.066804.

157. J. Liu and W. Zhang, *RSC Adv.* **3**, 21943 (2013), http://dx.doi.org/10.1039/C3RA44392B.

158. Y. Liu, H. Shu, P. Liang, D. Cao, X. Che, and W. Lu, *J. Appl. Phys.* **114**, 094308 (2013).

159. H.-X. Luan, C.-W. Zhang, F.-B. Zheng, and P.-J. Wang, *J. Phys. Chem. C* **117**, 13620 (2013), http://pubs.acs.org/doi/pdf/10.1021/jp4005357; http://pubs.acs.org/doi/abs/10.1021/jp4005357.

160. H. xing Luan, C. wen Zhang, F. Li, and P. ji Wang, *Phys. Lett. A* **377**, 2792 (2013), http://www.sciencedirect.com/science/article/pii/S037596011300724X.

161. L. Ma, J.-M. Zhang, K.-W. Xu, and V. Ji, *Physica B Condens. Matter* **425**, 66 (2013).

162. L. Matthes, P. Gori, O. Pulci, and F. Bechstedt, *Phys. Rev. B* **87**, 035438 (2013), http://link.aps.org/doi/10.1103/PhysRevB.87.035438.

163. L. Matthes, O. Pulci, and F. Bechstedt, *J. Phys. Condens. Matter* **25**, 395305 (2013).

164. N. Mehrotra, N. Kumar, and A. Sen, *AIP Conf. Proc.* **1512**, 1304 (2013).

165. B. Mohan, A. Kumar, and P. Ahluwalia, *AIP Conf. Proc.* **1512**, 378 (2013).

166. B. Mohan, A. Kumar, and P. Ahluwalia, *Physica E Low Dimens. Syst. Nanostruct. Res* **53**, 233 (2013), http://www.sciencedirect.com/science/article/pii/S1386947713001872.

167. A. Molle, C. Grazianetti, D. Chiappe, E. Cinquanta, E. Cianci, G. Tallarida, and M. Fanciulli, *Adv. Funct. Mater.* **23**, 4339 (2013), http://dx.doi.org/10.1002/adfm.201370179.

168. T. Morishita, M. J. S. Spencer, S. Kawamoto, and I. K. Snook, *J. Phys. Chem. C* **117**, 22142 (2013), http://pubs.acs.org/doi/pdf/10.1021/jp4080898; http://pubs.acs.org/doi/abs/10.1021/jp4080898.

169. M. Neek-Amal, A. Sadeghi, G. R. Berdiyorov, and F. M. Peeters, *Appl. Phys. Lett.* **103**, 261904 (2013), http://scitation.aip.org/content/aip/journal/apl/103/26/10.1063/1.4852636.

170. T. Ng, J. Yeo, and Z. Liu, *Int. J. Mech. Mater. Des.* **9**, 105 (2013), http://dx.doi.org/10.1007/s10999-013-9215-0.

171. V. O. Özçelik, H. H. Gurel, and S. Ciraci, *Phys. Rev. B* **88**, 045440 (2013), http://link.aps.org/doi/10.1103/PhysRevB.88.045440.

172. V. O. Özçelik and S. Ciraci, *J. Phys. Chem. C* **117**, 26305 (2013), http://pubs.acs.org/doi/pdf/10.1021/jp408647t; http://pubs.acs.org/doi/abs/10.1021/jp408647t.

173. Q.-X. Pei, Y.-W. Zhang, Z.-D. Sha, and V. B. Shenoy, *J. Appl. Phys.* **114**, 033526 (2013).

174. Q. Peng, X. Wen, and S. De, *RSC Adv.* **3**, 13772 (2013), http://dx.doi.org/10.1039/C3RA41347K.
175. Q. Peng, J. Crean, A. K. Dearden, C. Huang, X. Wen, S. P. A. Bordas, and S. De, *Mod. Phys. Lett. B* **27**, 1330017 (2013), http://www.worldscientific.com/doi/pdf/10.1142/S0217984913300172; http://www.worldscientific.com/doi/abs/10.1142/S0217984913300172.
176. P. Rubio-Pereda and N. Takeuchi, *J. Chem. Phys.* **138**, 194702 (2013).
177. F. Sánchez-Ochoa, J. Guerrero-Snchez, G. Canto, G. Cocoletzi, and N. Takeuchi, *J. Mol. Model.* **19**, 2925 (2013), http://dx.doi.org/10.1007/s00894-013-1873-1.
178. E. Scalise, M. Houssa, G. Pourtois, B. Broek, V. Afanasev, and A. Stesmans, *Nano Res.* **6**, 19 (2013).
179. J. Setiadi, M. D. Arnold, and M. J. Ford, *ACS Appl. Mater. Interfaces* **5**, 10690 (2013), http://pubs.acs.org/doi/pdf/10.1021/am402828k; http://pubs.acs.org/doi/abs/10.1021/am402828k.
180. Z.-G. Shao, X.-S. Ye, L. Yang, and C.-L. Wang, *J. Appl. Phys.* **114**, 093712 (2013).
181. E. F. Sheka, *International J. Quantum Chem.* **113**, 612 (2013), http://dx.doi.org/10.1002/qua.24081.
182. J. Sivek, H. Sahin, B. Partoens, and F. M. Peeters, *Phys. Rev. B* **87**, 085444 (2013), http://link.aps.org/doi/10.1103/PhysRevB.87.085444.
183. Y.-L. Song, J.-M. Zhang, D.-B. Lu, and K.-W. Xu, *Physica E Low Dimens. Syst. Nanostruct.* **53**, 173 (2013), http://www.sciencedirect.com/science/article/pii/S1386947713001586.
184. Y.-L. Song, S. Zhang, D.-B. Lu, H.-r. Xu, Z. Wang, Y. Zhang, and Z.-W. Lu, *Eur. Phys. J. B* **86**, 1 (2013), http://dx.doi.org/10.1140/epjb/e2013-31078-4.
185. M. J. S. Spencer, T. Morishita, and M. R. Bassett, *Proc. SPIE* **8923**, 89230D (2013), http://dx.doi.org/10.1117/12.2033776.
186. M. J. S. Spencer, M. R. Bassett, T. Morishita, I. K. Snook, and H. Nakano, *New J. Phys.* **15**, 125018 (2013), http://stacks.iop.org/1367-2630/15/i=12/a=125018.
187. C. J. Tabert and E. J. Nicol, *Phys. Rev. Lett.* **110**, 197402 (2013), http://link.aps.org/doi/10.1103/PhysRevLett.110.197402.
188. C. J. Tabert and E. J. Nicol, *Phys. Rev. B* **88**, 085434 (2013), http://link.aps.org/doi/10.1103/PhysRevB.88.085434.
189. M. Tahir and U. Schwingenschlögl, *Sci. Rep.* **3**, 1 (2013).
190. M. Tahir, A. Manchon, K. Sabeeh, and U. Schwingenschögl, *Appl. Phys. Lett.* **102**, 162412 (2013).
191. G. A. Tritsaris, E. Kaxiras, S. Meng, and E. Wang, *Nano Lett.* **13**, 2258 (2013), http://pubs.acs.org/doi/pdf/10.1021/nl400830u; http://pubs.acs.org/doi/abs/10.1021/nl400830u.
192. G. Wang, *Europhys. Lett.* **101**, 27005 (2013).
193. J. Wang, J. Li, S.-S. Li, and Y. Liu, *J. Appl. Phys.* **114**, 124309 (2013).
194. Y. Wang and Y. Ding, *Solid State Commun.* **155**, 6 (2013).
195. Y. Wang and Y. Ding, *Phys. Status Solidi Rapid Res. Lett.* **7**, 410 (2013), http://dx.doi.org/10.1002/pssr.201307110.
196. Y. Wang and Y. Lou, *J. Appl. Phys.* **114**, 183712 (2013), http://scitation.aip.org/content/aip/journal/jap/114/18/10.1063/1.4830020.
197. Y.-P. Wang and H.-P. Cheng, *Phys. Rev. B* **87**, 245430 (2013), http://link.aps.org/doi/10.1103/PhysRevB.87.245430.
198. Y.-P. Wang, J. N. Fry, and H.-P. Cheng, *Phys. Rev. B* **88**, 125428 (2013), http://link.aps.org/doi/10.1103/PhysRevB.88.125428.
199. W. Wei, Y. Dai, B. Huang, and T. Jacob, *Phys. Chem. Chem. Phys.* **15**, 8789 (2013), http://dx.doi.org/10.1039/C3CP51078F.
200. W. Wei and T. Jacob, *Phys. Rev. B* **88**, 045203 (2013), http://link.aps.org/doi/10.1103/PhysRevB.88.045203.
201. A. Yamakage, M. Ezawa, Y. Tanaka, and N. Nagaosa, *Phys. Rev. B* **88**, 085322 (2013), http://link.aps.org/doi/10.1103/PhysRevB.88.085322.
202. J.-A. Yan, R. Stein, D. M. Schaefer, X.-Q. Wang, and M. Y. Chou, *Phys. Rev. B* **88**, 121403 (2013), http://link.aps.org/doi/10.1103/PhysRevB.88.121403.
203. Y. Yan, H. Wu, F. Jiang, and H. Zhao, *Eur. Phys. J. B* **86**, 1 (2013), http://dx.doi.org/10.1140/epjb/e2013-40818-3.
204. K. Zberecki, M. Wierzbicki, J. Barnaś, and R. Swirkowicz, *Phys. Rev. B* **88**, 115404 (2013), http://link.aps.org/doi/10.1103/PhysRevB.88.115404.
205. X.-L. Zhang, L.-F. Liu, and W.-M. Liu, *Sci. Rep.* **3**, 2908 (2013).
206. F. bao Zheng, C. wen Zhang, P. ji Wang, and S. shi Li, *J. Appl. Phys.* **113**, 154302 (2013).
207. F.-B. Zheng, C.-W. Zhang, S.-S. Yan, and F. Li, *J. Mater. Chem. C* **1**, 2735 (2013), http://dx.doi.org/10.1039/C3TC30097H.
208. H. Zhou, M. Zhao, X. Zhang, W. Dong, X. Wang, H. Bu, and A. Wang, *J. Phys. Condens. Matter* **25**, 395501 (2013), http://stacks.iop.org/0953-8984/25/i=39/a=395501.
209. D. S. L. Abergel, J. M. Edge, and A. V. Balatsky, *New J. Phys.* **16**, 065012 (2014), http://stacks.iop.org/1367-2630/16/i=6/a=065012.

210. R.-L. An, X.-F. Wang, P. Vasilopoulos, Y.-S. Liu, A.-B. Chen, Y.-J. Dong, and M.-X. Zhai, *J. Phys. Chem. C* **118**, 21339 (2014), http://dx.doi.org/10.1021/jp506111a.
211. R. Ansari, S. Rouhi, and S. Ajori, *Superlattices Microstruct.* **65**, 64 (2014), http://www.sciencedirect.com/science/article/pii/S0749603613003765.
212. A. Atsalakis and L. Tsetseris, *J. Phys. Condens. Matter* **26**, 285301 (2014), http://stacks.iop.org/0953-8984/26/i=28/a=285301.
213. F. Bechstedt, P. Gori, S. Kokott, L. Matthes, P. Pflugradt, and O. Pulci, *AIP Conf. Proc.* **1618**, 56 (2014), http://scitation.aip.org/content/aip/proceeding/aipcp/10.1063/1.4897673.
214. G. Berdiyorov and F. Peeters, *RSC Adv.* **4**, 1133 (2014).
215. B. Bishnoi and B. Ghosh, *J. Comput. Electron.* **13**, 186 (2014), http://dx.doi.org/10.1007/s10825-013-0498-z.
216. G. R. Berdiyorov, M. Neek-Amal, F. M. Peeters, and A. C. T. van Duin, *Phys. Rev. B* **89**, 024107 (2014), http://link.aps.org/doi/10.1103/PhysRevB.89.024107.
217. B. Bishnoi and B. Ghosh, *Comput. Mater. Sci.* **85**, 16 (2014), http://www.sciencedirect.com/science/article/pii/S0927025613008008.
218. B. Bishnoi and B. Ghosh, *J. Comput. Theor. Nanosci.* **11**, 1271 (2014), http://www.ingentaconnect.com/content/asp/jctn/2014/00000011/00000005/art00008.
219. B. Bishnoi and B. Ghosh, *J. Comput. Electron.* **13**, 186 (2014), http://dx.doi.org/10.1007/s10825-013-0498-z.
220. V. Bocchetti, H. T. Diep, H. Enriquez, H. Oughaddou, and A. Kara, *J. Phys. Conf. Ser.* **491**, 012008 (2014), http://stacks.iop.org/1742-6596/491/i=1/a=012008.
221. S. Cahangirov, V. O. Özçelik, A. Rubio, and S. Ciraci, *Phys. Rev. B* **90**, 085426 (2014), http://link.aps.org/doi/10.1103/PhysRevB.90.085426.
222. H.-R. Chang, J. Zhou, H. Zhang, and Y. Yao, *Phys. Rev. B* **89**, 201411 (2014), http://link.aps.org/doi/10.1103/PhysRevB.89.201411.
223. A.-B. Chen, X.-F. Wang, P. Vasilopoulos, M.-X. Zhai, and Y.-S. Liu, *Phys. Chem. Chem. Phys.* **16**, 5113 (2014), http://dx.doi.org/10.1039/C3CP55447C.
224. J. Chen, X.-F. Wang, P. Vasilopoulos, A.-B. Chen, and J.-C. Wu, *Chemphyschem.* **15**, 2701 (2014), http://dx.doi.org/10.1002/cphc.201402171.
225. M. X. Chen and M. Weinert, *Nano Lett.* **14**, 5189 (2014), http://pubs.acs.org/doi/pdf/10.1021/nl502107v; http://pubs.acs.org/doi/abs/10.1021/nl502107v.
226. R. Das, S. Chowdhury, A. Majumdar, and D. Jana, *RSC Adv.* **5**, 41 (2015). http://dx.doi.org/10.1039/C4RA07976K.
227. Y.-J. Dong, X.-F. Wang, P. Vasilopoulos, M.-X. Zhai, and X.-M. Wu, *J. Phys. D Appl. Phys.* **47**, 105304 (2014), http://stacks.iop.org/0022-3727/47/i=10/a=105304.
228. T. Y. Du, J. Zhao, G. Liu, J. X. Le, and B. Xu, *Mod. Phys. Lett. B* **28**, 1450138 (2014), http://www.worldscientific.com/doi/pdf/10.1142/S0217984914501383; http://www.worldscientific.com/doi/abs/10.1142/S0217984914501383.
229. A. Durajski, D. Szcześniak, and R. Szcześniak, *Solid State Commun.* **200**, 17 (2014), http://www.sciencedirect.com/science/article/pii/S0038109814003664.
230. S. Dutta and K. Wakabayashi, *Jpn. J. Appl. Phys.* **53**, 06JD01 (2014), http://stacks.iop.org/1347-4065/53/i=6S/a=06JD01.
231. F. Ersan, O. Arslanalp, G. Gökoğlu, and E. Aktürk, *Appl. Surf. Sci.* **311**, 9 (2014), http://www.sciencedirect.com/science/article/pii/S0169433214009520.
232. M. Ezawa, *JPS Conf. Proc.* **1**, 012003 (2014).
233. M. Ezawa, *JPS Conf. Proc.* **3**, 012001 (2014).
234. D. Q. Fang, Y. Zhang, and S. L. Zhang, *New J. Phys.* **16**, 115006 (2014), http://stacks.iop.org/1367-2630/16/i=11/a=115006.
235. J.-W. Feng, Y.-J. Liu, H.-X. Wang, J.-X. Zhao, Q.-H. Cai, and X.-Z. Wang, *Comput. Mater. Sci.* **87**, 218 (2014).
236. F. Filippone, *J. Phys. Condens. Matter* **26**, 395009 (2014), http://stacks.iop.org/0953-8984/26/i=39/a=395009.
237. H. Fu, J. Zhang, Z. Ding, H. Li, and S. Meng, *Appl. Phys. Lett.* **104**, 131904 (2014), http://scitation.aip.org/content/aip/journal/apl/104/13/10.1063/1.4870534.
238. N. Gao, J. Li, and Q. Jiang, *Chem. Phys. Lett.* **592**, 222 (2014), http://www.sciencedirect.com/science/article/pii/S0009261413015170.
239. N. Gao, J. C. Li, and Q. Jiang, *Phys. Chem. Chem. Phys.* **16**, 11673 (2014), http://dx.doi.org/10.1039/C4CP00089G.
240. L. Gong, S. Xiu, M. Zheng, P. Zhao, Z. Zhang, Y. Liang, G. Chen, and Y. Kawazoe, *J. Mater. Chem. C* **2**, 8773 (2014), http://dx.doi.org/10.1039/C4TC01665C.
241. P. Gori, O. Pulci, R. de Lieto Vollaro, and C. Guattari, *Energy Procedia* **45**, 512 (2014), http://www.sciencedirect.com/science/article/pii/S1876610214000563.

242. C. Grazianetti, D. Chiappe, E. Cinquanta, G. Tallarida, M. Fanciulli, and A. Molle, *Appl. Surf. Sci.* **291**, 109 (2013), http://www.sciencedirect.com/science/article/pii/S0169433213016127.

243. G. Gupta, H. Lin, A. Bansil, M. B. Abdul Jalil, C.-Y. Huang, W.-F. Tsai, and G. Liang, *Appl. Phys. Lett.* **104**, 032410 (2014), http://scitation.aip.org/content/aip/journal/apl/104/3/10.1063/1.4863088.

244. M. Houssa, B. van den Broek, E. Scalise, B. Ealet, G. Pourtois, D. Chiappe, E. Cinquanta, et al., *Appl. Surf. Sci.* **291**, 98 (2013), http://www.sciencedirect.com/science/article/pii/S0169433213017030.

245. M. Houssa, E. Scalise, B. van den Broek, A. Lu, G. Pourtois, V. V. Afanas'ev, and A. Stesmans, *ECS Trans.* **64**, 111 (2014).

246. L. Hu, J. Zhao, and J. Yang, *J. Phys. Condens. Matter* **26**, 335302 (2014), http://stacks.iop.org/0953-8984/26/i=33/a=335302.

247. W. Hu, N. Xia, X. Wu, Z. Li, and J. Yang, *Phys. Chem. Chem. Phys.* **16**, 6957 (2014). http://dx.doi.org/10.1039/C3CP55250K.

248. B. Huang, H.-X. Deng, H. Lee, M. Yoon, B. G. Sumpter, F. Liu, S. C. Smith, and S.-H. Wei, *Phys. Rev. X* **4**, 021029 (2014), http://link.aps.org/doi/10.1103/PhysRevX.4.021029.

249. Z.-Q. Huang, B.-H. Chou, C.-H. Hsu, F.-C. Chuang, H. Lin, and A. Bansil, *Phys. Rev. B* **90**, 245433 (2014), http://link.aps.org/doi/10.1103/PhysRevB.90.245433.

250. T. Hussain, S. Chakraborty, A. De Sarkar, B. Johansson, and R. Ahuja, *Appl. Phys. Lett.* **105**, 123903 (2014), http://scitation.aip.org/content/aip/journal/apl/105/12/10.1063/1.4896503.

251. Y. Hwang, K.-H. Yun, and Y.-C. Chung, *J. Power Sources* **275**, 32 (2015), http://www.sciencedirect.com/science/article/pii/S0378775314018424.

252. H. Johll, M. D. K. Lee, S. P. N. Ng, H. C. Kang, and E. S. Tok, *Sci. Rep.* **4**, 7594 (2014).

253. T. P. Kaloni, N. Singh, and U. Schwingenschlögl, *Phys. Rev. B* **89**, 035409 (2014), http://link.aps.org/doi/10.1103/PhysRevB.89.035409.

254. T. P. Kaloni and U. Schwingenschlögl, *Phys. Status Solidi Rapid Res. Lett.* **8**, 685 (2014). http://dx.doi.org/10.1002/pssr.201409245.

255. D. Kaltsas, L. Tsetseris, and A. Dimoulas, *Appl. Surf. Sci.* **291**, 93 (2013), http://www.sciencedirect.com/science/article/pii/S016943321301756X.

256. S. Kaneko, H. Tsuchiya, Y. Kamakura, N. Mori, and M. Ogawa, *Appl. Phys. Express* **7**, 035102 (2014), http://stacks.iop.org/1882-0786/7/i=3/a=035102.

257. M. Kanno, R. Arafune, C. L. Lin, E. Minamitani, M. Kawai, and N. Takagi, *New J. Phys.* **16**, 105019 (2014), http://stacks.iop.org/1367-2630/16/i=10/a=105019.

258. Y. Kim, K. Choi, J. Ihm, and H. Jin, *Phys. Rev. B* **89**, 085429 (2014), http://link.aps.org/doi/10.1103/PhysRevB.89.085429.

259. S. Kokott, P. Pflugradt, L. Matthes, and F. Bechstedt, *J. Phys. Condens. Matter* **26**, 185002 (2014), http://stacks.iop.org/0953-8984/26/i=18/a=185002.

260. N. B. Le, T. D. Huan, and L. M. Woods, *Phys. Rev. Appl.* **1**, 054002 (2014), http://link.aps.org/doi/10.1103/PhysRevApplied.1.054002.

261. C.-C. Lee, A. Fleurence, R. Friedlein, Y. Yamada-Takamura, and T. Ozaki, *Phys. Rev. B* **90**, 241402 (2014), http://link.aps.org/doi/10.1103/PhysRevB.90.241402.

262. Y. Lee, K.-H. Yun, S. B. Cho, and Y.-C. Chung, *Chemphyschem* **15**, 4095 (2014). http://dx.doi.org/10.1002/cphc.201402613.

263. G. Li, J. Tan, X. Liu, X. Wang, F. Li, and M. Zhao, *Chem. Phys. Lett.* **595–596**, 20–24 (2014), http://www.sciencedirect.com/science/article/pii/S0009261414000517.

264. L. Li and M. Zhao, *J. Phys. Chem. C* **118**, 19129 (2014), http://dx.doi.org/10.1021/jp5043359.

265. R. Li, Y. Han, T. Hu, J. Dong, and Y. Kawazoe, *Phys. Rev. B* **90**, 045425 (2014), http://link.aps.org/doi/10.1103/PhysRevB.90.045425.

266. S. Li, Y. Wu, W. Liu, and Y. Zhao, *Chem. Phys. Lett.* **609**, 161 (2014), http://www.sciencedirect.com/science/article/pii/S0009261414005521.

267. S. hi Li, C.-W. Zhang, S.-S. Yan, S.-J. Hu, W.-X. Ji, P.-J. Wang, and P. Li, *J. Phys. Condens. Matter* **26**, 395003 (2014), http://stacks.iop.org/0953-8984/26/i=39/a=395003.

268. X. Lin, C. Lian, and J. Ni, *J. Phys. Conf. Ser.* **491**, 012005 (2014), http://stacks.iop.org/1742-6596/491/i=1/a=012005.

269. B. Liu, C. D. Reddy, J. Jiang, H. Zhu, J. A. Baimova, S. V. Dmitriev, and K. Zhou, *J. Phys. D Appl. Phys.* **47**, 165301 (2014), http://stacks.iop.org/0022-3727/47/i=16/a=165301.

270. B. Liu, J. A. Baimova, C. D. Reddy, S. V. Dmitriev, W. K. Law, X. Q. Feng, and K. Zhou, *Carbon* **79**, 236 (2014), http://www.sciencedirect.com/science/article/pii/S0008622314007027.

271. B. Liu, J. A. Baimova, C. D. Reddy, A. W.-K. Law, S. V. Dmitriev, H. Wu, and K. Zhou, *ACS Appl. Mater. Interfaces* **6**, 18180 (2014). http://dx.doi.org/10.1021/am505173s.

272. G. Liu, X. L. Lei, M. S. Wu, B. Xu, and C. Y. Ouyang, *Europhys. Lett.* **106**, 47001 (2014), http://stacks.iop.org/0295-5075/106/i=4/a=47001.

273. G. Liu, X. L. Lei, M. S. Wu, B. Xu, and C. Y. Ouyang, *J. Phys. Condens. Matter* **26**, 355007 (2014), http://stacks.iop.org/0953-8984/26/i=35/a=355007.

274. Y. Liu, X. Zhou, M. Zhou, M.-Q. Long, and G. Zhou, *J. Appl. Phys.* **116**, 244312 (2014), http://scitation.aip.org/content/aip/journal/jap/116/24/10.1063/1.4904751.

275. Y. Liu, X. Yang, X. Zhang, X. Hong, X.-F. Wang, J. Feng, and C. Zhang, *RSC Adv.* **4**, 48539 (2014), http://dx.doi.org/10.1039/C4RA07791A.

276. A. Lopez-Bezanilla, *J. Phys. Chem. C* **118**, 18788 (2014), http://dx.doi.org/10.1021/jp5060809.

277. K. L. Low, W. Huang, Y.-C. Yeo, and G. Liang, *IEEE Trans. Electron Devices* **61**, 1590 (2014).

278. L. Ma, J.-M. Zhang, K.-W. Xu, and V. Ji, *Phys. E Low Dimens. Syst. Nanostruct.* **60**, 112 (2014), http://www.sciencedirect.com/science/article/pii/S1386947714000642.

279. A. Majumdar, S. Chowdhury, P. Nath, and D. Jana, *RSC Adv.* **4**, 32221 (2014), http://dx.doi.org/10.1039/C4RA04174G.

280. A. Manjanath, V. Kumar, and A. K. Singh, *Phys. Chem. Chem. Phys.* **16**, 1667 (2014), http://dx.doi.org/10.1039/C3CP54655A.

281. A. Manjanath and A. K. Singh, *Chem. Phys. Lett.* **592**, 52 (2014), http://www.sciencedirect.com/science/article/pii/S0009261413014905.

282. L. Matthes, O. Pulci, and F. Bechstedt, *New J. Phys.* **16**, 105007 (2014), http://stacks.iop.org/1367-2630/16/i=10/a=105007.

283. B. Mohan, A. Kumar, and P. Ahluwalia, *Phys. E Low Dimens. Syst. Nanostruct.* **61**, 40 (2014), http://www.sciencedirect.com/science/article/pii/S1386947714001003.

284. B. Mohan, A. Kumar, and P. K. Ahluwalia, *AIP Conf. Proc.* **1591**, 1714 (2014), http://scitation.aip.org/content/aip/proceeding/aipcp/10.1063/1.4873087.

285. B. Mohan, A. Kumar, and P. K. Ahluwalia, *RSC Adv.* **4**, 31700 (2014), http://dx.doi.org/10.1039/C4RA02711F.

286. S. Naji, B. Khalil, H. Labrim, M. Bhihi, A. Belhaj, A. Benyoussef, M. Lakhal, and A. E. Kenz, *J. Phys. Conf. Ser.* **491**, 012006 (2014), http://stacks.iop.org/1742-6596/491/i=1/a=012006.

287. S. Naji, A. Belhaj, H. Labrim, M. Bhihi, A. Benyoussef, and A. E. Kenz, *Int. J. Mod. Phys. B* **28**, 1450086 (2014), http://www.worldscientific.com/doi/pdf/10.1142/S0217979214500866; http://www.worldscientific.com/doi/abs/10.1142/S0217979214500866.

288. Z. Ping Niu and S. Dong, *Appl. Phys. Lett.* **104**, 202401 (2014), http://scitation.aip.org/content/aip/journal/apl/104/20/10.1063/1.4876927.

289. V. O. Özçelik, E. Durgun, and S. Ciraci, *J. Phys. Chem. Lett.* **5**, 2694 (2014). http://dx.doi.org/10.1021/jz500977v.

290. H. Pan, X. Li, Z. Qiao, C.-C. Liu, Y. Yao, and S. A. Yang, *New J. Phys.* **16**, 123015 (2014), http://stacks.iop.org/1367-2630/16/i=12/a=123015.

291. Q.-X. Pei, Z.-D. Sha, Y.-Y. Zhang, and Y.-W. Zhang, *J. Appl. Phys.* **115**, 023519 (2014), http://scitation.aip.org/content/aip/journal/jap/115/2/10.1063/1.4861736.

292. Q. Peng and S. De, *Nanoscale* **6**, 12071 (2014), http://dx.doi.org/10.1039/C4NR01831A.

293. P. Pflugradt, L. Matthes, and F. Bechstedt, *Phys. Rev. B* **89**, 035403 (2014), http://link.aps.org/doi/10.1103/PhysRevB.89.035403.

294. P. Pflugradt, L. Matthes, and F. Bechstedt, *Phys. Rev. B* **89**, 205428 (2014), http://link.aps.org/doi/10.1103/PhysRevB.89.205428.

295. P. Pflugradt, L. Matthes, and F. Bechstedt, *New J. Phys.* **16**, 075004 (2014), http://stacks.iop.org/1367-2630/16/i=7/a=075004.

296. X. Pi, Z. Ni, Y. Liu, Z. Ruan, M. Xu, and D. Yang, *Phys. Chem. Chem. Phys.* **17**, 4146 (2015), http://dx.doi.org/10.1039/C4CP05196C.

297. R. Qin, W. Zhu, Y. Zhang, and X. Deng, *Nano Res. Lett.* **9**, 521 (2014).

298. S. Rachel and M. Ezawa, *Phys. Rev. B* **89**, 195303 (2014), http://link.aps.org/doi/10.1103/PhysRevB.89.195303.

299. O. D. Restrepo, R. Mishra, J. E. Goldberger, and W. Windl, *J. Appl. Phys.* **115**, 033711 (2014), http://scitation.aip.org/content/aip/journal/jap/115/3/10.1063/1.4860988.

300. E. Romera, J. Roldn, and F. de los Santos, *Phys. Lett. A* **378**, 2582 (2014), http://www.sciencedirect.com/science/article/pii/S0375960114006379.

301. N. J. Roome and J. D. Carey, *ACS Appl. Mater. Interfaces* **6**, 7743 (2014), http://pubs.acs.org/doi/pdf/10.1021/am501022x; http://pubs.acs.org/doi/abs/10.1021/am501022x.

302. T. Saari, C.-Y. Huang, J. Nieminen, W.-F. Tsai, H. Lin, and A. Bansil, *Appl. Phys. Lett.* **104**, 173104 (2014), http://scitation.aip.org/content/aip/journal/apl/104/17/10.1063/1.4873716.

303. H. Sadeghi, *J. Nanosci. Nanotechnol.* **14**, 4178 (2014).

304. H. Sadeghi, S. Bailey, and C. J. Lambert, *Appl. Phys. Lett.* **104**, 103104 (2014), http://scitation.aip.org/content/aip/journal/apl/104/10/10.1063/1.4868123.

Low-dimensional structures

305. M. A. Sadi, G. Gupta, and G. Liang, *J. Appl. Phys.* **116**, 153708 (2014), http://scitation.aip.org/content/aip/journal/jap/116/15/10.1063/1.4898357.

306. S. Sattar, R. Hoffmann, and U. Schwingenschlgl, *New J. Phys.* **16**, 065001 (2014), http://stacks.iop.org/1367-2630/16/i=6/a=065001.

307. E. Scalise, E. Cinquanta, M. Houssa, B. van den Broek, D. Chiappe, C. Grazianetti, G. Pourtois, et al., *Appl. Surf. Sci.* **291**, 113 (2013), http://www.sciencedirect.com/science/article/pii/S0169433213016048.

308. E. Scalise, M. Houssa, E. Cinquanta, C. Grazianetti, B. van den Broek, G. Pourtois, A. Stesmans, M. Fanciulli, and A. Molle, *2D Mater.* **1**, 011010 (2014), http://stacks.iop.org/2053-1583/1/i=1/a=011010.

309. K. Shakouri, P. Vasilopoulos, V. Vargiamidis, and F. M. Peeters, *Phys. Rev. B* **90**, 125444 (2014), http://link.aps.org/doi/10.1103/PhysRevB.90.125444.

310. K. Shakouri, P. Vasilopoulos, V. Vargiamidis, and F. M. Peeters, *Phys. Rev. B* **90**, 235423 (2014), http://link.aps.org/doi/10.1103/PhysRevB.90.235423.

311. M. Shen, Y.-Y. Zhang, X.-T. An, J.-J. Liu, and S.-S. Li, *J. Appl. Phys.* **115**, 233702 (2014), http://scitation.aip.org/content/aip/journal/jap/115/23/10.1063/1.4883193.

312. H. Shirkani and M. Golshan, *Physica E Low Dimens. Syst. Nanostruct.* **63**, 81 (2014), http://www.sciencedirect.com/science/article/pii/S1386947714001490.

313. N. Singh and U. Schwingenschlgl, *Phys. Status Solidi Rapid Res. Lett.* **8**, 353 (2014). http://dx.doi.org/10.1002/pssr.201409025.

314. E. H. Song, S. H. Yoo, J. J. Kim, S. W. Lai, Q. Jiang, and S. O. Cho, *Phys. Chem. Chem. Phys.* **16**, 23985 (2014). http://dx.doi.org/10.1039/C4CP02638A.

315. Y.-L. Song, J.-M. Zhang, D.-B. Lu, and K.-W. Xu, *Phys. E* **56**, 205 (2014).

316. H. Shu, D. Cao, P. Liang, X. Wang, X. Chen, and W. Lu, *Phys. Chem. Chem. Phys.* **16**, 304 (2014).

317. B. Soodchomshom, *J. Appl. Phys.* **115**, 023706 (2014), http://scitation.aip.org/content/aip/journal/jap/115/2/10.1063/1.4861644.

318. C. J. Tabert and E. J. Nicol, *Phys. Rev. B* **89**, 195410 (2014), http://link.aps.org/doi/10.1103/PhysRevB.89.195410.

319. X. Tan, F. Li, and Z. Chen, *J. Phys. Chem. C* **118**, 25825 (2014). http://dx.doi.org/10.1021/jp507011p.

320. X. Tan, C. R. Cabrera, and Z. Chen, *J. Phys. Chem. C* **118**, 25836 (2014). http://dx.doi.org/10.1021/jp503597n.

321. S. Thomas, A. C. Rajan, M. R. Rezapour, and K. S. Kim, *J. Phys. Chem. C* **118**, 10855 (2014). http://dx.doi.org/10.1021/jp501711d.

322. S. Trivedi, A. Srivastava, and R. Kurchania, *J. Comput. Theor. Nanosci.* **11**, 781 (2014).

323. S. Trivedi, A. Srivastava, and R. Kurchania, *J. Comput. Theor. Nanosci.* **11**, 789 (2014).

324. V. Y. Tsaran and S. G. Sharapov, *Phys. Rev. B* **90**, 205417 (2014), http://link.aps.org/doi/10.1103/PhysRevB.90.205417.

325. L. Tsetseris and D. Kaltsas, *Phys. Chem. Chem. Phys.* **16**, 5183 (2014). http://dx.doi.org/10.1039/C3CP55529A.

326. B. van den Broek, M. Houssa, E. Scalise, G. Pourtois, V. Afanas'ev, and A. Stesmans, *Appl. Surf. Sci.* **291**, 104 (2013).

327. B. Van Duppen, P. Vasilopoulos, and F. M. Peeters, *Phys. Rev. B* **90**, 035142 (2014), http://link.aps.org/doi/10.1103/PhysRevB.90.035142.

328. V. V. Hoang and H. T. C. Mi, *J. Phys. D Appl. Phys.* **47**, 495303 (2014), http://stacks.iop.org/0022-3727/47/i=49/a=495303.

329. V. Vargiamidis, P. Vasilopoulos, and G.-Q. Hai, *J. Phys. Condens. Matter* **26**, 345303 (2014), http://stacks.iop.org/0953-8984/26/i=34/a=345303.

330. V. Vargiamidis and P. Vasilopoulos, *Appl. Phys. Lett.* **105**, 223105 (2014), http://scitation.aip.org/content/aip/journal/apl/105/22/10.1063/1.4903248.

331. B. Wang, J. Wu, X. Gu, H. Yin, Y. Wei, R. Yang, and M. Dresselhaus, *Appl. Phys. Lett.* **104**, 081902 (2014), http://scitation.aip.org/content/aip/journal/apl/104/8/10.1063/1.4866415.

332. D. Wang and G. Jin, *Phys. Lett. A* **378**, 2557 (2014), http://www.sciencedirect.com/science/article/pii/S0375960114006410.

333. W. Rui, W. Shaofeng, and W. Xiaozhi, *J. Appl. Phys.* **116**, 024303 (2014), http://scitation.aip.org/content/aip/journal/jap/116/2/10.1063/1.4887353.

334. S. K. Wang, J. Wang, and K. S. Chan, *New J. Phys.* **16**, 045015 (2014), http://stacks.iop.org/1367-2630/16/i=4/a=045015.

335. Y. Wang, *Appl. Phys. Lett.* **104**, 032105 (2014), http://scitation.aip.org/content/aip/journal/apl/104/3/10.1063/1.4863091.

336. Y. Wang and Y. Lou, *Phys. Lett. A* **378**, 2627 (2014), http://www.sciencedirect.com/science/article/pii/S0375960114006367.

337. Y. Wang, R. Zheng, H. Gao, J. Zhang, B. Xu, Q. Sun, and Y. Jia, *Int. J. Hydrogen Energy* **39**, 14027 (2014), http://www.sciencedirect.com/science/article/pii/S0360319914019077.

338. J. Y. Wu, S. C. Chen, and M. F. Lin, *New J. Phys.* **16**, 125002 (2014), http://stacks.iop.org/1367-2630/16/i=12/a=125002.
339. Q. Wu, X.-H. Wang, T. Niehaus, and R.-Q. Zhang, *J. Phys. Chem. C* **0**, null (0), http://pubs.acs.org/doi/pdf/10.1021/jp501433t; http://pubs.acs.org/doi/abs/10.1021/jp501433t.
340. W. Wu, Z. Ao, T. Wang, C. Li, and S. Li, *Phys. Chem. Chem. Phys.* **16**, 16588 (2014), http://dx.doi.org/10.1039/C4CP01416B.
341. Y. Wu, K. Zhang, Y. Huang, S. Wu, H. Zhu, P. Cheng, and J. Ni, *Eur. Phys. J. B* **87**, 1 (2014), http://dx.doi.org/10.1140/epjb/e2014-41075-8.
342. P. Xiao, X.-L. Fan, and L.-M. Liu, *Comput. Mater. Sci.* **92**, 244 (2014), http://www.sciencedirect.com/science/article/pii/S0927025614003590.
343. X. Xiao, Y. Liu, and W. Wen, *J. Phys. Condens. Matter* **26**, 266001 (2014), http://stacks.iop.org/0953-8984/26/i=26/a=266001.
344. H. Xie, M. Hu, and H. Bao, *Appl. Phys. Lett.* **104**, 131906 (2014), http://scitation.aip.org/content/aip/journal/apl/104/13/10.1063/1.4870586.
345. P. Xu, Z. Yu, C. Yang, P. Lu, Y. Liu, H. Ye, and T. Gao, *Superlattices Microstruct.* **75**, 647 (2014), http://www.sciencedirect.com/science/article/pii/S0749603614003309.
346. S. Yamacli, *J. Nanopart. Res.* **16**, 2576 (2014), http://dx.doi.org/10.1007/s11051-014-2576-y.
347. C.-H. Yang, Z.-Y. Yu, P.-F. Lu, Y.-M. Liu, S. Manzoor, M. Li, and S. Zhou, *Proc. SPIE* **8975**, 89750K (2014), http://dx.doi.org/10.1117/12.2038401.
348. C. Yang, Z. Yu, P. Lu, Y. Liu, H. Ye, and T. Gao, *Comput. Mater. Sci.* **95**, 420 (2014), http://www.sciencedirect.com/science/article/pii/S0927025614005291.
349. K. Yang, S. Cahangirov, A. Cantarero, A. Rubio, and R. D'Agosta, *Phys. Rev. B* **89**, 125403 (2014), http://link.aps.org/doi/10.1103/PhysRevB.89.125403.
350. M. Yang, D.-H. Chen, R.-Q. Wang, and Y.-K. Bai, *Phys. Lett. A* **379**, 396 (2015), http://www.sciencedirect.com/science/article/pii/S0375960114012158.
351. X. F. Yang, Y. S. Liu, J. F. Feng, X. F. Wang, C. W. Zhang, and F. Chi, *J. Appl. Phys.* **116**, 124312 (2014), http://scitation.aip.org/content/aip/journal/jap/116/12/10.1063/1.4896630.
352. X.-S. Ye, Z.-G. Shao, H. Zhao, L. Yang, and C.-L. Wang, *RSC Adv.* **4**, 21216 (2014). http://dx.doi.org/10.1039/C4RA01802H.
353. X.-S. Ye, Z.-G. Shao, H. Zhao, L. Yang, and C.-L. Wang, *RSC Adv.* **4**, 37998 (2014), http://dx.doi.org/10.1039/C4RA03942D.
354. J. J. Yeo and Z. S. Liu, *J. Comput. Theor. Nanosci.* **11**, 1790 (2014), http://www.ingentaconnect.com/content/asp/jctn/2014/00000011/00000008/art00011.
355. S. Yu, X. Li, S. Wu, Y. Wen, S. Zhou, and Z. Zhu, *Mater. Res. Bull.* **50**, 268 (2014), http://www.sciencedirect.com/science/article/pii/S0025540813009239.
356. K. Zberecki, R. Swirkowicz, M. Wierzbicki, and J. Barnas, *Phys. Chem. Chem. Phys.* **16**, 12900 (2014), http://dx.doi.org/10.1039/C4CP01039F.
357. D. Zhang, M. Long, X. Zhang, C. Cao, H. Xu, M. Li, and K. Chan, *Chem. Phys. Lett.* **616–617**, 178 (2014), http://www.sciencedirect.com/science/article/pii/S000926141400894X.
358. J.-M. Zhang, W.-T. Song, K.-W. Xu, and V. Ji, *Comput. Mater. Sci.* **95**, 429 (2014), http://www.sciencedirect.com/science/article/pii/S0927025614005618.
359. R. Wu Zhang, C. Wen Zhang, S. Shi Li, W. Xiao Ji, P. Ji Wang, F. Li, P. Li, M. Juan Ren, and M. Yuan, *Solid State Commun.* **191**, 49 (2014), http://www.sciencedirect.com/science/article/pii/S0038109814001744.
360. R.-W. Zhang, C.-W. Zhang, W.-X. Ji, S.-J. Hu, S.-S. Yan, S.-S. Li, P. Li, P.-J. Wang, and Y.-S. Liu, *J. Phys. Chem. C* **118**, 25278 (2014), http://dx.doi.org/10.1021/jp508253x.
361. X.-L. Zhang, L.-F. Liu, and W.-M. Liu, *Sci. Rep.* **4**, 3801 (2014).
362. X. Zhang, H. Xie, M. Hu, H. Bao, S. Yue, G. Qin, and G. Su, *Phys. Rev. B* **89**, 054310 (2014), http://link.aps.org/doi/10.1103/PhysRevB.89.054310.
363. B. Zhao, J. Zhang, Y. Wang, and Z. Yang, *J. Chem. Phys.* **141**, 244701 (2014), http://scitation.aip.org/content/aip/journal/jcp/141/24/10.1063/1.4904285.
364. T. Zhao, S. Zhang, Q. Wang, Y. Kawazoe, and P. Jena, *Phys. Chem. Chem. Phys.* **16**, 22979 (2014), http://dx.doi.org/10.1039/C4CP02758B.
365. Y.-C. Zhao and J. Ni, *Phys. Chem. Chem. Phys.* **16**, 15477 (2014), http://dx.doi.org/10.1039/C4CP01549E.
366. R. Zheng, X. Lin, and J. Ni, *Appl. Phys. Lett.* **105**, 092410 (2014), http://scitation.aip.org/content/aip/journal/apl/105/9/10.1063/1.4895036.
367. J. Zhu and U. Schwingenschlögl, *ACS Appl. Mater. Interfaces* **6**, 11675 (2014), http://dx.doi.org/10.1021/am502469m.
368. J. Zhu and U. Schwingenschlögl, *ACS Appl. Mater. Interfaces* **6**, 19242 (2014), http://dx.doi.org/10.1021/am5052697.

Low-dimensional structures

369. V. Zólyomi, J. R. Wallbank, and V. I. Fal'ko, *2D Mater.* **1**, 011005 (2014), http://stacks.iop.org/2053-1583/1/i=1/a=011005.
370. H. Abdelsalam, T. Espinosa-Ortega, and I. Lukyanchuk, *Superlattices Microstruct.* **87**, 137 (2015), http://www.sciencedirect.com/science/article/pii/S0749603615002657.
371. S. Mehdi Aghaei and I. Calizo, *J. Appl. Phys.* **118**, 104304 (2015), http://scitation.aip.org/content/aip/journal/jap/118/10/10.1063/1.4930139.
372. X.-T. An, *Phys. Lett. A* **379**, 723 (2015), http://www.sciencedirect.com/science/article/pii/S0375960114012663.
373. H. Bao, J. Guo, W. Liao, and H. Zhao, *Appl. Phys. A* **118**, 431 (2015). http://dx.doi.org/10.1007/s00339-014-8837-x.
374. H. Bao, W. Liao, J. Guo, H. Zhao, and G. Zhou, *Laser Phys. Lett.* **12**, 095902 (2015), http://stacks.iop.org/1612-202X/12/i=9/a=095902.
375. H. Bao, W. Liao, J. Guo, X. Yang, H. Zhao, and G. Zhou, *J. Phys. D Appl. Phys.* **48**, 455306 (2015), http://stacks.iop.org/0022-3727/48/i=45/a=455306.
376. G. R. Berdiyorov, H. Bahlouli, and F. M. Peeters, *J. Appl. Phys.* **117**, 225101 (2015), http://scitation.aip.org/content/aip/journal/jap/117/22/10.1063/1.4921877.
377. Y. Borensztein, A. Curcella, S. Royer, and G. Prévot, *Phys. Rev. B* **92**, 155407 (2015), http://link.aps.org/doi/10.1103/PhysRevB.92.155407.
378. G. Cao, Y. Zhang, and J. Cao, *Phys. Lett. A* **379**, 1475 (2015), http://www.sciencedirect.com/science/article/pii/S0375960115002728.
379. R. Chandiramouli, A. Srivastava, and V. Nagarajan, *Appl. Surf. Sci.* **351**, 662 (2015), http://www.sciencedirect.com/science/article/pii/S0169433215013070.
380. S. Chowdhury, P. Nath, and D. Jana, *J. Phys. Chem. Solids* **83**, 32 (2015), http://www.sciencedirect.com/science/article/pii/S0022369715000724.
381. E. Cinquanta, G. Fratesi, S. dal Conte, C. Grazianetti, F. Scotognella, S. Stagira, C. Vozzi, G. Onida, and A. Molle, *Phys. Rev. B* **92**, 165427 (2015), http://link.aps.org/doi/10.1103/PhysRevB.92.165427.
382. C. Clendennen, N. Mori, and H. Tsuchiya, *J. Adv. Simulation Sci. Eng.* **2**, 171 (2015).
383. J. Dai and X. C. Zeng, *Phys. Chem. Chem. Phys.* **17**, 17957 (2015), http://dx.doi.org/10.1039/C4CP04953E.
384. D. Das and S. Sahoo, in *Intelligent Computing and Applications*, edited by D. Mandal, R. Kar, S. Das, and B. K. Panigrahi, vol. 343 *of Advances in Intelligent Systems and Computing*, Springer, New Delhi, India (2015), pp. 97–103.
385. Z. Deng, Z. Li, and W. Wang, in *28th International Vacuum Nanoelectronics Conference* (2015), pp. 50–51.
386. P. A. Denis, *Phys. Chem. Chem. Phys.* **17**, 5393 (2015). http://dx.doi.org/10.1039/C4CP05331A.
387. Y. Ding and Y. Wang, *Nanoscale Res. Lett.* **10**, 13 (2015).
388. H. Dong, D. Fang, B. Gong, Y. Zhang, E. Zhang, and S. Zhang, *J. Appl. Phys.* **117**, 064307 (2015), http://scitation.aip.org/content/aip/journal/jap/117/6/10.1063/1.4907582.
389. L. Drissi, K. Sadki, F. E. Yahyaoui, E. Saidi, M. Bousmina, and O. Fassi-Fehri, *Comput. Mater. Sci. A* **96**, 165 (2015), http://www.sciencedirect.com/science/article/pii/S0927025614006259.
390. L. Drissi and F. Ramadan, *Physica E Low Dimens. Syst. Nanostruct.* **68**, 38 (2015), http://www.sciencedirect.com/science/article/pii/S1386947714004378.
391. L. Drissi and K. Sadki, *Mech. Mater.* **89**, 151 (2015), http://www.sciencedirect.com/science/article/pii/S0167663615001416.
392. S. Dutta and K. Wakabayashi, *Phys. Rev. B* **91**, 201410 (2015), http://link.aps.org/doi/10.1103/PhysRevB.91.201410.
393. M. Ezawa, *JPS Conf. Proc.* **4**, 012001 (2015).
394. M. Ezawa, *J. Supercond. Nov. Magn.* **28**, 1249 (2015), http://dx.doi.org/10.1007/s10948-014-2900-x.
395. A. Esmailpour, M. Abdolmaleki, and M. Saadat, *Physica E Low Dimens. Syst. Nanostruct.* **77**, 144 (2015), http://www.sciencedirect.com/science/article/pii/S138694771530271X.
396. M. Farokhnezhad, M. Esmaeilzadeh, S. Ahmadi, and N. Pournaghavi, *J. Appl. Phys.* **117**, 173913 (2015), http://scitation.aip.org/content/aip/journal/jap/117/17/10.1063/1.4919659.
397. Y. Guo, S. Zhou, Y. Bai, and J. Zhao, *J. Supercond. Nov. Magn.* 29, 1–4 (2015). http://dx.doi.org/10.1007/s10948-015-3305-1.
398. G. Garcia, M. Atilhan, and S. Aparicio, *Phys.* 17, 16315–16326 *Chem. Chem. Phys.* (2015). http://dx.doi.org/10.1039/C5CP02432C.
399. A. Gert, M. Nestoklon, and I. Yassievich, *J. Exp. Theor. Phys.* **121**, 115 (2015). http://dx.doi.org/10.1134/S1063776115060072.
400. D. Ghosh, P. Parida, and S. K. Pati, *Phys. Rev. B* **92**, 195136 (2015), http://link.aps.org/doi/10.1103/PhysRevB.92.195136.
401. T. Gunst, M. Brandbyge, T. Markussen, and K. Stokbro, in *2015 International Conference on Simulation of Semiconductor Processes and Devices (SISPAD)*, (2015), pp. 32–35.

402. X. Guo, P. Guo, J. Zheng, L. Cao, and P. Zhao, *Appl. Surf. Sci.* **341**, 69 (2015), http://www.sciencedirect.com/science/article/pii/S0169433215005097.
403. Y.-H. Guo, J.-X. Cao, and B. Xu, *Chin. Phys. B* **25**, 017101 (2016), http://stacks.iop.org/1674-1056/25/i=1/a=017101.
404. T. Higuchi, C. Hu, and K. Watanabe, *e-J. Surf. Sci. Nanotechnol.* **13**, 115 (2015).
405. C. Hogan, S. Colonna, R. Flammini, A. Cricenti, and F. Ronci, *Phys. Rev. B* **92**, 115439 (2015), http://link.aps.org/doi/10.1103/PhysRevB.92.115439.
406. B. Hu, *J. Phys. Condens. Matter* **27**, 245301 (2015), http://stacks.iop.org/0953-8984/27/i=24/a=245301.
407. J. Hu, J. Zhang, S. Wu, and Z. Zhu, *Solid State Commun.* **209–210**, 59 (2015), http://www.sciencedirect.com/science/article/pii/S0038109815000733.
408. L.-F. Huang, P.-L. Gong, and Z. Zeng, *Phys. Rev. B* **91**, 205433 (2015), http://link.aps.org/doi/10.1103/PhysRevB.91.205433.
409. H. Ishida, Y. Hamamoto, Y. Morikawa, E. Minamitani, R. Arafune, and N. Takagi, *New J. Phys.* **17**, 015013 (2015), http://stacks.iop.org/1367-2630/17/i=1/a=015013.
410. P. Jamdagni, A. Kumar, M. Sharma, A. Thakur, and P. K. Ahluwalia, *AIP Conf. Proc.* **1661**, 080007 (2015), http://scitation.aip.org/content/aip/proceeding/aipcp/10.1063/1.4915398.
411. T.-T. Jia, M.-M. Zheng, X.-Y. Fan, Y. Su, S.-J. Li, H.-Y. Liu, G. Chen, and Y. Kawazoe, *J. Phys. Chem. C* **119**, 20747 (2015), http://dx.doi.org/10.1021/acs.jpcc.5b06626.
412. Q. G. Jiang, J. F. Zhang, Z. M. Ao, and Y. P. Wu, *J. Mater. Chem. C* **3**, 3954 (2015), http://dx.doi.org/10.1039/C4TC02829E.
413. Q. G. Jiang, J. F. Zhang, Z. M. Ao, and Y. P. Wu, *Sci. Rep.* **5**, 15734 (2015).
414. M. D. Kamatagi and N. S. Sankeshwar, *AIP Conf. Proc.* **1665**, 110036 (2015), http://scitation.aip.org/content/aip/proceeding/aipcp/10.1063/1.4918092.
415. L. Kou, Y. Ma, B. Yan, X. Tan, C. Chen, and S. C. Smith, *ACS Appl. Mater. Interfaces* **7**, 19226 (2015). http://dx.doi.org/10.1021/acsami.5b05063.
416. M. Krawiec and A. Podsiadly-Paszkowska, *Phys. Chem. Chem. Phys.* **17**, 2246 (2015). http://dx.doi.org/10.1039/C4CP05104A.
417. M.-Q. Le and D.-T. Nguyen, *Appl. Phys. A* **118**, 1437 (2015). http://dx.doi.org/10.1007/s00339-014-8904-3.
418. L. C. Lew Yan Voon, A. Lopez-Bezanilla, J. Wang, Y. Zhang, and M. Willatzen, *New J. Phys.* **17**, 025004 (2015), http://stacks.iop.org/1367-2630/17/i=2/a=025004.
419. F. Li, C. Zhang, W.-X. Ji, and M. Zhao, *Phys. Status Solidi B* **252**, 2072 (2015). http://dx.doi.org/10.1002/pssb.201552151.
420. R. Li, Y. Han, and J. Dong, *Phys. Chem. Chem. Phys.* **17**, 22969 (2015). http://dx.doi.org/10.1039/C5CP02538A.
421. S. Li, Y. Wu, Y. Tu, Y. Wang, T. Jiang, W. Liu, and Y. Zhao, *Sci. Rep.* **5**, 7881 (2015).
422. S. Shi Li, C. wen Zhang, and W. xiao Ji, *Mater. Chem. Phys.* **164**, 150 (2015), http://www.sciencedirect.com/science/article/pii/S0254058415302959.
423. C. Lian and J. Ni, *Phys. Chem. Chem. Phys.* **17**, 13366 (2015). http://dx.doi.org/10.1039/C5CP01557J.
424. X. Lin and J. Ni, *J. Appl. Phys.* **117**, 164305 (2015). http://scitation.aip.org/content/aip/journal/jap/117/16/10.1063/1.4919223.
425. S. Liu, H. Li, Y. He, X. Li, Y. Li, and X. Wang, *Mater. Des.* **85**, 60 (2015), http://www.sciencedirect.com/science/article/pii/S0264127515300745.
426. T. Ma, S. Wen, C.-X. Wu, L.-K. Yan, M. Zhang, Y. Kan, and Z.-M. Su, *J. Mater. Chem. C* **3**, 10085 (2015), http://dx.doi.org/10.1039/C5TC00792E.
427. F. Matusalem, M. Marques, L. K. Teles, and F. Bechstedt, *Phys. Rev. B* **92**, 045436 (2015), http://link.aps.org/doi/10.1103/PhysRevB.92.045436.
428. B. Mohan, Pooja, A. Kumar, and P. K. Ahluwalia, *AIP Conf. Proc.* **1665**, 140054 (2015), http://scitation.aip.org/content/aip/proceeding/aipcp/10.1063/1.4918263.
429. B. Molina, J. Soto, and J. Castro, *Chem. Phys.* **460**, 97 (2015), http://www.sciencedirect.com/science/article/pii/S0301010415001299.
430. S. Nigam, S. Gupta, D. Banyai, R. Pandey, and C. Majumder, *Phys. Chem. Chem. Phys.* **17**, 6705 (2015), http://dx.doi.org/10.1039/C4CP04861J.
431. S. Nigam, S. K. Gupta, C. Majumder, and R. Pandey, *Phys. Chem. Chem. Phys.* **17**, 1324 (2015), http://dx.doi.org/10.1039/C4CP05462H.
432. Z. P. Niu, Y. M. Zhang, and S. Dong, *New J. Phys.* **17**, 073026 (2015), http://stacks.iop.org/1367-2630/17/i=7/a=073026.
433. Z. P. Niu and S. Dong, *Europhys. Lett.* **111**, 37007 (2015), http://stacks.iop.org/0295-5075/111/i=3/a=37007.
434. M.-T. Nguyen, P. N. Phong, and N. D. Tuyen, *Chemphyschem* **14**, 1733 (2015). http://dx.doi.org/10.1002/cphc.201402902.

435. M. Noor-A-Alam, H. J. Kim, and Y.-H. Shin, *J. Appl. Phys.* **117**, 224304 (2015), http://scitation.aip.org/content/aip/journal/jap/117/22/10.1063/1.4922404.
436. V. O. Özçelik, D. Kecik, E. Durgun, and S. Ciraci, *J. Phys. Chem. C* **119**, 845 (2015). http://dx.doi.org/10.1021/jp5106554.
437. J. E. Padilha and R. B. Pontes, *J. Phys. Chem. C* **119**, 3818 (2015). http://dx.doi.org/10.1021/jp512489m.
438. A. Podsiady-Paszkowska and M. Krawiec, *Appl. Surf. Sci.* **373**, 45 (2015), http://www.sciencedirect.com/science/article/pii/S0169433215030196.
439. J. Prasongkit, R. G. Amorim, S. Chakraborty, R. Ahuja, R. H. Scheicher, and V. Amornkitbamrung, *J. Phys. Chem. C* **119**, 16934 (2015), http://dx.doi.org/10.1021/acs.jpcc.5b03635.
440. F. Peymanirad, M. Neek-Amal, J. Beheshtian, and F. M. Peeters, *Phys. Rev. B* **92**, 155113 (2015), http://link.aps.org/doi/10.1103/PhysRevB.92.155113.
441. N. Pournaghavi, M. Esmaeilzadeh, S. Ahmadi, and M. Farokhnezhad, *Solid State Commun.* **226**, 33 (2015), http://www.sciencedirect.com/science/article/pii/S0038109815003907.
442. O. Pulci, M. Marsili, V. Garbuio, P. Gori, I. Kupchak, and F. Bechstedt, *Phys. Status Solidi B* **252**, 72 (2015), http://dx.doi.org/10.1002/pssb.201350404.
443. Z. Qin, Z. Xu, and M. Buehler, *ASME. J. Appl. Mech.* **82**, 101003 (2015).
444. X. J. Qiu, Y. F. Cheng, Z. Z. Cao, and J. M. Lei, *J. Phys. D Appl. Phys.* **48**, 465105 (2015), http://stacks.iop.org/0022-3727/48/i=46/a=465105.
445. S. Rastgoo, H. Shirkani, and M. Golshan, *Phys. Lett. A* **379**, 1048 (2015), http://www.sciencedirect.com/science/article/pii/S0375960115001000.
446. J. Ribeiro-Soares, R. M. Almeida, L. G. Cançado, M. S. Dresselhaus, and A. Jorio, *Phys. Rev. B* **91**, 205421 (2015), http://link.aps.org/doi/10.1103/PhysRevB.91.205421.
447. E. Romera and M. Calixto, *Europhys. Lett.* **111**, 37006 (2015), http://stacks.iop.org/0295-5075/111/i=3/a=37006.
448. C. J. Rupp, S. Chakraborty, R. Ahuja, and R. J. Baierle, *Phys. Chem. Chem. Phys.* **17**, 22210 (2015). http://dx.doi.org/10.1039/C5CP03489B.
449. H. Sadeghi, S. Sangtarash, and C. Lambert, *Sci. Rep.* **5**, 9514 (2015).
450. R. Saxena, A. Saha, and S. Rao, *Phys. Rev. B* **92**, 245412 (2015), http://link.aps.org/doi/10.1103/PhysRevB.92.245412.
451. B. Sarebanha and S. Ahmadi, *Procedia Mater. Sci.* **11**, 259 (2015), http://www.sciencedirect.com/science/article/pii/S2211812815004770.
452. J. R. Soto, B. Molina, and J. J. Castro, *Phys. Chem. Chem. Phys.* **17**, 7624 (2015), http://dx.doi.org/10.1039/C4CP05912C.
453. R. Stephan, M.-C. Hanf, and P. Sonnet, *J. Phys. Condens. Matter* **27**, 015002 (2015), http://stacks.iop.org/0953-8984/27/i=1/a=015002.
454. R. Stephan, M.-C. Hanf, and P. Sonnet, *Phys. Chem. Chem. Phys.* **17**, 14495 (2015), http://dx.doi.org/10.1039/C5CP00613A.
455. R. Stephan, M.-C. Hanf, and P. Sonnet, *J. Chem. Phys.* **143**, 154706 (2015), http://scitation.aip.org/content/aip/journal/jcp/143/15/10.1063/1.4933369.
456. X. Sun, L. Wang, H. Lin, T. Hou, and Y. Li, *Appl. Phys. Lett.* **106**, 222401 (2015), http://scitation.aip.org/content/aip/journal/apl/106/22/10.1063/1.4921699.
457. A. Suwanvarangkoon and B. Soodchomshom, *J. Magn. Magn. Mater.* **374**, 479 (2015), http://www.sciencedirect.com/science/article/pii/S0304885314008154.
458. M. Syaputra, S. A. Wella, T. D. K. Wungu, A. Purqon, and Suprijadi, *AIP Conf. Proc.* **1677**, 080012 (2015), http://scitation.aip.org/content/aip/proceeding/aipcp/10.1063/1.4930743.
459. M. Syaputra, S. A. Wella, T. D. K. Wungu, A. Purqon, and Suprijadi, *AIP Conf. Proc.* **1677**, 080006 (2015), http://scitation.aip.org/content/aip/proceeding/aipcp/10.1063/1.4930737.
460. M. Tahir and U. Schwingenschlögl, *Eur. Phys. J. B* **88**, 285 (2015), http://dx.doi.org/10.1140/epjb/e2015-60719-7.
461. M. Vali, D. Dideban, and N. Moezi, *J. Comput. Electron.* **15**, 138 (2015), http://dx.doi.org/10.1007/s10825-015-0758-1.
462. N. Wang, H. Guo, Y.-J. Liu, J.-X. Zhao, Q.-H. Cai, and X.-Z. Wang, *Physica E Low Dimens. Syst. Nanostruct.* **73**, 21 (2015), http://www.sciencedirect.com/science/article/pii/S1386947715300503.
463. P. Wang, M. Zhou, G. Liu, Y. Liu, M.-Q. Long, and G. Zhou, *Eur. Phys. J. B* **88**, 243 (2015), http://dx.doi.org/10.1140/epjb/e2015-60316-x.
464. R. Wang, X. Pi, Z. Ni, Y. Liu, and D. Yang, *RSC Adv.* **5**, 33831 (2015), http://dx.doi.org/10.1039/C5RA05751E.
465. X. Wang, H. Liu, and S.-T. Tu, *RSC Adv.* **5**, 6238 (2015), http://dx.doi.org/10.1039/C4RA12257G.
466. X. Wang, H. Liu, and S.-T. Tu, *RSC Adv.* **5**, 65255 (2015), http://dx.doi.org/10.1039/C5RA12096A.

467. Z. Wang, T. Feng, and X. Ruan, *J. Appl. Phys.* **117**, 084317 (2015), http://scitation.aip.org/content/aip/journal/jap/117/8/10.1063/1.4913600.
468. W. Wei, Y. Dai, B. Huang, M.-H. Whangbo, and T. Jacob, *J. Phys. Chem. Lett.* **6**, 1065 (2015), http://dx.doi.org/10.1021/acs.jpclett.5b00106.
469. M. Wierzbicki, J. Barnaś, and R. Swirkowicz, *Phys. Rev. B* **91**, 165417 (2015), http://link.aps.org/doi/10.1103/PhysRevB.91.165417.
470. M. Wierzbicki, J. Barnaś, and R. Swirkowicz, *J. Phys. Condens. Matter* **27**, 485301 (2015), http://stacks.iop.org/0953-8984/27/i=48/a=485301.
471. H. Wu, Y. Qian, S. Lu, E. Kan, R. Lu, K. Deng, H. Wang, and Y. Ma, *Phys. Chem. Chem. Phys.* **17**, 15694 (2015), http://dx.doi.org/10.1039/C5CP01601K.
472. J.-Y. Wu, C.-Y. Lin, G. Gumbs, and M.-F. Lin, *RSC Adv.* **5**, 51912 (2015), http://dx.doi.org/10.1039/C5RA07721D.
473. W. C. Wu, Z. M. Ao, C. H. Yang, S. Li, G. X. Wang, C. M. Li, and S. Li, *J. Mater. Chem. C* **3**, 2593 (2015), http://dx.doi.org/10.1039/C4TC02095B.
474. X. Q. Wu and H. Meng, *J. Appl. Phys.* **117**, 203903 (2015), http://scitation.aip.org/content/aip/journal/jap/117/20/10.1063/1.4921799.
475. L. Xu, X.-F. Wang, L. Zhou, and Z.-Y. Yang, *J. Phys. D Appl. Phys.* **48**, 215306 (2015), http://stacks.iop.org/0022-3727/48/i=21/a=215306.
476. X. Xu, J. Li, X. Zhang, H. Xu, Z.-F. Ke, and C. Zhao, *RSC Adv.* **5**, 22135 (2015), http://dx.doi.org/10.1039/C4RA13754J.
477. J.-A. Yan, S.-P. Gao, R. Stein, and G. Coard, *Phys. Rev. B* **91**, 245403 (2015), http://link.aps.org/doi/10.1103/PhysRevB.91.245403.
478. J. Y. Yang and L. H. Liu, *Appl. Phys. Lett.* **107**, 091902 (2015), http://scitation.aip.org/content/aip/journal/apl/107/9/10.1063/1.4930025.
479. M. Yang, X.-L. Song, D.-H. Chen, and Y.-K. Bai, *Phys. Lett. A* **379**, 1149 (2015), http://www.sciencedirect.com/science/article/pii/S0375960115001784.
480. T. Yang, Q. Lin, and C. M. Wang, *Europhys. Lett.* **112**, 17009 (2015), http://stacks.iop.org/0295-5075/112/i=1/a=17009.
481. Y. Yao, S. Y. Liu, and X. L. Lei, *Phys. Rev. B* **91**, 115411 (2015), http://link.aps.org/doi/10.1103/PhysRevB.91.115411.
482. Z. Yu, H. Pan, and Y. Yao, *Phys. Rev. B* **92**, 155419 (2015), http://link.aps.org/doi/10.1103/PhysRevB.92.155419.
483. K. Zberecki, R. Swirkowicz, and J. Barna, *J. Magn. Magn. Mater.* **393**, 305 (2015), http://www.sciencedirect.com/science/article/pii/S030488531530192X.
484. D. Zha, C. Chen, and J. Wu, *Solid State Commun.* **219**, 21 (2015), http://www.sciencedirect.com/science/article/pii/S0038109815002264.
485. J. Zhang, Y. Hong, Z. Tong, Z. Xiao, H. Bao, and Y. Yue, *Phys. Chem. Chem. Phys.* **17**, 23704 (2015), http://dx.doi.org/10.1039/C5CP03323C.
486. R.-W. Zhang, C.-W. Zhang, W.-X. Ji, M.-J. Ren, F. Li, and M. Yuan, *Mater. Chem. Phys.* **156**, 89 (2015), http://www.sciencedirect.com/science/article/pii/S0254058415001248.
487. W. Zhang, Z. Song, and L. Dou, *J. Mater. Chem. C* **3**, 3087 (2015), http://dx.doi.org/10.1039/C4TC02758B.
488. X. Zhang, H. Bao, and M. Hu, *Nanoscale* **7**, 6014 (2015), http://dx.doi.org/10.1039/C4NR06523A.
489. H. Zhao, C. Zhang, S. Li, W. Ji, and P. Wang, *J. Appl. Phys.* **117**, 085306 (2015), http://scitation.aip.org/content/aip/journal/jap/117/8/10.1063/1.4913480.
490. R. Zheng, Y. Chen, and J. Ni, *Appl. Phys. Lett.* **107**, 263104 (2015), http://scitation.aip.org/content/aip/journal/apl/107/26/10.1063/1.4938755.
491. B. Zhou, B. Zhou, Y. Zeng, G. Zhou, and M. Duan, *Phys. Lett. A* **380**, 282 (2015), http://www.sciencedirect.com/science/article/pii/S0375960115008634.
492. B. Zhou, B. Zhou, X. Chen, W. Liao, and G. Zhou, *J. Phys. Condens. Matter* **27**, 465301 (2015), http://stacks.iop.org/0953-8984/27/i=46/a=465301.
493. B. Zhou, Y. Wang, and Y. Lou, *Phys. Lett. A* **380**, 502 (2015), http://www.sciencedirect.com/science/article/pii/S0375960115009548.
494. J. Zhu and U. Schwingenschlögl, *J. Mater. Chem. C* **3**, 3946 (2015), http://dx.doi.org/10.1039/C5TC00435G.
495. J. Zhu and U. Schwingenschlögl, *2D Mater.* **2**, 045004 (2015), http://stacks.iop.org/2053-1583/2/i=4/a=045004.
496. E. Akbari, Z. Buntat, A. Afroozeh, S. E. Pourmand, Y. Farhang, and P. Sanaati, *RSC Adv.* **6**, 81647–81653 (2016), http://dx.doi.org/10.1039/C6RA16736E.
497. M. Casuyac and R. Bantaculo, *MATEC Web Conf.* **40**, 02022 (2016), http://dx.doi.org/10.1051/matecconf/20164002022.
498. C. Chen, Z. Zhu, D. Zha, and J. Wu, *Chem. Phys. Lett.* **646**, 148 (2016), http://www.sciencedirect.com/science/article/pii/S0009261416000403.

499. M. X. Chen, Z. Zhong, and M. Weinert, *Phys. Rev. B* **94**, 075409 (2016), http://link.aps.org/doi/10.1103/PhysRevB.94.075409.

500. S.-H. Chen, *J. Magn. Magn. Mater.* **405**, 317 (2016), http://www.sciencedirect.com/science/article/pii/S0304885315309525.

501. X. Chen, J. Jiang, Q. Liang, R. Meng, C. Tan, Q. Yang, and X. Sun, *J. Mater. Chem. C* **4**, 7004 (2016), http://dx.doi.org/10.1039/C6TC01468B.

502. C. Cheng, H. Hu, Z. Zhaojin, and H. Zhang, *RSC Adv.* **6**, 7042 (2016), http://dx.doi.org/10.1039/C5RA18816D.

503. A. Debernardi and L. Marchetti, *Phys. Rev. B* **93**, 245426 (2016), http://link.aps.org/doi/10.1103/PhysRevB.93.245426.

504. A. Garcia-Fuente, L. J. Gallego, and A. Vega, *Phys. Chem. Chem. Phys.* **18**, 22606 (2016), http://dx.doi.org/10.1039/C6CP02961B.

505. E. Gkogkosi, A. Atsalakis, and L. Tsetseris, *J. Phys. Condens. Matter* **28**, 035304 (2016), http://stacks.iop.org/0953-8984/28/i=3/a=035304.

506. T. Gunst, T. Markussen, K. Stokbro, and M. Brandbyge, *Phys. Rev. B* **93**, 035414 (2016), http://link.aps.org/doi/10.1103/PhysRevB.93.035414.

507. S. Haldar, R. G. Amorim, B. Sanyal, R. Scheicher, and A. R. Rocha, *RSC Adv.* **6**, 6702 (2016), http://dx.doi.org/10.1039/C5RA23052G.

508. Y. Han, J. Dong, G. Qin, and M. Hu, *RSC Adv.* **6**, 69956 (2016), http://dx.doi.org/10.1039/C6RA14351B.

509. T. Hua, X. Zhai, Z. Yang, S. Wang, and B. Li, *Solid State Commun.* **244**, 43 (2016), http://www.sciencedirect.com/science/article/pii/S0038109816301454.

510. K. Iordanidou, M. Houssa, B. van den Broek, G. Pourtois, V. V. Afanasev, and A. Stesmans, *J. Phys. Condens. Matter* **28**, 035302 (2016), http://stacks.iop.org/0953-8984/28/i=3/a=035302.

511. Q. G. Jiang, W. C. Wu, J. F. Zhang, Z. M. Ao, Y. P. Wu, and H. J. Huang, *RSC Adv.* **6**, 69861 (2016), http://dx.doi.org/10.1039/C6RA11885B.

512. W. Ju, T. Li, Z. Hou, H. Wang, H. Cui, and X. Li, *Appl. Phys. Lett.* **108**, 212403 (2016), http://scitation.aip.org/content/aip/journal/apl/108/21/10.1063/1.4952770.

513. R. Li, Z.-L. Liu, Y. Gu, W. Zhang, and Y. Tan, *Phys. E Low Dimens. Syst. Nanostruct.* **79**, 152 (2016), http://www.sciencedirect.com/science/article/pii/S1386947715303404.

514. B. Mortazavi, A. Dianat, G. Cuniberti, and T. Rabczuk, *Electrochim. Acta* **213**, 865 (2016), http://www.sciencedirect.com/science/article/pii/S0013468616317224.

515. B. Peng, H. Zhang, H. Shao, Y. Xu, R. Zhang, H. Lu, D. W. Zhang, and H. Zhu, *ACS Appl. Mater. Interfaces* **8**, 20977 (2016), http://dx.doi.org/10.1021/acsami.6b04211.

516. A. P. Paszkowska and M. Krawiec, *J. Phys: Condens. Matter* **28**, 284004 (2016), http://stacks.iop.org/0953-8984/28/i=28/a=284004.

517. Y. G. Pogorelov and V. M. Loktev, *Phys. Rev. B* **93**, 045117 (2016), http://link.aps.org/doi/10.1103/PhysRevB.93.045117.

518. J. Qu, X. Peng, D. Xiao, and J. Zhong, *Phys. Rev. B* **94**, 075418 (2016), http://link.aps.org/doi/10.1103/PhysRevB.94.075418.

519. E. Romera, J. C. Bolívar, J. B. Roldán, and F. de los Santos, *EPL (Europhys. Lett.)* **115**, 20008 (2016), http://www.ingentaconnect.com/content/iop/epl/2016/00000115/00000002/art20008.

520. B. van den Broek, M. Houssa, K. Iordanidou, G. Pourtois, V. V. Afanasev, and A. Stesmans, *2D Mater.* **3**, 015001 (2016), http://stacks.iop.org/2053-1583/3/i=1/a=015001.

521. B. Xu, H.-S. Lu, B. Liu, G. Liu, M.-S. Wu, and C. Ouyang, *Chin. Phys. B* **25**, 067103 (2016), http://stacks.iop.org/1674-1056/25/i=6/a=067103.

522. X. Yuan, G. Lin, and Y. Wang, *Mol. Simul.* **42**, 1157 (2016), http://dx.doi.org/10.1080/08927022.2016.1148266.1148266.

523. M. Zare, F. Parhizgar, and R. Asgari, *Phys. Rev. B* **94**, 045443 (2016), http://link.aps.org/doi/10.1103/PhysRevB.94.045443.

524. Q. Zhang, K. S. Chan, and M. Long, *J. Phys. Condens. Matter* **28**, 055301 (2016), http://stacks.iop.org/0953-8984/28/i=5/a=055301.

525. W. Zhao, Z. Guo, Y. Zhang, J. Ding, and X. Zheng, *Sol. Stat. Commun.* **227**, 1 (2016), http://www.sciencedirect.com/science/article/pii/S0038109815004056.

526. B. Zhou, B. Zhou, G. Liu, D. Guo, and G. Zhou, *Physica B: Condens. Matter* **500**, 106 (2016), http://www.sciencedirect.com/science/article/pii/S0921452616303209.

527. P. Jamdagni, A. Kumar, M. Sharma, A. Thakur, and P. Ahluwalia, *Phys. E Low-Dimens. Syst. Nanostruct.* **85**, 65 (2017), http://www.sciencedirect.com/science/article/pii/S138694771630234X.

528. G. L. Lay, B. Aufray, C. Léandri, H. Oughaddou, J.-P. Biberian, P. D. Padova, M. Dávila, B. Ealet, and A. Kara, *Appl. Surf. Sci.* **256**, 524 (2009).

529. B. Aufray, A. Kara, S. Vizzini, H. Oughaddou, C. Léandri, B. Ealet, and G. L. Lay, *Appl. Phys. Lett.* **96**, 183102 (2010).
530. P. D. Padova, C. Quaresima, C. Ottaviani, P. M. Sheverdyaeva, P. Moras, C. Carbone, D. Topwal, et al., *Appl. Phys. Lett.* **96**, 261905 (2010).
531. B. Lalmi, H. Oughaddou, H. Enriquez, A. Kara, S. Vizzini, B. Ealet, and B. Aufray, *Appl. Phys. Lett.* **97**, 223109 (2010).
532. H. Okamoto, Y. Kumai, Y. Sugiyama, T. Mitsuoka, K. Nakanishi, T. Ohta, H. Nozaki, S. Yamaguchi, S. Shirai, and H. Nakano, *J. Am. Chem. Soc.* **132**, 2710 (2010).
533. Y. Sugiyama, H. Okamoto, T. Mitsuoka, T. Morikawa, K. Nakanishi, T. Ohta, and H. Nakano, *J. Am. Chem. Soc.* **132**, 5946 (2010).
534. P. D. Padova, C. Quaresima, B. Olivieri, P. Perfetti, and G. L. Lay, *Appl. Phys. Lett.* **98**, 081909 (2011).
535. P. D. Padova, C. Quaresima, B. Olivieri, P. Perfetti, and G. L. Lay, *J. Phys. D* **44**, 312001 (2011).
536. T. Ikuno, H. Okamoto, Y. Sugiyama, H. Nakano, F. Yamada, and I. Kamiya, *Appl. Phys. Lett.* **99**, 023107 (2011).
537. U. Kim, I. Kim, Y. Park, K.-Y. Lee, S.-Y. Yim, J.-G. Park, H.-G. Ahn, S.-H. Park, and H.-J. Choi, *ACS Nano* **5**, 2176 (2011).
538. B. Lalmi, J. P. Biberian, and B. Aufray, in *Physics, Chemistry and Applications of Nanostructures* (2011), pp. 430–432.
539. L. Chen, C.-C. Liu, B. Feng, X. He, P. Cheng, Z. Ding, S. Meng, Y. Yao, and K. Wu, *Phys. Rev. Lett.* **109**, 056804 (2012). http://link.aps.org/doi/10.1103/PhysRevLett.109.056804
540. D. Chiappe, C. Grazianetti, G. Tallarida, M. Fanciulli, and A. Molle, *Adv. Mat.* **24**, 5088 (2012).
541. M. E. Davila, A. Marele, P. De Padova, I. Montero, F. Hennies, A. Pietzsch, M. N. Shariati, J. M. Gómez-Rodriguez, and G. Le Lay, *Nanotechnology* **23**, 385703 (2012).
542. P. De Padova, O. Kubo, B. Olivieri, C. Quaresima, T. Nakayama, M. Aono, and G. Le Lay, *Nano Lett.* **12**, 5500 (2012), http://pubs.acs.org/doi/abs/10.1021/nl302598x.
543. B. Feng, Z. Ding, S. Meng, Y. Yao, X. He, P. Cheng, L. Chen, and K. Wu, *Nano Lett.* **12**, 3507 (2012), http://pubs.acs.org/doi/abs/10.1021/nl301047g.
544. A. Fleurence, R. Friedlein, T. Ozaki, H. Kawai, Y. Wang, and Y. Yamada-Takamura, *Phys. Rev. Lett.* **108**, 245501 (2012), http://link.aps.org/doi/10.1103/PhysRevLett.108.245501.
545. H. Jamgotchian, Y. Colignon, N. Hamzaoui, B. Ealet, J. Y. Hoarau, B. Aufray, and J. P. Bibérian, *J. Phys. Condens. Mat.* **24**, 172001 (2012).
546. G. L. Lay, P. D. Padova, A. Resta, T. Bruhn, and P. Vogt, *J. Phys. D: Appl. Phys.* **45**, 392001 (2012).
547. C.-L. Lin, R. Arafune, K. Kawahara, N. Tsukahara, E. Minamitani, Y. Kim, N. Takagi, and M. Kawai, *Appl. Phys. Express* **5**, 045802 (2012).
548. H. Nakano, M. Nakano, K. Nakanishi, D. Tanaka, Y. Sugiyama, T. Ikuno, H. Okamoto, and T. Ohta, *J Am. Chem. Soc.* **134**, 5452 (2012). http://pubs.acs.org/doi/pdf/10.1021/ja212086n.
549. P. Vogt, P. De Padova, C. Quaresima, J. Avila, E. Frantzeskakis, M. C. Asensio, A. Resta, B. Ealet, and G. Le Lay, *Phys. Rev. Lett.* **108**, 155501 (2012), http://link.aps.org/doi/10.1103/PhysRevLett.108.155501.
550. A. Acun, B. Poelsema, H. J. W. Zandvliet, and R. van Gastel, *Appl. Phys. Lett.* **103**, 263119 (2013). http://scitation.aip.org/content/aip/journal/apl/103/26/10.1063/1.4860964.
551. R. Arafune, C.-L. Lin, K. Kawahara, N. Tsukahara, E. Minamitani, Y. Kim, N. Takagi, and M. Kawai, *Surf. Sci.* **608**, 297 (2013), http://www.sciencedirect.com/science/article/pii/S0039602812003925.
552. J. Avila, P. D. Padova, S. Cho, I. Colambo, S. Lorcy, C. Quaresima, P. Vogt, A. Resta, G. L. Lay, and M. C. Asensio, *J. Phys. Condens. Matter* **25**, 262001 (2013).
553. E. Bianco, S. Butler, S. Jiang, O. D. Restrepo, W. Windl, and J. E. Goldberger, *ACS Nano* **7**, 4414 (2013). http://pubs.acs.org/doi/pdf/10.1021/nn4009406.
554. L. Chen, H. Li, B. Feng, Z. Ding, J. Qiu, P. Cheng, K. Wu, and S. Meng, *Phys. Rev. Lett.* **110**, 085504 (2013), http://link.aps.org/doi/10.1103/PhysRevLett.110.085504.
555. L. Chen, B. Feng, and K. Wu, *Appl. Phys. Lett.* **102**, 081602 (2013).
556. E. Cinquanta, E. Scalise, D. Chiappe, C. Grazianetti, B. van den Broek, M. Houssa, M. Fanciulli, and A. Molle, *J. Phys. Chem. C* **117**, 16719 (2013), http://pubs.acs.org/doi/pdf/10.1021/jp405642g.
557. E. Cinquanta, S. D. Conte, D. Chiappe, C. Grazianetti, M. Fanciulli, A. Molle, G. Cerullo, S. Stagira, F. Scotognella, and C. Vozzi, in *19th International Conference on Ultrafast Phenomena* (Optical Society of America, 2014), p. 09.Wed.P3.**42**, http://www.opticsinfobase.org/abstract.cfm?URI=UP-2014-09.Wed.P3.42.
558. S. Colonna, G. Serrano, P. Gori, A. Cricenti, and F. Ronci, *J. Phys.: Condens. Matter* **25**, 315301 (2013), http://stacks.iop.org/0953-8984/25/i=31/a=315301.
559. P. De Padova, C. Ottaviani, F. Ronci, S. Colonna, B. Olivieri, C. Quaresima, A. Cricenti, et al., *J. Phys. Condens. Matter* **25**, 014009 (2013).

560. P. D. Padova, P. Vogt, A. Resta, J. Avila, I. Razado-Colombo, C. Quaresima, C. Ottaviani, et al., *Appl. Phys. Lett.* **102**, 163106 (2013).

561. P. De Padova, J. Avila, A. Resta, I. Razado-Colombo, C. Quaresima, C. Ottaviani, B. Olivieri, et al., *J. Phys. Condens. Matter* **25**, 382202 (2013).

562. B. Feng, H. Li, C.-C. Liu, T.-N. Shao, P. Cheng, Y. Yao, S. Meng, L. Chen, and K. Wu, *ACS Nano* **7**, 9049 (2013), http://pubs.acs.org/doi/pdf/10.1021/nn403661h.

563. R. Friedlein, A. Fleurence, J. T. Sadowski, and Y. Yamada-Takamura, *Appl. Phys. Lett.* **102**, 221603 (2013).

564. C.-L. Lin, R. Arafune, K. Kawahara, M. Kanno, N. Tsukahara, E. Minamitani, Y. Kim, M. Kawai, and N. Takagi, *Phys. Rev. Lett.* **110**, 076801 (2013), http://link.aps.org/doi/10.1103/PhysRevLett.110.076801.

565. Z. Majzik, M. R. Tchalala, M. Svec, P. Hapala, H. Enriquez, A. Kara, A. J. Mayne, G. Dujardin, P. Jelnek, and H. Oughaddou, *J. Phys. Condens. Matter* **25**, 225301 (2013), http://stacks.iop.org/0953-8984/25/i=22/a=225301.

566. L. Meng, Y. Wang, L. Zhang, S. Du, R. Wu, L. Li, Y. Zhang, et al., *Nano Lett.* **13**, 685 (2013), http://pubs.acs.org/doi/pdf/10.1021/nl304347w.

567. A. Resta, T. Leoni, C. Barth, A. Ranguis, C. Becker, T. Bruhn, P. Vogt, and G. L. Lay, *Sci. Rep.* **3**, 2399 (2013).

568. M. Rachid Tchalala, H. Enriquez, A. J. Mayne, A. Kara, S. Roth, M. G. Silly, A. Bendounan, et al., *Appl. Phys. Lett.* **102**, 083107 (2013).

569. D. Tsoutsou, E. Xenogiannopoulou, E. Golias, P. Tsipas, and A. Dimoulas, *Appl. Phys. Lett.* **103**, 231604 (2013), http://scitation.aip.org/content/aip/journal/apl/103/23/10.1063/1.4841335.

570. T. Aizawa, S. Suehara, and S. Otani, *J. Phys. Chem. C* **118**, 23049 (2015), http://dx.doi.org/10.1021/jp505602c.

571. C. B. Azzouz, A. Akremi, M. Derivaz, J. L. Bischoff, M. Zanouni, and D. Dentel, *J. Phys. Conf. Ser.* **491**, 012003 (2014), http://stacks.iop.org/1742-6596/491/i=1/a=012003.

572. Y. Borensztein, G. Prévot, and L. Masson, *Phys. Rev. B* **89**, 245410 (2014), http://link.aps.org/doi/10.1103/PhysRevB.89.245410.

573. D. Chiappe, E. Scalise, E. Cinquanta, C. Grazianetti, B. van den Broek, M. Fanciulli, M. Houssa, and A. Molle, *Adv. Mat.* **26**, 2096 (2014), http://dx.doi.org/10.1002/adma.201304783.

574. Y. Du, J. Zhuang, H. Liu, X. Xu, S. Eilers, K. Wu, P. Cheng, et al., *ACS Nano* **8**, 10019 (2014). http://dx.doi.org/10.1021/nn504451t.

575. H. Enriquez, A. Kara, A. J. Mayne, G. Dujardin, H. Jamgotchian, B. Aufray, and H. Oughaddou, *J. Phys. Conf. Ser.* **491**, 012004 (2014), http://stacks.iop.org/1742-6596/491/i=1/a=012004.

576. A. Fleurence, Y. Yoshida, C.-C. Lee, T. Ozaki, Y. Yamada-Takamura, and Y. Hasegawa, *Appl. Phys. Lett.* **104**, 021605 (2014), http://scitation.aip.org/content/aip/journal/apl/104/2/10.1063/1.4862261.

577. R. Friedlein, A. Fleurence, K. Aoyagi, M. P. de Jong, H. Van Bui, F. B. Wiggers, S. Yoshimoto, et al., *J. Chem. Phys.* **140**, 184704 (2014), http://scitation.aip.org/content/aip/journal/jcp/140/18/10.1063/1.4875075.

578. R. Friedlein, H. Van Bui, F. B. Wiggers, Y. Yamada-Takamura, A. Y. Kovalgin, and M. P. de Jong, *J. Chem. Phys.* **140**, 204705 (2014), http://scitation.aip.org/content/aip/journal/jcp/140/20/10.1063/1.4878375.

579. C. Grazianetti, D. Chiappe, E. Cinquanta, M. Fanciulli, and A. Molle, *J. Phys. Condens. Matter* **27**, 255005 (2015), http://stacks.iop.org/0953-8984/27/i=25/a=255005.

580. H. Jamgotchian, Y. Colignon, B. Ealet, B. Parditka, J.-Y. Hoarau, C. Girardeaux, B. Aufray, and J.-P. Bibrian, *J. Phys. Conf. Ser.* **491**, 012001 (2014), http://stacks.iop.org/1742-6596/491/i=1/a=012001.

581. N. W. Johnson, P. Vogt, A. Resta, P. De Padova, I. Perez, D. Muir, E. Z. Kurmaev, G. Le Lay, and A. Moewes, *Adv. Funct. Mater.* **24**, 5253 (2014), http://dx.doi.org/10.1002/adfm.201400769.

582. K. Kawahara, T. Shirasawa, R. Arafune, C. Lin, N. Takahashi, M. Kawai, and N. Takagi, *Surf. Sci.* **623**, 25 (2014), http://www.sciencedirect.com/science/article/pii/S0039602814000053.

583. Z.-L. Liu, M.-X. Wang, J.-P. Xu, J.-F. Ge, G. L. Lay, P. Vogt, D. Qian, C.-L. Gao, C. Liu, and J.-F. Jia, *New J. Phys.* **16**, 075006 (2014), http://stacks.iop.org/1367-2630/16/i=7/a=075006.

584. S. K. Mahatha, P. Moras, V. Bellini, P. M. Sheverdyaeva, C. Struzzi, L. Petaccia, and C. Carbone, *Phys. Rev. B* **89**, 201416 (2014), http://link.aps.org/doi/10.1103/PhysRevB.89.201416.

585. A. J. Mannix, B. Kiraly, B. L. Fisher, M. C. Hersam, and N. P. Guisinger, *ACS Nano* **8**, 7538 (2014). http://dx.doi.org/10.1021/nn503000w.

586. P. Moras, T. O. Mentes, P. M. Sheverdyaeva, A. Locatelli, and C. Carbone, *J. Phys. Condens. Matter* **26**, 185001 (2014), http://stacks.iop.org/0953-8984/26/i=18/a=185001.

587. A. Orekhov, S. Savilov, V. Zakharov, A. Yatsenko, and L. Aslanov, *J. Nanopart. Res.* **16**, 1 (2014), http://dx.doi.org/10.1007/s11051-013-2190-4.

588. M. Rahman, T. Nakagawa, and S. Mizuno, in *Informatics, Electronics Vision (ICIEV), 2014 International Conference on* (2014), pp. 1–4.

589. M. S. Rahman, T. Nakagawa, and S. Mizuno, *Jpn. J. Appl. Phys.* **54**, 015502 (2015), http://stacks.iop.org/1347-4065/54/i=1/a=015502.

590. F. Ronci, G. Serrano, P. Gori, A. Cricenti, and S. Colonna, *Phys. Rev. B* **89**, 115437 (2014), http://link.aps.org/doi/10.1103/PhysRevB.89.115437.

591. E. Salomon, R. E. Ajjouri, G. L. Lay, and T. Angot, *J. Phys. Condens. Matter* **26**, 185003 (2014), http://stacks.iop.org/0953-8984/26/i=18/a=185003.

592. T. Shirai, T. Shirasawa, T. Hirahara, N. Fukui, T. Takahashi, and S. Hasegawa, *Phys. Rev. B* **89**, 241403 (2014), http://link.aps.org/doi/10.1103/PhysRevB.89.241403.

593. J. Sone, T. Yamagami, Y. Aoki, K. Nakatsuji, and H. Hirayama, *N. J. Phys.* **16**, 095004 (2014), http://stacks.iop.org/1367-2630/16/i=9/a=095004.

594. E. Speiser, B. Buick, N. Esser, W. Richter, S. Colonna, A. Cricenti, and F. Ronci, *Appl. Phys. Lett.* **104**, 161612 (2014), http://scitation.aip.org/content/aip/journal/apl/104/16/10.1063/1.4872460.

595. M. R. Tchalala, H. Enriquez, A. J. Mayne, A. Kara, G. Dujardin, M. A. Ali, and H. Oughaddou, *J Phys Conf. Ser.* **491**, 012002 (2014), http://stacks.iop.org/1742-6596/491/i=1/a=012002.

596. P. Vogt, P. Capiod, M. Berthe, A. Resta, P. De Padova, T. Bruhn, G. Le Lay, and B. Grandidier, *Appl. Phys. Lett.* **104**, 021602 (2014), http://scitation.aip.org/content/aip/journal/apl/104/2/10.1063/1.4861857.

597. X. Xu, J. Zhuang, Y. Du, H. Feng, N. Zhang, C. Liu, T. Lei, et al., *Sci. Rep.* **4**, 7543 (2014).

598. T. Aizawa, S. Suehara, and S. Otani, *J. Phys. Condens. Matter* **27**, 305002 (2015), http://stacks.iop.org/0953-8984/27/i=30/a=305002.

599. J. Chen, Y. Du, Z. Li, W. Li, B. Feng, J. Qiu, P. Cheng, S. X. Dou, L. Chen, and K. Wu, *Sci. Rep.* **5**, 13590 (2015).

600. N. W. Johnson, D. Muir, E. Z. Kurmaev, and A. Moewes, *Adv. Funct. Mater.* **25**, 4083 (2015), http://dx.doi.org/10.1002/adfm.201501029.

601. G.-W. Lee, H.-D. Chen, and D.-S. Lin, *Appl. Surf. Sci.* **354**, 187 (2015), http://www.sciencedirect.com/science/article/pii/S0169433215001920.

602. G.-W. Lee, H.-D. Chen, and D.-S. Lin, *Appl. Surf. Sci.* 354, 212–215 (2015), http://www.sciencedirect.com/science/article/pii/S0169433215000963.

603. L. Hui, H.-X. Fu, and M. Sheng, *Chin. Phys. B* **24**, 086102 (2015), http://stacks.iop.org/1674-1056/24/i=8/a=086102.

604. S. K. Mahatha, P. Moras, P. M. Sheverdyaeva, R. Flammini, K. Horn, and C. Carbone, *Phys. Rev. B* **92**, 245127 (2015), http://link.aps.org/doi/10.1103/PhysRevB.92.245127.

605. E. Noguchi, K. Sugawara, R. Yaokawa, T. Hitosugi, H. Nakano, and T. Takahashi, *Adv. Mater.* **27**, 856860 (2014), http://dx.doi.org/10.1002/adma.201403077.

606. M. Satta, S. Colonna, R. Flammini, A. Cricenti, and F. Ronci, *Phys. Rev. Lett.* **115**, 026102 (2015), http://link.aps.org/doi/10.1103/PhysRevLett.115.026102.

607. L. Tao, E. Cinquanta, D. Chiappe, C. Grazianetti, M. Fanciulli, M. Dubey, A. Molle, and D. Akinwande, *Nat. Nano.* **10**, 227 (2015).

608. H. Van Bui, F. B. Wiggers, R. Friedlein, Y. Yamada-Takamura, A. Y. Kovalgin, and M. P. de Jong, *J. Chem. Phys.* **142**, 064702 (2015), http://scitation.aip.org/content/aip/journal/jcp/142/6/10.1063/1.4907375.

609. W. Wang, W. Olovsson, and R. I. G. Uhrberg, *Phys. Rev. B* **92**, 205427 (2015), http://link.aps.org/doi/10.1103/PhysRevB.92.205427.

610. A. D. Alvarez, T. Zhu, J. Nys, M. Berthe, M. Empis, J. Schreiber, B. Grandidier, and T. Xu, *Surf. Sci.* **653**, 92 (2016), http://www.sciencedirect.com/science/article/pii/S0039602816302369.

611. P. Lagarde, M. Chorro, D. Roy, and N. Trcera, *J. Phys. Condens. Matter* **28**, 075002 (2016), http://stacks.iop.org/0953-8984/28/i=7/a=075002.

612. G. Prévot, R. Bernard, H. Cruguel, A. Curcella, M. Lazzeri, T. Leoni, L. Masson, A. Ranguis, and Y. Borensztein, *Phys Status Solidi B* **253**, 206 (2016), http://dx.doi.org/10.1002/pssb.201552524.

613. A. Kara, H. Enriquez, A. Seitsonen, L. C. Lew Yan Voon, S. Vizzini, and H. Oughaddou, *Surf. Sci. Rep.* **67**, 1 (2012).

614. H. Okamoto, Y. Sugiyama, and H. Nakano, *Chem. Eur. J.* **17**, 9864 (2011).

615. P. D. Padova, P. Perfetti, B. Olivieri, C. Quaresima, C. Ottaviani, and G. L. Lay, *J. Phys. Cond. Matter* **24**, 223001 (2012).

616. A. L. Ivanovskii, *Russ. Chem. Rev.* **81**, 571 (2012).

617. S. Z. Butler, S. M. Hollen, L. Cao, Y. Cui, J. A. Gupta, H. R. Gutirrez, T. F. Heinz, et al., *ACS Nano* 7, 2898 (2013), http://pubs.acs.org/doi/pdf/10.1021/nn400280c.

618. M. Xu, T. Liang, M. Shi, and H. Chen, *Chem. Rev.* **113**, 3766 (2013), http://pubs.acs.org/doi/pdf/10.1021/cr300263a.

619. Q. Tang and Z. Zhou, *Prog. Mat. Sci.* **58**, 1244 (2013), http://www.sciencedirect.com/science/article/pii/S0079642513000376.

620. M. A. Ali and M. R. Tchalala, *J. Phys. Conf. Ser.* **491**, 012009 (2014), http://stacks.iop.org/1742-6596/491/i=1/a=012009.

621. S. Das, M. Kim, J.-W. Lee, and W. Choi, *Crit. Rev. Solid State Mater. Sci.* **39**, 231 (2014).
622. D. Jose and A. Datta, *Acc. Chem. Res.* **47**, 593 (2014), http://pubs.acs.org/doi/pdf/10.1021/ar400180e.
623. M. Liu and F. Liu, *Nanotechnology* **25**, 135706 (2014), http://stacks.iop.org/0957-4484/25/i=13/a=135706.
624. L. Lew Yan Voon and G. Guzmán-Verri, *MRS Bull.* **39**, 366 (2014).
625. Y. Pan, L. Zhang, L. Huang, L. Li, L. Meng, M. Gao, Q. Huan, X. Lin, Y. Wang, S. Du, et al., *Small* **10**, 22152225 (2014), http://dx.doi.org/10.1002/smll.201303698.
626. Y. Yamada-Takamura and R. Friedlein, *Sci. Technol. Adv. Mat.* **15**, 064404 (2014), http://stacks.iop.org/1468-6996/15/i=6/a=064404.
627. H. J. Zandvliet, *Nano Today* **9**, 691 (2014), http://www.sciencedirect.com/science/article/pii/S1748013214001327.
628. A. Dimoulas, *Microelectron. Eng.* **131**, 68 (2015), http://www.sciencedirect.com/science/article/pii/S0167931714003542.
629. R. Friedlein and Y. Yamada-Takamura, *J. Phys. Condens. Matter* **27**, 203201 (2015), http://stacks.iop.org/0953-8984/27/i=20/a=203201.
630. M. Houssa, A. Dimoulas, and A. Molle, *J. Phys. Condens. Matter* **27**, 253002 (2015), http://stacks.iop.org/0953-8984/27/i=25/a=253002.
631. M. Inagaki, *Carbon* **87**, 462 (2015), http://www.sciencedirect.com/science/article/pii/S0008622315001189.
632. L. Geng, Z. Yin-Chang, Z. Rui, N. Jun, and W. Yan-Ning, *Chin. Phys. B* **24**, 087302 (2015), http://stacks.iop.org/1674-1056/24/i=8/a=087302.
633. L. Hong-Sheng, H. Nan-Nan, and Z. Ji-Jun, *Chin. Phys. B* **24**, 087303 (2015), http://stacks.iop.org/1674-1056/24/i=8/a=087303.
634. L. C. Lew Yan Voon, *Chin. Phys. B* **24**, 087309 (2015).
635. H. Oughaddou, H. Enriquez, M. R. Tchalala, H. Yildirim, A. J. Mayne, A. Bendounan, G. Dujardin, M. A. Ali, and A. Kara, *Prog. Surf. Sci.* **90**, 46 (2015), http://www.sciencedirect.com/science/article/pii/S0079681614000331.
636. N. Takagi, C.-L. Lin, K. Kawahara, E. Minamitani, N. Tsukahara, M. Kawai, and R. Arafune, *Prog. Surf. Sci.* **90**, 1 (2015), http://www.sciencedirect.com/science/article/pii/S0079681614000215.
637. W. Rong, X. Ming-Sheng, and P. Xiao-Dong, *Chin. Phys. B* **24**, 086807 (2015), http://stacks.iop.org/1674-1056/24/i=8/a=086807.
638. W. Yang-Yang, Q. Ru-Ge, Y. Da-Peng, and L. Jing, *Chin. Phys. B* **24**, 087201 (2015), http://stacks.iop.org/1674-1056/24/i=8/a=087201.
639. K.-H. Wu, *Chin. Phys. B* **24**, 086802 (2015), http://stacks.iop.org/1674-1056/24/i=8/a=086802.
640. J. Zhuang, X. Xu, H. Feng, Z. Li, X. Wang, and Y. Du, *Sci. Bull.* **69**, 1551–1562 (2015), http://dx.doi.org/10.1007/s11434-015-0880-2.
641. C. Grazianetti, E. Cinquanta, and A. Molle, *2D Mat.* **3**, 012001 (2016), http://stacks.iop.org/2053-1583/3/i=1/a=012001.
642. T. P. Kaloni, G. Schreckenbach, M. S. Freund, and U. Schwingenschlögl, *Phys. Status Solid. Rapid Res. Lett.* **10**, 133 (2016), http://dx.doi.org/10.1002/pssr.201510338.
643. L. C. Lew Yan Voon, J. Zhu, and U. Schwingenschlögl, *Appl. Phys. Rev.* **3**, 040802 (2016), http://scitation.aip.org/content/aip/journal/apr2/3/4/10.1063/1.4944631.
644. M. Dvila, L. L. Y. Voon, J. Zhao, and G. L. Lay, in *2D Materials*, edited by J. J. B. Francesca Iacopi and C. Jagadish, vol. 95 *of Semiconductors and Semimetals*, Elsevier, (2016), pp. 149–**188**, http://www.sciencedirect.com/science/article/pii/S0080878416300059.
645. L. C. Lew Yan Voon, in *Silicene*, edited by M. J. Spencer and T. Morishita, vol. 235 *of Springer Series in Materials Science*, Springer, (2016), pp. 3–33.
646. J. Zhao, H. Liu, Z. Yu, R. Quhe, S. Zhou, Y. Wang, C. C. Liu, et al., *Prog. Mater. Sci.* **83**, 24 (2016), http://www.sciencedirect.com/science/article/pii/S0079642516300068.
647. S. Iijima, *Nature* **354**, 56 (1991).
648. P. Wallace, *Phys. Rev.* **71**, 622 (1947).
649. M. Weinert, E. Wimmer, and A. J. Freeman, *Phys. Rev. B* **26**, 4571 (1982).
650. S. D. Sarma, S. Adam, E. H. Hwang, and E. Rossi, *Rev. Mod. Phys.* **83**, 407 (2011).
651. A. H. C. Neto, F. Guinea, N. M. R. Peres, K. S. Novoselov, and A. K. Geim, *Rev. Mod. Phys.* **81**, 109 (2009), http://link.aps.org/abstract/RMP/v81/p109.
652. C. Kittel, *Introduction to Solid State Physics*, 7th ed., Wiley, New York, (1996).
653. J. J. Ditri and J. L. Rose, *J. Comp. Mat.* **27**, 934 (1993).
654. E. P. Papadakis, T. Patton, Y. Tsai, D. O. Thompson, and R. B. Thompson, *J. Acoust. Soc. Am.* **89**, 2753 (1991).
655. J. L. Rose, J. J. Ditri, Y. Huang, D. Dandekar, and S.-C. Chou, *J. Nondestruc. Eval.* **10**, 159 (1991).
656. M. Klein, *IEEE J. Quant. Electron.* **QE-22**, 1760 (1986).
657. M. Cardona, *Superlatt. Microstr.* **5**, 27 (1989).

658. J. Menendez, *J. Lumin.* **44**, 285 (1989).
659. N. Ashcroft and N. Mermin, *Solid State Physics*, HRW Int. Eds., Philadelphia, PA, (1976).
660. R. E. Roman and S. W. Cranford, *Comput. Mat. Sci.* **82**, 50 (2014), http://www.sciencedirect.com/science/article/pii/S092702561300565X.
661. R. Arafune, C.-L. Lin, R. Nagao, M. Kawai, and N. Takagi, *Phys. Rev. Lett.* **110**, 229701 (2013), http://link.aps.org/doi/10.1103/PhysRevLett.110.229701.
662. J.-A. Yan, M. A. D. Cruz, S. Barraza-Lopez, and L. Yang, *Appl. Phys. Lett.* **106**, 183107 (2015), http://scitation.aip.org/content/aip/journal/apl/106/18/10.1063/1.4919885.
663. C. P. Chang, *J. Appl. Phys.* **110**, 013725 (2011), http://scitation.aip.org/content/aip/journal/jap/110/1/10.1063/1.3603040.
664. J. M. Luttinger and W. Kohn, *Phys. Rev.* **97**, 869 (1955).
665. V. P. Gusynin, S. G. Sharapov, and J. P. Carbotte, *Phys. Rev. Lett.* **98**, 157402 (2007), http://link.aps.org/doi/10.1103/PhysRevLett.98.157402.
666. Y. Zhang, Y.-W. Tan, H. L. Stormer, and P. Kim, *Nature* **438**, 201 (2005).
667. Y. Zhang, Z. Jiang, J. P. Small, M. S. Purewal, Y.-W. Tan, M. Fazlollahi, J. D. Chudow, J. A. Jaszczak, H. L. Stormer, and P. Kim, *Phys. Rev. Lett.* **96**, 136806 (2006), http://link.aps.org/doi/10.1103/PhysRevLett.96.136806.
668. Z. Jiang, E. A. Henriksen, L. C. Tung, Y.-J. Wang, M. E. Schwartz, M. Y. Han, P. Kim, and H. L. Stormer, *Phys. Rev. Lett.* **98**, 197403 (2007), http://link.aps.org/doi/10.1103/PhysRevLett.98.197403.
669. P. Yu and M. Cardona, *Fundamentals of Semiconductors*, Springer, Berlin, (1995).
670. P. Lautenschlager, M. Garriga, L. Vina, and M. Cardona, *Phys. Rev. B* **36**, 4821 (1987), http://link.aps.org/doi/10.1103/PhysRevB.36.4821.
671. R. R. Nair, P. Blake, A. N. Grigorenko, K. S. Novoselov, T. J. Booth, T. Stauber, N. M. R. Peres, and A. K. Geim, *Science* **320**, 1308 (2008).
672. F. Bechstedt, L. Matthes, P. Gori, and O. Pulci, *Appl. Phys. Lett.* **100**, 261906 (2012).
673. B. Adolph, J. Furthmüller, and F. Bechstedt, *Phys. Rev. B* **63**, 125108 (2011).
674. A. A. Balandin, S. Ghosh, W. Bao, I. Calizo, D. Teweldebrhan, F. Miao, and C. N. Lau, *Nano Lett.* **8**, 902 (2008).
675. M. Kamatagi, J. Elliott, N. Sankeshwar, and A. Lindsay Greer, in *Physics of Semiconductor Devices*, edited by V. K. Jain and A. Verma, Environmental Science and Engineering, Springer, Cham, Switzerland, **2014**, pp. 617–619. http://dx.doi.org/10.1007/978-3-319-03002-9_157.
676. R. G. Amorim, and R. H. Scheicher, *Nanotechnology* 26, 154002 (2015), https://doi.org/10.1088/0957-4484/26/15/154002.
677. S. M. Seyed-Talebi, I. Kazeminezhad, and J. Beheshtian, *Phys. Chem. Chem. Phys.* 17, 29689–29696 (2015). http://dx.doi.org/10.1039/C5CP04666A.
678. Y. Du, J. Zhuang, J. Wang, Z. Li, H. Liu, J. Zhao, X. Xu, et al., *Sci. Adv.* **2**, e1600067 (2016), http://advances.sciencemag.org/content/2/7/e1600.
679. S. Wei and M. Y. Chou, *Phys. Rev. B*, **50**, 2221 (1994).
680. Rohlfing, M., et al., *Phys. Rev. B*, **48**, 17791 (1993).

Silicene nanoribbons

6

Krzysztof Zberecki

Contents

6.1 INTRODUCTION

There is currently an increasing interest in two-dimensional (2D) materials such as graphene and 2D allotrope of carbon. In this planar structure, carbon atoms form a honeycomb lattice. Graphene exhibits unusual transport properties which follow from its famous electronic structure, in which electronic states around the K points of the Brillouin zone can be described by the Dirac model and the electrons behave like massless particles (Dirac cones). Graphene also has other promising properties, such as high electron mobility and long spin diffusion length, and is considered as a material for future nanoelectronic and spintronic devices.

The possibility of 2D structures made of silicon atoms (which belong to the same group in the periodic table as carbon) has been debated since the 1990s (Takeda and Shiraishi 1994). The main difference between graphene and silicene is that the silicon counterpart has a low-buckled structure (Cahangirov et al. 2009)—the two triangular sublattices are slightly displaced vertically. The electronic structure of silicene is similar to that of graphene. Spin–orbit (SO) interaction in the case of silicene opens an energy gap at the Fermi level but, like in graphene, this gap is rather small. But thanks to silicene's buckled atomic structure, an intrinsic SO interaction of Rashba form is present and plays a significant role in spin transport (Ezawa 2012b). Also, gate voltage can open an energy gap in silicene, but not in graphene.

From the application point of view, semiconducting transport properties are more desired than (semi)metallic, so opening a gap at the Fermi level in the electronic spectrum is one of the key challenges.

6.2 EXPERIMENTAL STATUS

The story of narrow silicon strips began in 2005, when Léandri et al. (2005) deposited Si atoms onto an Ag(110) surface, obtaining well-defined nanoribbons (NRs) for the first time (although these are called nanowires [NWs] in this work). Let us look more closely at this case, which will give us insight into the typical method of silicene nanoribbons (SiNRs) fabrication and characterization.

To obtain the sample, first the surface has to be prepared. Flat, Ag(110) surface is first cleaned by repeated cycles of Ar$^+$ bombardments, and then annealed at 700–750 K for few hours. Then, the deposition at room temperature of the Si atoms is conducted in an ultrahigh vacuum (UHV), using as the source of atoms a direct-current heated piece of silicon wafer. The obtained structures are then characterized by a variety of methods. First, scanning tunneling microscopy (STM) technique is used to visualize the structures (Figure 6.1). Figure 6.1a shows thin silicon structures, with length up to 30 nm, which are aligned along the (-110) direction of the Ag(110) surface. What is remarkable is their width, equal to 1.6 nm (which is four Ag–Ag atomic distances, a_1, where a_1 = 0.4 nm), is the same for all obtained NRs. As can be seen from Figure 6.1b and c, the height profile of each NR is rounded, with equally spaced protrusions. These protrusions are also periodically aligned, with a period corresponding to two distances between the Ag atoms (a_2 = 0.2885 nm). What is also important is that the NRs can be elongated up to 100 nm, by annealing at 230°C, keeping the width. The structural results were also confirmed by low-energy electron diffraction (LEED) measurements.

To get more insight into the atomic and electronic structure of the NRs, the advanced synchrotron radiation photoemission spectroscopy (PES) method has been applied. Angle-integrated measurements of Si 2p-core levels show that Si atoms of the NR belong to two different environments—the atoms (called Si$_1$) in contact with Ag surface atoms and the atoms (Si$_2$) in contact with atoms Si$_1$. The best fit to the Si 2p spectrum confirms the metallic nature of the nanostructures. (These findings were also confirmed later by STM measurements, conducted by Salomon and Kahn [2008]). The metallicity may be addressed to a proximity effect due to metal-induced gap states. The dispersion relations obtained with angle-resolved photoemission spectroscopy suggest the presence of electronic states confined within the NRs, leading to pronounced dispersion along the line Γ–X in the k-space.

The atomic structure of SiNW deposited on Ag(110) has been investigated more closely in Aufray et al. (2010) (Figure 6.2). New, high-resolution STM images revealed honeycomb-like patterns in the NWs.

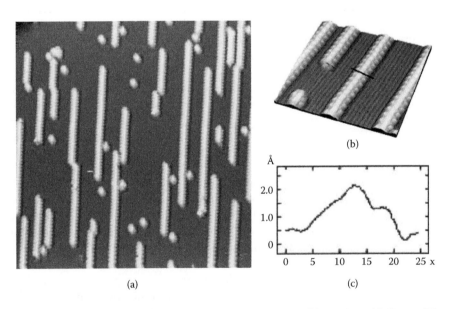

(a) (b)

Å

2.0

1.0

0

0   5   10   15   20   25 x

(c)

Figure 6.1 AFM images of SiNRs deposited on Ag(110) at room temperature: (a) overview with Si nanoribbons and nanodots (42 × 42 nm²), (b) zoom revealing the atomic structure of the NRs and of the bare substrate along the (110) direction, (c) height profile along the line in (b). (From Leandri, C., et al., *Surf. Sci.*, 574, L9–L15, 2005. With permission.)

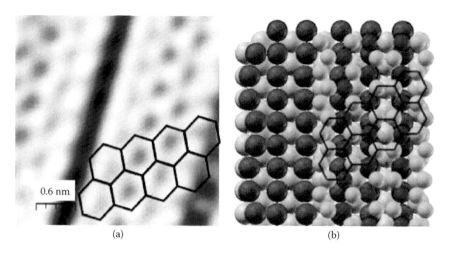

Figure 6.2 Atomic structure of SiNRs: (a) STM image of the structure with honeycombs drawn on the picture to guide the eye, (b) model of the calculated structure (dark blue balls—Ag atoms of the first layer, light blue—other Ag atoms, red—topmost Si atoms, green—other Si atoms). Hexagons formed by Si atoms are drawn with black line. (From Aufray, B., et al., *Appl. Phys. Lett.*, 96, 183102, 2010. With permission.)

Figure 6.3a clearly shows four silicon hexagons, forming a honeycomb arrangement. This hexagonal pattern, together with *ab initio* modeling (Aufray et al. 2010; Kara et al. 2010) helps to explain the bent structure of the NRs in the transverse direction (Figure 6.1b and c). Calculations, based on density-functional theory (DFT), reproduced both the hexagonal pattern and the buckling of the Si atoms on an Ag surface (Figure 6.3b). The model used in these calculations seems to work well, but it predicts different periodicity in the (-110) direction, x4, instead of x2 as observed in this direction. To overcome these problems, several new models have been proposed (see Aufray et al. [2015] and references therein), but none of them explains all spectroscopic data. The model of Tchalala et al. (Tchalala et al. 2014), based on STM and LEED measurements, predicts the armchair-edged NRs on an Ag surface (Figure 6.3) having higher buckling than the previous models, with proper periodicity, but this has not yet been fully confirmed experimentally (Aufray et al. 2015).

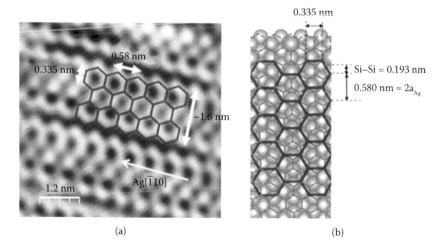

Figure 6.3 (a) STM image of silicene NRs grown on an Ag(110) surface (6 × 6 nm², V = −0.2 V, I = 1.9 nA); (b) model of the calculated structure (gray and yellow balls—Ag atoms, respectively, of the first and second layer, blue honeycombs correspond to the theoretical NRs and red ones to the observed structure). (From Tchalala, M. R., et al., *J. Phys. Conf. Ser.*, 491, 012002, 2014. With permission.)

Also very interesting is a study reported in Bernard et al. (2013), based on STM and grazing incidence X-ray diffraction (GIXD) measurements of the evolution of the Ag(110) surface during Si growth. The results suggest that during the NRs' growth, Ag atoms may be released, leading to surface reconstruction. This puts into question all previous models based on a nonreconstructed substrate. However, the most recent study (Feng et al. 2016) proposes the model of SiNR on Ag(110) which does not take into account the alloying with Ag atoms of the surface, but can describe the STM images well. To formulate this model, the authors used an interesting approach starting with over 50 different geometries of Si atoms deposited on Ag. Then, using *ab initio* results they chose two of them, for which the calculated STM image most resembled the experimental ones (Figure 6.4). Similar to the previous study, both NRs are periodic in the (-110) direction and have a honeycomb structure. The new model relates the width of the ribbon to a tensile stress between the lattices of silicene and Ag substrate, unlike in the previous models. The structure, resembling the $\sqrt{3} \times \sqrt{3}$ structure of monolayer silicene, is highly buckled, which stems from the reconstruction of NRs' edges.

So, in summary—silicene on Ag(110) surface forms narrow, one-dimensional (1D) structures of a metallic nature and quite complex structure, which is not fully understood yet.

It also turns out that these structures are extremely volatile to hydrogenation (Salomon et al. 2009), which leads to complete removal of Si atoms from the Ag surface. On the other hand, silicene on Ag(110) seems to be quite resistant toward oxidation (De Padova et al. 2011); the oxidation process may be started only with high oxygen exposure. What is more interesting is that the process of oxidation is enhanced in the presence of NRs' defects. These facts are very important from a technological point of view, giving a potential new tool for patterning the silicene 1D nanostructures.

There were also studies of Si atoms deposition on different substrates (Léandri et al. 2007). The deposition of Si atoms on an Ag(001) surface at RT leads to the formation of a p(3 × 3) superstructure, built by silicon tetramers. Then, after further deposition beyond 1 ML new structures appear, which are a mixture of chains of regular hexagons and local superstructures. Deposition of Si atoms under UHV conditions on

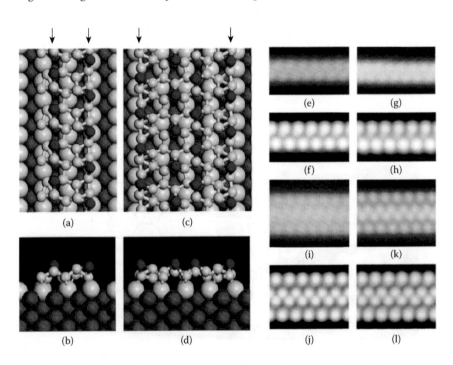

Figure 6.4 Ball and stick model of relaxed SiNRs on top of Ag(110): (a) and (b) top and side view of 1-nm wide SiNR, (c) and (d) top and side view of 2-nm wide SiNR. Light blue and dark blue balls are Ag atoms (topmost and underlying, respectively), red balls and yellow balls are Si, upper buckled (red) Si atoms can be probed by STM. (e) through (l)—simulated STM images versus experimental STM images at different bias voltages. (From Feng, B., et al., *Surf. Sci.*, 645, 74–79, 2016. With permission.)

an Ag(111) surface, which was undertaken by Lalmi et al. (2010), leads to the formation of a well-defined, 2D honeycomb silicene structure.

It is also possible to grow SiNR on an Au(110) substrate (Tchalala et al. 2013). The obtained structures are quite similar to those grown on Ag(110), with width equal to 1.6 nm and periodicity in the (-110) direction; the metallic nature of the NRs has been confirmed by high-resolution PES measurements.

Having established the experimental status of pristine SiNR, let us now come to more practical issues like modification of NRs with atoms or molecules. One can expect that such modification may change properties of nanostructures making them suitable for different applications. Since the SiNRs deposited on Ag(110) may uniformly cover the entire surface, one may try to functionalize these superstructures. One of the first experiments was to grow Co chains on previously prepared Si on Ag(110) (Sahaf et al. 2009). Co deposition in RT studied with STM technique revealed that Co atoms are adsorbed on the top of Si structures forming atomic chains ("nanolines"). Also, Co atoms were never adsorbed on the Ag surface. This shows that the deposition of Co atoms on SiNRs may be a good idea for building the future Si-based nanodevices (see also Masson et al. 2013). The magnetic properties of Co atoms on SiNRs have been also studied by Michez et al. (2015) with the use of X-ray magnetic circular dichroism.

The experimental results show that the magnetization is suppressed in the first Co layer deposited on the Si template due to decoupling of the Co from the metallic (Ag) substrate. Enhanced magnetization, with ferromagnetic (FM) ordering, exhibits only the second layer of Co atoms. Other magnetic atoms such as Mn grow quite differently on Si+Ag(110) structures (De Padova et al. 2013). They are adsorbed to the NRs' edges forming Si–Mn lines, which may be very interesting for the spintronic applications, although their magnetic properties have not yet been investigated.

It is known that attachment of organic molecules to graphene change its properties considerably. The study conducted by Salomon and Kahn (2008) is very interesting from the application point of view. There, the STM technique has been used to investigate the chemisorption of two organic molecules onto the SiNRs on Ag(110). The results show that the molecules are forming well-defined, 1D aggregation and what is more, changing the nature of SiNRs from metallic to semiconducting.

The problem with silicene is that, unlike graphene, it is unstable in air due to its mixed sp2/sp3 character. The deposition of silicene on an Ag(111) substrate studied with LEED (Acun et al. 2013) shows that the silicene layer is unstable against the formation of an sp3-like silicon structure. The formation of this structure seems to be triggered by thermal Si adatoms, which are created by the silicene layer itself. One way to overcome this problem would be to develop an encapsulation method preventing the silicene layer from degradation.

From the application point of view, it is most important to produce a silicene-based electronic device, for example, a field-effect transistor (FET). This was done for the first time by Tao et al. (2015), where silicene grown on Ag(110) was used to produce an FET operating at room temperature. The problem with silicene stability has been worked out by use of the novel fabrication silicene encapsulated delamination with native electrodes (SEDNE) method. This method allows the silicene properties to be preserved during transfer for device fabrication. The measured carrier mobility was equal to about 100 cm^{2}V^{-1}s^{-1}, which is a rather modest result.

6.3 THEORETICAL DEVELOPMENTS

6.3.1 TOOLS

Theoretical study of SiNRs is usually performed with use of one of two methods, which we will now describe briefly. The first one is the family of *ab initio* methods, based on DFT (Hohenberg and Kohn 1965; Kohn and Sham 1965). DFT describes all the electrons in the system using mean-field Hamiltonian. This approach leads to a system of equations, called Kohn–Sham equations. Solving them, one can obtain the electron density of the system as well as the electronic structure and other properties. In DFT, the total energy of the ground state is a functional of electron density ρ, namely

$$E[\rho] = T_s[\rho] + \int d\vec{r} v_{ext}(\vec{r})\rho[\vec{r}] + E_H[\rho] + E_{xc}[\rho] \tag{6.1}$$

where the first term is the kinetic energy of noninteracting electrons, the second is the energy of the external field (coming from the ions), the third is the Hartree term and the last is an exchange-correlation contribution. The method has two main disadvantages. The first is, this is a ground state theory so any property which stems from the excited states' spectrum must be determined by another approach (this is because Kohn–Sham equations are based on the variational principle). The second problem is that what differentiates this system from an interacting one is contained in the last part of the Hamiltonian (Equation 6.1). This so-called exchange-correlation term is usually parametrized and fitted to other theories (e.g., quantum Monte Carlo). So, the term *ab initio* may be a little misleading in this context. Nevertheless, the method is widely used and reliable, and its predictive power has been proven in countless cases (see Martin [2004]). The second method, commonly used to describe electronic properties of nanosystems, is the tight-binding (TB) theory (Slater and Koster 1954). In this theory, the Hamiltonian may written in the form

$$H = \sum_{ij;\alpha\beta} H_{ij;\alpha\beta} < i\alpha \,|\, j\beta >$$

(6.2)

where $|i\alpha>$ denotes the state that corresponds to the molecular orbital centered around $\overrightarrow{R_i}$, that is, $<\vec{r}\,|\,i\alpha> = \varphi_\alpha(\vec{r} - \overrightarrow{R_i})$. The wave function is expanded as a Bloch sum of orbitals of atomic nature, so the matrix element in Equation 6.2 has the form

$$H_{ij;\alpha\beta} = \int d\vec{r} \varphi_\alpha^*(\vec{r} - \overrightarrow{R_i})[T(\vec{r}) + V(\vec{r})]\varphi_\beta(\vec{r} - \overrightarrow{R_j})$$

(6.3)

where the T and V are kinetic energy operator and the potential, which describes the interaction between the electrons and ions, respectively. This leads to the system of equations of the Schrödinger form. In TB, unlike the DFT, the matrix elements (Equation 6.3) are not calculated but parametrized. These parametrizations are obtained by a fitting procedure, either to experimental data or to other theories (e.g., DFT). There may be a number of different parametrizations, each fitted to different data and suitable for different groups of nanosystems. For example, Papaconstantopoulos (2014) contains the complete set of parameters for all elements in building bulk materials. In the case of silicene, there exist a couple of parametrizations such as in Ezawa (2012b), which are fitted to *ab initio* data obtained for the 2D case.

6.3.2 STRUCTURAL AND ELECTRONIC PROPERTIES OF PRISTINE SiNRs

Before we reach the theoretical study of SiNRs, we should mention one important point. In the previous subsection, we have seen that up to now, silicene may have the form of well-established NRs on an Ag(110)/Au(110) surface. These structures are not easy to model as there is strong interplay between Si and Ag atoms. On the other hand, first theoretical works on silicene and its Ge-based relative, the germanene (Cahangirov et al. 2009; Lebegue and Eriksson 2009) consider this material (using the *ab initio* approach) as freestanding or suspended in a vacuum.

Such an approach, treating the structure as freestanding, has been used to study the 1D structures of silicene. Up to now there are only a few theoretical works (mentioned previously) on SiNRs deposited on a metallic surface. The reason for this is the fact that 1D free silicene structures have much more interesting electronic and transport properties and contain very rich physics. Such a model will now be used to discuss the electronic and magnetic features of SiNRs. So, from now on, by SiNR we mean freestanding SiNRs.

It is known that the atomic structure of 2D silicene is a little different from that of its carbon counterpart, graphene. The main difference is that silicene is not planar [as well as variety of existing and only postulated 2D materials (Sahaf et al. 2009)] but has a low-buckled structure. This means the structure consists of two sublattices, shifted vertically with respect to each other by about 0.4 Å (Cahangirov et al. 2009). The mechanism of buckling (or puckering) can be addressed to competition between sp^3 and sp^2 hybridization of Si orbitals causing the weakening of π–π bonding between p_z orbitals, perpendicular to the plane. This affects the stability of silicene, as its cohesive energy is two times lower than for graphene (Sahaf et al. 2009).

The structure of SiNRs, as considered in theoretical studies, is in analogy to NRs made of graphene. Generally, there are two kinds of NRs, zigzag (Z) and armchair (A), depending on the shape of the edges along the periodicity direction. Usually, their width is described by the number N of the chains in the ribbon (Figure 6.5).

Despite the small cohesive energy of silicone (Cahangirov et al. 2009), SiNRs can be surprisingly resistant to strain. Using the *ab initio* molecular dynamics (MD) method, Topsakal and Ciraci (2010)

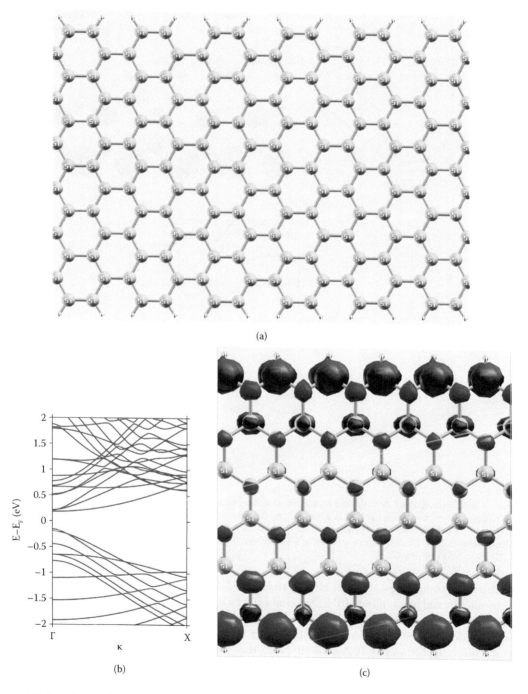

(a)

(b)

(c)

Figure 6.5 Atomic and electronic structure of SiNRs. (a) Model and (b) electronic structure of N = 13 ASiNR. Light blue balls Si atoms, small light blue balls H atoms. (c) Model and electronic structure. *(Continued)*

Low-dimensional structures

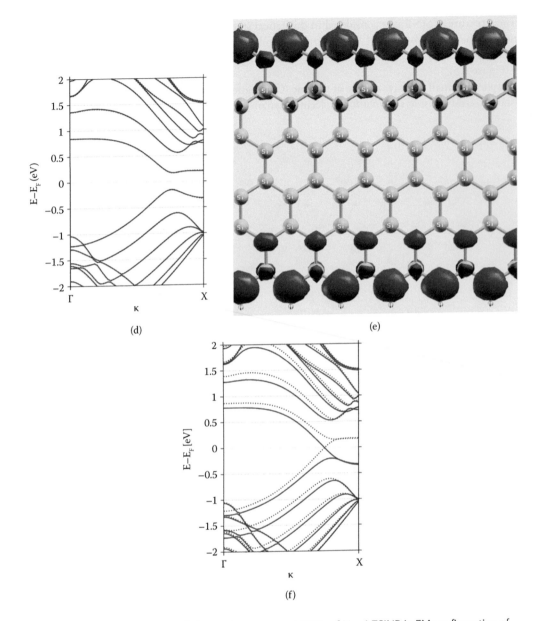

(d)

(e)

(f)

Figure 6.5 (Continued) (d) Atomic and electronic structure of SiNRs of N = 6 ZSiNR in FM configuration of edge magnetic moments. Red and dark blue balls on (c) are magnetic moments, spin-up (spin-down) bands are red (blue) on (d). (e) Model and (f) electronic structure of N = 6 ZSiNR in AFM configuration, same designations. Calculated with DFT, GGA, as implemented in Siesta code. (From Soler, J. M. E., et al., *J. Phys. Condens Matters.*, 14, 11, 2002. With permission.)

calculated the strain energy of several NRs (A and Z, ASiNRs, and ZSiNR, respectively) made of graphene, silicene, and boron nitride (BN). The strain energy is defined as the total energy at a given axial strain (ε) minus the total energy at zero strain. From this energy one may obtain a force constant κ, which is a second derivative of the strain energy with respect to the lattice constant (changed to induce a strain). The κ is much smaller for N = 10 ASiNR (30 N/m), than the κ for graphene nanoribbon (GNR) of the same type and width (176 N/m). Nevertheless, behavior under uniaxial tension of SiNRs is quite similar to that of GNRs. The electronic properties in the elastic regime ($\varepsilon < 0.2$) may be strongly modified, leading to closure of the band gap (equal to 1.1 eV for unmodified N = 10 ASiNR) in the case of A-type NRs. In the plastic regime ($\varepsilon = 0.2$ to 0.5 in the case of SiNRs), the structures may be elongated to form atomic chains between the patches, before they break [see also recent MD study of mechanical properties of silicene nanostructures in Fan et al. (2016)].

As in the case of graphene, the electronic structure of SiNRs is determined by the type (Z or A) and width. The armchair SiNRs are semiconducting, with a nonzero energy gap (E_g). The E_g width shows oscillatory behavior. For N = 3p + 2 (p – integer) Eg is small, but for N = 3p + 1 and for N = 3p it is much larger. The ground state is nonmagnetic. If one passivates the edges of the ribbon with H atoms (saturating the dangling bonds), the oscillating character persists, but for small N the gap is much wider. For zigzag NRs, the situation is even more interesting. Without the saturation of dangling bonds, the ground state is NM, but with H atoms on both edges, the ground state becomes magnetic. The magnetic moments are located mainly on the edges of the NRs, and there are two possible configurations—FM, having moments on both edges aligned in the same direction, and antiferromagnetic (AFM), where edge moments are aligned antiparallel (i.e., the moments on one edge are aligned in the same direction, antiparallel to the moments of the other edge). According to *ab initio* calculations (Cahangirov et al. 2009), AFM and FM configurations are very close in energy (1 meV), which is within the accuracy limits of DFT (so in fact one cannot determine which one is the ground state.) Other studies show that the AFM is the ground state (Pan et al. 2012; Zberecki et al. 2013) but the NM state lies much higher in energy. In FM configuration the band structure is metallic, while in the AFM case the structure is semiconducting with a narrow band gap.

The zigzag GNRs (ZGNRs) are famous for the presence of the special states localized on the edge atoms—so-called edge states (Nakada et al. 1996; Miyamoto et al. 1999). Studies conducted by Nakada et al. (1996) on the use of the TB method show that the wave functions which correspond to bands close to the Fermi level are confined to the edge atoms. These eigenstates may be expressed analytically and have the form of fast-decaying exponential functions. The study of the influence of the edge geometry on electronic structure for SiNRs has been conducted by Song (2010). *Ab initio* results confirm the presence of the edge states for H-terminated ZSiNRs. For ASiNRs, the edge states are not present (similar to the case of GNRs) due to their dimer Si–Si bond at edge.

The two presented shapes of NRs (Z and A) are not the only possible models of these 1D structures. There are other propositions, based on *ab initio* calculations. In Zhao and Ni (2014), the saw-shaped SiNRs are investigated. These NRs are predicted to be more stable than the ZSiNRs. They are all spin-semiconductors with the FM ground state (with a gap between majority and minority spin-polarized bands) and with the band gap tunable by strain. The magnetic moments, as in the ZSiNR, are located on the edges, forming the edge states. Although the saw-shaped SiNRs have never been observed experimentally, this may be an interesting direction to investigate by the experimentalists.

6.3.3 MODIFICATIONS

As in the experimental case, the zigzag and armchair NRs can be modified to obtain systems with more interesting features. Let us see how modifications like vacancies, doping atoms, or functionalizations may alter the electronic structure of SiNRs.

One of the earliest studies is the work of Song et al. (2011a), where structural and electronic properties of ASiNRs with H-passivated edges (like all SiNRs in this subsection, unless stated otherwise) with vacancies have been investigated by *ab initio* methods. The ASiNR (having N = 13 atomic chains) with mono- and divacancies have been considered, which may be created by removing one or two adjacent Si atoms. Vacancies located in the central region (far from the NR's edges) and in the non-central—close to the edges—have been considered. The total energy calculations predict that both locations of the structures with monovacancies are more stable than those with divacancies. A presence of mono- or divacancy leads to an indirect energy gap in ASiNRs, but ribbons remain nonmagnetic even for the cases (although not the most stable ones) where divacancy causes the system to turn metallic. Accordingly, the first study shows that the electronic properties of SiNRs are highly sensitive to structure modifications. The same group (Song et al. 2011b, 2013a) also investigated the modifications of a structure in the form of the monochains of carbon atoms, substituting the chain of Si atoms in the ZSiNR. These modifications, making the structure more stable than the pristine one, lead to FM ordering in the ground state. So, not only electronic, but also the magnetic structure of SiNRs can be changed by structure alterations.

Doping of ASiNRs with N and B atoms has been theoretically investigated in Ma et al. (2013). Firstly, the results confirm the oscillating nature of the band gap for pristine ASiNRs. Then, the formation energies have been calculated for N = 7 ASiNR, where N or B atoms have been substituted for Si atoms for different sites. It turned out that the most energetically favorable are the sites closest to the edge. The levels corresponding to the doping atoms are situated in the band gap, crossing the Fermi level, which makes the system metallic. Further doping with two (same or different) atoms leads to further modifications of the electronic structure. Namely, doping with two N atoms (in the most favorable atomic configuration) gives the indirect, very narrow (0.04 eV) band gap. Doping with two B atoms gives a metallic structure, while the pair N-B gives a semiconducting structure with two levels close to the Fermi level coming from the dopants and the band gap of 0.41 eV. The last value is two times smaller than for the pristine ribbon. So, doping ASiNR with N and B atoms gives ample opportunities for band gap engineering (see also Zheng et al. (2013a)). Doping of zigzag SiNRs with N and B atoms can also give interesting results (Luan 2013; Zheng et al. 2013a, 2013b). Compare also with Li et al. (2009) for an analogical study for ZGNRs. *Ab initio* calculations conducted analogously to that of Ma et al. (2013) show that doping may lead to several band structure modifications. To be more precise, the addition of one N or B atom leads to an atomic configuration in which dopant atoms are located in the sites closest to the edge. The ground state of such modified ZSiNR becomes FM as the edge with the substituted atom becomes nonmagnetic (which may be attributed to the perturbation of edge states). Doping with B atoms leads to an electronic structure in which one spin channel is insulating while the other is conducting, giving half-metallic (HM) behavior. In the case of N doping the situation is even more interesting, as both spin channels are semiconducting with band gaps 0.21 eV and 0.14 eV. The conduction band minimum (in minority spin channel) and the valence band maximum (in the majority spin channel) touch each other at E_F. This indicates that the band structure in the N doping case exhibits a typical semimetal and spin gapless semiconductor character. Doping with two N or two B atoms leads to metallic behavior, and to NM ground state due to the fact that now both edges are modified. Finally, for a pair of two atoms B and N, the GS is also NM, but this time a gap equal to 0.18 eV opens, making the ZSiNR a semiconductor. Similar results can be obtained by doping with Al and P atoms (Dong et al. 2014). Application of a single Al atom into the ZSiNR may result in half-metallic behavior and 100% spin polarization.

Doping with magnetic atoms can be also considered. As it has been shown in Lan et al. (2013), the addition of Co atoms as adsorbents may severely change the band structure of the ZSiNRs, as it raises the magnetic moments on the neighboring Si atoms and suppresses the edge magnetism if the edge is in the vicinity of the Co atom. In the most stable case the interesting magnetic configuration arises, in which the magnetic moment located on the Co atom is antiparallel to the moments located on the edges. Although the band structure is metallic, with Co d-states crossing the Fermi level, this opens the possibility of tailoring the magnetic properties of future silicene-based devices.

So, these results show that ZSiNRs are very susceptible to electronic and magnetic structure modifications making them promising candidates for building blocks of future electronic and spintronic devices.

Since the ASiNRs are semiconductors with rather wide band gap, it would be very beneficial if one could change their gap with some other means than the structure modifications. It turns out, that this may be possible. Results of the study in Song et al. (2013b) suggest that the width of the band gap can be tuned by the use of an external transverse electric field. For example, for N = 9 ASiNR, the gap may be (nonlinearly) constricted from 0.45 eV (no field) to 0.15 eV (E = 1.0 V/Å), and for N = 6 from 0.55 eV (E = 0) to 0.08 eV (1.0 V/Å).

Silicene nanoribbons are also supposed to be good materials for application in lithium (Li-ion batteries). In Deng et al. (2013), lithium adsorption and diffusion on SiNRs have been investigated by *ab initio* methods. The results suggest that SiNRs may be good electrode material, because of their good reactivity with charge-carrying Li ions that allow for their fast transport. The silicene edges provide the strongest binding of Li atoms from all the silicon structures. Also, the energy barriers for Li diffusion on SiNRs, which are very low (0.14–0.26 eV, depending on the NR's type and distance from the edge) suggest that SiNRs may provide much faster diffusion than a 2D silicene layer.

The fact that silicene has a relatively wide SO gap enables the possibility to observe the quantum spin Hall effect or topological insulator behavior (Liu et al. 2011a). These phenomena are possible due to a

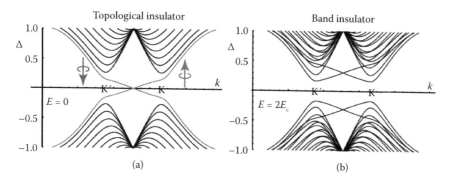

Figure 6.6 Band structure of SiNR, obtained by diagonalization of (4). (a) Topological insulator case—the bands (edge states) are crossing the gap, red (blue) line for spin-up (spin-down) states. (b) Band insulator case. (From Ezawa, M., *Phys. Rev. Lett.*, 109, 055502, 2012b. With permission.)

combination of time reversal symmetry and SO interaction. The effects of SO coupling including the Rashba effect and the influence of an external electric field can be described by the TB Hamiltonian, which has the general form (Liu 2011b; Ezawa 2012b):

$$H = t \cdot H_{NN} + \lambda_{SO} \cdot H_{so} + \lambda_R \cdot H_R + H_S \tag{6.4}$$

In the above, the first term represents the nearest-neighbor hopping on the honeycomb lattice with the transfer energy $t = 1.6$ eV. The second and third term represent, respectively, the effective SO coupling with constant $\lambda_{so} = 3.9$ meV and the Rashba SO coupling with $\lambda_R = 0.7$ meV. The last term is the staggered sublattice potential, proportional to the external electric field E_z, which is generated between silicon atoms of the two sublattices. As it has been shown in Ezawa (2012b), the application of this model to the monolayer silicene reveals that the band gap is a function of the electric field E_z: $\Delta(E_z) 2 |lE_z - \eta s_z \lambda_{so}|$, where l is the distance between the sublattices, $s_z = \pm 1$ is spin and $\eta = \pm 1$ is for K or K' special point in the first Brillouin zone. The gap is zero for $E_z = \eta s_z E_c$ where $E_c = \lambda_{so}/l$. But there is a big difference between situation with $|E_z| \le E_c$ and $|E_z| > E_c$. To see this, one may apply the Hamiltonian (Equation 6.4) to the silicene ZSiNR. The obtained spin-resolved band structure depicted for two points, $E_z = 0$ and $E_z = 2E_c$ can be seen in Figure 6.6 (Ezawa 2012b).

In the first case, $E_z = 0$, the band structure is that of topological insulator, as there are gapless modes coming from the two edges. For the second case, $E_z = 2E_c$, the band structure is of typical band insulator in the absence of gapless edge modes. So, in ZSiNRs, the transition from topological insulator to band insulator can be observed as the E_z changes.

Further study of the Hamiltonian with exchange term included shows that more phases are possible in ZSiNRs (Ezawa 2012a). Specially worth mentioning is the valley-polarized metal state, which appears for $ME_z \ne 0$, where M is the exchange field. This unique phase is characterized by the fact that it is a metallic state in which gaps are open at the K and K' points.

The role of an external electric field as well as effective SO coupling λ_{so} in the case of ASiNRs has been studied recently in Zhou et al. (2016). The results suggest that the band gap strongly depends on both, the λ_{so} and E_z.

6.4 TRANSPORT PROPERTIES

6.4.1 TOOLS

Theoretical study of electron transport in nanostructures is usually carried with use of a method which has become very popular in the last 20 years, because of its versatility and deep physical meaning. This is the (nonequilibrium) Green function method (NEGF) (Datta 1995), which is based on mathematical methods used in different areas of theoretical physics. In the case of electron transport description in the nanoscale systems, one usually starts with the approach presented briefly in Section 3.1—the system is described by a mean-field Hamiltonian. The charge carriers are provided by attached electrodes. These electrodes are modeled as semi-infinite, which breaks the translation invariance and because of applied voltage, the system is out of equilibrium. The situation can be dealt with by use of one-body Green's functions and the so-called partitioning technique. This technique can be briefly presented as follows. The system is divided into three regions—left electrode (L), right electrode (R), and the device region in between (D). The mean-field Hamiltonian of electrodes+device may have the matrix form:

$$H = \begin{pmatrix} H_L & H_{LD} & 0 \\ H_{DL} & H_D & H_{DR} \\ 0 & H_{RD} & H_R \end{pmatrix} \tag{6.5}$$

where H_{mn} is a part of the full Hamiltonian, which describes the interactions between m and n regions ($m,n = \{L,D,R\}$) and H_m is a part of the full Hamiltonian which describes isolated electrodes and device. No interaction between electrodes is assumed, so $H_{LR} = H_{RL} = 0$. The problem is that the electrodes are modeled as (semi)infinite, making the matrix (Equation 6.5) infinite. Use of the partitioning technique (described in detail in Jacob and Palacios [2011]) may overcome this problem by the concept of the self-energy. The self-energy $\Sigma_\alpha (E)$ can be expressed by the Hamiltonian and the Green's function of the isolated electrodes, $g_\alpha (E)$, as $\Sigma_\alpha (E) = (ES_{D\alpha} - H_{D\alpha}) g_\alpha (E)(ES_{D\alpha}^\dagger - H_{D\alpha}^\dagger)$, $\alpha = \{L,R\}$. S is a matrix which describes the overlap of the atomic orbitals, defined in analogy to Equation 6.5, which must be taken into account if the Hamiltonian matrix is defined with the use of a non-orthogonal basis.

The current through the nanoscopic conductor is given by the famous Landauer formula (Landauer 1957):

$$I(V) = \frac{2e}{h} \int dE \left[f(E - \mu_L) - f(E - \mu_L) \right] T(E) \tag{6.6}$$

where f is the Fermi distribution function, μ_L and μ_R are the left and right electrode chemical potentials, which are related to the applied bias voltage V by $eV = \mu_L - \mu_R$. The transmission function, $T(E)$, can then be calculated by the Caroli expression (Caroli et al. 1971):

$$T(E) = Tr \left[\Gamma_L(E) G_D^{(-)}(E) \Gamma_R(E) G_D^{(+)}(E) \right] \tag{6.7}$$

where the so-called coupling matrices $\Gamma_\alpha (E)$ are defined by the self-energies $\Sigma_\alpha (E)$ and $G_D^{(+/-)}(E)$ which are retarded (+) and advanced (-) Green's functions of the device region. This approach is widely used and one of its advantages is that the Hamiltonian in Equation 6.5 may be of Kohn–Sham type (Equation 6.1) defined on a molecular basis, as implemented, for example, in the Siesta code (Soler et al. 2002), the TB, or any mean-field-based code, which makes it possible to use different levels of theory to describe the studied system.

6.4.2 TRANSPORT PROPERTIES OF PRISTINE SiNRs

In the last chapter, the electronic structure of the SiNRs with various edge shapes was discussed, showing that the band structure of these silicene stripes is dependent on atomic structure. It can be also highly susceptible to various structure modifications. Before we check how these modifications may affect the transport properties, let us analyze the transport properties of pristine SiNRs.

As it is known (Li et al. 2008), the I(V) characteristics of ZGNRs may be very interesting due to their dependence on the symmetry of the ribbon. The current–voltage relationships can be very different for N-even and N-odd ZGNRs. For N-even, the NR has a mirror plane σ, which is situated exactly in the midplane between the edges, which is not the case if N is odd. As it has been shown in Li et al. (2008), the presence of such symmetry causes the π and π* bands to have opposite parity, which prevents the electron hopping between them. In the N-odd case the bands have no definite parity, so the electron transfer may actually exist. It turns out, this mechanism also works for ZSiNRs, as it has been shown by NEGF + *ab initio* calculations conducted by Kang et al. (2012). In ZSiNRs, the situation is different than in ZGNRs due to the fact that the silicene is not planar, and the two sublattices are shifted with respect to each other, making the presence of a σ plane not possible. However, the symmetry analysis of π and π^* orbitals reveals (Kang et al. 2012), that actually the N-dependence of I(V) also exists in the case of SiNRs with Z edges (Figure 6.7).

What is more interesting is that, under nonzero bias in the case of N-even, the gap opens in the conductance spectrum, which is growing with the bias. The gap is always slightly smaller than the bias window, which makes the very small current flow. In the magnetic case, the symmetry of the bands may also be used to obtain very strong magnetoresistance (MR) for N-even ZSiNRs. The MR is defined as $MR = (I_P - I_{AP})/I_{AP}$. In this formula, I_P is the current obtained in the case of parallel configuration of edge magnetic moments and the I_{AP} is the current for antiparallel configuration, realized as in Figure 6.8c. In the antiparallel case the current is almost zero (Figure 6.8d), due to opposite parity of π and π^* bands under the c_2 symmetry operation (Kang et al. 2012). The resulting MR reaches very high values (Figure 6.8e), up to $10^6\%$.

To further assess the MR value for the SiNRs, the NEGF + *ab initio* calculations have been conducted for the N = 3–8 ZSiNR in Xu et al. (2012). The system has been modeled as an NR with zigzag edges connected to armchair-edged electrodes of greater width than the device ZGNR. The I(V) characteristics have been calculated for two cases—FM, where the magnetic moments of the edge atoms of the zigzag region were parallel to each other and AFM, where moments of the edges were antiparallel to each other. MR has

been defined as $MR = \dfrac{I_{FM} - I_{AFM}}{I_{AFM}}$, where I_{FM} (I_{AFM}) is the total current for the FM(AFM) edge magnetic

moment configuration of the ZSiNR. The calculated MR takes values up to 2000% (for N = 3 and low V_{bias}), although much smaller than for the system studied in Kang et al. (2012) still suggests that the GMR may actually occur in the pristine ZSiNRs.

6.4.3 MODIFICATIONS

The discussion in Subsection 6.3.3 shows that the structure modifications may affect the electronic properties of SiNRs. Since the transmission function is in close relation with the band structure it would be instructive to check whether doping or strain affects the transport properties.

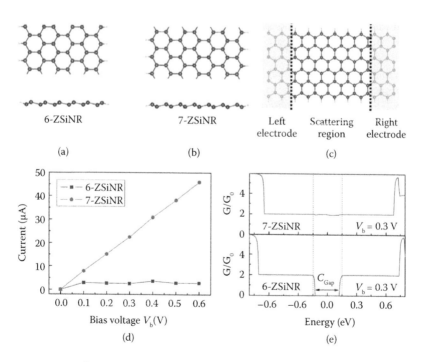

Figure 6.7 Atomic structure of (a) N = 6 ZSiNR and (b) N = 7 ZSiNR from top view (upper) and side view (lower). Blue balls Si atoms, white balls H atoms. (c) Schematic view of the system. (d) Calculated I(V) characteristics for N = 6,7 ZSiNRs. (e) Transmission function for N = 6,7 ZSiNR under a bias voltage of 0.3 V. G_0 equals e^2/h. (From Kang, J., et al., *Appl. Phys. Lett.*, 100, 233122, 2012. With permission.)

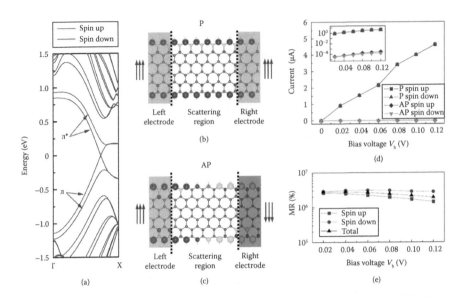

Figure 6.8 (a) Band structure of N = 6 ZSiNR in FM configuration, red (blue) lines correspond to spin-up (down). Spin density for (b) N = 6 SiNR in P and (b) AP (c) configuration. Pink and light blue balls correspond to spin-up and down components, respectively. (d) I(V) characteristics for P and AP configurations. The inset is semilogarithmic scale plot. (e) The spin-up, spin-down, and total magnetoresistance on semilogarithmic scale. (From Kang, J., et al., *Appl. Phys. Lett.*, 100, 233122, 2012. With permission.)

The doping with single Al or P atoms of ZSiNRs has been studied in Zberecki et al. (2014a) by an *ab initio* + NEGF approach. The calculations show that the ground state of pristine NR is magnetic with AFM edge moments configuration, while the FM state is higher in energy only by 0.02 eV. So, both configurations have been studied for the doping case. The impurity atoms (Al or P) were distributed periodically along the chain. Three localizations have been taken into consideration: at one of the edges (PE configuration), at the NR center (PC), and in the middle between the edge and central atoms of the NR (PM). The results confirm findings described in Subsection 6.3.3 (B/N impurity studies)—atoms located on the edges suppress the magnetic moments in the vicinity. The low-energy state is no longer AFM for the PE case, but small net magnetization can be observed. So, the configuration is FM in NRs with impurities (in the case of impurities located at one edge and of a sufficiently large concentration, the configuration becomes FM because of the suppression of magnetic moments of the whole edge modified with impurities). The energy gap at the Fermi level, which exists in pristine ZSiNRs in the AFM state, survives in the presence of impurities (Figure 6.9a). Also, the transmission function now depends on the spin orientation, because the magnetic moments of the two edges do not fully compensate each other, so it is significantly modified with respect to the pristine case. There are three types of modifications—the new gaps are opening above the Fermi level (Figure 6.9a through c), there are observed impurity states in the gap close to the Fermi level, and also in PC and PM cases one can observe the Fano antiresonances (Figure 6.9a, b, and d) coming from the quantum interference effects. In the FM state, slightly higher in energy than the FM one, the impurities also modify the transmission function. Especially interesting may be the situation for the PM case, where one spin channel then remains conductive in the vicinity of the Fermi level, whereas the second channel becomes semiconducting due to a wide dip in the transmission at the Fermi level.

The doping with single Al atoms has been investigated with the same method as in Deng et al. (2016), where the edges were asymmetrically passivated with hydrogen atoms. The results show that this kind of edge modification may cause interesting bipolar magnetic semiconducting behavior, and Al substitutions close to the edge can tune the band structure to be HM. The Al atom as a dopant introduces one hole to the NR and this makes the Fermi level shift down to the valence bands. If the Al atoms are doped symmetrically to both electrodes, one can obtain 100% spin filtering and also, the junction becomes a rectifier with maximum rectification ratios that can reach the order of 10^5. This is consistent with the study

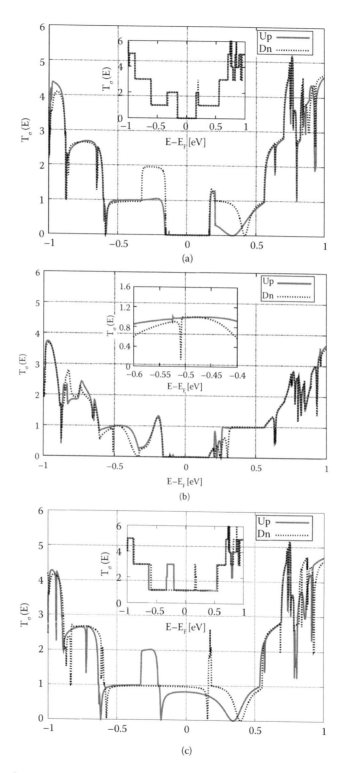

Figure 6.9 Spin-dependent transmission function for N = 6 ZSiNR. Red (black) lines are for spin-up (down). (a) Transmission for low-energy state with Al impurity in PE location, in inset $T_\sigma(E)$ for pristine ZSiNR in AFM configuration, (b) transmission for low-energy state with Al impurity in PC location, in inset part of the curve in a narrow energy range, where the Fano antiresonances are well resolved, (c) transmission for FM configuration with Al impurity in PE location, in inset $T_\sigma(E)$ for pristine ZSiNR in FM configuration. (From Zberecki, K., et al., *Phys. Rev. B*, 89, 165419, 2014a. With permission.)

(Continued)

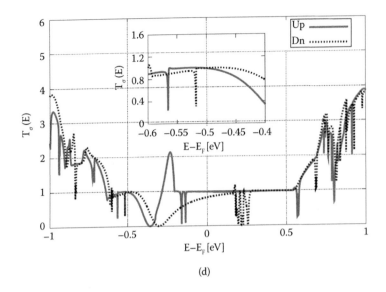

(d)

Figure 6.9 (Continued) Spin-dependent transmission function for N = 6 ZSiNR. Red (black) lines are for spin-up (down). (d) transmission for FM configuration with Al impurity in PC location, in inset $T_\sigma(E)$ for pristine ZSiNR in FM configuration. (From Zberecki, K., et al., *Phys. Rev. B*, 89, 165419, 2014a. With permission.)

of Zhang et al. (2014), where 100% bipolar spin-filtering behavior in the ZSiNRs with asymmetric edge hydrogenation has been obtained by the DFT + NEGF study.

6.4.4 GATED SILICENE NANORIBBONS

One of the most important issues from the application point of view is the influence of the gate potential on the transport properties of a nanostructure. In Section 3.3, we have seen that one of the silicene's key features is its buckled structure. The two sublattices are shifted and under an external field become inequivalent, so opening a gap in the bandstucture may become possible in the case of the 2D layer (unlike the graphene case). Theoretical study of the influence of a vertical electric field on a band gap of silicene monolayer has been undertaken in Ni et al. (2012). The results show that the gap may actually be opened with the width of 0.17 eV under the field of 1 V/Å and twice more if the silicene is inserted between two monolayers of BN. Let us now see how the properties of the 1-D silicene can be tuned by the external electric field.

Since the H-edge passivated ASiNRs are semiconductors, it would be interesting to investigate the properties of an ASiNR-based FET. This has been done by Li (2012) by *ab initio* methods, where the model of a two-probe FET has been proposed. The model consists of a previously optimized ASiNR connected to two ZSiNRs acting as electrodes. The gate voltage is applied to the armchair area. The structural model of the FET can be seen in Figure 6.10a.

The obtained transfer characteristics show that for the N=6 and N=9 ASiNR FETs, the perfect bipolar gate effects may be observed. The threshold voltage is estimated to be $V_{th} = \pm 1V$. Figure 6.10b shows the output characteristics of 6-ASiNR FET with L=9.89nm. The I(V) curve plotted for different V_{gate} is at first linear (for $V_{bias} < 0.1V$), then becomes constant ($V_{bias} > 0.1V$) for the $|V_{gate}| > |Vth|$. This current saturation together with calculated I_{on}/I_{off} ratio of over 10^6 suggests that this SiNR-based FET would be very interesting for applications . The high I_{on}/I_{off} ratio has been confirmed by Pan (2014), where the semi-hydrogenated ASiNR-based FET model has been analyzed. The results suggest that the spin-polarized current can be generated in this system for certain values of the gate voltage. Application of positive gate voltage causes shifting of the Fermi level to the valence bands of the minority spin, making the device HM, which gives the possibility of spin-filter behavior. The spin-filter efficiency of the device reaches 100% at Vg = 1.9 V, suggesting that semi-hydrogenated SiNRs may serve as a HM spintronic device. The spin-filter behavior has been also studied by Tsai et al. (2013). Using the Li Hamiltonian (Li 2012), they proposed and analyzed the SiNR-based device with spin-polarized output current, which can be switched electrically

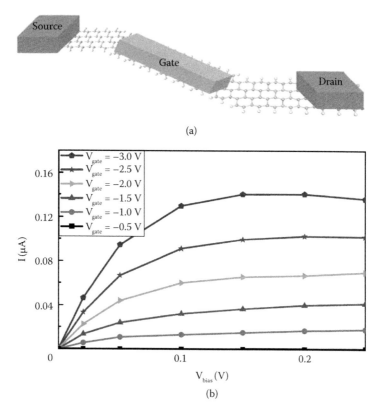

Figure 6.10 (a) Ball and stick model of a device—N = 6 ASiNR (gated) connected to N = 4 ASiNR electrodes. Yellow balls Si atoms, white balls h atoms. (b) I(V) characteristics of a FET for different gate voltages. (From Li, H., *Eur. Phys. J. B*, 85, 274, 2012. With permission.)

without switching external magnetic fields. The proposed spin filter would be highly efficient (nearly 100% spin polarization) and robust against edge imperfections. Using similar Hamiltonian with AF exchange field (Pournaghavi et al. 2016) has shown that ZSiNRs in the presence of gate voltage can also be used as spin-filter devices.

6.4.5 THERMOELECTRIC PROPERTIES

The thermoelectric properties of nanostructure are currently of great interest because of the possibility of heat to electrical energy conversion at nanoscale (Seebeck effect), which may be important for applications.

To study the thermoelectric phenomena, the linear response regime (Mahan 2000) is normally used. In this case, the electric I and heat I_Q currents flowing through the system from left to right can be related to the difference between the electric potentials of the left and right electrodes $\Delta V = V_L - V_R$ and the temperature difference, $\Delta T = T_L - T_R$ via the matrix equation:

$$
\begin{pmatrix} I \\ I_Q \end{pmatrix} = \begin{pmatrix} e^2 L_0 & \dfrac{e}{T} L_1 \\ eL_1 & \dfrac{1}{T} L_2 \end{pmatrix} \begin{pmatrix} \Delta V \\ \Delta T \end{pmatrix}
\tag{6.9}
$$

Here, e is the electron charge, L_n is defined as $L_n = -\dfrac{1}{h} \int dE\, T(E)(e - \mu)^n \dfrac{\partial f}{\partial E}$, $n = 0,1,2$ and $f(E)$ is the F–D distribution corresponding to chemical potential μ. $T(E)$ is the transmission function, which may be calculated with a variety of methods, but *ab initio* + NEGF formalism is mostly used. Having calculated $T(E)$, one can obtain all the moments L_n and from them the quantities, which may be used to analyze thermoelectric properties of the system. So, electrical conductance G can be given by $= e^2 L_0$, while the electronic contribution to the thermal conductance κ_e

$$\kappa_e = \frac{1}{T}\left(L_2 - \frac{L_1^2}{L_0}\right) \tag{6.10}$$

where T is the temperature. The thermopower $S = -\Delta V/\Delta T$ can be obtained from

$$S = -\frac{L_1}{|e|TL_0} \tag{6.11}$$

Another important quantity is thermoelectric efficiency, given by

$$ZT = \frac{S^2 GT}{\kappa} \tag{6.12}$$

where $\kappa = \kappa_e + \kappa_{ph}$ is the sum of electronic and phonon contributions to thermal conductance.

The assessment of thermoelectric properties of pristine SiNRs was made by Pan et al. (2012) and Zberecki et al. (2013). In the first work, the properties of both armchair and zigzag NRs have been calculated by use of the DFT + NEGF method for different widths and temperatures. Calculated band structures for ASiNRs show oscillations of the gap width as mentioned before. S obtained for ASiNRs is also oscillating, reaching a maximum value of about 850 μV/K for N = 7 and minimum for N = 8 (less than 50 μV/K). A similar tendency is present for ZT, where phonon contribution has been calculated with two methods—nonequilibrium molecular dynamic and *ab initio* with the use of perturbation theory. Again, the highest value, close to 3 (which is quite a high value compared to graphene in which ZT may be up to 0.8 for NRs of chevron type [Liang et al. 2012]), has been obtained for N = 7 and it oscillates as the width increases. Calculations performed in both papers for ZSiNRs predict that the ground state of the system will have AFM configuration. The calculations in both works are generally consistent, although the maximum values of S in both works differ by about a factor of 2 due to usage of different parametrizations of DFT exchange-correlation term (LDA vs. GGA).

Let us now analyze briefly the magnitude of S as a function of chemical potential μ (Figure 6.11a), using as an example N = 5 ZSiNR in AFM configuration (Zberecki et al. 2013).

The chemical potential μ can be changed in the vicinity of the Fermi level by p-type doping (μ<0) or n-type doping (μ>0) (Fermi level of undoped system corresponds to μ=0). As can be seen, for a μ close to zero the thermopower vanishes, since the currents (via electrons and holes) compensate each other. For μ>0, but inside the band gap, S is negative. It gains maximum absolute value when μ is at the distance of an order of several kT from the upper edge of the gap, because the hole current is blocked and the charge current is mediated by electrons. The electrons flow from left to right (the thermocurrent is flowing from right to left), so a positive voltage is needed to block the current and the thermopower is negative (cf. the definition). For μ<0, but inside the band gap, the current is dominated by holes, and the situation is reversed—the current is flowing from left to right, and now the S is positive. Thermopower is also temperature dependent. As can be seen from Figure 6.11b, the maxima of |S|(when μ is still inside the band gap) move toward the gap edges with decreasing T. Moreover, S almost vanishes in a wide region of μ between the two peaks in |S|, and this plateau of zero S value becomes broader with decreasing T. This behavior can be linked to the facts that the Fermi–Dirac distribution function has a quite long tail and the functions L1 and L2 vary differently with temperature (Equation 6.11).

In a situation when two spin channels can be treated as independent in the whole system (i.e., they are not mixed by spin-flip transitions), the temperature gradient may lead not only to charge accumulation at the ends of an open system, but also to spin accumulation, which means that it also gives rise to the spin voltage. This spin voltage can be observed, if the length of the sample is smaller than the spin-flip length. The above conditions are fulfilled for the SiNRs, so, for the FM configuration one can introduce the spin thermopower $S_s = -\dfrac{\Delta V_s}{\Delta T}$ which corresponds to a spin voltage V_s generated by a temperature gradient (spin Seebeck effect). The spin thermopower S_S calculated in Zberecki et al. (2013) for ZSiNRS for different widths and temperatures suggests that the spin thermoelectric effects may play a significant role in the pristine ribbons. Recently has been suggested (Fu et al. 2015), that ZSiNR may also be used to design a spin Seebeck diode that allows the thermal-induced spin current to flow only in one direction.

Modifications of the structure may significantly change the thermoelectric properties. Results in Zberecki et al. (2014a) suggest that in Al- and P-doped ZSiNRs, the spin thermopower can be considerably enhanced by the impurities. In Zberecki et al. (2015), the doping of ZSiNRs with Co atoms has been studied. The results show that the most stable configuration is the one with the Co atom in the edge position.

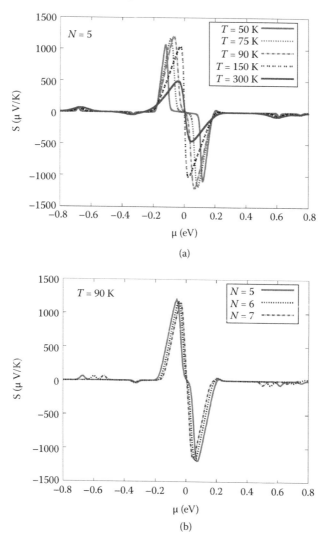

Figure 6.11 (a) Thermopower S for $N = 5$ ZSiNR in the AFM configuration (b) S for $N = 5$–7 ZSiNRs under temperature $T = 90$ K if the AFM configuration. (From Zberecki, K., et al., *Phys. Rev. B*, 88, 115404, 2013. With permission.)

In this case, the low-energy state is FM with relatively high-spin thermopower. The thermoelectric efficiency may be also considerably enhanced in the case of different edge passivation by hydrogen atoms (Zberecki et al. 2014b) as well as by introducing a vacancy (An et al. 2014) or gate voltage (Yan et al. 2013). Thermoelectric performance of defected SiNRs is studied in Zhao et al. (2016). The significant enhancement of thermoelectric efficiency ZT is obtained by blocking the phonons with the introduction of a large number of defects, which leads to the lowering of phonon thermal conductance κ_{ph} (see also Wierzbicki et al. (2013) for a similar discussion, on ZGNRs).

6.5 CONCLUSIONS

In this chapter, we reviewed recent progress in investigating properties of SiNRs. Theoretical results suggest that SiNRs may have very interesting properties and may be considered as building blocks of future electronic devices. Due to its buckled structure, silicene has unique electronic and transport properties. Also, calculations suggest that silicene may be very good thermoelectric material. Experimental results are also encouraging. Recently, huge progress in the growth and characterization of silicene on different metallic substrates has been achieved, not to mention the working silicene transistor device.

REFERENCES

A. Acun, B. Poelsema, H. J. W. Zandvliet, and R. van Gastel, The instability of silicene on Ag(111), *Appl. Phys. Lett.* 103, 263119 (2013).

R.-L. An, Vacancy effects on electric and thermoelectric properties of zigzag silicene nanoribbons, *J. Phys. Chem. C* 118, 21339–21346 (2014).

B. Aufray, B. Ealet, H. Jamgotchian, H. Maradj, J.-Y. Hoarau, and J.-P. Biberian, Growth of silicon nano-ribbons on Ag(110): State of the Art, In *Silicene—Structure, properties and applications*, Eds. M. J. S. Spencer and T. Morishita, Cham, Switzerland: Springer, 2016. pp. 183–202.

B. Aufray, A. Kara, S. V., H. Oughaddou, C. Leandri, B. Ealet, and G. Le Lay, Graphene-like silicon nanoribbons on Ag(110): A possible formation of silicene, *Appl. Phys. Lett.* 96, 183102 (2010).

R. Bernard et al., Growth of Si ultrathin films on silver surfaces: Evidence of an Ag(110) reconstruction induced by Si, *Phys. Rev B* 88, 121411(R) (2013).

S. Cahangirov, M. Topsakal, E. Akturk, H. Sahin, and S. Ciraci, Two- and one-dimensional honeycomb structures of silicon and germanium, *Phys. Rev. Lett.* 102, 236804 (2009).

C. Caroli, R. Combescot, P. Nozieres, and D. Saint-James, Direct calculation of the tunneling current, *J. Phys. C* 4, 916 (1971).

S. Datta, *Electronic Transport in Mesoscopic Systems*, New York: Cambridge University Press, 1995.

J. Deng, J. Z. Liu, and N. V. Medhekar, Enhanced lithium adsorption and diffusion on silicene nanoribbons, *RSC Adv.* 3, 20338 (2013).

X.Q. Deng, Z.H. Zhang, Z.Q. Fan, G.P. Tang, L. Sun, and C.X. Li, Spin-filtering and rectifying effects for Al-doped zigzag-edged silicene nanoribbons with asymmetric edge hydrogenation, *Org. Electron.* 32, 41–46 (2016).

P. De Padova, C. Quaresima, B. Olivieri, P. Perfetti, and G. Le Lay, Strong resistance of silicene nanoribbons towards oxidation, *J. Phys. D Appl. Phys.* 44, 312001 (2011).

P. De Padova et al., Mn-silicide nanostructures aligned on massively parallel silicon nanoribbons, *J. Phys. Condens Matter* 25, 014009 (2013).

Y.-J. Dong, X.-F. Wang, P. Vasilopoulos, M.-X. Zhai, and X.-M. Wu, Half-metallicity in aluminum-doped zigzag silicene nanoribbons, *J. Phys. D Appl. Phys.* 47, 105304 (2014).

M. Ezawa, A topological insulator and helical zero mode in silicene under an inhomogeneous electric field, *New J. Phys.* 14, 033003 (2012a).

M. Ezawa, Valley-polarized metals and quantum anomalous hall effect in silicene, *Phys. Rev. Lett.* 109, 055502 (2012b).

Y.-C. Fan, T.-H. Fang, and T.-H. Chen, Stress waves and characteristics of zigzag and armchair silicene nanoribbons, *Nanomaterials* 6, 120 (2016).

B. Feng, H. Li, S. Meng, L. Chen, and K. Wu, Structure and quantum well states in silicene nanoribbons on Ag(110), *Surf. Sci.* 645, 74–79 (2016).

H.-H. Fu, D.-D. Wu, L. Gu, M. Wu, and R. Wu, Design for a spin-Seebeck diode based on two-dimensional materials, *Phys. Rev. B* 92, 045418 (2015).

P. Hohenberg and W. Kohn, Inhomogeneous electron gas, *Phys. Rev.* 136, B864 (1965).

D. Jacob and J. J. Palacios, Critical comparison of electrode models in density functional theory based quantum transport calculations, *J. Chem. Phys.* 134, 044118 (2011).

J. Kang, F. Wu, and J. Li, Symmetry-dependent transport properties and magnetoresistance in zigzag silicene nanoribbons, *Appl. Phys. Lett.* 100, 233122 (2012).

A. Kara et al., Silicon nano-ribbons on Ag(110): A computational investigation, *J. Phys. Condens Matter* 22, 045004 (2010).

W. Kohn and L. J. Sham, Self-consistent equations including exchange and correlation effects, *Phys. Rev.* 140, A1133 (1965).

B. Lalmi et al., Epitaxial growth of a silicene sheet, *Appl. Phys. Lett.* 97, 223109 (2010).

M. Lan, G. Xiang, C. Zhang, and X. Zhang, Vacancy dependent structural, electronic, and magnetic properties of zigzag silicene nanoribbons:Co, *J. Appl. Phys.* 114, 163711 (2013).

R. Landauer, Spatial variation of currents and fields due to localized scatterers in metallic conduction, *IBM J. Res. Dev.* 1, 223–231 (1957).

C. Leandri, H. Oughaddou, B. Aufray, J.M. Gay, G. Le Lay, A. Ranguis, and Y. Garreau, Growth of Si nanostructures on Ag(001), *Surf. Sci.* 601, 262–267 (2007).

C. Leandri et al., Self-aligned silicon quantum wires on Ag(110), *Surf. Sci.* 574, L9–L15 (2005).

S. Lebegue and O. Eriksson, Electronic structure of two-dimensional crystals from ab initio theory, *Phys. Rev. B* 79, 115409 (2009).

H. Li, High performance silicene nanoribbon field effect transistors with current saturation, *Eur. Phys. J. B* 85, 274 (2012).

Z. Li, H. Qian, J. Wu, B.-L. Gu, and W. Duan, Role of symmetry in the transport properties of graphene nanoribbons under bias, *Phys. Rev. Lett.* 100, 206802 (2008).

Y. Li, Z. Zhou, P. Shen, and Z. Chen, Spin gapless semiconductor metal half-metal properties in nitrogen-doped zigzag graphene nanoribbons, *ACS Nano* 3(7), 1952–1958 (2009).

L. Liang, E. Cruz-Silva, E. Costa Girao, and V. Meunier, Enhanced thermoelectric figure of merit in assembled graphene nanoribbons, *Phys. Rev. B* 86, 115438 (2012).

C.-C. Liu, W. Feng, and Y. Yao, Low-energy effective Hamiltonian involving spin-orbit coupling in silicene and two-dimensional germanium and tin, *Phys. Rev. B* 84, 195430 (2011b).

C.-C. Liu, W. Feng, and Y. Yao, Quantum spin hall effect in silicene and two-dimensional germanium, *Phys. Rev. Lett.* 107, 076802 (2011a).

H.-X. Luan, C.-W. Zhang, F.-B. Zheng, and P.-J. Wang, First-principles study of the electronic properties of B/N atom doped silicene nanoribbons, *J. Phys. Chem. C* 117, 13620–13626 (2013).

L. Ma, J.-M. Zhang, K.-W. Xu, and V. Ji, Structural and electronic properties of substitutionally doped armchair silicene nanoribbons, *Physica B* 425, 66–71 (2013).

G. D. Mahan, *Many-Particle Physics*, 3rd ed., New York: Kluwer Academic/Plenum Publishers, 2000.

R. Martin, *Electronic structure*, Cambridge, UK: Cambridge University Press, 2004.

L. Masson, H. Sahaf, P. Amsalem, F. Dettoni, E. Moyen, N. Koch, and M. Hanbucken, Nanoscale Si template for the growth of self-organized one-dimensional nanostructures, *Appl. Surf. Sci.* 267, 192–195 (2013).

L. Michez et al., Magnetic properties of self-organized Co dimer nanolines on Si/Ag(110), *Beilstein J. Nanotechnol.* 6, 777–784 (2015).

Y. Miyamoto, K. Nakada, and M. Fujita, First-principles study of edge states of H-terminated graphitic ribbons, *Phys. Rev. B* 59, 9858 (1999).

K. Nakada, M. Fujita, G. Dresselhaus, and M. S. Dresselhaus, Edge state in graphene ribbons: Nanometer size effect and edge shape dependence, *Phys. Rev. B* 54(17), 954 (1996).

Z. Ni et al., Tunable bandgap in silicene and germanene, *Nano Lett.* 12, 113–118 (2012).

F. Pan, Gate-induced half-metallicity in semi-hydrogenated silicene, *Physica E* 56, 43–47 (2014).

L. Pan et al., Thermoelectric properties of armchair and zigzag silicene nanoribbons, *Phys. Chem. Chem. Phys.* 14, 13588–13593 (2012).

D. A. Papaconstantopoulos, *Handbook of the Band Structure of Elemental Solids*, New York: Springer, 2014.

N. Pournaghavi, M. Esmaeilzadeh, S. Ahmadi, and M. Farokhnezhad, Electrically controllable spin conductance of zigzag silicene nanoribbons in the presence of anti-ferromagnetic exchange field, *Solid State Comm.* 226, 33–38 (2016).

H. Sahaf, C. Leandri, E. Moyen, M. Mace, L. Masson, and M. Hanbucken, Growth of Co nanolines on self-assembled Si nanostripes, *EPL* 86, 28006 (2009).

E. Salomon, T. Angot, C. Thomas, J.-M. Layet, P. Palmgren, C.I. Nlebedim, and M. Gothelid, Etching of silicon nanowires on Ag(1 1 0) by atomic hydrogen, *Surf. Sci.* 603, 3350–3354 (2009).

E. Salomon and A. Kahn, One-dimensional organic nanostructures: A novel approach based on the selective adsorption of organic molecules on silicon nanowires, *Surf. Sci.* 602, L79–L83 (2008).

J. Slater and G. Koster, Simplified LCAO method for the periodic potential problem, *Phys. Rev.* 94, 1498 (1954).

J. M. Soler, E. Artacho, J. D. Gale, A. Garcia, J. Junquera, P. Ordejon, and D. Sanchez-Portal, The SIESTA method for ab initio order-N materials simulation, *J. Phys. Condens Matters* 14, 11 (2002).

Y.-L. Song, J.-M. Zhang, D.-B. Lu, and K.-W. Xu, Structural and electronic properties of a single C chain doped zigzag silicene nanoribbon, *Physica E* 53, 173–177 (2013a).

Y.-L. Song, Y. Zhang, J.-M. Zhang, and D.-B. Lu, Effects of the edge shape and the width on the structural and electronic properties of silicene nanoribbons, *Appl. Surf. Sci.* 256, 6313–6317 (2010).

Y.-L. Song, Y. Zhang, J.-M. Zhang, D.-B. Lu, and K.-W. Xu, First-principles study of the structural and electronic properties of armchair silicene nanoribbons with vacancies, *J. Mol. Struct.* 990, 75–78 (2011a).

Y.-L. Song, Y. Zhang, J.-M. Zhang, D.-B. Lu, and K.-W. Xu, Modulation of the electronic and magnetic properties of the silicene nanoribbons by a single C chain, *Eur. Phys. J. B* 79, 197–202 (2011b).

Y.-L. Song et al., Band-gap modulations of armchair silicene nanoribbons by transverse electric fields, *Eur. Phys. J. B* 86, 488 (2013b).

K. Takeda and K. Shiraishi, Theoretical possibility of stage corrugation in Si and Ge analogs of graphite, *Phys. Rev. B* 50, 14916 (1994).

L. Tao et al., Silicene field-effect transistors operating at room temperature, *Nat Nanotechnol* 10, 227–231 (2015).

M. Topsakal and S. Ciraci, Elastic and plastic deformation of graphene, silicene, and boron nitride honeycomb nanoribbons under uniaxial tension: A first-principles density-functional theory study, *Phys. Rev. B* 81, 024107 (2010).

M. R. Tchalala et al., Atomic structure of silicene nanoribbons on Ag(110), *J. Phys Conf. Ser.* 491, 012002, (2014).

M. R. Tchalala et al., Formation of one-dimensional self-assembled silicon nanoribbons on Au(110)-(2 × 1), *App. Phys. Lett.* 102, 083107 (2013).

W.-F. Tsai, C.-Y Huang, T.-R. Chang, H. Lin, H.-T. Jeng, and A. Bansil, Gated silicene as a tunable source of nearly 100% spin-polarized electrons, *Nat. Comm.* 4, 1500 (2013).

M. Wierzbicki, J. Barnaś, and R. Swirkowicz, Giant spin thermoelectric efficiency in ferromagnetic graphene nanoribbons with antidots, *Phys. Rev. B* 88, 235434 (2013).

C. Xu et al., Giant magnetoresistance in silicene nanoribbons, *Nanoscale* 4, 3111 (2012).

Y. Yan, H. Wu, F. Jiang, and H. Zhao, Enhanced thermopower of gated silicene, *Eur. Phys. J. B* 86, 457 (2013).

K. Zberecki, R. Swirkowicz, and J. Barnaś, Spin effects in thermoelectric properties of Al- and P-doped zigzag silicene nanoribbons, *Phys. Rev. B* 89, 165419 (2014a).

K. Zberecki, R. Swirkowicz, and J. Barnaś, Thermoelectric properties of zigzag silicene nanoribbons doped with Co impurity atoms, *JMMM* 393, 305–309 (2015).

K. Zberecki, M. Wierzbicki, J. Barnaś, and R. Swirkowicz, Thermoelectric effects in silicene nanoribbons, *Phys. Rev. B* 88, 115404 (2013).

K. Zberecki, M. Wierzbicki, J. Barnaś, and R. Swirkowicz, Enhanced thermoelectric efficiency in ferromagnetic silicene nanoribbons terminated with hydrogen atoms, *Phys. Chem. Chem. Phys.* 16, 12900 (2014b).

D. Zhang et al., Bipolar spin-filtering, rectifying and giant magnetoresistance effects in zigzag silicene nanoribbons with asymmetric edge hydrogenation, *Chem. Phys. Lett.* 616–617, 178–183 (2014).

Y.-C. Zhao and J. Ni, Spin-semiconducting properties in silicene nanoribbons, *Phys. Chem. Chem. Phys.* 16, 15477 (2014).

W. Zhao, Z. X. Guo, Y. Zhang, J. W. Ding, and X. J. Zheng, Enhanced thermoelectric performance of defected silicene nanoribbons, *Solid State Comm.* 227, 1–8 (2016).

F.-B. Zheng, C.-W. Zhang, P.-J. Wang, and S.-S. Li, Novel half-metal and spin gapless semiconductor properties in N-doped silicene nanoribbons, *J. Appl. Phys.* 113, 154302 (2013).

F.-B. Zheng, C.-W. Zhang, S-S. Yan, and F. Li, Novel electronic and magnetic properties in N or B doped silicene nanoribbons, *J. Mater. Chem. C* 1, 2735 (2013).

B. Zhou, B. Zhou, Y. Zeng, G. Zhou, and M. Duana, Tunable electronic and transport properties for ultranarrow armchair-edge silicene nanoribbons under spin–orbit coupling and perpendicular electric field, *Phys. Lett. A* 380, 282–287 (2016).

7

Hexagonal honeycomb silicon: Silicene

Xin Tan, Sean C. Smith, and Zhongfang Chen

Contents

7.1 INTRODUCTION

The successful exfoliation of graphene from graphite in 2004 has shown that it is possible to create stable, single and few-atom-thick layers of van der Waals materials (Novoselov et al. 2004), and these two-dimensional (2D) layered materials can exhibit fascinating and technologically useful properties different from their bulk counterparts. The most extensively studied 2D material is graphene, which is an extended honeycomb network of sp^2-hybridized carbon atoms, due to its exceptional electronic, optoelectronic, electrochemical, and biomedical applications (Geim and Novoselov 2007; Geim 2009). Following graphene, there are many 2D materials that have been explored, ranging from hexagonal boron nitride (h-BN), transition metal chalcogenides (MoO_3, MoS_2, $MoSe_2$, WS_2, WSe_2, etc.), MXenes, to monoelemental 2D semiconductors (silicene, germanene, stanene, borophene, phosphorene, etc.) (Butler et al. 2013; Koski and Cui 2013; Song et al. 2013; Tang and Zhou 2013; Xu et al. 2013; Miró et al. 2014; Gupta et al. 2015; Kou et al. 2015; Zhao et al. 2016).

Silicene, the silicon analogue and the bigger cousin of graphene, has recently attracted significant attention because of its one-atom-thick honeycomb structure and unique physical and chemical properties

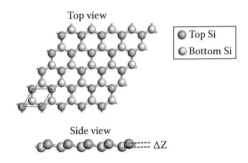

Figure 7.1 Top and side views of buckled crystal structure of freestanding silicene.

(Kara et al. 2012; Balendhran et al. 2015; Houssa et al. 2015; Oughaddou et al. 2015). A natural advantage of silicene over graphene is obviously its better compatibility with mature Si-based electronics. There are notable and important differences and similarities between silicene and graphene. For instance, the atomic arrangement of freestanding silicene follows the hexagonal honeycomb structure of graphene, but unlike C atoms in graphene, Si atoms prefer to adopt an sp^2/sp^3-hybridized state in silicene (Röthlisberger et al. 1994). In its most stable state, freestanding silicene was predicted to be slightly buckled with two sublattices displaced from each other in the out-of-plane direction, which is contrary to the perfectly planar graphene (Jose and Datta 2012). The density functional theory (DFT) computations revealed that the atomic configuration of freestanding silicene consists of top and bottom Si atoms, separated by a vertical distance ΔZ of about 0.44 Å, as shown in Figure 7.1 (Cahangirov et al. 2009). The lattice parameters of freestanding silicene are about $a = b = 3.87$ Å, $\alpha = \beta = 90°$, $\gamma = 120°$, and the Si–Si bond length is about 2.28 Å. The unique structural characteristics distinguished from graphene enhance the chemical reactivity on the silicene surface and allow tunable electronic states by chemical functionalization. In addition, the interlayer coupling in silicenes is very strong, which is significantly different from graphite that consists of weakly stacked graphene layers through dispersion forces.

In recent years, tremendous theoretical efforts have been devoted to assessing the fundamental properties of freestanding silicene, and various exciting, rich, physical and chemical properties have been predicted (Cahangirov et al. 2009; Chen et al. 2012; Feng et al. 2012). For example, the electronic structures of silicene and graphene are similar: both have a Dirac cone and linear electronic dispersion around K points (Cahangirov et al. 2009). Experimentally observable quantum spin Hall effect (QSHE) and quantum anomalous Hall effect (QAHE) were also predicted in silicene (Liu et al. 2011).

However, the experimental analyses are still in their infancy because of the big challenge to synthesize silicene. It was not until 2010 that silicene was proved to exist (Lalmi et al. 2010). The epitaxial silicene was successfully synthesized on Ag(111) (Lalmi et al. 2010; Lin et al. 2012), Ir (111) (Meng et al. 2013), ZrB_2(0001) (Fleurence et al. 2012), and $LaAlO_3$(111) substrates (Azzouz et al. 2014), but freestanding silicene has not been achieved so far. Note that the electronic properties of silicene could be modulated by the effects of substrate, defect, and interlayer coupling. Thus, the exploration of the fundamental science and practical synthesis routes associated with silicene remains far from being exhausted. As the quality of silicene-based materials and devices continues to improve, more breakthroughs can be expected.

In this chapter, we focus on the experimental and theoretical studies in silicene fabrication, properties, and applications. We first introduce the current experimental methods for fabrication silicene. Then, we discuss the fundamental electronic, mechanical, and thermal properties of silicene. Finally, the potential applications of silicene are proposed.

7.2 FABRICATION OF 2D SILICENE

Si and C are similar atoms: they lie next to each other in the same group on the periodic table and have similar electronic configurations. However, C atoms tend to adopt sp² hybridization over sp³ hybridization, but for Si the situation is the reverse (Röthlisberger et al. 1994). Thus, it is not energetically favorable to spontaneously form 2D silicene from Si atoms. Although theoretical investigations have shown that freestanding silicene has the potential to be realized, clear experimental evidence for the formation of 2D silicene honeycomb structures was achieved only recently through epitaxial growth of silicene as ribbons on Ag(110) (De Padova et al. 2011; Davila et al. 2012), and sheets on Ag(111) (Lalmi et al. 2010; Lin et al. 2012). Since then, the epitaxial 2D silicene has been successfully synthesized on other substrates such as Ir(111) (Meng et al. 2013), and ZrB$_2$(0001) (Fleurence et al. 2012), but the freestanding silicene has not been achieved so far. Recently, a soft synthetic method for fabricating graphitic-like silicon sheets in gram-scale quantities has been proposed by redox-assisted chemical exfoliation (RACE) of calcium disilicide (CaSi$_2$) (Tchalala et al. 2013; Ali and Tchalala 2014).

7.2.1 SILICENE EPITAXIAL GROWTH ON METAL SUBSTRATES

7.2.1.1 One-dimensional silicene nanoribbons growth on Ag(110)

A submonolayer deposition of silicon on Ag(110) under ultrahigh vacuum (UHV) conditions at room temperature leads to the formation of silicene NRs (Figure 7.2) (Leandri and Lay 2005). Scanning tunneling microscopy (STM) measurements revealed that these 1D silicene structures are all oriented along the (−110) direction of the Ag(110) surface with the same characteristic width of 16 Å (equal to $4a_{Ag(100)}$), while the lengths are highly disparate (1.5–30 nm). All these NRs show rounded protrusions with a height of 0.2 nm, and 5.77 Å periodicity along their lengths. Moreover, the structural ordering of the silicene NRs could be improved by annealing at 230°C for about 10 minutes: a significant elongation of the NRs well beyond 100 nm, while their widths are kept the same (Leandri and Lay 2005). The complex structures and properties of these 1D silicene NRs have been extensively investigated both by different experimental techniques (Lay et al. 2009; Kara et al. 2009; Ronci et al. 2010) and DFT calculations (He 2006; Kara et al. 2010; Ding and Wang 2014).

7.2.1.2 2D silicene growth on Ag(111)

On bare Ag(111) substrate, the deposition of 1ML of silicon results in the formation of a $\left(2\sqrt{3} \times 2\sqrt{3}\right)R30°$ 2D silicene sheet (Figure 7.3) (Lalmi et al. 2010). The atomic resolved STM images and low-energy electron diffraction (LEED) pattern revealed that the honeycomb structure presents two silicon sublattices occupying positions at different heights, with a height difference of about 0.02 nm (Lalmi et al. 2010). Following this study, other ordered phases of 2D silicene growth on Ag(111) substrate have been obtained by varying the substrate temperature during the growth, such as (4×4) and $\left(\sqrt{13} \times \sqrt{13}\right)R13.9°$ superstructures (Feng et al. 2012; Gao and Zhao 2012; Jamgotchian et al. 2012; Lin et al. 2012; Acun et al. 2013; Majzik et al. 2013). Detailed studies have shown that they correspond to different orientations of the silicene sheet relative to the Ag(111) surface (Majzik et al. 2013; Tchalala et al. 2014). On the other hand, several theoretical investigations have also been performed to study the different structures and properties of these silicenes on Ag(111) (Cahangirov et al. 2009; Liu et al. 2011; Kawahara et al. 2014; Mahatha et al. 2014;

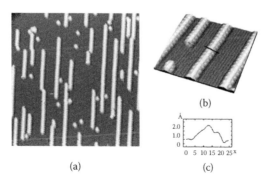

(a)　　　　(b)　　　　(c)

Figure 7.2 Topographic images of ~0.25 monolayer (ML) of Si deposited on Ag(110) at room temperature: (a) 42 × 42 nm² overview with Si NRs and nanodots, (b) 12.1 × 12.1 nm² zoom revealing the atomic rows of the bare substrate along the (−110) direction and the profile of the NRs, (c) height profile along the black line in (b). (Reprinted from Leandri, C., and Lay, G. L., 2005, *Surface Science*, 574, 9–15, Copyright 2005, with permission from Elsevier BV.)

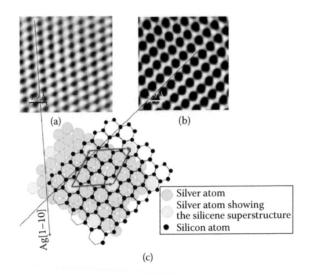

Figure 7.3 (a) Filled-state atomically resolved STM image of the clean Ag(111) surface, (b) filled-state atomically resolved STM image of the same sample (without any rotation) after deposition of 1M silicon, (c) proposed ball model of silicene on Ag(111) derived from both STM images (a) and (b) and from the observed $\left(2\sqrt{3} \times 2\sqrt{3}\right)R30°$ LEED pattern. (Reprinted with permission from Lalmi, B., et al., 2010, Epitaxial growth of a silicene sheet, *Appl. Phys. Lett.*, 97, 223109, Copyright 2010, American Institute of Physics.)

Pan et al. 2014). These DFT calculations showed that the epitaxial silicenes are slightly incommensurable with the Ag(111) substrate, which induce strongly corrugated silicenes compared to freestanding silicene. Moreover, these DFT calculations indicated that the electronic structures of these epitaxial silicenes are strongly affected by the substrate interaction compared to that of freestanding silicene.

7.2.1.3 2D silicene growth on Ir(111)

Iridium is another metal substrate for 2D silicene epitaxial growth. The deposition and annealing of 1ML of silicon on Ir(111) was shown to produce a $\left(\sqrt{7} \times \sqrt{7}\right)R19°$ superstructure, as characterized by both the LEED pattern and the STM images (Figure 7.4) (Meng et al. 2013). The STM images revealed the 2D silicene on Ir(111) exhibits the corrugation, and the height and periodicity of the protrusions are 0.83 and 7.2 Å, respectively. The subsequent detailed DFT calculations and simulated STM image showed that the structure of $\left(\sqrt{7} \times \sqrt{7}\right)R19°$ superstructure coincides with the $\left(\sqrt{3} \times \sqrt{3}\right)$ superlattice of 2D silicene.

7.2.2 SILICENE EPITAXIAL GROWTH ON NONMETALLIC SUBSTRATES

One of the key challenges for silicene applications is the fabrication of silicene on nonmetallic substrates, which could retain its intrinsic properties and avoid the strongly coupling effect between silicene and the metal substrate. The first nonmetallic material for silicene epitaxial growth is zirconium diboride (ZrB_2). The 2D silicene spontaneously forms when the Si atoms segregate on the (0001) surface of ZrB_2 thin films grown on an Si(111) substrate. The detailed STM measurements and DFT computations showed that this epitaxial silicene has a $\left(\sqrt{3} \times \sqrt{3}\right)R30°$ periodicity with respect to the (2 × 2) lattice of the ZrB_2 substrate (Fleurence et al. 2012). Subsequently, reflection high energy electron diffraction (RHEED) and X-ray photoelectron spectroscopy (XPS) investigations indicated that 2D silicene can also epitaxially grow on the insulating $LaAlO_3$(111) substrate at a temperature between 300°C and 500°C (Azzouz et al. 2014). Besides, 2D silicene epitaxy on clean MoS_2(0002) crystal surface at 200°C has also been observed in the STM topography and RHEED pattern (Chiappe et al. 2014).

7.2.3 CHEMICAL SYNTHESIS OF SILICENE

The chemical processes provide an alternative route to synthesis of silicene in gram-scale quantities. However, the synthesis of silicene without using conventional vacuum processes and vapor deposition is rather challenging.

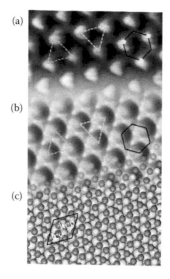

Figure 7.4 (a) Zoomed-in STM image of the epitaxial silicene on Ir(111) substrate. Besides the brightest protrusions, two other regions showing different contrast are indicated by the upward and downward triangles. The honeycomb feature is indicated by the black hexagon. (b) Simulated STM image, showing features identical with the experimental results in the same triangles and hexagons. (c) Top view of the relaxed atomic model of the $(\sqrt{3} \times \sqrt{3})$silicene/$(\sqrt{7} \times \sqrt{7})$Ir(111) configuration. (Reprinted with permission from Meng, L., et al., 2013. 13, 685–690. Copyright 2013 American Chemical Society.)

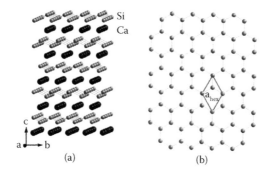

Figure 7.5 (a) Crystal structure of $CaSi_2$ (cell parameters: $a = 3.855$ Å and $c = 30.60$ Å with space group R-3m). The large circles represent Ca atoms and the small circles represent Si atoms. (b) Silicon sheet in $CaSi_2$ crystal structure. (The source of the material Ali, M.A., and Tchalala, M.R., 2014, *J. Phys. Conf. Ser.*, 491, 012009, IOP Publishing, is acknowledged.)

Recently, a soft synthetic method for large-scale fabrication of silicene was developed by RACE of calcium disilicide ($CaSi_2$) (Tchalala et al. 2013; Ali and Tchalala 2014). Detailed XPS, transmission electron microscopy (TEM) and energy-dispersive X-ray spectroscopy (XEDS) revealed that the obtained crystalline silicon sheets show a 2D hexagonal graphitic structure (Figure 7.5).

7.3 FUNDAMENTAL PROPERTIES OF SILICENE

7.3.1 ELECTRONIC PROPERTIES

7.3.1.1 Electronic band structure

Many theoretical investigations have predicted the possible existence and properties of silicene. By mean of DFT computations on the structure and phonon modes, as well as finite temperature molecular dynamics, Cahangirov et al. (2009) predicted that silicon can have stable, one-atom-thick, low-buckled, honeycomb

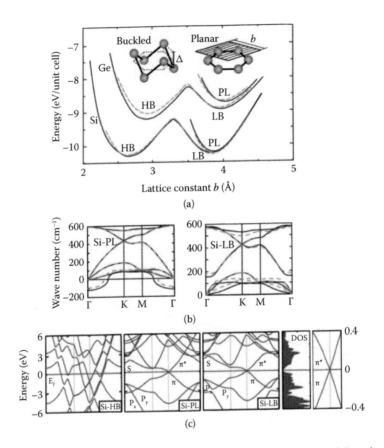

Figure 7.6 (a) The variation of the energy versus hexagonal lattice constant of 2D Si and Ge calculated for various honeycomb structures. HB = high-buckled, PL = planar, and LB = low-buckled. (b) Phonon dispersion curves obtained by force-constant and linear response theory are presented by black (dark) and dashed green (dashed light) curves, respectively. (c) The silicene band structures for HB, PL, and LB structures. The density of states for LB structure is also presented. (Reprinted with permission from Cahangirov, S., et al., 2009, *Phys. Rev. Lett.*, 102, 236804. Copyright 2009 by the American Physical Society.)

structures, which should be dynamically stable based on the computed phonons (Figure 7.6). Similar to graphene, the electronic structure of freestanding silicene has a Dirac cone, and its charge carriers can behave like a massless Dirac fermion due to their π and π* bands, which are crossed linearly at the Fermi level. Experimental evidences of silicene on Ag substrate showed the linear energy–momentum dispersion and a large Fermi velocity (1.2×10^6 m/s), which prove that quasiparticles in silicene behave as massless Dirac fermions (Chen et al. 2012; Feng et al. 2012). However, in such a case, it is believed that the electronic properties of the silicene are substantially affected by the metal substrate.

7.3.1.2 Quantum spin Hall effect

The QSH effect is caused by the spin–orbit coupling (SOC) opened energy gap at the Dirac point, which facilitates the transition of 2D material from semimetallic to a QSH insulator (Lin et al. 2013). The existence of the QSH effect was first proposed by Kane et al. in graphene (Kane and Mele 2005). However, graphene has extremely weak SOC, which means it is very unlikely to support a QSH state at temperatures achievable with today's technologies. So far, 2D HgTe/CdTe semiconductor quantum wells are the only materials able to demonstrate the QSH effect in a realistic system (Bernevig et al. 2006; König et al. 2007). Since Si atoms have greater intrinsic SOC strength than C atoms, recent DFT computations showed that an appreciable band gap of 1.55 meV (in comparison to that of graphene) can be opened at the Dirac point (Figure 7.7), which results in a significant QSH effect in an experimentally accessible low temperature regime (Liu et al. 2011). This property makes silicene particularly interesting for applications as QSH-effect devices.

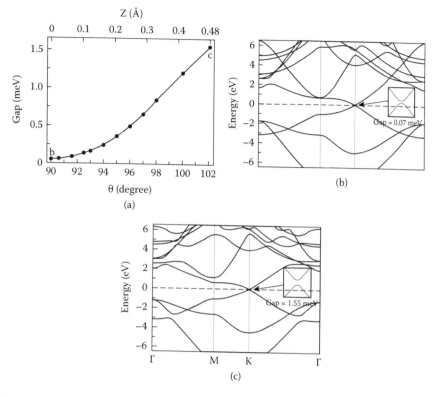

Figure 7.7 The adiabatic evolution of the gap and the calculated relativistic band structure of silicene. (a) The evolution of the gap opened by SOC for the π orbital at the Dirac point K from the planar honeycomb geometry to the low-buckled honeycomb geometry with keeping the Si–Si bond length constant. The top and bottom abscissas correspond to the difference of vertical height between A sublattice and B sublattice and the θ angle aforementioned, respectively, during evolution. (b) and (c) are the relativistic band structures with the corresponding geometries in (a). (Reprinted with permission from Liu, C.C., et al., 2011, *Phys. Rev. Lett.*, 107, 076802. Copyright 2011 by the American Physical Society.)

7.3.1.3 Tunability of electronic properties

Chemical functionalization, such as hydrogenation, fluorination, and surface adsorption of dopants, has been proposed as an effective method to tune the electronic properties of silicene.

DFT computations showed that silicanes, the hydrogenated derivative of 2D silicene, have an indirect gap of 3.8–4.0 eV for a chair-like structure and a direct gap of 2.9–3.3 eV for a boat-like structure (Figure 7.8) (Houssa et al. 2011). Thus, these materials are potentially interesting for optoelectronic applications in the blue/violet spectral range. The electronic properties of silicene with halogen functionalization (F, Cl, or Br) were also investigated, and the resulting gap value (1~2 eV) was predicted to be smaller than silicane (Lu et al. 2009; Gao et al. 2012). Besides, Wang et al. investigated the fully oxidized silicene structure (Si:O = 1:1) based on DFT calculations, and found that the zigzag ether-like conformation (z-sSiO) is energetically the most favorable (Wang and Ding 2013a). After oxidation, the semimetallic silicene is transformed into semiconductors with a narrow direct band gap, and the calculated charge mobility of z-sSiO is in the order of 10^4 cm²/Vs, which endows this nanostructure with potential applications in nanoelectronics and devices.

The electronic structures of silicene can also be tuned by other methods, such as the electric field (E-field) and the strain. By using DFT computations, Drummond et al. (2012) predicted that the vertical E-field generates a tunable gap in the Dirac-type electronic structure of silicene, and Ni et al. (2012) also revealed that the band gap opens in semimetallic silicene by when vertical E-field is applied, and the size of the band gap increases linearly with the E-field strength. Besides, Yan et al. (2015) studied the effects of small biaxial strain (|ε| ≤ 5%) and a vertical E-field on the electronic and

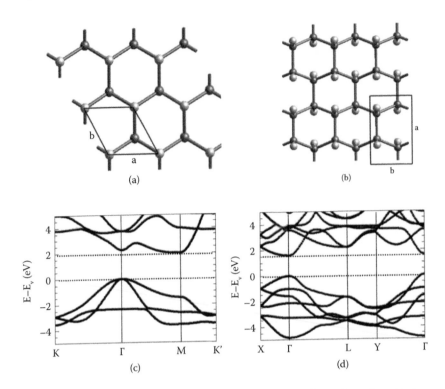

Figure 7.8 Two different atomic configurations of silicane: (a) Chair-like configuration, where H atoms are alternating on both sides of the plane, and (b) boat-like configuration, where the H atoms are alternating in pairs. Si and H atoms are represented by gray and white spheres, respectively. The corresponding energy band structures, calculated using local-density approximation functional, for (c) chair-like silicane, and (d) boat-like silicane. (Reprinted with permission from Houssa, M., et al., *Appl. Phys. Lett.*, 2011, 98, 223107. Copyright 2011 by the American Physical Society.)

phonon properties of silicene (Figure 7.9). The conduction bands of silicene at Γ and M dramatically change under a small biaxial strain, while the E-field mainly affects the band dispersions near the Γ and opens a small band gap at the K point in silicene. In addition, they also found that the field-induced gap opening in silicene could be enhanced by a compressive strain while mitigated by a tensile strain, indicating the possibility to tune the electronic properties of silicene by combining mechanical strain and an E-field.

7.3.2 SURFACE CHEMICAL REACTIVITY

Since Si atoms tend to adopt sp^3 hybridization compared to C atoms, the chemical reactivity on a silicene surface is significantly enhanced. Recently, the stability of silicene in O_2 was studied by DFT computations (Figure 7.10a), and the results showed that the O_2 molecule easily dissociates on the silicene surface from both thermodynamic and kinetic points of view, which is dramatically different from the case of graphene (Liu et al. 2014a). Moreover, the dissociation processes of H_2 on silicene were also investigated by DFT calculations. Though thermodynamically the H_2 dissociation releases an energy of about 0.25 eV, this reaction has a high activation energy barrier of about 1.75 eV (Figure 7.10b) (Liu et al. 2014b; Wu et al. 2014). However, compared with the H_2 dissociation barrier on graphene (Figure 7.10c), the energy barrier of H_2 dissociation on silicene decreases significantly (McKay et al. 2010; Tozzini and Pellegrini 2013).

7.3.3 MECHANICAL PROPERTIES

The mechanical properties of silicene are of utmost importance for its potential applications in miniaturized devices, and a few theoretical studies focused on this issue. Hu et al. (2013) reported that silicene has a

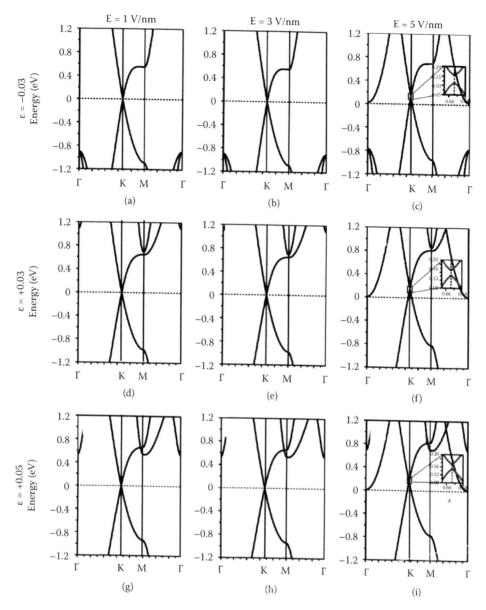

Figure 7.9 Band structures of silicene under various E-fields for typical strains. ε = −0.03 with (a) E = 1V/nm, (b) 3V/nm, and (c) 5V/nm. ε = +0.03 with (d) E = 1V/nm (e) 3V/nm, and (f) 5V/nm. ε = +0.05 with (g) E = 1V/nm, (h) 3V/nm, and (i) 5V/nm. Insets of (c), (f), and (i) show the zoom-in of the band gap opening near the K point. (Reprinted with permission from Yan, J.A., et al., 2015, 91, 245403. Copyright 2015 American Chemical Society.)

superior mechanical flexibility compared with graphene. Poisson's ratio, known as the coefficient of expansion on the transverse axial, of silicene increases remarkably, while the Poisson's ratio of graphene decreases with an increase in the uniaxial stretching strain (Figure 7.11). This difference indicates that silicene is much more flexible for the deformation in the transverse direction relative to graphene. Manjanath et al. (2014) investigated the in-plane stiffness of pristine silicene and transition metal-doped (such as Ni, Fe, and Co) silicene and found that the flat planar structures have higher stiffness values than the buckled structures for both pristine and doped silicenes, which can be explained by the fact that the flat planar structure has much stronger sp^2 hybridization than the buckled structure (sp^2-sp^3 hybridization). Moreover, they also found that transition metal doping can increase the in-plane stiffness in comparison to that of the pristine silicene.

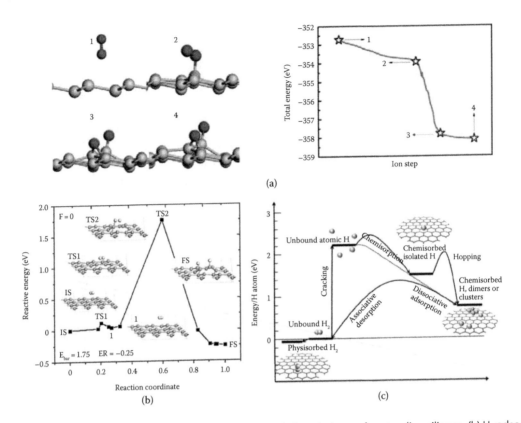

(a)

(b)

(c)

Figure 7.10 The diagrammatic sketch of (a) O_2 adsorption and dissociation on freestanding silicene, (b) H_2 adsorption and dissociation on freestanding silicene, and (c) H_2 adsorption and dissociation on freestanding graphene. Note that the energy in (c) is in eV per H atom, that is, to obtain the values per H_2 each energy level and barrier value must be doubled. (The source of the material Liu, G., et al., 2014a, *J. Phys. Condens Matter*, 26, 355007, IOP Publishing, is acknowledged; Wu, W., et al., 2014, *Phys. Chem. Chem. Phys.*, 16, 16588–16594, reproduced by permission of The Royal Society of Chemistry; Tozzini, V., and Pellegrini, V., 2013, *Phys. Chem. Chem. Phys.*, 15, 80–89, reproduced by permission of The Royal Society of Chemistry.)

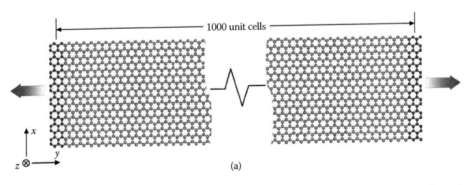

(a)

Figure 7.11 (a) Snapshot of a typical zigzag silicene sheet used as model system in the molecular dynamic simulations. The x axis (transverse direction) is along the armchair edge; the y axis (longitudinal direction) is along the zigzag edge, and the uniaxial tension along the y direction is indicated by blue arrows; the z axis is the out-of-plane flexural direction.

(Continued)

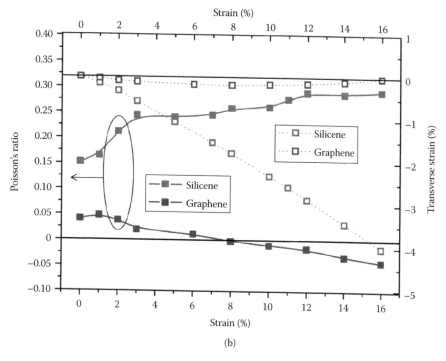

Figure 7.11 (Continued) (b) Comparison of Poisson's ratio (left axis) and transverse strain (right axis) as a function of uniaxial stretching strain between single-layer silicene and graphene sheet. (Reprinted with permission from Hu, M., et al., 2013, *Phys. Rev. B*, 87, 195417. Copyright 2013 American Chemical Society.)

Wang et al. examined the strain-induced self-doping phenomenon in silicene (Wang and Ding 2013b) and found that the Dirac point is shifted below and above the Fermi level under the compressive and tensile strains, respectively, which means that the compressive and tensile strains can introduce n-type and p-type doping to silicene.

7.3.4 THERMAL AND THERMOELECTRIC PROPERTIES

Thermal conductivity is the property of a material to conduct heat. Zhang et al. (2014) studied the thermal conductivity of silicene using equilibrium and nonequilibrium molecular dynamics simulations and anharmonic lattice dynamics calculations with an optimized Stillinger–Weber potential. They predicted that silicene has an extremely low thermal conductivity and a short phonon mean-free path in comparison to graphene, which suggests silicene as a potential candidate for high-efficiency thermoelectric materials. Moreover, Hu et al. (2013) discovered the anomalous effect of strain on the thermal conductivity of silicene. Contrary to its counterpart of graphene, the thermal conductivity of silicene and silicene nanoribbons increases significantly with applied tensile strain.

Recently, the thermoelectric property of silicene was also investigated by DFT calculations (Yang et al. 2014). The thermoelectric effect is the direct conversion of temperature differences to electric voltage and vice versa, and the efficiency of the thermoelectric conversion is characterized by a dimensionless parameter, called the figure of merit:

$$ZT = \frac{\sigma S^2 T}{\kappa}$$

where σ is the electric conductance, S is the Seebeck coefficient, T is the absolute temperature, and $\kappa = \kappa_e + \kappa_p$ is the total thermal conductance that is usually split into electron and phonon contributions, respectively (Yang et al. 2014). The calculations predicted that the figure of merit of silicene is quite high, ranging from 1 to 2 at room temperature.

Low-dimensional structures

7.4 APPLICATIONS OF SILICENE

7.4.1 FIELD-EFFECT TRANSISTOR

Silicene has the unique band structure with a "massless Dirac cone" and the surface of silicene is sensitive, which offers the tunable electronic states by chemical functionalization, external fields, and interface interactions. Moreover, silicene is considered as a feasible 2D nanomaterial beyond graphene because of its better compatibility with ubiquitous silicon semiconductor technology. All these features make silicene promising for use in high-speed electronic devices.

Recently, Tao et al. (2015) reported the silicene field-effect transistor (FET) operating at room temperature, which is the first proof-of-concept silicene device that is in agreement with theoretical predictions of Dirac-like ambipolar charge transport. The fabrication process of silicene FET includes four key steps (Figure 7.12a): (1) epitaxial growth of silicene on crystallized Ag(111) thin film, (2) *in situ* Al_2O_3 capping, (3) encapsulated delamination transfer of silicene, and (4) native contact electrode formation to enable back-gated silicene transistors. The observed linear relationship between drain current output (I_d) and drain voltage (V_d) in silicene FETs with native Ag electrodes (Figure 7.12c) indicates an ohmic contact under ambient conditions. The extracted low residual carrier density and high gate modulation compared with graphene suggest a small band gap opening in the experimental silicene FET devices. Moreover, the measured carrier mobility of silicene FET is ~100 $cm^2 V^{-1} s^{-1}$ (Figure 7.12d), which is within the estimated range of 10–1,000 $cm^2 V^{-1} s^{-1}$ for supported silicene from recent theoretical calculations (Li et al. 2013; Wang et al. 2013).

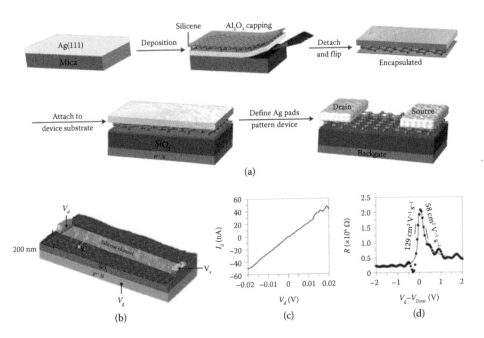

Figure 7.12 (a) Schematics of silicene-encapsulated delamination with native electrode (SEDNE) process. (b) Three-dimensional rendering of AFM image on a silicene FET on 90-nm-thick SiO_2/p^{++} Si substrate, including the channel (false-colored for visual guide) and source/drain contacts (~100-nm-thick) defined in native Ag film. V_g, V_s, and V_d are the gate, source, and drain voltages, respectively, in electrical measurements. (c) Low-field linear I_d versus V_d response at $V_g = 0$. (d) R versus ($V_g - V_{dirac}$) of silicene device. Measured transfer characteristics (dots) are in good agreement with a widely used ambipolar diffusive transport model (line) (Kim et al. 2009), which yields extracted low-field hole and electron carrier mobilities of 129 and 99 $cm^2 V^{-1} s^{-1}$, with similar residual carrier concentration of ~3–7 × 10⁹ cm^{-2}, more than an order of magnitude lower than in graphene transistors. (Reprinted by permission from Macmillan Publishers Ltd., Tao, L., et al., 2015, *Nat. Nanotechnol.*, 10, 227–231, copyright 2015.)

7.4.2 LITHIUM-ION BATTERIES (LIBs)

Lithium-ion batteries (LIBs) are ubiquitously used in portable and telecommunication electronic devices and are also promising for electric vehicles and electric grid applications. The most commonly used anode in current commercial batteries is graphite (Jeong et al. 2011). Recently, several theoretical studies suggested that silicene is a better anode material that can replace graphene for higher capacity and better cycling performance of LIBs.

Tritsaris et al. (2013) investigated the adsorption and diffusion of lithium on single-layer (SL) and double-layer (DL) silicene for Li-ion storage, and they predicted the significantly high charge capacity of 954 and 715 mAh/g for the SL and DL silicene, respectively, which are 2–3 times higher than for graphite (372 mAh/g) (Figure 7.13a and b). Moreover, the barriers of Li diffusion on silicene are relatively low, typically less than 0.6 eV (Figure 7.13c). These theoretical results indicate that silicene could serve as a promising high capacity and fast charge/discharge rate anode material for LIBs.

Considering the low defect formation energy for silicene compared with graphene, silicene is more likely to contain defects. Using DFT computations, Setiadi et al. (2013) calculated the Li-ion adsorption and diffusion on silicene with defects and found that the porous silicene also has large Li adsorption energy and high Li mobility, suggesting porous silicene is also a suitable material for use in LIBs.

Besides, Tan et al. (2014) studied the possibility of utilization of B-doped silicene (BSi_3) as high-capacity anode material for LIBs and predicted that the metallic BSi_3 silicene has good electronic conductivity, very high theoretical charge capacity, fast Li diffusion, and low open circuit voltage.

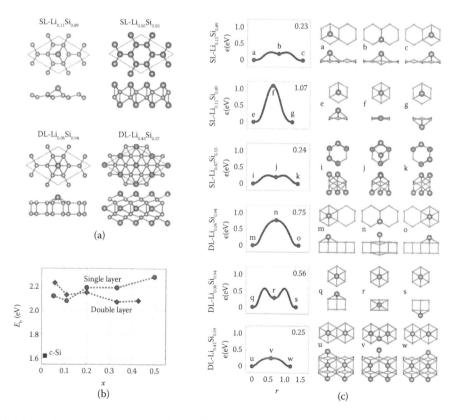

Figure 7.13 (a) Atomic structures corresponding to lithiated single-layer (SL) and double-layer (DL) silicene at low and high Li contents. (b) Binding energy, E_b, of Li as a function of Li content, x, for SL (red) and DL (green) silicene, Li_xSi_{1-x}. (c) Pathways for the diffusion of atomic Li on SL and DL silicene at low ($x = 0.11$) and high ($x > 0.40$) Li content. (Reprinted with permission from Tritsaris, G.A., et al., 2013, 13, 2258–2263. Copyright 2013 American Chemical Society.)

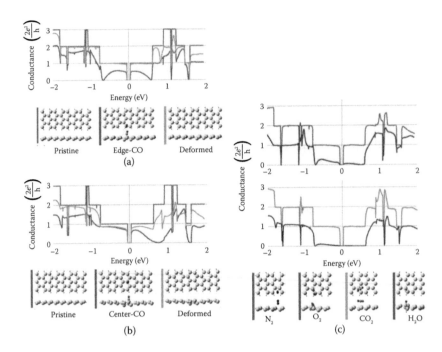

Figure 7.14 (a) and (b), Quantum conductance of pristine silicene nanoribbon (blue), nanoribbon with CO adsorption (red), and deformed nanoribbon with CO removed (green). (c) Quantum conductance modulation resulting from environmental gas molecule adsorption on nanoribbon: N_2 (blue), O_2 (red), CO_2 (green), and H_2O (purple). (Reprinted with kind permission from Springer Science+Business Media: Osborn, T.H., and Farajian, A.A., 2014, *Nano Res.*, 7, 945–952, Copyright 2014.)

7.4.3 GAS SENSORS

Many 2D materials, such as graphene, MoS_2, phosphorene, etc., have shown great promise for gas-sensing applications due to their large specific surface area, low Johnson noise, and sensitivity of electronic properties to the changes in the surroundings. Recently, Osborn and Farajian (2014) investigated the effect of CO, N_2, O_2, CO_2, and H_2O adsorption on the quantum conduction of pristine silicene nanoribbons and found that the adsorption of CO, O_2, and H_2O cause the detectable changes in conductance of silicene nanoribbons, while the adsorption of N_2 and CO_2 essentially do not affect conductance (Figure 7.14). The origin of this modification is the charge transfer from CO, O_2, or H_2O to the silicene nanoribbon and the deformation induced by these molecules' chemisorption. Moreover, the moderate binding energies of CO on silicene nanoribbons provide an optimal mix of high detectability and recoverability, thus the pristine silicene nanoribbons are considered as promising carbon monoxide nanosensors with molecular resolution. Similarly, Osborn and Farajian (2014) also considered the silver-contact-coupled sensors, and found that the short silicene segments (~1 nm) do not exhibit any systematic change in conductance due to the effects of silver contacts in close proximity, which reveals the length effects of silicene nanoribbons for sensor applications.

7.5 CONCLUSION

The recent discovery of graphene has initiated tremendous interest in exploring graphene-like materials. Silicene, the silicon analogue of graphene, is a one-atom-thick silicon sheet arranged in a 2D honeycomb lattice. Although theoretical investigations have shown that freestanding silicene has the potential to be realized, freestanding silicene has not been achieved experimentally so far. Nevertheless, the epitaxial silicene has been successfully synthesized on Ag(111), Ir(111), ZrB_2(0001), $LaAlO_3$(111), and MoS_2(0002) substrates.

There are notable and important differences and similarities between silicene and graphene. For example, both silicene and graphene have honeycomb structures, while silicene prefers to be slightly buckled with two sublattices displaced from each other in the out-of-plane direction, which is contrary to the perfectly planar graphene. The unique structural characteristics of silicene distinguished from graphene enhance its chemical reactivity on the surface, and allow tunable electronic states by chemical functionalizations. Besides, the electronic structures of silicene and graphene are similar: both have a Dirac cone and linear electronic dispersion around K points. However, Si atoms have greater intrinsic SOC strength than C atoms, accordingly, an appreciable band gap of 1.55 meV can be opened at the Dirac point for silicene, which leads to a significant QSH effect in an experimentally accessible low-temperature regime. All these fascinating and technologically useful properties of silicene and the natural advantage of silicene's better compatibility with mature Si-based electronics, make silicene particularly interesting for applications as FET, LIBs, gas sensors, etc.

However, the fundamental investigations and technological applications of silicene are still in their infancy. We strongly believe that deeper understandings can be obtained by further theoretical studies, and the quality of silicene-based materials and devices will continue to improve. No doubt, more breakthroughs for silicene can be achieved in the near future.

REFERENCES

Acun A, Poelsema B, Zandvliet HJW, van Gastel R (2013). The instability of silicene on Ag(1 1 1). *Appl Phys Lett* 103:263119.

Ali MA, Tchalala MR (2014). Chemical synthesis of silicon nanosheets from layered calcium disilicide. *J Phys Conf Ser* 491:012009.

Azzouz CB, Akremi A, Derivaz M, Bischoff J-L, Zanouni M, Dentel D (2014). Two dimensional Si layer epitaxied on LaAlO$_3$(1 1 1) substrate: RHEED and XPS investigations. *J Phys Conf Ser* 491:012009.

Balendhran S, Walia S, Nili H, Sriram S, Bhaskaran M (2015). Elemental analogues of graphene: Silicene, germanene, stanene, and phosphorene. *Small* 11:640–652.

Bernevig BA, Hughes TL, Zhang S-C (2006). Quantum spin hall effect and topological phase transition in HgTe quantum wells. *Science* 314:1757–1761.

Butler SZ, Hollen SM, et al. (2013). Progress, challenges, and opportunities in two-dimensional materials beyond graphene. *ACS Nano* 7:2898–2926.

Cahangirov S, Topsakal M, Aktürk E, Şahin H, Ciraci S (2009). Two- and one-dimensional honeycomb structures of silicon and germanium. *Phys Rev Lett* 102:236804.

Chen L, Liu C-C, et al. (2012). Evidence for dirac fermions in a honeycomb lattice based on silicon. *Phys Rev Lett* 109:056804.

Chiappe D, Scalise E, et al. (2014). Two-dimensional Si nanosheets with local hexagonal structure on a MoS$_2$ surface. *Adv Mater* 26:2096–2101.

Davila ME, Marele A, Padova PD, Montero I, Hennies F, Pietzsch A, Shariati MN, Gomez-Rodrıguez JM, Lay GL (2012). Comparative structural and electronic studies of hydrogen interaction with isolated versus ordered silicon nanoribbons grown on Ag(110). *Nanotechnology* 23:385703.

De Padova P, Quaresima C, Olivieri B, Perfetti P, Lay GL (2011). Strong resistance of silicene nanoribbons towards oxidation. *J Phys D Appl Phys* 44:312001.

Ding Y, Wang Y (2014). Electronic structures of reconstructed zigzag silicene NRs. *Appl Phys Lett* 104:083111.

Drummond ND, Zolyomi V, Fal'ko VI (2012). Electrically tunable band gap in silicene. *Phys Rev B* 85:075423.

Feng B, Ding Z, et al. (2012). Evidence of silicene in honeycomb structures of silicon on Ag(1 1 1). *Nano Lett* 12:3507–3511.

Fleurence A, Friedlein R, Ozaki T, Kawai H, Wang Y, Yamada-Takamura Y (2012). Experimental evidence for epitaxial silicene on diboride thin films. *Phys Rev Lett* 108:245501

Gao J, Zhao J (2012). Initial geometries, interaction mechanism and high stability of silicene on Ag(1 1 1) surface. *Sci Rep* 2:861.

Gao N, Zheng WT, Jiang Q (2012). Density functional theory calculations for two-dimensional silicene with halogen functionalization. *Phys Chem Chem Phys* 14:257–261.

Geim AK (2009). Graphene: Status and prospects. *Science* 324:1530–1534.

Geim AK, Novoselov KS (2007). The rise of graphene. *Nat Mater* 6:183–191.

Gupta A, Sakthivel T, Seal S (2015). Recent development in 2D materials beyond graphene. *Prog Mater Sci* 73:44–126.

He GM (2006). Atomic structure of Si nanowires on Ag(1 1 0): A density-functional theory study. *Phys Rev B* 73:035311.

Houssa M, Dimoulas A, Molle A (2015). Silicene: A review of recent experimental and theoretical investigations. *J Phys Condens Matter* 27:253002.

Houssa M, Scalise E, Sankaran K, Pourtois G, Afanas'ev VV, Stesmans A (2011). Electronic properties of hydrogenated silicene and germanene. *Appl Phys Lett* 98:223107.

Hu M, Zhang X, Poulikakos D (2013). Anomalous thermal response of silicene to uniaxial stretching. *Phys Rev B* 87:195417.

Jamgotchian H, Colignon Y, et al. (2012). Growth of silicene layers on Ag(1 1 1): Unexpected effect of the substrate temperature. *J Phys Condens Matter* 24:172001.

Jeong G, Kim Y-U, Kim H, Kim Y-J, Sohn H-J (2011). Prospective materials and applications for Li secondary batteries. *Energ Environ Sci* 4:1986–2002.

Jose D, Datta A (2012). An understanding of the buckling distortions in silicene. *J Phys Chem C* 116:24639–24648.

Kane CL, Mele EJ (2005). Quantum spin hall effect in graphene. *Phys Rev Lett* 95:226801.

Kara A, Enriquez H, et al. (2012). A review on silicene—New candidate for electronics. *Surf Sci Rep* 67:1–18.

Kara A, Léandri C, et al. (2009). Physics of silicene stripes. *J Supercond Nov Magn* 22:259–263.

Kara A, Vizzini S, et al. (2010). Silicon nano-ribbons on Ag(1 1 0): A computational investigation. *J Phys Condens Matter* 22:045004.

Kawahara K, Shirasawa T, et al. (2014). Determination of atomic positions in silicene on Ag(1 1 1) by low energy electron diffraction. *Surf Sci* 623:25–28.

Kim S, Nah J, et al. (2009). Realization of a high mobility dual-gated graphene field-effect transistor with Al_2O_3 dielectric. *Appl Phys Lett* 94:062107.

König M, Wiedmann S, et al. (2007). Quantum spin hall insulator state in HgTe quantum wells. *Science* 318:766–770.

Koski KJ, Cui Y (2013). The new skinny in two-dimensional nanomaterials. *ACS Nano* 7:3739–3743.

Kou L, Chen C, Smith SC (2015). Phosphorene: Fabrication, properties, and applications. *J Phys Chem Lett* 6:2794–2805.

Lalmi B, Oughaddou H, et al. (2010). Epitaxial growth of a silicene sheet. *Appl Phys Lett* 97:223109.

Lay GL, Aufray B, et al. (2009). Physics and chemistry of silicene nano-ribbons. *Appl Surf Sci* 256:524–529.

Leandri C, Lay GL (2005). Self-aligned silicon quantum wires on Ag(110). *Surf Sci* 574:9–15.

Li X, Mullen JT, Jin Z, Borysenko KM, Nardelli MB, Kim KW (2013). Intrinsic electrical transport properties of monolayer silicene and MoS_2 from first principles. *Phys Rev B* 87:115418.

Lin CL, Arafune R, et al. (2012). Structure of silicene grown on Ag(1 1 1). *Appl Phys Express* 5:045802.

Lin CL, Arafune R, et al. (2013). Substrate-induced symmetry breaking in silicene. *Phys Rev Lett* 110:076801.

Liu CC, Feng W, Yao Y (2011). Quantum spin hall effect in silicene and two-dimensional germanium. *Phys Rev Lett* 107:076802.

Liu G, Lei XL, Wu MS, Xu B, Ouyang CY (2014a). Comparison of the stability of free-standing silicene and hydrogenated silicene in oxygen: A first principles investigation. *J Phys Condens Matter* 26:355007.

Liu G, Lei XL, Wu MS, Xu B, Ouyang CY (2014b). Is silicene stable in O_2?—First-principles study of O_2 dissociation and O_2-dissociation–induced oxygen atoms adsorption on free-standing silicene. *EPL* 106:47001.

Lu N, Li Z, Yang J (2009). Electronic structure engineering via on-plane chemical functionalization: A comparison study on two-dimensional polysilane and graphane. *J Phys Chem C* 113:16741–16746.

Mahatha SK, Moras P, et al. (2014). Silicene on Ag(1 1 1): A honeycomb lattice without Dirac bands. *Phys Rev B* 89:201416.

Majzik Z, Tchalala MR, et al. (2013). Combined AFM and STM measurements of a silicene sheet grown on the Ag(1 1 1) surface. *J Phys Condens Matter* 25:225301.

Manjanath A, Kumar V, Singh AK (2014). Mechanical and electronic properties of pristine and Ni-doped Si, Ge, and Sn sheets. *Phys Chem Chem Phys* 16:1667–1671.

McKay H, Wales DJ, Jenkins SJ, Verges JA, de Andres PL (2010). Hydrogen on graphene under stress: Molecular dissociation and gap opening. *Phys Rev B* 81:075425.

Meng L, Wang Y, et al. (2013). Buckled silicene formation on Ir(1 1 1). *Nano Lett* 13:685–690.

Miró P, Audiffred M, Heine T (2014). An atlas of two-dimensional materials. *Chem Soc Rev* 43:6537–6554.

Ni Z, Liu Q, et al. (2012). Tunable bandgap in silicene and germanene. *Nano Lett* 12:113–118.

Novoselov KS, Geim AK, et al. (2004). Electric field effect in atomically thin carbon films. *Science* 306:666–669.

Osborn TH, Farajian AA (2014). Silicene nanoribbons as carbon monoxide nanosensors with molecular resolution. *Nano Res* 7:945–952.

Oughaddou H, Enriquez H, et al. (2015). Silicene, a promising new 2D material. *Prog Surf Sci* 90:46–83.

Pan H, Li Z, Liu CC, Zhu G, Qiao Z, Yao Y (2014). Valley-polarized quantum anomalous Hall effect in silicene. *Phys Rev Lett* 112:106802.

Ronci F, Colonna S, et al. (2010). Low temperature STM/STS study of silicon NW grown on the Ag(1 1 0) surface. *Phys Status Solidi C* 7:2716–2719.

Röthlisberger U, Andreoni W, Parrinello M (1994). Structure of nanoscale silicon clusters. *Phys Rev Lett* 72:665–668.

Setiadi J, Arnold MD, Ford MJ (2013). Li-ion adsorption and diffusion on two-dimensional silicon with defects: A first principles study. *ACS Appl Mater Interfaces* 5:10690–10695.

Song X, Hu J, Zeng H (2013). Two-dimensional semiconductors: Recent progress and future perspectives. *J Mater Chem C* 1:2952–2969.

Tan X, Cabrera CR, Chen Z (2014). Metallic BSi_3 silicene: A promising high capacity anode material for lithium-ion batteries. *J Phys Chem C* 118:25836–25843.

Tang Q, Zhou Z (2013). Graphene-analogous low-dimensional materials. *Prog Mater Sci* 58:1244–1315.

Tao L, Cinquanta E, et al. (2015). Silicene field-effect transistors operating at room temperature. *Nat Nanotechnol* 10:227–231.

Tchalala MR, Ali MA, et al. (2013). Silicon sheets by redox assisted chemical exfoliation. *J Phys Condens Matter* 25:442001.

Tchalala MR, Enriquez H, et al. (2014). Atomic and electronic structures of the ($\sqrt{13} \times \sqrt{13}$)R13.9° of silicene sheet on Ag(1 1 1). *Appl Surf Sci* 303:61–66.

Tozzini V, Pellegrini V (2013). Prospects for hydrogen storage in graphene. *Phys Chem Chem Phys* 15:80–89.

Tritsaris GA, Kaxiras E, Meng S, Wang E (2013). Adsorption and diffusion of lithium on layered silicon for Li-ion storage. *Nano Lett* 13:2258–2263.

Wang R, Pi X, et al. (2013). Silicene oxides: Formation, structures and electronic properties. *Sci Rep* 3:3507.

Wang Y, Ding Y (2013a). Mechanical and electronic properties of stoichiometric silicene and germanene oxides from first-principles. *Phys Status Solidi-Rapid Res Lett* 7:410–413.

Wang Y, Ding Y (2013b). Strain-induced self-doping in silicene and germanene from first-principles. *Solid State Commun* 155:6–11.

Wu W, Ao Z, Wang T, Li C, Li S (2014). Electric field induced hydrogenation of silicene. *Phys Chem Chem Phys* 16:16588–16594.

Xu M, Liang T, Shi M, Chen H (2013). Graphene-like two-dimensional materials. *Chem Rev* 113:3766–3798.

Yan JA, Gao SP, Stein R, Coard G (2015). Tuning the electronic structure of silicene and germanene by biaxial strain and electric field. *Phys Rev B* 91:245403.

Yang K, Cahangirov S, Cantarero A, Rubio A, D'Agosta R (2014). Thermoelectric properties of atomically thin silicene and germanene nanostructures. *Phys Rev B* 89:125403.

Zhang X, Xie H, et al. (2014). Thermal conductivity of silicene calculated using an optimized Stillinger-Weber potential. *Phys Rev B* 89:054310.

Zhao J, Liu H, et al. (2016). Rise of silicene: A competitive 2D material. *Prog Mater Sci* 83:24–151. doi: http://dx.doi.org/10.1016/j.pmatsci.2016.04.001.

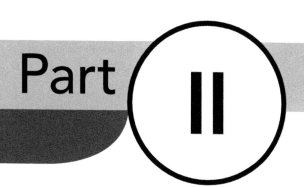

Part II

Clusters, nanoparticles, and quantum dots

Fluorescent silicon clusters and nanoparticles

Klaus von Haeften

Contents

8.1 INTRODUCTION

Clusters, consisting of a small number of atoms, have been in the focus of physical and chemical research for several decades. They often show dramatic size effects. The addition of a single atoms can change their properties rather abruptly because of, for example, the discreteness of shell filling (Knight et al. 1984) or sphere-packing effects (Echt et al. 1981). When clusters become larger and reach the nanometer scale, other effects are observed, such as quantum confinement; the intense red fluorescence observed for nanostructured silicon (Canham 1990; Cullis and Canham 1991; Wilson et al. 1993; Lockwood 1994; Cullis et al. 1997) is a popular and frequently cited example of this effect. The discovery of fluorescent nanoscale silicon at room temperature by Canham (Canham 1990) increased the already quite intense research into silicon clusters further, and to date numerous examples of nanostructured forms of fluorescent silicon have been reported (Takagi et al. 1990; Brus et al. 1995; Hirschman et al. 1996; Borsella et al. 1997; Ehbrecht et al. 1997; Cullis et al. 1997; Huisken et al. 1999; Pavesi et al. 2000; Belomoin et al. 2000, 2002; Falconieri et al. 2005; Mangolini et al. 2005; Brewer and von Haeften 2009; Vincent et al. 2010; He et al. 2011; Dasog et al. 2014; Li et al. 2016). Hence, we have a rich set of data available on electronic and structural properties that underpin our understanding of the fluorescence of silicon clusters and nanoparticles.

The strong appeal of fluorescent silicon clusters and nanoparticles arises due to a veritable multitude of applications, for example, in electronic circuits (Pavesi 2003; Švrček et al. 2004; Stupca et al. 2007) and biomedicine (Nel et al. 2009; Chinnathambi et al. 2014). Silicon is the most frequently used semiconductor material in electronics. It has been suggested that by combining electric and optical signal transmission, higher performance can be achieved (Canham 2000; Pavesi 2003). Issues directly related to the decreasing size of electronic units, such as signal delay caused by increasingly longer interconnects, can be addressed by optically transmitted refresh pulses (Pavesi 2003). Fluorescent silicon clusters and nanoparticles are also attractive in biomedical applications because nanoscale silicon and silicon dioxide are considered to be nontoxic, or at least considerably less harmful than fluorescent nanoparticles of other materials (Warner et al. 2005; Erogbogbo and Swihart 2007; Erogbogbo et al. 2008; Choi et al. 2009;

Erogbogbo et al. 2010; Cheng et al. 2014; McVey and Tilley 2014; Peng et al. 2014) as well as biodegradable (Park et al. 2009). Fluorescent silicon nanoparticles play an important role as biological markers (Nel et al. 2009; Montalti et al. 2015).

In this chapter, the foundations of silicon cluster experiments and cluster production will be discussed. The underlying principles behind the fluorescence of silicon clusters and nanoparticles, such as quantum confinement and surface passivation, are introduced, and contrasted with the current state of research on fluorescent silicon clusters and nanoparticles. Owing to the vast number of publications on this subject that can already be found in the literature, this book chapter cannot be exhaustive. Rather, it complements recent review articles on fluorescent nanoscale silicon (Cheng et al. 2014; Dohnalová et al. 2014; McVey and Tilley 2014; Peng et al. 2014; Priolo et al. 2014; Montalti et al. 2015; Dasog et al. 2016).

8.2 FUNDAMENTAL CONCEPTS FOR PRODUCING FLUORESCENT NANOSCALE SILICON

Bulk crystalline silicon is known as a poor light emitter because of its indirect band gap. Fluorescence is a relatively inefficient relaxation process in electronically excited indirect band gap semiconductors because fluorescence has to simultaneously occur with the absorption of a phonon of matching momentum. This mechanism is illustrated in more detail in Figure 8.1. The entire electronic excitation and relaxation/fluorescence cycle is shown for direct and indirect band gap semiconductors.

The left-hand side of Figure 8.1 shows an energy band schematic of a direct band gap semiconductor, characterized by the conduction and valence band maxima and minima being on top of each other. Photoexcitation of an electron follows an energetic pathway indicated by the vertical arrow, reaching from the top of the valence band and into the conduction band, from where it returns to recombine with the hole, inducing fluorescence. Photoexcitation using higher energies is possible, but less likely because of the decreasing density of states along the parabola, and indeed the fluorescence wavelength will remain unaltered because of the relatively short timescale on which electronic relaxation occurs in the conduction and valence bands. The fluorescence intensity will be unchanged because both electron and hole have the same momentum.

This situation changes in indirect band gap semiconductors, as shown on the right-hand side in Figure 8.1. In indirect band gap semiconductors, the minima of the conduction band and maxima of the valence band are shifted. As a consequence, electrons that are photoexcited into the conduction bands quickly undergo electronic relaxation to the minimum of the conduction band. However, any subsequent direct, vertical decay to the valence band is not possible because all states with similar momenta, $\hbar k$, are populated with electrons; in other words, hole states for direct recombination with the excited electrons are not available. "Diagonal" recombination with the available, original hole state at the maximum of the

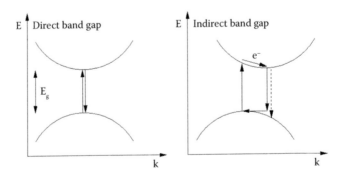

Figure 8.1 Schematic of fluorescence emission in direct and indirect band gap semiconductors. The scheme shows the valence (bottom) and conduction bands (top) of a direct (left) and an indirect band gap semiconductor. The arrows indicate the magnitude of the energy band gap, E_g, the pathways of excitation of an electron from the valence band to the conduction band, electronic relaxation, fluorescence, and absorption of phonons (see text).

valence band is not an option because momentum would no longer be conserved. So, in order for diagonal recombination to happen, a phonon with matching momentum has to be absorbed during fluorescence, but because these two processes would have to happen simultaneously, this scenario is clearly a rare occurrence. As a consequence, the fluorescence lifetime from bulk crystalline silicon is very long and fluorescence intensity is very weak.

When bulk crystalline silicon is reduced in size to a scale approaching that of the nanoscale regime, translational symmetry is gradually lost. A consequence of small size of nanocrystals is, therefore, that materials that are indirect band gap semiconductors in the bulk phase become quasi-direct semiconductors at the nanoscale. For silicon, this means that high fluorescence intensities are possible if nanocrystallites are only small enough to display a quasi-direct band gap.

Another feature of such reductions in crystal size is quantum confinement. Quantum confinement relates to the shift of energy levels with size, and neatly explains the energy spectrum of, for example, color-center defects in crystals (Hayes and Stoneham 2004; Fox 2010), and electrons confined in nanoscale bubbles in liquid helium (Fowler and Dexter 1968; Grimes and Adams 1990, 1992).

These energy shifts can be understood using the popular particle-in-a-box model that is almost perennially reviewed in quantum mechanical text books. The energy spectrum, (n), of an electron confined in a one-dimensional box of length l with infinitely high box walls can be straightforwardly derived to give the following equation:

$$E(n) = \frac{n^2 \pi^2 \hbar^2}{2 m_e l^2} \tag{8.1}$$

where n is the principal quantum number, $\hbar$ the reduced Planck (or Dirac) constant, and m_e the electron mass. The quantum number, n, is indexed from 1, and the energy difference $E(n = 2)$ and $E(n = 1)$ would be equivalent to the fluorescence energy from the first excited state to the ground state.

The analogy of the one-dimensional model can be straightforwardly extended to three dimensions. A more realistic model, using the work function rather than infinitely high potential walls, requires solving transcendental equations. This latter, more rigorous, treatment lowers the energy values by not more than 10%, for which decreases in the energy difference and effective magnitude of the box diameter for a given transition energy are implicit. Furthermore, a more realistic potential surface than a square well would also lower the energy levels as the constraints imposed by the former must, by their very nature, be more relaxed.

Lockwood and coworkers observed that the fluorescence of an Si/SiO_2 superlattice depended on the silicon film thickness (Lockwood et al. 1996). They attributed this behavior to one-dimensional confinement of the excited electron within the silicon film. To explain the shift in the fluorescence energy, they adopted the particle-in-a-box model and showed that the peak energy of the observed red fluorescence band followed Equation 8.2.

$$E(n) = E_g + \frac{\pi^2 \hbar^2}{2 d^2} \left(\frac{1}{m_e^*} + \frac{1}{m_h^*} \right) \tag{8.2}$$

Here, d is the silicon layer thickness, m_e^* and m_h^* are the "effective masses" of electron and hole, although the authors acknowledge that, strictly speaking, the concept of effective electron and hole masses has no physical meaning in nanoscale systems that do not exhibit the translational symmetry of crystals. A good fit was reported for $E \, (eV) = 1.60 + 0.72 d^{-2}$ eV, with d given in nm, which implies effective masses $m_e^* \approx m_h^* \approx 1 \, m_e$ in good agreement with bulk crystalline silicon were $m_e^* (Si_{bulk}) = 1.18 \, m_e$ and $m_h^* (Si_{bulk}) = 0.81 \, m_e$. The fit shows that the band gap energy, at $E_g = 1.60$ eV, is considerably larger than that of bulk crystalline silicon (E_g (c-Si) = 1.12 eV at 295 K), and is, in fact, more similar to that of amorphous silicon (E_g (a-Si) = 1.5 - 1.6 eV at 295 K). The good fit with experimental data and the similarity between the confinement term in Equation 8.2 and the original particle-in-a-box Equation 8.1 is remarkable, and strongly supports the presence of quantum confinement.

Clusters, nanoparticles, and quantum dots

Park and coworkers investigated amorphous silicon quantum dots embedded in silicon nitride. They observed fluorescence in the form of a single band whose maximum shifted with average quantum dot size. The size dependence of fluorescence energy fit to the equation was found as $E\ (eV) = 1.56 + 2.40d^{-2}$ eV, which confirmed the earlier observed band gap energy of amorphous silicon (Lockwood et al. 1996) in the limit of large dot sizes. However, the dependence on quantum confinement, $2.40d^{-2}$ eV, was much larger. The discrepancy with work conducted on Si/SiO$_2$ superlattices was attributed to the three-dimensional confinement of the quantum dots in silicon nitride (Park et al. 2001).

Summarizing, we have so far seen that with reduction in size of a crystal, translational symmetry is gradually lost with the consequence that indirect band gap semiconductors become quasi-direct semiconductors at the nanoscale. At the same time, the band gap energy increases because of quantum confinement.

A further important factor determining the ability to fluoresce is the electronic structure at the surface of nanocrystals. Surfaces break translational symmetry, a consequence of which is that one cannot expect the same energy band structure as might be observed for "infinite" crystals. Band gap energies may be smaller, or may not even exist. For nanocrystals, this means that surface effects may compete with quantum confinement. In the following, nonradiative decay at nanocrystal surfaces will be discussed.

To prevent nonradiative decay at its surface, a nanocrystal may be embedded in another semiconductor or insulator of larger band gap energy. This effect is illustrated in Figure 8.2.

In broadest terms, one can expect that surface effects on the band gap energy can be minimized for such a system. It is assumed that the nanocrystal structure fits well with that of the host and that a minimum of additional interface states are produced. Ideally, this would mean that an electron promoted across E_{gl} from the valence band of the nanocrystal to its conduction band would remain confined within the nanocrystal. The electron would have no other choice than to fluoresce to the ground state because no discrete states are available within the gap.

To better account for nonradiative decay in real systems, vibrational relaxation is often considered. Because of the possibility of surface reconstruction, vibrational relaxation at surfaces is particularly important for free nanoscale crystals and clusters. Figure 8.3 illustrates the cycle of excitation, electronic migration, and relaxation within the conduction band and vibrational relaxation at the surface. In this simple picture, a high density of vibrational states at the surface is assumed to exist. The electronically excited electron "finds" the surface on a subfemtosecond timescale, and relaxes by jumping down the energy ladder of vibrational states.

Another simplification implicit in this picture of a dense spectrum of vibrational states is that the short timescales of nonradiative decay are ignored. The energy level scheme shown in Figure 8.3 implies that the vibrational levels are time averages.

Under closer inspection, the consideration of relaxation pathways along such eigenstates in nanocrystals does not appear to be justified. Particularly in free clusters, the atoms at the surface can perform large amplitude vibrations. These large amplitude vibrations give rise to electronic–vibrational coupling, and open other, nonadiabatic routes for relaxation from electronically excited states to the ground state. The simple example of the triatomic, homonuclear molecule shown in Figure 8.4 illustrates this mechanism.

In this example, the molecule performs a bend vibration. At the classical turning points of this motion, structures belonging to three different point groups, $D_{\infty h}$, C_{2v}, and D_{3h}, can be identified, with the actual point group depending on the amplitude of the vibration. In other words, by performing a bend

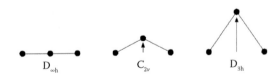

Figure 8.2 Simple schematic illustrating how vibrational–electronic coupling can lead to nonadiabatic relaxation from electronically excited states to the ground state. A linear, homonuclear triatomic molecule performing a bending vibration assumes, at various points during one such oscillation, structures that belong to three different point groups, and hence three different electronic states.

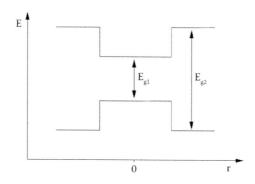

Figure 8.3 Schematic of a nanoscale crystal (quantum dot) embedded in a material of larger band gap energy. E_{g1} and E_{g2} refer to the band gap energies of the nanocrystal and host material, respectively.

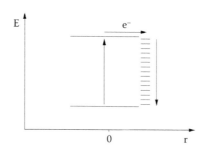

Figure 8.4 Simplified energy level diagram of a nanoscale crystal with a high density of vibrational states at its surface. The small size of the crystal means that electrons delocalized over the conduction band quickly localizing at the surface. They then relax nonradiatively by hopping across the vibrational states.

vibration, the molecule is able to intersect the different electronic states associated with these point groups. These intersection points provide passages to lower lying states.

These lower states may be dissociative. Therefore, a tendency toward isomerization and dissociation may be expected, particularly for small homonuclear clusters. This corroborates the relevance of caging through a rigid shell of atoms, moieties, or solvent molecules. In fact, deposition of metal clusters into argon matrices and argon droplets has made the observation of photoluminescence possible (Felix et al. 2001; Sieber et al. 2004; Conus et al. 2006; Harbich et al. 2007). Argon shells have also been used to cage oxygen molecules; the observed luminescence was attributed to oxygen atoms that had first dissociated but were then forced, by the cage, to recombine (Laarmann et al. 2008).

In summary, nonadiabatic decay at the surface of homonuclear clusters is an important relaxation channel in terms of its competition with fluorescence. The design and engineering of fluorescent clusters therefore requires that this possibility be suppressed. In broadest terms, the passivation of dangling bonds at a surface with atoms such as hydrogen (hydrides), halogens (halides), oxygen (oxides), or molecular groups such as alkenes, amines, or polymers can be understood to confine the electronically excited electron to the core region of the cluster. The distinctly higher fluorescence intensities that arise from this confinement can be observed. This was confirmed by Seraphin and coworkers (Seraphin et al. 1996), who investigated silicon nanoparticles produced by laser ablation with and without passivation, in vacuo. They found that passivation increased fluorescence intensity, but left the fluorescence energy itself unchanged. Embedding clusters and nanocrystals within a lattice-matching host of a larger band gap energy fulfills the same purpose. Also, in the ideal case of a lattice-matching interface, large amplitude vibrations would be successfully suppressed purely as, though not necessarily limited to, a matter of sterics. Passivated or core-shell clusters embedded in a solvent can also effectively transfer and absorb excess vibrational energy.

Clusters, nanoparticles, and quantum dots

8.3 CLUSTERS

Clusters are understood as being particles consisting of as few as two, three, or four, or as having as many as a few thousand, atoms. The term "cluster" is used alongside the term "nanoparticles," but clusters are commonly understood to represent smaller particles. The study of clusters is motivated by the desire to understand the often dramatic changes of material properties that accompany changes in size. Material properties also depend on structure and dimensionality, and similarly accompany changes in size. All such characteristics are relevant in cluster science research.

8.3.1 PRODUCTION OF SILICON CLUSTERS

Free clusters are frequently studied in supersonic beams. A gas under high pressure is expanded through a tiny aperture into vacuum. During expansion, it cools and nucleates via removal of excess energy through three-body collisions, forming clusters. They fly through vacuum where they can, for instance, be investigated free from external interactions, or can otherwise be deposited on a substrate. For silicon clusters, this production method is unsuitable because it requires atomic vapor of significant pressure. Silicon has a comparably low vapor pressure, even at high temperatures, rendering supersonic expansion of thermally produced silicon vapor through a nozzle into vacuum an unrealistic mechanism for silicon cluster production.

Supersonic beams have been employed in sources where silicon has been generated by decomposition of precursor gases which are then mixed with a carrier/aggregation gas. Silane diluted in helium and exposed to a discharge has been expanded through a pulsed piezoelectric valve (Hoops et al. 2001). Ehbrecht and Huisken (1999) used a pulsed CO_2 laser to decompose silane diluted in helium. The products were expanded into vacuum and subsequently investigated by time-of-flight mass spectrometry or deposited onto CF_2 or LiF substrates (Ehbrecht et al. 1997).

Silicon clusters have been produced using the principle of gas aggregation (Sattler et al. 1980). Silicon vapor is mixed with a "seeding" gas, which induces the three-body collisions required for nucleation. It also facilitates growth by mediating collisions between silicon atoms and removing the latent heat that is released during any subsequent growth of clusters from these collisions.

To generate silicon vapor within a seed gas, laser vaporization, sputtering, and pulsed arcs have been used. Laser vaporization sources employ pulsed laser sources that are fired at a rotating silicon rod (Bloomfield et al. 1985; Jarrold and Constant 1991; Bower and Jarrold 1992; Honea et al. 1993a, 1999). Sputter gas aggregation sources employ a modified sputter electrode configuration, allowing higher operating gas pressures than are usually needed in sputtering for thin film production (Haberland et al. 1991; Hoffmann et al. 2001b; Pratontep et al. 2005; Wegner et al. 2006; von Haeften et al. 2009). Argon and helium are used as seed gases. Helium produces smaller clusters because it is much lighter than silicon and energy transfer is less efficient. The pulsed arc cluster ion source is another variation of a gas aggregation-type cluster source that has been used for silicon cluster production (Maus et al. 2000).

These sources produce neutral and ionic clusters, including both cations and anions; charged clusters can be easily manipulated in a spatial sense via electrostatics, and can therefore be mass selected. Kitsopoulos and coworkers size-selected silicon cluster anions, which were then photoexcited into the neutral ground state as well as other low-lying electronically excited states of the neutral clusters by the process of electron detachment (Kitsopoulos et al. 1990). Honea and coworkers used arrangements of linear and perpendicular quadrupole mass selectors to size-select silicon cluster ions and deposit them onto a surface (Honea et al. 1999). To soften the impact, rare gas buffer layers were codeposited onto a cold, predeposited rare gas matrix (Honea et al. 1999). A similar scheme was used by Grass and coworkers, who soft-landed Si_4 on highly oriented pyrolytic graphite (HOPG) and performed X-ray photoelectron spectroscopy (Grass et al. 2002). Astruc Hoffmann and coworkers used a reflectron time-of-flight (RETOF) mass spectrometer in combination with a multiwire mass gate to size-select silicon anion clusters on which to subsequently perform photoelectron spectroscopy. The size-selected clusters were irradiated with UV light from an ArF excimer laser and analyzed in a magnetic bottle photoelectron spectrometer (Hoffmann et al. 2001b).

Hirsch and coworkers introduced size-selected silicon clusters into an ion trap (Hirsch et al. 2009; Lau et al. 2011; Vogel et al. 2012; Kasigkeit et al. 2015). The trapped clusters were excited by monochromatic synchrotron radiation, allowing the determination of core-level shifts for specific cluster sizes.

8.3.2 GEOMETRIC STRUCTURE

A great variety of geometric structures have been reported for silicon clusters using both theoretical and experimental methods. In general, it has been found that the structures of neutral, cation, and anion cluster species differ considerably. Anions have frequently been used to elucidate electronic and geometric features. Their structures are affected by Jahn–Teller distortions. Care must be taken when compared to neutral clusters.

In a number of early studies, silicon clusters produced using supersonic beam techniques were deposited and investigated spectroscopically. Honea and coworkers deposited size-selected silicon clusters into argon, krypton, and nitrogen matrices using codeposition onto a liquid helium-cooled substrate (Honea et al. 1993b, 1999). Using surface plasmon polariton-enhanced Raman spectroscopy, sharp features characteristic of Si_4, Si_6, and Si_7 structures were identified in their spectra, which were in good agreement with earlier *ab initio* calculations (Raghavachari and Logovinsky 1985; Raghavachari 1986; Tománek and Schlüter 1986; Raghavachari and Rohlfing 1988; Ballone et al. 1988). The spectra also revealed evidence for the presence of cluster–cluster aggregation within the rare gas matrix. The spectra of the aggregates of Si_4, Si_6, and Si_7 bore considerable similarities to those of larger clusters, such as Si_{25-35}, as well as those of amorphous silicon (Honea et al. 1999).

The structure of free silicon clusters in the size range from n = 10 to n = 100 was addressed using drift mobility measurements, the results of which revealed a prevalence for prolate-shaped structures. A structural transition occurs at sizes around n = 27; however, larger clusters were found to prefer more spherical configurations (Jarrold and Constant 1991). The preference for prolate shapes was attributed to a tendency of the silicon atoms to minimize coordination. This trend competes with the surface energy of the clusters, ultimately leading to a preference for spherical structural motifs in larger clusters (Jarrold and Constant 1991). The experiments were repeated at higher resolution (Hudgins et al. 1999), and calculations confirm these experimental findings (Ho et al. 1998; Jackson et al. 2004).

Vibrational spectroscopy of anions has been performed by photoexcitation spectroscopy into the neutral state using electron detachment (Kitsopoulos et al. 1990; Xu et al. 1998). Comparison of the resultant spectral features with calculations reveals the structures of the anions. Recent work on larger silicon cluster anions reveals vibrational spectra indicative of the previously observed transition from prolate to spherical shapes (Meloni et al. 2004).

The FELIX free electron laser source provides intense and tunable infrared radiation including in the spectral range from 166 to 600 cm^{-1} where silicon clusters absorb. Vibrational spectroscopy of small silicon cluster cations was performed using multiple photon dissociation spectroscopy. The ions were analyzed in a time-of-flight mass spectrometer. Also, isotopically selected ^{129}Xe atoms were attached to the clusters. Absorption of multiple IR photons would lead to a depletion of the ion signal, allowing the requisition of spectra of size-selected clusters. Comparison of the spectra with calculations using density functional theory (DFT) revealed novel structures and a growth motif that started with a pentagonal bipyramid building block and changed to a trigonal prism for larger clusters (Lyon et al. 2009).

Related work provided information on the structure of small neutral silicon clusters (Fielicke et al. 2009). Using a combination of tunable far-infrared radiation from the FELIX free-electron laser source and vacuum-ultraviolet two-color ionization, it was possible to scan the spectrum of homonuclear neutral silicon clusters in the spectral range from 200 to 550 cm^{-1}. The use of vacuum-ultraviolet two-color ionization provided the advantage of detection of the initially neutral clusters with mass selectivity. An increase of the ionization rate was observed when IR photons had been absorbed. Comparison with DFT and Møller–Plesset (MP2) perturbation theory calculations revealed that the ground-state potential energy surface was very flat. Therefore, rapid interconversion between different structures might well be expected, as well as the presence of higher energy isomers in real-world experimental samples (Fielicke et al. 2009).

Furthermore, theoretical work suggested that, at intermediate sizes, around 20 atoms silicon clusters tend to build irregular cages stabilized by a small number of encapsulated silicon atoms (Mitas et al. 2000). Silicon cages can also be stabilized by encapsulated foreign atoms (Kumar and Kawazoe 2001, 2003a, 2003b). Stable hollow structure similar to C_{60} buckminsterfullerene (buckyballs), can nevertheless be excluded (Sun et al. 2003).

Clusters, nanoparticles, and quantum dots

In summary, the structures of silicon clusters grown in the gas phase differ from the structures of silicon nanocrystals that have been produced, for example, by etching of bulk crystalline silicon. The structures of small silicon clusters are characterized by the tendency of the atoms to minimize coordination, thereby favoring growth of prolate shapes. Therefore, their bond angles are smaller than their counterparts in sp^3 bonded, cubic diamond-structured bulk silicon. The consequences of this behavior are shorter internuclear separations and higher atomic densities. An exemption from these principles is shown in the work of Laguna and coworkers, however, who deposited silicon clusters produced by pyrolysis of silane onto holey carbon films. High-resolution transmission electron images clearly show nanoparticles with lattice planes surrounded by an amorphous oxide shell (Hofmeister et al. 1999; Laguna et al. 1999).

8.3.3 ELECTRONIC STRUCTURE

Photoelectron spectra of silicon cluster anions have frequently been reported in the literature. These spectra exhibit rich features for small clusters, which become smoother as the clusters become larger. Maus et al. (2000) assign the low-binding energy features to the extra electron occupying the conduction bands. The higher-energy features observed are attributed to the valence electrons. The energy difference between the two corresponds to the band gap in the bulk picture. Small clusters between n = 3 and 13 were found to have band gap energies smaller than those typically seen for bulk crystals (Maus et al. 2000). This is incommensurate with the idea of quantum confinement, which would predict larger band gap energies for anything smaller than the bulk. The results were attributed to the entire geometric and electronic structure being affected by surface effects, similar to the reconstruction of the surface of bulk silicon crystals. While such an effect must clearly dominate the electronic structure of small clusters, the trend continues for larger clusters as well; Hoffmann et al., for instance, report the absence of a band gap for clusters up to 1000 atoms (Hoffmann et al. 2001a). For large Si cluster anions, the photoelectron spectrum is dominated by a single smooth and broad feature. The onset of photoemission shifts with size toward larger binding energies, a trend that is incommensurate with a bulk band gap picture and contrary to what one would expect from quantum confinement.

The observations made through photoelectron spectroscopy of cluster anions in free beams match measurements of band gap energies of silicon clusters deposited on HOPG (Marsen et al. 2000). Such band gap energies were always smaller than those of their bulk counterparts, which agrees with a different geometric structure than the diamond cubic structure of the bulk. Also, a transition region was found for sizes around 1.5 nm, corresponding to 44 atoms; larger clusters had no band gap.

Band gap energies of cationic VSi_n^+ were determined by X-ray spectroscopy using monochromatic synchrotron radiation and an ion trap to store the size-selected clusters (Lau et al. 2011). By measuring the ion yield of specific ion decay channels, it was possible to record a direct 2p photoionization spectrum separately from the resonant 2p photoionization spectrum, yielding the energy difference E_{XAS} between the core level and the lowest unoccupied molecular orbital (LUMO), and E_{CL}, the energy difference between the core and the vacuum level, E_{VAC}. Measurement of the valence state photoionization spectrum E_{VB} yields the difference between the highest occupied molecular orbital (HOMO) and the the vacuum level, E_{VAC}. For the band gap energy, E_g, or, more precisely, the HOMO–LUMO energy difference, it follows $E_g = E_{VB} - E_{CL} + E_{XAS}$.

Photoionization thresholds were measured for silicon clusters in the size range from n = 2 to 200 using laser photoionization with RETOF mass spectrometry detection (Fuke et al. 1993). The ionization potential revealed features that were ascribed to a structural transition for sizes around n = 20. Measurements of the 2p core level and valence electron binding energies using monochromatic synchrotron radiation and an ion trap show a similar size dependence (Vogel et al. 2012). Both 2p binding energy and ionization potential show a linear dependence on the inverse cluster radius $n^{-1/3}$. Such a dependence is expected from the size-dependent charging energy, similar to metal clusters (Halder and Kresin 2015). Furthermore, core-level shifts were derived and compared to *ab initio* calculations (Vogel et al. 2012).

8.4 FLUORESCENT SILICON CLUSTERS

The electronic level structure associated with dense packing, suggesting very small band gap energies for small- and medium-sized silicon clusters and even the absence of band gaps, is unfavorable toward fluorescence. Indeed, fluorescence from free silicon clusters in traps or in molecular beams has not yet been reported in the literature. The work on neutral and cationic silver clusters in argon droplets and solid matrices (Felix et al. 2001; Sieber et al. 2004; Conus et al. 2006; Harbich et al. 2007) suggests that fluorescence might be possible if silicon clusters are deposited and embedded in rare gas matrices. While such deposition experiments have been carried out (Honea et al. 1993b, 1999), attempts to observe fluorescence with this setup are not known to the author.

It appears that fluorescence reported for nanoscale silicon can be attributed to the effects of quantum confinement, passivation, and the presence of defects. To achieve confinement, passivation, or defects due to other materials, molecules, or atoms are actively introduced. In the vast majority of bottom-up methods used to produce fluorescent silicon nanoparticles, chemical methods are employed. Exceptions are laser vaporization of silicon targets in liquids (Švrček et al. 2009a; Švrček et al. 2009b; Alkis et al. 2012; Intartaglia et al. 2012a; Švrček et al. 2016; Rodio et al. 2016), pyrolysis of silane in gas-flow reactors (Ehbrecht et al. 1997) and silicon cluster codeposition with water vapor (von Haeften et al. 2009).

8.5 RED–ORANGE LUMINESCENCE

Silicon clusters produced by CO_2 laser-induced decomposition of silane were found to show red photoluminescence after they were deposited onto LiF or CaF_2 substrates, and transferred to ambient air (Ehbrecht et al. 1997). Owing to a continuous supersonic beam and a pulsed CO_2 laser, the part of the beam containing clusters was also pulsed. A velocity selector was employed to select velocity segments of the cluster pulse, and the cluster mass was measured by time-of-flight mass spectrometry (Ledoux et al. 2002). The average number of atoms in the clusters, $\bar{N}$, was found to vary from $\bar{N} = 395$, corresponding to an average diameter, $\bar{D}$, of 2.47 nm to $\bar{N} = 9070$, corresponding to $\bar{D} = 7.03$ nm. The diameters were deduced using a spherical particle model,

$$D(N) = \left(\frac{3N}{4\pi} V_{unit} \right)^{1/3}$$

(8.3)

where $V_{unit} = 0.1601$ nm³ is the volume of the unit cell of bulk crystalline silicon. Equation 8.3 takes into account the fact that the unit cell has a diamond cubic structure and contains eight atoms. It is assumed that bulk and nanoparticle densities are the same.

These samples were photoexcited at 488 nm using continuous laser radiation, and the resultant fluorescence spectrum was measured. Each sample showed an almost symmetric fluorescent band whose peak wavelength decreased with particle diameter. The size-dependent fluorescence wavelength shifts agreed with the results one might anticipate from quantum confinement. Deviations were attributed to partial oxidation of the surface layer, which could also be seen in high-resolution transmission electron microscopy (HRTEM) images (Laguna et al. 1999). The oxide shell thickness could be correlated linearly to the cluster diameter. The smallest clusters of 6 nm in diameter had an oxide shell with a thickness of 0.81 nm, while the largest particles had a diameter of 34 nm and a 3 nm thick oxide shell (Hofmeister et al. 1999). The HRTEM images showed that the nanoparticles had a crystalline core. A number of different silicon lattice planes were identified from diffraction rings.

Hofmeister and coworkers also investigated the spacing of a (111) lattice as a function of cluster size. They found that large silicon clusters exhibited compressed (111) lattices compared to bulk crystalline silicon. However, clusters smaller than 3 nm in diameter were found to be dilated. The dependence of the (111) lattice spacing on the cluster diameter followed:

$$d(111) = \frac{0.023}{D} + 0.307 \, [\text{nm}]$$

(8.4)

where D is the cluster diameter.

Because the photoluminescence of silicon was found to depend on pressure, the authors concluded that the size-dependent lattice separation must be taken into account in a modified equation for the photoluminescence energy caused by quantum confinement (Ledoux et al. 2000).

$$E_{PL}^{corr} = E_0 + \frac{3.73}{D^{1.39}} + \frac{0.881}{D} - 0.245 \tag{8.5}$$

Here, E_{PL}^{corr} is the energy of the photoluminescent band, as corrected for size-dependent lattice separation, in eV. E_0 is the band gap energy of bulk crystalline silicon at room temperature (1.17 eV) (Ledoux et al. 2000).

It is important to note that all samples had been transferred to air testing for photoluminescence. The samples were kept in an argon atmosphere during transfer (Ehbrecht et al. 1997). After production and exposure to air, the crystalline core section of the particles was found to reduce in diameter (Ledoux et al. 2000, 2001). Also, the photoluminescence energy was found to blue shift, which was attributed to the smaller sizes of the silicon clusters, supporting the assertion that quantum confinement was controlling fluorescence properties (Ledoux et al. 2001). The effect of quantum confinement was also investigated by etching the oxide layer using hydrofluoric acid (HF). This was found to narrow the spectral band width of the luminescence but not the peak wavelength, since the silicon crystalline core itself would clearly not be affected by such treatment with HF. The effect of passivation on the fluorescence intensity, but not on the fluorescence wavelength, is in line with earlier work by Seraphin and coworkers (Seraphin et al. 1996).

Pyrolysis of silane in vacuum was also used by Li and coworkers to produce silicon clusters (Li et al. 2004a, 2004b), who post-processed the samples by etching with HF and nitric acid (HNO$_3$). This was found to reduce the cluster size and the intensity of visible luminescence.

8.6 BLUE FLUORESCENCE

To investigate the effect of passivation of silicon clusters *in situ*, von Haeften and coworkers used a molecular beam codeposition scheme (von Haeften et al. 2009). They produced silicon clusters by gas aggregation using ion sputtering in an argon–helium atmosphere, codepositing them with a beam of water vapor onto a liquid nitrogen-cooled target. After a deposition time of 30 minutes, the target was warmed up, whereupon the ice-silicon mixture melted and a few milliliters of liquid sample was collected. A schematic of the apparatus used is shown in Figure 8.5.

When photoexcited with 308 nm UV light, all liquid samples showed a blue fluorescence that peaked at 420 nm (von Haeften et al. 2009). The fluorescence intensity was stable over several months (Brewer and von Haeften 2009). When the photoexcitation wavelength was decreased from 310 to 240 nm, the wavelength of the fluorescent band remained at 420 nm; however, additional fluorescence bands appeared in the UV region (von Haeften et al. 2010a, 2010b; Torricelli et al. 2011). When the clusters were embedded

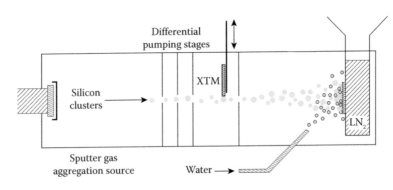

Figure 8.5 Schematic of experimental setup used by von Haeften et al. (2009); Brewer and von Haeften (2009); Torricelli et al. (2011).

in liquid ethanol and isopropanol, the fluorescence wavelength shifted to 365 and 380 nm, respectively (Galinis et al. 2012b). The fluorescence lifetime was measured using monochromatic synchrotron radiation. For an excitation wavelength at 195 nm and fluorescence at 300 nm, a lifetime of 3.7 ns was determined (Yazdanfar et al. 2012).

Time-correlated fluorescence spectroscopy showed that the blue fluorescence consisted of at least two components: the spectral component with a long fluorescent lifetime that peaked at 2.7 eV (Yazdanfar et al. 2012), which is a perfect match with the fluorescence of defect-rich silica (Skuja et al. 1984), and a second short-lived, and much more intense, component, which peaked at 3.0 eV (Yazdanfar et al. 2012). The good energy match and the long lifetime of the first of these bands suggested it arose due to the spin-forbidden $T_1 \rightarrow S_0$ transition observed for twofold coordinated Si in SiO_2 (O-Si-O) (Skuja et al. 1984; Skuja 1992; Nishikawa et al. 1992; Fitting et al. 2004).

Short lifetimes have frequently been reported for the blue fluorescence of nanoscale silicon (Kovalev et al. 1994). Tsybeskov and coworkers investigated the lifetime of blue fluorescence emitted from thermally and chemically oxidized porous silicon (Tsybeskov et al. 1994). The decay was multiexponential with a lifetime of ~1 ns, which was independent of the excitation energy. Furthermore, the appearance of blue fluorescence was correlated to the presence of silicon oxide. Silicon-hydrogen bonds were absent.

Harris and coworkers prepared blue light-emitting silicon samples by electrochemical etching. They found that the samples rapidly degraded, but were able to measure photoluminescence spectra at a sample temperature of 120 K; both red- and blue-emitting components were observed. Fluorescence decay of the blue fluorescence was measured at 77 K as having a time constant of 0.86 ns. The decay was monoexponential (Harris et al. 1994).

By using different post-processing chemical treatments, porous silicon can be prepared to emit either blue or red fluorescence. Žídek and coworkers used such methods to measure the luminescence decay time separately for the two different wavelength ranges. They found that both types of fluorescence have distinct characteristics in their ultrafast decay time. The blue fluorescent band is attributed to several underlying bands which vary further, depending on sample preparation (Žídek et al. 2011).

Light emission in the blue spectral range is a known phenomenon for colloidal suspensions of silicon nanoparticles (Kimura 1999; Belomoin et al. 2000; Valenta et al. 2008) and silicon nanocrystal films (Loni et al. 1995; Canham et al. 1996; de Boer et al. 2010; Ondič et al. 2014). At the present time, there seems to be a consensus that the vast majority of reported fluorescent bands in the blue spectral range are due to localized transitions, rather than being caused by quantum confinement (Dasog et al. 2013, 2016). The precise nature of this fluorescence is nevertheless debated, though it is nevertheless reasonable to attribute it to a range of different transitions which have similar transition energies.

To elucidate the nature of the blue fluorescence, various sample preparation techniques have been explored. Responses to various chemical treatments and correlation to fluorescence performance, as well as chemical analysis, have been employed. The results are not always consistent.

Konkievicz and coworkers prepared porous silicon films and investigated photoluminescence over a wide spectral range. Under photoexcitation with 193 nm excimer laser radiation, they observed red luminescence. After annealing in an oxygen atmosphere with 2% organochlorine, blue luminescence around 400 nm appeared. The blue band appeared at annealing temperatures of 750°C and increased in intensity up to temperatures of 150°C, after which no further increases were seen (Kontkiewicz et al. 1994). Fourier transform infrared (FT-IR) measurements showed the presence of silicon oxide. This preparation was also repeated in a nitrogen atmosphere. Annealing in nitrogen did not produce blue luminescence. This led to the conclusion that the blue luminescence originates from oxidized nanostrutured silicon, although later work showed a correlation between blue fluorescence intensity and nitrogen content (Dasog et al. 2013). Results similar to those of Konkievicz and coworkers have also been reported from porous Si that was oxidized and annealed at 880°C (Kanemitsu et al. 1994). Depending on the production method, the band maxima range from 400 to 460 nm (Yu et al. 1998). The specific response of red and blue luminescence intensity to repeated etching and oxidation was investigated by Lockwood and coworkers, leading them to the conclusion that, at least, quantum confinement cannot be responsible for the blue luminescence (Lockwood et al. 1996).

An important aspect of chemical treatment is how silicon nanoparticles interact with an aqueous environment. This is because water is a strong oxidizing agent, but also because of the relevance of nanoparticles to biomedical applications. Water may also be expected to quench fluorescence because of its dense vibrational spectrum. The interaction with water has been found to chemically modify silicon nanoparticles, and as a consequence its fluorescence activity has been seen to "degrade" (Li and Ruckenstein 2004; Erogbogbo and Swihart 2007; Erogbogbo et al. 2008). However, blue fluorescence from nanoscale silicon has been frequently observed, specifically in connection with treatment with water (Švrček et al. 2009a; Švrček et al. 2009b, 2016; Alkis et al. 2012; Intartaglia et al. 2012a, 2012b; Rodio et al. 2016).

Hou and coworkers treated light-emitting porous silicon with boiling water (Hou et al. 1993) and observed a large blue shift in the fluorescence toward the green–blue spectral range. Infrared spectroscopy was performed with the specific goal to assess whether the formation of silicon monohydride (2080 cm^{-1}) and silicon dihydride (2120 cm^{-1}) was correlated with the appearance of blue fluorescence. Prior to treatment with water, both lines were present in the spectrum and decreased in intensity after boiling water was added. Instead, a band appeared at 1105 cm^{-1}, showing that treatment with water had caused oxidation (Hou et al. 1993).

Koyama and coworkers annealed oxidized porous silicon in water vapor and observed a drastic enhancement in the blue fluorescence intensity (Koyama et al. 1998). Infrared absorption spectroscopy indicated that this annealing increased the absorption peaks related to OH vibrations except for those of free silanol, which disappeared completely. No traces of carbon-related signals were observed, contrary to the previously suspected involvement of carbon (Kontkiewicz et al. 1994; Canham et al. 1996).

Many authors report that the emission of blue fluorescence is correlated with very small structures, perhaps only a few nanometers in size. Akcakir and coworkers etched p-type boron-doped silicon using H_2O_2 and HF (Yamani et al. 1997; Akcakir et al. 2000). The combined effect of the two chemicals produced exceptionally small structures, which were then dispersed in acetone. Under photoexcitation at 355 nm, blue fluorescence was observed. Using two-photon excitation with 780 nm light pulses of 150 fs duration, fluorescence correlation spectroscopy (FCS) was performed, which suggested a hydrodynamic radius of 0.9 nm (Akcakir et al. 2000). The same research group presented TEM images of graphite films coated with this colloidal solution. The images showed agglomerated particles of 1 nm in diameter, in very good agreement with the FCS work (Belomoin et al. 2000). IR spectra showed various Si–H bands of the freshly prepared samples: 520–750 cm^{-1} (SiH$_2$ scissors or SiH$_3$), 880–900 cm^{-1} (Si–H wagging), and 2070–2090 cm^{-1} (SiH stretch and coupled H–Si–Si–H). The 1070 cm^{-1} Si–O stretch was also observed. Treatment with H_2O_2 and subsequent IR spectroscopy was found to replace first the di- and trihydrogen bonds and then the Si–H with Si–O. The coupled H–Si–Si–H bonds showed somewhat greater resilience. The blue fluorescence intensity changed by no more than a factor of two after H_2O_2 treatment (Belomoin et al. 2000).

Fluorescent silicon clusters produced by codeposition with water vapor onto a cold target showed similar sizes. Atomic force microscopy in noncontact and constant force mode of cluster films produced by drop-casting colloidal solution onto freshly cleaved HOPG showed uncovered regions of graphite, and agglomerated monolayers, as well as double layers, of clusters (Torricelli et al. 2011). The height of the monolayers reflected the difference of the tip-HOPG and tip-cluster forces, and hence cannot be taken as a measure of cluster height. However, measuring the differences between the tip-first cluster and tip-second cluster layers was expected to give a fair estimate of the height of the clusters in the film. Values between 0.92 and 1.62 nm were found (Galinis et al. 2012a), in very good agreement with Belomoin and coworkers (Belomoin et al. 2000).

Further studies using chemically produced silicon nanoparticles confirm the relation between blue luminescence, small cluster sizes, and localized transitions. Zhong and coworkers report XRD diffraction peaks similar to the diamond structure of bulk crystalline silicon (Zhong et al. 2013). Size distributions were measured by TEM, for which an average size of 2.2 nm was found (Zhong et al. 2013). Li and coworkers observed very high quantum yields of blue fluorescence, up to 75%, which were attributed to surface treatment with nitrogen-containing agents (Li et al. 2013). Furthermore, Li showed that the fluorescence wavelength can be tuned by different ligands attached to nitrogen-capped silicon clusters (Li et al. 2016). Also, the quantum yield could be increased further, up to 90%, and the emission bandwidth could be narrowed. They attribute their observations to localized transitions at the cluster surface and propose a "ligand law" controlling the photoluminescence (Li et al. 2016).

8.7 CONCLUSIONS

Silicon clusters consisting of a small number of atoms are fascinating objects through which one can study the evolution of material properties with complexity and size. Free clusters produced in molecular beams have properties that are unfavorable for light emission. However, when passivated or embedded in a suitable host, they may emit fluorescence. The current available data show that both quantum confinement and localized transitions, often at the surface, are responsible for fluorescence. By building silicon clusters atom by atom, and by embedding them in shells atom by atom, new insights into the microscopic origins of fluorescence from nanoscale silicon can be expected.

The methods needed to perform such experiments, such as spectroscopy in droplets of argon (Felix et al. 2001) and helium (Feng et al. 2015; Katzy et al. 2016), have recently been developed. It can be hoped that they will be used for the study of fluorescence of silicon clusters. In view of the rising number of studies of fluorescent silicon nanostructures for biomedical and other applications (Dasog et al. 2014; McVey and Tilley 2014), the value and importance of such studies is clear.

REFERENCES

Akcakir, O., Therrien, J., Belomoin, G., Barry, N., Muller, J. D., Gratton, E., and Nayfeh, M. (2000). Detection of luminescent single ultrasmall silicon nanoparticles using fluctuation correlation spectroscopy. *Appl. Phys. Lett.*, 76(14):1857–1859.

Alkis, S., Okyay, A. K., and Ortaç, B. (2012). Post-treatment of silicon nanocrystals produced by ultra-short pulsed laser ablation in liquid: Toward blue luminescent nanocrystal generation. *J. Phys. Chem. C*, 116(5):3432–3436.

Ballone, P., Andreoni, W., Car, R., and Parrinello, M. (1988). Equilibrium structures and finite temperature properties of silicon microclusters from ab initio molecular-dynamics calculations. *Phys. Rev. Lett.*, 60(4):271.

Belomoin, G., Therrien, J., and Nayfeh, M. (2000). Oxide and hydrogen capped ultrasmall blue luminescent Si nanoparticles. *Appl. Phys. Lett.*, 77:779.

Belomoin, G., Therrien, J., Smith, A., Rao, S., Twesten, R., Chaieb, S., Nayfeh, M., Wagner, L., and Mitas, L. (2002). Observation of a magic discrete family of ultrabright Si nanoparticles. *Appl. Phys. Lett.*, 80(5):841–843.

Bloomfield, L. A., Freeman, R. R., and Brown, W. L. (1985). Photofragmentation of mass-resolved clusters. *Phys. Rev. Lett.*, 54(20):2246–2249.

Borsella, E., Botti, S., Cremona, M., Martelli, S., Montereali, R., and Nesterenko, A. (1997). Photoluminescence from oxidised Si nanoparticles produced by CW CO_2 laser synthesis in a continuous-flow reactor. *J. Mater. Sci. Lett.*, 16(3):221–223.

Bower, J. E., and Jarrold, M. F. (1992). Properties of deposited size-selected clusters: Reactivity of deposited silicon clusters. *J. Chem. Phys.*, 97(11):8312–8321.

Brewer, A., and von Haeften, K. (2009). In-situ passivation and blue luminescence of silicon clusters using a cluster beam/H_2O co-deposition production method. *Appl. Phys. Lett.*, 94:261102.

Brus, L., Szajowski, P., Wilson, W., Harris, T., Schuppler, S., and Citrin, P. (1995). Electronic spectroscopy and photophysics of Si nanocrystals: Relationship to bulk c-Si and porous Si. *J. Am. Chem. Soc.*, 117(10):2915–2922.

Canham, L. (2000). Gaining light from silicon. *Nature*, 408:411–412.

Canham, L., Loni, A., Calcott, P., Simons, A., Reeves, C., Houlton, M., Newey, J., Nash, K., and Cox, T. (1996). On the origin of blue luminescence arising from atmospheric impregnation of oxidized porous silicon. *Thin Solid Films*, 276(1):112–115.

Canham, L. T. (1990). Silicon quantum wire array fabrication by electrochemical and chemical dissolution of wafers. *Appl. Phys. Lett.*, 57:1046–1048.

Cheng, X., Lowe, S. B., Reece, P. J., and Gooding, J. J. (2014). Colloidal silicon quantum dots: From preparation to the modification of self-assembled monolayers (SAMs) for bio-applications. *Chem. Soc. Rev.*, 43(8):2680–2700.

Chinnathambi, S., Chen, S., Ganesan, S., and Hanagata, N. (2014). Silicon quantum dots for biological applications. *Adv. Health. Mat.*, 3(1):10–29.

Choi, J., Zhang, Q., Reipa, V., Wang, N. S., Stratmeyer, M. E., Hitchins, V. M., and Goering, P. L. (2009). Comparison of cytotoxic and inflammatory responses of photoluminescent silicon nanoparticles with silicon micron-sized particles in RAW 264.7 macrophages. *J. Appl. Toxicol.*, 29(1):52–60.

Conus, F., Rodrigues, V., Lecoultre, S., Rydlo, A., and Félix, C. (2006). Matrix effects on the optical response of silver nanoclusters. *J. Chem. Phys.*, 125:024511.

Cullis, A., and Canham, L. (1991). Visible light emission due to quantum size effects in highly porous crystalline silicon. *Nature*, 353(6342):335–338.

Cullis, A., Canham, L., and Calcott, P. (1997). The structural and luminescence properties of porous silicon. *J. Appl. Phys.*, 82:909–965.

Dasog, M., De los Reyes, G. B., Titova, L. V., Hegmann, F. A., and Veinot, J. G. (2014). Size vs surface: Tuning the photoluminescence of freestanding silicon nanocrystals across the visible spectrum via surface groups. *ACS Nano*, 8(9):9636–9648.

Dasog, M., Kehrle, J., Rieger, B., and Veinot, J. G. (2016). Silicon nanocrystals and silicon-polymer hybrids: Synthesis, surface engineering, and applications. *Angew. Chem. Int. Ed.*, 55(7):2322–2339.

Dasog, M., Yang, Z., Regli, S., Atkins, T. M., Faramus, A., Singh, M. P., Muthuswamy, E., Kauzlarich, S. M., Tilley, R. D., and Veinot, J. G. (2013). Chemical insight into the origin of red and blue photoluminescence arising from free-standing silicon nanocrystals. *ACS Nano*, 7(3):2676–2685.

de Boer, W., Timmerman, D., Dohnalová, K., Yassievich, I., Zhang, H., Buma, W., and Gregorkiewicz, T. (2010). Red spectral shift and enhanced quantum efficiency in phonon-free photoluminescence from silicon nanocrystals. *Nat. Nanotechnol.*, 5:878.

Dohnalová, K., Gregorkiewicz, T., and Kusová, K. (2014). Silicon quantum dots: Surface matters. *J. Phys.*, 26(17):173201.

Echt, O., Sattler, K., and Recknagel, E. (1981). Magic numbers for sphere packings: Experimental verification in free xenon clusters. *Phys. Rev. Lett.*, 47(16):1121–1124.

Ehbrecht, M., and Huisken, F. (1999). Gas-phase characterization of silicon nanoclusters produced by laser pyrolysis of silane. *Phys. Rev. B*, 59(4):2975–2985.

Ehbrecht, M., Kohn, B., Huisken, F., Laguna, M. A., and Paillard, V. (1997). Photoluminescence and resonant Raman spectra of silicon films produced by size-selected cluster beam deposition. *Phys. Rev. B*, 56(11):6958–6964.

Erogbogbo, F., and Swihart, M. T. (2007). Photoluminescent silicon nanocrystals with mixed surface functionalization for biophotonics. *Mater. Res. Soc. Symp. Proc.*, 958:239.

Erogbogbo, F., Yong, K., Roy, I., Xu, G., Prasad, P., and Swihart, M. (2008). Biocompatible luminescent silicon quantum dots for imaging of cancer cells. *ACS Nano*, 2(5):873–878.

Erogbogbo, F., Yong, K.-T., Roy, I., Hu, R., Law, W.-C., Zhao, W., Ding, H., et al. (2010). In vivo targeted cancer imaging, sentinel lymph node mapping and multi-channel imaging with biocompatible silicon nanocrystals. *ACS Nano*, 5(1):413–423.

Falconieri, M., Borsella, E., De Dominicis, L., Enrichi, F., Franzò, G., Priolo, F., Iacona, F., Gourbilleau, F., and Rizk, R. (2005). Probe of the Si nanoclusters to Er energy transfer dynamics by double-pulse excitation. *Appl. Phys. Lett.*, 87:061109.

Felix, C., Sieber, C., Harbich, W., Buttet, J., Rabin, I., Schulze, W., and Ertl, G. (2001). Ag$_8$ fluorescence in argon. *Phys. Rev. Lett.*, 86(14):2992–2995.

Feng, C., Latimer, E., Spence, D., Al Hindawi, A. M., Bullen, S., Boatwright, A., Ellis, A. M., and Yang, S. (2015). Formation of Au and tetrapyridyl porphyrin complexes in superfluid helium. *Phys. Chem. Chem. Phys.*, 17(26):16699–16704.

Fielicke, A., Lyon, J. T., Haertelt, M., Meijer, G., Claes, P., De Haeck, J., and Lievens, P. (2009). Vibrational spectroscopy of neutral silicon clusters via far-IR-VUV two color ionization. *J. Chem. Phys.*, 131(17):171105.

Fitting, H., Ziems, T., von Czarnowski, A., and Schmidt, B. (2004). Luminescence center transformation in wet and dry SiO$_2$. *Radiat. Meas.*, 38(4–6):649–653.

Fowler, W., and Dexter, D. (1968). Electronic bubble states in liquid helium. *Phys. Rev.*, 176(1):337–343.

Fox, M. (2010). *Optical Properties of Solids*, vol. 3. Oxford University Press, Oxford, UK.

Fuke, K., Tsukamoto, K., Misaizu, F., and Sanekata, M. (1993). Near threshold photoionization of silicon clusters in the 248–146 nm region: Ionization potentials for Sin. *J. Chem. Phys.*, 99(10):7807–7812.

Galinis, G., Torricelli, G., Akraiam, A., and von Haeften, K. (2012a). Measurement of cluster-cluster interaction in liquids by deposition of silicon clusters onto HOPG surfaces. *J. Nanopart. Res.*, 14:1057.

Galinis, G., Yazdanfar, H., Bayliss, M., Watkins, M., and von Haeften, K. (2012b). Towards biosensing via fluorescent surface sites of nanoparticles. *J. Nanopart. Res.*, 14(8):1019.

Grass, M., Fischer, D., Mathes, M., Ganteför, G., and Nielaba, P. (2002). A form of bulk silicon consisting of "magic" clusters. *Appl. Phys. Lett.*, 81(20):3810–3812.

Grimes, C. C., and Adams, G. (1990). Infrared spectrum of the electron bubble in liquid helium. *Phys. Rev. B*, 41(10):6366–6371.

Grimes, C. C., and Adams, G. (1992). Infrared-absorption spectrum of the electron bubble in liquid helium. *Phys. Rev. B*, 45(5):2305.

Haberland, H., Karrais, M., and Mall, M. (1991). A new type of cluster and cluster ion source. *Z. Phys. D*, 20(1):413–415.

Halder, A., and Kresin, V. V. (2015). Nanocluster ionization energies and work function of aluminum, and their temperature dependence. *J. Chem. Phys.*, 143(16):164313.

Harbich, W., Sieber, C., Meiwes-Broer, K., and Félix, C. (2007). Electronic excitations induced by the impact of coinage metal ions and clusters on a rare gas matrix: Neutralization and luminescence. *Phys. Rev. B*, 76(10):104306.

Harris, C., Syväjärvi, M., Bergman, J., Kordina, O., Henry, A., Monemar, B., and Janzen, E. (1994). Time-resolved decay of the blue emission in porous silicon. *Appl. Phys. Lett.*, 65(19):2451–2453.

Hayes, W., and Stoneham, A. M. (2004). *Defects and Defect Processes in Nonmetallic Solids*. Dover, Mineloa, NY.

He, Y., Zhong, Y., Peng, F., Wei, X., Su, Y., Lu, Y., Su, S., Gu, W., Liao, L., and Lee, S.-T. (2011). One-pot microwave synthesis of water-dispersible, ultraphoto-and pH-stable, and highly fluorescent silicon quantum dots. *J. Angew. Chem. Int. Ed.*, 133(36):14192–14195.

Hirsch, K., Lau, J., Klar, P., Langenberg, A., Probst, J., Rittmann, J., Vogel, M., Zamudio-Bayer, V., Möller, T., and Issendorff, B. (2009). X-ray spectroscopy on size-selected clusters in an ion trap: From the molecular limit to bulk properties. *J. Phys. B*, 42:154029.

Hirschman, K., Tsybeskov, L., Duttagupta, S., and Fauchet, P. (1996). Silicon-based visible light-emitting devices integrated into microelectronic circuits. *Nature*, 384(6607):338–341.

Ho, K., Shvartsburg, A., Pan, B., Lu, Z., Wang, C., Wacker, J., Fye, J., and Jarrold, M. (1998). Structures of medium-sized silicon clusters. *Nature*, 392(6676):582–585.

Hoffmann, G., Kliewer, J., and Berndt, R. (2001a). Luminescence from metallic quantum wells in a scanning tunneling microscope. *Phys. Rev. Lett.*, 87(17):176803.

Hoffmann, M. A., Wrigge, G., Issendorff, B. V., Müller, J., Ganteför, G., and Haberland, H. (2001b). Ultraviolet photoelectron spectroscopy of Si 4- to Si 1000. *Eur. J. Phys. D Atmos. Mol. Clusters*, 16(1):9–11.

Hofmeister, H., Huisken, F., and Kohn, B. (1999). Lattice contraction in nanosized silicon particles produced by laser pyrolysis of silane. *Eur. J. Phys. D Atmos. Mol. Clusters Atomic Mol., Opt. Plasm. Phys.*, 9(1):137–140.

Honea, E. C., Kraus, J., Bower, J., and Jarrold, M. (1993a). Optical spectra of size-selected matrix-isolated silicon clusters. *Z. Phys. D Atmos. Mol. Clusters*, 26(1):141–143.

Honea, E. C., Ogura, A., Murray, C. A., Raghavachari, K., Sprenger, W. O., Jarrold, M. F., and Brown, W. L. (1993b). Raman spectra of size-selected silicon clusters and comparison with calculated structures. *Nature*, 366:40–44.

Honea, E.C, Ogura, A., Peale, D., Félix, C., Murray, C., Raghavachari, K., Sprenger, W., Jarrold, M., and Brown, W. (1999). Structures and coalescence behavior of size-selected silicon nanoclusters studied by surface-plasmon-polariton enhanced Raman spectroscopy. *J. Chem. Phys.*, 110:24.

Hoops, A. A., Bise, R.T., Choi, H., and Neumark, D. M. (2001). Photodissociation spectroscopy and dynamics of Si_4. *Chem. Phys. Lett.*, 346(1):89–96.

Hou, X., Shi, G., Wang, W., Zhang, F., Hao, P., Huang, D., and Wang, X. (1993). Large blue shift of light emitting porous silicon by boiling water treatment. *Appl. Phys. Lett.*, 62(10):1097–1098.

Hudgins, R. R., Imai, M., Jarrold, M. F., and Dugourd, P. (1999). High-resolution ion mobility measurements for silicon cluster anions and cations. *J. Chem. Phys.*, 111(17):7865–7870.

Huisken, F., Kohn, B., Alexandrescu, R., Cojocaru, S., Crunteanu, A., Ledoux, G., and Reynaud, C. (1999). Silicon carbide nanoparticles produced by CO2 laser pyrolysis of SiH_4/C_2H_2 gas mixtures in a flow reactor. *J. Nanopart. Res.*, 1(2):293–303.

Intartaglia, R., Bagga, K., Genovese, A., Athanassiou, A., Cingolani, R., Diaspro, A., and Brandi, F. (2012a). Influence of organic solvent on optical and structural properties of ultra-small silicon dots synthesized by UV laser ablation in liquid. *Phys. Chem. Chem. Phys.*, 14(44):15406–15411.

Intartaglia, R., Bagga, K., Scotto, M., Diaspro, A., and Brandi, F. (2012b). Luminescent silicon nanoparticles prepared by ultra short pulsed laser ablation in liquid for imaging applications. *Opt. Mat. Exp.*, 2(5):510–518.

Jackson, K. A., Horoi, M., Chaudhuri, I., Frauenheim, T., and Shvartsburg, A. A. (2004). Unraveling the shape transformation in silicon clusters. *Phys. Rev. Lett.*, 93(1):013401.

Jarrold, M. F., and Constant, V. A. (1991). Silicon cluster ions: Evidence for a structural transition. *Phys. Rev. Lett.*, 67(21):2994–2997.

Kanemitsu, Y., Futagi, T., Matsumoto, T., and Mimura, H. (1994). Origin of the blue and red photoluminescence from oxidized porous silicon. *Phys. Rev. B*, 49(20):14732–14735.

Kasigkeit, C., Hirsch, K., Langenberg, A., Moöller, T., Probst, J., Rittmann, J., Vogel, M., et al. (2015). Higher ionization energies from sequential vacuum-ultraviolet multiphoton ionization of size-selected silicon cluster cations. *J. Phys. Chem. C*, 119(20):11148–11152.

Katzy, R., Singer, M., Izadnia, S., LaForge, A., and Stienkemeier, F. (2016). Doping He droplets by laser ablation with a pulsed supersonic jet source. *Rev. Sci. Instrum.*, 87(1):013105.

Kimura, K. (1999). Blue luminescence from silicon nanoparticles suspended in organic liquids. *J. Cluster Sci.*, 10(2):359–380.

Kitsopoulos, T. N., Chick, C. J., Weaver, A., and Neumark, D. M. (1990). Vibrationally resolved photoelectron spectra of and. *J. Chem. Phys.*, 93(8):6108–6110.

Knight, W., Clemenger, K., de Heer, W., Saunders, W., Chou, M., and Cohen, M. (1984). Electronic shell structure and abundances of sodium clusters. *Phys. Rev. Lett.*, 52(24):2141–2143.

Kontkiewicz, A. J., Kontkiewicz, A. M., Siejka, J., Sen, S., Nowak, G., Hoff, A., Sakthivel, P., et al. (1994). Evidence that blue luminescence of oxidized porous silicon originates from SiO_{2n}. *Appl. Phys. Lett.*, 65(11):1436–1438.

Kovalev, D., Yaroshetzkii, I., Muschik, T., Petrova-Koch, V., and Koch, F. (1994). Fast and slow visible luminescence bands of oxidized porous Si. *Appl. Phys. Lett.*, 64(2):214–216.

Koyama, H., Matsushita, Y., and Koshida, N. (1998). Activation of blue emission from oxidized porous silicon by annealing in water vapor. *J. Appl. Phys.*, 83(3):1776–1778.

Kumar, V., and Kawazoe, Y. (2001). Metal-encapsulated fullerenelike and cubic caged clusters of silicon. *Phys. Rev. Lett.*, 87(4):045503.

Kumar, V., and Kawazoe, Y. (2003a). Hydrogenated silicon fullerenes: Effects of H on the stability of metal-encapsulated silicon clusters. *Phys. Rev. Lett.*, 90(5):055502.

Kumar, V., and Kawazoe, Y. (2003b). Metal-doped magic clusters of Si, Ge, and Sn: The finding of a magnetic superatom. *Appl. Phys. Lett.*, 83(13):2677–2679.

Laarmann, T., Wabnitz, H., von Haeften, K., and Möller, T. (2008). Photochemical processes in doped argon-neon core-shell clusters: The effect of cage size on the dissociation of molecular oxygen. *J. Chem. Phys.*, 128(1):014502.

Laguna, M., Paillard, V., Kohn, B., Ehbrecht, M., Huisken, F., Ledoux, G., Papoular, R., and Hofmeister, H. (1999). Optical properties of nanocrystalline silicon thin films produced by size-selected cluster beam deposition. *J. Lumin.*, 80(1–4):223–228.

Lau, J., Vogel, M., Langenberg, A., Hirsch, K., Rittmann, J., Zamudio-Bayer, V., Möller, T., and von Issendorff, B. (2011). Communication: Highest occupied molecular orbital–lowest unoccupied molecular orbital gaps of doped silicon clusters from core level spectroscopy. *J. Chem. Phys.*, 134(4):041102.

Ledoux, G., Amans, D., Gong, J., Huisken, F., Cichos, F., and Martin, J. (2002). Nanostructured films composed of silicon nanocrystals. *Mater. Sci. Eng. C*, 19(1–2):215–218.

Ledoux, G., Gong, J., and Huisken, F. (2001). Effect of passivation and aging on the photoluminescence of silicon nanocrystals. *Appl. Phys. Lett.*, 79(24):4028–4030.

Ledoux, G., Guillois, O., Porterat, D., Reynaud, C., Huisken, F., Kohn, B., and Paillard, V. (2000). Photoluminescence properties of silicon nanocrystals as a function of their size. *Phys. Rev. B*, 62(23):15942–15951.

Li, Q., He, Y., Chang, J., Wang, L., Chen, H., Tan, Y.-W., Wang, H., and Shao, Z. (2013). Surface-modified silicon nanoparticles with ultrabright photoluminescence and single-exponential decay for nanoscale fluorescence lifetime imaging of temperature. *J. Am. Chem. Soc.*, 135(40):14924–14927.

Li, Q., Luo, T.-Y., Zhou, M., Abroshan, H., Huang, J., Kim, H. J., Rosi, N. L., Shao, Z., and Jin, R. (2016). Silicon nanoparticles with surface nitrogen: 90% quantum yield with narrow luminescence bandwidth and the ligand structure based energy law. *ACS Nano*, 10(9):8385–8393.

Li, X., He, Y., and Swihart, M. T. (2004a). Surface functionalization of silicon nanoparticles produced by laser-driven pyrolysis of silane followed by $HF-HNO_3$ etching. *Langmuir*, 20(11):4720–4727.

Li, X., He, Y., Talukdar, S., and Swihart, M. (2004b). Preparation of luminescent silicon nanoparticles by photothermal aerosol synthesis followed by acid etching. *Phase Transitions*, 77(1–2):131–137.

Li, Z., and Ruckenstein, E. (2004). Water-soluble poly (acrylic acid) grafted luminescent silicon nanoparticles and their use as fluorescent biological staining labels. *Nano Lett.*, 4(8):1463–1467.

Lockwood, D. (1994). Optical properties of porous silicon. *Solid State Commun.*, 92(1–2):101–112.

Lockwood, D. J., Lu, Z. H., and Baribeau, J.-M. (1996). Quantum confined luminescence in Si/SiO_2 superlattices. *Phys. Rev. Lett.*, 76(3):539.

Loni, A., Simons, A., Calcott, P., and Canham, L. (1995). Blue photoluminescence from rapid thermally oxidized porous silicon following storage in ambient air. *J. Appl. Phys.*, 77(7):3557–3559.

Lyon, J. T., Gruene, P., Fielicke, A., Meijer, G., Janssens, E., Claes, P., and Lievens, P. (2009). Structures of silicon cluster cations in the gas phase. *J. Angew. Chem. Int. Ed.*, 131(3):1115–1121.

Mangolini, L., Thimsen, E., and Kortshagen, U. (2005). High-yield plasma synthesis of luminescent silicon nanocrystals. *Nano Lett.*, 5(4):655–659.

Marsen, B., Lonfat, M., Scheier, P., and Sattler, K. (2000). Energy gap of silicon clusters studied by scanning tunneling spectroscopy. *Phys. Rev. B*, 62(11):6892–6895.

Maus, M., Ganteför, G., and Eberhardt, W. (2000). The electronic structure and the band gap of nano-sized Si particles: Competition between quantum confinement and surface reconstruction. *Appl. Phys. A Mater. Sci. Process.*, 70(5):535–539.

McVey, B. F., and Tilley, R. D. (2014). Solution synthesis, optical properties, and bioimaging applications of silicon nanocrystals. *Acc. Chem. Res.*, 47(10):3045–3051.

Meloni, G., Ferguson, M. J., Sheehan, S. M., and Neumark, D. M. (2004). Probing structural transitions of nanosize silicon clusters via anion photoelectron spectroscopy at 7.9 eV. *Chem. Phys. Lett.*, 399(4):389–391.

Mitas, L., Grossman, J. C., Stich, I., and Tobik, J. (2000). Silicon clusters of intermediate size: Energetics, dynamics, and thermal effects. *Phys. Rev. Lett.*, 84(7):1479–1482.

Montalti, M., Cantelli, A., and Battistelli, G. (2015). Nanodiamonds and silicon quantum dots: Ultrastable and biocompatible luminescent nanoprobes for long-term bioimaging. *Chem. Soc. Rev.*, 44(14):4853–4921.

Nel, A., Madler, L., Velegol, D., Xia, T., Hoek, E., Somasundaran, P., Klaessig, F., Castranova, V., and Thompson, M. (2009). Understanding biophysicochemical interactions at the nano-bio interface. *Nat. Mater.*, 8(7):543–557.

Nishikawa, H., Shiroyama, T., Nakamura, R., Ohki, Y., Nagasawa, K., and Hama, Y. (1992). Photoluminescence from defect centers in high-purity silica glasses observed under 7.9-eV excitation. *Phys. Rev. B*, 45(2):586–591.

Ondič, L., Kusová, K., Ziegler, M., Fekete, L., Gärtnerová, V., Cháb, V., Holý, V., et al. (2014). A complex study of the fast blue luminescence of oxidized silicon nanocrystals: The role of the core. *Nanoscale*, 6(7):3837–3845.

Park, J., Gu, L., von Maltzahn, G., Ruoslahti, E., Bhatia, S., and Sailor, M. (2009). Biodegradable luminescent porous silicon nanoparticles for in vivo applications. *Nat. Mater.*, 8(4):331–336.

Park, N., Kim, T., and Park, S. (2001). Band gap engineering of amorphous silicon quantum dots for light-emitting diodes. *Appl. Phys. Lett.*, 78:2575.

Pavesi, L. (2003). Will silicon be the photonic material of the third millenium? *J. Phys. C*, 15:1169.

Pavesi, L., Negro, L. D., Mazzoleni, C., Franzò, G., and Priolo, F. (2000). Optical gain in silicon nanocrystals. *Nature*, 408:440–444.

Peng, F., Su, Y., Zhong, Y., Fan, C., Lee, S.-T., and He, Y. (2014). Silicon nanomaterials platform for bioimaging, biosensing, and cancer therapy. *Acc. Chem. Res.*, 47(2):612–623.

Pratontep, S., Carroll, S. J., Xirouchaki, C., Streun, M., and Palmer, R. E. (2005). Size-selected cluster beam source based on radio frequency magnetron plasma sputtering and gas condensation. *Rev. Sci. Instrum.*, 76:045103.

Priolo, F., Gregorkiewicz, T., Galli, M., and Krauss, T. F. (2014). Silicon nanostructures for photonics and photovoltaics. *Nat. Nanotechnol.*, 9(1):19–32.

Raghavachari, K. (1986). Theoretical study of small silicon clusters: Equilibrium geometries and electronic structures of Si_n (n= 2–7, 10). *J. Chem. Phys.*, 84(10):5672–5686.

Raghavachari, K., and Logovinsky, V. (1985). Structure and bonding in small silicon clusters. *Phys. Rev. Lett.*, 55(26):2853.

Raghavachari, K., and Rohlfing, C. M. (1988). Bonding and stabilities of small silicon clusters: A theoretical study of Si_7 Si_{10}. *J. Chem. Phys.*, 89(4):2219–2234.

Rodio, M., Brescia, R., Diaspro, A., and Intartaglia, R. (2016). Direct surface modification of ligand-free silicon quantum dots prepared by femtosecond laser ablation in deionized water. *J. Colloid Interface Sci.*, 465:242–248.

Sattler, K., Mühlbach, J., and Recknagel, E. (1980). Generation of metal clusters containing from 2 to 500 atoms. *Phys. Rev. Lett.*, 45(10):821–824.

Seraphin, A., Ngiam, S.-T., and Kolenbrander, K. (1996). Surface control of luminescence in silicon nanoparticles. *J. Appl. Phys.*, 80(11):6429–6433.

Sieber, C., Buttet, J., Harbich, W., Felix, C., Mitric, R., and Bonacic-Koutecky, V. (2004). Isomer-specific spectroscopy of metal clusters trapped in a matrix: Ag_9. *Phys. Rev. A*, 70(4):041201.

Skuja, L. (1992). Isoelectronic series of twofold coordinated Si, Ge, and Sn atoms in glassy SiO_2: A luminescence study. *J. Non-Cryst. Solids*, 149(1/2):77–95.

Skuja, L., Streletsky, A., and Pakovich, A. (1984). A new intrinsic defect in amorphous SiO_2: Twofold coordinated silicon. *Solid State Commun.*, 50(12):1069–1072.

Stupca, M., Alsalhi, M., Al Saud, T., Almuhanna, A., and Nayfeh, M. (2007). Enhancement of polycrystalline silicon solar cells using ultrathin films of silicon nanoparticle. *Appl. Phys. Lett.*, 91:063107.

Sun, Q., Wang, Q., Jena, P., Rao, B. K., and Kawazoe, Y. (2003). Stabilization of SiO_{60} cage structure. *J Phys. Rev. Lett.*, 90(13):135503.

Švrček, V., Mariotti, D., Cvelbar, U., Filipič, G., Lozach, M., McDonald, C., Tayagaki, T., and Matsubara, K. (2016). Environmentally friendly processing technology for engineering silicon nanocrystals in water with laser pulses. *J. Phys. Chem. C*, 120(33):18822–18830.

Švrček,, V., Mariotti, D., and Kondo, M. (2009a). Ambient-stable blue luminescent silicon nanocrystals prepared by nanosecond-pulsed laser ablation in water. *Opt. Express.*, 17(2):520–527.

Švrček, V., Sasaki, T., Katoh, R., Shimizu, Y., and Koshizaki, N. (2009b). Aging effect on blue luminescent silicon nanocrystals prepared byápulsed laser ablation of silicon wafer in de-ionized water. *Appl. Phys. B*, 94(1):133–139.

Švrček, V., Slaoui, A., and Muller, J. (2004). Silicon nanocrystals as light converter for solar cells. *Thin Solid Films*, 451:384–388.

Takagi, H., Ogawa, H., Yamazaki, Y., Ishizaki, A., and Nakagiri, T. (1990). Quantum size effects on photoluminescence in ultrafine Si particles. *Appl. Phys. Lett.*, 56:2379.

Tománek, D., and Schlüter, M. (1986). Calculation of magic numbers and the stability of small Si clusters. *Phys. Rev. Lett.*, 56(10):1055–1058.

Torricelli, G., Akraiam, A., and von Haeften, K. (2011). Size-selecting effect of water on fluorescent silicon clusters. *Nanotechnology*, 22:315711.

Tsybeskov, L., Vandyshev, J. V., and Fauchet, P. (1994). Blue emission in porous silicon: Oxygen-related photoluminescence. *Phys. Rev. B*, 49(11):7821.

Valenta, J., Fucikova, A., Pelant, I., Kusová, K., Dohnalová, K., Aleknavičius, A., Cibulka, O., Fojtík, A., and Kada, G. (2008). On the origin of the fast photoluminescence band in small silicon nanoparticles. *New J. Phys.*, 10:073022.

Vincent, J., Maurice, V., Paquez, X., Sublemontier, O., Leconte, Y., Guillois, O., Reynaud, C., Herlin-Boime, N., Raccurt, O., and Tardif, F. (2010). Effect of water and UV passivation on the luminescence of suspensions of silicon quantum dots. *J. Nanopart. Res.*, 12(1):39–46.

Vogel, M., Kasigkeit, C., Hirsch, K., Langenberg, A., Rittmann, J., Zamudio-Bayer, V., Kulesza, A., et al. (2012). 2p core-level binding energies of size-selected free silicon clusters: Chemical shifts and cluster structure. *Phys. Rev. B*, 85(19):195454.

von Haeften, K., Akraiam, A., Torricelli, G., and Brewer, A. (2010a). Fluorescence of silicon nanoparticles suspended in water: Reactive co-deposition for the control of surface properties of clusters. *AIP Conf. Proc.*, 1275:40.

von Haeften, K., Binns, C., Brewer, A., Crisan, O., Howes, P., Lowe, M., Sibbley-Allen, C., and Thornton, S. C. (2009). A novel approach towards the production of luminescent silicon nanoparticles: Sputtering, gas aggregation and co-deposition with H_2O. *Eur. Phys. J. D*, 52(1–3):11–14.

von Haeften, K., Lowe, M., and Brewer, A. (2010b). *Fluorescence of Silicon Nanoparticles Suspended in Water*. Innsbruck University Press, Innsbruck, Austria.

Warner, J., Hoshino, A., Yamamoto, K., and Tilley, R. (2005). Water-soluble photoluminescent silicon quantum dots. *Angew. Chem. Int. Ed.*, 44(29):4550–4553.

Wegner, K., Piseri, P., Tafreshi, H., and Milani, P. (2006). Cluster beam deposition: A tool for nanoscale science and technology. *J. Phys. D Appl. Phys.*, 39(22):R439–R459.

Wilson, W., Szajowski, P., and Brus, L. (1993). Quantum confinement in size-selected, surface-oxidized silicon nanocrystals. *Science*, 262(5137):1242–1244.

Xu, C., Taylor, T. R., Burton, G. R., and Neumark, D. M. (1998). Vibrationally resolved photoelectron spectroscopy of silicon cluster anions Si − n(n = 3–7). *J. Chem. Phys.*, 108(4):1395–1406.

Yamani, Z., Thompson, W. H., AbuHassan, L., and Nayfeh, M. H. (1997). Ideal anodization of silicon. *Appl. Phys. Lett.*, 70(25):3404–3406.

Yazdanfar, H., Kotlov, A., and von Haeften, K. (2012). *Blue Photoluminescence of Oxidised Silicon Nanoparticles in Solution and on Surfaces*. Technical report. DESY Photon Science, Annual Report, Hamburg, Germany.

Yu, D., Hang, Q., Ding, Y., Zhang, H., Bai, Z., Wang, J., Zou, Y., Qian, W., Xiong, G., and Feng, S. (1998). Amorphous silica nanowires: Intensive blue light emitters. *Appl. Phys. Lett.*, 73:3076.

Zhong, Y., Peng, F., Bao, F., Wang, S., Ji, X., Yang, L., Su, Y., Lee, S.-T., and He, Y. (2013). Large-scale aqueous synthesis of fluorescent and biocompatible silicon nanoparticles and their use as highly photostable biological probes. *J. Am. Chem. Soc.*, 135(22):8350–8356.

Žídek, K., Trojanek, F., Malý, P., Pelant, I., Gilliot, P., and Hönerlage, B. (2011). Ultrafast photoluminescence dynamics of blue-emitting silicon nanostructures. *Phys. Stat. Sol. (c)*, 8(3):979–984.

9 Silicon nanoparticles from pulsed laser ablation

Canan Kurşungöz, Elif Uzcengiz Şimşek, and Bülend Ortaç

Contents

9.1 INTRODUCTION AND MOTIVATION

The research on silicon nanocrystals (SiNCs) has been considerably attractive for the scientist in the last decades due to the SiNCs' significant size-dependent optical properties, and their potential for stimulated emission (Pavesi et al. 2000). The bright photoluminescence (PL) properties of SiNCs is mainly observed in visible regions of the spectrum (Wolkin et al. 1999). On the other hand, blue-luminescent SiNCs would provide a decrease in free carrier absorption and an increase in the stimulated emission (Švrček et al. 2006). These excellent properties and their biocompatible nature make SiNCs a promising candidate for applications such as light-emitting devices or energy sources, in biomedicine and photodynamic therapy (Walters et al. 2005; Stupca et al. 2007; Wang et al. 2008). Moreover, it was suggested that the production of SiNCs in colloidal suspension provides considerable advantages in optoelectronics applications due to the increased stimulated emission process (Luterova et al. 2016).

A number of chemical (Zhang et al. 2007; Rosso-Vasic et al. 2009) and physical (Knipping et al. 2004; Khang and Lee 2010) methods were developed for the production of SiNCs. Chemical methods were much more favorable due to their potential for controlling both the particle size and surface properties (Intartaglia et al. 2011; Intartaglia et al. 2012a). These chemical methods included reduction reactions (Warner et al. 2005; Zhang et al. 2007; Rosso-Vasic et al. 2009), electrochemical etching (Belomoin et al. 2002; Kůsová et al. 2010), gaseous phase decomposition of silane (Li et al. 2004; Mangolini et al. 2005). However, the use of chemical products and reducing agents in chemical methods results in by-product contamination in the nanocrystal solution, which would affect the downstream procedures (Warner et al. 2005; Kůsová et al. 2010). Furthermore, these methods are time consuming and costly since they consist of multistep procedures (Kabashin and Meunier 2006; Intartaglia et al. 2012a).

Pulsed laser ablation in liquids (PLAL) is a promising method for the production of pure colloidal nanoparticle solutions (Rioux et al. 2009; Intartaglia et al. 2011). PLAL offers a number of advantages over other nanoparticle production methods. First, the colloidal nanoparticle solution is produced in a chemically pure environment since there is no need for any chemical precursors for the nanoparticles to be produced. Thus, at the end of the production, there are no residual byproducts or contamination which would affect the downstream application processes in the solution. Second, the PLAL method is a relatively simple one when compared to other nanoparticle production methods, since there is no requirement for extreme temperature and pressure conditions. Moreover, it can be applied to the production of a number of different nanoparticles. Last but not least, PLAL is a versatile method allowing for further nanoscale functionalization as the nanoparticles are in a colloidal solution (Yang et al. 2008; Yang et al. 2009a; Intartaglia et al. 2011).

The (111) crystalline structure of SiNCs could be achieved by using different types of lasers with varying pulse energies and pulse durations. In general, 800 nm femtosecond lasers, 532, 1064, and 355 nm nanosecond (ns) lasers and also picosecond lasers are used in the PLAL process for SiNC production. Although different solvents, such as hexane, chloroform, ethanol, and a variety of surfactants, are used for SiNC generation with PLAL, deionized water is the commonly preferred and successful solvent for this process. The laser parameters (wavelength, pulse duration, pulse energy, and repetition rate), the liquid environment, and ablation duration are the key parameters for the production of SiNC with PLAL mechanism. The overall literature search about these parameters and resulting SiNC features are summarized in Table 9.1.

In this chapter, the PLAL mechanism and a general setup will be introduced. The effects of laser ablation parameters (pulse energy, pulse duration, laser wavelength, and liquid environment) will be discussed in detail with a comprehensive literature search. The physical and chemical properties of produced SiNCs and the effect of posttreatment methods will be reviewed. The chapter will be concluded with the different applications of SiNCs produced with the PLAL mechanism.

9.2 PLAL MECHANISM AND SETUP

In the ablation of a solid target in a liquid environment, a series of processes are observed. First, the plasma plume is generated when the laser light reaches the solid target. Then, the generated plasma plume is transformed and condensed; it is strongly affected by the liquid confinement when considered in terms of its thermodynamic and kinetic properties.

The liquid confinement leads to the formation of a shock wave due to the adiabatic expansion of the laser-induced plasma at a supersonic velocity in the plasma plume. Due to the incoming laser pulse, the continuous material removal from the solid target takes place with vaporization. Then, extra pressure is produced in the plasma due to the shock wave, which is called plasma-induced pressure, and the temperature in the plasma increases. Thus, the thermodynamic state of the plasma changes to the state with higher temperature, higher pressure, and higher density, which allows the formation of the metastable phases (Yang 2007).

There are four types of chemical reactions observed in the laser-induced plasma and the interface between the liquid and the plasma during its transformation process. The first reaction takes place inside the plasma due to the high temperature and high-pressure state of the laser-induced, high-density plasma, and a new phase can form by these chemical reactions between the ablations from the target. The second chemical reaction again takes place inside the plasma. In this reaction, the liquid molecules at the interface are excited and evaporated due to the high temperature and high pressure in the plasma. Thus, a new plasma from the liquid molecules is generated at the interface, which is called plasma-induced plasma. The laser-induced plasma and plasma-induced plasma are mixed and the laser-induced plasma includes a number of species from the plasma-induced plasma inside. The species generated from the laser ablation of the target and the species formed due to the excitation of the liquid molecules are engaged in the chemical reactions in this part. The third chemical reaction takes place at the interface of the laser-induced plasma and the liquid due to the high temperature, pressure, and density. The fourth reaction is observed inside the liquid when the ablated species from the solid target is affected by the high pressure of the laser-induced plasma. In the chemical reactions where two species, namely the ones from the solid target and the ones from the confining liquid, are

Table 9.1 Detailed summary of the literature indicating the effects of laser parameters and liquid environment on the SiNC produced by PLAL

LASER WAVELENGTH	LASER PULSE DURATION	PULSE ENERGY	REPETITION RATE	ABLATION TIME	TARGET MATERIAL	LIQUID ENVIRONMENT	POSTTREATMENT	CRYSTAL STRUCTURE	DIAMETER	OPTICAL PROPERTIES AND BAND GAP	REFERENCE
800 nm	120 fs	0.1–1 mJ	1 kHz	30 min	<100> Si wafer	Deionized water		<111> crystalline state with 3.14 Å lattice constant. Amorphous structures	2.4 nm	1270 nm emission (PL)	Rioux et al. 2009
532 nm		10 mJ/cm²			Single-crystal silicon wafer	Water, hexane			<10 nm	2.9 eV gap energy in water 3.5 eV gap energy in hexane	Umezu et al. 2007
800 nm	100 fs	0.40 mJ; 0.27 mJ; 0.16 mJ	1 kHz	60 min	Bulk Si target	Deionized water		<111> crystalline state with 3.12 Å lattice constant	60 nm; 3.5 nm; 2.5 nm	1.21 eV; 1.56 eV; 1.67 eV	Intartaglia et al. 2011
355 nm	40 ns	40 J/cm²	5 kHz		Si wafer	Chloroform	Physiochemical posttreatment (isopropanol/HF/hexane 3:1:3 and ultrasonic posttreatment)	Monocrystaline and polycrystalline. <111> crystalline state with 3.1 Å lattice constant	50 nm diameter decreased to 4 nm	500 nm emission (PL)	Abderrafi et al. 2011
800 nm	110 fs	0.15 mJ; 0.4 mJ	1 kHz		Bulk Si target	Deionized water		<111> crystalline state with 3.12 Å lattice constant	1–8 nm (mean 5.5 nm); 10–120 nm (mean 65 nm)	460 nm absorption, 475 nm emission; 485 nm absorption, 575 nm emission	Intartaglia et al. 2012

(Continued)

Clusters, nanoparticles, and quantum dots

Table 9.1 (*Continued*) Detailed summary of the literature indicating the effects of laser parameters and liquid environment on the SiNC produced by PLAL

LASER WAVELENGTH	LASER PULSE DURATION	PULSE ENERGY	REPETITION RATE	ABLATION TIME	TARGET MATERIAL	LIQUID ENVIRONMENT	POSTTREATMENT	CRYSTAL STRUCTURE	DIAMETER	OPTICAL PROPERTIES AND BAND GAP	REFERENCE
355 nm	8 ns	0.07 mJ	30 Hz	30 min	<100> Crystalline Si wafer	Deionized water		<111> crystalline state with 3 Å lattice constant	2–100 nm (mean 60 nm)	427 nm emission	Švrček et al. 2006
		1.1 mJ									
		2.4 mJ									
		6.0 mJ							2–50 nm (mean 20 nm)	399 nm emission	
1064 nm	10 ns	150 mJ	10 Hz	60 min	Bulk Si target	0.03 M SDS	Centrifuge + addition of ethanol to the pellet + ultrasonic rinsing	<111> crystalline state with 3 Å lattice constant	10–20 nm	415 nm and 435 nm emission	Yang et al. 2008
387 nm	180 fs	3.5 µJ	1 kHz	70 min	<001> Si wafer	Deionized water		Crystalline structure	5–200 nm (median 20 nm)	1033 nm absorption	Semaltianos et al. 2010
800 nm	35 fs	4 J/cm²			Single-crystal silicon wafer	Ethanol		<111> crystalline state	~40 nm mean	635 nm emission	Kuzmin et al. 2010
	100 fs								~25 nm mean		
	200 fs								~50 nm mean		
	600 fs								~60 nm mean		
	900 fs								~50 nm mean		
800 nm	100 fs	1 mJ	1 kHz	60 min	Bulk Si target	Deionized water			60 nm	485 nm absorption	Intartaglia et al. 2012
						Single-stranded oligonucleotide solution			5 nm	460 nm absorption, 450 nm emission	

(*Continued*)

Table 9.1 (Continued) Detailed summary of the literature indicating the effects of laser parameters and liquid environment on the SiNC produced by PLAL

LASER WAVELENGTH	LASER PULSE DURATION	PULSE ENERGY	REPETITION RATE	ABLATION TIME	TARGET MATERIAL	LIQUID ENVIRONMENT	POSTTREATMENT	CRYSTAL STRUCTURE	DIAMETER	OPTICAL PROPERTIES AND BAND GAP	REFERENCE
1064 nm	10 ns	50 mJ	10 Hz	30 min	<111> Si wafer	Water	The samples prepared in ethanol with 100 mJ were centrifuged as follows:		19 nm	Blue shift in NPs when compared to bulk Si in absorption. More blue shift with SiNC in water. Two emission peaks for both samples, at 415 and 435 nm	Yang et al. 2009b
						Ethanol			6.8 nm		
		100 mJ				Water			16 nm		
						Ethanol			5.4 nm		
						Water:ethanol 1:1			6.1 nm		
		150 mJ				Water:ethanol 3:1			7.1 nm		
						Water	4000 r/min—9.7 nm		14 nm		
						Ethanol	8000 r/min—8.3 nm		3.6 nm		
		200 mJ				Water	14000 r/min—6.8 nm		21 nm		
						Ethanol	left colloid: 4.4 nm		3.1 nm		

involved, the generation of new materials in nanoscale by the combination of the target and the liquid takes place. Finally, the plasma plume cools down and condenses in the confining liquid in two different ways. It can either condenses and deposits back on the target solid or condenses and become dispersed in the liquid by forming nanoscale materials (Yang 2007). When we explain the nanoscale material formation mechanism in detail, we should first assume the clusters and the surrounding plasma have the same temperature T, then isothermal nucleation time is given by

$$\tau = \sqrt{2\pi mkT}\,\frac{kT\gamma}{p_s(T)(\Delta\mu)^2} \tag{9.1}$$

where m, k, T, γ, $P_s(T)$, $\Delta\mu$ denote the mass of a single atom, the Boltzmann constant, the absolute temperature, the surface energy density of the material, saturated vapor pressure of nuclei at the temperature of T, and atom chemical potential difference, respectively. When the nucleation time decreases, the pressure increases. However, the temperature and the nucleation time are adversely related to each other due to effect of the saturated vapor pressure of nuclei. Consequently, the diameter of nanomaterials could be expressed as

$$d = V\left(2\tau_d - \tau\right) + 2r^* \tag{9.2}$$

where τ_d and r^* are the laser pulse duration and the size of critical nuclei of diamond, respectively. These theoretical calculations were also shown to be in good agreement with the experimental results (Yang et al. 1999; Yang 2007).

Yang et al. (2009b) also investigated the effect of a liquid environment and the laser energy on the structure of the produced SiNCs, specifically, and they proposed a mechanism for these effects similar to the nucleation and growth theory (Yang et al. 2009b). As mentioned above, the interaction between pulsed laser light and the target leads to an instant local high-temperature and high-pressure plasma plume at the interface of the target and the liquid environment. Then, such plasma will ultrasonically and adiabatically expand, resulting in the cooling of the plume region and formation of clusters. With the annihilation of the plasma, the adjacent formed clusters aggregate quickly into nanoparticles. They suggested that at low laser energies, the generation of Si atoms filled in the plasma plume is less and the pressure and temperature of the plasma plume is low. Hence, nucleation is relatively difficult. When the pressure and temperature of the plasma plume suddenly decrease, further nucleation becomes almost unattainable. The small number of nuclei formed uses the surrounding Si clusters, thus leading to large-sized nanoparticles. Under high laser energy, due to relatively high temperature and pressure of the formed plasma plume, nucleation takes place almost simultaneously all around the plume region. Therefore, a number of supplied Si atoms at high laser energy are shared by more nuclei, resulting in small-sized nanoparticles (Yang et al. 2009b).

Besides the effect of laser energy, the liquid environment is also important in the PLAL process. Different transparencies are suggested to directly affect the peak value of the pressure and temperature of the plasma plume after laser irradiation. Moreover, different thermal conductivities of the liquids have an effect on the decay process of the temperature and pressure of the plume. Furthermore, various density and viscosity values have an impact on the expansion of the plasma plume. All these liquid parameters have an effect together on the nucleation and the following growth process of the nanoparticles. This is why different-sized SiNCs are obtained with different liquid environments (Yang et al. 2009b).

A general setup scheme for the generation of SiNC with PLAL mechanism is given in Figure 9.1. As shown in the figure, there are key parameters for PLAL to produce SiNCs successfully. These parameters are the laser wavelength, laser pulse duration, pulse energy, repetition rate, ablation time, and liquid environment. The effect of those parameters on SiNC will be discussed in detail in the following section.

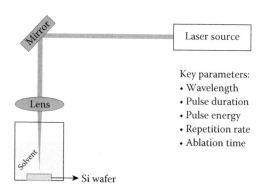

Figure 9.1 A representative PLAL setup for the formation of SiNC. The laser light is transferred with mirror(s) to the solvent in which an Si wafer target is placed. By focusing the laser light with the lens on the target, the ablation process takes place and SiNCs are formed.

Not only the solid targets in the liquids but also the powder targets, which are considered as colloids, can also be subjected to PLAL for the production of nanoscale materials. In this case, pulsed laser melting in liquid (PLML) and pulsed laser fragmentation in liquid (PLFL) processes are observed.

9.3 EFFECT OF LASER ABLATION PARAMETERS ON THE SILICON NANOPARTICLE GENERATION

Laser parameters and liquid environment (solvents) are effective on the nanoparticle formation and the physical properties of nanoparticles such as size, shape, crystallinity, and so on. In this section, the effects of different laser pulse durations, pulse energies, laser wavelength, and the liquid environment will be discussed in terms of their effects on nanoparticle properties. The summary of the literature search mentioned below in this section is given as a summary in Table 9.1.

9.3.1 PULSE ENERGY EFFECT

Rioux et al. demonstrated the production of SiNCs by ablating an Si wafer for 30 minutes with a femtosecond laser operating at 0.1–1 mJ pulse energy and 1kHz repetition rate, with 120 fs pulse duration and an 800 nm laser wavelength. SiNCs were produced in deionized water and they had a 2.4 nm average diameter with crystalline structure. Amorphous SiNCs were also observed. When the laser energy increased, they showed that the average diameter of the nanoparticles increased together with an increase in the size dispersion (Rioux et al. 2009). Similarly, in another study conducted using an 800 nm femtosecond laser operating at 100 fs pulse duration and 1kHz repetition rate, increasing laser energies (0.16, 0.27, and 0.40 mJ) were used to control the nanoparticle size. SiNCs were produced by ablating a bulk Si target for 60 minutes in deionized water. It was revealed that by decreasing the laser energy, smaller SiNCs were obtained. The average diameter of SiNCs were determined as 60 nm, 3.5 nm, and 2.5 nm for the laser energies 0.40 mJ, 0.27 mJ, and 0.16 mJ, respectively (Figure 9.2) (Intartaglia et al. 2011).

The ablation of a bulk Si target in deionized water by using a laser operating at 800 nm laser wavelength, 110 fs pulse duration, and 1kHz repetition rate was also shown. In this study, two different laser pulse energies were used. While 0.15 mJ laser pulse energy resulted in SiNCs with 5.5 nm average size, 0.4 mJ laser energy led to SiNCs having 65 nm average diameter (Intartaglia et al. 2012). Besides femtosecond lasers, the effect of laser energy was also analyzed in studies conducted by using nanosecond lasers. The laser operating at 355 nm wavelength, 8 ns pulse duration, and 30 Hz repetition rate was utilized to obtain SiNCs by ablating a bulk Si wafer in deionized water for 30 minutes. It was revealed that SiNCs with 60 nm average diameter were obtained with 0.07 mJ laser energy, but 6 mJ laser energy resulted in SiNCs with 20 nm average diameter (Švrček et al. 2006). Moreover, by using deionized water and ethanol as the liquid media, an Si wafer was ablated for 30 minutes with a nanosecond laser having the wavelength of 1064 nm, 10 ns pulse duration, and 10 Hz repetition rate. Higher laser energies resulted in much smaller SiNCs in both water and ethanol, which was demonstrated as a good liquid environment for the production of smaller and more stable SiNCs (Figure 9.3) (Yang et al. 2009b). Although there is tendency to produce smaller SiNCs by using lower energy with a femtosecond laser and higher energy with a nanosecond laser, the effect of the laser repetition rate should be taken into account and further studies should be performed to make a better interpretation.

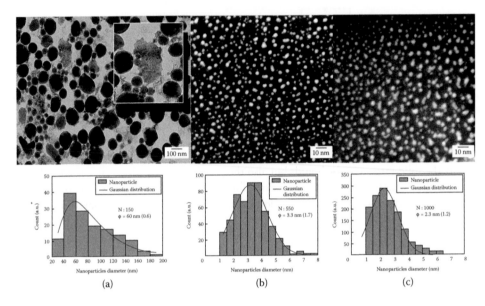

Figure 9.2 TEM and STEM analyses of the obtained solution by femtosecond laser ablation using different pulse energies: (a) 0.40 mJ TEM image, (b) 0.27 mJ, and (c) 0.16 mJ STEM images. (From Intartaglia, R., et al., *Opt. Mater. Exp.*, 2, 510, 2012. With permission.)

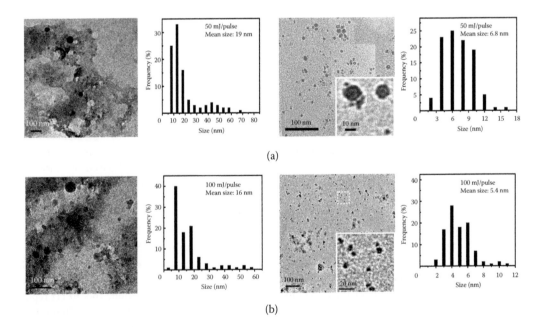

Figure 9.3 TEM results and the size distribution of the SiNCs produced in water (left) and in ethanol (right) at different laser fluences: (a) 50 and (b) 100.   *(Continued)*

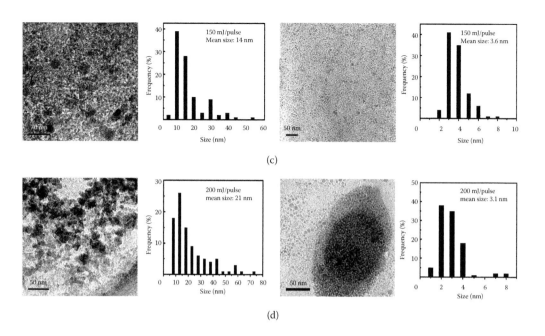

Figure 9.3 (Continued) TEM results and the size distribution of the SiNCs produced in water (left) and in ethanol (right) at different laser fluences: (c) 150 and (d) 200 mJ/pulse (From Yang, S., et al., *J. Phys. Chem. C*, 113, 19091–19095, 2009b. With permission.).

9.3.2 PULSE DURATION EFFECT

When we consider the effect of laser pulse duration, Kuzmin et al. investigated the effect of laser pulse duration by using an 800 nm femtosecond laser in the range of 35, 100, 200, 600, and 900 fs pulse duration. The laser had 4J/cm² fluence and the ablation was performed in ethanol. It was revealed that 40 nm, 25 nm, 50 nm, 60 nm, and 50 nm average nanoparticle sizes were obtained by starting from the lowest pulse duration to the longest pulse duration, respectively (Figure 9.4). The authors concluded that the average diameter of SiNCs and the overall size distribution decreases when the laser pulse duration decreases. They suggested that SiNCs interact with the white light continuum due to low pulse duration and this results in smaller SiNCs at lower pulse durations (Kuzmin et al. 2010).

9.3.3 LIQUID ENVIRONMENT EFFECT

The liquid environment in which SiNCs are produced has a great importance on the size of the nanoparticles. Yang et al. produced SiNCs in both ethanol and deionized water by using the same laser parameters with the PLAL technique, and they showed that the average diameter of SiNCs was smaller in ethanol when compared to water (Yang et al. 2009). Moreover, the same group demonstrated the effect of a surfactant, namely 0.03 M sodium dodecyl sulfate (SDS), on SiNC size and it was indicated that the average size of SiNCs was similar to that of the nanoparticles in water (Yang et al. 2008). Another study showed that SiNCs produced in single-stranded oligonucleotide solution, as a surfactant, had an average size of 5 nm but the ones produced in deionized water by using the same parameters had 60 nm average size (Intartaglia et al. 2012). SiNCs were produced in water and a polymer matrix (a mixture of ethylpolysilicate ethanol and ethyl acetate) by using a nanosecond laser system by the PLAL method. It was demonstrated that SiNCs produced in deionized water were much smaller when compared to the ones in polymer (Švrček and Kondo 2009). Mansour et al. produced SiNCs with average size 3.5 nm by PLAL by using dimethyl sulfoxide (DMSO) as the liquid medium (Mansour et al. 2012). These studies indicate that the liquid medium is quite important for the size of SiNCs, and the other suitable solvents for the production of SiNCs should be investigated in detail.

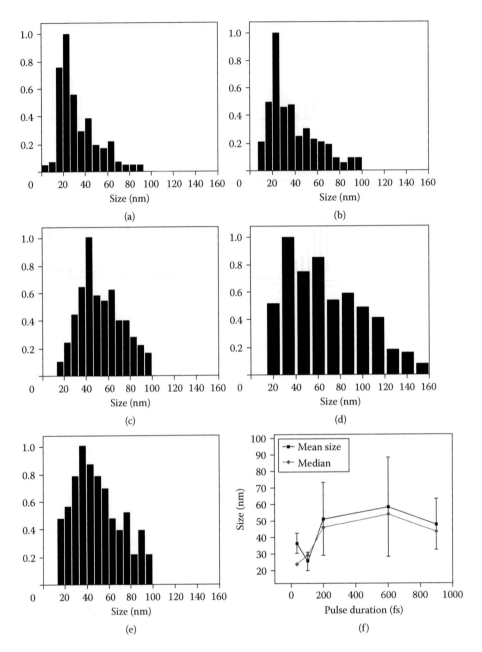

Figure 9.4 Size distribution calculated from TEM images of SiNCs obtained by laser ablation in ethanol. Pulse duration of (a) 35, (b) 100, (c) 200, (d) 600, and (e) 900 fs. Dependence on the laser pulse duration of the nanoparticle (NP) mean size with the vertical bars being the full width at half-maximum of the corresponding distribution functions and NP median (f). (From Kuzmin, P.G., et al., *J. Phys. Chem. C*, 114, 15266–15273, 2010. With permission.)

In conclusion, the laser parameters such as the laser pulse duration and the pulse energy, and the liquid environment are significant determinants for the production of SiNCs by the PLAL method. We can summarize that there is a tendency to produce smaller SiNCs by using lower energy with a femtosecond laser and higher energy with a nanosecond laser but we believe this issue needs further investigation. Moreover, the average diameter of SiNCs and the overall size distribution decreases with lower laser pulse durations and the liquid medium is quite important for the size of SiNCs.

9.4 PHYSICAL AND CHEMICAL PROPERTIES OF SILICON NANOPARTICLES

The physical and chemical properties of silicon nanoparticles produced by the pulsed laser ablation method will be discussed in detail by introducing different characterization methods to explore the different characteristics of the nanoparticles. SiNCs are characterized in terms of their physical and chemical properties after they are produced by the pulsed laser ablation method. The physical properties such as size, crystal structure, absorption, material content, and PL of nanocrystals are examined by certain characterization methods. The methods used for determining physical properties are scanning electron microscope (SEM) images, transmission electron microscope (TEM) or high-resolution TEM (HRTEM) images, selected area electron diffraction (SAED), photoluminescence emission (PLE) spectrum, absorption spectrum, Fourier transform infrared (FTIR), and Raman shift (spectroscopy). The size of the SiNCs is highly related with the parameters of the pulsed laser, type of liquid environment, and ablation time. These effects are mentioned in the "Effect of laser ablation parameters on the silicon nanoparticle generation," (Section 9.3). The size of SiNCs is determined by SEM for tens of nanometer ranges. For the smaller NCs, TEM or a high-resolution transmission microscope are generally used for several angstrom (Å) ranges. The diameter of the SiNCs could be approximately in the range of 2–200 nm. The size diameter of SiNCs produced by the PLAL method are shown in Table 9.1 (Literature details). SiNCs are produced in the liquid environment by laser ablation. The silicon volume fraction, ϕ_V inside the liquid that is ablated inside is calculated by following Equation (9.3) (Mansour et al. 2012):

$$\phi_V = \frac{V_s}{V_s + V_L} \tag{9.3}$$

where V_s is the volume of the particles, V_L is the volume of the liquid. The size and presence of the SiNCs are also determined by Raman spectroscopy. The Raman spectrum represents a sharp peak around at 520 cm^{-1} which has the asymmetricity toward the lower frequency side which proves the presence of the SiNCs in the sample (Rioux et al. 2009). The size of the SiNCs and the size distribution are evaluated by deconvoluting the Raman spectrum. The functions which are used for devolution are Lorentzian, and the phonon confinement method (Richter et al. 1981; dos Santos and Torriani 1993), and Gaussian distribution function. The peak at 520 cm^{-1} represents the bulk Si which has a crystalline size larger than 9 nm; the peak in the 510–520 cm^{-1} range is coming from the small Si crystallites such as Si nanoparticles, and the 480 cm^{-1} peak stands for an amorphous Si (Mchedlidze et al. 2008). SiNCs are produced using the Si wafer targets having <100> or <001> states and the crystalline structure is investigated by HRTEM. The crystalline structure of nanosized material displays (111) lattice sets with an interplanary spacing of approximately 3–3.14 Å, which is a characteristic property of Si bulk (ablated target material) (see Table 9.1). The FTIR method is used to find the presence of SiNCs, and 1080 cm^{-1} shows the Si–O vibration modes in the FTIR spectra (Umezu et al. 2007). The optical properties of SiNCs are determined by absorption spectra. The absorption spectra cover a large and continuous band between 200 and 800 nm and have a typical shoulder with minima at around 400 nm (Intartaglia et al. 2012). The peaks at 485 nm and 460 nm in the absorption spectra confirm the successful production of small-sized SiNCs (Intartaglia et al. 2011). The pulsed laser influence affects the PL intensity; as the laser influence increases, the PL intensity increases, and a blue shift of PL is observed as shown in Figure 9.5 (Švrček et al. 2006).

The band gap energy of the Si bulk is E_o = 1.12 eV. Silicon bulk and nanoparticles have different band gap energies due to the quantum confinement effect. The band gap of SiNCs can be found as a function of the particle diameter, d, using the PL experimental results and the linear combination of atomic orbitals technique calculations in Equation 9.4 (Delerue et al. 1993; Ledoux et al. 2000; Meier et al. 2007):

$$E_g = E_0 + \left(\frac{3.73}{d^{1.39}}\right) \tag{9.4}$$

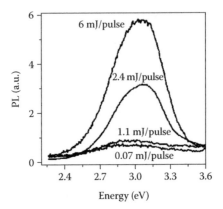

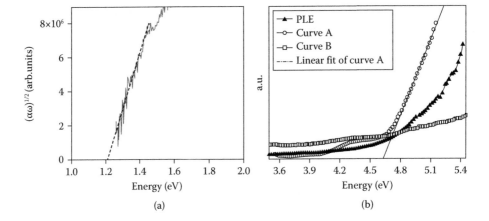

Figure 9.5 (a) Photoluminescence spectra of SiNCs in water prepared by laser ablation of silicon wafer in deionized water with different laser fluences and aged for 6 months. (From Švrček, V., et al., *Appl. Phys. Lett.*, 89, 213113, 2006. With permission.)

Figure 9.6 PLE measurements of SiNCs which have (a) indirect band gap (from Semaltianos, N.G., et al., *J. Nanopart. Res.*, 12(2), 573–580, 2010) and (b) direct band gap (from Du, X.W., et al., *J. Appl. Phys.*, 102, 013518–013518-4, 2007. With permission.).

Bulk silicon has indirect band gap and it is reported that silicon nanoparticles produced by PLA method or other methods have an indirect band gap (Makino et al. 2005; Meier et al. 2007; Umezu et al. 2007). The band gap energy of the nanoparticles are calculated by the following Equation (9.5) (Semaltianos et al. 2010):

$$\alpha(\hbar\omega) \propto \omega^{-1}(\hbar\omega - E_g)^2 \tag{9.5}$$

α is the absorption coefficient, $\hbar$ is Planck constant, ω is angular frequency, and E_g is the band gap energy. Equation 9.5 is valid for indirect band gap semiconductors. In Figure 9.6a, the linear fitting is represented which exposes E_g = 1.2 eV, implying a blue shift of 0.08 eV from the band gap of Si bulk and it has an indirect band gap. The Si nanoparticle size can be determined using Equation 9.5. For example, given E_g and E_o values above, the size is calculated as d≈15.8 nm.

The intrinsic character of SiNCs having indirect band gap limits efficient applications. There are theoretical studies that explain the band structure can be altered by controlling the atomic arrangement in Si nanostructures (Tomašek and Schluter 1987; Delley and Steigmeier 1993; Zhou et al. 2003). SiNCs having face-centered cubic crystal structure have been produced by the PLA method with a microsecond pulse width, and light emission in the visible range was obtained (Du et al. 2007). The PLE graph is shown in Figure 9.6b. If the PLE of SiNCs is fitted to Equation 9.6, it represents direct band gap semi-conductor (Kux and Chorin 1996):

$$(\alpha h\nu)^2 \propto \left(h\nu - E_g\right) \tag{9.6}$$

Only curve A represents a linear behavior in Figure 9.6b, thus Equation 9.6 is fitted to curve A, and the band gap energy is calculated as 4.6 eV. The band gap energy is higher than the PL emission peak, 2.6 eV, which proposes that excitation and emission have different routes.

In relation to redox behavior, which is an important chemical property, it is reported that SiNCs represent reductive properties to noble metal ions (Yang et al. 2009b). SiNCs display reductive properties to Au^{3+} ions because of the overall positive redox potential for the metal ions—SiNCs system, whereas Si microsized powders do not exhibit such reductive behavior.

9.5 POSTTREATMENT METHODS

It is quite important to obtain the silicon nanoparticles with sizes under the quantum confinement limits to display their unique properties. Thus, to obtain much smaller silicon nanoparticles, post-treatment methods are also applied and these methods will be introduced in this section. In order to obtain smaller nanoparticles, ultrasonic chemical posttreatment in HF after producing larger-sized nanoparticles by using ns pulsed laser ablation is applied (Alkis et al. 2012). The posttreatment can also be applied by a chemical-free process which has two stages: first, the particles in the range of 5–100 nm produced by the PLA method; and second, chemical-free ultrasonic and filtering posttreatment is applied to obtain smaller-sized particles in the range of 1–5.5 nm (Alkis et al. 2012). It is possible to obtain smaller SiNCs after posttreatment. The size of the nanocrystals affects the optical properties. When silicon crystallite becomes small in diameter for $D \leq 10$ nm, the Raman phonon band becomes larger and shifts down. The Raman shift is related with the crystallite size in the quantitative phonon confinement model in Equation 9.7 (Mafuné et al. 2001):

$$\Delta\nu = -52.3\left(\frac{0.543}{D}\right)^{1.586} \tag{9.7}$$

where $\Delta\nu$ (cm^{-1}) is the shift of the Raman peak of SiNCs compared to bulk Si, D (nm) is the diameter of the SiNCs. For example, if the Raman shift is observed as 2 cm^{-1}, then such a shift stands for 4.3 nm size of SiNCs (Abderrafi et al. 2011).

9.6 APPLICATIONS OF SILICON NANOPARTICLES PRODUCED BY PULSED LASER ABLATION METHOD

In the past decades, SiNCs have attracted attention in various application areas due to their unique electrical, optical, and chemical properties (Hirschman et al. 1996; Bruchez et al. 1998). Due to the quantum confinement effect, SiNCs can emit blue light (Švrček et al. 2006; Yang et al. 2008) which can be used in biomedical tagging, silicon-based full color display, and flash memories (Ding et al. 2002; Larson et al. 2003; Wang et al. 2004). SiNCs confocal microscopy studies confirm the possible use of biocompatible SiNCs for imaging applications. SiNCs having a diameter size less than 10 nm have been identified by fluorescence imaging microscopy technique (Figure 9.7) (Intartaglia et al. 2012). SiNCs are a very good candidate for fluorescence imaging of biological samples since they are biocompatible.

Reductive SiNCs have shown possible applications in pollution remediation on heavy metal ions and in wastewater or soil which are coming from industrial waste, and these are toxic to bacteria, plants, and humans (Browning 1969; Smith and Lec 1972; Brauer and Wetterhahn 1991).

Clusters, nanoparticles, and quantum dots

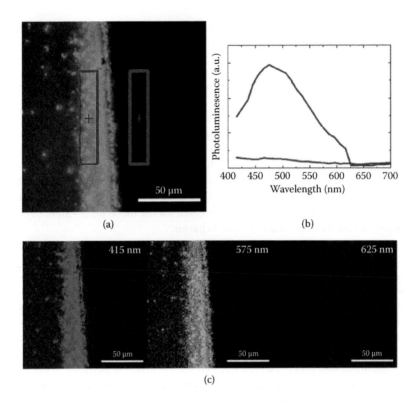

Figure 9.7 Fluorescence imaging microscopy of silicon nanoparticles prepared via the PLAL method in the low energy regime. (a) Spectral image acquired on the ring border and (b) the PL spectra relative to two different regions of interest. Color of the PL spectra corresponds to the color used in the covered area in the optical image; (c) Selection of channels centered at three different wavelengths. (From Intartaglia, R., et al., *Opt. Mater. Exp.*, 2(5), 510, 2012.)

9.7 CONCLUSION

The PLAL mechanism was shown to be a suitable method for SiNCs generation. The effects of laser ablation parameters (pulse energy, pulse duration, laser wavelength, and liquid environment) were discussed in detail with a comprehensive literature search throughout the chapter. One can easily manipulate the physical and chemical properties of produced SiNCs by altering the laser parameters and liquid environment. Moreover, the posttreatment approach was reviewed to obtain much smaller SiNCs with superior quantum confinement effects. Finally, various applications of SiNCs produced by PLAL were discussed in detail.

REFERENCES

Abderrafi, K., G. Raul, M. B. Gongalsky, I. Suarez, R. Abarques, V. S. Chirvony, V. Y. Timoshenko, R. Ibanez, and J. P. Martínez-Pastor. 2011. Silicon Nanocrystals Produced by Nanosecond Laser Ablation in an Organic Liquid. *Journal of Physical Chemistry C* 115: 5147–51. doi:10.1021/jp109400v.

Alkis, S., A. K. Okyay, and B. Ortaç. 2012. Post-Treatment of Silicon Nanocrystals Produced by Ultra-Short Pulsed Laser Ablation in Liquid: Toward Blue Luminescent Nanocrystal Generation. *Journal of Physical Chemistry C* 116 (5): 3432–6. doi:10.1021/jp211521k.

Belomoin, G., J. Therrien, A. Smith, S. Rao, R. Twesten, S. Chaieb, M. H. Nayfeh, L. Wagner, and L. Mitas. 2002. Observation of a Magic Discrete Family of Ultrabright Si Nanoparticles. *Applied Physics Letters* 80 (5): 841–3. doi:10.1063/1.1435802.

Brauer, S. L., and K. E. Wetterhahn. 1991. Chromium(VI) Forms a Thiolate Complex with Glutathione. *Journal of American Chemical Society* 113 (8): 3001–7. doi:10.1021/ja00008a031.

Browning, Y. E. 1969. Chromium. In *Toxicity of Industrial Metals*, 2nd ed., Browning, Y.E. (editor), Butterworths, London, UK, pp. 119–31.

Bruchez, M., Jr., M. Moronne, P. Gin, S. Weiss, and A. P. Alivisatos. 1998. Semiconductor Nanocrystals as Fluorescent Biological Labels. *Science* 281 (5385): 2013–16. doi:10.1126/science.281.5385.2013.

Delerue, C., G. Allan, and M. Lannoo. 1993. Theoretical Aspects of the Luminescence of Porous Silicon. *Physical Review B* 48 (15): 11024–36. doi:10.1103/PhysRevB.48.11024.

Delley, B., and E. F. Steigmeier. 1993. Quantum Confinement in Si Nanocrystals. *Physical Review B* 47 (3): 1397–400. doi:10.1103/PhysRevB.47.1397.

Ding, Z., B. M. Quinn, S. K. Haram, L. E. Pell, B. A. Korgel, and A. J. Bard. 2002. Electrochemistry and Electrogenerated Chemiluminescence from Silicon Nanocrystal Quantum Dots. *Science (New York, N.Y)* 296 (5571): 1293–7. doi:10.1126/science.1069336.

dos Santos, D. R., and I. L. Torriani. 1993. Crystallite Size Determination in Mc-Ge Films by X-Ray Diffraction and Raman Line Profile Analysis. *Solid State Communications* 85 (4): 307–10. doi:10.1016/0038-1098(93)90021-E.

Du, X. W., W. J. Qin, Y. W. Lu, X. Han, Y. S. Fu, and S. L. Hu. 2007. Face-Centered-Cubic Si Nanocrystals Prepared by Microsecond Pulsed Laser Ablation. *Journal of Applied Physics* 102 (1), 013518–013518-4. doi:10.1063/1.2752785.

Hirschman, K. D., L. Tsybeskov, S. P. Duttagupta, and P. M. Fauchet. 1996. Silicon-Based Visible Light-Emitting Devices Integrated into Microelectronic Circuits. *Nature* 384 (6607): 338–41. doi:10.1038/384338a0.

Intartaglia, R., K. Bagga, F. Brandi, G. Das, A. Genovese, E. Di Fabrizio, and A. Diaspro. 2011. Optical Properties of Femtosecond Laser-Synthesized Silicon Nanoparticles in Deionized Water. *Journal of Physical Chemistry C* 115 (12): 5102–7. doi:10.1021/jp109351t.

Intartaglia, R., K. Bagga, M. Scotto, A. Diaspro, and F. Brandi. 2012. Luminescent Silicon Nanoparticles Prepared by Ultra Short Pulsed Laser Ablation in Liquid for Imaging Applications. *Optical Materials Express* 2 (5): 510. doi:10.1364/OME.2.000510.

Intartaglia, R., A. Barchanski, K. Bagga, A. Genovese, G. Das, P. Wagener, E. D. Fabrizio, A. Diaspro, F. Brandi, and S. Barcikowski. 2012. Bioconjugated Silicon Quantum Dots from One-Step Green Synthesis. *Nanoscale* 4 (4): 1271. doi:10.1039/c2nr11763k.

Kabashin, A. V., and Meunier, M. 2006. Laser ablation-based synthesis of functionalized colloidal nanomaterials in biocompatible solutions. *Journal of Photochemistry and Photobiology A: Chemistry*, 182(3): 330–4.

Khang, Y., and J. Lee. 2010. Synthesis of Si Nanoparticles with Narrow Size Distribution by Pulsed Laser Ablation. *Journal of Nanoparticle Research* 12 (4): 1349–54. doi:10.1007/s11051-009-9669-z.

Knipping, J., Wiggers, H., Rellinghaus, B., Roth, P., Konjhodzic, D., and Meier, C. 2004. Synthesis of high purity silicon nanoparticles in a low pressure microwave reactor. *Journal of nanoscience and nanotechnology*, 4(8): 1039-44.

Kůsová, K., O. Cibulka, K. Dohnalová, I. Pelant, J. Valenta, A. Fůcíková, K. Žídek, et al. 2010. Brightly Luminescent Organically Capped Silicon Nanocrystals Fabricated at Room Temperature and Atmospheric Pressure. *ACS Nano* 4 (8): 4495–504. doi:10.1021/nn1005182.

Kux, A., and M. B. Chorin. 1996. Red, Green and Blue Luminescence in Porous Silicon Excitation Spectra. *Thin Solid Films* 276: 272–5.

Kuzmin, P. G., G. A. Shafeev, V. V. Bukin, S. V. Garnov, C. Farcau, R. Carles, B. Warot-Fontrose, V. Guieu, and G. Viau. (2010). Silicon Nanoparticles Produced by Femtosecond Laser Ablation in Ethanol: Size Control, Structural Characterization, and Optical Properties. *The Journal of Physical Chemistry C* 114(36): 15266–73.

Larson, D. R., W. R. Zipfel, R. M. Williams, S. W. Clark, M. P. Bruchez, F. W. Wise, and W. W. Webb. 2003. Water-Soluble Quantum Dots for Multiphoton Fluorescence Imaging in Vivo. *Science (New York, NY)* 300 (5624): 1434–6. doi:10.1126/science.1083780.

Ledoux, G., O. Guillois, D. Porterat, C. Reynaud, F. Huisken, B. Kohn, and V. Paillard. 2000. Photoluminescence Properties of Silicon Nanocrystals as a Function of Their Size. *Physical Review B* 62 (23): 15942–51. doi:10.1103/PhysRevB.62.15942.

Li, X., Y. He, and M. T. Swihart. 2004. Surface Functionalization of Silicon Nanoparticles Produced by Laser-Driven Pyrolysis of Silane Followed by HF-HNO 3 Etching. *Langmuir* 20 (11): 4720–7. doi:10.1021/la036219j.

Luterova, K., K. Dohnalova, V. Svrcek, I. Pelant, J.-P. Likforman, O. Cregut, P. Gilliot, and B. Hönerlage. 2016. Optical Gain in Porous Silicon Grains Embedded in Sol-Gel Derived SiO2 Matrix under Femtosecond Excitation. *Applied Physics Letters* 84 (17): 3280–2.

Mafuné, F., J. Kohno, and Y. Takeda. 2001. Formation of Gold Nanoparticles by Laser Ablation in Aqueous Solution of Surfactant. *The Journal of Physical Chemistry B* 105 (22): 5114–20. http://pubs.acs.org/doi/abs/10.1021/jp0037091.

Makino, T., M. Inada, I. Umezu, and A. Sugimura. 2005. Structural and Optical Properties of Surface-Hydrogenated Silicon Nanocrystallites Prepared by Reactive Pulsed Laser Ablation. *Journal of Physics D: Applied Physics* 38 (18): 3507–11. doi:10.1088/0022-3727/38/18/028.

Mangolini, L., E. Thimsen, and U. Kortshagen. 2005. High-Yield Plasma Synthesis of Luminescent Silicon Nanocrystals. *Nano Letters* 5 (4): 655–9. doi:10.1021/nl050066y.

Mansour, N., A. Momeni, R. Karimzadeh, and M. Amini. 2012. Blue-Green Luminescent Silicon Nanocrystals Fabricated by Nanosecond Pulsed Laser Ablation in Dimethyl Sulfoxide. *Optical Materials Express* 2 (6): 740–8. doi:10.1364/OME.2.000740.

Mchedlidze, T., T. Arguirov, S. Kouteva-Arguirova, M. Kittler, R. Rölver, B. Berghoff, D. Bätzner, and B. Spangenberg. 2008. Light-Induced Solid-to-Solid Phase Transformation in Si Nanolayers of Si-SiO2 Multiple Quantum Wells. *Physical Review B* 77: 2–5. doi:10.1103/PhysRevB.77.161304.

Meier, C., A. Gondorf, S. Lüttjohann, A. Lorke, and H. Wiggers. 2007. Silicon Nanoparticles: Absorption, Emission, and the Nature of the Electronic Bandgap. *Journal of Applied Physics* 101 (10): 103112. doi:10.1063/1.2720095.

Pavesi, L., L. D. Negro, C. Mazzoleni, G. Franzò, and F. Priolo. 2000. Optical Gain in Silicon Nanocrystals. *Nature* 408 (6811): 440–4. doi:10.1038/35044012.

Richter, H., Z. P. Wang, and L. Ley. 1981. The One Phonon Raman Spectrum in Microcrystalline Silicon. *Solid State Communications* 39 (5): 625–9. doi:10.1016/0038-1098(81)90337-9.

Rioux, D., M. Laferrière, A. Douplik, D. Shah, L. Lilge, A. V. Kabashin, and M. M. Meunier. 2009. Silicon Nanoparticles Produced by Femtosecond Laser Ablation in Water as Novel Contamination-Free Photosensitizers. *Journal of Biomedical Optics* 14 (2): 021010. doi:10.1117/1.3086608.

Rosso-Vasic, M., E. Spruijt, Z. Popovic, K. Overgaag, B. van Lagen, B. Grandidier, D. Vanmaekelbergh, D. Dominguez-Gutierrez, L. De Cola, and H. Zuilhof. 2009. Amine-Terminated Silicon Nanoparticles: Synthesis, Optical Properties and Their Use in Bioimaging. *Journal of Materials Chemistry* 19 (33): 5926–33. doi:10.1039/b902671a.

Semaltianos, N. G., S. Logothetidis, W. Perrie, S. Romani, R. J. Potter, S. P. Edwardson, P. French, M. Sharp, G. Dearden, and K. G. Watkins. 2010. Silicon Nanoparticles Generated by Femtosecond Laser Ablation in a Liquid Environment. *Journal of Nanoparticle Research* 12 (2): 573–80. doi:10.1007/s11051-009-9625-y.

Smith, R. G., and Lec, D. H. K. 1972. *Chromium in Metallic Contaminants and Human Health.* Academic Press, New York.

Stupca, M., M. Alsalhi, T. Al Saud, A. Almuhanna, and M. H. Nayfeh. 2007. Enhancement of Polycrystalline Silicon Solar Cells Using Ultrathin Films of Silicon Nanoparticle. *Applied Physics Letters* 91 (6): 1–4. doi:10.1063/1.2766958.

Švrček, V., and M. Kondo. 2009. Blue Luminescent Silicon Nanocrystals Prepared by Short Pulsed Laser Ablation in Liquid Media. *Applied Surface Science* 255 (24): 9643–6. doi:10.1016/j.apsusc.2009.04.126.

Švrček, V., T. Sasaki, Y. Shimizu, and N. Koshizaki. 2006. Blue Luminescent Silicon Nanocrystals Prepared by Ns Pulsed Laser Ablation in Water. *Applied Physics Letters* 89 (21): 213113. doi:10.1063/1.2397014.

Tomańek, D., and M. A. Schluter. 1987. Structure and Bonding of Small Semiconductor Clusters. *Physical Review B* 36 (2): 1208–17. doi:10.1103/PhysRevB.36.1208.

Umezu, I., H. Minami, H. Senoo, and A. Sugimura. 2007. Synthesis of Photoluminescent Colloidal Silicon Nanoparticles by Pulsed Laser Ablation in Liquids. *Journal of Physics: Conference Series* 59 (1): 392–5. doi:10.1088/1742-6596/59/1/083.

Walters, R. J., G. I. Bourianoff, and H. A. Atwater. 2005. Field-Effect Electroluminescence in Silicon Nanocrystals. *Nature Materials* 4 (2): 143–6. doi:10.1038/nmat1307.

Wang, G., S. T. Yau, K. Mantey, and M. H. Nayfeh. 2008. Fluorescent Si Nanoparticle-Based Electrode for Sensing Biomedical Substances. *Optics Communications* 281 (7): 1765–70. doi:10.1016/j.optcom.2007.07.070.

Wang, L., V. Reipa, and J. Blasic. 2004. Silicon Nanoparticles as a Luminescent Label to DNA. *Bioconjugate Chemistry* 15 (2): 409–12. doi:10.1021/bc030047k.

Warner, J. H., H. Rubinsztein-Dunlop, and R. D. Tilley. 2005. Surface Morphology Dependent Photoluminescence from Colloidal Silicon Nanocrystals. *Journal of Physical Chemistry B* 109 (41): 19064–7. doi:10.1021/jp054565z.

Wolkin, M., J. Jorne, P. Fauchet, G. Allan, and C. Delerue. 1999. Electronic States and Luminescence in Porous Silicon Quantum Dots: The Role of Oxygen. *Physical Review Letters* 82: 197–200. doi:10.1103/PhysRevLett.82.197.

Yang, G. W. 2007. Laser Ablation in Liquids: Applications in the Synthesis of Nanocrystals. *Progress in Materials Science* 52 (4): 648–98. doi:10.1016/j.pmatsci.2006.10.016.

Yang, G.-W., J.-B Wang, and Q.-X. Liu. 1999. Preparation of Nano-Crystalline Diamonds Using Pulsed Laser Induced Reactive Quenching. *Journal of Physics: Condensed Matter* 10 (35): 7923–7. doi:10.1088/0953-8984/10/35/024.

Yang, S., W. Cai, G. Liu, H. Zeng, and P. Liu. 2009a. Optical Study of Redox Behavior of Silicon Induced by Laser Ablation in Liquid. *Journal of Physical Chemistry C* 113 (16): 6480–4. doi:10.1021/jp810787d.

Yang, S., W. Cai, H. Zeng, and Z. Li. 2008. Polycrystalline Si Nanoparticles and Their Strong Aging Enhancement of Blue Photoluminescence. *Journal of Applied Physics* 104 (2): 023516. doi:10.1063/1.2957053.

Yang, S., W. Cai, H. Zhang, X. Xu, and H. Zeng. 2009b. Size and Structure Control of Si Nanoparticles by Laser Ablation in Different Liquid Media and Further Centrifugation Classification. *The Journal of Physical Chemistry C* 113 (44): 19091–5. doi:10.1021/jp907285f.

Zhang, X., D. Neiner, S. Wang, A. Y. Louie, and S. M. Kauzlarich. 2007. A New Solution Route to Hydrogen-Terminated Silicon Nanoparticles: Synthesis, Functionalization and Water Stability. *Nanotechnology* 18 (9): 095601. doi:10.1088/0957-4484/18/9/095601.

Zhou, Z., L. Brus, and R. Friesner. 2003. Electronic Structure and Luminescence of 1 . 1- and 1 . 4-Nm Silicon Nanocrystals : Oxide Shell versus Hydrogen Passivation. *Nano Letters* 3 (2): 163–7. doi:10.1021/nl025890q.

10 Silicon nanoparticles via pulsed laser ablation in liquid

Fernando Brandi

Contents

10.1 INTRODUCTION TO PULSED LASER ABLATION IN LIQUID

Since its establishment as a production technology, nanotechnology has revolutionized the everyday life of billions of people worldwide. The main merit of nanotechnology is the miniaturization, which brings many advantages like the possibility to create more compact devices, to achieve better energy usage, and higher surface-to-volume ratio which enhances physicochemical reactivity, to mention a few. All these effects are related to the mere scaling down in the physical dimension. The nanometric length scale, however, also brings out new physical phenomena related to the finite size of the constituents of matter—atoms and electrons. Among these effects, the quantum confinement phenomenon appearing for dimensions smaller than the exciton Bohr radius is one of the most striking since it reveals itself through a drastic change in the optical properties of semiconductor materials. For example, an indirect band gap semiconductor like silicon switches to direct band gap behavior when the physical size is smaller than about 5–10 nm (Barbagiovanni et al. 2014). This transition substantially modifies the optical absorption and emission properties of the silicon nanomaterial compared to the bulk: the absorption edge shifts to the blue region of the optical spectrum, and bright photoluminescence (PL) appears with wavelength in the optical spectrum depending on the actual size of the nanocrystals.

The possibility to finely control the synthesis of nanomaterials allows to fabricate quantum-confined nanoparticles (NPs), so-called quantum dots, with well-defined optical properties. The most widely known quantum dots are based on binary or ternary compounds which enables a great versatility in tuning the optical properties. However, such compounds relay on heavy metal elements (e.g., lead and cadmium) that are highly toxic, which raises issues on the chemical safety of their synthesis, usage, and disposal.

In this respect, heavy-metal-free NPs have been deeply investigated in recent years as alternative quantum dots. Silicon nanoparticles (SiNPs) offer a great opportunity to tackle and solve basic safety issues of heavy metal quantum dots given their high level of biocompatibility (Peng et al. 2014; Baati et al. 2016), and are readily compatible with standard silicon technology for easy integration in existing devices. Moreover, when used as photoluminescent nanomarkers for bioimaging, the SiNPs offer the advantage of much lower photobleaching when compared to standard dye molecules (Zhong et al. 2015).

SiNPs can be fabricated using a variety of approaches such as solution chemistry routes (Rosso-Vasic et al. 2009; Xiao et al. 2011; Zhong et al. 2013), as well as thermal decomposition and laser pyrolysis of silane, laser ablation of a solid target in a vacuum or a controlled atmosphere, and nonthermal plasma synthesis (Mangolini 2013).

Another efficient and versatile emerging method for the production of stable SiNP colloidal solutions is pulsed laser ablation in liquid (PLAL), which is a simple approach consisting basically of the process of laser ablation of solid materials, either in bulk or powder form, confined in a liquid environment. Since its introduction at the beginning of the 1990s (Lida et al. 1991; Neddersen et al. 1993), the PLAL process has been investigated mostly for the production of metallic colloids, but in recent years it has been applied to a plethora of materials from semiconductors to ceramics and from alloys to polymers.

The basic principles of the PLAL process are schematically illustrated in Figure 10.1: the solid material is immersed in a liquid and irradiated by a pulsed laser beam that is focused to locally reach a fluence, i.e., the energy delivered per unit area, higher than the threshold for ablation of the solid material. As schematically shown in the highlight of Figure 10.1, the ablation process in a liquid environment creates a plasma plume and a subsequent bubble in which the actual nucleation and growth of NPs takes place. In Figure 10.2, an example of the actual PLAL process is shown: the left image shows the starting condition with the bulk material (in this case a cylindrical silicon rod) immersed in the liquid (water in this case) with the laser turned off; on the right the same image is shown when the laser is turned on and PLAL is taking place. It is interesting to note the typical NP "spiral" cloud caused by the liquid convection motion due to the temperature gradient inside the liquid, with a higher temperature on the surface of the solid material due to the ablation process.

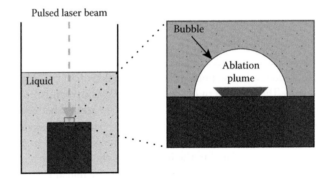

Figure 10.1 Schematic of the PLAL process: the pulsed laser beam (in green for better visualization) is focused on the surface of the material immersed in the liquid. Highlighted is the ablation region showing the ablation plume (in red) and the subsequent bubble confined in the liquid, where the nanoparticles are generated.

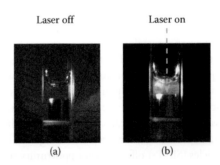

Figure 10.2 Actual pictures of the PLAL process: (a) the starting condition with the bulk material immersed in the liquid in a glass cuvette with the laser turned off; (b) the laser is turned on and PLAL is taking place, with the typical nanoparticle cloud "spiral" caused by the liquid convection motion due to the temperature gradient inside the liquid. The ablation plume plasma emission is not visible in this image due to limited time resolution of the camera used.

In general, the PLAL method relies on a very fast and repetitive process in which the ablation plume and the bubble interacting with the surrounding liquid constitute a kind of microreactor where the NPs are formed and eventually functionalized. The synthesis is not only limited in space within the microreactor (which can actually have up to mm size depending on the laser spot diameter and laser pulse fluence) but also limited to a very short time duration in the order of microseconds, which is the time it takes for the plume to cool down and for the bubble to collapse within the liquid. The actual time duration of each microreaction depends on the specific parameters used and mainly on the laser fluence that dictates the amount of ablated material as well as the energy stored in the plume. During PLAL, each laser pulse triggers a microreaction process which can be described as follows:

1. First stage—laser–matter interaction, during which the energy is delivered to the "microreactor" triggering the ablation process, typically in the time scale of femtosecond to nanoseconds depending on the laser pulse duration.
2. Second stage—ablation plume evolution and bubble formation, during which gas/plasma phase reactions occur among the target material constituents and eventually the vaporized/ionized constituents of the liquid; in this stage some gas–liquid reaction can take place at the interface of the bubble with the liquid.
3. Third stage—the plasma cools and the bubble collapses, while the produced NPs are released in the liquid; during this stage liquid phase reactions occur between the produced NP and the molecules present in the liquid (e.g., surface capping or functionalization).
4. Fourth stage—this may be present, during which the formed NPs interact with the subsequent laser pulses.

This last stage is governed by laser–matter interactions at the nanoscale, which can be drastically different compared with laser–matter interaction at the bulk or micron scale. Specifically, the light absorption properties of the NPs are of importance at this stage. For example, it is well known that in metallic NPs the plasmonic response related to the collective free electron motion induces an absorption resonance band characteristic of the element with little dependence on the actual nanoparticles size (e.g., at ~520 nm for Au and ~400 nm for Ag). Instead, in semiconductor laser–matter interaction at the nanoscale strongly depends on the actual size of the NPs. For example, in silicon the transition from indirect to direct band gap behavior induces a large shift of the band edge to the blue, which is reflected in a wider transmission band (e.g., near-infrared radiation is basically absorbed in bulk silicon, while it is transmitted by SiNPs colloids).

All in all, PLAL is a kind of hybrid fabrication process between the top-down and bottom-up approaches: a) it is not as atomically precise as other bottom-up approaches but indeed relies on nucleation of atoms/ions present in the ablation plume; b) it starts from bulk or microsized material that is machined (ablated), by externally controlled tools (the laser) as in top-down approaches. In this respect, PLAL has the advantage of efficiently using the raw material which is fully consumed, and with minimal reagents being used (except for the liquid used for ablation) therefore minimizing the production of toxic waste. In fact, PLAL possesses several advantages compared with other fabrication methods:

1. It is a clean synthesis without the need of chemical precursors, reducing agents, and surface ligands, that fulfills the principles of green chemistry (Anastas and Warner 1998) resulting in highly pure and stable colloidal solutions.
2. Its implementation is relatively simple and economical, since it is performed at ambient conditions.
3. It is adaptable and versatile, since the obtained NPs are naturally in colloidal solution form, giving the opportunity of further nanoscale manipulations such as biofunctionalization (Intartaglia et al. 2012c; Bagga et al. 2013).
4. It is a scalable production technique, since it relies on a repeated very fast process on a microsecond time scale for every single laser pulse.

One of the main drawbacks of PLAL is that the size distribution of the produced NPs is often much wider than what is achievable by other fabrication methods. This is reflected to some extend in a relatively broad optical response of the colloids, in terms of both absorption and emission properties. Even if methods to reduce the wide-size distribution have been investigated [e.g., photofragmentation (Bagga et al. 2013)], more systematic studies are still necessary to mitigate this problem.

Clusters, nanoparticles, and quantum dots

Another critical issue of PLAL is related to the many parameters that play a crucial role in the process, from the liquid and the solid material physicochemical properties (e.g., viscosity, composition, and optical absorption of the liquid and the solid material at the laser wavelength) to the laser and irradiation characteristics (e.g., wavelength, fluence, laser pulse time duration and repetition rate, total irradiation time, flowing, or static liquid). The exact mechanism that leads to the formation of a stable colloid depends on the interplay of all these parameters as well as on the very fast ablation plume dynamics (femtosecond to microsecond timescale) often far from thermodynamic equilibrium.

Although many experimental studies on PLAL have been performed and theoretical descriptions attempted in specific cases, the investigation of the PLAL process is basically phenomenological due to the very complex interplay of all the parameters involved. In this scenario, a standardization of the PLAL procedure both in terms of its implementation and sample characterization is necessary in order to bring PLAL out from the research field to the actual (large scale) production arena. Nonetheless, PLAL technique is sufficiently evolved that it is already an established approach for the synthesis and commercialization of (mainly metallic) colloidal solutions*.

Very good overviews on PLAL and its application for NP generation can be found in Tan et al. (2013), Zeng et al. (2012), Amendola and Meneghetti (2013), and Semaltianos (2016).

In this chapter, a specific overview of the status and future needs in PLAL of silicon for the production of SiNPs colloids is given, with emphasis on the synthesis of photoluminescent SiNPs.

10.2 SILICON NANOPARTICLES PRODUCED BY PLAL

10.2.1 OPTICAL PROPERTIES

Photoluminescent SiNPs are very good candidates for applications ranging from light-emitting devices (Belomoin et al. 2002; Walters et al. 2005) to photovoltaic solar cell technology and optoelectronics (Stupca et al. 2007; Alkis et al. 2012). Biocompatibility and stability against photobleaching make of these NPs ideal substitute to dyes as fluorescent labels for imaging (Li and Ruckenstein 2004; Borsella et al. 2010; Erogbogbo et al. 2011; Intartaglia et al. 2012a; Zhong et al. 2013; Chinnathambi et al. 2014), as well as efficient photosensitizer for photo-therapy treatments (Rioux et al. 2009; Xiao et al. 2011).

The large-size distribution typical of SiNPs produced by PLAL broadens the emission spectrum related to the quantum confinement effect. However, the emission properties of SiNPs produced by PLAL are actually mainly dictated by the surface/defects chemistry with some possible contribution from the quantum confinement effect. In fact, the rapid process of ablation and nucleation leads to the formation of unavoidable defects in the nanocrystals (e.g., dangling bonds and P_b centers), and in parallel the presence of the surrounding liquid and vapor inside the bubble creates surface-functionalized particles: (a) mainly partially oxidized and hydroxyl-capped in water and (b) covered with carbonyl, amorphous carbon, and graphitic structures for PLAL performed in carbon-containing solvents. However, for practical purposes, the SiNPs surface reactivity can be exploited for further functionalization of the NPs (Intartaglia et al. 2012c; Bagga et al. 2013).

The quantum confinement effect clearly manifests itself in the absorption properties of the PLAL-produced SiNP colloids. The band gap is markedly blue shifted compared to bulk silicon (1.1 eV) resulting in an almost colorless colloid absorbing in the near UV. This is common to almost all reported experiments and demonstrates the presence of small (<10 nm) and ultrasmall (<5 nm) silicon quantum dots which can be in the form of isolated nanocrystals, aggregated polynanocrystals, or embedded in a silicon oxide matrix.

Figure 10.3 shows an example of the optical properties of SiNPs colloid produced by PLAL with picosecond laser in water. With decreasing SiNPs size, the absorption edge shifts to the blue (i.e., higher photon energy). For comparison, the absorption coefficients in bulk silicon at 300 K at the harmonic wavelengths of the Nd:YAG laser (often used in PLAL) are ~10^6 cm^{-1} at 355 nm, ~10^4 cm^{-1} at 532 nm, and 10 cm^{-1} at 1064 nm (Green 2008), with orders of magnitude variation within the optical region of the electromagnetic spectrum. Importantly, ultrasmall SiNPs produced by PLAL also exhibit a marked blue PL, reported as the black curve in Figure 10.3.

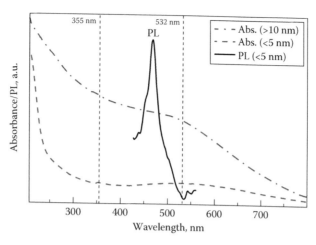

Figure 10.3 Example of typical optical properties of SiNPs colloids generated by PLAL: UV–Vis absorption spectrum of SiNPs with size >10 nm (blue dotted-dashed curve) and <5 nm (red dashed curve); photoluminescence (PL) spectrum from colloids with SiNPs with size <5 nm (black continuous curve). The vertical dashed lines indicate the second-harmonic and third-harmonic wavelengths of the Nd:YAG laser, 532 nm and 355 nm respectively, often used for PLAL. (Adapted from Intartaglia, R., et al., *Opt. Express*, 22, 3117–3127, 2014.)

10.2.2 PHOTOFRAGMENTATION

In Figure 10.4, the transmission electron microscopy (TEM) images of the SiNPs obtained by PLAL in water using a picosecond laser are reported. It is found that the SiNP size changes drastically with laser wavelength and ablation time. Specifically:

1. The larger SiNPs (~40 nm) shown in Figure 10.4a are obtained using the fundamental wavelength of the Nd:YAG laser at 1064 nm;
2. The small SiNPs (~10 nm) shown in Figure 10.4b are obtained by a short ablation time with the third-harmonic of the Nd:YAG laser at 355 nm;
3. The ultrasmall SiNPs (<5 nm) reported in Figure 10.4c result from longer ablation time with 355 nm laser pulses.

The smaller size of the latter SiNPs is due to the interplay of ablation and photofragmentation processes in liquid, as schematically shown in Figure 10.5. The as ablated SiNPs have a large-size and a wide-size distribution. Further irradiation of such SiNPs with 1064 nm laser light does not change the size distribution significantly since large SiNPs do not absorb near-infrared photons (see absorption spectra in Figure 10.3). Instead, when using UV laser light at 355 nm, the larger SiNPs do absorb photons resulting in the photofragmentation effect due to the high fluence used, while the ultrasmall SiNPs do not absorb such photons (see absorption spectra in Figure 10.3) resulting in an accumulation of smaller SiNPs in the colloid.

10.2.3 PRODUCTION CAPABILITY

An overview of the production capability of SiNPs via PLAL, is given in Tables 10.1 and 10.2. Specifically, Table 10.1 lists a collection of the outcome from experiments on PLAL of silicon in which the PL is not reported, while Table 10.2 also reports PL data. The data reported in the tables are:

- **Liquid** used for ablation, comprising mainly water and carbon-based solvents.
- λ is the wavelength of the laser used for ablation/photofragmentation, which ranges from UV to IR.
- **PD** is the laser pulse time duration, which spans from fs to sub-ms. Typically, the PD is indicated in ns, and other units are explicitly reported when necessary.
- **RR** is the repetition rate of the laser—the number of laser pulses delivered per second, which ranges from sub-Hz to few-kHz.
- **F** is the fluence applied for ablation—the energy per unit surface impinging on the solid target for each laser pulse, which ranges from mJ/cm^{-2} to 10s of J/cm^{-2}.

- **AT** is the total ablation time—the period of time the ablation/fragmentation process has been carried out, which spans from a few minutes to few hours.
- **Size** is the peak or median value of the distribution of the physical diameter of the produced NPs as reported in the publications. As noted before, the size distribution on SiNPs produced by PLAL is quite broad compared with other fabrication methods, and it is typically 10s of nm for NP size in the 20–100 nm range, while it is a few nm in the range 1–20 nm. The size values reported in the

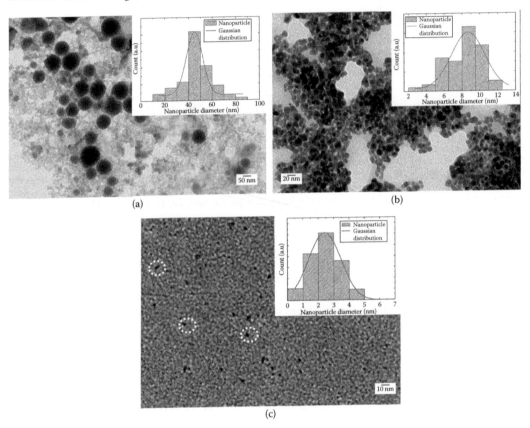

Figure 10.4 TEM images of SiNPs by PLAL in water: (a) large nanoparticles produced with 1064 nm ps laser; (b) small nanoparticles produced with 355 nm ps laser and short ablation time; (c) ultrasmall nanoparticles produced with 355 nm ps laser and long ablation time. (Adapted from Intartaglia, R., et al., *Opt. Express*, 22, 3117–3127, 2014.)

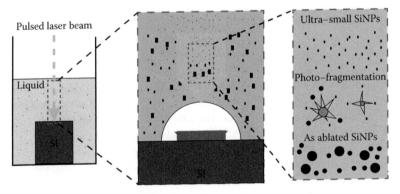

Figure 10.5 Schematic of PLAL and photofragmentation of silicon: larger SiNPs are generated by the ablation process (as ablated SiNPs, size >10 nm) and subsequently photofragmented by further irradiation leading to smaller nanoparticle formation (ultrasmall SiNPs, size <5 nm). The laser beam is shown in green for better visualization, while the ablation plume is shown in red.

tables correspond to the diameter of the NPs produced, which may be composed of isolated nano-crystals, nanopolycrystals with nanocrystal size usually below 10 nm, or a combination of such NPs embedded in a silicon oxide matrix. In some cases when the distribution is not clearly peaked, a size range is reported.

- **Exc.** is the wavelength of the excitation radiation used for PL measurements. In some cases, a systematic PL study has been performed varying the excitation wavelength in a selected range, and in these cases the investigated range is reported. The excitation wavelength spans from deep UV to green, with the vast majority of cases in the UV blue range.

- **Emis.** is the peak wavelength of the reported PL emission spectrum. It has to be noted that the emission spectrum of SiNPs produced by PLAL is always quite broad, with a width up to 100 nm. In some cases multiple peaks are observed, and then a range value is reported. Similarly, when a systematic PL study has been performed varying the excitation wavelength, the resulting emission range is reported rather than the one-by-one correspondence between excitation and emission wavelengths. The emission wave-length spans from UV to red, with the vast majority of cases in the blue-green range.

- **Ref.** is the bibliographic reference of the reported data.

The items in Tables 10.1 and 10.2 are grouped by the liquid used and ordered with decreasing laser wavelength and pulse duration.

From Table 10.1, it is noted that the first work on PLAL of silicon was reported at the beginning of the century (Dolgaev et al. 2002), and most of the activity about the production and characterization of SiNPs by PLAL has been conducted in the last decade, with a growing interest and research activity on this topic. This fact is due to the relative simplicity of the PLAL method and availability of pulsed lasers in many labo-ratories, as well as the versatility of the synthesis method, which allows plenty of room to perform phenom-enological and systematic studies given the large variety of parameters to be investigated.

As evidenced by the data reported in Table 10.2, the PL of SiNPs produced via PLAL is excited mainly in the wavelength range 250–400 nm with emission wavelength in the range 350–550 nm. A schematic of the light absorption and emission processes in SiNPs colloids produced by PLAL is shown in Figure 10.6. It is generally well established and accepted that the quantum confinement effect is responsible for the pho-ton absorption in SiNPs colloids produced by PLAL by electron–hole pair formation. Instead, the photon emission mechanism is still debated and is not related to a unique process, but it is most probably due to both quantum confinement effect and other radiative recombination processes on defects in the Si/SiO_x ($1 < x < 2$) interface or on surface states. The former case is typically found for PLAL in water, resulting in partially oxidized SiNPs. For example, a systematic study on aging and annealing of the produced SiNPs revealed that blue PL comes from electron–hole recombination at the near-interface traps located between the silicon core and surrounding SiO_x layer (Yang et al. 2011a). Another detailed study on the oxidation state of SiNPs produced by near-infrared nanosecond PLAL of silicon in water revealed the presence of two distinct SiNPs groups, one composed of smaller size fully oxidized NPs (< 40 nm) with PL bands in the UV blue range and another group of larger NPs (> 40 nm) which are formed by polynanocrystallites and exhibit a PL emission in the green–red region, possibly related to quantum confinement (Vaccaro et al. 2016). Instead, the emission from surface states is dominant when PLAL is performed in carbon-containing liquids, and it is due to carbon-based functional groups on the surface of the SiNPs (e.g., alkyl groups). As example, a study of PLAL of Si in hexane revealed SiNPs colloid where light absorption is due to quantum confinement effect at the X and Γ points, while PL emission originates from surface state of the Si–C and Si–C–H_2 bonds (Hao et al. 2016).

A characteristic aspect of photoluminescent SiNPs produced by PLAL is the lifetime of the light emis-sion, which may help in understanding the origin of the PL. Moreover, the control of the PL lifetime is important from an application point of view. For example, the relatively long lifetime of tens of ns to μs can be exploited for time-gated fluorescence imaging, in order to rule out the very fast (few ns) autofluorescence typical of bio and polymeric materials (Dahan et al. 2001; Gu et al. 2013). Up to the present time, the life-time of the PL from SiNPs produced by PLAL has not been extensively investigated with only few works reporting data on PL lifetime. Even if the green-NIR PL is ascribed to the quantum confinement effect, the indirect band gap nature of bulk silicon is retained in the long lifetime of the size-dependent PL emission, μs, while the strongest blue–green PL emission in SiNPs produced via PLAL has a typical lifetime on the

Clusters, nanoparticles, and quantum dots

Table 10.1 **A collection of works on SiNPs by PLAL without PL reported**

LIQUID	SIZE (nm)	LASER PARAMETERS λ (nm)/PD(ns) RR(Hz)/F(J/cm²)	ABL. TIME (min.)	REFERENCES
Water	>100	10600/2μs 1.5–3/1.5–7.5	30–60	Popovic et al. (2013)
Water	2	1064/18 1/70	–	Karimzadeh and Mansour (2013)
Water	70	1064/0.15 10/4.4	15	Popovic et al. (2014)
Water	56	1064/0.15 10/15.7	15	Popovic et al. (2014)
Water	77	1064/0.15 10/15.7	15	Popovic et al. (2014)
Water	64	1064/0.15 10/15.7	15	Popovic et al. (2014)
Water	>100	1064/1.5 60000/0.8	6	Chen et al. (2014)
Water	10–30	1064/34 ps 10/12.7	120	Eroshova et al. (2012)
Water	25	1064/30 ps 10/0.7	30	Perminov et al. (2011)
Water	50	1025/480 fs 1–5k/1	2–5	Baati et al. (2016)
Water	40	800/2 ps 1000/~19	30	Hamad et al. (2014)
Water	1–5	800/200 fs 1000/–	5	Alkis et al. (2012)
Water	2.4	800/120 fs 1000/0.05	30	Rioux et al. (2009)
Water	84	511/20 15000/0.7	–	Dolgaev et al. (2002)
Water	74	511/20 15000/1.2	–	Dolgaev et al. (2002)
Water	20–40	387/180 fs 1000/0.8	70	Semaltianos et al. (2010)
Water	20–150	387/180 fs 1000/0.8	45	Semaltianos et al. (2010)
Water	10	355/40 5000/40	2	Jiménez et al. (2010)

(Continued)

Table 10.1 *(Continued)* A collection of works on SiNPs by PLAL without PL reported

LIQUID	SIZE (nm)	LASER PARAMETERS λ (nm)/PD(ns) RR(Hz)/F(J/cm²)	ABL. TIME (min.)	REFERENCES
Ethanol	<10	1064/10 10/03	30	Yang et al. (2009)
Ethanol	10–50	800/50 fs 1k/~3	70	Li et al. (2015)
Ethanol	6	532/13 10/0.17–0.63	30 to 360	Kobayashi et al. (2013)
Ethanol	6	532/13 10/0.44	30	Chewchinda et al. (2014)
Ethanol	76	511/20 15000/0.7	–	Dolgaev et al. (2002)
Ethanol	78	511/20 15000/1.2	–	Dolgaev et al. (2002)
EDC	60	511/20 15000/0.7	–	Dolgaev et al. (2002)
EDC	66	511/20 15000/1.5	–	Dolgaev et al. (2002)
TCM	42	800/2 ps 1000/~19	30	Hamad et al. (2014)
TCM	20–120	355/40 5000/40	60	Abderrafi et al. (2013)
N_2	<40	1064/30 ps 02/06	60	Perminov et al. (2011)
Glycerol	<40	1064/30 ps 10/06	60	Perminov et al. (2011)
Acetone	10	800/2 ps 1000/~19	30	Hamad et al. (2014)
DCM	45	800/2 ps 1000/~19	30	Hamad et al. (2014)
Octene	2.5	532/– 20/0.8	60	Kitasako and Saitow (2013)
DMSO	3.5	1064/18 1/70	–	Karimzadeh and Mansour (2013)

Table 10.2 **A collection of works on SiNPs by PLAL with PL reported**

LIQUID SIZE (nm)	EXC. (nm) EMIS. (nm)	ABLATION PARAMETERS λ (nm)/PD(ns)/F(J/cm²) RR(Hz)/AT(min)	REFERENCES
Water 4	201 365–477	1064/0.4 ms/20 3.6k/240	Du et al. (2007)
Water 20	360 415–435	1064/10/1.6–6.4 10/30	Yang et al. (2009)
Water –	400 415–435	1064/10/3.2–6.4 10/30	Yang et al. (2011a)
Water 10–20	400–450 600	1064/9/5.5 10/25	Mahdieh and Momeni (2015)
Water 7	– 420	1064/8/~10 20/–	Svetlichnyi et al. (2015)
Water >40	300 580	1064/5/5.1 10/45	Vaccaro et al. (2016)
Water <40	250 300–464	1064/5/5.1 10/45	Vaccaro et al. (2016)
Water 3	354 652	1064/5/0.6 10/30	Vaccaro et al. (2014b)
Water <10 nm	248 281–460	1064/5/0.6–5 10/30–45	Vaccaro et al. (2014a), Vaccaro et al. (2014b)
Water 2.5	364 450–900	1025/480 fs/1 1k/45	Blandin et al. (2013)
Water 20	364 450–900	1025/480 fs/1 1k/300	Blandin et al. (2013)
Water <5	370 425	800/200 fs/– 1k/5	Alkis et al. (2012)
Water 5	400 460	800/110 fs/0.075 1k/60	Intartaglia et al. (2012b)
Water 65	400 580	800/110 fs/0.2 1k/60	Intartaglia et al. (2012b)
Water <100	270 300–440	800/90 fs/0.14 1k/120	Li et al. (2013)
Water <1 μm	230 367	532/10/1000 2.5/120	Liu et al. (2013)
Water 70–400	e–25 KeV 354 to 652	532/6/136 10/210	Alima et al. (2012)
Water 5–60	300 450–750	532/5/1.5 10/10	Fazio et al. (2011)

(Continued)

Table 10.2 *(Continued)* **A collection of works on SiNPs by PLAL with PL reported**

LIQUID SIZE (nm)	EXC. (nm) EMIS. (nm)	ABLATION PARAMETERS λ (nm)/PD(ns)/F(J/cm²) RR(Hz)/AT(min)	REFERENCES
Water 3	325 445	532/–/0.01 –/–	Umezu et al. (2007)
Water 60	300 428	355/8/0.009 30/30	Švrček et al. (2006)
Water 20	300 400	355/8/0.76 30/30	Švrček et al. (2006)
Water –	342 420	355/8/0.69 30/30	Švrček et al. (2009b)
Water 2	290 470	355/60 ps/3.2 20/60	Intartaglia et al. (2012a)
Water 2	325 410–525	248/10/0.024 20/240	Švrček et al. (2009a)
Water 2–70	400 660	248/10/0.024 20/10	Švrček et al. (2013)
H_2O_2 <100	300 450–750	532/5/1.5 10/10	Fazio et al. (2011)
Acetone 4–135	355 440–515	800/40 fs/0.8–40 1k/40	Hamad et al. (2014)
Acetone 7	350 430	800/40 fs/14 1k/40	Vendamani et al. (2015)
Ethanol 5	360 415–435	1064/10/1.6–6.4 10/30	Yang et al. (2009)
Ethanol –??	400 415–435	1064/10/3.2–6.4 10/30	Yang et al. (2011a)
Ethanol 7	350 423–460	1064/10/1.2 10/30	Chewchinda et al. (2013)
Ethanol 10–20	300–450 350–500	1064-10-07 10/50	Momeni and Mahdieh (2016)
Ethanol 20–50	400 640	800/<1 ps/4 1k/10	Kuzmin et al. (2010)
Ethanol 5	– 360–470	532/10/0.3 10/–	Kobayashi et al. (2014)
Ethanol 3	350 413	532/10/1.2 10/30	Chewchinda et al. (2013)
Ethanol <10	300 450–750	532/5/1.5 10/10	Fazio et al. (2011)

(Continued)

Table 10.2 *(Continued)* **A collection of works on SiNPs by PLAL with PL reported**

LIQUID SIZE (nm)	EXC. (nm) EMIS. (nm)	ABLATION PARAMETERS λ (nm)/PD(ns)/F(J/cm^2) RR(Hz)/AT(min)	REFERENCES
Octene 3	280–400 350–500	790/–/33 1k/1.3	Dewan et al. (2016)
Octene 2–10	266 380	532/10/0.5 10/30	Shirahata et al. (2009)
Octene 2	266 415	532/5/0.07 15/360	Nakamura et al. (2014)
Octene 2–10	266 360–480	532/5/0.2 10/30	Shirahata et al. (2010)
Toluene 3	350 420	355/60 ps/3.2 20/60	Intartaglia et al. (2012a)
Toluene 1	340–520 440–570	308/15/2 –/120	Lu et al. (2012)
Toluene 2	375 450	308/15/1 –/–	Lu et al. (2010)
N$_2$ <10	337 750	1250/120 fs/3.9 10/120	Eroshova et al. (2012)
Hexane 3	320–380 400–450	800/120 fs/10–20 1k/60	Tan et al. (2011)
Hexane 3	300–380 412	800/100 fs/0.15 1k/120	Hao et al. (2016)
Hexane 3	325 400	532/–/0.01 –/–	Umezu et al. (2007)
SDS 20	360 410–450	1064/10/3.2 10/30	Yang et al. (2008)
Acrylic 2	320–560 400–600	800/120 fs/10–20 1k/60	Tan et al. (2012)
DCM 2.4	375 470	308/20/2 10/30	Xu and Han (2014)
DMSO 3	350–475 455–520	1064/18/70 1/120	Mansour et al. (2012)
TCM 50	337 500	355/40/40 5k/–	Abderrafi et al. (2011)
CO$_2$ 6	325–633 427–653	532/9/0.8 20/20	Wei et al. (2012), Saitow and Yamamura (2009)

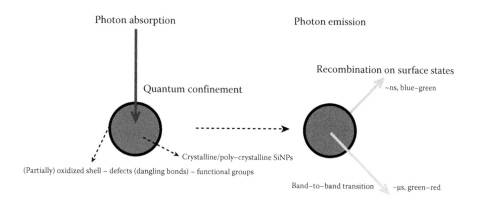

Figure 10.6 Basic schematic of the light absorption and emission processes in photoluminescent SiNPs generated by PLAL: photon absorption is due to the quantum confinement effect of the crystalline or polycrystalline SiNPs, while photon emission—PL—is due to recombination within the nanocrystals by band-to-band transition (lifetime ~µs, green–red range) and/or from the SiNP's surface due to radiative recombination at defects or functional groups (lifetime ~ns, blue–green range).

ns time scale (Ghosh and Shirahata 2014). However, the specific origin of such fast PL is not unique since it is related to the surface state and/or defects.

The blue PL in SiNPs colloids fabricated by femtosecond PLAL in hexane has a lifetime in the ns range and it is attributed both to radiative electron–hole recombination at surface state related to the Si–C or Si–C–H$_2$ bonds (Hao et al. 2016), and to radiative recombination of electron–hole pairs in oxide-related centers on the NP's surface (Tan et al. 2011). The surface-originated PL is further confirmed by the vibrational modulation of the PL spectrum ascribed to coupling between excitons and the Si–C scissoring vibration mode on the SiNP surface (Tan et al. 2011).

A detailed analysis on PL of SiNPs colloids produced by nanosecond PLAL in water reveals various bands (Vaccaro et al. 2014b, 2016): (i) a red PL band with µs lifetime ascribed to quantum confinement effect; (ii) PL bands in the UV with ns lifetime and in the blue with ms lifetime associated with allowed singlet-to-singlet and forbidden triplet-to-singlet transitions in oxygen deficient centers of amorphous SiO$_2$, respectively.

When performing PLAL of silicon in carbon-containing solvents, C-related chemical bonds are present on the SiNP surface, which induce excitonic recombination leading to fast PL in the blue–green region. For example, SiNP colloids produced by PLAL in dichloromethane (DCM) (Xu and Han 2014) have a excitation wavelength-dependent PL in the blue–green region with ns lifetime. Further, it was also observed that after natural oxidation by exposure to air of the dried SiNPs the PL shifts to the red and its lifetime increases to the µs range. This fact is attributed to the presence of a disordered oxide layer which acts as a carrier recombination center. Therefore, a different mechanism is at the origin of the PL of the SiNP colloid compared to the dried SiNPs.

A very important issue when studying the PL characteristics of SiNP colloids produced by PLAL is the possible contribution from the irradiated liquid or from the by-product originating from the ablation process, particularly when performing PLAL in carbon-containing solvents. In fact the work by Intartaglia et al. (2012a) on PLAL of silicon in toluene highlighted the dramatic need to also investigate the PL properties of the irradiated liquid alone as a control sample, demonstrating that carbon-based by-product was at the origin of the observed PL. Indeed, Yang et al. (2011b) studied the composition of the carbon-based by-product obtained by PLAL of silicon in toluene, evidencing the presence of graphitic structures being formed during laser ablation.

From these findings, the importance of the PL properties of the irradiated liquid becomes clear, and other works focused on the systematic investigation of the liquid effect on the PL. In Dewan et al. (2016), a study on SiNPs produced by PLAL in octane results in alkyl-capped SiNPs with PL emission at 335 nm when excited at 280 nm, possibly originating from electron–hole recombination at the direct Γ point transition in SiNPs shifted due to the quantum confinement effect. Besides, it was found that the carbon-based by-product from the PLAL process, and separated by the SiNPs by ultracentrifugation of the ablated colloidal solution, has a excitation-dependent PL in the range 370–450 nm with excitation wavelength in the range 300–400 nm. It is claimed that the source of this PL is amorphous and graphitic-like carbon compounds.

In some cases, the effect of the liquid itself must also be considered when studying light emission from PLAL-generated SiNPs colloids. In Momeni and Mahdieh (2016), dual pulse ablation of silicon in water and ethanol is performed, and after a detailed analysis of the PL the authors claim that the light emission comes from Raman scattering by hydroxy groups in the solvent with some contribution from hydroxy groups on the SiNP surface.

10.3 CONCLUSIONS

10.3.1 SUMMARY

PLAL is a facile and very versatile method for the production of NP colloids. The synthesis mechanism of the NPs in PLAL relies on the laser-induced ablation of solid materials in a liquid environment. A large variety of solid materials and liquids can be used in PLAL, from metallic to semiconductor and from water to carbon-containing solvents, with a corresponding large variety of different colloids being produced. The specific case of silicon is very interesting since it allows the production of photoluminescent colloids with SiNP size ranging from few to tens of nm, with potential applications as nanomarkers for bioimaging. The optical properties of SiNPs produced by PLAL are found to be determined by the quantum confinement effect and by the presence of defects, surface states, and (partially) oxidized sites. In fact, while the photon absorption mechanism is generally ascribed to the quantum confinement effect in the crystalline or polycrystalline SiNPs, with the absorption edge shifting to the blue with decreasing crystallite size, the photoluminescent mechanism is still debated with the main contribution due to the presence of defects, surface states, and (partially) oxidized sites with a smaller contribution from quantum confinement effect. The PL of SiNPs produced by PLAL is efficiently excited in the range 250–400 nm with emission mainly in the range 350–550 nm, and with some tunability within this range.

10.3.2 OUTLOOK

Even if the potentiality of SiNPs produced by PLAL has been clearly demonstrated, their actual implementation into practical application needs further investigation directed mainly toward the enhancement of the PL efficiency as well as the increase in productivity. In order to enhance the PL is necessary to determine precisely the origin of photon emission. In this respect, a lot of studies have been carried out and it is now well established that the main contribution to PL comes from the surface of the SiNPs, and that functionalization/decoration of the NP surface can lead to an efficient PL emission. On the contrary, not much work has been performed to study and optimize the productivity of SiNPs via PLAL, and much more effort is necessary in order to bring PLAL of silicon from the research field to the actual production arena. A productivity up to several grams per hour is foreseen using high-power laser sources (Intartaglia et al. 2014), which is actually similar to what is achieved by other methods (Zhong et al. 2015). However, more experimental work is necessary to reach such a productivity level.

The specific thermodynamic and atomic pathway that leads to the nucleation of the ablated material and the formation of stable colloidal solutions is still not fully understood. The many parameters involved in the PLAL process are at the origin of the basically phenomenological approach used so far to investigate the process. Many studies are reporting systematic investigations varying a few of these parameters (laser wavelength, laser pulse fluence, ablation time, liquid etc.). In order to fully integrate the results from the many experimental works a standardization of both the PLAL procedure and the methods used to characterize the produced SiNPs would be highly beneficial. In this respect, a possible list of the parameters to be specified in order to have a standardized approach is as follows:

- For the PLAL procedure: laser fluence, wavelength, pulse time duration, beam profile (typically either Gaussian or flat-top), amount of liquid used and height of the liquid surface on top of the bulk material surface, ablation time;

- For the characterization of the colloids and SiNPs: UV-Vis absorption spectrum of the colloid, PL emission and excitation wavelengths, PL quantum efficiency and lifetime (if possible also on the dry powder after ultracentrifugation), TEM and scanning electronic microscopy (SEM) imaging, dynamic light scattering (DLS), and Inductively Coupled Plasma-Mass Spectrometry (ICP-MS).

Online monitoring of the colloidal solution during PLAL is also very useful to analyze and investigate the process. As an example, online spectroscopy allows a monitoring of the size of the produced NPs and optimizing the PLAL process (Giorgetti et al. 2007).

Regarding the characterization of the colloidal solution obtained by PLAL, it is of paramount importance to perform ultracentrifugation to separate the possibly multiple components in solution, specifically the SiNPs and the by-product originating from the ablation process. The pellet and supernatant thus obtained should be characterized separately using the methods listed above. It is also important to characterize the control sample constituted by the liquid irradiated without the solid material, in order to evaluate the effect on the liquid itself of the pulsed laser irradiation. Techniques useful to identify and characterize species generated during PLAL are Fourier Transformed Infrared Spectroscopy (FTIR), X-ray diffraction (XRD), X-ray Photoelectron Spectroscopy (XPS), and Raman spectroscopy.

Another significant issue to be studied more systematically is the effect of aging on the optical properties, and stability of the colloid, which is very important toward the practical use of PLAL-generated SiNPs colloids.

In conclusion, PLAL of silicon is a relatively young technology to fabricate SiNPs colloids, which holds great potentiality to become a production technology. However in order to bring PLAL of silicon out from its infancy much experimental work is still needed with a standardization in the procedures, which will help in defining a unique picture of the methodology thus allowing to optimize and engineer the fabrication process.

REFERENCES

Abderrafi, K., Calzada, R. G., Gongalsky, M. B., Suarez, I., Abarques, R., Chirvony, V. S., Timoshenko, V. Y., Ibáñez, R., and Martínez-Pastor, J. P. (2011). Silicon nanocrystals produced by nanosecond laser ablation in an organic liquid. *J. Phys. Chem. C*, 115:51475151.

Abderrafi, K., García-Calzada, R., Sanchez-Royo, J., Chirvony, V., Agouram, S., Abargues, R., Ibáñez, R., and Martínez-Pastor, J. (2013). Laser ablation of a silicon target in chloroform: Formation of multilayer graphite nanostructures. *J. Phys. D: Appl. Phys.*, 46:135301(9).

Alima, D., Estrin, Y., Rich, D. H., and Bar, I. (2012). The structural and optical properties of supercontinuum emitting Si nanocrystals prepared by laser ablation in water. *J. Appl. Phys.*, 112:114312.

Alkis, S., Okyay, A. K., and Ortaç, B. (2012). Post-treatment of silicon nanocrystals produced by ultra-short pulsed laser ablation in liquid: Toward blue luminescent nanocrystal generation. *J. Phys. Chem. C*, 116:34323436.

Amendola, V., and Meneghetti, M. (2013). What controls the composition and the structure of nanomaterials generated by laser ablation in liquid solution? *Phys. Chem. Chem. Phys.*, 15:3027–3046.

Anastas, P. T., and Warner, J. C. (1998). *Green Chemistry: Theory and Practice*. Oxford University Press, New York.

Baati, T., Al-Kattan, A., Esteve, M.-A., Njim, L., Ryabchikov, Y., Chaspoul, F., Hammami, M., Sentis, M., Kabashin, A. V., and Braguer, D. (2016). Ultrapure laser-synthesized Si-based nanomaterials for biomedical applications: In vivo assessment of safety and biodistribution. *Sci. Rep.*, 6:25400.

Bagga, K., Barchanski, A., Intartaglia, R., Dante, S., Marotta, R., Diaspro, A., Sajti, C. L., and Brandi, F. (2013). Laser-assisted synthesis of *Staphylococcus aureus* protein-capped silicon quantum dots as bio-functional nanoprobes. *Laser Phys. Lett.*, 10:065603.

Barbagiovanni, E. G., Lockwood, D. J., Simpson, P. J., and Goncharova, L. V. (2014). Quantum confinement in Si and Ge nanostructures: Theory and experiment. *Appl. Phys. Rev.*, 1:011302.

Belomoin, G., Therrien, J., Smith, A., Rao, S., Twesten, R., Chaieb, S., Nayfeh, M. H., Wagner, L., and Mitas, L. (2002). Observation of a magic discrete family of ultrabright Si nanoparticles. *Appl. Phys. Lett.*, 80:841–843.

Blandin, P., Maximova, K. A., Gongalsky, M. B., Sanchez-Royo, J. F., Chirvony, V. S., Sentis, M., Timoshenko, V. Y., and Kabashin, A. V. (2013). Femtosecond laser fragmentation from water-dispersed microcolloids: Toward fast controllable growth of ultrapure Si-based nanomaterials for biologicalapplications. *J. Mater. Chem. B*, 1:2489.

Borsella, E., Falconieri, M., Herlin, N., Loschenov, V., Miserocchi, G., Nie, Y., Rivolta, I., Ryabova, A., and Wang, D. (2010). Biomedical and sensor applications of silicon nanoparticles. In: *Silicon Nanocrystals: Fundamentals, Synthesis and Applications* (Pavesi, L., and Turan, R., eds.), pp. 507–536. Wiley-VCH Verlag GmbH, KGaA, Weinheim, Germany.

Chen, L., Jiang, X.-F., Guo, Z., Zhu, H., Kao, T.-S., Xu, Q.-H., Ho, G., and Hong, M. (2014). Tuning optical nonlinearity of laser-ablation-synthesized silicon nanoparticles via doping concentration. *J. Nanomater.*, 2014(7):652829.

Chewchinda, P., Hayashi, K., Ichida, D., Seo, H., Uchida, G., Shiratani, M., Odawara, O., and Wada, H. (2014). Preparation of Si nanoparticles by laser ablation in liquid and their application as photovoltaic material in quantum dot sensitized solar cell. *J. Phys. Conf. Ser.* 518:012023.

Chewchinda, P., Tsuge, T., Funakubo, H., Odawara, O., and Wada, H. (2013). Laser wavelength effect on size and morphology of silicon nanoparticles prepared by laser ablation in liquid. *Jpn. J. Appl. Phys.*, 52:025001.

Chinnathambi, S., Chen, S., Ganesan, S., and Hanagata, N. (2014). Silicon quantum dots for biological applications. *Adv. Healthcare Mater.*, 3:10–29.

Dahan, M., Laurence, T., Pinaud, F., Chemla, D. S., Alivisatos, A. P., Sauer, M., and Weiss, S. (2001). Time-gated biological imaging by use of colloidal quantum dots. *Opt. Lett.*, 26:825–827.

Dewan, S., Odhner, J. H., Tibbetts, K. M., Afsari, S., Levis, R. J., and Borguet, E. (2016). Resolving the source of blue luminescence from alkyl-capped silicon nanoparticles synthesized by laser pulse ablation. *J. Mater. Chem. C*, 4:6894–6899. doi: 10.1039/C6TC02283A.

Dolgaev, S., Simakin, A., Voronov, V., Shafeev, G., and Bozon-Verduraz, F. (2002). Nanoparticles produced by laser ablation of solids in liquid environment. *App. Surf. Sci.*, 28(1–4):546–551.

Du, X.-W., Qin, W.-J., Lu, Y.-W., Han, X., Fu, Y.-S., and Hu, S.-L. (2007). Face-centered-cubic Si nanocrystals prepared by microsecond pulsed laser ablation. *J. Appl. Phys.*, 102:013518.

Erogbogbo, F., Yong, K. T., Roy, I., Hu, R., Law, W. C., Zhao, W. W., Ding, H., et al. (2011). In vivo targeted cancer imaging, sentinel lymph node mapping and multi-channel imaging with biocompatible silicon nanocrystals. *ACS Nano*, 5:413–423.

Eroshova, O. I., Perminov, P. A., Zabotnov, S. V., Gongal'skii, M. G., Ezhov, A. A., Golovan, L. A., and Kashkarov, P. K. (2012). Structural properties of silicon nanoparticles formed by pulsed laser ablation in liquid media. *Crystallogr. Rep.*, 57:831–835.

Fazio, E., Barreca, F., Spadaro, S., Curra, G., and Neri, F. (2011). Preparation of luminescent and optical limiting silicon nanostructures by nanosecond-pulsed laser ablation in liquids. *Mater. Chem. Phys.*, 130:418–424.

Ghosh, B., and Shirahata, N. (2014). Colloidal silicon quantum dots: Synthesis and luminescence tuning from the near-UV to the near-IR range. *Sci. Technol. Adv. Mater.*, 15:014207.

Giorgetti, E., Giusti, A., Laza, S. C., Marsili, P., and Giammanco, F. (2007). Production of colloidal gold nanoparticles by picosecond laser ablation in liquids. *Phys. Status Solidi (A)*, 204:1693–1698.

Green, M. A. (2008). Self-consistent optical parameters of intrinsic silicon at 300 K including temperature coefficients. *Solar Energy Mater. Solar Cells*, 92:1305–1310.

Gu, L., Hall, D. J., Qin, Z., Anglin, E., Joo, J., Mooney, D. J., Howell, S. B., and Sailor, M. J. (2013). In vivo time-gated fluorescence imaging with biodegradable luminescent porous silicon nanoparticles. *Nat. Commun.*, 4:2326.

Hamad, S., Podagatlapalli, G. K., Vendamani, V. S., Nageswara Rao, S. V. S., Pathak, A. P., Tewari, S. P., and Rao Soma, V. (2014). Femtosecond ablation of silicon in acetone: Tunable photoluminescence from generated nanoparticles and fabrication of surface nanostructures. *J. Phys. Chem. C*, 118:7139–7151.

Hao, H. L., Wu, W. S., Zhang, Y., Wu, L. K., and Shen, Z. (2016). Origin of blue photoluminescence from colloidal silicon nanocrystals fabricated by femotosecond laser ablation in solution. *Nanotechnology*, 27:325702.

Intartaglia, R., Bagga, K., and Brandi, F. (2014). Study on the productivity of silicon nanoparticles by picosecond laser ablation in water: Towards gram per hour yield. *Opt. Express*, 22:3117–3127.

Intartaglia, R., Bagga, K., Genovese, A., Athanassiou, A., Cingolani, R., Diaspro, A., and Brandi, F. (2012a). Influence of organic solvent on optical and structural properties of ultra-small silicon dots synthesized by UV laser ablation in liquid. *Phys. Chem. Chem. Phys.*, 14:15406–15411.

Intartaglia, R., Bagga, K., Scotto, M., Diaspro, A., and Brandi, F. (2012b). Luminescent silicon nanoparticles prepared by ultra short pulsed laser ablation in liquid for imaging applications. *Opt. Mater. Express*, 2:510–518.

Intartaglia, R., Barchanski, A., Bagga, K., Genovese, A., Das, G., Wagener, P., Di Fabrizio, E., Diaspro, A., Brandi, F., and Barcikowski, S. (2012c). Bioconjugated silicon quantum dots from one-step green synthesis. *Nanoscale*, 4:1271–1274.

Jiménez, E., Abderrafi, K., Abargues, R., Valdes, J., and Martínez-Pastor, J. (2010). Laser-ablation-induced synthesis of SiO$_2$-capped noble metal nanoparticles in a single step. *Langmuir*, 26(10):7458–7463.

Karimzadeh, R., and Mansour, N. (2013). Solvent dependent optical limiting and thermo-optical response of silicon nanoparticle colloids. *Optik*, 124:7032–7035.

Kitasako, T., and Saitow, K. (2013). Si quantum dots with a high absorption coefficient: Analysis based on both intensive and extensive variables. *Appl. Phys. Lett.*, 103:151912.

Kobayashi, H., Chewchinda, P., Inoue, Y., Funakubo, H., Hara, M., Fujino, M., Odawara, O., and Wada, H. (2014). Photovoltaic properties of Si-based quantum-dot-sensitized solar cells prepared using laser plasma in liquid. *Jpn. J. Appl. Phys.*, 53:010208.

Kobayashi, H., Chewchinda, P., Ohtani, H., Odawara, O., and Wada, H. (2013). Effects of laser energy density on silicon nanoparticles produced using laser ablation in liquid. In *Journal of Physics: Conference Series, Proceedings of the 11th Asia Pacific Conference on Plasma Science and Technology (APCPST) and 25th Symposium on Plasma Science for Materials (SPSM)*, 441:012035.

Kuzmin, P. G., Shafeev, G. A., Bukin, V. V., Garnov, S. V., Farcau, C., Carles, R., Warot-Fontrose, B., Guieu, V., and Viau, G. (2010). Silicon nanoparticles produced by femtosecond laser ablation in ethanol: Size control, structural characterization, and optical properties. *J. Phys. Chem. C*, 114:15266–15273.

Li, C.-Q., Zhang, C.-Y., Huang, Z.-S., Li, X.-F., Dai, Q.-F., Lan, S., and Tie, S.-L. (2013). Assembling of silicon nanoflowers with significantly enhanced second harmonic generation using silicon nanospheres fabricated by femtosecond laser ablation. *J. Phys. Chem. C*, 117:2462524631.

Li, X., Zhang, G., Jiang, L., Shi, X., Zhang, K., Rong, W., Duan, J., and Lu, Y. (2015). Production rate enhancement of size-tunable silicon nanoparticles by temporally shaping femtosecond laser pulses in ethanol. *Opt. Express*, 23:4226–4232.

Li, Z. F., and Ruckenstein, E. (2004). Water-soluble poly(acrylic acid) grafted luminescent silicon nanoparticles and their use as fluorescent biological staining labels. *Nano Lett.*, 4:1463–1467.

Lida, Y., Tsuge, A., Uwanimo, Y., Morikawa, H., and Ishizuka, T. (1991). Laser ablation in a liquid-medium as a technique for solid sampling. *J. Anal. At. Spectrom.*, 6:541–544.

Liu, P., Liang, Y., Li, H. B., Xiao, J., He, T., and Yang, G. W. (2013). Violet-blue photoluminescence from Si nanoparticles with zinc-blende structure synthesized by laser ablation in liquids. *AIP Adv.*, 3:022127.

Lu, W., Bian, Y., Liu, H., Han, L., Yu, W., and Fu, G. (2010). Preparation and luminescence properties of spin-coated PMMA films containing Si nanocrystallites. *Mater. Lett.*, 64:1073–1076.

Lu, W., Wu, L., Yu, W., Wang, X., Li, X., and Fu, G. (2012). Blue photoluminescence from ultrasmall silicon nanocrystals produced by nanosecond pulsed laser ablation in toluene. *Micro Nano Lett.*, 7:1125–1128.

Mahdieh, M. H., and Momeni, A. (2015). From single pulse to double pulse ns laser ablation of silicon in water: Photoluminescence enhancement of silicon nanocrystals. *Laser Phys.*, 25:015901.

Mangolini, L. (2013). Synthesis, properties, and applications of silicon nanocrystals. *J. Vac. Sci. Technol. B*, 31:020801.

Mansour, N., Momeni, A., Karimzadeh, R., and Amini, M. (2012). Blue-green luminescent silicon nanocrystals fabricated by nanosecond pulsed laser ablation in dimethyl sulfoxide. *Opt. Mater. Express*, 2:740.

Momeni, A., and Mahdieh, M. H. (2016). Photoluminescence analysis of colloidal silicon nanoparticles in ethanol produced by double-pulse ns laser ablation. *J. Lumin.*, 176:136–143.

Nakamura, T., Yuan, Z., and Adachi, S. (2014). High-yield preparation of blue-emitting colloidal Si nanocrystals by selective laser ablation of porous silicon in liquid. *Nanotechnology*, 25:275602.

Neddersen, J., Chumanov, G., and Cotton, T. M. (1993). Laser ablation of metals: A new method for preparing SERS active colloids. *Appl. Spectros.*, 47:1959–1964.

Peng, F., Su, Y., Zhong, Y., Fan, C., Lee, S.-T., and He, Y. (2014). Silicon nanomaterials platform for bioimagin, biosensing, and cancer therapy. *Acc. Chem. Res.*, 47:612–623.

Perminov, P., Dzhun, I., Ezhov, A., Zabotnov, S., Golovan, L., Ivlev, G., Gatskevich, E.I. Malevich, V., and Kashkarov, P. (2011). Creation of silicon nanocrystals using the laser ablation in liquid. *Laser Phys.*, 21(4):801–804.

Popovic, D., Chai, J., Zekic, A., Trtica, M., Momcilovic, M., and Maletic, S. (2013). Synthesis of silicon-based nanoparticles by 10.6 μm nanosecond CO_2 laser ablation in liquid. *Laser Phys. Lett.*, 10:026001(7).

Popovic, D., Chai, J., Zekic, A., Trtica, M., Stasic, J., and Sarvan, M. (2014). The influence of applying the additional continuous laser on the synthesis of silicon-based nanoparticles by picosecond laser ablation in liquid. *Laser Phys. Lett.*, 11:116101(6).

Rioux, D., Laferrière, M., Douplik, A., Shah, D., Lilge, L., Kabashin, A., and Meunier, M. (2009). Silicon nanoparticles produced by femtosecond laser ablation in water as novel contamination-free photosensitizers. *J. Biomed. Opt.*, 14(2):021010.

Rosso-Vasic, M., Spruijt, E., Popovic, Z., Overgaag, K., Van Lagen, B., Grandidier, B., Vanmaekelbergh, D., Dominguez-Gutierrez, D., De Cola, L., and Zuilhof, H. (2009). Amine-terminated silicon nanoparticles: Synthesis, optical properties and their use in bioimaging. *J. Mater. Chem*, 19:5926–5933.

Saitow, K., and Yamamura, T. (2009). Effective cooling generates efficient emission: Blue, green, and red light-emitting Si nanocrystals. *J. Phys. Chem. C*, 113:8465.

Semaltianos, N., Logothetidis, S., Perrie, W., Romani, S., Potter, R., Edwardson, S., French, P., Sharp, M., Dearden, G., and Watkins, K. (2010). Silicon nanoparticles generated by femtosecond laser ablation in a liquid environment. *J. Nanopart. Res.*, 12(2):573–580.

Semaltianos, N. G. (2016). Nanoparticles by laser ablation of bulk target materials in liquids. In: *Handbook of Nanoparticles* (Aliofkhazraei, M., ed.), pp. 67–92. Springer International, Cham, Switzerland.

Shirahata, N., Hirakawa, D., and Sakka, Y. (2010). Interfacial-related color tuning of colloidal Si nanocrystals. *Green Chem.*, 12:2139–2141.

Shirahata, N., Linford, M. R., Furumi, S., Pei, L., Sakka, Y., Gates, R. J., and Asplund, M. C. (2009). Laser-derived one-pot synthesis of silicon nanocrystals terminated with organic monolayers. *Chem. Commun.*, 2009:4684–4686.

Stupca, M., Alsalhi, M., Al Saud, T., Almuhanna, A., and Nayfeh, H. M. (2007). Enhancement of polychrystalline silicon solar cells using ultra thin films of silicon nanoparticle. *Appl. Phys. Lett.*, 91:063107.

Svetlichnyi, V. A., Izaak, T. I., Lapin, I. N., Martynova, D. O., Stonkus, O. A., Stadnichenko, A. I., and Boronin, A. I. (2015). Physicochemical investigation of nanopowders prepared by laser ablation of crystalline silicon in water. *Adv. Powder Technol.*, 26:478–486.

Švrček, V., Mariotti, D., Blackley, R. A., Zhou, W. Z., Nagai, T., Matsubar, K., and Kondo, M. (2013). Semiconducting quantum confined silicon–tin alloyed nanocrystals prepared by ns pulsed laser ablation in water. *Nanoscale*, 5:6725–6730.

Švrček, V., Mariotti, D., and Kondo, M. (2009a). Ambient-stable blue luminescent silicon nanocrystals prepared by nanosecond-pulsed laser ablation in water. *Opt. Express*, 17:520–527.

Švrček, V., Sasaki, T., Katoh, R., Shimizu, Y., and Koshizaki, N. (2009b). Aging effect on blue luminescent silicon nanocrystals prepared by pulsed laser ablation of silicon wafer in de-ionized water. *Appl. Phys. B*, 94:133–139.

Švrček, V., Sasaki, T., Shimizu, Y., and Koshizaki, N. (2006). Blue luminescent silicon nanocrystals prepared by ns pulsed laser ablation in water. *Appl. Phys. Lett.*, 89:213113.

Tan, D., Ma, Z., Xu, B., Dai, Y., Ma, G., He, M., Jin, Z., and Qiu, J. (2011). Surface passivated silicon nanocrystals with stable luminescence synthesized by femtosecond laser ablation in solution. *Phys. Chem. Chem. Phys.*, 13:20255–20261.

Tan, D., Xu, B., Chen, P., Dai, Y., Zhou, S., Ma, G., and Qiu, J. (2012). One-pot synthesis of luminescent hydrophilic silicon nanocrystals. *RSC Adv.*, 2:8254–8257.

Tan, D., Zhou, S., Qiu, J., and Khusroa, N. (2013). Preparation of functional nanomaterials with femtosecond laser ablation in solution. *J. Photochem. Photobiol. C-Photochem. Rev.*, 17:50–68.

Umezu, I., Minami, H., Senoo, H., and Sugimura, A. (2007). Synthesis of photoluminescent colloidal silicon nanoparticles by pulsed laser ablation in liquids. *J. Phys. Conf. Ser.*, 59:392–395.

Vaccaro, L., Camarda, P., Messina, F., Buscarino, G., Agnello, S., Gelardi, F. M., Cannas, M., and Boscaino, R. (2014a). Oxidation of silicon nanoparticles produced by nanosecond laser ablation in liquids. *AIP Conf. Proc.*, 1624:174.

Vaccaro, L., Popescu, R., Messina, F., Camarda, P., Schneider, R., Gerthsen, D., Gelardi, F. M., and Cannas, M. (2016). Self-limiting and complete oxidation of silicon nanostructures produced by laser ablation in water. *J. Appl. Phys.*, 120:024303.

Vaccaro, L., Sciortino, L., Messina, F., Buscarino, G., Agnello, S., and Cannas, M. (2014b). Luminescent silicon nanocrystals produced by near-infrared nanosecond pulsed laser ablation in water. *Appl. Surf. Sci.*, 302:62–65.

Vendamani, V. S., Hamad, S., Saikiran, V., Pathak, A. P., Venugopal Rao, S., Ravi Kanth Kumar, V. V., and Nageswara Rao, S. V. S. (2015). Synthesis of ultra-small silicon nanoparticles by femtosecond laserablation of porous silicon. *J. Mater. Sci.*, 50:1666–1672.

Walters, R., Bourianoff, G., and Atwater, H. (2005). Field-effect electroluminescence in silicon nanocrystals. *Nat. Mater.*, 4:143–146.

Wei, S., Yamamura, T., Kajiya, D., and Saitow, K.-I. (2012). White-light-emitting silicon nanocrystal generated by pulsed laser ablation in supercritical fluid: Investigation of spectral components as a function of excitation wavelengths and aging time. *J. Phys. Chem. C*, 116:3928–3934.

Xiao, L., Gu, L., Howell, S. B., and Sailor, M. J. (2011). Porous silicon nanoparticle photosensitizers for singlet oxygen and their phototoxicity against cancer cells. *ACS Nano*, 5:3651–3659.

Xu, Y. and Han, Y. (2014). Effects of natural oxidation on the photoluminescence properties of Si nanocrystals prepared by pulsed laser ablation. *Appl. Phys. A*, 117:1557–1562.

Yang, S., Cai, W., Zeng, H., and Li, Z. (2008). Polycrystalline Si nanoparticles and their strong aging enhancement of blue photoluminescence. *J. Appl. Phys.*, 104:023516.

Yang, S., Cai, W., Zhang, H., Xu, X., and Zeng, H. (2009). Size and structure control of Si nanoparticles by laser ablation in different liquid media and further centrifugation classification. *J. Phys. Chem. C*, 113:19091–19095.

Yang, S., Li, W., Cao, B., Zeng, H., and Cai, W. (2011a). Origin of blue emission from silicon nanoparticles: Direct transition and interface recombination. *J. Phys. Chem. C*, 115:21056–21062.

Yang, S., Zeng, H., Zhao, H., Zhang, H., and Cai, W. (2011b). Luminescent hollow carbon shells and fullerene-like carbon spheres produced by laser ablation with toluene. *J. Mater. Chem.*, 21:4432–4436.

Zeng, H., Du, X.-W., Singh, S. C., Kulinich, S. A., Yang, S., He, J., and Cai, W. (2012). Nanomaterials via laser ablation/irradiation in liquid: A review. *Adv. Funct. Mater.*, 22:1333–1353.

Zhong, Y., Peng, F., Bao, F., Wang, S., Ji, X., Yang, L., Su, Y., Lee, S.-T., and He, Y. (2013). Large-scale aqueous synthesis of fluorescent and biocompatible silicon nanoparticles and their use as highly photostable biological probes. *JACS*, 135:8350–8356.

Zhong, Y., Sun, X., Wang, S., Peng, F., Bao, F., Su, Y., Li, Y., Lee, S.-T., and He, Y. (2015). Facile, large-quantity synthesis of stable, tunable-color silicon nanoparticles and their application for long-term cellular imaging. *ACS Nano*, 9:5958–5967.

11 Silicon nanoparticles with zincblende structure

Pu Liu, Jun Xiao, and Guowei Yang

Contents

11.1 INTRODUCTION

Silicon (Si) products have been extensively investigated during the last 60 years and are currently the most important and dominant semiconductors used in communication transmission, microelectronics, and solar cells. With the development of nanotechnology, numerous low-dimensional Si nanomaterials, such as nanowires, nanobelts, and nanoparticles, have been synthesized in the last 20 years.[1-4] Si nanomaterials possess unique physical and chemical properties, such as size confinement and high surface area, and are of growing interest to researchers. On the other hand, along with the research of Si products, the complex phase diagram of Si has also attracted intense research interest whether in theoretical or in experimental studies for more than 50 years, because of its key position for technological application and fundamental condensed matter physics study.[5-8] Moreover, the phase transformation research of Si nanomaterials and unique phase synthesis of Si nanocrystals (NCs) have also been of high concern from 20 years ago,[9-12] because of the unique Si nanomaterials, especially their special nanophase which may be applied in field-emission devices,[106,107] biomedical imaging, biosensors, and energy storage.[5-7] For example, by using high-pressure X-ray diffraction (XRD) and resistivity measurements in a diamond anvil cell (DAC), or micro-Raman spectroscopy associate indentation experiments with precisely controlled strain rates, researchers have found that Si undergoes a series of phase transitions during compression. Si can be transformed from the diamond cubic structure (Si-I) at ambient condition to the metallic β-Sn phase (Si-II) at about 12 GPa, then to Imma (Si-XI) at about 13 GPa, and to a primitive hexagonal structure (Si-V) at about 16G Pa.[10] Moreover, these phase transitions can be reversible upon decompression. Instead of a reversible transition to the original Si-I phase, the Si-II phase would transform to a semimetallic R8 (Si-XII) phase at approximately 9.3 GPa and further to a metastable body-centered BC8 (Si-III) phase when pressure is completely released.[13] However, it should be pointed out that the abundant and unexpected mestable phase of Si has not been explored to the end, just utilizing DAC or indentation experiment (including nanoindentation research) or pressure-induced transformation methods.[14–17] Some of the predicted metastable phase—zincblende structure of Si—has not been achieved by the traditional high-pressure compression method, however, it may be trapped utilizing a far-from-thermodynamic equilibrium process.[18] In addition, the knowledge of the phase and stability of these metastable structures are therefore critical for developing practical applications.

In recent years, understanding the luminescent properties of Si nanomaterials and their underlying mechanisms has been a major challenge that is currently being investigated because of their potential use in fundamental physics and optoelectronics applications.[19–21] Many researchers have focused their attention on the emission characteristics of various Si nanostructures,[12,19,22,23] and luminescent Si nanoparticles have received particular interest.[24–26] Research of Si nanoparticles exhibiting visible emission will greatly benefit Si-based optoelectronics, full color displays, and biotechnology.[27–30] However, there have been few in-depth studies investigating the origin of luminescence from Si nanomaterials, and reports of violet–blue light emission from Si NCs are scarce.[31,32] For example, the emission mechanism and detailed origin of the photoluminescence (PL) in Si nanoparticles are controversial, and at least four main pathways for emission have been proposed, including (i) quantum size confinement effects,[1,33] (ii) direct band gap emission related to the ultrasmall size of Si,[34] (iii) oxide-related effects,[35,36] and (iv) surface state or stacking fault-induced emission effects.[37,38] A few years ago, Zeng et al.[32] proposed a mechanism where some excitons form and are trapped by nonradiative Pb centers within Si NCs, and others recombine in the area between Si and the surrounding SiO_x, to explain the blue emission from their Si nanoparticles. However, are there other mechanisms apart from those mentioned above that can explain the origin of violet–blue luminescence in Si nanomaterials? It is really a challenging question when it comes to the metastable phase of Si, for instance, the zincblende structure of Si.

Therefore, aiming at an overview of the synthesis and property of SiNCs with zincblende structure that was achieved under laser ablation in liquids (LAL), in this chapter, the formation and characterized properties of unique micro- and nanocubes of single-crystalline Si with zincblende structure and the zincblende structure SiNCs that emit violet–blue light are primarily introduced, utilizing the synthesis method of pulse laser-induced liquid–solid interface reaction-assisted low-concentration inorganic salt solutions (PLIIR), and electric field–assisted laser ablation in liquids (EFLAL). This chapter is organized into four sections. After the brief introduction about the motivation and sporadic research of metastable structure of Si in Section 11.1, we will first introduce the synthesis process that is based on LAL and its growth mechanism upon LAL—the basic physical processes involved in the metastable zincblende structure of SiNCs synthesis and nanocubes fabrication upon LAL—in Section 11.2. Through Section 11.3, we introduce a series of detailed properties characterization of the micro- and nanocubes and the NCs of Si with zincblende structure using scanning electron microscopy (SEM), transmission electron microscopy (TEM), associated energy-dispersive X-ray spectroscopy (EDS), XRD, UV absorption spectroscopy, PL measurements, and Raman spectroscopy. Further, annealing treatment studies of the Si zincblende structure are also performed to probe the thermal stability of this metastable phase and the relationship between the metastable structure and PL in the SiNCs with zincblende structure. Finally, a summary of the further research and its potential applications will be discussed in Section 11.4, along with the challenges that remain in the field.

11.2 FORMATION OF METASTABLE ZINCBLENDE Si NANOSTRUCTURE UPON LASER ABLATION IN A CONFINING LIQUID ENVIRONMENT

In the past decades, extensive research on inorganic nanostructures has been extensively carried out because of the unique applications of these inorganic nanostructures in mesoscopic chemistry and physics, and in microelectronic and optoelectronic devices. The phase- and morphology-controlled synthesis of nanostructures has been one of the frontier fields in nanostructure fabrication because the shape and phase of nanostructures play crucial roles in their chemical and physical properties.[39,40] For instance, as building blocks for fabricating nanodevices, numerous nanoconfigurations with interesting morphologies or specific structures have been synthesized. Similarly, nanocubes with wide applications have attracted special interest from both theoretical and applicable perspectives because of their shape-dependent properties.[41–46] However, compared with the synthesis of metal or metal oxide nanostructure,[47–49] still few researches are involved in the novel Si cubic morphology in present securable literatures.[50] Therefore, it is a great challenge to develop new methods to achieve the shape and phase-controlled synthesis of Si nanocubes or novel Si nanoparticles. Aiming at the controlled synthesis of micro- and nanocubes of Si NCs, Liu et al. developed unique techniques, that is, PLIIR-assisted low-concentration inorganic salt solutions and EFLAL, and successfully used them to synthesize a kind of micro- and nanocubes of Si with zincblende structure as well as a novel kind of metastable SiNCs with zincblende structures.[51]

11.2.1 SYNTHESIS OF MICRO- AND NANOCUBES OF SILICON WITH ZINCBLENDE STRUCTURE

Briefly, for the detailed PLIIR synthesis of micro- and nanocubes of Si, the solid target is first set as a single-crystalline Si substrate with purity of 99.99% and deposited thin amorphous carbon layer with the thickness of 100–300 nm, which has been prepared prior by a filtered cathode vacuum arc (FCVA) technique with the pressure of 3×10^{-5} Torr and the substrate temperature of 300 K. The liquid, which is used as assistant agents in the synthesis, is selected to be a mixture of twice-distilled water, ethanol, acetone, and very low-concentration inorganic salts solutions (below < 8 mM) such as KCl and NaCl solutions. Note that the motivation lies specifically in utilizing simple inorganic ions as additives instead of surfactant molecules or organic molecules. In general, inorganic ions have a more pronounced influence on the nucleation process of crystals than that of organic surfactants or polymers.[52,53] All aqueous solutions that are used as the laser ablation environments are prepared from the twice-distilled water, KCl (99.5+%), and NaCl (99.5+%), and the concentration ration of the twice-distilled water, ethanol, acetone, and inorganic salt is about 5:2.5:1.5:1. A second-harmonic laser is produced by a Q-switched Nd:YAG laser with a wavelength of 532 nm, pulse width of 10 ns, power density of 10^{10} W/cm^2 and repetition rate of 10 Hz.

The formation process can be described as follows. The solid target is first fixed on the bottom of a quartz chamber. Then, the liquid is poured slowly into the chamber until the target is covered by 2–3 mm. Finally, the pulsed laser is focused onto the surface of the solid target (an ablation rate of about 10–20 nm/pulse can be obtained in the previous case[54]). During the laser ablating, the target and liquid are maintained at room temperature. Meanwhile, the target rotates or moves horizontally at a slow speed. After the pulsed laser has interacted with the target for 15 to 20 min, the solid target is taken out and dried from the liquid. Note that, in this case, the surface of Si targets are covered by the synthesized products, which is the sample to be used in the further characterizations. Before measuring, these samples are washed and dialyzed carefully with the deionized water for 1 hour to remove the remaining inorganic salts, and then dried at 60°C again in an oven.

It is well known that PLIIR is a very fast and far-from-thermodynamic equilibrium process. Thus, all stable and metastable phases forming at the initial, the intermediate, and the final stages of the synthesis process may be reserved in the final products, especially the metastable intermediate phase.[55–57] For instance, two structures, the metastable hexagonal and the stable cubic structure of diamonds, have been observed simultaneously in the same sample synthesized by PLIIR.[58,59] Following Fabbro's studies, at the initial stage of LAL, a lot of species having large initial kinetic energy eject from the solid target surface and form a dense region—a laser-induced plasma plume—in the vicinity of the solid–liquid interface due to the confinement effect of the liquid environment.[55–57] Then, the laser energy should be greatly spent in mechanical or dynamical effects in the LAL condition.[60,61] Since the plasma plume is strongly confined in the liquid, the liquid would stop the plasma plume expansion to form an adiabatic region,[62] in which a shock wave will be created at supersonic velocity in front and inducing an extra pressure, called laser-induced pressure, in the plasma plume. Further, the laser-induced pressure leads to a temperature increase in the plasma plume. Therefore, the plasma plume created in the duration of pulsed-laser ablation at liquid–solid interface was in the higher temperature, higher pressures, and higher density (HTHPHD) state. For example, in the case of laser ablation of graphite in water, the temperature is about 4000–5000 K and the pressure is about 10–15 GPa.[63] Thus, the high amplitude stress waves would be applied on the solid target, while the phase transition between species in the plasma plume would happen.[64] As a result of the strong confinement effect of liquid, the quenching time of the plasma plume in the liquid becomes so short that the metastable phase forming at the intermediate stage of the synthesis could be frozen in the synthesized final products. In addition, the laser fluence plays an important role in the controlling phase formation upon PLIIR, so the different laser fluence could result in forming the different metastable phases.[65,66]

In terms of the detailed formation process of micro- and nanocubes of Si upon PLIIR-assisted low-concentration inorganic salt solution, the laser-induced plasma plume was first generated at the liquid–solid interface when the pulsed laser was ablating the interface between the solid target and the liquid. Since there are amorphous carbons covering the Si target, the plasma plume contains Si and carbon species from the target, and the salt ions from the liquid. Then, the expansion of the plasma plume will be delayed due to the confinement of liquid. Thus, the laser-induced pressure would be an order of magnitude greater and the shock wave duration would be 2–3 times longer than that in the direct regime at the same power density.[55] Therefore, the incident laser radiation and the laser-induced pressure would drive the plasma plume into the HTHPHD state,[64] in which Si species will impact with each other drastically to form the stable and mestable phase. The phase transition from the diamond to zincblende structures of Si may take place in the plasma plume with the HTHPHD state. As known, the Si with the zincblende structure is a metastable phase compared with the diamond structural Si. Therefore, the diamond structural Si is energetically preferable to the zincblende Si in thermodynamic. However, PLIIR is a far-from-thermodynamic equilibrium process, which provides many opportunities to the formation of metastable phases.[54,67,68] For example, the intermediate rhombohedral graphite is observed in the pulsed-laser-induced phase transformation, and the geometric path of the phase transition of graphite-to-rhombohedral graphite is proposed in previous work.[67] Therefore, some Si species with the diamond structure from the laser ablation of Si targets may transform into those with the zincblende structure in the plasma plume with the HTHPHD state by series crystal basal planes sliding. Moreover, during cooling down and condensation of the plasma plume in the confining liquid, the nucleation and growth of the zincblende Si would also occur. Meanwhile, the rapid quenching and growth time (about 20 ns in this case[62]) of the plasma plume leads to the synthesized metastable and stable phase frozen in the final products.[62] Additionally, the size of the final products is usually in the range of micro- and nanometer scale due to the short quenching time.

There is a great influence of the inorganic salt ions in the plasma plume on the forming morphology of the synthesized cubes upon PLIIR-assisted low-concentration inorganic salt solution. The inorganic salt ions in the plasma plume could act as an oriented agent to induce the growth of the Si nuclei due to the anisotropy in the adsorption stability of the salt ions. For the metastable nuclei with the cubic shape, the preferential adsorption lowers the surface energy of the bound plane and hinders the crystallite growth perpendicular to the plane, which results in the final morphology of the synthesized products. Importantly, in the case of Si micro- and nanocubes research, the relevant experimental studies show that the usual single-crystalline nuclei with the stable phase only develop into the typical Si balls upon PLIIR without the inorganic salts in the liquid.[51] As a comparison, in inorganic salt-assisted PLIIR, the metastable nuclei with cubic morphology will benefit from the preferential adsorption and result in the final morphology of the synthesized micro- and nanocubes. Also, some of the usual single crystallite nuclei with the stable phase would develop into the typical Si micro- and nanoballs. Moreover, the experiments also confirm that the amorphous carbon layer covering the Si target has influence on the shape formation of Si cubes in the synthesis process, though its detailed mechanisms are still unclear. An illustrative summary of the synthesis mechanism is depicted schematically in Figure 11.1.

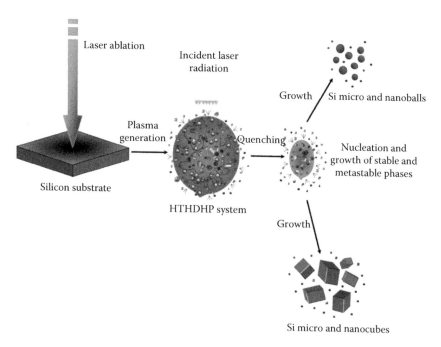

Figure 11.1 Schematic illustration of the synthesis mechanisms of silicon micro- and nanocubes upon LALs technology. (From Liu, P., et al., *Chem. Mater.*, 20, 494–502, 2008. With permission.)

11.2.2 CONSTRUCTION OF SPHERICAL Si MICRO- AND NANOPARTICLES WITH ZINCBLENDE STRUCTURE

Techniques based on the interaction of a pulsed laser with materials in a liquid environment have attracted substantial interest in the past few years and have been applied in diverse fields from micromachining to nanoparticles synthesis.[58,59,67–77] Moreover, there have been several experimental reports on using LAL to trap metastable phases.[78–81] Therefore, based on this method, a facile technique called EFLAL has been developed by Liu et al. which enables controlled fabrication of micro- and nanostructures with metastable configuration without using any catalyst or organic additives (i.e., it is a "green" synthesis method).[82]

Here, details of the improved synthetic procedure based on EFLAL are provided, which is depicted schematically in Figure 11.2.[83] A single-crystalline Si target with diamond-type structure and 99.999% purity is used as the starting material. Firstly, the target is held in a polytetrafluoroethylene fitting and suspended at one end of a rectangular quartz chamber with dimensions of $10.0 \times 4.0 \times 3.0$ cm. Two parallel square graphite electrodes separated by a distance of 3.3 cm are placed next to the target to provide a dc electric field of 120 V. The chamber is filled with ultrapure water (resistance 18.2 MΩ) that is maintained at room temperature during the synthesis process. High-power laser pulses from two harmonic Q-switched Nd:YAG lasers (wavelength 532 nm, pulse width 10 ns, power density 10^{11} W/cm^2, and repetition frequency 2.5 Hz) are finely focused on the target surface. A magnetic stirrer is used to disperse the fine particles generated during the process, and the whole chamber system is placed on an E-Motion surface, which moved the Si target horizontally at slow speed during the laser ablation process. After the laser pulse was applied to the target for 120 min, a gray–white powder generated was collected from the chamber and dried at 50°C in a vacuum oven. Note that the mild treatment condition in the drying process did not change the crystalline structure of the product.

Similar to the formation process of micro- and nanocubes of Si using the PLIIR method, the fabrication mechanism of LAL is a rapid process using far-from-thermodynamic equilibrium.[84] Thus, all stable and metastable phases formed during the initial, intermediate, and final stages of the synthesis process can be preserved in the final products.[57,79–81,85] According to our previous studies,[56,57] a large

Clusters, nanoparticles, and quantum dots

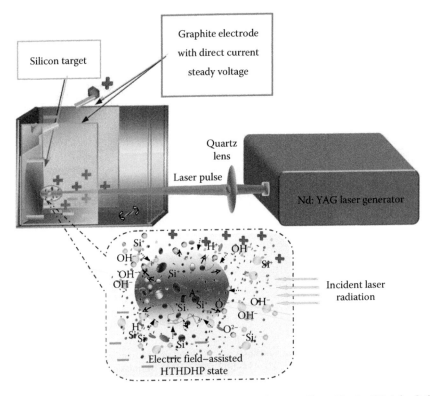

Figure 11.2 Schematic illustration of EFLAL and its fabrication mechanism. (From Liu, P., *AIP Adv.*, 3, 022127, 2013. With permission.)

number of Si species with high initial kinetic energy will be generated from the solid target surface and form a dense region during the initial stage of LAL. This is called a laser-induced plasma plume, and it forms in the vicinity of the solid–liquid interface because of the confinement effect of liquid. Then, because the plasma plume is strongly confined in the liquid, the expansion of the plasma plume is delayed to form an adiabatic region.[62] A shock wave is created in this region at supersonic velocity, and this induces extra pressure, called laser-induced pressure. Subsequently, the temperature increases because of the extra pressure in the plasma plume. Therefore, the plasma plume created during LAL at the solid–liquid interface is in an HTHPHD state, in which the Si species collide with each other to form stable and metastable phases. As a result of the strong confinement effect of liquids, the quenching time of the plasma plume in the liquid is short, so the metastable phase formed during the intermediate stage of the synthesis can be found in the final products.

In the detailed EFLAL case, the phase transition of Si from diamond to zincblende structure is taken place in the plasma plume within the HTHPHD state. Because LAL is a totally far-from-thermodynamic equilibrium process, metastable phases can form,[39–41] and thus some Si species from laser ablation may readily transform into Si zincblende structures in the plasma plume within the HTHPHD state by sliding of the crystal basal planes. The particles cool down as the plasma plume condenses in the confining liquid, causing nucleation and growth of the zincblende SiNCs. The small size of the final products is because of the short quenching time. The equilibrium of crystalline phase and structure can be determined by the ambient energy distribution. During LAL, the applied electrical field influences the plasma plume and plays an important role in the growth of defect-containing structures. Specifically, the applied electrical field could disturb the nonequilibrium process during metastable nucleation, which would influence the final morphology of the products. Importantly, the relevant experimental studies (Figure 11.2) definitely show that the applied electrical field affects the formation of defect-containing structures in the products.[86] However, the detailed formation mechanisms of such configurations are still unclear; further studies on this topic are underway.

11.3 STRUCTURE CHARACTERIZATION AND PROPERTIES ANALYSIS OF THE ZINCBLENDE Si NANOCRYSTALS

Structure characterization should be the important route to understand the properties of zincblende phase Si nanoparticles. Both the micro- and nanocubes of Si and the SiNCs with zincblende structure synthesized based on the improved LAL environment are examined carefully by SEM (JSM-6330F, 150 kV) and TEM (JEOL JEM-2010H, 200 KV), with corresponding selected area electron diffraction (SAED) patterns are taken. For detailed high-resolution transmission electron microscopy (HRTEM) and phase characterization of the micro- and nanocubes, a JEM-2010F TEM operating at 200 kV is used to measure the HRTEM images and the electron energy loss spectrum (EELS) spectra of the Si nanocubes. Raman scattering measurements are carried out under $\lambda = 514.5$ nm excitation from an Ar-ion laser line using a Renishaw inVia +Plus laser microRaman system (maximum laser power 20 mW, 500 × objective) for all the samples. Moreover, the SiNCs are deposited on pure quartz substrates for XRD, PL, and UV absorption measurements. All analysis measurements are obtained at room temperature.

11.3.1 SEM AND XRD PROPERTIES OF THE SiNCs WITH ZINCBLENDE STRUCTURE

Typical SEM images of the as-synthesized Si micro- and nanocubes, and SiNCs are shown in Figure 11.3. Interestingly, for the PLIIR result, it can be seen clearly that the as-synthesized samples are cubes in the high-magnification SEM image of Figure 11.3b, and these cubes (shown in Figure 11.3a) have perpendicular fringes that are either parallel to or orthogonal to the upright facet of cubes, with the side length in the range of 200–500 nm. Further, in terms of the EFLAL case, most of the spherical micro- and nanoparticles are uniform in size and possess rough surfaces, with an average diameter of 40–60 nm (Figure 11.3d). Interestingly, the high-magnification SEM image shown in the inset of Figure 11.3c reveals both micro- and nanosized crystals have an average diameter of 200 nm and rough surfaces. Figure 11.3e further shows the larger SiNCs consist of a main body in which many smaller NCs are embedded. A recent report indicated that the bimodal size distribution of nanoparticles formed by LAL is governed by a sudden rise in solution concentration caused by plasma erosion of the surface upon bubble collapse.[87] However, the rough, NC-embedded morphologies differ considerably to others published,[88–90] which implies that EFLAL produces NCs with unique morphologies.

In addition, an XRD pattern and the EDS analysis (Figure 11.3f and its inset) of the synthesized NCs clearly show that the fabricated micro- and nanoparticles contain a crystalline phase of pure Si. The three XRD peaks located at 28.64, 47.62, and 56.51° correspond to the (111), (220), and (311) crystalline planes of the zincblende structure of Si, respectively (JCPDS Card File No. 800018). A weak, broad band located at about 23° is attributed to the quartz substrate and the presence of a small amount of amorphous Si. Because there are no diffraction peaks consistent with oxide phases in the XRD pattern, and the small Cu and C signals in the EDS spectrum originate from the underlying copper grid, we conclude that both the synthesized Si samples are crystalline and of high purity.

11.3.2 TEM ANALYSIS OF THE STRUCTURE PROPERTIES OF SiNCs WITH ZINCBLENDE STRUCTURE

For systematically expounding of the crystalline phase and zincblende structure of the micro- and nanocubes of Si and SiNCs in detail, TEM and HRTEM images and corresponding SAED patterns are obtained and shown in Figures 11.4 through 11.6. Specifically, as shown in Figure 11.4, the bright-field images and the corresponding SAED patterns of a nanocube are obtained under a JEOL JEM-2010H TEM operated at 200kV, and the corresponding EDS spectrum within the measurement error of 2% is shown in Figure 11.4b. Thus, these results undoubtedly indicate that the synthesized cubes are pure Si. In Figure 11.4b, the Cu, Cr, and C peaks originate from the copper grid and the amorphous carbon film support, respectively, while the very slight Cl peak comes from the impurity of salt ion and the weak O peak is supposed to come from the amorphous carbon film support or the exterior oxide layer of the Si cubes.[51]

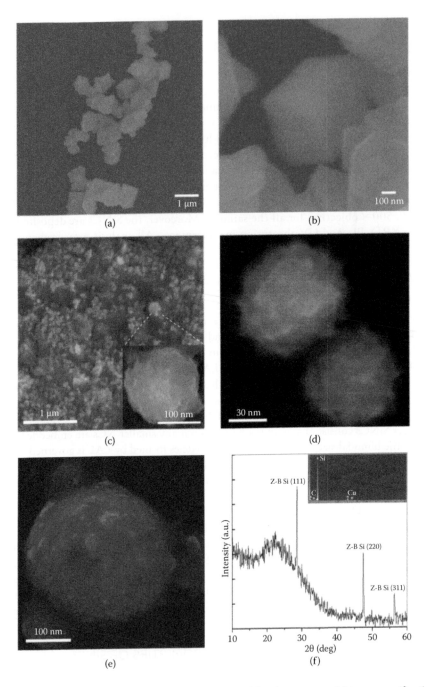

Figure 11.3 SEM images of the synthesized silicon NCs with zincblende structure. (a) Low-magnification of micro- and nanocubes. (b) High-magnification of an Si nanocube. (c) Low-magnification SEM image of the synthesized SiNCs. (d) High-magnification SEM image of two SiNCs. (e) High-magnification SEM image of SiNCs containing smaller embedded NCs. (f) XRD pattern of the synthesized SiNCs.

Three corresponding SAED patterns of the Si cubes are shown in Figure 11.4c through e. The representative SAED patterns are obtained by directing the electron beam perpendicular to the different crystal facets of the cube that shown in Figure 11.4a. These SAED patterns reveal that the synthesized Si cubes are single crystal and are bounded mainly by 2 {111} facets, 2 {220} facets, and 2 {112} facets. These results are schematically illustrated in Figure 11.4f. Additionally, the d_{exp} values of Si cubes can

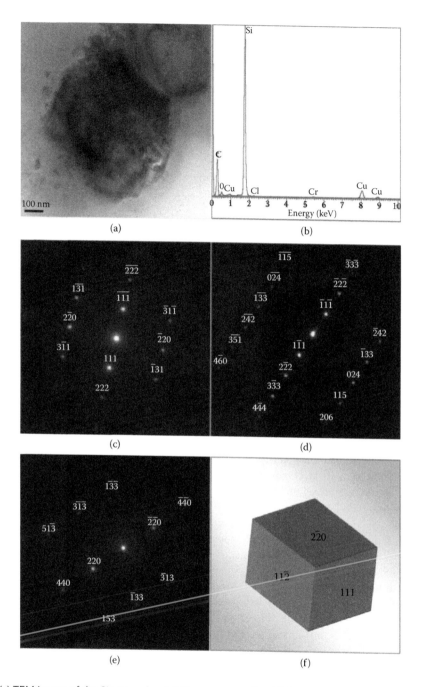

Figure 11.4 (a) TEM image of the Si nanocube. (b) The corresponding EDS spectrum (c through e) The corresponding SAD patterns of the single Si cube. (f) A model of the silicon cube that was rebuilt based on the SAED analysis. (From Liu, P., et al., *Chem. Mater.*, 20, 494–502, 2008. With permission.)

be calculate from the SAED patterns. Table 11.1 lists a comparison of the experimental data of Si cubes, the calculated value of the zincblende structure of Si (JCPDS Card File No. 800018) and the value of the single-crystalline Si with diamond structure (JCPDS Card File No. 895012).[18,91] Clearly, the d_{exp} values of Si cubes are highly consistent with the theoretical d_{calc} value of the zincblende structure of Si. Therefore, these results show that the synthesized cubes are single-crystalline Si with the zincblende structure.

Table 11.1 **The indexing results of SAD patterns of micro- and nanocubes, with the comparison of the experimental data of zincblende and diamond-type structures of silicon**

(HKL)	D_{EXP} (Å)	D_{CALC}^{*} (Å)	D_{CALC}^{**} (Å)
111	3.1063	3.1130	3.1355
131	1.6291	1.6257	1.6374
133	1.2366	1.2370	1.2458
220	1.9002	1.9063	1.9201
222	1.5537	1.5565	1.5677
242	1.0998	1.1006	1.1085
311	1.6304	1.6257	1.6374
422	1.1002	1.1006	1.1085

Source: C. Y. Yeh, Z. W. Lu, S. Froyen and A. Zunger, *Phys. Rev. B,* **46**: 10086, 1992; L. Pavesi, L. D. Negro, C. Mazzoleni, G. Franzo and F. Priolo, *Nature,* **408**: 440–444, 2000.

Note: D_{exp} refers to our experimental values, D_{calc}^{*} refers to the predicted values of zincblende structure, and D_{calc}^{**} refers to the predicted values of diamond structure.

*JCPDS Card File No. 800018;**JCPDS Card File No. 895012.

A JEM-2010F TEM operated at 200kV is employed to measure the HRTEM image and the corresponding EEL spectra of the nanocubes. Figure 11.5a shows an HRTEM image of one facet of the Si cube. A careful examination indicates that the interplanar spacings of the cube are 0.310 and 0.189 nm, which are both smaller than the d values of the diamond structural crystalline Si, but in good agreement with the d_{calc} value of (111) and (220) of the zincblende structure of Si. A fast Fourier transform (FFT) analysis (see inset in Figure 11.3) shows the square spot array of the 2D lattice fringes, which can just be indexed to the (111) and (220) directions. Moreover, a thin amorphous layer can be seen on the outer surface of the Si cube indicated by a dashed square. This result shows that the Si cube is covered with a thin amorphous oxide layer, which leads to the O peak in the EDS spectrum (Figure 11.2b). Note that there are some saturated regions in the lattice planes (marked out with white circles). This result implies that there should be some asymmetrical lattice structures in the cube,[92] which is suspected to be caused by inner defects or slight dislocations. Figure 11.5b through e shows the EELS of the Si nanocube, in which Figure 11.5b and c shows a low-loss spectrum and a high-loss spectrum of the samples, respectively. Meanwhile, a standard low-loss spectrum and a high-loss spectrum of the diamond structure crystalline Si are exhibited in Figure 11.5d and e. In the low-loss spectra, we can see that the main plasmon peak in Figure 11.5b is located at around 17 eV, which is in good agreement with that shown in Figure 11.5d. This result indicates that the synthesized Si cubes are single-crystalline Si. A stretching mode after 17 eV shown in Figure 11.5b seems different from that shown in Figure 11.5d, because the standard spectrum exhibits only one single and broad plasmon peak at 17 eV in Figure 11.5d. Thus, this stretching mode at about 21 eV shown in Figure 11.5b is suggested to be attributed to the zincblende Si. In the high-loss spectra, we confirm that the plasmon peak at 101.5 eV shown in Figure 11.5c is in agreement with the peak at 101 eV shown in Figure 11.5e, which is the L-edge absorption peak of crystalline Si. Then, the second plasmon peak shown in Figure 11.5c is at 104 eV, which is different to the second plasmon peak at 108 eV in Figure 11.5e.

Considering that the fine structure of a sample can be inspected in the high-loss spectrum, the peak shift of Figure 11.5c indicates that there is a difference between the crystalline structures of Si cube and single-crystalline Si with diamond structure. To clarify this issue, the first-principle calculations were carried out to theoretically determine the EEL spectrum (low-loss) of the zincblende structure and diamond structure of Si, and the calculated results are shown in Figure 11.5f. In Figure 11.5f, the black solid line represents the zincblende structure, and the red dotted line represents the diamond structure. In the calculations, an accurate full-potential, linearized, augmented plane wave method is used, and the exchange and correlation effects are treated within the generalized

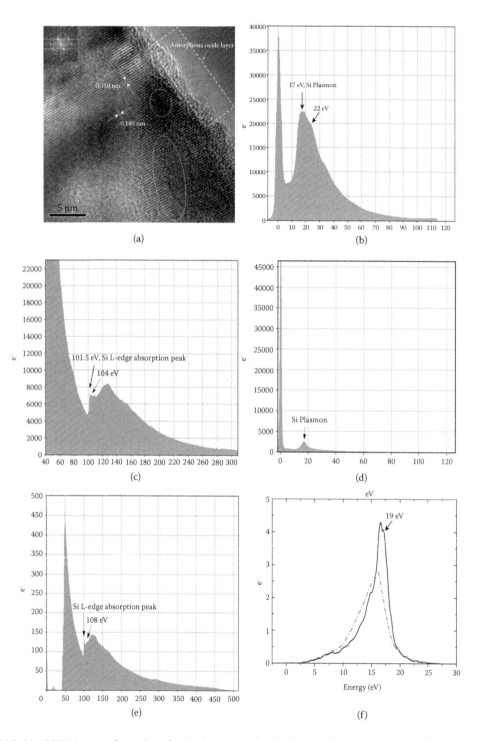

Figure 11.5 (a) HRTEM image of an edge of a single silicon cube, the inset is the corresponding FFT analysis. (b) Low-loss and (c) high-loss spectra of EELS of samples. (d) Low-loss and (e) high-loss spectra of the standard diamond structure of silicon. (f) Theoretical calculated EELS (low-loss) of the silicon with the zincblende and diamond structures through the first-principles method. (From Liu, P., et al., *Chem. Mater.*, 20, 494–502, 2008. With permission.)

gradient approximation (GGA).[51] The muffin-tin sphere of Si is set to R = 2.1 for zincblende structure and R = 1.25 for diamond structure. Note that, for the zincblende structure, a mesh of 10 × 10 × 10 is used to represent 47 k-points in the irreducible wedge of Brillouin zone.[51] Comparing Figure 11.5b and d with Figure 11.5f, we can see there is indeed a difference between the zincblende and diamond structures of Si in the low-loss spectrum. In detail, the red dotted line spectrum in our calculations exhibits only one single plasmon peak at 17 eV, which is in good agreement with the standard spectrum shown in Figure 11.5d. Meanwhile, the black solid line spectrum depicted in Figure 11.5f indicates not only a plasmon peak at 17 eV, but also a skew-stretching mode after 17 eV, which is at about 19 eV and is very similar to that shown in Figure 11.5b. Since the calculated value would have little difference to the actual value, these results confirm that the synthesized cubes exactly having zincblende structure of Si.

In terms of the SiNCs that are synthesized upon EFLAL, Figure 11.6a and c shows two different morphologies are present in the samples (cf. Figure 11.6). The HRTEM image of a single nanoparticle in Figure 11.6a indicates a multitude of saturated, thin regions in the lattice planes (marked with white circles). This result not only clarifies the structure shown in the SEM analysis of Figure 11.3d, it also implies that there should be many stacking fault defects or asymmetrical lattice structures existing in the NC.[51,92,93] From the magnified HRTEM image (shown in the inset of Figure 11.6a), the interplanar spacings are 0.309 and 0.189 nm, which correspond to the d values of the (111) and (220) crystallographic planes of the zincblende structure of Si, respectively. The corresponding SAED pattern (Figure 11.6b)

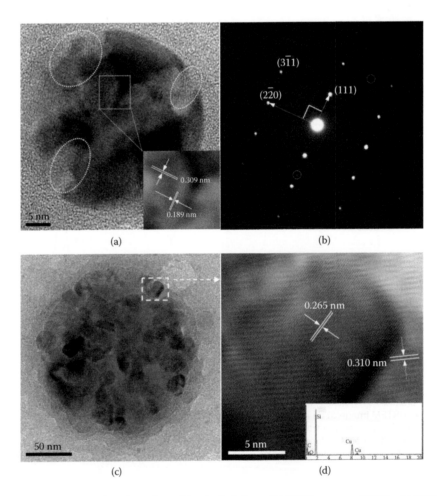

Figure 11.6 (a) HRTEM image of SiNCs with stacking fault defects. (b) SAED pattern of SiNCs. (c) TEM and (d) HRTEM images of an SiNC. (From Liu, P., *AIP Adv.*, 3, 022127, 2013. With permission.)

confirms the HRTEM result and also identifies some very weak diffraction points (marked with white circles). It is believe that these weak diffraction points are caused by stacking fault defects or asymmetrical lattice structures in the NC. Figure 11.6c shows a bright-field image of a large microcrystal, which has similar geometry to that of the particle in Figure 11.3e. An HRTEM image of the region indicated by a white rectangle in Figure 11.6c reveals that the particle consists of several smaller NCs with a size of 7–14 nm. The interplanar spacings shown in Figure 11.6d give d values of the lattice fringe of 0.265 and 0.310 nm. Note that the lattice spacing of 0.265 nm is just the d value of the (200) crystallographic plane of the zincblende structure of Si, and cannot be seen in crystalline Si with a diamond-type structure. In the corresponding EDS (inset of Figure 11.6d), the Si component is also obtained with a measurement error of 2%, and the weak Cu and C peaks originate from the copper grid and amorphous carbon film support, respectively. Because almost no oxide layer is present on the Si samples (cf. HRTEM analysis above), the very weak O peak in the spectrum originates from the amorphous carbon film support. Therefore, from all the above TEM analysis discussion, one can conclude that the synthesized NCs consist of pure Si with a metastable zincblende structure.

11.3.3 RAMAN CHARACTERIZATION OF THE SiNCs WITH ZINCBLENDE STRUCTURE

Raman spectrum analysis should be the most powerful technique for determining crystalline structure of semiconductor. Therefore, the corresponding Raman spectrum of the SiNCs is both obtained from the synthesized micro- and nanocubes of Si and metastable Si nanoparticles. Figure 11.7b shows the Raman spectrum of the Si micro- and nanocubes, while a contrasting Raman spectrum of single-crystalline Si substrate with diamond structure is shown in Figure 11.7a. There are definitely two different Raman peaks in Figure 11.7; one peak at 504 cm^{-1} is from the as-synthesized Si micro- and nanocubes, and the other peak at 520 cm^{-1} is from the single-crystalline Si substrate. Thus, there is an obvious Raman shift between the Si cubes and the single-crystalline Si substrate. Usually, Si NCs have a Raman shift because of the quantum size effect or the heating by the visible laser[43,44]. However, in this case, the size of Si cubes seems far beyond the quantum size-effect regime. In addition, the energy of the detected laser is only set to be 1 mW. Therefore, the Raman shift of Si cubes is not attributed to the quantum size effect or laser heating-up effect, but seems to originate from the crystalline defects, so there is inner stress in the Si cubes. First, the cubic shape is not a natural morphology of single-crystalline Si particle, thus, there would be some inner energy or lattice stress in the Si cube, which may be caused by the lattices mismatched or distorted, and not be released completely. Second, the zincblende structure of Si is a metastable phase. Therefore, some defects can form in the synthesized Si micro- and nanocubes as shown in Figure 11.5a. Moreover, some theoretical calculations also indicate that the cubic morphology would induce the spectral shifts.[94] Accordingly, the Raman peak at 504 cm^{-1} seems the intrinsic Raman peak of the Si cubes with the zincblende structure.

For a deeper analysis of the structure properties that are contained in the SiNCs with zincblende structure, a series of Raman spectra of the SiNCs are further recorded under different laser power densities over a period of 10 s (Figure 11.8). In the Raman spectrum obtained with a laser power density of 0.2 mW (Figure 11.8a, pattern i), the strong peak at 520.17 cm^{-1} is consistent with the first-order Raman scattering band of crystalline Si, while the weak peak at 298.86 cm^{-1} is attributed to the 2TA(X) mode.[95] To the best of our knowledge, the Raman peak located at 298.86 cm^{-1} is very weak and hard to detect compared with the signal at 520.17 cm^{-1}. Therefore, it is difficult to simultaneously observe both the peaks at 520.17 and 298.86 cm^{-1} in the same spectrum.[49–51] The Raman spectrum recorded at high laser power density (1 mW, Figure 11.8a, pattern ii) exhibits peaks at 510.5 and 294.34 cm^{-1}. These peaks are both shifted to higher energy compared with those observed at low laser power density (Figure 11.8a, pattern i), and the peak at 294.34 cm^{-1} became very weak. Based on previous reports,[86,95,96] the Raman spectra can be interpreted as follows: SiNCs normally exhibit a Raman shift because of the quantum size effect or heating effects from the detecting laser. In this case, however, most of the SiNCs are far too large to exhibit a quantum size effect, which is usually observed for nanoparticles of ~8 nm or smaller (cf. SEM and TEM analysis, and no obvious band broadening is found in the Raman scattering bands[86,95]). In addition, the laser power density used in Figure 11.4a (pattern i) was very low (0.2 mW) and for a short period (10 s), so heating effects should be negligible. No Raman shift of the first-order Raman scattering band at 520.17 cm^{-1} can be seen in pattern i, so the band at 298.86 cm^{-1} is assigned to the zincblende structure of the SiNCs. The NCs

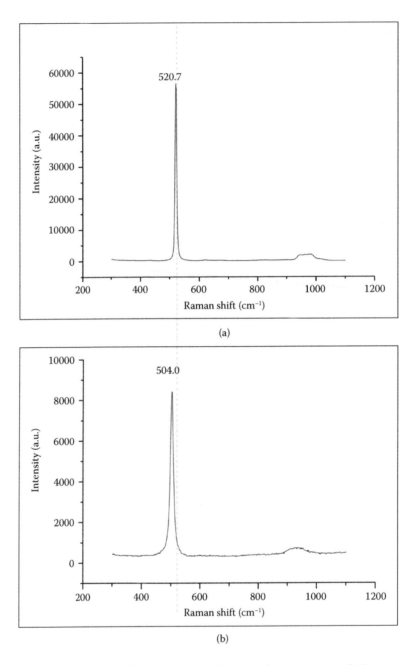

Figure 11.7 (a) Raman spectrum of the silicon substrate with diamond-type structure, (b) Raman spectrum of Si cubes with detected laser energy of 1 mW.

contain lattice mismatches or distortions, and the zincblende structure of Si is metastable, so it is reasonable to expect that the crystalline structure is sensitive to laser heating at high-power densities (1 mW, Figure 11.4a, pattern ii). That is, the crystalline defects should contribute to the appearance and small shift of the Raman peak from 298.86 to 294.34 cm^{-1}.[51,95] Moreover, in the Raman analysis of Si micro- and nanocube (Figure 11.7), the Raman spectra show similar behavior to that in Figure 11.8a (pattern ii). Therefore, it can be believed that the Raman peak shifts to 510.5 cm^{-1} because of laser heating and scattering effects from the Si zincblende crystalline NCs containing defects.[95] For comparison, Raman spectra of the original single-crystalline Si target with a diamond-type structure were recorded under the same conditions as that in Figure 11.8a, and are shown in Figure 11.8b. These spectra are identical and the peaks

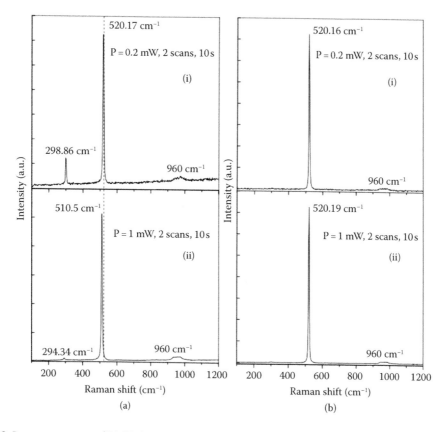

Figure 11.8 Raman spectrum of (a) SiNCs and (b) single-crystalline Si target with diamond structure using laser power densities of (i) 0.2 mW, and (ii) 1 mW. (From Liu, P., *AIP Adv.*, 3, 022127, 2013. With permission.)

are consistent with single-crystalline Si with diamond structure, indicating that the typical diamond-type crystal structure of Si does not show obvious heating effects. Therefore, these results of the Raman spectra shown in Figures 11.7 and 11.8 further clarify the structure properties that originate from the metastable zincblende crystalline structure of Si.

11.3.4 VIOLET–BLUE LUMINESCENCE PROPERTY AND UV ABSORPTION ANALYSIS OF THE SiNCs WITH ZINCBLENDE STRUCTURE

Violet–blue luminescence from Si nanostructures has been widely investigated, because of its potential use in optoelectronic and bioimaging devices. For the synthesized SiNCs with zincblende structure, its PL spectra of three typical samples are shown in Figure 11.9a. A dominant broad peak at 366 nm is observed in all spectra, along with a broad shoulder at 395 nm. In addition, a peak is observed at 324 nm in spectrum (ii) and a broad shoulder appears at 326 nm in spectrum (iii). No similar peak is observed in spectrum (i). Therefore, the emission peaks at about 324 and 395 nm are probably both caused by stacking fault defects or asymmetrical lattice structures present in the samples, because of their weak, variable emission behavior. Therefore, the SiNCs with zincblende structure exhibit violet–blue emission at 366 nm, which has not been reported previously for SiNCs.[26,27,31,35,88,97–100] Although each PL spectrum exhibits a dominant peak at about 366 nm, the variable presence of minor peaks implies there is possibly more than one origin of violet–blue emission in these SiNCs.

A UV absorption spectrum of the synthesized SiNCs is shown in Figure 11.9b, and does not contain any obvious absorption peak. However, analysis of the UV curve revealed a weak absorption band that can be assigned to the intersection of two tangent line equations fitted to the data gained in the UV absorption spectrum. The intersection of the two tangent lines is at about 240 nm, which implies that a seriate absorption appears at a higher wavelength than 240 nm in the UV spectrum. Because the PL emission spectra

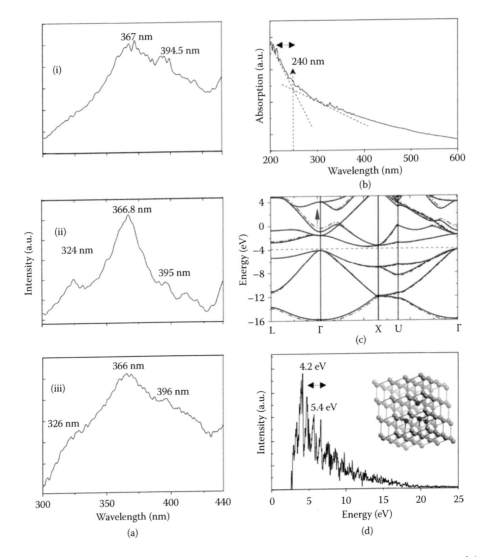

Figure 11.9 (a) PL spectra of SiNCs produced by EFLAL (i through iii) and (b) UV absorption spectrum of the synthesized SiNCs. The calculated energy band structure of the SiNCs with a zincblende structure is shown in (c), and the corresponding absorption coefficient is calculated in (d). (From Liu, P., *AIP Adv.*, 3, 022127, 2013. With permission.)

were measured with an excitation wavelength of 230 nm, the deduced absorption region is consistent with the PL optical measurement.

To investigate the band gap structures and corresponding absorption coefficient of the SiNCs, a band states of the SiNCs with both zincblende and ordinary diamond-type structures are calculated using first-principles density functional theory with the SIESTA code.[101–103] Electron wave functions are expanded using a double-ζ basis set plus polarization functions.[104] The numerical integrals are performed on a real space grid with an equivalent cutoff of 150 Ry, while the Brillouin zone is sampled in a 10 × 10 × 10 k-mesh following the Monkhorst–Pack scheme for bulk Si. The structure geometry is based on the obtained crystalline data (Figure 11.6, JCPDS Card File No. 800018).[18] The results of the calculations are shown in Figure 11.9c and d. Figure 11.9c shows that both the zincblende and diamond-type structures of Si have indirect band gaps, which are indicated by the blue solid line and red dotted line, respectively. Interestingly, the calculation reveals band gap broadening, indicated by a red arrow in Figure 11.9c, which could be the origin of violet–blue emission from the SiNCs with zincblende structure. The absorption coefficient of the

Si zincblende structure, which is calculated from the real and imaginary parts of the dielectric constant, is shown in Figure 11.9d, and a schematic diagram of a zincblende crystal framework is shown in its inset. Based on first-principles theory, the main absorption appears around 4.2–5.4 eV, which is less than the result deduced directly from the UV absorption spectrum. Considering that the SiNCs contain numerous stacking fault defects that would influence the absorption behavior, this calculated result confirms that the violet–blue PL emission is induced by the zincblende structure of Si. It is well known that the optical properties of materials are determined by their energy bands. The surface states, stacking fault defects and size of nanoparticles influence their optical properties by affecting polarization (extrinsic) and the energy band structure (intrinsic). Thus, we conjecture that the PL peaks at 395 and 354 nm (Figure 11.9a) arise from an unstable exciton transition that is influenced by the distribution of surface states in the sample, which could be caused by the rough or defect-containing surface of the NCs. Accordingly, on the basis of the UV spectra and first-principles calculations, the prominent peak at about 367 nm is assigned as the intrinsic PL peak of SiNCs with zincblende structure.

To be able to investigate the thermal stability property of the zincblende structure of Si and clarify if the metastable SiNCs is the origin of their violet–blue emission, the samples synthesized by EFLAL are annealed in a horizontal high-vacuum stove for 90 min at 600°C. SEM, XRD, TEM, Raman, and PL measurements are recorded under the same conditions as for the unannealed samples; the results of these measurements are shown in Figure 11.10. SEM and TEM analysis (Figure 11.10a and c, respectively) reveal that the surfaces of the products are glazed, and the size of the particles is in the range of 300–400 nm. These results indicate that annealing removes crystalline defects and increases the size of the SiNCs, producing a stable crystalline structure. The XRD pattern of the annealed sample (Figure 11.10b) can be completely indexed to an Si phase with diamond-type crystalline structure (JCPDS card 271402). It is speculated that the increase in peak intensity compared with that of the unannealed sample is caused by the growth and/or consumption of the small SiNCs during the heating process. The TEM image shown in Figure 11.10c shows the ripening SiNCs, while the HRTEM image in Figure 11.10d indicates that the annealed SiNCs possess a diamond structure, because the d value of 0.314 nm is consistent with the typical lattice spacing of (111) crystallographic planes of crystalline Si with a diamond-type structure. Similarly, compared with the Raman spectrum in Figure 11.8b, the spectrum in Figure 11.10e of the annealed SiNCs indicates that crystalline Si with diamond structure has formed. Obviously, annealing converts the metastable zincblende structure in the defect-containing SiNCs to a typical diamond structure. Therefore, the PL spectrum of the annealed SiNCs (Figure 11.10f) does not contain any peak. Finally, the annealing of mestastable Si Cs and its corresponding structure analysis and PL measurements all suggest that the origin of the violet–blue luminescence is indeed from the SiNCs with a metastable zincblende structure that contains defects.

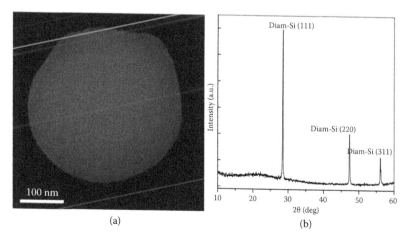

Figure 11.10 (a) SEM image, (b) XRD pattern.

(Continued)

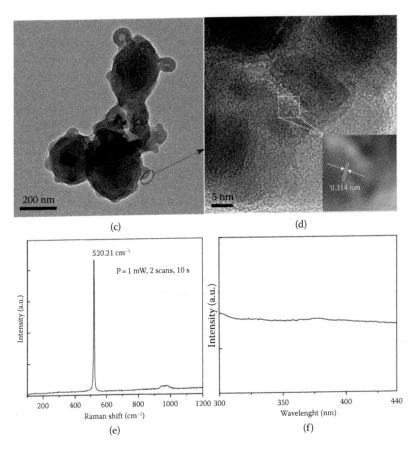

(c)   (d)

(e)   (f)

Figure 11.10 (Continued) (c) TEM image of the annealed sample. (d) HRTEM image of the (111) facet of SiNCs. (e) Raman spectrum of the annealed SiNCs with a laser power density 2 mW. (f) PL spectrum of the annealed SiNCs. (From Liu, P., *AIP Adv.*, 3, 022127, 2013. With permission.)

11.4 SUMMARY AND PERSPECTIVE

In this chapter, single-crystalline micro- and nanocubes of Si and metastable SiNCs with zincblende structure that are synthesized utilizing the formation process based on improved LALs is introduced, including the PLIIR assisted low-concentration inorganic salts solutions and EFLAL methods. They are both proved to be effective and a general strategy for nanostructuring, that is, from phase- and shape-controlled synthesis of NCs to functional nanostructures fabrication, especially for the synthesis of nanostructures with metastable phases and shapes.

For the properties exploration, through the SEM, XRD, and TEM analysis as well as the calculated band gaps of the SiNCs with zincblende structure, it can safely exclude the possibility of the violet–blue luminescence arising from either the SiO_2 component or band-to-band recombination within ordinary SiNCs,[32] and the broad emission peak at 366 nm (3.38 eV) is consistent with direct electron–hole recombination at the previously reported Γ point ($\Gamma_{25} \rightarrow \Gamma_{15}$).[105] Moreover, from the UV absorption spectrum (Figure 11.9b), one can determine that the optical absorption of the synthesized SiNCs for an indirect transition is in a broad UV region, which indicates the existence of the stacking fault effect. According to previous studies, the stacking fault effect not only influences the indirect band gap but also the direct transition energies at Γ or X points.[65] Therefore, the seriate absorption at <240 nm implies that the violet–blue emission from the SiNCs is caused by absorption around the UV region, corresponding to direct transitions at the Γ or X points (Figure 11.9c).

Considering the metastable phase of the zincblende structure, the fate of the photoexcited excitons (electron–hole pairs) in Si can be determined by traps and intrinsic recombination processes within the

NCs. Generally, the phonon-assisted indirect band gap electron–hole recombination in Si is very weak.[65] However, when a metastable structure is present, the optical band gap is large (Figure 11.9c), and direct electron–hole recombination at the Γ point ($\Gamma_{25} \rightarrow \Gamma_{15}$) is facilitated, especially because there are numerous stacking fault defects that act as luminescence centers. Subsequent annealing at 600°C removes a large proportion of crystalline defects in the SiNCs, inducing a crystalline structure, and weakening the blue PL intensity. However, there may be a small amount of amorphous Si in the product, so the effect of these amorphous components on the outer layer of the SiNCs is still not clear and further investigation is needed.

In summary, metastable SiNCs with zincblende structures are prepared in both micro- and nanocube morphology, and novel embedded spherical nanoparticle forms. These SiNCs emit violet–blue light, which disappears upon subsequent annealing of the NCs. However, the violet–blue luminescence of the initial SiNCs can remain stable even after aging for 6 months at room temperature. Analysis of the structure and PL measurements of various SiNCs indicated that the violet–blue PL cannot be attributed to a Si oxide species or quantum confinement effects as in typical SiNCs. We infer that excitons form inside the stacking fault defects in the SiNCs from transitions at the Γ or X points, and then transfer to and recombine at the Γ or X points to emit violet–blue light, which is induced by the metastable phase of Si with a zincblende structure. This mechanism readily explains the disappearance of the violet–blue PL from SiNCs upon thermal annealing. Moreover, understanding the mechanism of violet–blue PL from defect-containing SiNCs with a zincblende structure may be useful in applications such as Si-based full color displays and biomedical imaging. Furthermore, due to the metastable phase-induced unique shape and property, the cube or embedded spherical morphology SiNCs could be used as building blocks for fabrications of new nanodevices in future.

ACKNOWLEDGMENT

The National Basic Research Program of China under Grant No. 2014CB931700, the National Natural Science Foundation of China No. 91233203, and the State Key Laboratory of Optoelectronic Materials and Technologies funded this work. The author is grateful to Professor G. W. Yang and Professor C. X. Wang of Sun Yat-sen University, who gave much critical advice and made important contributions to the research field covered by this work. In addition, the author is grateful to Dr. J. Xiao of Sun Yat-sen University, who worked in the author's group, for helpful stimulating discussions.

REFERENCES

1. H. Takagi, H. Ogawa, Y. Yamazaki, A. Ishizaki and T. Nakagiri, Quantum size effects on photoluminescence in ultrafine Si particles, *Appl. Phys. Lett.*, **56**(1990): 2379–2380.
2. Y. Cui and C. M. Lieber, Functional nanoscale electronic devices assembled using silicon nanowire building blocks, *Science*, **291**(2001): 851–853.
3. F. Erogbogbo, K.-T. Yong, I. Roy, G. Xu, P. N. Prasad and M. T. Swihart, Biocompatible luminescent silicon quantum dots for imaging of cancer cells, *ACS Nano*, **2**(2008): 873–878.
4. U. Kim, I. Kim, Y. Park, K. Y. Lee, S. Y. Yim, J. G. Park, H. G. Ahn, S. H. Park and H. J. Choi, Synthesis of Si nanosheets by a chemical vapor deposition process and their blue emissions, *ACS Nano*, **5**(2011): 2176–2181.
5. B. N. Dutta, Lattice constants and thermal expansion of silicon up to 900°C by X-ray method, *Phys. Status Solidi*, **2**(1962): 984–987.
6. R. J. Needs and A. Mujica, First-principles pseudopotential study of the structure phase of silicon, *Phys. Rev. B*, **51**(1995): 9652–9660.
7. J. Crain, G. J. Ackland, J. R. Maclean, R. O. Piltz, P. D. Hatton and G. S. Pawley, Reversible pressure-induced structural transitions between metastable phase of silicon, *Phys. Rev. B*, **50**(1994): 13043–13046.
8. Z. D. Zeng, Q. S. Zeng, W. L. Mao and S. X. Qu, Phase transitions in metastable phases of silicon, *J. Appl. Phys.*, **115**(2014): 103514.
9. Y. Zhang, Z. Iqbal, S. Vijayalakshmi and H. Grebel, Stable hexagonal-wurtzite silicon phase by laser ablation, *Appl. Phys. Lett.*, **75**(1999): 2758–2760.
10. A. San-Miguel, P. Kéghélian, X. Blase, P. Mélinon, A. Perez, J. P. Itié, A. Polian, E. Reny, C. Cros and M. Pouchard, High pressure behavior of silicon clathrates: A new class of low compressibility materials, *Phys. Rev. Lett.*, **83**(1999): 5290–5393.

11. D. C. Mahon, P. J. Mahon and D. C. Creagh, The effect of laser excitation on the Raman microspectroscopy of nanoindentation-induced silicon phase transformation, *Nucl. Instrum. Methods Phys. Res.*, **580**(2007): 430–433.

12. S. Ruffell, B. Haberl, S. Koenig, J. E. Bradby and J. S. Williams, Annealing of nanoindentation-induced high pressure crystalline phases created in crystalline and amorphous silicon, *J. Appl. Phys.*, **105**(2009): 093513.

13. R. O. Piltz, J. R. Maclean, S. J. Clark, G. J. Ackland, P. D. Hatton and J. Crain, Structure and properties of silicon XII: A complex tetrahedrally bonded phase, *Phys. Rev. B*, **52**(1995): 4072–4085.

14. S. Ruffell, J. E. Bradby, J. S. Williams and P. Munroe, Formation and growth of nanoindentation-induced high pressure phases in crystalline and amorphous silicon, *J. Appl. Phys.*, **102**(2007): 063521.

15. Y. B. Gerbig, S. J. Stranick, and R. F. Cook, Direct observation of phase transformation anisotropy in indented silicon studied by confocal Raman spectroscopy, *Phys. Rev. B*, **83**(2011): 205209.

16. S. Wong, B. Haberl, J. S. Williams, and J. E. Bradby, Phase transformation as the single-mode mechanical deformation of silicon, *Appl. Phys. Lett.*, **106**(2015): 252103.

17. Y. B. Gerbig, C. A. Michaels, and R. F. Cook, In situ observation of the spatial distribution of crystalline phases during pressure-induced transformations of indented silicon thin films, *J. Mater. Res.*, **30**(2015): 390–406.

18. C. Y. Yeh, Z. W. Lu, S. Froyen and A. Zunger, Zinc-blende-wurtzite polytypism in semiconductors, *Phys. Rev. B*, **46**(1992): 10086.

19. L. Pavesi, L. D. Negro, C. Mazzoleni, G. Franzo and F. Priolo, Optical gain in silicon nanocrystals, *Nature*, **408**(2000): 440–444.

20. Y. Kanemitsu and K. Suzuki, Luminescence properties of a cubic silicon cluster octasilacubane, *Phys. Rev. B*, **51**(1995): 10666–10670.

21. J. R. Siekierzycka, M. R. Vasic, H. Zuihof and A. Brouwer, Photophysics of n-butyl-capped silicon nanoparticles, *J. Phys. Chem. C*, **115**(2011): 20888–20895.

22. Z. H. Cen, J. Xu, Y. S. Liu, W. Li, L. Xu, Z. Y. Ma, X. F. Huang and K. J. Chen, Visible light emission from single layer Si nanodots fabricated by laser irradiation method, *Appl. Phys. Lett.*, **89**(2006): 163107.

23. S. S. Walavalkar, C. E. Hofmann, A. P. Homyk, M. D. Henry, H. A. Atwater and A. Scherer, Tunable visible and near-IR emission from sub-10 nm etched single-crystal Si nanopillars, *Nano Lett.*, **10**(2010): 4423–4428.

24. T. Orii, M. Hirasawa and T. Seto, Tunable, narrow-band light emission from size-selected Si nanoparticles produced by pulsed-laser ablation, *Appl. Phys. Lett.*, **83**(2003): 3395–3397.

25. V. Švrček, T. Sasaki, Y. Shimizu and N. Koshizaki, Blue luminescent silicon nanocrystals prepared by ns pulsed laser ablation in water, *Appl. Phys. Lett.*, **89**(2006): 213113.

26. R. Intartaglia, K. Bagga, F. Brandi, G. Das, A. Genovese, E. Di Fabrizio and A. Diaspro, Optical properties of femtosecond laser-synthesized silicon nanoparticles in deionized water, *J. Phys. Chem. C*, **115**(2011): 5102–5107.

27. T. Yoshida, Y. Yamada and T. Orii, Electroluminescence of silicon nanocrystallites prepared by pulsed laser ablation in reduced pressure inert gas, *J. Appl. Phys.*, **83**(1998): 5427–5432.

28. S.W. Lin and D.H. Chen, Synthesis of water-soluble blue photoluminescent silicon nanocrystals with oxide surface passivation, *Small*, **5**(2009): 72–76.

29. J. Y. Fan and P. K. Chu, Group IV nanoparticles: Synthesis, properties, and biological applications, *Small*, **6**(2010): 2080–2098.

30. H. Föll, H. Hartz, E. Ossei-Wusu, J. Carstensen and O. Riemenschneider, Si nanowire arrays as anodes in Li ion batteries, *Phys. Status Solidi RRL*, **4**(2010): 4–6.

31. R. M. Sankaran, D. Holunga, R. C. Flagan and K. P. Giapis, Synthesis of blue luminescent Si nanoparticles using atmospheric-pressure microdischarges, *Nano Lett.*, **5**(2005): 537–534.

32. S. K. Yang, W. Z. Li, B. Q. Cao, H. B. Zeng and W. P. Cai, Origin of blue emission from silicon nanoparticles: Direct transition and interface recombination, *J. Phys. Chem. C*, **115**(2011): 21056–21062.

33. S. Furukawa and T. Miyasato, Quantum size effects on the optical band gap of microcrystalline Si:H, *Phys. Rev. B*, **38**(1988): 5726–5729.

34. G. Belomoin, J. Therrien, A. Smith, S. Rao, R. Twesten, S. Chaieb, M. H. Nayfeh, L. Wagner and L. Mitas, Observation of a magic discrete family of ultrabright Si nanoparticles, *Appl. Phys. Lett.*, **80**(2002): 841–843.

35. Z. Y. Zhou, L. Brus and R. Friesner, Electronic structure and luminescence of 1.1 and 1.4-nm silicon nanocrystals: Oxide shell versus hydrogen passivation, *Nano Lett.*, **3**(2003): 163–167.

36. G. G. Qin and Y. J. Li, Photoluminescence mechanism model for oxidized porous silicon and nanoscale-silicon-particle-embedded silicon oxide, *Phys. Rev. B*, **68**(2003): 085309.

37. E. Rogozhina, G. Belomoin, A. Smith, L. Abuhassan, N. Barry, O. Akcakir, P. V. Braun and M. H. Nayfeh, Si-N linkage in ultrabright, ultrasmall Si nanoparticles, *Appl. Phys. Lett.*, **78**(2001): 3711–3713.

38. Y. Q. Wang, R. Smirani, and G. G. Ross, Stacking faults in Si nanocrystals, *Appl. Phys. Lett.*, **86**(2005): 221920.

39. F. Li, F. Qiang, J. Xiang, and C. M. Lieber, Nanowire electronic and optoelectronic devices, *Mater. Today*, **9**(2006): 18–27.

40. V. F. Puntes, K. M. Krishnan and A. P. Alivisatos, Colloidal nanocrystal shape and size control: The case of cobalt, *Science*, **291**(2001): 2115–2117.

41. C. J. Murphy, Nanocubes and nanoboxes, *Science*, **298**(2002): 2139–2141.
42. Y. Sun and Y. Xia, Shape-controlled synthesis of gold and silver nanoparticles, *Science*, **298**(2002): 2176–2179.
43. I. O. Sosa, C. Noguez and R. G. Barrera, Optical properties of metal nanoparticles with arbitrary shapes, *J. Phys. Chem. B*, **107**(2003): 6269–6275.
44. F. Dumestre, B. Chaudret, C. Amiens, P. Renaud and P. Fejes, Superlattices of iron nanocubes synthesized from Fe[N(SiMe$_3$)$_2$]$_2$, *Science*, **303**(2004): 821–823.
45. E. R. Chan, X. Zhang, C. Y. Lee, M. Neurock and S. C. Glotzer, Simulations of tetra-tethered organic/inorganic nanocube–polymer assemblies, *Macromolecules*, **38**(2005): 6168–6180.
46. L. J. Sherry, S. H. Chang, G. C. Schatz, R. P. Van Duyne, B. J. Wiley and Y. Xia, Localized surface plasmon resonance spectroscopy of single silver nanocubes, *Nano Lett.*, **5**(2005): 2034–2038.
47. D. Yu and V. W. Yam, Controlled synthesis of monodisperse silver nanocubes in water, *J. Am. Chem. Soc.*, **126**(2004): 13200–13201.
48. R. Liu, F. Oba, E. W. Bohannan, F. Ernst and J. A. Switzer, Shape control in epitaxial electrodeposition: Cu$_2$O nanocubes on InP(001), *Chem. Mater.*, **15**(2003): 4882–4885.
49. R. Xu and H. C. Zeng, Self-generation of tiered surfactant superstructures for one-pot synthesis of Co$_3$O$_4$ nanocubes and their close- and non-close-packed organizations, *Langmuir*, **20**(2004): 9780–9790.
50. H. Matsumoto, K. Higuchi, S. Kyushin and M. Goto, Octakis (1,1,2-trimethylpropyl) octasilacubane: Synthesis, molecular structure, and unusual properties, *Angew. Chem. Int. Ed. Engl.*, **31**(1992): 1354–1356.
51. P. Liu, Y. L. Cao, H. Cui, X. Y. Chen and G. W. Yang, Micro-and nanocubes of silicon with zinc-blende structure, *Chem. Mater.* **20**(2008): 494–502.
52. A. Pileni and M. P. Filankembo, Is the template of self-colloidal assemblies the only factor that controls nanocrystal shapes? *J. Phys. Chem. B*, **104**(2000): 5865–5868.
53. B. Wiley, T. Herricks, Y. Sun and Y. Xia, Polyol synthesis of silver nanoparticles: Use of chloride and oxygen to promote the formation of single-crystal, truncated cubes and tetrahedrons, *Nano Lett.*, **4**(2004): 1733–1739.
54. G. W. Yang, Laser ablation in liquids: Applications in the synthesis of nanocrystals, *Prog. Mater. Sci.*, **52**(2007): 648–698.
55. L. Berthe, R. Fabbro, P. Peyer, L. Tollier and E. Bartnicki, Shock waves from a water-con°ned laser-generated plasma, *J. Appl. Phys.*, **82**(1997): 2826–2832.
56. L. Berthe, R. Fabbro, P. Peyer and E. Bartnicki, Wavelength dependent of laser shock-wave generation in the water-confinement regime, *J. Appl. Phys.*, **85**(1999): 7552–7555.
57. A. Sollier, L. Berthe and R. Fabbro, Numerical modeling of the transmission of breakdown plasma generated in water during laser shock processing, *Eur. Phys. J. Appl. Phys.*, **16**(2001): 131–139.
58. G. W. Yang, J. B. Wang and Q. X. Liu, Preparation of nano-crystalline diamonds using pulsed laser induced reactive quenching, *J. Phys. Condens. Matter*, **10**(1998): 7923–7927.
59. J. B. Wang, C. Y. Zhang, X. L. Zhong and G. W. Yang, Cubic and hexagonal structures of diamond nanocrystals formed upon pulsed laser induced liquid–solid interfacial reaction, *Chem. Phys. Lett.*, **361**(2002): 86–90.
60. S. Zhu, Y. F. Lu, M. H. Hong and X. Y. Chen, Laser ablation of solid substrates in water and ambient air, *J. Appl. Phys.*, **89**(2001): 2400–2403.
61. A. D. Giacomo, M. Dell'Aglio, F. Colao and R. Fantoni, Double pulse laser produced plasma on metallic target in seawater: Basic aspects and analytical approach, *Spectrochimica Acta Part B*, **59**(2004): 1431–1438.
62. C. X. Wang, P. Liu, H. Cui and G. W. Yang, Nucleation and growth kinetics of nanocrystals formed upon pulsed-laser ablation in liquid, *Appl. Phys. Lett.*, **87**(2005): 201913.
63. J. B. Wang and G. W. Yang, Phase transformation between diamond and graphite in preparation of diamonds by pulsed-laser induced liquid-solid interface reaction, *J. Phys. Condens. Matter*, **11**(1999): 7089–7094.
64. M. N. R. Ashfold, F. Claeyssens, G. M. Fuge and S. J. Henley, Pulsed laser ablation and deposition of thin films, *Chem. Soc. Rev.*, **33**(2004): 23–31.
65. P. P. Patil, D. M. Phase, S. A. Kulkarni, S. V. Ghaisas, S. K. Kulkarni, S. M. Kanetkar and S. B. Ogale, Pulsed-laser-induced reactive quenching at a liquid-solid interface: Aqueous oxide of iron, *Phys. Rev. Lett.*, **58**(1987): 238–241.
66. S. B. Ogale, P. P. Patil, D. M. Phase, Y. V. Bhandarkar, S. K. Kulkarni, S. Kulkarni, S. V. Ghaisas and S. M. Kanetkar, Synthesis of metastable phase via pulse-laser-induced reactive quenching at liquid-solid interface, *Phys. Rev. B*, **36**(1987): 8237–8250.
67. G. W. Yang and J. B.Wang, Pulsed-laser-induced transformation path of graphite to diamond via an intermediate rhombohedral graphite, *Appl. Phys. A*, **72**(2001): 475–479.
68. Q. X. Liu, C. X. Wang, W. Zhang and G.. W. Yang, Immiscible silver-nickel alloying nanorods growth upon pulsed-laser induced liquid/solid interfacial reaction, *Chem. Phys. Lett.*, **382**(2003): 1–5.
69. G. W. Yang and J. B. Wang, Carbon nitride nanocrystals having cubic structure using pulsed laser induced liquid–solid interfacial reaction, *Appl. Phys. A*, **71**(2000): 343–344.

70. G. W. Yang and J. B. Wang, Pulsed-laser-induced transformation path of graphite to diamond via an intermediate rhombohedral graphite, *Appl. Phys. A*, **72**(2001): 475–479.

71. J. B. Wang, G. W. Yang and C. Y. Zhang, Cubic-BN nanocrystals synthesis by pulsed laser induced liquid-solid interfacial reaction, *Chem. Phys. Lett.*, **367**(2003): 10–14.

72. V. Amendola, P. Riello and M. Meneghetti, Magnetic nanoparticles of iron carbide, iron oxide, iron@iron oxide, and metal iron synthesized by laser ablation in organic solvents, *J. Phys. Chem. C*, **115**(2011): 5140–5146.

73. X. Y. Li, A. Pyatenko, Y. Shimizu, H. Q. Wang, K. Koga and N. Koshizaki, Fabrication of crystalline silicon spheres by selective laser heating in liquid medium, *Langmuir*, **27**(2011): 5076–5080.

74. F. Lin, J. Yang, S. H. Lu, K. Y. Niu, Y. Liu, J. Sun and X. W. Du, Laser synthesis of gold/oxide nanocomposites, *J. Mater. Chem.*, **20**(2010): 1103–1106.

75. P. Liu, Y. Liang, X. Z. Lin, C. X. Wang and G. W. Yang, A general strategy to fabricate simple polyoxometalate nanostructures: Electrochemistry-assisted laser ablation in liquid, *ACS Nano*, **5**(2011): 4748–4755.

76. S. Besner, A.V. Kabashin, F. M. Winnik and M. Meunier, Ultrafast laser based "green" synthesis of non-toxic nanoparticles in aqueous solutions, *Appl. Phys. A*, **93**(2008): 955–959.

77. S. K. Yang, W. P. Cai, H. W. Zhang, X. X. Xu and H. B. Zeng, Size and structure control of Si nanoparticles by laser ablation in different liquid media and further centrifugation classification, *J. Phys. Chem. C*, **113**(2009): 19091–19095.

78. X. Y. Chen, H. Cui, P. Liu and G. W. Yang, Double-layer hexagonal Fe nanocrystals and magnetism, *Chem. Mater.*, **20**(2008): 2035–2038.

79. P. Liu, H. Cui and G. W. Yang, Synthesis of body-centered cubic carbon nanocrystals, *Cryst. Growth Des.*, **8**(2008): 581–586.

80. P. Liu, Y. L. Cao, C. X. Wang, X. Y. Chen and G. W. Yang, Micro- and nanocubes of carbon with C8-like and blue luminescence, *Nano Lett.*, **8**(2008): 2570–2575.

81. P Liu, Y. L. Cao, X. Y. Chen and G. W. Yang, Trapping high-pressure nanophase of Ge upon laser ablation in liquid, *Cryst. Growth. Des.*, **9**(2009): 1390–1393.

82. P. Liu, C. X. Wang, X. Y. Chen and G. W. Yang, Controllable fabrication and cathodoluminescence performance of high-index facets GeO$_2$ micro- and nanocubes and spindles upon electrical-field-assisted laser ablation in liquid, *J. Phys. Chem. C*, **112**(2008): 13450–13456.

83. P. Liu, Y. Liang, H. B. Li, J. Xiao, T. He, and G. W. Yang, Violet-blue photoluminescence from Si nanoparticles with zinc-blende structure synthesized by laser ablation in liquids, *AIP Adv.*, **3**(2013): 022127.

84. P. Liu, H. Cui, C. X. Wang and G. W. Yang, From nanocrystal synthesis to functional nanostructure fabrication: Laser ablation in liquid, *Phys. Chem. Chem. Phys.*, **12**(2010): 3942–3952.

85. P. G. Kuzmin, G. A. Shafeev, V. V. Bukin, S. V. Garnov, C. Farcau, R. Carles, B. Warot-Fontrose, V. Guieu and G. Viau, Silicon nanoparticles produced by femtosecond laser ablation in ethanol: Size control structural characterization, and optical properties, *J. Phys. Chem. C*, **114**(2010): 15266–15273.

86. G. Faraci, S. Gibilisco, A. R. Pennisi and C. Faraci, Quantum size effects in Raman spectra of Si nanocrystals, *J. Appl. Phys.*, **109**(2011): 074311.

87. T. E. Itina, On nanoparticle formation by laser ablation in liquids, *J. Phys. Chem. C*, 115(2011): 5044–5048.

88. K. Hata, S. Yoshida, M. Fujita, S. Yasuda, T. Makimura, K. Murakami and H. Shigekawa, Self-assembled monolayer as a template to deposit silicon nanoparticles fabricated by laser ablation, *J. Phys. Chem. B*, **105**(2001): 10842–10846.

89. J. Zou, R. K. Baldwin, K. A. Pettigrew and S. M. Kauzlarich, Solution synthesis of ultrastable luminescent siloxane-coated silicon nanoparticles, *Nano Lett.*, **4**(2004): 1181–1186.

90. K. Abderrafi, R. G. Calzada, M. B. Gongalsky, I. Suarez, R. Abarques, V. S. Chirvony, V. Y. Timoshenko, R. Ibanez and J. P. Martínez-Pastor, Silicon nanocrystals produced by nanosecond laser ablation in an organic liquid, *J. Phys. Chem. C*, **115**(2011): 5147–5151.

91. M. E. Straumanis and E. Z. Aka, Lattice parameters, coefficients of thermal expansion, and atomic weights of purest silicon and germanium, *J. Appl. Phys.*, **23**(1952): 330–334.

92. Z. L. Wang, Characterization of nanophase materials, *Part. Part. Syst. Charact.*, **18**(2001): 142–165.

93. Z. L. Wang, New developments in transmission electron microscopy for nanotechnology, *Adv. Mater.*, **15**(2003): 1497–1514.

94. B. J. Wiley, S. H. Im, Z. Y. Li, J. McLellan, A. Siekkinen and Y. Xia, Maneuvering the surface plasmon resonance of silver nanostructures through shape-controlled synthesis, *J. Phys. Chem. B*, **110**(2006): 15666–15675.

95. J. Khajehpour, W. A. Daoud, T. Williams and L. Bourgeois, Laser-induced reversible and irreversible changes in silicon nanostructures: One- and multi-phonon Raman scattering study, *J. Phys. Chem. C*, **115**(2011): 22131–22137.

96. C. Georgi, M. Hecker and E. Zschech, Effects of laser-induced heating on Raman stress measurements of silicon and silicon-germanium structures, *J. Appl. Phys.*, **101**(2007): 123104.

97. H. Ow, D. R. Larson, M. Srivastava, B. A. Baird, W. W. Webb and U. Wiesner, Bright and stable core–shell fluorescent silica nanoparticles, *Nano Lett.*, **5**(2005): 113–117.

98. V. Svrcek, D. Mariotti, T. Nagai, Y. Shibata, I. Turkevych and M. Kondo, Photovoltaic applications of silicon nanocrystal based nanostructures induced by nanosecond laser fragmentation in liquid media, *J. Phys. Chem. C*, **115**(2011): 5084–5093.

99. H. Morisaki, F. W. Ping, H. One and K. Yazawa, Above-band-gap photoluminescence from Si fine particles with oxide shell, *J. Appl. Phys.*, **70**(1991): 1869.

100. X. L. Wu, S. J. Xiong, G. G. Siu, G. S. Huang, Y. F. Mei, Z. Y. Zhang, S. S. Deng and C. Tan, Optical emission from excess Si defect centers in Si nanostructures, *Phys. Rev. Lett.*, **91**(2003): 157402.

101. P. Hohenberg and W. Kohn, Inhomogeneous electron gas, *Phys. Rev.*, **136**(1964): B864–B871.

102. D. Sanchez-Portal, P. Ordejon, E. Artacho and J. M. Soler, Density-functional method for very large systems with LCAO basis sets, *Int. J. Quantum Chem.*, **65**(1997): 453–461.

103. J. M. Soler, E. Artacho, J. D. Gale, A. Garcia, J. Junquera, P. Ordejon and D. Sanchez-Portal, The SIESTA method for ab initio order-N materials simulation, *J. Phys. Condens Matter*, **14**(2002): 2745–2779.

104. P. Ordejon and J. M. Soler, Self-consistent order-N density-functional calculations for very large systems, *Phys. Rev. B*, **53**(1996): R10441–R10444.

105. J. P. Wilcoxon, G. A. Samara and P. N. Provencio, Optical and electronic properties of Si nanoclusters synthesized in inverse micelles, *Phys. Rev. B*, **60**(1999): 2704–2714.

106. J. C. She, S. Z. Deng, N. S. Xu, R. H. Yao and J. Chen, Fabrication of vertically aligned Si nanowires and their application in a gated field emission device, *Appl. Phys. Lett.*, **88**(2006): 013112.

107. L. Fletcher and P. Mitchell, Silicon light up imaging, *Nat. Biotechnol.*, **20**(2002): 351.

12 Silicon nanocrystals from plasma synthesis

Samantha K. Ehrenberg, Katharine I. Hunter, and Uwe R. Kortshagen

Contents

12.1 INTRODUCTION

In this chapter, we will explore the bottom-up growth of freestanding (unembedded) Si nanocrystals in the gas phase using plasmas. Si nanocrystals can be unique nanostructures due to the effect of quantum confinement operating in all three spatial dimensions (Efros and Rosen 2000). This three dimensional (3D) quantum confinement affords the ability to modify the accessible energy levels within the material, leading to size-dependent optical and electronic properties (Brus 1986). Additionally, as the surface area to volume ratio increases with reduced nanocrystal size, the surface plays an increasingly important role for small, freestanding crystallites. This allows for the use of surface and strain engineering as additional parameters to tune material properties to a greater extent than is accessible in bulk materials or even in 1- or 2-D nanostructures. Freestanding Si nanocrystals are forward compatible with bottom-up manufacturing, such as self-assembly and 3-D printing, which adds new degrees of freedom in materials and device design compared to traditional top-down processing. As we discuss in this chapter, plasmas provide a growth environment that is well-tailored to the processing of high-quality Si nanocrystals for optoelectronic applications.

Plasmas—partially ionized gases created by application of an electric field—have become ubiquitous in the field of semiconductor processing. Commonly, low-pressure plasmas operated at pressures of 10^{-4}–10^{-2} times the atmospheric pressure are used for the deposition of Si-based thin films from the gas phase onto solid substrates, often using silane (SiH_4) as the precursor gas. However, under "unfavorable" conditions, these plasmas are prone to the nucleation and growth of Si nanoparticles in the gas phase, resulting in a "dusty" plasma. These undesired Si nanoparticles can contaminate thin film growth, resulting in defective devices. Consequently, the early history of dusty plasma research focused on methods for suppressing the formation of these Si nanoparticle contaminants to improve film quality.

However, interest in quantum-confined Si took off with the first observations of bright, visible photoluminescence, notably in porous Si in 1990 (Canham 1990). It became clear that Si nanoparticles possess useful properties in their own right, and methods were developed specifically for the synthesis of high-quality Si nanocrystals. The high crystallization temperature of Si demands that most synthesis methods expend high amounts of energy to achieve crystalline nanoparticles. Initial efforts to controllably grow Si nanoparticles used flow-through hot wall aerosol reactors to thermally decompose silane (Alam and Flagan 1986) or disilane (Littau et al. 1993). However, particle growth rate and the resulting size distribution proved difficult to control in neutral aerosols due to gas-phase agglomeration of the uncharged particles.

The growth of Si nanocrystals in plasmas overcomes many of the challenges seen in these other gas-phase processes. Nanoparticles grown in plasmas are highly crystalline and monodisperse in size, as shown in Figure 12.1 (Kortshagen et al. 2008). When these nanocrystals are made small enough—around a few nanometers—photoluminescence is observed, ranging from near-infrared into visible wavelengths with decreasing nanocrystal size. Figure 12.2 shows this bright, visible luminescence from 2–5 nm oxidized Si nanocrystals dispersed in ethanol under ultraviolet (UV) illumination (Pi et al. 2008b).

The first report of size-dependent photoluminescence (PL) from plasma-produced Si nanocrystals came in 1990 from nanocrystals grown in a microwave discharge around the same time as the observation of luminescence from porous Si (Takagi et al. 1990). The origin of this visible PL from Si nanocrystals has been a contentious subject and often requires careful spectroscopic investigation to reveal. Size-dependent, quantum-confined photoluminescence was originally observed from colloidally synthesized cadmium sulfide (CdS) nanocrystals in 1983 (Rossetti et al. 1983) and famously described through an effective mass model (Brus 1986)

$$E_g \sim E_{g,bulk} + \frac{h^2}{8R_{NC}^2}\left(\frac{1}{m_e} + \frac{1}{m_h}\right) - \frac{1.8e^2}{\epsilon R}$$

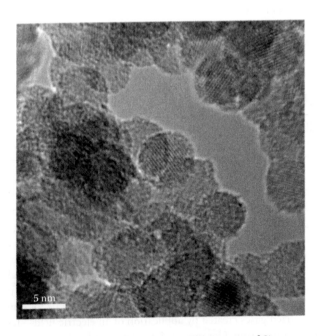

5 nm

Figure 12.1 Representative transmission electron microscopy (TEM) image of Si nanocrystals formed in a non-thermal plasma. (Reprinted with permission from Kortshagen, U., R. et al., *Pure and Applied Chemistry* 80 (9): 1901–8, 2008.)

Figure 12.2 Visible photoluminescence from oxidized Si nanocrystals dispersed in ethanol. (Reprinted from Pi, X. D., et al., *Nanotechnology*, 19, 245603, 2008b. With permission.)

which estimates the dependence of the band gap energy E_g in a semiconductor nanocrystal on the nanocrystal radius, R_{NC}, and carrier effective masses, m_e and m_h. The deviation of the nanocrystal band gap from the band gap energy of the bulk materials, $E_{g,bulk}$, is approximated by the quantum confinement of the exciton wavefunction, described by the particle-in-a-box model. Accordingly, experimental evidence of this size-dependent band gap widening has been documented in Si nanocrystals. Figure 12.3 summarizes the experimental evidence for this quantum-confined shift in band gap energy compared to this simple model (Wheeler et al. 2015).

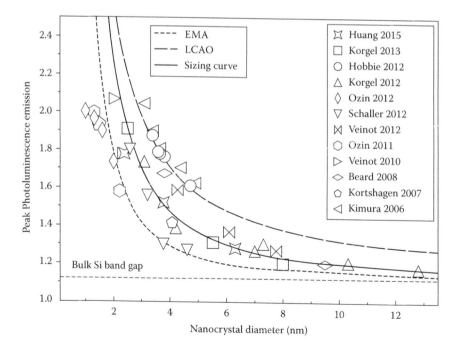

Figure 12.3 Peak photoluminescence energy of quantum-confined Si nanocrystals versus crystal diameter as measured by transmission electron microscopy. Effective mass approximation model without Coulomb interaction (short dashed line) and linear combination of atomic orbitals model (long dashed line) also plotted with a power fit law of the experimental data, $E(d) = 1.12 + 3.73d^{-1.69}$ (solid line). (Adapted with permission from Wheeler, L. M., et al., 2015, Silyl radical abstraction in the functionalization of plasma-synthesized silicon nanocrystals, *Chem. Mater.*, 27, 6869–6878. Copyright 2015 American Chemical Society.)

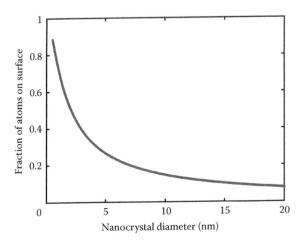

Figure 12.4 Approximate fraction of Si atoms at the surface of a Si nanocrystal for a range of crystallite sizes. Surface atoms are considered to be within one unit cell of the nanocrystal outer radius.

While nanocrystals are often depicted as perfect crystals, it is important to be mindful that the crystal lattice may contain defects and imperfections, particularly at the nanoparticle surface, where an amorphous or disordered layer is often observed. High-energy X-ray diffraction (HEXRD) measurements have shown that the crystal structure of freestanding, ligand-passivated Si nanocrystals becomes significantly distorted at diameters smaller than 3 nm (Petkov et al. 2013). For nanoparticles, the atoms at the surface represent a significant number of atoms in the whole particle. Therefore, surface structure can have a significant impact on nanocrystal electronic and optical properties. Figure 12.4 shows an estimate of the fraction of atoms at the nanocrystal surface for a range of sizes, assuming that all Si atoms within one unit cell of the surface are effectively surface atoms. This high fraction of surface atoms for small nanocrystals requires careful attention and surface atoms must be appropriately terminated or functionalized if defects are of concern for the desired application. As Si readily oxidizes, nanocrystal termination and exposure to air and water significantly affects nanocrystal oxidation rate (Buriak 2002). Therefore, it is important to consider the environmental stability of Si nanocrystals and the effect of oxidation on nanocrystal properties.

The quality of the nanocrystal, and thus its technological applicability, depends strongly on growth conditions. For Si nanocrystals, photoluminescence is a key metric for assessing nanocrystal quality as it is dependent on the density of defect states in the Si nanocrystals. While robust solution-phase synthesis routes have been developed, these techniques require specialized processing of highly reactive reagents and produce nanocrystals often exhibiting weak, defect-related photoluminescence (Bley and Kauzlarich 1996; Atkins et al. 2012). In contrast, gas-phase processing of Si nanocrystals has produced high-quality nanocrystals exhibiting desirable properties, including bright, size-tunable photoluminescence (Jurbergs et al. 2006). These gas-phase approaches generate little chemical waste and do not necessarily require the use of organic ligands, which would otherwise limit synthesis temperatures.

The use of plasmas has emerged as a leading technique for the synthesis of Si nanocrystals because of the unique set of advantages which set this approach apart from other gas-phase approaches. Plasma properties, specifically the temperature and density of neutral and charged species (electrons and ions), dictate the reactive environment within which Si particles are shaped. Therefore, it is important to introduce basics of these characteristic plasma properties to understand their effect on resulting nanoparticle properties.

12.2 PROCESSING OF Si NANOCRYSTALS USING PLASMAS

The term plasma refers to a partly or fully ionized gas. In nature, plasmas occur in stars, interstellar nebulae, the aurora borealis, and lightning, to mention just a few examples. In laboratory settings, plasmas are usually produced by applying electric fields to a gas at low or atmospheric pressure. The electric field accelerates some initial free electrons that may stem from natural radioactivity or cosmic radiation to energies at which they can ionize some of the gas atoms to produce more free electrons and gaseous ions. The electrons absorb the vast majority of the electrical power from the electric field, as their mobility is by 3 to 4 orders of magnitude higher than that of the more massive ions. Accordingly, the plasma electrons typically have an average kinetic energy in the range of 1–5 electronvolts (eV), which nominally corresponds to an electron temperature of ~10,000–50,000 K. The electrons then transfer some of their energy to the "heavy" neutral gas atoms and gaseous ions through elastic collisions or Coulombic interactions. The degree to which energy is transferred from the electrons to the heavy gas species determines how much the electron temperature T_e, will equilibrate with the temperature of the heavy species, T_h. In so-called thermal plasmas, often encountered at atmospheric pressure in arc discharges at very high-power densities, the electron and heavy species temperatures almost equilibrate, so $T_e \approx T_h$. By contrast, in nonthermal plasmas, the energy transfer from electrons to the heavy species is weak and $T_e \gg T_h$. In the laboratory, nonthermal plasmas are produced at low or atmospheric pressure but at significantly lower energy densities than thermal plasmas.

12.2.1 THERMAL VS. NONTHERMAL PLASMAS

While thermal plasmas have been employed for nanoparticle synthesis (Rao et al. 1995; Shigeta and Murphy 2011), they suffer from some disadvantages associated with the high gas temperature environment. Due to the high-power density used to produce thermal plasmas, both the electron and heavy species temperatures are in a range of ~10,000–30,000 K, which exceeds the melting and boiling point temperatures of common materials. Thermal plasmas are thus very efficient at dissociating gaseous precursors for nanocrystal growth. However, the temperatures within the thermal plasma are too high to facilitate the nucleation of nanomaterials. Therefore, nanoparticle nucleation in thermal plasmas usually happens at the fringes or in the plasma effluent, which are characterized by extreme gradients in temperatures and plasma species densities, which makes the nanoparticle nucleation process hard to control.

Nonthermal plasmas, which have dominated research on plasma-synthesized nanocrystals, offer a number of advantages for the synthesis of nanoparticles. Most importantly, due to their low gas temperature, which is often close to room temperature, nonthermal plasmas enable the generation and growth of nanoparticles in the main body of the plasmas, where spatial gradients are much smaller than in the plasma fringes. This enables delicate control over the nanoparticle growth conditions.

As most reactive nanoparticle growth species in the nonthermal plasma are produced through electron collision-induced dissociation of precursor molecules, the nanoparticle growth kinetics in nonthermal plasmas are highly dependent on the density of free electrons, often simply called the "electron density" or "plasma density." The electron density in a nonthermal plasma can be adjusted through the electrical power coupled into the plasma. In steady state, the electrical power provided to the plasma balances the energy loss of the electrons through various loss channels, including excitation and ionization of atoms and molecules, dissociation of precursor molecules, elastic collisions with gaseous species, and the loss of electrons (and other plasmas species to the walls). A simplified power balance is:

$$P_{elec} = N_e \varepsilon_T$$

Here N_e is the total number of electrons in the plasma and ε_T is the "energy price," the energy acquired and lost again by one average electron during its lifetime in the plasma (Lieberman and Lichtenberg 1994). This power balance suggests that the average electron density $n_{e,ave} = N_e/V_{plasma}$ is roughly proportional to the provided electrical power, if the plasma volume V_{plasma} remains constant. In most nanoparticle synthesis plasmas operated at low pressure, the electron density is in the order of 10^9–10^{11} cm^{-3}, which corresponds to a fractional ionization of the plasma gas of only 10^{-4}–10^{-5}.

The plasma electron temperature, T_e, is an equally important parameter, as it determines the rate constants for most electron-induced reactions. It can be estimated from a simple particle balance for the electrons. In steady state, the rate at which electrons are created in the bulk of the plasma by ionization,

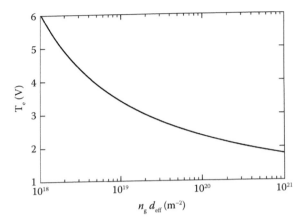

Figure 12.5 Electron temperature T_e versus the product of gas density and effective plasma diameter, $n_g d_{eff}$, for Maxwellian electrons in argon. (Reprinted with permission from Lieberman M. A., and A.J. Lichtenberg. 1994. *Discharges and Materials Processing Principles of Plasma Discharges and Materials*. New York: Wiley.)

$R_{iz} e^-/cm^3$, must be equal to the rate at which they are lost by diffusion to the walls of the plasma container, $R_{e^- loss} e^-/cm^2$:

$$R_{e^- loss} \times A_s = R_{iz} \times V_p$$

The surface area of the reactor, A_s, and the plasma volume, V_p, are both dependent on the system geometry. Both the rate of loss by diffusion, $R_{e^- loss}$, and the rate of ionization, R_{iz}, are a function of the electron temperature, T_e. This simple balance equation reveals that the electron temperature, among other factors, can be adjusted by the design of the reactor geometry. An estimation of these rates, as presented in many plasma textbooks (Lieberman and Lichtenberg 1994) yields that T_e is inversely related to the product of gas density and effective plasma diameter, $n_g d_{eff}$ (Figure 12.5). This simple analysis shows that for typical gas pressures of 10^{-3}–10^{-2} times the atmospheric pressure and characteristic plasma dimensions in the order of 1 cm, the electron temperature is a few eV (10^4 K).

It is interesting to note that the electron temperature is largely independent of the electron density, and thus essentially independent of the provided electrical power. Accordingly, the plasma reactor geometry is the primary determinant for the process conditions because it dictates the plasma electron temperature for a given pressure. Furthermore, the electron temperature in nonthermal plasmas is typically much lower than the ionization threshold energies for most common plasma gases, which are around 13–25 eV. This implies that only the electrons in the high energy tail of the electron energy distribution perform ionizing collisions. Electrons in the distribution tail are also important for many plasma chemical reactions that produce nanoparticle growth species such as ions and radicals. Thus, designing the plasma geometry to achieve a certain electron temperature and adjusting the electrical power to achieve an appropriate plasma density are important parameters in determining suitable conditions for nanoparticle growth.

12.2.2 PLASMA SYNTHESIS REACTORS FOR Si NANOCRYSTALS

Laboratory plasmas are generated by the application of an electric field to a confined gas under well-defined conditions. All plasma reactors designed for the synthesis of Si nanocrystals have several common components: (1) a reaction vessel for containment of the plasma, (2) a gas delivery system supplying a plasma carrier gas, typically argon or helium, and the process gas, typically silane, (3) a power supply (AC or DC) and a means of transferring the power to the plasma, and (4) a method for collecting the Si nanoparticles. The design and selection of these components, however, depends on the operating regime of the plasma. For example, the plasma can be operated at a high or low pressure, electric power can be coupled capacitively or inductively, and nanocrystals can be produced in a batch or continuous flow-through mode. All of these characteristics together dictate reactor design.

Clusters, nanoparticles, and quantum dots

Additionally, nonthermal plasmas are often described by a similar product of pressure (p) and plasma dimension (d), often simply referred to as the pd product, in order to achieve similar electron temperatures. Hence, nonthermal plasmas at low pressure can be operated in larger volumes while atmospheric pressure operation usually requires small plasma dimensions. For this reason, it is often convenient to discuss practical reactor implementations based on pressure regime—low or high (atmospheric) pressure.

12.2.2.1 Low-pressure nonthermal plasma synthesis

A wide variety of nanoparticle synthesis reactors have been designed to exploit the formation of nanoparticles that can occur in dusty plasmas for nanoparticle synthesis. Different schemes for producing nonthermal plasmas at low pressure are shown in Figure 12.6, and will be discussed below.

(a) Parallel-plate reactors. A low-pressure discharge can be generated by the application of an AC voltage across two parallel plates—a geometry commonly used for the growth of Si thin films by plasma-enhanced chemical vapor deposition. Several groups have successfully demonstrated the controlled growth of nanometer-scale Si particles in capacitively coupled parallel-plate reactors and have achieved high-quality Si crystals with them. However, as both the nanocrystals as well as the reactor walls are negatively charged by the high mobility electrons, nanocrystals tend to be trapped in the reactor, which makes parallel-plate geometries more suitable for batch synthesis of nanocrystals. Oda and colleagues devised a "digital process" of precursor injection and nanocrystal growth followed by pulsed discharge to control the nanocrystal growth process (Oda 1997).

(b) Capacitively coupled flow-through reactors. Over the last decade, capacitively coupled plasma (CCP) flow-through reactors have grown in popularity (Mangolini et al. 2005; Gresback et al. 2011; Yasar-Inceoglu et al. 2012; Sagar et al. 2015) for the production of nanocrystals due to their ability to continuously produce nanocrystals in steady state operation. In this design, process gases are passed through a cylindrical dielectric tube (e.g., glass, quartz, ceramic). Nanocrystals form through nucleation and grow while residing in the plasma, and are carried through the plasma reactor by the gas flow. In this geometry, power is applied to a pair of ring electrodes external to the reactor tube, which separates the electrodes from contact with the plasma.

A basic diagram of the flow-through CCP reactor is shown in Figure 12.7, adapted from Mangolini et al. 2005. The carrier gas, typically argon or helium, and precursor gases, typically silane (SiH_4) or silicon tetrachloride ($SiCl_4$), flow into the top of a glass or quartz reactor tube. A CCP is generated by applying a radio frequency (RF) voltage through an impedance matching network connected to the ring electrodes. Particle nucleation and growth occurs in the diffuse plasma above the powered electrode. Nanoparticles initially grow through a combination of the coagulation of small clusters and surface attachment of molecular precursors. As the particles grow larger than a few nanometers, the particle concentration drops below the positive ion density which allows the nanocrystals to acquire enough negative charge to prevent further coagulation of nanoparticles and clusters (Kortshagen and Bhandarkar 1999). Particles continue to grow through surface deposition. In the bright discharge region below the ring electrodes, the nanoparticles are heated above the crystallization temperature. After transiting through the plasma, the nanocrystals can be collected by diffusion onto a filter or can be accelerated through an orifice and collected by impaction onto a substrate (Holman and Kortshagen 2010).

(c) Inductively coupled flow-through reactors. Inductively coupled plasmas (ICPs) use an external coil in place of ring electrodes to couple power into the plasma through a dielectric (Gorla et al. 1997).

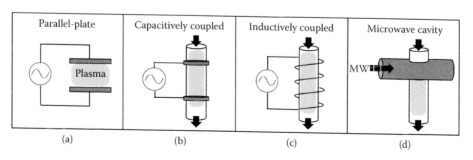

Figure 12.6 Sketches depicting various plasma geometries commonly used for the low-pressure plasma synthesis of Si nanocrystals.

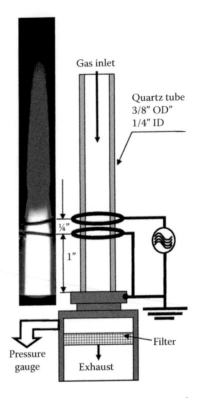

Gas inlet

Quartz tube
3/8" OD"
1/4" ID

¼"

1"

Filter

Pressure
gauge

Exhaust

Figure 12.7 Schematic of capacitively coupled nonthermal plasma reactor for the synthesis of Si nanocrystals with accompanying photograph of argon-silane discharge. (Reprinted from Mangolini, L, E., et al., *Nano Letters* 5 (4): 655–59, 2005. With permission.)

These ICP flow-through reactors have similar advantages to their CCP counterparts; however, due to high plasma densities in the bulk of the plasma, nanocrystal growth is often suppressed by intense heating and cluster disintegration, making the synthesis of small, quantum-confined Si nanocrystals challenging.

(d) Microwave flow-through reactors. Microwave reactors operate similarly to capacitively coupled flow-through reactors, but power is coupled to the plasma through a microwave chamber rather than through ring or plate electrodes (Knipping et al. 2004). The plasmas often have an elevated plasma density, leading to increased particle heating and elevated gas temperatures, which must be managed appropriately. These reactors are promising for high throughput, commercial applications (Hülser et al. 2011).

12.2.2.2 High-pressure nonthermal plasma synthesis

Reactors designed to synthesize Si nanocrystals at or above atmospheric pressure are a logistically enticing approach. These reactors do not require the use of vacuum equipment and are directly compatible with atmospheric pressure post-processing of the resulting nanocrystals. However, any such implementation must address the key challenge of maintaining a diffuse and uniform nonthermal plasma environment at high pressure while suppressing thermalization. Nonthermal glow discharges are preferred for their stability and safety compared to arc discharges or thermal plasmas.

A popular approach for generating nonthermal plasmas at atmospheric pressure is to reduce the overall plasma size (Sankaran et al. 2005; Nozaki et al. 2007). Termed "microplasmas," the small scale of these discharges increases the local electric field to promote breakdown at high pressures, according to Paschen's Law

$$V_b = \frac{B(pd)}{\ln(A\ pd) - \ln\left[\ln\left(1 + \frac{1}{\gamma_{se}}\right)\right]}$$

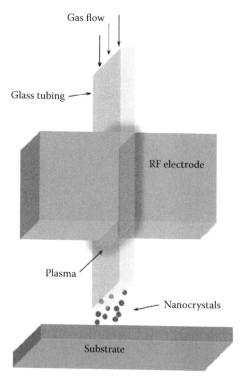

Figure 12.8 Schematic of cross-flow plasma reactor operated at atmospheric pressure. The plasma container is a rectangular quartz tube with an internal cross-section of 0.5 mm × 5 mm and 0.3 mm wall thickness. The plasma is created within by applying RF-power (13.56 MHz) through two rectangular copper electrodes with a gap spacing of 0.5 mm. (Reprinted with permission from Askari, S., et al., *Journal of Physics D: Applied Physics* 48 (31): 314002, 2015.)

which says that the breakdown voltage V_b is dependent on the product of pressure and typical plasma dimension (pd), where A and B are constants specific to the carrier gas, and γ_{se} is the secondary electron emission coefficient of the electrode. If necessary, a third "triggering" electrode is added to initiate breakdown. Importantly, the high surface area to volume ratio of these reactors leads to efficient cooling of the gas through heat transport from the reactor vessel, and may even be enhanced by the addition of a cooling fluid surrounding the chamber. The reactor geometries presented previously for low-pressure plasmas can be miniaturized to function at atmospheric pressure (reviewed by Mariotti and Sankaran 2010). This modification relies on the small volume of the discharge to minimize gas heating and thermalization.

The clear challenge to this approach is the low throughput inherent in the miniaturized dimensions of microplasma reactors. One approach has been to keep one dimension small while scaling up the dimensions perpendicular to the gas flow, as demonstrated in the cross-flow reactor in Figure 12.8. This design uses two parallel plates at a sub-millimeter spacing to generate and maintain a nonthermal glow, but the width of the reactor is scaled up to increase throughput (Askari et al. 2015).

12.2.3 PARTICLE PROPERTIES' DEPENDENCES ON PLASMA ENVIRONMENT

Nonthermal plasmas provide a unique environment for the controlled growth of Si nanocrystals. The properties of the generated particles depend on the plasma parameters and geometry of the reactor. However, these parameters are often coupled and require careful consideration when tuning a desired nanocrystal property.

Nanoparticle Nucleation and Growth. Si nanocrystals are synthesized from Si-containing precursors, most commonly silane (SiH_4), followed by other precursors such as silicon tetrachloride. The nanocrystal growth is initiated by the electron collision-induced dissociation of the precursor, followed by chemical clustering of precursor radicals. Depending on the plasma conditions, the clustering can be dominated by a negative ion-induced reaction pathway or a neutral radical dominated pathway. Watanabe (2006) provided an excellent review of this topic. Following the nucleation of initial clusters, larger clusters and protoparticles grow quickly through the coagulation of the initial clusters. As clusters coagulate, their density drops to a point at which the resulting protoparticles become uniformly negatively charged, due to the higher mobility of electrons compared to ions in the plasma. Coagulation then ceases and nanoparticles continue to grow through surface reactions with the remaining precursor radicals (Boufendi and Bouchoule 1994; Watanabe et al. 1996; Ravi and Girshick 2009).

Size. The average size of the resulting nanocrystal distribution is most directly dictated by nanocrystal residence time within the plasma, which can be estimated by

$$t_{res} = \frac{A_c \, L_{plasma}}{Q}$$

where A_c is the cross-sectional area of the reactor, L_{plasma} is the length of the plasma, and Q is the total flow rate of gas through the reactor.

In addition to this ability to tune average nanoparticle size, low-pressure plasma often exhibits a smaller dispersity in nanoparticle size than other gas-phase approaches. While nanoparticles in neutral aerosols continue to grow by Brownian coagulation, approaching a geometric standard deviation of 1.355 in the free-molecular regime regardless of the initial size distribution (Lee et al. 1984), nanocrystals grown in low-pressure plasmas often exhibit much narrower distributions due to their charging. Negative charging leads to electrostatic repulsion between nanoparticles and suppresses particle aggregation.

Crystallinity. Considering that the melting temperature of bulk Si is more than 1600 K, it may be surprising that it is possible to produce crystalline Si nanoparticles from an environment in which the background gas temperature does not significantly deviate from room temperature. However, nanoparticles in low-pressure, nonthermal plasma can be heated to temperatures well above that of the background gas through energetic surface reactions, such as electron-ion recombination. Relevant surface reactions for Si nanoparticle crystallization in nonthermal plasmas are shown in Figure 12.9 (Kramer et al. 2014).

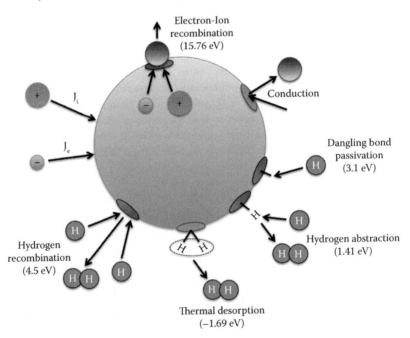

Figure 12.9 Diagram of surface events responsible for the heating and cooling of the Si nanoparticle surface in the nonthermal plasma environment. Heating events: Hydrogen recombination, hydrogen abstraction, dangling bond passivation, electron-ion recombination. Cooling events: heat conduction through collisions with colder gas atoms, hydrogen thermal desorption. (Reprinted from Kramer, N. J., et al., *J. Phys. D. Appl. Phys.* **47**, 075202, 2014. With permission.)

These surface reactions selectively heat the nanoparticles while the main cooling mechanism is heat conduction to the background gas, a processes which is highly dependent on gas density. At low pressure, this process is inefficient, leading to poor thermal coupling of the nanoparticles to the background gas. Models suggest that this poor thermal coupling allows nanoparticles to reach temperatures more than a few hundred Kelvin above room temperature, which is sufficient for crystallizing nanomaterials as they have a lower crystallization temperature than their bulk counterparts (Kramer et al. 2014). At high pressure, nanoparticle cooling by transfer of thermal energy to the background gas is much more efficient and requires more frequent surface reactions to heat the nanoparticles. The rate of these energetic surface reactions is dependent on plasma and radical densities and is often controlled by the power input to the plasma (Askari et al. 2014).

Doping. The intentional introduction of impurities into an intrinsic semiconductor is a critical tool for controlling electronic behavior in bulk materials. Similarly, doping of semiconducting nanocrystals has been explored as a way to tune their optoelectronic properties. For solution-synthesized nanocrystals, doping has been challenging due to a process known as self-purification, in which the impurities are excluded to the surface to lower the free energy of the nanocrystal (Norris et al. 2008). However, the use of a nonthermal plasma to create a highly nonequilibrium environment has allowed for much easier gas-phase synthesis of doped semiconductor nanocrystals (Pi et al. 2008a). Doping of these Si nanocrystals is accomplished by the introduction of gaseous dopant precursors like phosphine (PH_3) or diborane (B_2H_6), into the reactor along with the silane and argon carrier gas. The doping fraction of the nanocrystals is controlled by varying the ratio of precursors in the gas feed. In bulk Si, typical doping concentrations range anywhere from 10^{13} cm^{-3} to 10^{18} cm^{-3}; in this case, less than 0.1% of Si atoms have been replaced by dopants. In contrast, doped Si nanocrystals have doping fractions up to the order of 10%—much higher than in bulk. However, in this case, only a small fraction (10^{-4}–10^{-2}) of the dopants may be electrically activated (Gresback et al. 2014).

12.2.4 ADVANTAGES OF NONTHERMAL PLASMAS FOR Si NANOCRYSTAL SYNTHESIS

Nonthermal plasmas possess key advantages which make them attractive for Si nanocrystal synthesis including:

1. Selective heating—Si nanocrystals are heated above their crystallization temperature by energetic surface reactions. The background gas and reactor vessel remain near room temperature, reducing the total input power requirements.
2. Particle charging—Si nanocrystals are negatively charged by collisions with electrons. Coulombic repulsion prevents nanocrystals from agglomerating in flight, reducing size dispersion.
3. Nanocrystal confinement—The large mobility of the electrons compared to that of the ions also leads to negative charging of the dielectric reactor walls. This negative charge repels the negatively charged particles and keeps them confined within the bulk of the plasma. This nanocrystal confinement increases yield compared to other gas-phase processes.

12.3 PROPERTIES

Si nanoparticles have been produced on the scale of tens of nanometers to just a few nanometers in plasmas. Larger crystals can be highly facetted or even cubic shaped. For example, Bapat et al. (2007) reported 35 ± 4.7 nm cube-shaped Si nanocrystals synthesized in a two-stage plasma reactor. Smaller nanocrystals have been the focus of most studies, however, and are spherical when produced in plasmas due to fast thermal quenching upon exiting the reactive plasma environment. Surface effects, quantum confinement, and doping have led to a wide range of interesting and tunable properties of Si nanocrystals, which we review here.

12.3.1 ABSORPTION

One of the most fundamental properties of plasma-synthesized Si nanocrystals is the size-dependent band gap which strongly emerges for crystal sizes below ~6 nm. As the diameter of the crystal falls below the average distance of electron-hole pairs (also known as the Bohr exciton radius), the band gap is modified and recombination rates change as a result of the 3D spatial confinement of the excitons.

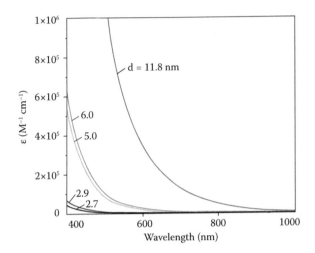

Figure 12.10 Molar absorption coefficients for various diameters of Si nanocrystals. (Adapted with permission from Hessel, C. M., et al., 2012, Synthesis of ligand-stabilized Si nanocrystals with size-dependent photoluminescence spanning visible to near-infrared wavelengths. *Chem. Mater.*, 24, 393–401. Copyright 2012 American Chemical Society.)

This quantum-confinement effect results in a widening band gap as the crystal size shrinks and is demonstrated in Si by the absorption shifting to higher energies (blueshift) as seen in Figure 12.10. Frequently, band gaps are estimated via Tauc plots extracted from absorption curves. The optical absorbance strength depends on the difference between the photon energy and band gap as follows:

$$(\alpha h v)^n = A(h v - E_g)$$

where α is the absorption coefficient, h is Planck's constant, v is the photon's frequency, E_g is the band gap and A is a proportionality constant. The value of the exponent depends on the nature of the band gap with n = ½ for allowed direct transitions and n = 2 for allowed indirect transitions. By plotting $(\alpha h v)^n$ versus $h v$, one can estimate E_g by extrapolating for the intercept and also understand the nature of the transition by comparing the linear fits for different exponents (Viezbicke et al. 2015). Si has a very gradual absorption onset, however, which makes finding the linear portion on which to perform Tauc analysis challenging, and high uncertainties in both the exponent and estimated band gap arise from using this method.

12.3.2 EMISSION

Another effect of quantum confinement is brightened photoluminescence. The efficiency of emission is described by the photoluminescence quantum yield (PLQY), which depends on the relative rates of radiative (k_r) and nonradiative (k_{nr}) recombination:

$$PLQY = \frac{k_r}{k_{nr} + k_r}$$

The rate of nonradiative recombination is determined primarily by Auger recombination and trap-assisted recombination. As illustrated in Figure 12.11, Auger recombination results in the energy being passed off to a third carrier. During trap-assisted recombination, an electron-hole pair recombines through a mid-gap state which usually derives from a surface dangling bond or defect. As will be discussed later, eliminating trap states through surface modifications is one of the most effective tools for improving Si nanocrystal PLQY.

The emission and absorption of plasma-synthesized Si nanocrystals are essentially similar to that of porous Si which has been studied since the 1990s (Hybertsen 1994). However, the detailed mechanism

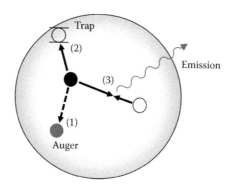

Figure 12.11 Electron-hole recombination pathways within a nanocrystal: (1) Auger recombination, wherein the energy is passed off to a third carrier, (2) recombination to a trap state, and (3) radiative recombination.

of photoluminescence in Si, no matter the synthesis method, remains disputed. As an indirect semiconductor, the transition from the bottom of the conduction band to the top of the valence band in Si does not conserve momentum and requires the assistance of phonons to occur. Radiative recombination is believed to increase in confined nanocrystals due to reduced importance of nonradiative pathways and also an increased overlap of the electron and hole wave functions in k-space. Quantum confinement theory predicts that decreasing crystal size could overlap the electron and hole wave functions enough to cause an indirect-to-direct transition of the band gap and a sharpening of absorption and emission features. However, size-dependent Si emission seems to remain phonon-mediated and absorption features remain characteristic of an indirect band gap despite the size-dependent blueshift (Kovalev et al. 1999).

Studies of porous Si demonstrated that "bands" of emission arise through distinct recombination pathways. The most common bands exhibited by plasma-synthesized Si are the so-called F-band, Y-band, and S-band, also referred to as the blue, yellow, and red bands respectively (Cullis et al. 1997). The fast F-band exhibits emission in the blue–green (~470 nm or 2.6 eV) with recombination timescales in the order of nanoseconds and is believed to arise from surface oxide trap states. The quantum yield is reportedly low, less than 20%. The yellow Y-band also has fast, nanosecond recombination rates and low quantum yields, but is demonstrated by unoxidized nanocrystals with Si–H and Si–C surface bonds (Wen et al. 2015). Neither the blue nor yellow bands exhibit emission tunability with size, only the red S-band does. Here, emission can be tuned with particle size from the near-infrared to about 2.1 eV as detailed in Figure 12.3.

Combined with its high emission efficiency, the red band has generated the most research interest. Emission has long lifetimes on the order of microseconds and is broad, as seen in Figure 12.12. Emission linewidth has been narrowed by ~30% using size selection schemes like density gradient ultracentrifugation, but still does not match the narrow linewidth of direct band gap materials like cadmium selenide (Miller et al. 2012). Near-infrared emission has been reported to have the highest quantum yields for 3–4 nm Si nanocrystals, with the quantum yield dropping significantly for larger or smaller Si crystals. While some research on red emission has also pointed to surface-related origins, its source is more frequently believed to arise from the characteristically indirect nanocrystal core (Mangolini 2013). However, the importance of the surface to this emission cannot be overstated since surface defects can drastically affect recombination rates and pathways.

Hydrogen-terminated Si produced in plasma reactors typically has weak emission which increases up to ~45% once a stable surface oxide grows (Anthony and Kortshagen 2009; Botas et al. 2014). Peak emission wavelength can often be tunable with size for oxidized particles produced in plasmas and tends to blueshift with time, since the crystal core shrinks as oxidation progresses. For Si nanocrystals passivated with fluorine in the plasma, *ex situ* oxidation grows the emission efficiency up to ~50% (Yang et al. 2014). The very highest quantum yields of up to ~70% are reported for oxide-free, ligand-terminated Si nanocrystals (Jurbergs et al. 2006). Unfortunately, the long-term stability of this surface depends on remaining oxide free (Yang et al. 2014).

Clusters, nanoparticles, and quantum dots

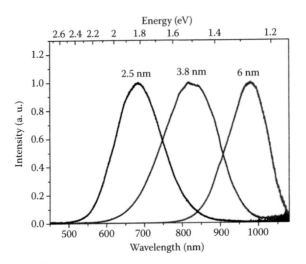

Figure 12.12 Photoluminescent spectra of plasma-synthesized and solution passivated Si particles 2.5, 3.8, and 6 nm in diameter. (Adapted from Wen, X., et al., *Sci. Rep.*, 5, 12469, 2015. With permission.)

12.3.3 CONDUCTIVITY

With the advent of nanocrystal inks, the possibility of printed electronics holds more promise than ever before for reducing manufacturing costs while improving flexibility. As a result, Si nanocrystals are of high interest as conductive thin films. The actual implementation of highly conductive Si nanocrystal films is not trivial, however, and their electron mobilities remain significantly lower than that of bulk Si films. Much work has been done to increase film conductivity via the introduction of additional free carriers, improved particle-to-particle contact, and increased film density and uniformity. Poor particle contact is perhaps a tunneling barrier easiest to understand since one can imagine similar problems also in bulk materials from cracks or insulating contaminations. Indeed, nanocrystals provide ample surface area for undesirable adsorbates and low packing fractions provide many gaps between particles. Additionally, surface ligands are often required to produce a uniform film from solution but they are usually less than ideal for electron transport. In some cases, however, adsorbates or ligands may actually temporarily improve film conductivities, including water which was studied by Rastgar and Rowe in 2013.

Transport in nanocrystal films is governed by disorder which arises in some unique ways. Nonuniformity in particle size, doping, and particle-to-particle contact leads to "hopping" conduction, the details of which depend on the amount and type of disorder. In hopping transport, conductance, G, generally decreases with decreasing temperature, T, according to

$$G \propto \exp\left(-\left(\frac{T_o}{T}\right)^{\gamma}\right)$$

where γ depends on the transport mechanism and T_o is a characteristic temperature whose form is also dependent upon the transport mechanism (Chen et al. 2014). The same system may exhibit different conduction mechanisms for different temperature regimes. For $\gamma = 1$, the mechanism is nearest to neighbor hopping (NNH). For $\gamma = 0.5$, the mechanism is Efros-Shklovskii variable range hopping (ES-VRH) and for $\gamma = 0.25$ it is Mott variable range hopping (M-VRH) in three dimensions. A key feature of hopping conduction is the emergence of an activation energy required to move carriers. For NNH, the energy barrier is low, allowing transport locally between neighboring crystals. For VRH mechanisms, the required activation energy varies from crystal to crystal and is often great enough that carriers may travel further spatially in order to minimize their activation energy, leading to a "variable range" behavior.

Undoped Si nanocrystal films have been reported to exhibit NNH conduction (Chen et al. 2014) but heavily doped films have been reported to exhibit ES-VRH (Chen et al. 2015). VRH is in part preferred for highly doped films since increased doping tends to lead to a wider distribution of free carriers per nanocrystal. If the crystals have no net charge as synthesized, once in a film, carriers will collect in crystals which started with a lower net carrier concentration in order to level out the film's Fermi level. The subsequent particle charging develops an uneven Coulomb potential and more variable activation energy, leading to VRH.

Theoretically, increasing the number of free carriers sufficiently should lead to metallic behavior even in nanocrystalline films. For a bulk semiconductor, doping can be used to transition the material from insulating to metallic behavior if the free electron concentration, n_M, reaches the Mott criterion:

$$n_M a_b^3 \geq 0.02$$

The transition is dependent on the Bohr radius, a_b. Nanocrystalline films are not expected to behave according to the Mott criterion, however, due to their unique hopping conduction mechanism. Recently, Chen et al. (2015) derived a new criterion, showing that for a ligand-free nanocrystalline film with interparticle contact facets of radius ρ, the critical free electron density, n_c, is instead determined by:

$$n_c \rho^3 \cong 0.3g$$

where g is the number of equivalent minima in the conduction band. As expected, this criterion predicts a much higher critical free electron density than the Mott criterion. The facet radius of spherical nanocrystals can be approximated as

$$\rho_a = \sqrt{da/2}$$

where a is the lattice constant and d is the crystal diameter. For a film of 8 nm diameter Si nanocrystals, it is found that $n_c \cong 5 \times 10^{20}$ cm^{-3} while $n_M = 3 \times 10^{18}$ cm^{-3}. The work by Chen et al. also showed that for plasma-synthesized and doped Si nanocrystals, the metal–insulator transition was approached (but not yet reached) for doping concentrations up to $n = 2.8 \times 10^{20}$ cm^{-3}, further proving the increased challenge in producing highly conductive Si nanocrystal films compared to the bulk counterpart.

12.3.4 LOCALIZED SURFACE PLASMON RESONANCE

Significantly doped Si nanocrystals can also exhibit a localized surface plasmon resonance (LSPR). An LSPR is a coherent, localized oscillation of free electrons induced by the electric field of incident photons having a wavelength larger than the particle size. Size, shape, composition, dielectric environment, and interparticle spacing can all impact the LSPR, but it is the free electron concentration which is most important in defining the LSPR intensity and wavelength, particularly for nonmetals (Luther et al. 2011). The free electron concentration, n, can be determined from the plasmon peak position as described by

$$\omega = \sqrt{\frac{4\pi n e^2}{m^*(\varepsilon + 2\varepsilon_m)}}$$

where ω is the LSPR resonance frequency, ε is the dielectric constant for bulk Si (11.7), ε_m is the dielectric constant for the surrounding medium (~1 for nitrogen), e is the electron constant, and m^* is the effective electron mass. Plasmonic Si nanocrystals typically exhibit resonance in the mid-infrared region, about 600 to 3000 cm^{-1} (0.07 to 0.37 eV). Rowe et al. first produced plasmonic Si nanocrystals via plasma synthesis through *in situ* substitutional phosphorous doping in 2013. Following studies soon unveiled an intriguing role of oxidation in the development of LSPRs in Si. In 2015, Zhou et al. first showed an LSPR from boron doped particles, but it did not emerge until significant oxidation had occurred. Furthermore, oxidation greatly suppresses and even eliminates LSPR behavior from phosphorus doped Si (Kramer et al. 2015; Zhou et al. 2015). The emergence of plasmon resonance in Si is not yet well understood but it does provide a useful benchmark for estimating free electron densities in extrinsically doped Si nanocrystals.

12.4 SURFACE ENGINEERING

As nanocrystals shrink in size, the fraction of atoms lying at the surface increases drastically, scaling with the inverse of the particle radius (see Figure 12.4). For Si nanoparticles 4 nm in diameter, more than 30% of atoms are likely located on the surface. Simply put, the surface matters (Roduner 2006; Mariotti et al. 2013). Much research has been dedicated to improving Si quantum dot surfaces precisely because the surface characteristics of nanocrystals dominate their properties and range of potential applications. As will be described, studies of plasma-synthesized Si nanocrystals point to a unique surface structure which can affect the performance of the various surface passivation schemes tested on Si synthesized by other means (Jariwala et al. 2011). For this reason, we exclude discussion of surface treatments which have not yet been reported with plasma-synthesized crystals.

In plasmas, Si nanocrystals can be produced directly with a variety of surface terminations and with defect densities as low as in the order of one dangling bond per 200 nanocrystals (Pereira et al. 2012). Silane is typically used to produce hydrogen-terminated Si. Halogens are easily attached to the surface during synthesis, using precursors like silicon tetrachloride and silicon tetrafluoride, or via the addition of another halogenated gas like carbon tetrafluoride or sulfur hexafluoride. Nitridation of the Si surface has also been reported through the addition of nitrogen gas (Uchida et al. 2011). *In situ* ligand attachment has been achieved by injecting alkenes or alkynes into the plasma (Mangolini and Kortshagen 2007; Anthony et al. 2012; Weeks et al. 2012). A wide range of surface modifications, from simple oxidation to silanization, can also be obtained *ex situ*, which are briefly reviewed here.

12.4.1 LIGAND FUNCTIONALIZATION

Hydrosilylation is perhaps the most studied surface passivation technique for improving luminescence efficiency and the dispersion of Si nanocrystals in nonpolar solvents. Hydrosilylation covalently attaches organic chains, ligands, or brushes to the hydrogenated Si surface (see Figure 12.13). Thermal initiation is typically employed, but photochemical activation, catalyst mediation, and other means may also be used (Buriak 2013). The process had long been thought to primarily propagate through Si–H surface bonds, but recent studies show that this functionalization method is more dependent on silyl (SiH_3) abstraction for Si synthesized in the plasma environment (Shu et al. 2015; Wheeler et al. 2015). It is more accurately termed silylsilylation, in this case.

The amount of more weakly bonded silyl species on the nanocrystal surface can be effectively increased through hydrogen injection at the end or "afterglow" region of the plasma, and when coupled with thermal silylsilylation in solution, it yields the highest quantum yields for plasma-synthesized Si (Anthony et al. 2011). As previously mentioned, these nanocrystals are not immune to degradation through oxidation as some $Si–H_x$ bonds remain, even after passivation. Energetic ultraviolet light has also been shown to cleave weak surface bonds, dropping the ensemble quantum yield to about one half of its initial value, similar to the trend which is seen for oxidation (Yang et al. 2014). Photodegradation alone has been proved reversible, however, and may be eliminated altogether through passivation schemes which include photochemical activation (Wu and Kortshagen 2015).

Hydrosilylation may even be achieved directly within the plasma. Gas-phase ligand attachment is attractive as it eliminates time-consuming, waste-generating liquid passivation techniques, and can also shorten

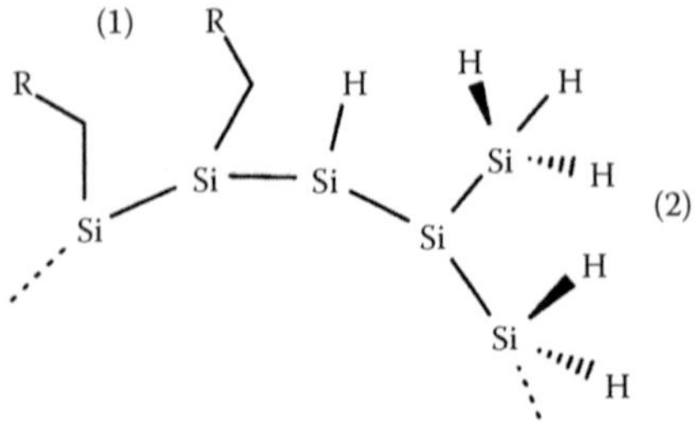

Figure 12.13 Depiction of a hydrosilylated/silylsilylated Si surface with (1) long organic ligands and (2) various silyl species.

the length of attaching ligands. Short ligands are desirable for applications where electrical conductivity is important but they can be less practical to attach in the liquid phase as their low boiling points limit thermal energy inputs. To achieve *in situ* functionalization with organic ligands, the ligands can be injected into the plasma afterglow region similarly to hydrogen injection schemes or through a secondary plasma (Mangolini and Kortshagen 2007; Anthony et al. 2012). Unfortunately, these particles reportedly exhibit weaker photoluminescence and lower stability, indicating that passivation is not as complete as is achieved through liquid phase routes. However, some work has been done to thermally attach vaporized short- and long-chain alkynes within the plasma chamber directly, showing high improvement of the ligand surface coverage. The work by Weeks et al. (2012) demonstrated hydrocarbon surface coverage of about 58%, just below the theoretical 60% limit. The improved surface functionalization and resulting reduction of Si-H$_x$ moieties slowed and reduced the total oxidation of the Si nanocrystals, improving stability.

Beyond hydrosilylation and/or silylsilylation, several other unique surface passivation schemes have been attempted. Yasar-Inceoglu et al. (2015) employed a series of two low-pressure, RF plasmas to synthesize ~12 nm Si nanocrystals and then coated them with ~4 nm shell of polyaniline. Very low input powers for the secondary shell forming plasma were used in order to prevent the formation of silicon carbide at the core/shell interface. Interestingly, they also found that the surface structure of the polyaniline could be tuned by the presence of hydrogen in the plasma.

Silanization is a technique similar to hydrosilylation that instead acts upon surface hydroxyl groups to attach siloxane groups of silane coupling agents (SCAs) with various functional groups to improve particle solubility. Anderson et al. silanized plasma-synthesized Si nanocrystals *ex situ* by etching with HNO$_3$ to produce an SiOH surface to which dodecyldimethylchlorosilane ligands were thermally attached. The resulting particles exhibited size-tunable emission with peak wavelengths as low as ~600 nm and low quantum yields, ~2%. The emission was stable upon further air exposure, however, with no further changes measured after 60 days of exposure (Anderson et al. 2012).

Recently, attention has turned to using plasmas for engineering the Si nanocrystal surface while the particles are in a liquid suspension. Laser-induced plasmas as well as the application of atmospheric pressure microdischarges have been considered (Askari et al. 2015). Early studies have indicated that such treatments might improve environmental stability of the nanocrystals as well as their dispersion in water. The chemistry in this case depends greatly upon the liquid suspension under study and research is ongoing (Mariotti et al. 2013).

12.4.2 LIGAND-FREE NANOCRYSTALS

Some ligand-free Si nanocrystals can be stabilized in certain solutions and are generally preferred for conductive film applications. In particular, chlorinated particles exhibit hypervalent interactions with hard donor (Lewis base) molecules enabling single-particle dispersions (Wheeler et al. 2013). The unbound solvating molecules are then easily removed under vacuum for device fabrication. Furthermore, the interaction of the particle with the hard donor can also produce a surface-induced doping effect with resonance in the mid-infrared. The free carrier concentration was estimated to be in the order of 10^{20} cm^{-3} or less, as evaporation of the solution molecules suppresses the plasmonic behavior. Unfortunately, chlorinated Si nanocrystals are significantly less robust against oxidation compared to hydrogen or ligand-terminated Si (Yasar-Inceoglu et al. 2012).

On the other hand, oxidation itself is a form of surface passivation which can lead to higher quantum yields compared to as-produced crystals. Anthony and Kortshagen (2009) reported quantum yields of up to ~45% after 80 days of oxidation of initially hydrogen-terminated Si nanocrystals in air. The rate of oxidation was found to increase drastically for halogenated surfaces and reach similar quantum yields (Liptak et al. 2009). However, particles with native oxides do not singly disperse in solvents, nor does the addition of an insulating silicon oxide layer generally benefit transport properties in films.

12.5 APPLICATIONS

The research done to understand, passivate, and stabilize the Si nanocrystal surface has enabled Si nanocrystals to be considered for a diverse list of applications. Most applications roughly fall into two distinct categories: (1) nanocrystal analogues to bulk devices and (2) photon management devices. Since bulk Si

Clusters, nanoparticles, and quantum dots

dominates the semiconductor industry today, interest in Si nanocrystal electronic devices is a natural extension. Bottom-up processing techniques through Si inks or direct, gas-phase deposition holds promise for tunable, more flexible devices and low-cost, roll-to-roll manufacturing.

Investigations of plasma-synthesized Si nanocrystal devices include diodes (Becker et al. 2012; Meseth et al. 2012), field-effect transistors (Zhou et al. 2009; Gresback et al. 2014; Pereira et al. 2014), and battery anode materials (Yasar-Inceoglu et al. 2015). Alternatively, Si nanocrystals may be used to enhance bulk film properties. Examples of this include mixed phase films (Cabarrocas et al. 2007; Adjallah et al. 2010; Fields et al. 2014) or the use of heavily doped nanocrystals as the dopant source for doped bulk crystalline films (Meseth et al. 2013). Plasma-synthesized nanocrystals are attractively compatible for these technologies as they may be easily integrated with the plasma-enhanced chemical vapor deposition (PECVD) techniques already used for Si thin film deposition.

Beyond these desirable processing traits, Si nanocrystals have generated interest simply due to their nanocrystal nature for thermoelectric applications. Good thermoelectric materials exhibit high electrical conductivity combined with low thermal conductivity to leverage the Seebeck effect and generate electricity from heat. Grain boundaries between crystals impede thermal conduction and can improve the Seebeck coefficient. Combined with high doping fractions, plasma-synthesized Si nanocrystals have been used in films whose thermoelectric performance exceeds that of the bulk Si counterpart (Petermann et al. 2011).

Quantum confinement is not a requirement for producing high-quality thermoelectric materials from Si nanocrystals, but it has led to high interest for Si in optical applications. The optical applications studied with plasma-synthesized Si to date include quantum dot-based light-emitting diodes (QLEDs) and luminescent downshifters. Si's tunable, efficient, visible emission has been harnessed to produce a QLED with an external quantum efficiency as high as 8.6% out of plasma-synthesized Si quantum dots, the highest reported efficiency for QLEDs at the time (Cheng et al. 2011). Furthermore, the fabrication of a QLED was demonstrated completely in the gas phase (Anthony et al. 2012). These QLEDs emitted in the red to near-infrared as luminescent efficiencies are reduced on the blue side of the visible spectrum. While this limits use in visible lighting applications, red luminescence is actually ideal for the application of luminescent downshifters. In this application, the crystals are deposited on top of a conventional solar cell. There, they absorb high-energy photons and shift them to lower energies which are better absorbed by the underlying cell, producing a small increase in power conversion efficiency, ~0.4% (Yuan et al. 2011).

12.6 CONCLUSION AND OUTLOOK

The synthesis of Si nanocrystals in plasmas has come a long way since its humble beginnings as an unwanted problem for the microelectronics industry. Researchers in this field have uncovered many unique advantages for Si produced in plasmas including good size control, low defect densities, and facile doping, among others. Si is the most studied nanocrystalline material synthesized in plasmas, yet opportunities for continued progress abound (Kortshagen et al. 2016). Research is expected to proceed in a diverse range of areas, such as particle morphology control, heterostructure fabrication, and surface engineering. Further developing plasma synthesis techniques in these areas may enable Si quantum dots to displace traditional quantum dot materials in future applications. One drawback of well-studied, solution processed quantum dots like CdSe, PbS, or CuInS, is the toxicity and/or scarcity inherent to their elemental components. Synthesized from an extremely abundant and nontoxic element, Si nanocrystals are an attractive solution.

REFERENCES

Adjallah, Y., C. Anderson, U. Kortshagen, and J. Kakalios. 2010. Structural and Electronic Properties of Dual Plasma Codeposited Mixed-Phase Amorphous/Nanocrystalline Thin Films. *Journal of Applied Physics* 107 (4): 043704. doi:10.1063/1.3285416.

Alam, M. K., and R. C. Flagan. 1986. Controlled Nucleation Aerosol Reactors: Production of Bulk Silicon. *Aerosol Science and Technology* 5 (2): 237–48. doi:10.1080/02786828608959090.

Anderson, I. E., R. A. Shircliff, C. MacAuley, D. K. Smith, B. G. Lee, S. Agarwal, P. Stradins, and R. T. Collins. 2012. Silanization of Low-Temperature-Plasma Synthesized Silicon Quantum Dots for Production of a Tunable, Stable, Colloidal Solution. *The Journal of Physical Chemistry C* 116 (6): 3979–87. doi:10.1021/jp211569a.

Anthony, R., and U. Kortshagen. 2009. Photoluminescence Quantum Yields of Amorphous and Crystalline Silicon Nanoparticles. *Physical Review B* 80 (11): 1–6. doi:10.1103/PhysRevB.80.115407.

Anthony, R. J., K. Cheng, Z. C. Holman, R. J. Holmes, and U. R. Kortshagen. 2012. An All-Gas-Phase Approach for the Fabrication of Silicon Quantum Dot Light-Emitting Devices. *Nano Letters* 12: 2822–5. doi:/10.1021/nl300164z.

Anthony, R. J., D. J. Rowe, M. Stein, J. Yang, and U. Kortshagen. 2011. Routes to Achieving High Quantum Yield Luminescence from Gas-Phase-Produced Silicon Nanocrystals. *Advanced Functional Materials* 21 (21): 4042–6. doi:10.1002/adfm.201100784.

Askari, S., I. Levchenko, K. Ostrikov, P. Maguire, and D. Mariotti. 2014. Crystalline Si Nanoparticles below Crystallization Threshold: Effects of Collisional Heating in Non-Thermal Atmospheric-Pressure Microplasmas. *Applied Physics Letters* 104 (16): 163103. doi:10.1063/1.4872254.

Askari, S., M. Macias-Montero, T. Velusamy, P. Maguire, V. Svrcek, and D. Mariotti. 2015. Silicon-Based Quantum Dots: Synthesis, Surface and Composition Tuning with Atmospheric Pressure Plasmas. *Journal of Physics D: Applied Physics* 48 (31): 314002. doi:10.1088/0022-3727/48/31/314002.

Atkins, T. M., A. Y. Louie, and S. M. Kauzlarich. 2012. An Efficient Microwave-Assisted Synthesis Method for the Production of Water Soluble Amine-Terminated Si Nanoparticles. *Nanotechnology* 23 (29): 294006. doi:10.1088/0957-4484/23/29/294006.

Bapat, A., M. Gatti, Y.-P. Ding, S. A. Campbell, and U. Kortshagen. 2007. A Plasma Process for the Synthesis of Cubic-Shaped Silicon Nanocrystals for Nanoelectronic Devices. *Journal of Physics D: Applied Physics* 40 (8): 2247–57. doi:10.1088/0022-3727/40/8/S03.

Becker, A., G. Schierning, R. Theissmann, M. Meseth, N. Benson, R. Schmechel, D. Schwesig, N. Petermann, H. Wiggers, and P. Ziolkowski. 2012. A Sintered Nanoparticle P-N Junction Observed by a Seebeck Microscan. *Journal of Applied Physics* 111 (5): 054320. doi:10.1063/1.3693609.

Bley, R. A., and S. M. Kauzlarich. 1996. A Low-Temperature Solution Phase Route for the Synthesis of Silicon Nanoclusters. *Journal of the American Chemical Society* 118 (49): 12461–2. doi:10.1021/ja962787s.

Botas, A. M. P., R. A. S. Ferreira, R. N. Pereira, R. J. Anthony, T. Moura, D. J. Rowe, and U. R. Kortshagen. 2014. High Quantum Yield Dual-Emission from Gas Phase Grown Crystalline Si Nanoparticles. *The Journal of Physical Chemistry C* 118 (19): 10375–83. doi:10.1021/jp5000683.

Boufendi, L., and A. Bouchoule. 1994. Particle Nucleation and Growth in a Low-Pressure Argon-Silane Discharge. *Plasma Sources Science and Technology* 3 (3): 262–7. doi:10.1088/0963-0252/3/3/004.

Brus, L. 1986. Electronic Wave Functions in Semiconductor Clusters: Experiment and Theory. *The Journal of Physical Chemistry* 90 (12): 2555–60. doi:10.1021/j100403a003.

Buriak, J. M. 2002. Organometallic Chemistry on Silicon and Germanium Surfaces. *Chemical Reviews* 102 (5): 1271–308. doi:10.1021/cr000064s.

Buriak, J. M. 2014. Illuminating Silicon Surface Hydrosilylation: An Unexpected Plurality of Mechanisms. *Chemistry of Materials*. 26: 762–72. doi:10.1021/cm402120f.

Cabarrocas, P. R. I., T. Nguyen-Tran, Y. Djeridane, A. Abramov, E. Johnson, and G. Patriarche. 2007. Synthesis of Silicon Nanocrystals in Silane Plasmas for Nanoelectronics and Large Area Electronic Devices. *Journal of Physics D: Applied Physics* 40: 2258–66. doi:10.1088/0022-3727/40/8/S04.

Canham, L. T. 1990. Silicon Quantum Wire Array Fabrication by Electrochemical and Chemical Dissolution of Wafers. *Applied Physics Letters* 57: 1046. doi:10.1063/1.103561.

Chen, T., K. V. Reich, N. J. Kramer, H. Fu, U. R. Kortshagen, and B. I. Shklovskii. 2015. Metal-Insulator Transition in Films of Doped Semiconductor Nanocrystals. *Nature Materials* 15: 1–6. doi:10.1038/nmat4486.

Chen, T., B. Skinner, W. Xie, B. I. Shklovskii, and U. R. Kortshagen. 2014. Carrier Transport in Films of Alkyl Ligand-Terminated Silicon Nanocrystals. *The Journal of Physical Chemistry C* 118: 19580–8. doi:10.1021/jp5051723.

Cheng, K.-Y., R. Anthony, U. R. Kortshagen, and R. J. Holmes. 2011. High-Efficiency Silicon Nanocrystal Light-Emitting Devices. *Nano Letters* 11 (5): 1952–6. doi:10.1021/nl2001692.

Cullis, A.G., L. T. Canham, and P. D. J. Calcott. 1997. The Structural and Luminescence Properties of Porous Silicon. *Journal of Applied Physics* 82 (3): 909–65. doi:10.1063/1.366536.

Efros, A. L., and M. Rosen. 2000. The Electronic Structure of Semiconductor Nanocrystals. *Annual Review of Materials Science* 30 (1): 475–521. doi:10.1146/annurev.matsci.30.1.475.

Fields, J. D., S. McMurray, L. R. Wienkes, J. Trask, C. Anderson, P. L. Miller, B. J. Simonds, et al. 2014. Quantum Confinement in Mixed Phase Silicon Thin Films Grown by Co-Deposition Plasma Processing. *Solar Energy Materials and Solar Cells* 129: 7–12. doi:10.1016/j.solmat.2013.10.028.

Gorla, C. R., S. Liang, G. S. Tompa, W. E. Mayo, and Y. Lu. 1997. Silicon and Germanium Nanoparticle Formation in an Inductively Coupled Plasma Reactor. *Journal of Vacuum Science & Technology A: Vacuum, Surfaces, and Films* 15 (3): 860. doi:10.1116/1.580721.

Gresback, R., N. J. Kramer, Y. Ding, T. Chen, U. R. Kortshagen, and T. Nozaki. 2014. Controlled Doping of Silicon Nanocrystals Investigated by Solution-Processed Field Effect Transistors. *ACS Nano* 8 (6): 5650–6. doi:10.1021/nn500182b.

Gresback, R., T. Nozaki, and K. Okazaki. 2011. Synthesis and Oxidation of Luminescent Silicon Nanocrystals from Silicon Tetrachloride by Very High Frequency Nonthermal Plasma. *Nanotechnology* 22 (30): 305605. doi:10.1088/0957-4484/22/30/305605.

Hessel, C. M., et al. 2012. Synthesis of Ligand-Stabilized Silicon Nanocrystals with Size-Dependent Photoluminescence Spanning Visible to Near-Infrared Wavelengths. *Chemistry of Materials* 24 (2): 393–401. doi:10.1021/cm2032866.

Holman, Z. C., and U. R. Kortshagen. 2010. A Flexible Method for Depositing Dense Nanocrystal Thin Films: Impaction of Germanium Nanocrystals. *Nanotechnology* 21 (33): 335302. doi:10.1088/0957-4484/21/33/335302.

Hülser, T., S. M. Schnurre, H. Wiggers, and C. Schulz. 2011. Gas-Phase Synthesis of Nanoscale Silicon as an Economical Route towards Sustainable Energy Technology. *KONA Powder and Particle Journal* 29 (29): 191–207. doi:10.14356/kona.2011021.

Hybertsen, M. S. 1994. Absorption and Emission of Light in Nanoscale Silicon Structures. *Physical Review Letters* 72 (10): 1514–17. doi:10.1103/PhysRevLett.72.1514.

Jariwala, B. N., N. J. Kramer, M. C. Petcu, D. C. Bobela, M. C. M. Van De Sanden, P. Stradins, C. V. Ciobanu, and S. Agarwal. 2011. Surface Hydride Composition of Plasma-Synthesized Si Nanoparticles. *The Journal of Physical Chemistry C* 115 (42): 20375–9. doi:10.1021/jp2028005.

Jurbergs, D., E. Rogojina, L. Mangolini, and U. Kortshagen. 2006. Silicon Nanocrystals with Ensemble Quantum Yields Exceeding 60%. *Applied Physics Letters* 88 (23): 233116. doi:10.1063/1.2210788.

Knipping, J., H. Wiggers, B. Rellinghaus, P. Roth, D. Konjhodzic, and C. Meier. 2004. Synthesis of High Purity Silicon Nanoparticles in a Low Pressure Microwave Reactor. *Journal of Nanoscience and Nanotechnology* 4 (8): 1039–44. doi:10.1166/jnn.2004.149.

Kortshagen, U., R. Anthony, R. Gresback, Z. Holman, R. Ligman, C.-Y. Liu, L. Mangolini, and S. A. Campbell. 2008. Plasma Synthesis of Group IV Quantum Dots for Luminescence and Photovoltaic Applications. *Pure and Applied Chemistry* 80 (9): 1901–8. doi:10.1351/pac200880091901.

Kortshagen, U., and U. Bhandarkar. 1999. Modeling of Particulate Coagulation in Low Pressure Plasmas. *Physical Review E, Statistical Physics, Plasmas, Fluids, and Related Interdisciplinary Topics* 60 (1): 887–98. doi:10.1103/PhysRevE.60.887.

Kortshagen, U. R., R. M. Sankaran, R. N. Pereira, S. L. Girshick, J. J. Wu, and E. S. Aydil. 2016. Nonthermal Plasma Synthesis of Nanocrystals: Fundamental Principles, Materials, and Applications. *Chemical Reviews* 116: 11061–127. doi:10.1021/acs.chemrev.6b00039.

Kovalev, D., H. Heckler, G. Polisski, and F. Koch. 1999. Optical Properties of Si Nanocrystals. *Physica Status Solidi (B)* 215 (2): 871–932. doi:10.1002/(SICI)1521-3951(199910)215:2%3C871::AID-PSSB871%3E3.0.CO;2-9.

Kramer, N. J., R. J. Anthony, M. Mamunuru, E. S. Aydil, and U. R. Kortshagen. 2014. Plasma-Induced Crystallization of Silicon Nanoparticles. *Journal of Physics D: Applied Physics* 47 (7): 075202. doi:10.1088/0022-3727/47/7/075202.

Kramer, N. J., K. S. Schramke, and U. R. Kortshagen. 2015. Plasmonic Properties of Silicon Nanocrystals Doped with Boron and Phosphorus. *Nano Letters* 15 (8): 5597–603. doi:10.1021/acs.nanolett.5b02287.

Lee, K. W., J. Chen, and J. A. Gieseke. 1984. Log-Normally Preserving Size Distribution for Brownian Coagulation in the Free-Molecule Regime. *Aerosol Science and Technology* 3 (1): 53–62. doi:10.1080/02786828408958993.

Lieberman, M. A., and A. J. Lichtenberg. 1994. *Discharges and Materials Processing Principles of Plasma Discharges and Materials.* New York: Wiley.

Liptak, R. W., U. Kortshagen, and S. A. Campbell. 2009. Surface Chemistry Dependence of Native Oxidation Formation on Silicon Nanocrystals. *Journal of Applied Physics* 106 (6): 064313. doi:10.1063/1.3225570.

Littau, K. A., P. J. Szajowski, A. J. Muller, A. R. Kortan, and L. E. Brus. 1993. A Luminescent Silicon Nanocrystal Colloid via a High-Temperature Aerosol Reaction. *The Journal of Physical Chemistry* 97 (6): 1224–30. doi:10.1017/CBO9781107415324.004.

Luther, J. M., P. K. Jain, T. Ewers, and A. P. Alivisatos. 2011. Localized Surface Plasmon Resonances Arising from Free Carriers in Doped Quantum Dots. *Nature Materials* 10 (5): 361–6. doi:10.1038/nmat3004.

Mangolini, L. 2013. Synthesis, Properties, and Applications of Silicon Nanocrystals. *Journal of Vacuum Science & Technology B: Microelectronics and Nanometer Structures* 31 (2): 20801. doi:10.1116/1.4794789.

Mangolini, L., and U. Kortshagen. 2007. Plasma-Assisted Synthesis of Silicon Nanocrystal Inks. *Advanced Materials* 19 (18): 2513–19. doi:10.1002/adma.200700595.

Mangolini, L., E. Thimsen, and U. Kortshagen. 2005. High-Yield Plasma Synthesis of Luminescent Silicon Nanocrystals. *Nano Letters* 5 (4): 655–9. doi:10.1021/nl050066y.

Mariotti, D., S. Mitra, and V. Švrček. 2013. Surface-Engineered Silicon Nanocrystals. *Nanoscale* 5 (4): 1385. doi:10.1039/c2nr33170e.

Mariotti, D., and R. M. Sankaran. 2010. Microplasmas for Nanomaterials Synthesis. *Journal of Physics D: Applied Physics* 43 (32): 323001. doi:10.1088/0022-3727/43/32/323001.

Meseth, M., B. C. Kunert, L. Bitzer, F. Kunze, S. Meyer, F. Kiefer, M. Dehnen, et al. 2013. Excimer Laser Doping Using Highly Doped Silicon Nanoparticles. *Physica Status Solidi A* 210 (11): 2456–62. doi:10.1002/pssa.201329012.

Meseth, M., P. Ziolkowski, G. Schierning, R. Theissmann, N. Petermann, H. Wiggers, N. Benson, and R. Schmechel. 2012. The Realization of a Pn-Diode Using Only Silicon Nanoparticles. *Scripta Materialia* 67 (3): 265–8. doi:10.1016/j.scriptamat.2012.04.039.

Miller, J. B., A. R. Van Sickle, R. J. Anthony, D. M. Kroll, U. R. Kortshagen, and E. K. Hobbie. 2012. Ensemble Brightening and Enhanced Quantum Yield in Size-Purified Silicon Nanocrystals. *ACS Nano* 6 (8): 7389–96. doi:10.1021/nn302524k.

Norris, D. J., A. L. Efros, and S. C. Erwin. 2008. Doped Nanocrystals. *Science* 319 (5871): 1776–9. doi:10.1126/science.1143802.

Nozaki, T., K. Sasaki, T. Ogino, D. Asahi, and K. Okazaki. 2007. Microplasma Synthesis of Tunable Photoluminescent Silicon Nanocrystals. *Nanotechnology* 18: 235603. doi:10.1088/0957-4484/18/23/235603.

Oda, S. 1997. Preparation of Nanocrystalline Silicon Quantum Dot Structure by a Digital Plasma Process. *Advances in Colloid and Interface Science* 71–72: 31–47. doi:10.1016/S0001-8686(97)90008-7.

Pereira, R. N., J. Coutinho, S. Niesar, T. A. Oliveira, W. Aigner, H. Wiggers, M. J. Rayson, P. R. Briddon, M. S. Brandt, and M. Stutzmann. 2014. Resonant Electronic Coupling Enabled by Small Molecules in Nanocrystal Solids. *Nano Letters* 14 (7): 3817–26. doi:10.1021/nl500932q.

Pereira, R. N., D. J. Rowe, R. J. Anthony, and U. Kortshagen. 2012. Freestanding Silicon Nanocrystals with Extremely Low Defect Content. *Physical Review B* 86 (8): 85449. doi:10.1103/PhysRevB.86.085449.

Petermann, N., N. Stein, G. Schierning, R. Theissmann, B. Stoib, M. S. Brandt, C. Hecht, C. Schulz, and H. Wiggers. 2011. Plasma Synthesis of Nanostructures for Improved Thermoelectric Properties. *Journal of Physics D: Applied Physics* 44 (17): 174034. doi:10.1088/0022-3727/44/17/174034.

Petkov, V., C. M. Hessel, J. Ovtchinnikoff, A. Guillaussier, B. A. Korgel, X. Liu, and C. Giordano. 2013. Structure-Properties Correlation in Si Nanoparticles by Total Scattering and Computer Simulations. *Chemistry of Materials* 25 (11): 2365–71. doi:10.1021/cm401099q.

Pi, X.D., R. Gresback, R. W. Liptak, S. A. Campbell, and U. Kortshagen. 2008a. Doping Efficiency, Dopant Location, and Oxidation of Si Nanocrystals. *Applied Physics Letters* 92 (12): 2–5. doi:10.1063/1.2897291.

Pi, X. D., R. W. Liptak, J. Deneen Nowak, N. P. Wells, C. B. Carter, S. A. Campbell, and U. Kortshagen. 2008b. Air-Stable Full-Visible-Spectrum Emission from Silicon Nanocrystals Synthesized by an All-Gas-Phase Plasma Approach. *Nanotechnology* 19: 245603. doi:10.1088/0957-4484/19/24/245603.

Rao, N., S. Girshick, J. Heberlein, P. McMurry, S. Jones, D. Hansen, and B. Micheel. 1995. Nanoparticle Formation Using a Plasma Expansion Process. *Plasma Chemistry and Plasma Processing* 15 (4): 581–606. doi:10.1007/BF01447062.

Rastgar, N., and D. J. Rowe. 2013. Effects of Water Adsorption and Surface Oxidation on the Electrical Conductivity of Silicon Nanocrystal Films. *The Journal of Physical Chemistry C* 117: 4211–18. doi:10.1021/jp308279m.

Ravi, L., and S. L. Girshick. 2009. Coagulation of Nanoparticles in a Plasma. *Physical Review E—Statistical, Nonlinear, and Soft Matter Physics* 79 (2): 1–9. doi:10.1103/PhysRevE.79.026408.

Roduner, E. 2006. Size Matters: Why Nanomaterials Are Different. *Chemical Society Reviews* 35 (7): 583–92. doi:10.1039/B502142c.

Rossetti, R., S. Nakahara, and L. E. Brus. 1983. Quantum Size Effects in the Redox Potentials, Resonance Raman Spectra, and Electronic Spectra of CdS Crystallites in Aqueous Solution. *Journal of Chemical Physics* 79 (1983): 1086–8. doi:10.1063/1.445834.

Rowe, D. J., J. S. Jeong, K. A. Mkhoyan, and U. R. Kortshagen. 2013. Phosphorus-Doped Silicon Nanocrystals Exhibiting Mid-Infrared Localized Surface Plasmon Resonance. *Nano Letters* 13: 1317–22. doi:10.1021/nl4001184.

Sagar, D. M., J. M. Atkin, P. K. B. Palomaki, N. R. Neale, J. L. Blackburn, J. C. Johnson, A. J. Nozik, M. B. Raschke, and M. C. Beard. 2015. Quantum Confined Electron-Phonon Interaction in Silicon Nanocrystals. *Nano Letters* 15 (3): 1511–16. doi:10.1021/nl503671n.

Sankaran, R. M., D. Holunga, R. C. Flagan, and K. P. Giapis. 2005. Synthesis of Blue Luminescent Si Nanoparticles Using Atmospheric-Pressure Microdischarges. *Nano Letters* 5 (3): 537–41.

Shigeta, M., and A. B. Murphy. 2011. Thermal Plasmas for Nanofabrication. *Journal of Physics D: Applied Physics* 44 (17): 174025. doi:10.1088/0022-3727/44/17/174025.

Shu, Y., U. Kortshagen, B. G. Levine, and R. J. Anthony. 2015. Surface Structure and Silicon Nanocrystal Photoluminescence: The Role of Hypervalent Silyl Groups. *The Journal of Physical Chemistry C* 119 (47): 26683–91. doi:10.1021/acs.jpcc.5b08578.

Takagi, H., H. Ogawa, Y. Yamazaki, A. Ishizaki, and T. Nakagiri. 1990. Quantum Size Effects on Photoluminescence in Ultrafine Si Particles. *Applied Physics Letters* 56 (24): 2379–80. doi:10.1063/1.102921.

Uchida, G., K. Yamamoto, Y. Kawashima, M. Sato, K. Nakahara, K. Kamataki, N. Itagaki, K. Koga, M. Kondo, and M. Shiratani. 2011. Surface Nitridation of Silicon Nano-Particles Using Double Multi-Hollow Discharge Plasma CVD. *Physica Status Solidi C* 8 (10): 3017–20. doi:10.1002/pssc.201001230.

Viezbicke, B. D., S. Patel, B. E. Davis, and D. P. Birnie. 2015. Evaluation of the Tauc Method for Optical Absorption Edge Determination: ZnO Thin Films as a Model System. *Physica Status Solidi B* 11 (8): 1700–1710. doi:10.1002/pssb.201552007.

Watanabe, Y. L. B. 2006. Formation and Behaviour of Nano/Micro-Particles in Low Pressure Plasmas. *Journal of Physics D: Applied Physics* 39 (19): R329. doi:10.1088/0022-3727/39/19/R01.

Watanabe, Y., M. Shiratani, H. Kawasaki, S. Singh, T. Fukuzawa, Y. Ueda, and H. Ohkura. 1996. Growth Processes of Particles in High Frequency Silane Plasmas. *Dusty Plasmas—'95 Workshop on Generation, Transport, and Removal of Particles in Plasmas* 14 (2): 540. doi:10.1116/1.580141.

Weeks, S. L., B. Macco, M. C. M. Van De Sanden, and S. Agarwal. 2012. Gas-Phase Hydrosilylation of Plasma-Synthesized Silicon Nanocrystals with Short- and Long-Chain Alkynes. *Langmuir* 28 (50): 17295–301. doi:10.1021/la3030952.

Wen, X., P. Zhang, T. A. Smith, R. J. Anthony, U. R. Kortshagen, P. Yu, Y. Feng, S. Shrestha, G. Coniber, and S. Huang. 2015. Tunability Limit of Photoluminescence in Colloidal Silicon Nanocrystals. *Scientific Reports* 5: 12469. doi:10.1038/srep12469.

Wheeler, L. M., N. C. Anderson, P. K. B. Palomaki, J. L. Blackburn, J. C. Johnson, and N. R. Neale. 2015. Silyl Radical Abstraction in the Functionalization of Plasma-Synthesized Silicon Nanocrystals. *Chemistry of Materials* 27 (19): 6869–78. doi:10.1021/acs.chemmater.5b03309.

Wheeler, L. M., N. R. Neale, T. Chen, and U. R. Kortshagen. 2013. Hypervalent Surface Interactions for Colloidal Stability and Doping of Silicon Nanocrystals. *Nature Communications* 4: 2197. doi:10.1038/ncomms3197.

Wu, J. J., and U. Kortshagen. 2015. Photostability of Thermally-Hydrosilylated Silicon Quantum Dots. *RSC Advances* 5 (126): 103822–8. doi:10.1039/C5RA22827A.

Yang, J., R. Liptak, D. Rowe, J. Wu, J. Casey, D. Witker, S. A. Campbell, and U. Kortshagen. 2014. UV and Air Stability of High-Efficiency Photoluminescent Silicon Nanocrystals. *Applied Surface Science* 323: 54–8. doi:10.1016/j.apsusc.2014.08.027.

Yasar-Inceoglu, O., T. Lopez, E. Farshihagro, and L. Mangolini. 2012. Silicon Nanocrystal Production through Non-Thermal Plasma Synthesis: A Comparative Study between Silicon Tetrachloride and Silane Precursors. *Nanotechnology* 23 (25): 255604. doi:10.1088/0957-4484/23/25/255604.

Yasar-Inceoglu, O., L. Zhong, and L. Mangolini. 2015. Core/shell Silicon/Polyaniline Particles via in-Flight Plasma-Induced Polymerization. *Journal of Physics D: Applied Physics* 48 (31): 314009. doi:10.1088/0022-3727/48/31/314009.

Yuan, Z., G. Pucker, A. Marconi, F. Sgrignuoli, A. Anopchenko, Y. Jestin, L. Ferrario, P. Bellutti, and L. Pavesi. 2011. Silicon Nanocrystals as a Photoluminescence Down Shifter for Solar Cells. *Solar Energy Materials and Solar Cells* 95 (4): 1224–7. doi:10.1016/j.solmat.2010.10.035.

Zhou, S., X. Pi, Z. Ni, Y. Ding, Y. Jiang, C. Jin, and C. Delerue. 2015. Comparative Study on the Localized Surface Plasmon Resonance Silicon Nanocrystals. *ACS Nano* 9 (1): 378–86. doi:10.1021/nn505416r.

Zhou, X., K. Uchida, H. Mizuta, and S. Oda. 2009. Electron Transport in Surface Oxidized Si Nanocrystal Ensembles with Thin Film Transistor Structure. *Journal of Applied Physics* 106 (4): 044511. doi:10.1063/1.3204669.

13 Silicon nanocrystals in water

Nastaran Mansour and Ashkan Momeni Bidzard

Contents

13.1 INTRODUCTION

Silicon nanomaterials are being increasingly used for optoelectronic devices as well as biomedical applications due to excellent biocompatibility, unique electronic, optical, and mechanical properties, and compatibility with conventional silicon technology (Cheng et al. 2011; Tamarov et al. 2014; Dutta et al. 2015; Song et al. 2015; Bae et al. 2016). In recent years, observation of size-tunable visible to near-infrared (NIR) emissions from quantum confined silicon nanocrystals (SiNCs) has stimulated worldwide interest in obtaining efficient nanocrystal Si-based light-emitting devices (LEDs) for low-cost applications (Cheng et al. 2010; Maier-Flaig et al. 2013a; Xin et al. 2015). Moreover, surface modification of the biocompatible SiNCs with functional ligands can enhance their optical properties for potential application in optoelectronics and biological fields (Anderson et al. 2012; Intartaglia et al. 2012b; Ding et al. 2014; Premnath et al. 2015). More importantly, water-dispersible SiNCs have emerged as a novel kind of promising material for biological imaging and diagnostic applications due to strong fluorescence in the visible to NIR region, photostability, low toxicity, and available versatile preparation techniques (Erogbogbo et al. 2008; Zhong et al. 2013, 2015; Gongalsky et al. 2016). Thus, it is of great scientific effort to study the optical, thermal, and electronic properties of the SiNCs for a variety of device applications.

In this chapter, we present a review of the recent research achievements in synthesis and device fabrications based on the colloidal SiNCs in water. The most commonly used methods for fabrication of the SiNCs including pulsed laser ablation (Mansour et al. 2013; Rodio et al. 2016), electrical discharge (Mardanian et al. 2012), and plasma-assisted synthesis approaches (Mariotti et al. 2012; Wu et al. 2016) result in production of the nanocrystals with oxide and hydroxyl surface modifications. Therefore, this chapter mainly focuses on the optical, thermal, and electronic properties of SiNCs considering size and oxide-related surface effects which are important for the nanocrystal device applications. It is hoped that the presented material will stimulate a general interest in further developing silicon materials-based nanotechnology.

13.2 SYNTHESIS METHODS, FUNCTIONALIZATION, AND CHARACTERIZATION OF COLLOIDAL SILICON NANOCRYSTALS IN WATER

The conventional physical/chemical methods commonly used for fabrication of colloidal SiNCs are pulsed laser ablation (Mansour et al. 2013; Momeni and Mahdieh 2015; Rodio et al. 2016), chemical synthesis preparations (He et al. 2009b; Dohnalová et al. 2014) and electrochemical etching (Hwang et al. 2015), electrical discharge (Liu et al. 2005; Mardanian et al. 2012, 2014), and plasma and microwave-assisted synthesis approaches (Atkins et al. 2011; Mariotti et al. 2012; Mitra et al. 2014; Wu et al. 2016).

13.2.1 PULSED LASER ABLATION IN LIQUID

Pulsed laser ablation synthesis in liquids is a simple and green technique for preparing the ultrasmall SiNCs with average size <10 nm. A typical experimental setup for the laser ablation is shown in Figure 13.1; a high-intense laser beam is irradiated on a silicon target immersed in a liquid medium. This technique allows control of the SiNC size distribution as well as their surface chemistry via changing materials and laser parameters (Yang et al. 2009; Intartaglia et al. 2011; Amendola and Meneghetti 2013; Bagga et al. 2013; Popovic et al. 2014; Mahdieh and Momeni 2015). Laser pulse duration has an especially strong influence on the SiNC size distribution and surface characteristics engineering which plays an important role in applied device fabrication. Tables 13.1 and 13.2 summarize characteristics of sizes and surface properties

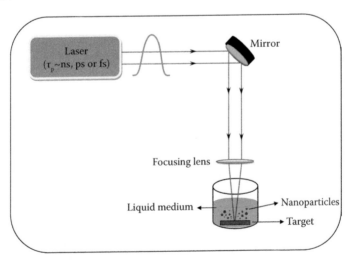

Figure 13.1 Experimental setup of the laser ablation process in liquids.

Clusters, nanoparticles, and quantum dots

Table 13.1 **Pulsed laser ablation of silicon target in water**

TARGET	LASER TYPE, WAVELENGTH, PULSE WIDTH, REPETITION RATE, FLUENCE (OR ENERGY PER PULSE), ABLATION TIME	PRODUCT (SIZE AND NOTE)	REFERENCES
Si p-type	Nd:YAG, 1064 nm, 10 ns, 10 Hz, 1.2–3.5 J/cm², 25 min	Single-pulse and double-pulse laser ablation processes result in production of silicon nanoparticles with different average sizes of 15–26 nm.	Momeni and Mahdieh 2015
Si	Ti:sapphire, 800 nm, 15 fs, 1 KHz, 150 μJ/pulse, 30 min	Silicon nanoparticles with mean size of 37 nm.	Hamad et al. 2015
Si	Nd:YAG, 1064 and 355 nm, 60 ps, 20 Hz, 0.4–4.45 J/cm², 30 min	In case of laser ablation at 1064 nm, silicon nanoparticles with a mean diameter of 40 nm have been produced. While ablation at 355 nm has resulted in smaller nanoparticles with a mean diameter of 3 nm due to *in situ* ablation/photo-fragmentation physical process.	Intartaglia et al. 2014
Si	Nd:YAG, 1064 nm, 150 ps, 10 Hz, 4.4 and 15.7 J/cm², 15 min	Application of the continuous laser changes the NP size distribution, reducing the maximum distribution by about 13 nm, from 69.4 ± 1.2 nm to 56.1 ± 0.4 nm.	Popovic et al. 2014
Silicon–tin (SiSn) thin film	Excimer laser KrF, 245 nm, 10 ns, 20 Hz, 23.5 J/cm², 10 min	Semiconducting alloyed SiSn nanocrystals with diameter ranging from 2 nm to 70 nm and a mean value of about 17 nm.	Švrček et al. 2013
Si	CO_2, 10.6 μm, 100 ns, 1.5 and 3 Hz 1.5–7.5 J/cm², 30 and 60 min	Colloidal silicon nanoparticles with particle diameters of 50–320 nm	Popovic et al. 2013
Sidiamond-type	Nd:YAG, 532 nm, 10 ns, 2.5 Hz, 1000 J/cm², 120 min	SiNC with size distribution of 7–14 nm	Liu et al. 2013a
Si	Ti:sapphire, 800 nm, 200 fs, 1 KHz, 1.6 mJ/pulse, 5 min	Colloidal silicon nanocrystals with spherical shape and the particle diameters of 5–100 nm.	Alkis et al. 2012a
Si p-type	Nd:YAG, 1064 nm, 34 ps, 10 Hz, 1 mJ/pulse, 120 min	Colloidal silicon nanocrystals with spherical shape and the particle diameters of 18–23 nm.	Eroshova et al. 2012
Si	Ti:sapphire, 800 nm, 110 fs, 1 KHz, 0.15–0.4 mJ/pulse, 15–90 min	SiNCs with the size distributions of 10–120 and 1–8 nm were prepared at 0.4 and 0.15 mJ/pulse, respectively	Intartaglia et al. 2012a

(Continued)

Clusters, nanoparticles, and quantum dots

Table 13.1 *(Continued)* **Pulsed laser ablation of silicon target in water**

TARGET	LASER TYPE, WAVELENGTH, PULSE WIDTH, REPETITION RATE, FLUENCE (OR ENERGY PER PULSE), ABLATION TIME	PRODUCT (SIZE AND NOTE)	REFERENCES
Si p-type	Nd:YLF, 527 nm, 100 ns, 1 KHz, 16 mJ/pulse, 5 min	Colloidal SiNC with size distribution of 5–100 nm were prepared.	Alkis et al. 2012b
Si	Nd:YAG, 532 nm, 6 ns, 10 Hz, 136 J/cm², 210 min	SiNC with size distribution of 1.75–18.09 nm—peak at 3 nm.	Alima et al. 2012
Si	Ti:sapphire, 800 nm, 100 fs, 1 KHz, 0.05–0.5 mJ/pulse, 60 min	The mean size is found to vary from 60 to 2.5 nm by decreasing the pulse energy value.	Intartaglia et al. 2011
Si	Ti:sapphire, 387.5 nm, 180 fs, 1 KHz, 350 μJ/pulse, 70 min	Silicon nanoparticles with diameters of 5–200 nm were formed in the colloidal solution.	Semaltianos et al. 2010
Si n-type & p-type	Excimer laser KrF, 245 nm, 20 ns, 20 Hz, 20 mJ/cm², 120 min	SiNCs with size <3 nm.	Švrcčk and Kondo 2010
Si p-type	Excimer laser KrF, 245 nm, 20 ns, 10 Hz, 23.5 mJ/cm², 120 min	SiNCs with size <2 nm.	Švrček et al. 2009a
Si	Nd:YAG, 1064 nm, 10 ns, 10 Hz, 50–200 mJ/pulse, 30 min	SiNCs with size distribution of 14 to 21 nm were produced.	Yang et al. 2009
Si	Nd:YAG, 355 nm, 8 ns, 30 Hz, <7 mJ/pulse, 30–120 min	SiNCs with size distribution of 1 to 6 nm were synthesized.	Švrček et al. 2009b
Si	KrF & Nd:YAG, 245&355 nm, 10&8 ns, 20&30 Hz, <6 mJ/pulse, 120 min	SiNCs with size < 2 nm.	Švrček and Kondo 2009
Si	KrF, 245 nm, 10 ns, 20 Hz, 5.3 mJ/pulse, 120 min	Spherical silicon nanoparticles with an average size of 40 nm.	Švrček et al. 2008
Si p-type	Nd:YAG, 355 nm, 8 ns, 30 Hz, 0.07–6 mJ/pulse, 30 min	SiNCs with the size distributions of 2–10 and 2–8 nm were prepared at 0.07 and 6 mJ/pulse, respectively.	Švrček et al. 2006

of the colloidal SiNCs in water produced by laser ablation or fragmentation process. As it is shown in Table 13.2, laser ablation in water is an effective technique for production of well-dispersed colloidal SiNCs with unique surface characteristics such as oxide and hydroxyl surface passivation.

13.2.2 CHEMICAL PROCESSES AND ELECTROCHEMICAL ETCHING

Various chemical solution-based and electrochemical etching methods are reported in the literature for the SiNCs production (Belomoin et al. 2002; Veinot 2006; He et al. 2009b; Lin and Chen 2009). Figure 13.2 shows different chemical strategies used to synthesize the ultrasmall (<10 nm) matrix-embedded and

Table 13.2 Summary of colloidal silicon nanoparticles with unique surface characteristics such as oxide and hydroxyl surface passivation, obtained using the ablation techniques in water and reported by different groups

TARGET	LASER PARAMETERS (λ, τ_{PULSE}, FLUENCE)	SiNPs MEAN SIZE (nm)	SURFACE BONDING	PRODUCT/PROPERTIES	REFERENCES
Si (99.999%)	Ti:sapphire, 800 nm, 100 fs, 0.09 mJ/pulse	4	Si–O–Si, Si–OH and Si–H	Amine-terminated silicon nanoparticles were prepared by formation of interfacial Si–O bonds between the organosilane and hydroxyl (OH) groups of the nanoparticles.	Rodio et al. 2016
Si p-type	Nd:YAG, 1064 nm, 10 ns, 7 J/cm²	<10	Si–O–Si, Si–OH and Si-H	Orange–red (550–650 nm) emission enhancement has been observed due to a different type of Si–OH surface terminations.	Mahdieh and Momeni 2015
Si n-type	Nd:YAG, 1064 nm, 5 ns, 0.6 J/cm²	3	Si–O–Si, Si–H	Silicon nanocrystal surrounded by a silica layer with the UV (4.4 eV) and blue (2.7 eV) emissions originating from oxygen-deficient centers in the SiO₂ shell.	Vaccaro et al. 2014
Si n-type	Nd:YAG, 1064 nm, 18 ns, 70 J/cm²	3	Si–O–Si, Si–OH and Si–H	Prominent red emissions (675 nm) have been observed due to quantum confinement size effect of SiNCs and radiative centers associated with the Si=O and Si–OH surface states.	Mansour et al. 2013
SiNC aggregates of different sizes up to micrometer	Excimer laser KrF, 245 nm, 10 ns, 2 J/cm²	<10	Si–O–Si, Si–OH and Si–H	PL intensity enhancement at 600 nm due to the OH terminations surface engineering of the SiNCs.	Švrček et al. 2011b
Si	Nd:YAG, 1064 nm, 10 ns, 3 J/cm²	<10	Si–O–Si	Blue–green light emission due to the recombination at the near-interface states between Si and SiOₓ.	Yang et al. 2011
Si p-type	Excimer laser KrF, 245 nm, 10 ns, 23.5 J/cm²	<10	Si–O–Si, Si–OH and Si–H	SiNCs with unique surface characteristics have shown PL emission at 550 nm due to the self-trapped excitonic surface states.	Švrček et al. 2011a
Si	Nd:YAG, 532 nm, 10 mJ/cm²	<3	Si–O–Si	Blue light emission has been observed due to the quantum confinement size effect of SiNCs.	Umezu et al. 2007

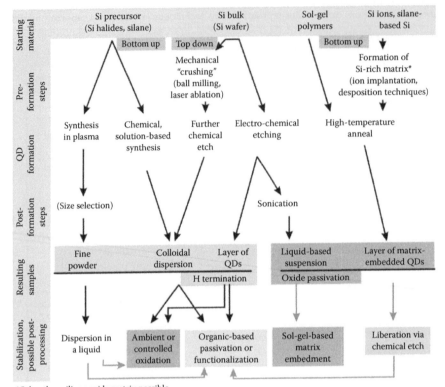

Figure 13.2 Different chemical strategies that have been developed for production of the matrix-embedded and freestanding silicon nanoparticles. (From Dohnalova, K., et al., *J. Phys. Condens. Matter*, 26, 173201, 2014. With permission.)

freestanding silicon nanoparticles (Dohnalová et al. 2014). The water-dispersible SiNCs are commonly prepared by electrochemical etching and ultrasonication of Si bulk/powders in a mixture of HNO_3 and HF (He et al. 2009a; Hwang et al. 2015). An experimental setup of this process is shown in Figure 13.3.

13.2.3 ELECTRICAL DISCHARGE

In this method, the colloidal SiNCs are produced via electrical spark discharge between two bulk silicon electrodes immersed in water as shown in Figure 13.4. By the electrodes distance and discharge voltage, large amounts of the nanoparticles can be prepared. The formation of the ultrasmall SiNCs occurs based on the condensation of silicon vapor coming out of the melting surface of electrodes resulting from the cooling process in the water interface (Liu et al. 2005; Mardanian et al. 2012, 2014). In addition, the SiNCs surface reaction with water may induce the Si–O, Si–OH, or Si–H bonds on the nanoparticle surface.

13.2.4 PLASMA-ASSISTED SYNTHESIS

Plasma treatment stabilizes the optoelectronic properties of SiNC by surface engineering in water (Mariotti et al. 2012; Mitra et al. 2014; Wu et al. 2016). This technique induces nonequilibrium liquid chemistry that passivates the SiNC surface with oxygen or hydroxyl terminations. Figure 13.5 shows the change in the surface chemistry of the SiNCs using the microplasma treatment, resulting in different Si–O and Si–OH bonding arrangements.

13.2.5 MICROWAVE-ASSISTED SYNTHESIS

A microwave-assisted reaction has been developed to produce water-dispersible SiNCs with different surface chemistry. This technique allows easy large-scale production of the highly photoluminescence (PL)

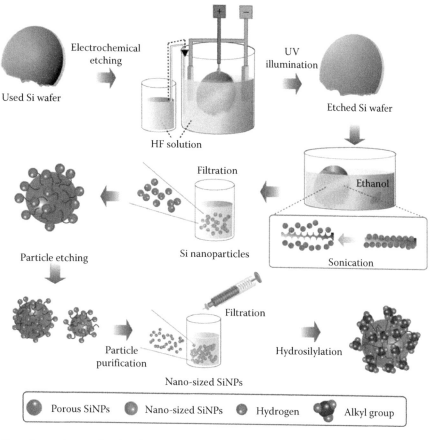

Figure 13.3 Experimental setup for the electrochemical etching of silicon. (From Hwang, J., et al., *Ind. Eng. Chem. Res.*, 54, 5982–5989, 2015. With permission.)

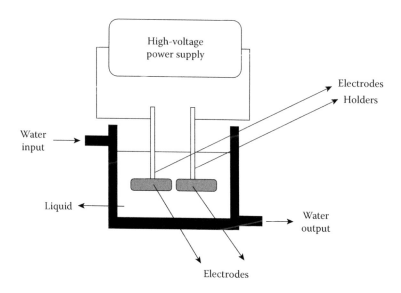

Figure 13.4 Schematic diagram of electrical spark discharge between two bulk silicon electrodes immersed in water. (From Mardanian, M., et al., *Appl. Phys. A*, 112, 437–442, 2012. With permission.)

Clusters, nanoparticles, and quantum dots

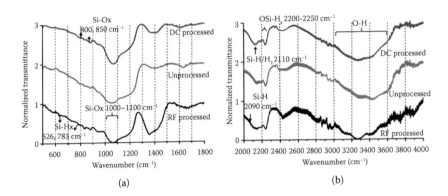

Figure 13.5 Comparison of the Fourier transform infrared (FT-IR) spectra of unprocessed silicon nanocrystals in water with those of plasma processed by RF or direct-current (DC) microplasma. (a) and (b) show the FT-IR spectra from 500 to 1800 cm⁻¹ and 2000 to 4000 cm⁻¹, respectively. (From Mitra, S., et al., *Plasma Process. Polym.*, 11, 158, 2014. With permission.)

SiNCs in water with high quantum yields (QYs) (Wu et al. 2015a), ultrasmall amine-terminated SiNCs with blue emission (Atkins et al. 2012), protein-covered silicon nanoparticles (Zhong et al. 2012), and biocompatible SiNCs with ultrahigh photo- and pH-stability (He et al. 2011).

13.3 OPTICAL, THERMAL, AND ELECTRONIC PROPERTIES OF SILICON NANOCRYSTAL

Bulk silicon crystal has extensive industrial applications and has been used for decades in microelectronic, photovoltaic, and MEMS technologies (Yates 1998; Pavesi and Lockwood 2004; Muller et al. 2006; Tilli et al. 2015). However, its use in optoelectronic applications is hampered by the inability of silicon to efficiently emit light due to its indirect band gap (Pavesi and Turan 2010). The observation of strong photoluminescence PL from porous silicon at room temperature in 1990 (Canham 1990) has stimulated extensive scientific researches in the field of preparation of nanoscale silicon structures. In addition, SiNCs are more attractive than other kind of semiconductor nanoparticles because they are fully compatible with existing technologies. Thus, it is of great scientific interest to understand the optical, thermal, and electronic properties of silicon nanocrytals for a variety of device applications.

13.3.1 NANOSILICON STRUCTURE SIZE EFFECT ON THE OPTICAL PROPERTIES

It has been reported that the quantum and spatial confinement effects results in wide range of the visible to NIR emissions from SiNCs due to a widening of their electronic band gaps (Shirahata et al. 2010; Mastronardi et al. 2012; Maier-Flaig et al. 2013a; Chen et al. 2015). For silicon nanoparticles with sizes less than 10 nm, wavelength-tunable emissions in the UV, visible, and IR regions can be achieved by precise nanocrystals size selection as shown in Figures 13.6 and 13.7. Based on the quantum confinement theory, the band gap of SiNCs increases compared to silicon bulk, resulting in a blue shift in PL emission and obeying the following relation as a function of particle diameters (Ledoux et al. 2002):

$$\Delta E_g = \frac{3.73}{d^{1.39}}.$$

Including the SiNCs lattice parameter changes, the following modified quantum size effect is proposed to determine the maximum of PL emissions as a function of the nanocrystal size (Ledoux et al. 2000):

$$E_{PL} = E_0 + \frac{3.73}{d^{1.39}} + \frac{0.881}{d} - 0.245$$

where E_0 is the band gap energy of bulk silicon (1.17 eV) and d is the size of SiNCs.

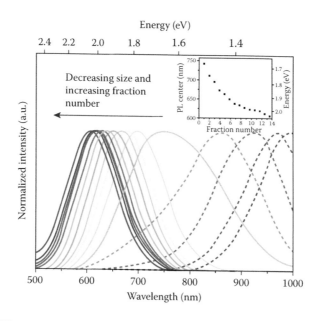

Figure 13.6 Visible to NIR emitting allylbenzene-capped silicon nanocrystals fractions obtained using size-selective precipitation. The PL blue shift with decreasing size is consistent with quantum size effects expected for the SiNCs. (From Mastronardi, M.L., et al., *Nano Lett.*, 12, 337–342, 2012. With permission.)

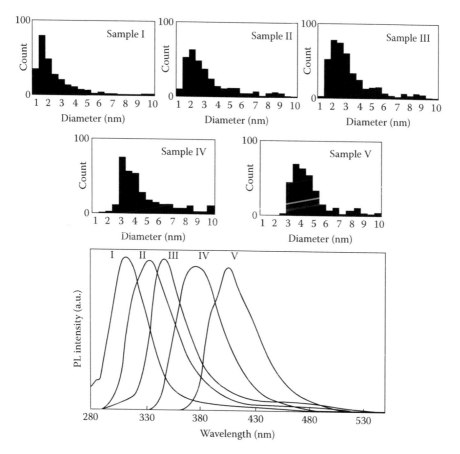

Figure 13.7 Wavelength-tunable UV-blue luminescence from SiNCs with different size distributions. (From Shirahata, N., et al., *Small*, 6, 915–921, 2010. With permission.)

Clusters, nanoparticles, and quantum dots

13.3.2 OPTICAL ABSORPTION

Bulk crystalline silicon has an indirect band gap of 1.1 eV and a direct band gap transition of 3.4 eV (Wilcoxon and Samara 1999). Therefore, fundamental band gap transition is momentum forbidden (indirect) and the optical absorption is of low efficiency because of phonon-assisted transitions. After the first observation of strong PL from porous silicon in 1990 (Canham 1990), extensive studies have been undertaken to control the intrinsic material properties to yield stronger optical absorption and consequently emission for silicon-based photonics applications (Wilcoxon and Samara 1999; Meier et al. 2007; Gresback et al. 2013; Singh et al. 2014; Lee et al. 2016; Sychugov et al. 2016). It has been reported that the colloidal SiNCs in water show a broad absorbance with visible onset (up to ~2.2 eV) and increasing absorption cross section at higher energies as a result of quantum confinement size effect (Intartaglia et al. 2012a; Mardanian et al. 2012; Mansour et al. 2013; Popovic et al. 2014). Figure 13.8 indicates the optical absorption spectra of the colloidal SiNCs in water with different sizes which shows blue shifts in the absorption edge with reduction of their size (Intartaglia et al. 2011). Furthermore, in SiNCs the momentum conservation law of optical transition partially relaxes due to the Heisenberg uncertainty principle, resulting in the enhancement of optical absorption in SiNCs (Sychugov et al. 2016). Indeed, it has been shown that the joint effects of both the Heisenberg uncertainty principle and interface scattering-induced Γ-X coupling can result in quasi-direct optical transitions in the SiNCs, at energies between the quantum-confined band gap in nanocrystals and the bulk crystalline silicon direct band gap at 3.4 eV (see Figure 13.9) (Lee et al. 2016). In addition, highly oxidized surface modification of the SiNCs may lead to drastically different optical properties if they were associated with surface states from interfaces or ligands.

13.3.3 PHOTOLUMINESCENCE PROPERTIES

Colloidal SiNCs in water are being extensively investigated as a means of improving the PL properties of silicon for different applications (Švrček et al. 2011b; Intartaglia et al. 2012a; Tamarov et al. 2014; Mahdieh and Momeni 2015). The physical mechanisms underlying the PL of the SiNC colloids are mainly due to the quantum confinement size effect and surface features of the nanocrystals. Control of surface characteristics is one of the main remaining challenges, having important effects on the optical and electronic properties of the SiNCs (Švrček et al. 2011b; Mariotti et al. 2012; Mitra et al. 2014). For different synthesis methods of the colloidal SiNCs, oxide and hydroxyl surface modification are observed (Mardanian et al. 2012; Mansour et al. 2013; Rodio et al. 2016; Wu et al. 2016). The PL emission properties of the SiNCs are highly sensitive to the oxide-related surface effects, so that oxygen passivation at the surface of the SiNCs

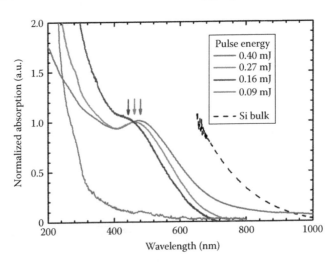

Figure 13.8 Optical absorption spectra of the SiNCs with different size distributions produced via femtosecond laser ablation of a silicon target in deionized water. (From Intartaglia, R., et al., *J. Phys. Chem. C*, 115, 5102–5107, 2011. With permission.)

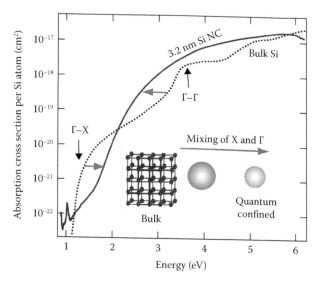

Figure 13.9 Optical absorption behavior of SiNCs. (From Lee, B.G., *Nano Lett.*, 16, 1583–1589, 2016. With permission.)

can result in changing the density of states, thereby changing the band gap. Many experimental and theoretical studies have shown that localized surface states can form in the oxygen-passivated SiNCs which may affect the recombination process in light emission from the nanocrystals (Godefroo et al. 2008; Dohnalová et al. 2013, 2014; Xu and Han 2014). Although the mechanisms of carrier relaxation in the SiNCs is not yet fully understood, the optical transitions from different energy states including core energy states (transition across the direct and indirect band gap) and oxygen-based surface/defect states have been suggested as the main responsible mechanisms of the PL emissions (Alkis et al. 2012a; Mariotti et al. 2012; Mansour et al. 2013; Mitra et al. 2014; Mahdieh and Momeni 2015). It should be pointed out that the PL emission spectrum of the SiNCs usually contains two emission bands referred to as "S-band" and "F-band" due to the slow and fast decay times, respectively (Dohnalová et al. 2010). As Figure 13.10 shows, the S-band is peaked at the spectral range of 600–850 nm and has a slow decay in the range of 10–100 μs. However, the F-band has a very fast decay of nanoseconds generally observed in the 400–500 nm wavelength range (Pelant 2011). The origin of the S-band is still controversial and it has been ascribed either to the quantum confinement effect (core recombination) or to a recombination at the surface oxide bonds which produce stable states within the band gap. Mansour et al. have shown that the oxidized SiNCs (~3 nm) exhibit green emission due to the confinement size effect and intense red emission associated with oxide radiative centers (Mansour et al. 2013). Moreover, it has been shown that atmospheric pressure microplasma/nanosecond laser processing can improve the stability and red PL emission of the colloidal SiNCs through the OH-related surfaces engineering (see Figure 13.11) (Švrček et al. 2011b; Mariotti et al. 2012; Mitra et al. 2014). Furthermore, observation of blue emissions peaking at the wavelength range of 420–450 nm are reported for the SiNCs colloids (Švrček and Kondo 2009; Švrček et al. 2009a; Yang et al. 2011; Alkis et al. 2012a; Mardanian et al. 2012; Vaccaro et al. 2014). In literature on the blue PL emission, different views are proposed including quantum-confined ultrasmall SiNCs (<2 nm) and an oxide-related surface effect. Since the emission spectra blue shifts by filtering of the SiNCs size, confinement size effect has been suggested to be the responsible mechanism for the blue emission band peaked at 420 nm (Alkis et al. 2012a). On the other hand, Yang et al. have reported blue emission due to the recombination at near-interface traps located at the interface areas between silicon and the surrounding SiO_x (0 < x < 2) layer (Yang et al. 2011). Figure 13.12 shows this process—transfer of excitons to the near-interface traps, and blue light is emitted due to recombination. It should be noticed that controlling the SiNCs sizes and surface oxide layer characteristics may effectively influence the PL properties including emission tunability (Kang et al. 2009; Mansour et al. 2013), luminescence intensity (Švrček et al. 2009b; Mahdieh and Momeni 2015), and emission linewidths (Sychugov et al. 2014).

Clusters, nanoparticles, and quantum dots

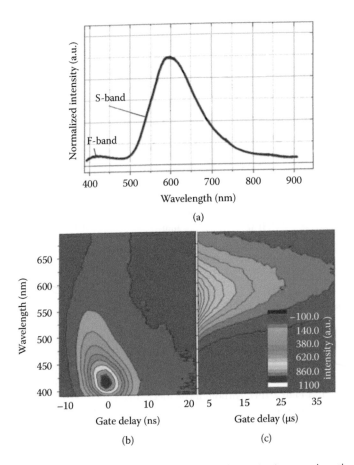

Figure 13.10 (a) The room temperature PL spectra of SiNCs under the excitation wavelength of 325 nm contains two emission bands, referred to as S-band and F-band. The temporal evolutions of the F and S bands are shown in (b) and (c), respectively. (From Pelant, I., *Phys. Status Solidi A*, 208, 625–630, 2011. With permission.)

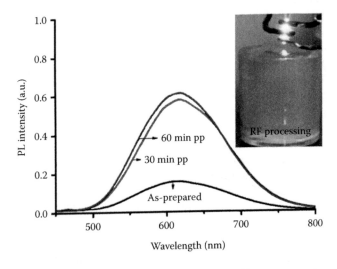

Figure 13.11 The RF plasma treatment of the SiNCs/water colloid for different processing times which results in the PL emission enhancement through the OH-related surfaces engineering. (From Mitra, S., et al., *Plasma Process. Polym.*, 11, 158, 2014. With permission.)

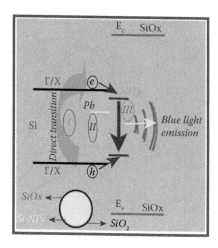

Figure 13.12 Schematic of the blue PL from oxidized SiNCs. This emission originated from the recombination at near-interface traps (NITs) located at the interface areas between silicon and the surrounding SiO_x ($0 < x < 2$) layer of the SiNCs. (From Yang, S., et al., *J. Phys. Chem. C*, 115, 21056–21062, 2011. With permission.)

13.3.4 ELECTROLUMINESCENCE

The observation of size-tunable visible emission from quantum-confined SiNCs has stimulated interest in obtaining efficient nanocrystal Si-based LEDs for low-cost applications (Cheng et al. 2010; Maier-Flaig et al. 2013a; Xin et al. 2015). The first silicon nanoparticle electroluminescent displays were obtained by embedding the SiNCs in nonconductive matrices, especially silicon oxide, which improves the quantum efficiency of the nanoparticles by reducing the number of nonradiative decay pathways (Photopoulos and Nassiopoulou 2000; Jambois et al. 2005). Jambois et al. have shown the visible to NIR EL emissions from a SiNCs/SiO$_2$ sample for a bias voltage in the range of 2–4 V (Jambois et al. 2005). They have suggested that the observation of a more energetic EL signal as a result of the increase of electric field is related to injection of carriers in smaller clusters (see Figure 13.13). Furthermore, the PL and EL signals with the similar peak behaviors near 750 nm were observed for the SiNCs of approximately 2–4 nm in diameter due to the radiative recombination of excitons within the nanocrystals, as shown in Figure 13.14 (Walters et al. 2005). It should be noted that optimization of the SiNC shell could be a promising way to improve the EL signal in oxidized SiNCs to consider the fabrication of Si-based electroluminescent devices. Ligman et al. have reported stable electrically induced red emission from the surface-oxidized SiNCs which can improve by the addition of the nanocrystals to the poly (9-vinyl carbazole) polymer (PVK) matrix (Ligman et al. 2007). Moreover, De La Torre et al. have shown the stable red EL from 3 nm SiNCs in an SiO$_2$ matrix due to the recombination at the surface states (De La Torre et al. 2003).

13.3.5 SiNCs SURFACE FUNCTIONALIZATION

In addition to the quantum confinement size effect, surface modification of nanocrystals with functional ligands can be an alternative strategy to enhance the optical properties of the SiNCs for potential application in optoelectronics and biological fields. In the past decade, the SiNCs surface functionalization has attracted increasing attention for achieving highly stable and tunable PL emissions from the nanocrystals (Zou and Kauzlarich 2008; Gupta et al. 2009; Anderson et al. 2012; Intartaglia et al. 2012b; Bagga et al. 2013; Wang et al. 2014a; Premnath et al. 2015). It has been indicated that the surface functionalization of SiNCs with different alkene molecules by UV-induced hydrosilylation can increase the stability of their PL substantially (Gupta et al. 2009). Figure 13.15 exhibits the functionalization of SiNCs with alkene molecules that allows them to form a stable and optically clear dispersion in many organic solvents. Wang et al. have shown that the SiNCs surface modification by 9-ethylanthracene resulted in dual-emissive hybrid nanomaterial with peaks centered at the blue and NIR regions that can be applicable in LEDs or bioimaging (Wang et al. 2014a). It has been reported that the surface modification of silicon nanoparticles with

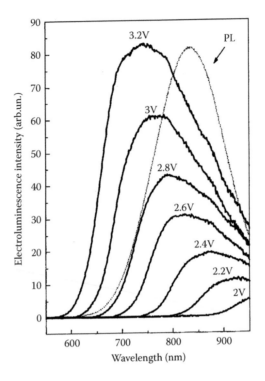

Figure 13.13 The EL spectra of the SiNCs/SiO$_2$ sample containing the nanocrystals with a mean size of 2.8 nm as a function of the bias voltage. (From Jambois, O., *J. Appl. Phys.* 98, 046105, 2005. With permission.)

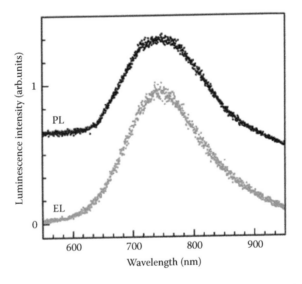

Figure 13.14 Photoluminescence (PL) and electroluminescence (EL) emission spectra of an array of SiNCs. (From Walters, R.J., *Nat. Mater.*, 4, 143–146, 2005. With permission.)

different organic compounds promotes the light harvesting ability for the luminescence through a charge transfer transition between the protective molecules and the nanoparticles, and also enhances the radiative rate of the silicon nanoparticles (Miyano et al. 2016). Moreover, it has been reported that the ultrashort pulsed laser synthesis can result in production of luminescent protein-functionalized and silicon–DNA conjugate nanoparticles which are promising for fluorescence imaging and probing of biological systems (see Figure 13.16) (Intartaglia et al. 2012b; Bagga et al. 2013). This process has been proposed as a suitable

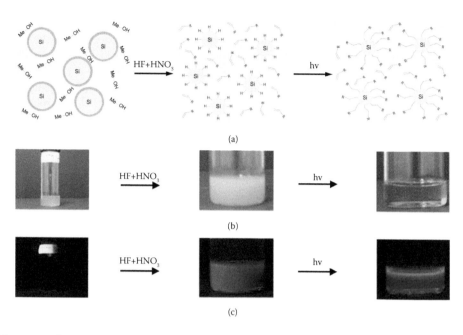

Figure 13.15 (a) Surface functionalization of SiNCs with alkene molecules. (b) and (c) show the photographs of the nanoparticles at various stages of the functionalization process under the ambient and UV illuminations, respectively. (From Gupta, A., et al., *Adv. Funct. Mater.*, 19, 696–703, 2009. With permission.)

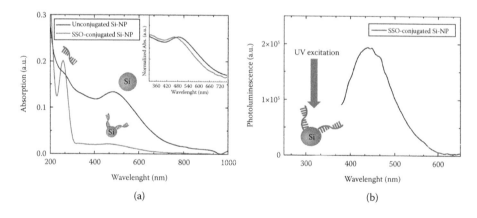

Figure 13.16 (a) Absorption spectra of the unconjugated (black line) and SSO (single-stranded oligonucleotides)-conjugated silicon nanoparticles (red line). (b) PL emission property of the conjugated silicon nanoparticles under UV excitation. (From Intartaglia, R., et al., *Nanoscale*, 4, 1271–1274, 2012b. With permission.)

method for generation of functionalized multiphased silicon/silicon oxide nanobiomaterials which exhibit remarkable drug-like properties for cancer therapy applications (Premnath et al. 2015).

13.3.6 BIOLOGICAL FLUORESCENCE IMAGING

Fluorescent biological imaging has been known as one the most powerful noninvasive tools for biomedical investigations (Peng et al. 2014). Because of favorable biocompatibility and low toxicity, luminescent SiNCs have great potential for use in biological imaging and diagnostic applications (Wang et al. 2011, 2014b; Premnath et al. 2015; Zhong et al. 2015). In order to exploit this potential, it is essential that the SiNCs have a substantial PL quantum yield in the visible to NIR regions and remain water dispersible and hydrophilic to prevent aggregation and precipitation in biological media (Wang et al. 2011). In recent years, there has been considerable interest in producing hydrophilic SiNCs or functionalizing them with hydrophilic species (e.g., hydrophilic molecule, polymer, and micelle), to make the water-dispersible nanocrystals suitable for bioimaging applications (Warner et al. 2005; Erogbogbo et al. 2008; Zhong et al. 2012, 2013). However, there have been many challenges in fabricating colloidally and optically stable water-dispersible SiNCs, including instability of luminescence and the difficulty of attaching hydrophilic molecules to the SiNC surfaces. Manhat et al. have reported successful biofunctional ligand grafting onto the oxidized luminescent SiNCs which allows for tuning of the particle surface charge, solubility, and functionality (Manhat et al. 2011). The nanocrytals show biological stability when used to examine cellular uptake and distribution in live N2a cells (Manhat et al. 2011). Fujioka et al. have shown that luminescent stable passive-oxidized SiNCs were more than 10 times safer than CdSe quantum dots, demonstrating the suitability of the nanocrystals for bioimaging (Fujioka et al. 2008). In addition, it has been reported that laser-synthesized oxide-passivated SiNCs exhibit strong PL with QY of several percent (Gongalsky et al. 2016). These nanocrystals do not show any sign of toxicity, and demonstrate biodegradability and excellent cellular uptake, which makes them extremely promising candidates for biological imaging tasks (see Figure 13.17).

13.3.7 THERMO-OPTICAL RESPONSE

Because of size-dependent PL properties and biocompatible features, knowledge of thermo-optical properties of the SiNCs is important for a variety of applications in optoelectronic and photothermal (PT) therapy. It has been reported that the SiNCs exhibit a notable PT response while still maintaining

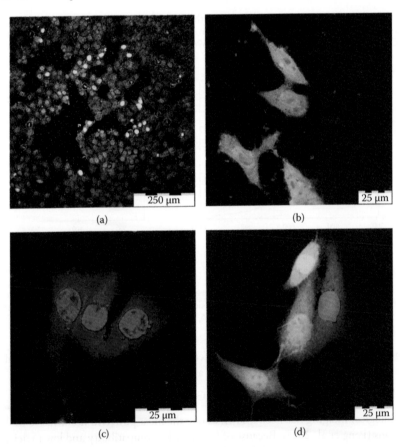

Figure 13.17 Confocal fluorescent microscopy images of CF2Th cancer cells with added SiNCs (colored red, pink, and partially violet) under different magnifications scales (a through c) and that of a control sample without the nanocrystals (d). Panel (c) represents detailed views of cells after washing out the SiNCs from extracellular space. Cell nuclei and their cytoplasm are colored blue and green, respectively, in panels (b through d). (From Gongalsky, M., et al., *Sci. Rep.*, 6, 24732, 2016. With permission.)

relatively high NIR photoluminescent QYs, which can be applicable for combined *in vivo* PL imaging and PT therapy (Regli et al. 2012). Since the PT effect increases with the SiNC size, defect concentration, and irradiating energy, a combination of carrier thermalization and defect-mediated heating have been suggested as the origin of the PT response (Regli et al. 2012). Recently, Afshani et al. have shown a significant enhancement of the efficiency of PT energy conversion by a new hybrid nanostructured material consisting of the surface-oxidized SiNCs anchored to the surface of reduced graphene oxide nanosheets dispersed in water (see Figure 13.18) (Afshani et al. 2016). They have proposed the SiNCs–graphene oxide nanocomposites as promising materials with potential applications in PT therapy, heating and evaporation of liquids by solar energy, ignition of solid fuels, and welding of composite materials (Afshani et al. 2016). Furthermore, Ishii et al. have experimentally demonstrated that an ensemble of silicon nanoparticles with different sizes dispersed in water can efficiently harvest sunlight to accelerate heating and vaporization of water by nanoscale local heating (Ishii et al. 2016). As shown in Figure 13.19, the vaporization and heating speeds of the colloidal silicon nanoparticle in water can increase by up to twofold compared to those of pure water.

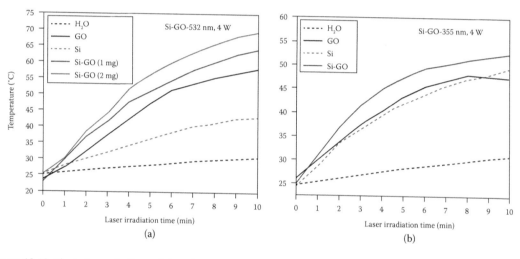

Figure 13.18 Photothermal effect of the silicon (Si) nanocrystals, graphene oxide (GO) and the Si-GO nanocomposites in water measured by using laser irradiation at the wavelengths of (a) 532 and (b) 355 nm. (From Afshani, P., et al., *Chem. Phys. Lett.*, 650, 148–153, 2016. With permission.)

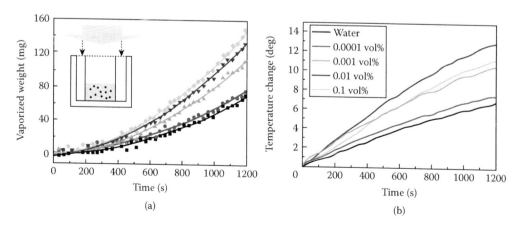

Figure 13.19 Comparison of the (a) weight losses and (b) temperature changes of the colloidal silicon nanoparticle in water to those of pure water upon irradiation by a solar simulator. (From Ishii, S., *Opt. Mater. Express*, 6, 640–648, 2016. With permission.)

Clusters, nanoparticles, and quantum dots

13.3.8 OPTICAL GAIN

As a first report in this field, Pavesi et al. have shown the light amplification using silicon quantum dots dispersed in a silicon dioxide matrix with the material gain being of the same order as that of direct band gap quantum dots (Pavesi et al. 2000). They have been reported the stimulated emission and light amplification in the S-band of SiNCs (~ 800 nm) based on the population inversion of radiative states associated with the nanocrystal oxide interface (Pavesi et al. 2000). Figure 13.20 shows the SiNCs optical gain in a single pass configuration and a three-level model that proposed to explain the observed gain. Up to now, many investigations have been focusing on the occurrence of single-passage light amplification and optical gain in the S-band (red-emission band) of the SiNCs/SiO₂-based samples (Ruan et al. 2003; Luterova et al. 2004; Pelant 2011). Furthermore, observation of the optical gain at ~450 nm (F-band) has been also reported on the oxidized SiNCs (Dohnalová et al. 2009). Recently, Wang et al. have found white light emission (spans the blue to red region) from a single SiNC thin film due to radiative transitions from different energy levels of Si=O interfacial states to the ground states of SiNCs (Wang et al. 2015). They have shown positive optical gain (around 10^2 cm^{-1}) for the red, green, and blue components of the white emission which can be applicable for multicolor silicon lasing (Wang et al. 2015). The measured PL lifetimes and optical gains of the different emission components are listed in Table 13.3.

13.3.9 PHOTOVOLTAIC PROPERTIES

Due to the desirable electronic and optical properties of semiconductor nanocrystals, the incorporation of nanocrystals to form hybrid solar cells are being increasingly introduced to improve the performance of organic solar cells (Wright and Uddin 2012). Recently, remarkable accomplishments have been achieved

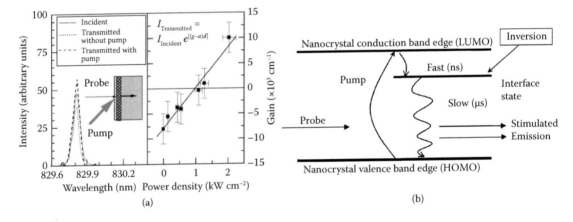

Figure 13.20 (a) Optical gain of the SiNCs medium measured by considering the transmission of a probe beam in presence (dashed line) or in absence (dotted line) of a pump beam. Right panel shows the dependence of the material gain value on the pump power density. (b) Schematic energy diagram for the nanocrystal showing a three-level model for population inversion and the observed optical gain. (From Pavesi, L., et al., *Nature*, 408, 440–444, 2000. With permission.)

Table 13.3 Summary of the net optical gains, total optical losses, real optical gains, and PL lifetimes for the four components of the SiNCs thin film white light emission

WAVELENGTH (nm)	400	490	560	620
G (cm^{-1})	72.26	83.96	54.51	46.41
α_{Tot} (cm^{-1})	45.25	39.24	40.16	43.42
g (cm^{-1})	117.49	123.20	94.57	89.83
τ (ns)	0.44	1.09	1.31	1.63

Source: Wang, D.C., et al., *Nanotechnology*, 26, 475203, 2015.

Clusters, nanoparticles, and quantum dots

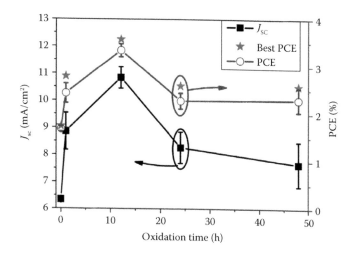

Figure 13.21 Short circuit current (J_{SC}) and power conversion efficiency (PCE) of the SiNC-based HSCs are changed depending on SiNC oxidation time. For each condition, best PCEs of devices are also marked with asterisk. (From Ding, Y., et al., *Nano Energ.*, 10, 322–328, 2014. With permission.)

in nanocrystals/polymer hybrid solar cells including the cadmium sulfide- (CdS) (Ren et al. 2011), lead sulfide- (PbS) (Liu et al. 2013b), and cadmium selenide- (CdSe) (Zhou et al. 2010) based HSCs with power conversion efficiency (PCE) of 2–5.5%. Among all kinds of semiconductor nanocrystals, the SiNCs have been known as an attractive candidate because of the nontoxicity and abundance of silicon. The SiNCs have been extensively studied and shown feasibility as a proper material that could be used to boost the efficiency of photovoltaic energy conversion due to the quantum confinement and surface-induced effects (Liu et al. 2008; Ingenhoven et al. 2013; Ding et al. 2014; Dutta et al. 2015). Švrček et al. have investigated the potential use of SiNCs in photovoltaics technology as a down-converter of high-energy photons to improve the silicon solar efficiency at low cost (Švrček et al. 2004). They have shown that the high-energetic photons are absorbed within the luminescence converter SiNCs, resulting an increase in carrier collections which can improve the efficiency of a silicon solar cell up to approximately 1.2% (Švrček et al. 2004). Furthermore, it has been reported that the controlled SiNC oxidization can effectively improve their electron mobility, and as a result, SiNC-based HSCs achieved an efficiency of 3.6%, which is more than twice the efficiency of devices fabricated with fresh SiNCs (without passivation) (Ding et al. 2014). As Figure 13.21 shows, an increase in the short circuit current (J_{SC}) and PCE was observed until the oxidation time of SiNCs exceeded 12 h, so the average PCE has been improved considerably due to effective oxygen passivation which can reduce carrier traps easily and effectively (Ding et al. 2014). In addition to those mentioned above, SiNCs offer interesting features of importance for photovoltaics including the reduction of the density of states and discretization of energy levels that affect hot carrier cooling processes, and the enhancement of the Coulomb interaction between carriers enclosed in small volumes, promoting collective effects such as multiple exciton generation (MEG) (Beard et al. 2007; Priolo et al. 2014).

13.4 SILICON NANOCRYSTALS DEVICE APPLICATIONS

13.4.1 LED

Fabrication of LEDs using fluorescent SiNCs as an emitting layer with utilizing their EL properties has received worldwide attention (Cheng et al. 2010; Maier-Flaig et al. 2013b; Xin et al. 2015; Yang et al. 2015). It has been reported that the EL of the silicon nanoparticle-based LEDs is sufficiently bright in the visible to NIR region. Puzzo et al. have presented two classes of SiNCs featuring size-tunable EL (i.e., 3.2 nm for 685 nm EL and 3.0 nm for 645 nm EL) as emitter layers for the construction of LED devices (Puzzo et al. 2011). Figure 13.22 illustrates a fabrication design for a highly efficient and widely

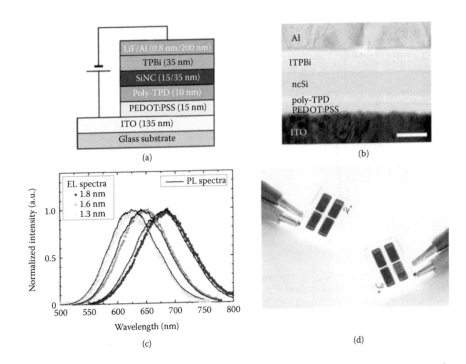

Figure 13.22 (a) Schematic view of silicon-based NCs-LED stack of the devices. (b) BF-TEM image of a cross section of a real device showing the layer structure of the device (scale bar: 50 nm). (c) Comparison of EL and PL of devices built with SiNCs emitters of three different sizes. (d) SiNCs-LEDs connected to a 9 V battery in series to an ohmic resistor limiting the voltage to 6 V, show bright, stable, and homogeneous strong red and orange luminescence emissions. (From Maier-Flaig, F., et al., *Nano Lett.*, 13, 475–480, 2013a. With permission.)

color-tunable silicon-based nanocrystals-LED. Maier-Flaig and coworkers have employed bright, long-term, stable, and color-tunable silicon LED featuring intense EL from the NIR down to the yellow spectral region by using size-separated SiNCs (Maier-Flaig et al. 2013a). Cheng et al. have reported infrared-emitting LED devices with a wavelength of 853 nm by using small-sized (diameters: 3–5 nm) SiNCs as the emitting layer, with external quantum efficiency of 8.6% (Cheng et al. 2011).

13.4.2 SENSOR

Recently, SiNCs, as an emerging type of luminescent nanomaterial, have received considerable attention for fluorescent sensors in the field of chemical analysis (Feng et al. 2014; Liao et al. 2016), biomedicine (Chen et al. 2014; Zhang et al. 2015), environmental monitoring (Guo et al. 2016), high-energy compounds detection (Gonzalez et al. 2014), and so on. The sensing mechanism is based on the adsorption of test materials in the SiNC assembly which can result in a loss of its PL intensity with a recovery property of the original PL as shown in Figure 13.23 (Sailor and Wu 2009). Feng et al. have shown that the label-free SiNCs with an average size of 2.5 nm can be used as a highly sensitive pH sensor (Feng et al. 2014). Zhang and his coworkers have reported highly efficient detection of the dopamine (DA) using water-soluble silicon nanoparticles (Zhang et al. 2015). This allows selective detection of the DA molecule based on a fluorescence quenching effect. Hypochlorite in tap water has been detected utilizing blue-emissive SiNC with relatively good selectivity and sensitivity (Guo et al. 2016). High-energy compounds such as nitrobenzene, nitrotoluene, and dinitrotoluene are easily detected using an SiNC luminescent paper sensor made by coating the nanocrystals onto a filter paper (Gonzalez et al. 2014). Figure 13.24 illustrates the sensing of explosive solutions of TNT, RDX, and PETN by the SiNCs paper sensor through the rapid fluorescence quenching by all the compounds (Gonzalez et al. 2014). The SiNCs paper sensor may be applicable for detection of solution, solid, and vapor phases.

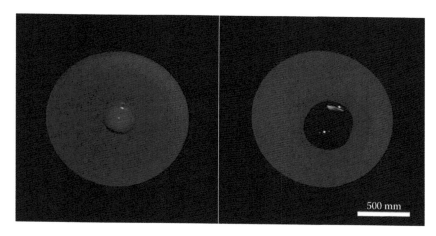

Figure 13.23 Adsorbate-induced quenching of photoluminescence from the SiNC assembly. Both samples contain a 3 mL drop of liquid on their surface; the drop on the left is pure water, the one on the right is 40% ethanol in water. (From Sailor, M.J., and Wu, E.C., *Adv. Funct. Mater.*, 19, 3195–3208, 2009. With permission.)

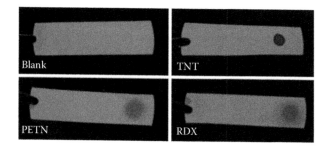

Figure 13.24 Images of SiNC-coated filter paper under a handheld UV-lamp without the presence of nitrocompounds and in the presence of solutions of TNT, PETN, and RDX, as indicated. (Gonzalez, C.M., et al., *Nanoscale*, 6, 2608–2612, 2014. With permission.)

13.4.3 BIOLOGICAL LABEL

Water-dispersible SiNCs have emerged as a novel kind of promising material for luminescent biological labels due to strong fluorescence in the visible to NIR region, photostability, low toxicity, and available versatile preparation techniques. Various SiNCs-based fluorescent biological labels have been reported for nanomedicine applications (Erogbogbo et al. 2008; Wang et al. 2011; Chinnathambi et al. 2013; Das et al. 2013; Zhong et al. 2013; Wu et al. 2015b). The red-emitting hydrophobic SiNCs have been transformed into water-soluble folate functionalized nanoprobes and used as fluorescent labels for cancer cells and tissue, as shown in Figure 13.25 (Das et al. 2013). Peptides-conjugated SiNCs probes are highly suitable for real-time immunofluorescence imaging of cancer cells (Song et al. 2015). The luminescent micelle-encapsulated SiNCs have been used as nontoxic optical probes for pancreatic cancer cells (Erogbogbo et al. 2008). Very stable water-soluble poly (acrylic acid) grafted luminescent silicon nanoparticles have been used as fluorescent biological staining labels for bioimaging (Li and Ruckenstein 2004).

13.4.4 CANCER THERAPY

Silicon nanoparticles have the potential to be used as radiosensitizers for improving the therapeutic efficacy and reducing toxic side effects of radiotherapy treatment of cancers (Gara et al. 2012). It has been shown that, at clinically relevant doses of X-ray irradiation, the surface-oxidized silicon nanoparticle can enhance the yield of reactive oxygen species (ROS) formation which could result in the oxidative damage to lipids, proteins, and DNA at excessive amounts (Gara et al. 2012; Klein et al. 2014). Furthermore, amino-functionalized oxidized silicon nanoparticles have been used as radiosensitizers

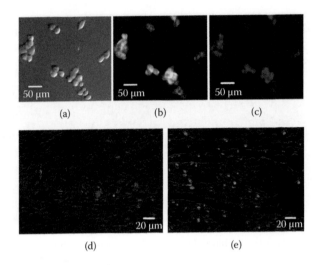

Figure 13.25 Folate functionalized silicon nanoparticles selectively label HeLa cells (top row a through c) and cervical cancer positive tissue (d and e). Fluorescence microscopic images of cervical cancer positive tissue after labeling with silicon nanoparticles with and without folate functionalization are shown in (d) and (e), respectively. The advantage of the silicon nanoprobe is that the cells/tissues can be imaged under blue (b), green (c and d), or UV excitation (e) to obtain yellow or red emission. In tissue imaging (d and e) blue-emitting Hoechst dye is used to stain the cell nucleus. (From Das, P., et al., *Nanoscale*, 5, 5732–5737, 2013. With permission.)

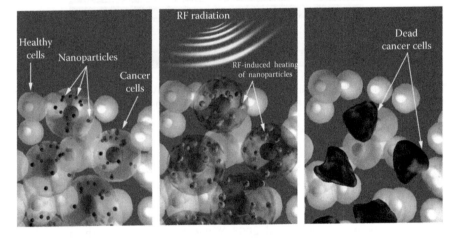

Figure 13.26 Schematic representation of the RF radiation-induced hyperthermia treatment procedure using RF-absorbing Si nanoparticle-based sensitizers. The nanoparticles are targeted into a tumor area (actively or passively), accumulate in it, and then absorb main RF radiation power to heat cancer cells resulting a local increase of temperature and selective destruction of cancer cells. (From Tamarov, K.P., et al., *Sci. Rep.*, 4, 7034, 2014. With permission.)

for X-ray treatment of human breast cancer and mouse fibroblast cells (Klein et al. 2013). Recently, a new modality for mild cancer therapy has been developed based on silicon nanomaterials, with the possibility of the involvement of parallel imaging and treatment channels due to their unique optical properties. This allows mild, noninvasive, and deep cancer therapy using low-energy sources. Tamarov et al. have introduced a novel cancer treatment modality using SiNCs as sensitizers of radio frequency (RF) radiation-induced hyperthermia as schematically indicated in Figure 13.26 (Tamarov et al. 2014). They have shown that the excitation of silicon nanoparticles by the low-intensity RF radiation not only strongly inhibited the growth of carcinoma tumors but also resulted in a decrease of the tumor volume (Tamarov et al. 2014).

13.4.5 LITHIUM-ION BATTERIES

SiNC-embedded SiO$_x$ nanocomposite has attracted a great deal of attention as a high-capacity lithium storage material for lithium-ion batteries (Park et al. 2014; Park et al. 2015; Bae et al. 2016). High-performance anode materials for lithium-ion batteries are suggested since the Si/SiO$_x$ nanocomposite provides facile electrochemical kinetics by offering a large surface area and short Li$^+$ pathways. It has been shown that incorporating the SiNCs into amorphous SiO$_x$ provides a high capacity with a notably improved initial efficiency (~74%) and stable cycle performance over 100 cycles (Park et al. 2015). Excellent cycle performance and rate capability with high-dimensional stability have been observed for Si/SiO$_x$ nanocomposite materials with a uniform carbon coating layer (Park et al. 2014). Furthermore, the electroconductive black TiO$_{2-x}$ coating highly improved the performance of the nanocomposite anode materials as shown in Figure 13.27 (Bae et al. 2016).

13.4.6 HYBRID SOLAR CELL

SiNCs have attracted considerable interest in the field of silicon solar cells thanks to their desirable electronic and optical properties arising from the quantum confinement and surface-induced effects (Cho et al. 2008; Ingenhoven et al. 2013; Ding et al. 2014, 2016; Dutta et al. 2015; Zhao et al. 2016). TheSiNCs-incorporated hybrid solar cells have been mainly explored as binary systems by coupling SiNCs and different materials. It has been reported that the hybrid solar cells with power conversion efficiencies (PCE) of up to ~3.6 % are fabricated by blending SiNCs with different materials such as P3HT (Liu et al. 2010), fullerenes (Švrček et al. 2010), and low band gap polymer of PTB7 (Ding et al. 2014). Figure 13.28 shows the

Figure 13.27 (a) Schematic illustration of Si/SiOx nanosphere coated with a very thin TiO$_{2-x}$ shell. (b) TEM image of the nanocomposite combined with cycle performance of TiO$_{2-x}$ coated Si/SiOx nanospheres at a current density of 200 mA g^{-1} for 100 cycles. (From Bae, J., et al., *ACS Appl. Mater. Interfaces*, 8, 4541–4547, 2016. With permission.)

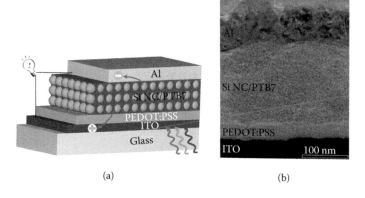

Figure 13.28 (a) Schematic illustration, and (b) cross-sectional TEM image of the SiNC/PTB7 hybrid solar cells. (From Ding, Y., et al., *Nano Energ.*, 10, 322–328, 2014. With permission.)

Clusters, nanoparticles, and quantum dots

schematic diagram of the SiNC/PTB7 hybrid solar cell with PCE of 3.6% that has been designed by Ding and coworkers (Ding et al. 2014). Dutta et al. have reported an efficient hybrid solar cell utilizing SiNCs and silicon nanowires with a high cell efficiency of 12.9% (Dutta et al. 2015). Recently, it has been found that incorporating SiNCs into the P3HT:PCBM binary system resulted in novel ternary hybrid solar cells with maximum PCE of 4.11% (Zhao et al. 2016). It has been shown that the efficiency of the P3HT:PCBM solar cell may increase by ~40% when ~5% of PCBM in the original blend is replaced with SiNCs (Zhao et al. 2016).

13.5 SUMMARY

SiNCs have attracted considerable interest for optoelectronic devices as well as biomedical applications due to desirable electronic and optical properties, favorable biocompatibility, and low toxicity. Luminescent water-dispersible SiNCs have especially great potential for use in biological imaging and diagnostic applications due to strong fluorescence in the visible to NIR region, photostability, low toxicity, and available versatile preparation techniques. Thus, it is of great scientific interest to study the optical, thermal, and electronic properties of the SiNCs for a variety of device applications. This chapter presents a review of the recent research achievements in synthesis and device fabrications based on the colloidal SiNCs dispersible in water. First, the conventional physical/chemical methods commonly used for fabrication of the SiNCs including pulsed laser ablation, chemical synthesis preparations and electrochemical etching, electrical discharge, plasma, and microwave-assisted synthesis approaches have been presented. Then, the optical absorption and emissions, thermo-optical response, and photovoltaic properties of the SiNCs were studied and discussed by mainly focusing on the size and oxide-related surface effects. The SiNC device applications at the end of this chapter present some of the recent achievements in fabrication of LEDs, fluorescent sensors, luminescent biological labels, radiosensitizers for cancer treatment, high-capacity lithium storage materials for lithium-ion batteries, and silicon solar cells using SiNCs. It is hoped that the presented material will stimulate a general interest in further developing silicon materials-based nanotechnology.

REFERENCES

Afshani P, Moussa S, Atkinson G, Kisurin VY, Samy El-Shall M. (2016). Enhanced photothermal effect of surface oxidized silicon nanocrystals anchored to reduced graphene oxide nanosheets. *Chem. Phys. Lett.* 650:148–153.

Alima D, Estrin Y, Rich DH, Bar I. (2012). The structural and optical properties of supercontinuum emitting Si nanocrystals prepared by laser ablation in water. *J. Appl. Phys.* 112:114312.

Alkis S, Okyay AK, Ortac B. (2012a). Post-treatment of silicon nanocrystals produced by ultra-short pulsed laser ablation in liquid: Toward blue luminescent nanocrystal generation. *J. Phys. Chem. C* 116:3432–3436.

Alkis S, Oruc FB, Ortac B, Kosger AC, Okyay AK. (2012b). A plasmonic enhanced photodetector based on silicon nanocrystals obtained through laser ablation. *J. Opt.* 14:125001.

Amendola V, Meneghetti M. (2013). What controls the composition and the structure of nanomaterials generated by laser ablation in liquid solution? *Phys. Chem. Chem. Phys.* 15:3027–3046.

Anderson I, Shircliff R, Macauley C, Smith D, Lee B, Agarwal S, Stradins P, Collins R. (2012). Silanization of low-temperature-plasma synthesized silicon quantum dots for production of a tunable, stable, colloidal solution. *J. Phys. Chem. C* 116:3979–3987.

Atkins TM, Louie AY, Kauzlarich SM. (2012). An efficient microwave-assisted synthesis method for the production of water soluble amine-terminated Si nanoparticles. *Nanotechnology* 23:294006.

Atkins TM, Thibert A, Larsen DS, Dey S, Browning ND, Kauzlarich SM. (2011). Femtosecond ligand/core dynamics of microwave-assisted synthesized silicon quantum dots in aqueous solution. *J. Am. Chem. Soc.* 133:20664–20667.

Bae J, Kim DS, Yoo H, Park E, Lim Y-G, Park M-S, Kim Y-J, Kim H. (2016). High-performance Si/SiOx nanosphere anode material by multipurpose interfacial engineering with black TiO$_{2-x}$. *ACS Appl. Mater. Interfaces* 8:4541–4547.

Bagga K, Barchanski A, Intartaglia R, Dante S, Marotta R, Diaspro A, Sajti CL, Brandi F. (2013). Laser-assisted synthesis of *Staphylococcus aureus* protein-capped silicon quantum dots as bio-functional nanoprobes. *Laser Phys. Lett.* 10:065603.

Beard MC, Knutsen KP, Yu P, Luther JM, Song Q, Metzger WK, Ellingson RJ, Nozik AJ. (2007). Multiple exciton generation in colloidal silicon nanocrystals. *Nano Lett.* 7:2506–2512.

Belomoin G, Therrien J, Smith A, Rao S, Twesten R, Chaieb S, Nayfeh M, Wagner L, Mitas L. (2002). Observation of a magic discrete family of ultrabright Si nanoparticles. *Appl. Phys. Lett.* 80:841–843.

Canham LT. (1990). Silicon quantum wire array fabrication by electrochemical and chemical dissolution of wafers. *Appl. Phys. Lett.* 57:1046–1048.

Chen KK, Mastronardi ML, Kübel C, Ozin GA. (2015). Size-selective separation and purification of "water-soluble" organically capped brightly photoluminescent silicon nanocrystals. *Part. Part. Syst. Charact.* 32:301–306.

Chen Q, Liu M, Zhao J, Peng X, Chen X, Mi N, Yin B, Li H, Zhang Y, Yao S. (2014). Water-dispersible silicon dots as a peroxidase mimetic for the highly-sensitive colorimetric detection of glucose. *Chem. Commun.* 50:6771–6774.

Cheng K-Y, Anthony R, Kortshagen UR, Holmes RJ. (2010). Hybrid silicon nanocrystal—Organic light-emitting devices for infrared electroluminescence. *Nano Lett.* 10:1154–1157.

Cheng K-Y, Anthony R, Kortshagen UR, Holmes RJ. (2011). High-efficiency silicon nanocrystal light-emitting devices. *Nano Lett.* 11:1952–1956.

Chinnathambi S, Chen S, Ganesan S, Hanagata N. (2013). Silicon quantum dots for biological applications. *Adv. Healthcare Mater.* 3:10–29.

Cho E-C, Park S, Hao X, Song D, Conibeer G, Park S-C, Green MA. (2008). Silicon quantum dot/crystalline silicon solar cells. *Nanotechnology* 19:245201.

Das P, Saha A, Maity AR, Ray SC, Jana NR. (2013). Silicon nanoparticle based fluorescent biological label via low temperature thermal degradation of chloroalkylsilane. *Nanoscale* 5:5732–5737.

De La Torre J, Souifi A, Poncet A, Busseret C, Lemiti M, Bremond G, Guillot G, Gonzalez O, Garrido B, Morante J. (2003). Optical properties of silicon nanocrystal LEDs. *Physica E* 16:326–330.

Ding Y, Sugaya M, Liu Q, Zhou S, Nozaki T. (2014). Oxygen passivation of silicon nanocrystals: Influences on trap states, electron mobility, and hybrid solar cell performance. *Nano Energ.* 10:322–328.

Ding Y, Zhou S, Juangsa FB, Sugaya M, Zhang X, Zhao Y, Nozaki T. (2016). Double-parallel-junction hybrid solar cells based on silicon nanocrystals. *Org. Electron.* 30:99–104.

Dohnalova K, Gregorkiewicz T, Kusova K. (2014). Silicon quantum dots: Surface matters. *J. Phys. Condens. Matter* 26:173201.

Dohnalova K, Ondic L, Kusova K, Pelant I, Rehspringer J, Mafouana R-R. (2010). White-emitting oxidized silicon nanocrystals: Discontinuity in spectral development with reducing size. *J. Appl. Phys.* 107:053102.

Dohnalova K, Poddubny AN, Prokofiev AA, de Boer WDAM, Umesh CP, Paulusse JMJ, Zuilhof H, Gregorkiewicz T. (2013). Surface brightens up Si quantum dots: Direct bandgap-like size-tunable emission. *Light Sci. Appl.* 2:e47.

Dohnalova K, Zidek K, Ondic L, Kusova K, Cibulka O, Pelant I. (2009). Optical gain at the F-band of oxidized silicon nanocrystals. *J. Phys. D Appl. Phys.* 42:135102.

Dutta M, Thirugnanam L, Trinh PV, Fukata N. (2015). High efficiency hybrid solar cells using nanocrystalline Si quantum dots and Si nanowires. *ACS Nano* 9:6891–6899.

Erogbogbo F, Yong K-T, Roy I, Xu G, Prasad PN, Swihart MT. (2008). Biocompatible luminescent silicon quantum dots for imaging of cancer cells. *ACS Nano* 2:873–878.

Eroshova OI, Perminov PA, Zabotnov SV, Gongal'skii MB, Ezhov AA, Golovan' LA, Kashkarov PK. (2012). Structural properties of silicon nanoparticles formed by pulsed laser ablation in liquid media. *Crystallogr. Rep.* 57:831–835.

Feng Y, Liu Y, Su C, Ji X, He Z. (2014). New fluorescent pH sensor based on label-free silicon nanodots. *Sens. Actuators B* 203:795–801.

Fujioka K, Hiruoka M, Sato K, Manabe N, Miyasaka R, Hanada S, Hoshino A, Tilley RD, Manome Y, Hirakuri K. (2008). Luminescent passive-oxidized silicon quantum dots as biological staining labels and their cytotoxicity effects at high concentration. *Nanotechnology* 19:415102.

Gara PMD, Garabano NI, Portoles MJL, Moreno MS, Dodat D, Casas OR, Gonzalez MC, Kotler ML. (2012). ROS enhancement by silicon nanoparticles in X-ray irradiated aqueous suspensions and in glioma C6 cells. *J Nanopart. Res.* 14:1–13.

Godefroo S, Hayne M, Jivanescu M, Stesmans A, Zacharias M, Lebedev O, Van Tendeloo G, Moshchalkov VV. (2008). Classification and control of the origin of photoluminescence from Si nanocrystals. *Nat. Nanotech.* 3:174–178.

Gongalsky M, Osminkina L, Pereira A, Manankov A, Fedorenko A, Vasiliev A, Solovyev V, Kudryavtsev A, Sentis M, Kabashin A. (2016). Laser-synthesized oxide-passivated bright Si quantum dots for bioimaging. *Sci. Rep.* 6:24732.

Gonzalez CM, Iqbal M, Dasog M, Piercey DG, Lockwood R, Klapotke TM, Veinot JGC. (2014). Detection of high-energy compounds using photoluminescent silicon nanocrystal paper based sensors. *Nanoscale* 6:2608–2612.

Gresback R, Murakami Y, Ding Y, Yamada R, Okazaki K, Nozaki T. (2013). Optical extinction spectra of silicon nanocrystals: Size dependence upon the lowest direct transition. *Langmuir* 29:1802–1807.

Guo Y, Zhang L, Cao F, Mang L, Lei X, Cheng S, Song J. (2016). Hydrothermal synthesis of blue-emitting silicon quantum dots for fluorescent detection of hypochlorite in tap water. *Anal. Methods* 8:2723–2728.

Gupta A, Swihart MT, Wiggers H. (2009). Luminescent colloidal dispersion of silicon quantum dots from microwave plasma synthesis: Exploring the photoluminescence behavior across the visible spectrum. *Adv. Funct. Mater.* 19:696–703.

Hamad S, Podagatlapalli GK, Mounika R, Rao SN, Pathak A, Rao SV. (2015). Studies on linear, nonlinear optical and excited state dynamics of silicon nanoparticles prepared by picosecond laser ablation. *AIP Adv.* 5:127127.

He Y, Kang ZH, Li QS, Tsang CHA, Fan CH, Lee ST. (2009a). Ultrastable, highly fluorescent, and water-dispersed silicon-based nanospheres as cellular probes. *Angew. Chem. Int. Ed.* 121:128–132.

He Y, Su Y, Yang X, Kang Z, Xu T, Zhang R, Fan C, Lee S-T. (2009b). Photo and pH stable, highly-luminescent silicon nanospheres and their bioconjugates for immunofluorescent cell imaging. *J. Am. Chem. Soc.* 131:4434–4438.

He Y, Zhong Y, Peng F, Wei X, Su Y, Lu Y, Su S, Gu W, Liao L, Lee S-T. (2011). One-pot microwave synthesis of water-dispersible, ultraphoto- and pH-stable, and highly fluorescent silicon quantum dots. *J. Am. Chem. Soc.* 133:14192–14195.

Hwang J, Jeong Y, Lee KH, Seo Y, Kim J, Hong JW, Kamaloo E, Camesano TA, Choi J. (2015). Simple preparation of fluorescent silicon nanoparticles from used Si wafers. *Ind. Eng. Chem. Res.* 54:5982–5989.

Ingenhoven P, Anopchenko A, Tengattini A, Gandolfi D, Sgrignuoli F, Pucker G, Jestin Y, Pavesi L, Balboni R. (2013). Quantum effects in silicon for photovoltaic applications. *Phys. Status Solidi A* 210:1071–1075.

Intartaglia R, Bagga K, Brandi F. (2014). Study on the productivity of silicon nanoparticles by picosecond laser ablation in water: Towards gram per hour yield. *Opt. Express* 22:3117–3127.

Intartaglia R, Bagga K, Brandi F, Das G, Genovese A, Di Fabrizio E, Diaspro A. (2011). Optical properties of femtosecond laser-synthesized silicon nanoparticles in deionized water. *J. Phys. Chem. C* 115:5102–5107.

Intartaglia R, Bagga K, Scotto M, Diaspro A, Brandi F. (2012a). Luminescent silicon nanoparticles prepared by ultra short pulsed laser ablation in liquid for imaging applications. *Opt. Mater. Express* 2:510–518.

Intartaglia R, Barchanski A, Bagga K, Genovese A, Das G, Wagener P, Di Fabrizio E, Diaspro A, Brandi F, Barcikowski S. (2012b). Bioconjugated silicon quantum dots from one-step green synthesis. *Nanoscale* 4:1271–1274.

Ishii S, Sugavaneshwar RP, Chen K, Dao TD, Nagao T. (2016). Solar water heating and vaporization with silicon nanoparticles at mie resonances. *Opt. Mater. Express* 6:640–648.

Jambois O, Rinnert H, Devaux X, Vergnat M. (2005). Photoluminescence and electroluminescence of size-controlled silicon nanocrystallites embedded in SiO_2 thin films. *J. Appl. Phys.* 98:046105.

Kang Z, Liu Y, Tsang CHA, Ma DDD, Fan X, Wong NB, Lee ST. (2009). Water-soluble silicon quantum dots with wavelength-tunable photoluminescence. *Adv. Mater.* 21:661–664.

Klein S, Dell'Arciprete ML, Wegmann M, Distel LV, Neuhuber W, Gonzalez MC, Kryschi C. (2013). Oxidized silicon nanoparticles for radiosensitization of cancer and tissue cells. *Biochem. Biophys. Res. Commun.* 434:217–222.

Klein S, Sommer A, Dell'Arciprete ML, Wegmann M, Ott SV, Distel LV, Neuhuber W, Gonzalez MC, Kryschi C. (2014). Oxidized silicon nanoparticles and iron oxide nanoparticles for radiation therapy. *J. Nanomater. Mol. Nanotechnol.* S2:002.

Ledoux G, Gong J, Huisken F, Guillois O, Reynaud C. (2002). Photoluminescence of size-separated silicon nanocrystals: Confirmation of quantum confinement. *Appl. Phys. Lett.* 80:4834–4836.

Ledoux G, Guillois O, Porterat D, Reynaud C, Huisken F, Kohn B, Paillard V. (2000). Photoluminescence properties of silicon nanocrystals as a function of their size. *Phys. Rev. B* 62:15942.

Lee BG, Luo J-W, Neale NR, Beard MC, Hiller D, Zacharias M, Stradins P, Zunger A. (2016). Quasi-direct optical transitions in silicon nanocrystals with intensity exceeding the bulk. *Nano Lett.* 16:1583–1589.

Li ZF, Ruckenstein E. (2004). Water-soluble poly (acrylic acid) grafted luminescent silicon nanoparticles and their use as fluorescent biological staining labels. *Nano Lett.* 4:1463–1467.

Liao B, Wang W, Deng X, He B, Zeng W, Tang Z, Liu Q. (2016). A facile one-step synthesis of fluorescent silicon quantum dots and their application for detecting Cu^{2+}. *RSC Adv.* 6:14465–14467.

Ligman RK, Mangolini L, Kortshagen UR, Campbell SA. (2007). Electroluminescence from surface oxidized silicon nanoparticles dispersed within a polymer matrix. *Appl. Phys. Lett.* 90:061116.

Lin SW, Chen DH. (2009). Synthesis of water-soluble blue photoluminescent silicon nanocrystals with oxide surface passivation. *Small* 5:72–76.

Liu CY, Holman ZC, Kortshagen UR. (2008). Hybrid solar cells from P3HT and silicon nanocrystals. *Nano Lett.* 9:449–452.

Liu CY, Holman ZC, Kortshagen UR. (2010). Optimization of SiNC/P3HT hybrid solar cells. *Adv. Funct. Mater.* 20:2157–2164.

Liu P, Liang Y, Li H, Xiao J, He T, Yang G. (2013a). Violet-blue photoluminescence from Si nanoparticles with zinc-blende structure synthesized by laser ablation in liquids. *AIP Adv.* 3:022127.

Liu SM, Kobayashi M, Sato S, Kimura K. (2005). Synthesis of silicon nanowires and nanoparticles by arc-discharge in water. *Chem. Commun.* 2005:4690–4692.

Liu Z, Sun Y, Yuan J, Wei H, Huang X, Han L, Wang W, Wang H, Ma W. (2013b). High-efficiency hybrid solar cells based on polymer/PbSxSe1-x nanocrystals benefiting from vertical phase segregation. *Adv. Mater.* 25:5772–5778.

Luterova K, Dohnalova K, Svrcek V, Pelant I, Likforman J-P, Cregut O, Gilliot P, Honerlage B. (2004). Optical gain in porous silicon grains embedded in sol-gel derived SiO_2 matrix under femtosecond excitation. *Appl. Phys. Lett.* 84:3280–3282.

Mahdieh MH, Momeni A. (2015). From single pulse to double pulse ns laser ablation of silicon in water: Photoluminescence enhancement of silicon nanocrystals. *Laser Phys.* 25:015901.

Maier-Flaig F, Kubel C, Rinck J, Bocksrocker T, Scherer T, Prang R, Powell AK, Ozin GA, Lemmer U. (2013b). Looking inside a working SiLED. *Nano Lett.* 13:3539–3545.

Maier-Flaig F, Rinck J, Stephan M, Bocksrocker T, Bruns M, Kubel C, Powell AK, Ozin GA, Lemmer U. (2013a). Multicolor silicon light-emitting diodes (SiLEDs). *Nano Lett.* 13:475–480.

Manhat BA, Brown AL, Black LA, Ross JA, Fichter K, Vu T, Richman E, Goforth AM. (2011). One-step melt synthesis of water-soluble, photoluminescent, surface-oxidized silicon nanoparticles for cellular imaging applications. *Chem. Mater.* 23:2407–2418.

Mansour N, Momeni A, Karimzadeh R, Amini M. (2013). Surface effects on the luminescence properties of colloidal silicon nanocrystals in water. *Phys. Scr.* 87:035701.

Mardanian M, Nevar AA, Tarasenko NV. (2012). Optical properties of silicon nanoparticles synthesized via electrical spark discharge in water. *Appl. Phys. A* 112: 437–442.

Mardanian M, Tarasenko NV, Nevar AA. (2014). Influence of liquid medium on optical characteristics of the Si nanoparticles prepared by submerged electrical spark discharge. *Braz. J. Phys.* 44:240–246.

Mariotti D, Svrcek V, Hamilton JWJ, Schmidt M, Kondo M. (2012). Silicon nanocrystals in liquid media: Optical properties and surface stabilization by microplasma-induced non-equilibrium liquid chemistry. *Adv. Funct. Mater.* 22:954–964.

Mastronardi ML, Maier-Flaig F, Faulkner D, Henderson EJ, Kübel C, Lemmer U, Ozin GA. (2012). Size-dependent absolute quantum yields for size-separated colloidally-stable silicon nanocrystals. *Nano Lett.* 12:337–342.

Meier C, Gondorf A, Luttjohann S, Lorke A, Wiggers H. (2007). Silicon nanoparticles: Absorption, emission, and the nature of the electronic bandgap. *J. Appl. Phys.* 101:103112.

Mitra S, Svrcek V, Mariotti D, Velusamy T, Matsubara K, Kondo M. (2014). Microplasma-induce liquid chemistry for stabilizing of silicon nanocrystals optical properties in water. *Plasma Process. Polym.* 11:158.

Miyano M, Kitagawa Y, Wada S, Kawashima A, Nakajima A, Nakanishi T, Ishioka J, Shibayama T, Watanabe S, Hasegawa Y. (2016). Photophysical properties of luminescent silicon nanoparticles surface-modified with organic molecules via hydrosilylation. *Photochem. Photobiol. Sci.* 15:99–104.

Momeni A, Mahdieh MH. (2015). Double-pulse nanosecond laser ablation of silicon in water. *Laser Phys. Lett.* 12:076102.

Muller A, Ghosh M, Sonnenschein R, Woditsch P. (2006). Silicon for photovoltaic applications. *Mater. Sci. Engl. B* 134:257–262.

Park E, Yoo H, Lee J, Park M-S, Kim Y-J, Kim H. (2015). Dual-size silicon nanocrystal-embedded SiO_x nanocomposite as a high-capacity lithium storage material. *ACS Nano* 9:7690–7696.

Park MS, Park E, Lee J, Jeong G, Kim KJ, Kim JH, Kim Y-J, Kim H. (2014). Hydrogen silsequioxane-derived Si/SiOx nanospheres for high-capacity lithium storage materials. *ACS Appl. Mater. Interfaces* 6:9608–9613.

Pavesi L, Dal Negro L, Mazzoleni C, Franzo G, Priolo F. (2000). Optical gain in silicon nanocrystals. *Nature* 408:440–444.

Pavesi L, Lockwood DJ. (2004). *Silicon photonics*. Springer-Verlag, Heidelberg, Berlin.

Pavesi L, Turan R. (2010). *Silicon nanocrystals: Fundamentals, synthesis and applications*. Weinheim: Wiley.

Pelant I. (2011). Optical gain in silicon nanocrystals: Current status and perspectives. *Phys. Status Solidi A* 208:625–630.

Peng F, Su Y, Zhong Y, Fan C, Lee S-T, He Y. (2014). Silicon nanomaterials platform for bioimaging, biosensing, and cancer therapy. *Acc. Chem. Res.* 47:612–623.

Photopoulos P, Nassiopoulou AG. (2000). Room- and low-temperature voltage tunable electroluminescence from a single layer of silicon quantum dots in between two thin SiO_2 layers. *Appl. Phys. Lett.* 77:1816–1818.

Popovic DM, Chai JS, Zekic AA, Trtica M, Momcilovic M, Maletic S. (2013). Synthesis of silicon-based nanoparticles by 10.6 μm nanosecond CO_2 laser ablation in liquid. *Laser Phys. Lett.* 10:026001.

Popovic DM, Chai JS, Zekic AA, Trtica M, Stasic J, Sarvan MZ. (2014). The influence of applying the additional continuous laser on the synthesis of silicon-based nanoparticles by picosecond laser ablation in liquid. *Laser Phys. Lett.* 11:116101.

Premnath P, Tan B, Venkatakrishnan K. (2015). Engineering functionalized multi-phased silicon/silicon oxide nano-biomaterials to passivate the aggressive proliferation of cancer. *Sci. Rep.* 5:12141.

Priolo F, Gregorkiewicz T, Galli M, Krauss TF. (2014). Silicon nanostructures for photonics and photovoltaics. *Nat. Nanotechnol.* 9:19–32.

Puzzo DP, Henderson EJ, Helander MG, Wang Z, Ozin GA, Lu Z. (2011). Visible colloidal nanocrystal silicon light-emitting diode. *Nano Lett.* 11:1585–1590.

Regli S, Kelly JA, Shukaliak AM, Veinot JGC. (2012). Photothermal response of photoluminescent silicon nanocrystals. *J. Phys. Chem. Lett.* 3:1793–1797.

Ren S, Chang LY, Lim SK, Zhao J, Smith M, Zhao N, Bulovic V, Bawendi M, Gradecak S. (2011). Inorganic–organic hybrid solar cell: Bridging quantum dots to conjugated polymer nanowires. *Nano Lett.* 11:3998–4002.

Rodio M, Brescia R, Diaspro A, Intartaglia R. (2016). Direct surface modification of ligand-free silicon quantum dots prepared by femtosecond laser ablation in deionized water. *J. Colloid Interface Sci.* 465:242–248.

Ruan J, Fauchet PM, Dal Negro L, Cazzanelli M, Pavesi L. (2003). Stimulated emission in nanocrystalline silicon superlattices. *Appl. Phys. Lett.* 83:5479–5481.

Sailor MJ, Wu EC. (2009). Photoluminescence-based sensing with porous silicon films, microparticles, and nanoparticles. *Adv. Funct. Mater.* 19:3195–3208.

Semaltianos N, Logothetidis S, Perrie W, Romani S, Potter R, Edwardson S, French P, Sharp M, Dearden G, Watkins K. (2010). Silicon nanoparticles generated by femtosecond laser ablation in a liquid environment. *J Nanopart. Res.* 12:573–580.

Shirahata N, Hasegawa T, Sakka Y, Tsuruoka T. (2010). Size-tunable UV-luminescent silicon nanocrystals. *Small* 6:915–921.

Singh V, Yu Y, Sun QC, Korgel B, Nagpal P. (2014). Pseudo-direct bandgap transitions in silicon nanocrystals: Effects on optoelectronics and thermoelectrics. *Nanoscale* 6:14643–14647.

Song C, Zhong Y, Jiang X, Peng F, Lu Y, Ji X, Su Y, He Y. (2015). Peptide-conjugated fluorescent silicon nanoparticles enabling simultaneous tracking and specific destruction of cancer cells. *Anal. Chem.* 87:6718–6723.

Švrček V, Kondo M. (2009). Blue luminescent silicon nanocrystals prepared by short pulsed laser ablation in liquid media. *Appl. Surf. Sci.* 255:9643–9646.

Švrček V, Kondo M. (2010). Blue light emitting silicon nanocrystals prepared by laser ablation of doped Si wafers in water. *J. Laser Micro Nanoen.* 5:103–108.

Švrček V, Mariotti D, Blackley RA, Zhou WZ, Nagai T, Matsubara K, Kondo M. (2013). Semiconducting quantum confined silicon-tin alloyed nanocrystals prepared by ns pulsed laser ablation in water. *Nanoscale* 5:6725–6730.

Švrček V, Mariotti D, Hailstone R, Fujiwara H, Kondo M. (2008). Luminescent colloidal silicon nanocrystals prepared by nanoseconds laser fragmentation and laser ablation in water. *Mater. Res. Soc. Symp. Proc.* 2066:A18.

Švrček V, Mariotti D, Kalia K, Dickinson C, Kondo M. (2011a). Formation of single-crystal spherical particle architectures by plasma-induced low-temperature coalescence of silicon nanocrystals synthesized by laser ablation in water. *J. Phys. Chem. C* 115:6235–6242.

Švrček V, Mariotti D, Kondo M. (2009a). Ambient-stable blue luminescent silicon nanocrystals prepared by nanosecond-pulsed laser ablation in water. *Opt. Express* 17:520–527.

Švrček V, Mariotti D, Nagai T, Shibata Y, Turkevych I, Kondo M. (2011b). Photovoltaic applications of silicon nanocrystal based nanostructures induced by nanosecond laser fragmentation in liquid media. *J. Phys. Chem. C* 115:5084–5093.

Švrček V, Mariotti D, Shibata Y, Kondo M. (2010). A hybrid heterojunction based on fullerenes and surfactant-free, self-assembled, closely packed silicon nanocrystals. *J. Phys. D Appl. Phys.* 43:415402.

Švrček V, Sasaki T, Katoh R, Shimizu Y, Koshizaki N. (2009b). Aging effect on blue luminescent silicon nanocrystals prepared by pulsed laser ablation of silicon wafer in de-ionized water. *Appl. Phys. B* 94:133–139.

Švrček V, Sasaki T, Shimizu Y, Koshizaki N. (2006). Blue luminescent silicon nanocrystals prepared by ns pulsed laser ablation in water. *Appl. Phys. Lett.* 89:213113.

Švrček V, Slaoui A, Muller JC. (2004). Silicon nanocrystals as light converter for solar cells. *Thin Solid Films* 451:384–388.

Sychugov I, Fucikova A, Pevere F, Yang Z, Veinot JGC, Linnros J. (2014). Ultra-narrow luminescence linewidth of silicon nanocrystals and influence of matrix. *ACS Photonics* 1:998–1005.

Sychugov I, Pevere F, Luo J W, Zunger A, Linnros J. (2016). Single-dot absorption spectroscopy and theory of silicon nanocrystals. *Phys. Rev. B* 93:161413.

Tamarov KP, Osminkina LA, Zinovyev SV, Maximova KA, Kargina JV, Gongalsky MB, Ryabchikov Y, et al. (2014). Radio frequency radiation-induced hyperthermia using Si nanoparticle-based sensitizers for mild cancer therapy. *Sci. Rep.* 4:7034.

Tilli M, Motooka T, Airaksinen V-M, Franssila S, Paulasto-Krockel M, Lindroos V. (2015). *Handbook of silicon based MEMS materials and technologies.* William Andrew, Elsevier, London.

Umezu I, Minami H, Senoo H, Sugimura A. (2007). Synthesis of photoluminescent colloidal silicon nanoparticles by pulsed laser ablation in liquids. *J. Phys. Conf. Ser.* 59:392–395.

Vaccaro L, Sciortino L, Messina F, Buscarino G, Agnello S, Cannas M. (2014). Luminescent silicon nanocrystals produced by near-infrared nanosecond pulsed laser ablation in water. *Appl. Surf. Sci.* 302:62–65.

Veinot JGC. (2006). Synthesis, surface functionalization, and properties of freestanding silicon nanocrystals. *Chem. Commun.* 2006:4160–4168.

Walters RJ, Bourianoff GI, Atwater HA. (2005). Field-effect electroluminescence in silicon nanocrystals. *Nat. Mater.* 4:143–146.

Wang DC, Hao HC, Chen JR, Zhang C, Zhou J, Sun J, Lu M. (2015). White light emission and optical gains from a Si nanocrystal thin film. *Nanotechnology* 26:475203.

Wang G, Ji J, Xu X. (2014a). Dual-emission of silicon quantum dots modified by 9-ethylanthracene. *J. Mater. Chem. C* 2:1977–1981.

Wang J, Ye D-X, Liang G-H, Chang J, Kong J-L, Chen J-Y. (2014b). One-step synthesis of water-dispersible silicon nanoparticles and their use in fluorescence lifetime imaging of living cells. *J. Mater. Chem. B* 2:4338–4345.

Wang Q, Ni H, Pietzsch A, Hennies F, Bao Y, Chao Y. (2011). Synthesis of water-dispersible photoluminescent silicon nanoparticles and their use in biological fluorescent imaging. *J. Nanopart. Res.* 13:405–413.

Warner JH, Hoshino A, Yamamoto K, Tilley R. (2005). Water-soluble photoluminescent silicon quantum dots. *Angew. Chem. Int. Ed.* 117:4626–4630.

Wilcoxon J, Samara G. (1999). Tailorable, visible light emission from silicon nanocrystals. *Appl. Phys. Lett.* 74:3164–3166.

Wright M, Uddin A. (2012). Organic—Inorganic hybrid solar cells: A comparative review. *Sol. Energ. Mater. Sol. Cells* 107:87–111.

Wu FG, Zhang X, Kai S, Zhang M, Wang HY, Myers JN, Weng Y, Liu P, Gu N, Chen Z. (2015a). One-step synthesis of superbright water-soluble silicon nanoparticles with photoluminescence quantum yield exceeding 80%. *Adv. Mater. Interfaces* 2:1500360.

Wu JJ, Kondeti VSSK, Bruggeman PJ, Kortshagen UR. (2016). Luminescent, water-soluble silicon quantum dots via micro-plasma surface treatment. *J. Phys. D Appl. Phys.* 49:08LT02.

Wu S, Zhong Y, Zhou Y, Song B, Chu B, Ji X, Wu Y, Su Y, He Y. (2015b). Biomimetic preparation and dual-color bioimaging of fluorescent silicon nanoparticles. *J. Am. Chem. Soc.* 137:14726–14732.

Xin Y, Nishio K, Saitow K-I. (2015). White-blue electroluminescence from a Si quantum dot hybrid light-emitting diode. *Appl. Phys. Lett.* 106:201102.

Xu Y, Han Y. (2014). Effects of natural oxidation on the photoluminescence properties of Si nanocrystals prepared by pulsed laser ablation. *Appl. Phys. A* 117:1557.

Yang L, Liu Y, Zhong Y-L, Jiang X-X, Song B, Ji X-Y, Su Y-Y, Liao L-S, He Y. (2015). Fluorescent silicon nanoparticles utilized as stable color converters for white light-emitting diodes. *Appl. Phys. Lett.* 106:173109.

Yang S, Cai W, Zhang H, Xu X, Zeng H. (2009). Size and structure control of Si nanoparticles by laser ablation in different liquid media and further centrifugation classification. *J. Phys. Chem. C* 113:19091–19095.

Yang S, Li W, Cao B, Zeng H, Cai W. (2011). Origin of blue emission from silicon nanoparticles: Direct transition and interface recombination. *J. Phys. Chem. C* 115:21056–21062.

Yates JT. (1998). A new opportunity in silicon-based microelectronics. *Science* 279:335–336.

Zhang X, Chen X, Kai S, Wang H-Y, Yang J, Wu F-G, Chen Z. (2015). Highly sensitive and selective detection of dopamine using one-pot synthesized highly photoluminescent silicon nanoparticles. *Anal. Chem.* 87:3360–3365.

Zhao S, Pi X, Mercier C, Yuan Z, Sun B, Yang D. (2016). Silicon-nanocrystal-incorporated ternary hybrid solar cells. *Nano Energ.* 26:305–312.

Zhong Y, Peng F, Bao F, Wang S, Ji X, Yang L, Su Y, Lee S-T, He Y. (2013). Large-scale aqueous synthesis of fluorescent and biocompatible silicon nanoparticles and their use as highly photostable biological probes. *J. Am. Chem. Soc.* 135:8350–8356.

Zhong Y, Peng F, Wei X, Zhou Y, Wang J, Jiang X, Su Y, Su S, Lee ST, He Y. (2012). Microwave-assisted synthesis of biofunctional and fluorescent silicon nanoparticles using proteins as hydrophilic ligands. *Angew. Chem. Int. Ed.* 51:8485–8489.

Zhong Y, Sun X, Wang S, Peng F, Bao F, Su Y, Li Y, Lee S-T, He Y. (2015). Facile, large-quantity synthesis of stable, tunable-color silicon nanoparticles and their application for long-term cellular imaging. *ACS Nano* 9:5958–5967.

Zhou Y, Riehle FS, Yuan Y, Schleiermacher H-F, Niggemann M, Urban GA, Krüger M. (2010). Improved efficiency of hybrid solar cells based on non-ligand-exchanged CdSe quantum dots and poly(3-hexylthiophene). *Appl. Phys. Lett.* 96:013304.

Zou J, Kauzlarich SM. (2008). Functionalization of silicon nanoparticles via silanization: Alkyl, halide and ester. *J. Clust. Sci.* 19:341–355.

14 Surface-engineered silicon nanocrystals

Calum McDonald, Tamilselvan Velusamy, Davide Mariotti, and Vladimir Svrcek

Contents

14.1 INTRODUCTION

14.1.1 THE IMPORTANCE OF SILICON WITH QUANTUM CONFINEMENT

When the radius of a nanoparticle is reduced below its Bohr exciton radius, the nanoparticle electronic energy structure becomes confined, leading to a set of unique properties which enable a range of new applications. Quantum confinement is generally observed in the widening of the band gap, where the band gap energy becomes inversely proportional to particle size. The possibility to tune the energy gap by controlling particle size is an intriguing and exciting aspect of quantum-confined materials, which allows us to fine-tune material properties as per application requirements. In addition to an increase in the band gap energy, other changes following quantum confinement of the energy structure are as follows: a change in the oscillator strength; an enhancement of carrier life times, necessary for carrier multiplication or hot-carrier extraction; and a change in transition dynamics, for example, a shift toward direct band gap behavior for silicon.

As well as the particle size, the surface properties require important considerations for quantum-confined nanocrystals. Due to the high surface-to-volume ratio, which exists in zero-dimensional nanostructures, the influence of surface states on the overall properties becomes significant. At the surface of a nanocrystal, atoms experience vastly different conditions than at the core, such as increased strain, changes to crystal structure, and variations in surface terminations. Surface terminations play a vital role in nanocrystal properties, and therefore surface termination of an SiNC must be considered. SiNCs exhibit a wavefunction which overlaps with the surface states and thus the surface terminations influence the wavefunction of the core and therefore the energy structure. The surface termination of SiNCs has also been well demonstrated to affect both the photoluminescence (PL) intensity and energy, and the absorption properties of the nanocrystal. The PL and absorption properties of a nanocrystal are usually a good indicator of the surface properties; however, it is important to note that in the case of SiNCs it is sometimes insufficient to distinguish between Si–C and Si–H terminations. PL emission from an indirect gap semiconductor such as Si is an interesting property in itself, given that a radiative recombination via an indirect transition requires the absorption or emission of a phonon in order to conserve crystal momentum. Certainly, different surface terminations are well observed to vary PL energy, while reducing particle size is also observed to strongly influence PL emission.

Following most synthesis techniques, SiNCs typically exhibit a hydride, halogen, or oxide surface, depending on the synthesis route. H-terminated SiNCs have large spectral tunability and are well-studied, but are chemically unstable and readily oxidize. In general, H terminations on SiNCs will be replaced by a hydroxyl/oxide termination in the presence of water. OH terminations provide a higher degree of passivation than H-terminated surfaces, are more stable than H terminated SiNCs, and display an enhanced PL intensity. However, over time, OH-terminated SiNCs will oxidize.

Oxidized SiNCs, while optically and chemically stable, have a red-shifted PL, a low PL quantum yield, low radiative rates, are difficult to surface engineer, and completely oxidized SiNCs have been shown to possess delocalized electronic states.[1] However, surface oxidation of SiNCs has been shown to improve the performance of SiNCs in solar cells.[2] As-prepared SiNCs often contain dangling bonds which act as carrier traps, while an oxidized surface favorably replaces dangling bonds with Si–O–Si bridge bonds and reduces carrier traps. It was well-noted by Ding et al.[2] that the amount of oxidation played a vital role in determining the effectiveness of the SiNCs in a hybrid solar cell; an oxidized SiNC surface was shown to improve solar cell device performance, while an excessively oxidized surface deteriorated the electrical properties and hampered device performance. Therefore, techniques for inducing oxide layer growth in a controlled manner are desirable.

In this chapter, we will see that SiNC properties are strongly dependent on the surface, and methods to modify surface terminations will be outlined. In particular, freestanding SiNCs will be discussed, where "freestanding" refers to unsupported SiNCs which are neither grown within a solid matrix nor on a solid substrate. While both freestanding nanocrystals and those supported by a matrix or on a substrate have individual merits, the latter two types do not have their surfaces exposed, and therefore access to the surface is not convenient which makes it difficult to tune their properties by surface engineering. Freestanding SiNCs refer to nanocrystals which are produced in liquid/gas phases or have been released from solid matrices, and are preferential for studies regarding surface engineering due to the ease of access to their surfaces. Therefore, since this chapter is concerned with the surface engineering of SiNCs, we will focus on freestanding nanocrystals.

Different types of surface terminations for SiNCs include short-chain terminations, ligands, and inorganic semiconductor coatings. Various wet-chemical methods exist, particularly for ligand attachment and termination with halides, however, these types of terminations are either unsuitable for electronic devices or highly unstable. The story emerging is that a controlled oxide layer on an SiNC is desirable for both passivating SiNCs while also maintaining the excellent optoelectronic properties of SiNCs. Probably the most effective and simple method to achieve surface passivation without long-chain ligands is the liquid–plasma chemistry method. This method can be used to produce conductive and stable SiNCs, modify the Fermi level, and enhance PL intensity.

14.1.2 COMPATIBILITY WITHIN PHOTOVOLTAICS

SiNCs are of particular interest due to the nature of the preexisting electronics industry, which is largely based on silicon wafer technology. The ability to combine SiNCs with Si-wafer technology is of paramount interest and poses the most significant application opportunity and development route for SiNCs in the near future. Since a well-established infrastructure for Si-based technologies already exists, it is unlikely that manufacturers will be persuaded to move away from this industry given their commitment and investment. Therefore, while silicon wafer technology still dominates, industrial investment is more likely to be available for technologies which expand on what is already available, without requiring any significant change. In addition, there is already a plethora of knowledge regarding the industrial production of high-quality crystalline silicon and the processing of raw materials, precursors, storage, and waste; to change to a new material would be a substantial move and require huge commitment across all steps of production. Currently, while silicon is so dominant and well established, integration of SiNCs in existing silicon wafer-based devices is most practical.

Examples of possible applications of SiNCs on Si-wafer technology are light-emitting devices,[3] flash memory,[4] solar cells,[5,6] single-electron devices,[7,8] and spintronics.[9] Many of these examples regard SiNCs grown in a solid matrix and therefore cannot be surface engineered, therefore the properties of the SiNCs cannot be optimized. One of the benefits of using SiNCs with bulk silicon photovoltaics (PV) is entailed in the larger band gap of SiNCs. Bulk silicon PV has a band gap of 1.1 eV and a wide absorption spectrum across the visible range of the solar spectrum. Photons absorbed with energy in excess of 1.1 eV generate hot carriers, and excess energy is lost through thermalization (i.e., phonon scattering) as carriers relax to the conduction band edge. Because of this, the optimum band gap energy for a single gap solar cell is roughly 1.5 eV. By employing a wider band gap material such as SiNCs on top of the Si wafer, it is possible to capture and make use of the high-energy light which is not used efficiently by bulk Si solar cells.

Another benefit of quantum-confined SiNCs is carrier multiplication: the process by which photons with energy equal or greater than twice the band gap of the nanocrystal can generate a second carrier via impact ionization. This is an effect observed minutely in all semiconductors under high-intensity illumination of at least twice the band gap energy; however, the process is profoundly inefficient such that there is no noticeable benefit. The effect is far more pronounced in semiconductors with quantum confinement due to the greater separation in the energy of available states, meaning that the relaxation of hot carriers must occur via a multiphonon scattering process, the probability for which is very low. Carrier lifetimes in quantum dots is therefore much higher than in bulk materials, which can theoretically increase the efficiency of solar cells significantly via the process of carrier multiplication.

In the context of photovoltaic applications, one of the drawbacks of SiNCs is the large band gap in the strong quantum confinement regime. Small particle sizes are highly desirable as this increases absorption cross section, carrier lifetimes, and exciton dissociation. However, these small SiNCs with diameters around 2–3 nm have a very large band gap close to or above 3 eV, which is not suitable for third generation photovoltaic concepts.[10] SiNCs with such a large band gap would capture a very small portion of the solar spectrum and carrier multiplication would be irrelevant, as this would require the absorption of photons with energy >6 eV for SiNCs with a band gap of 3 eV. If SiNCs are to be implemented into a photovoltaic device, a smaller band gap is required while maintaining the favorable properties intrinsic to quantum confinement. A smaller band gap is achieved for SiNC with larger diameters, however, this introduces surface defects and dangling bonds which must be well passivated through surface engineering. Therefore, tailoring absorption properties of SiNCs while retaining strong electronic transport and surface passivation is certainly a major challenge for the materials community.

14.1.3 ON THE ORIGIN OF PL IN SiNCs

Interest in SiNCs increased significantly following the report of red PL in electrochemically etched silicon by Canham et al. in 1990.[11] The discovery of bright PL in SiNCs is particularly interesting given that bulk silicon is highly inefficient at emitting light. As silicon is an indirect band gap semiconductor, light absorption and emission requires the simultaneous absorption or emission of a phonon in order to preserve crystal momentum[12]; therefore recombination in bulk silicon is largely nonradiative.

The exact origin of bright PL in quantum-confined Si has been the subject of discussion for the past two decades,[13–18] and is typically assigned to two models; (1) the quantum confinement model and (2) the surface chemistry model. In the quantum confinement model, the PL is attributed to the quantization of the electronic energy structure within the core of the nanocrystals. Evidence supporting the quantum confinement model has been reported.[16,19–28] In the surface chemistry model, the PL is attributed to highly localized surface defect states; excited carriers in the core states relax into lower lying defect states that slowly radiate. This model is supported experimentally in the literature.[16,24,29–31] Nonetheless, both size and surface termination certainly contribute to the energy and intensity of the PL.

14.1.4 QUANTUM-CONFINED SILICON

The Bohr exciton radius is about 4.2–4.6 nm for Si,[32] and quantum confinement has been shown to increase the indirect band gap from 1.1 eV in bulk silicon up to in excess of 3 eV for particles of ~2 nm diameter.[29,33] Quantum efficiencies approaching that of direct transitions have been observed[33,34]; as the indirect gap is increased, the direct gap is simultaneously decreased (from 3.32 eV) with increasing quantum confinement. In bulk silicon, the direct gap is skipped and electrons relax into the conduction band edge at the indirect gap very quickly. For quantum-confined SiNCs, as the particle size is reduced the indirect gap increases and direct gap decreases, leading to a change in the transition dynamics as direct transitions become more favorable.

Calculated band gap energies for SiNCs embedded in a lattice with varying sizes of the quantum structure are shown in Figure 14.1.[35–38] For structures with diameters <3 nm, there is an exponential increase in the band gap energy as the diameter is reduced further, and SiNCs with diameters of 2 nm are expected to have band gap as large as >3 eV, in agreement with the experimental results discussed above.[29,33] The increase in band gap energy must be accompanied by a shift in the position of valence and conduction band edges; this can be experimentally observed using X-ray and ultraviolet photoelectron spectroscopy techniques, and can be confirmed with effective mass approximation for the band edges.[39]

14.1.5 SURFACE PROPERTIES

While particle size has been demonstrated to determine the energy structure of a nanocrystal, the optical properties, particularly the PL, are also strongly influenced by the surface functionalization process. The optical properties of quantum-confined SiNCs are highly influenced by atoms/molecules/radicals, or

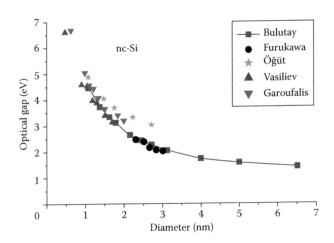

Figure 14.1 Theoretical and experimental band gap trend of quantum-confined SiNCs with H terminations in a matrix reported. (Reproduced from Bulutay, C., *Phys. Rev. B*, 76, 205321, 2007; initial data are reported elsewhere Öğüt, S., et al., *Phys. Rev. Lett.*, 79, 1770–1773, 1997; Vasiliev, I., et al., *Phys. Rev. Lett.*, 86, 1813–1816, 2001; Furukawa, S., and Miyasato, T., *Phys. Rev. B*, 38, 5726–5729, 1988.)

defects at or close to the surface. It is also worth noting that certain surface terminations play an important role in the stability, dispersity, and biocompatibility of the SiNCs in various media. Hydride-terminated SiNCs are not soluble in water; for water-soluble SiNCs, the nanocrystals must be functionalized with polar molecules. Hydride-terminated SiNCs are typically difficult to disperse in common solvents such as water. Toxicity and biocompatibility are also largely determined by the surface terminations; since certain ligand terminations are cytotoxic, they are not compatible in biological applications. Hence, the importance of surface functionalization goes well beyond electronic and optical properties. Surface engineering becomes highly important for integration in devices as it allows: (1) the passivation of the nanocrystal, (2) the tailoring of the optical and electronic properties, and (3) control of mixing/integration/compatibility. However, for the most part, the stability and the electrical and optical properties of the nanocrystal are considered first, while the challenges of integration and compatibility are addressed later. Therefore, this chapter will focus on (1) and (2).

Surface effects in nanocrystals concerning optical properties arise from two fundamental mechanisms:
1. The wavefunctions of carriers in the nanocrystals are delocalized over the whole nanocrystal to include its surface, and therefore the surface wavefunctions interact with the core nanocrystal wavefunction.
2. Due to the small volume of the SiNCs, carriers can easily diffuse to/from the core/surface of the nanocrystal.

The change in PL can arise from changes to surface chemistry, allowing for the tuning of particle properties not only through SiNC size. For example, PL tuning of SiNCs has been demonstrated by the attachment of surface ligands.[29,31,40,41] Differences in surface passivation have been reported to shift the PL emission of similar-sized SiNCs by up to 400 nm in wavelength.[29,31,41] Figure 14.2 shows the change in PL position for SiNCs terminated with different halides. Despite the large spectral tunability available with halides, this type of functionalization is highly unstable due to the reactivity of halides.

While varying surface terminations have been demonstrated to control the PL energy for SiNCs, PL tuning via particle size is still possible even for surfaces functionalized with long organic chains. K. Cheng et al. demonstrated that, for SiNCs synthesized via nonthermal plasma and with 1-dodecene surface terminations, the typical blue shift in PL is still observed as particle size is reduced.[42] The peak PL emission wavelengths were reported as 853 nm and 777 nm for SiNCs with diameters of 5 nm and 3 nm, respectively. The observed PL energy is far smaller than what is expected for H-terminated SiNCs, and the lower PL energy is attributed to recombination via surface ligands. This demonstrates that it is still possible to tune the band gap via particle size even when the surface is highly functionalized. Comparatively, for

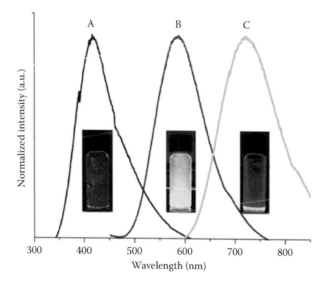

Figure 14.2 The photoluminescence spectra of hexyl-functionalized SiNCs in toluene with (A) chloride, (B) iodide, and (C) bromide surface terminations with particle diameters of approximately 3 nm. (From Dasog, M., et al., *Chem. Mater.*, 27, 1153–1156, 2015.)

H-terminated SiNCs, a reduction in particle size also leads to an increase and blue shift of the excitonic emission intensity.

14.2 SiNC SURFACE ENGINEERING TECHNIQUES

The initial surface termination plays an important role in the surface engineering process. The surface termination present on an SiNC is determined both by the synthesis route and the precursors used. Various methods exist for the synthesis of SiNCs; reduction of silicon halides,[43] steric stabilization,[33] oxidation of Zintl phases,[44] plasma-assisted magnetron sputtering,[45] atmospheric pressure plasma,[46–49] hydrosilylation,[50] and electrochemical etching.[51] Of the widely employed synthesis methods for SiNCs, the majority yield particles with a surface terminated by a hydride, halogen, or oxide. Therefore, the surface engineering technique should be compatible with these surfaces. In this section, several prominent methods for surface engineering will be discussed.

14.2.1 LIGAND SURFACE ATTACHMENT

The functionalization of SiNCs with ligands is a popular method for surface engineering as there is a vast array of possible ligand attachments, which allows for a significant amount of control over surface properties. Ligand attachment can stabilize the surface of the nanocrystals and provide steric hindrance. Alkanes, which are the simplest organic ligands consisting of only carbon and hydrogen, are the most commonly investigated ligand for surface functionalization, and can be easily attached to the surface of SiNCs via the hydrosilylation of an alkyl radical. Alkyl-terminated SiNCs are stable due to strong Si–C bonds which prevent photooxidation, provide steric protection, and also avert aggregation when in solution.

Alkyl chains tend to have a minimum effect on the optical gap of SiNCs because of the often type-I alignment of the energy levels. In type-I alignment, both the electron and hole are confined in the core, while in type-II alignment, one carrier is mostly confined to the shell and the other mostly confined to the core. Type-I alignment occurs when the highest occupied molecular orbital (HOMO) and lowest unoccupied molecular orbital (LUMO) of the core exist within the band gap of the shell energy levels, and have a minimal effect on the optical gap of the SiNCs.[40] Type-II band alignment is achieved, for example, through an energy band alignment in which both the HOMO and LUMO of the core are lower than their respective shell energy levels. This is shown schematically in Figure 14.3. Both type-I and type-II

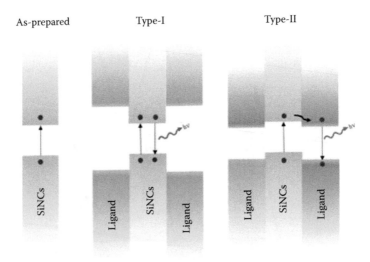

Figure 14.3 Band alignment for as-prepared silicon nanocrystals (SiNCs) and SiNCs with type-I alignment and type-II alignment via surface ligand functionalization. In a type-I alignment, the carriers are confined to the SiNCs and recombination occurs within the core, while in a type-II alignment the carriers can relax into the LUMO of the ligand and recombine.

alignments are possible with SiNCs, and type-II alignment can be used to modify the optical properties of the SiNC, which has been previously demonstrated.[52] In this manner, it is possible to control the optical properties without changing particle size and by only changing the ligand.

14.2.1.1 Hydrosilylation

Hydrosilylation is one of the most widely used methods for the attachment of alkyls to SiNC surfaces due to its simplicity, and involves the replacement of a typically H-terminated SiNC surface with Si–C, Si–O, Si–N, or Si–S bonds. Si–C bonds are predominantly favorable due to strong resistance to oxidation and compatibility with organic compounds, allowing the formation of surface-functionalized SiNCs with ligands.

The two most common types of hydrosilylation employed are thermal and photochemical hydrosilylation. Thermally induced hydrosilylation occurs via the thermally initiated cleavage of Si–H bonds followed by the formation of an Si–C bond. Thermal hydrosilylation is independent of particle size and does not require a catalyst, however, high reaction temperatures are necessary which limits the selection of alkenes and alkynes to those which can withstand such temperatures. Photochemical hydrosilylation occurs via UV-light-induced cleavage of Si–H bonds. Photochemical hydrosilylation can be performed at room temperature and in various solvents, which determines the type of surface functionalization obtained.

14.2.1.2 Advantages and disadvantages

The main advantage of both thermally and photochemically induced hydrosilylation is the ease and simplicity of the procedure. These techniques do not require much equipment and are therefore inexpensive and simple to set up. There is also a wide array of possible surface functionalization available owing to a high degree of tunability in terms of optical properties, solubility, and biocompatibility. This presents many opportunities for applications in biolabeling, for example, and thus a large amount of research interest exists in the hydrosilylation of SiNCs.

One of the major issues with hydrosilylation is that the attachment of long-chain alkyls inhibits the performance of SiNCs in electronic devices. Long-chain organic terminations are a hindrance particularly for electronic devices which require good transport properties. Ligands are not suitable for electronic devices as they inhibit certain electronic processes, such as carrier transport properties and exciton dissociation in photovoltaic devices. In order to produce useful electronic devices from ligand-terminated SiNCs, the electrically insulating native ligands must be exchanged in order to produce conductive SiNC arrays, introducing significant complexity into the fabrication process.

In addition to this, it is difficult to cover the entire surface of the particle uniformly with ligands, leading to irregularities in properties and stability; any untreated surfaces offer opportunities for inward oxidation. Thermal hydrosilylation is not effective for short-chain alkyls and is limited to alkyls which can withstand the high reaction temperatures necessary for thermal hydrosilylation, while the reaction dynamics and outcome of photochemical hydrosilylation is dependent on the initial particle size, and is not effective for particles with diameters >5 nm. Photochemical hydrosilylation reaction times are long, typically around a few days, and result in a decrease in initial particle size which must be factored in when designing surface engineering experiments.

14.2.2 INORGANIC SHELLS

The coating of SiNCs with inorganic shells is a remarkable yet highly challenging method for the control of SiNC surface properties. In this method, semiconductor quantum dots are typically contained within the shell of a semiconductor material. Like ligand attachment, this can be achieved in a type-I or type-II alignment, determined by the energy structure of the shell material relative to the SiNC. The type of alignment selected will depend on the properties of the SiNC that are desired.

SiNCs coated with CdS have been reported,[53] forming a type-II nanocrystal; absorption occurs via a transition within the core SiNC and recombination occurs via a transition in the CdS shell. In this work, SiNCs were synthesized initially by hydrogen reduction of $HSiO_{1.5}$ forming an Si–SiO_2 composite. The SiNCs were then removed by HF etching to yield H- and F-terminated SiNCs. Following extraction reflux, the SiNCs were capped with 1-decene to improve dispersion. The CdS shell was then grown by a successive ion layer adhesion and reaction (SILAR) technique. This work showed that the spectroscopic properties and the electrical conductivity of the Si–CdS nanocrystals can be controlled by varying the thickness of the CdS shell. Specifically, the absorption onset for Si–CdS NCs was reduced to a lower energy for a thicker CdS shell, though this is highly likely due to increased CdS absorption. PL decay times still indicated that the transition for thicker CdS shells remained within the shell.

14.2.2.1 Advantages and disadvantages

Inorganic shell coatings present an excellent method to passivate the SiNC surface. Since SiNCs easily oxidize, it is possible to select shell materials which are less prone to oxidation than Si. However, while the SILAR technique is quite simplistic in itself, it is rather limited with regard to which shell materials can be grown; typically, binary II–VI compounds such as CdS and ZnS. Currently, only type-II core-shell SiNCs have been reported using CdS, likely due to the limited compatibility of the SILAR method. In addition, the use of the toxic element cadmium defeats the ecological benefits of using silicon. The growth of other types of semiconductor shells is challenging and requires the development of many new techniques and understanding for growing uniform shells on nanocrystal surfaces with complete coverage.

For solar energy applications, the large band gap of SiNCs is unattractive as it does not match well with the solar spectrum. Since in type-II alignment the absorption occurs in the SiNC, the amount of energy harvestable will be very low, greatly reducing the efficiency of the device. As stated earlier, small SiNCs (1–3 nm) are desirable due to high-absorption cross sections, enhanced exciton transport, and higher carrier lifetimes. However, we have also seen that such small SiNCs have very large band gaps, unsuitable for solar energy applications. To address this, a transition from the HOMO of a surface state to the LUMO of the SiNC can generate an absorption edge lower than the band gap of SiNCs. This can be achieved via type-II alignment, and would require careful materials selection in order to correctly align the band energies of the materials; however, these materials may not necessarily be compatible with the SILAR method.

14.2.3 SURFACE ENGINEERING BY PLASMA–LIQUID CHEMISTRY

In this section, the surface functionalization of freestanding SiNCs in liquid media by atmospheric pressure microplasma will be presented. This type of surface engineering can be carried out during or post synthesis, while wet-chemical methods like hydrosilylation are mostly performed post synthesis. In this technique, the surface engineering of nanocrystals is carried out directly in a colloid, where the liquid media can be a wide range of solvents yielding different surface chemistries; however, this chapter will focus on the surface engineering in ethanol and water. A chemically active surface is required for plasma–liquid surface engineering, for example, Si–H, Si–OH, Si–Cl terminations, and the starting surface termination can determine the type of engineering achieved. Si–H terminations can be readily achieved following synthesis via electrochemical etching in hydrofluoric acid. Three main methods have been developed for surface engineering by atmospheric pressure plasma in liquid media; pulsed laser processes, direct-current (DC) plasma, and ultrahigh frequency (UHF) plasma.[54]

14.2.3.1 Plasma generated by laser

The laser-based surface engineering method uses a pulsed laser to generate a plasma plume in the colloidal solution. This technique is a combined fragmentation and surface engineering process which can be achieved with many types of lasers and is compatible with a wide range of solvents. This process is performed at room temperature for different time scales (from a few minutes to a few hours). The focal spot size of the laser used is typically 3 mm in diameter, with a focal length of 250 mm. Photothermal heating and Coulomb explosion processes occur in the colloidal solution which fragment the SiNCs and generate the plasma. The resulting surface chemistry obtained depends highly on the solvent used and is sensitive to the processing time. For water, this process transforms Si–H bonds to Si–OH after 15 min processing, and following 45 min processing, inward oxide growth commences at Si–Si back bonds and Si dimers observed by a decrease in PL intensity. Due to the stable oxygen-based terminations on the SiNCs after 15 min processing, oxide growth must be facilitated by strained bonds due to Si-dimers present on the SiNC surface. This undesirable slow oxide growth introduces further defects into the SiNC surface which reduces the SiNC core size and deteriorates PL properties. For surface engineering in ethanol, the process favors

Si–OH formation and yields a higher PL intensity and red shift. The growth of an oxide layer is restricted for surface engineering in ethanol and thus the PL properties are more stable.[47]

14.2.3.2 Plasmas generated in gas

This section will discuss both DC and UHF methods of microplasma-induced surface engineering. The resulting chemistries and surface engineering are very similar for both DC and UHF plasma treatment.[55–57] Both plasma processes stabilize the optical properties of the SiNCs and prevent degradation of the PL intensity even for SiNCs processed in water. For surface engineering in water, these methods produce hydrogen peroxide (H_2O_2) which contributes to the removal of surface defects and surface oxidation in combination with other chemical processes. In water, there is no observed shift in PL position for both DC and UHF processes.

In the DC atmospheric pressure plasma method, a plasma is generated between a metal tube and the surface of the colloid. The metal tube is made of either nickel or stainless steel, with an internal diameter of typically either 0.7 mm or 0.25 mm, and the plasma is generated between the metal tube and the surface of the colloid. A counter electrode is also used, which is made of either a carbon rod or metal wire, and is inserted approximately 5 mm in the solution, approximately 2 cm from the metal tube. The plasma is achieved using either pure He or Ar gas at a flow rate of 25–100 sccm with a constant current of 0.5–5 mA applied. The distance from the metal tube to the colloid is 0.5–1 mm, and this distance is adjusted throughout the reaction to account for solvent evaporation. The SiNC concentration in the colloidal solution, the type of solvent used, and the processing current all affect the outcome of the surface engineering process. Since the plasma is created between the metal tubing and the grounded solution, it is necessary for the colloid to be somewhat conductive in order to achieve a plasma. This is a slight disadvantage to this method and limits its applicability and flexibility, and therefore the UHF method was developed in which the plasma is generated within a quartz tube independent of the solution.[55–58]

In the UHF method, also referred to as the radio frequency (RF) method, a plasma is generated within a quartz capillary tube. The plasma can be generated independent of a counter electrode, and the process can be carried out with non-conductive colloids. The distance from the capillary tube to the surface of the colloid is 2 mm initially, which was also adjusted according to solvent evaporation. Although the resultant surface engineering is effectively the same, the physical processes involved in DC and UHF plasma processing are quite different.

14.2.3.3 Advantages and disadvantages

As these methods can be carried out in colloidal solution, they have the benefit of easy storage, and can be deposited at low cost via spin coating, screen printing, and spray coating. However, the experimental setup is more complicated than the wet-chemical methods, and therefore requires expertise in plasma configuration. The DC and UHF methods are not sufficient at fragmenting SiNCs, while the laser method is capable of highly fragmenting the SiNCs into small spherical particles, which is particularly useful for applications in electronic devices. However, the type of surface engineering achieved through the laser method is highly susceptible to reaction times and must be optimized accordingly.

14.3 SURFACE ENGINEERING OF DOPED SiNCs IN LIQUID PLASMA

In this section, we will discuss the surface engineering of p- and n-type SiNCs by atmospheric pressure plasma in liquid.

14.3.1 p-TYPE SiNCs

The surface engineering of p-type SiNCs has been studied in DC and RF plasma, and it is observed that both processes result in the same surface engineering.[55–58] DC and RF plasma processes in water produce mostly OH⁻ ions and OH radicals, and a small concentration of H_2O_2. SiNCs synthesized via electrochemical etching are H terminated with occasional Si dimers and dangling bonds. The surface engineering of H-terminated p-type SiNCs in water leads to the transformation of H terminations to OH terminations

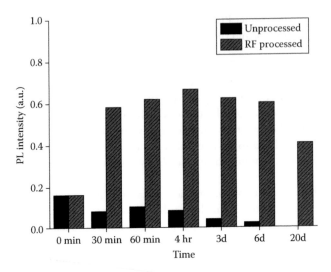

Figure 14.4 Enhanced photoluminescence stability of RF microplasma surface-engineered p-type silicon nanocrystals in water. (From Mitra, S., et al., *Plasma Process. Polym.*, 11, 158–163, 2014.)

via the highly reactive OH⁻ and OH species. These OH terminations in turn condensate into a thin surface oxide layer, preventing further inward oxidation. It was found that H_2O_2 can cleave Si back bonds of H-terminated SiNCs leading to inward oxidation; however, the rate of H_2O_2 formation is expected to be much lower than the formation of OH⁻ and OH species. The effect of surface engineering on p-type SiNCs in water is shown in Figure 14.4. The PL intensity is significantly increased owed to the formation of Si–O terminations and is stable over a period of 20 days.

In the case of plasma processing in ethanol, CH_3CH_2OH is dissociated into $CH_3CH_2O^-$ + H•. For p-type SiNCs with H terminations, these $CH_3CH_2O^-$ radicals can replace H terminations on SiNCs with Si–O–C_2H_5. Water and ethanol processing both yield SiNCs with similar peak PL, with a higher peak PL intensity observed for surface engineering carried out in water. The higher PL intensity is attributed to the formation of R–O terminations for surface engineering in water which presents a recombination pathway. This oxidation occurs very slowly in ethanol, possibly because of a small presence of water in the ethanol.

14.3.2 n-TYPE SiNCs

While the surface engineering of p-type SiNCs is established, n-type engineering is not well understood. Both p- and n-type SiNCs are required for particular electronic devices, and therefore an understanding of the engineering of both types is necessary. Velusamy et al. reported drastically different properties for the surface engineering of p- and n-type SiNCs under the same conditions.[59]

The differences in n-type and p-type SiNC surface engineering is shown schematically in Figure 14.5. The dissimilarities in surface termination can be attributed to the different types of allowed reactions which can take place at the surface of the SiNC. In short, a p-type surface exhibits surface holes (i.e., positively charged) while an n-type surface is electron rich. Thus, in the case of p-type SiNCs, the positively charged surface can react with negatively charged ions present in the solution (e.g., $CH_3CH_2O^-$ in the case of ethanol) formed due to the plasma. These Si–H terminations are replaced with more stable Si–O–R terminations. For n-type SiNCs, due to the electron-rich surface, the reaction of negatively charged ions is prohibited. Instead, slower reactions with OH, H_2O_2, and H_2O likely lead to the formation of Si-H_x, O_2SiH_2, and a surface oxide network of Si–O–Si bonds.

The effect of surface engineering on the PL spectrum of n- and p-type SiNCs is shown in Figure 14.5. As-prepared n- and p-type SiNCs both exhibit low PL intensity with the PL peak observed at roughly 600 nm and 550 nm, respectively. Following surface engineering in ethanol, the PL intensity of both n- and p-type SiNCs was enhanced dramatically. For surface-engineered n-type SiNCs, the peak PL position was red shifted by approximately 105 nm to 700 nm, while the peak PL of p-type SiNCs was red shifted by

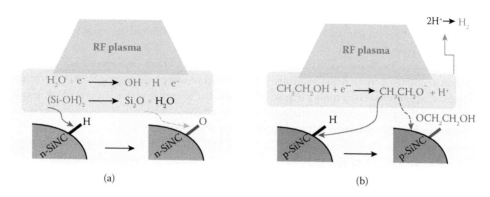

Figure 14.5 Comparison between the surface engineering of (a) n-type and (b) p-type silicon nanocrystals in ethanol. (From Velusamy, T., et al., *ACS Appl. Mater. Interfaces* 7, 28207–28214, 2015.)

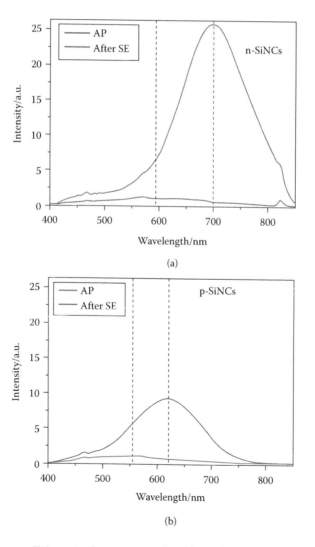

Figure 14.6 Photoluminescene (PL) spectra for as-prepared and for surface-engineered (SE) (a) n-type and (b) p-type silicon nanocrystals (SiNCs) in ethanol. The dotted lines indicate the PL maximum. (From Velusamy, T., et al., *ACS Appl. Mater. Interfaces* 7, 28207–28214, 2015.)

Clusters, nanoparticles, and quantum dots

approximately 70 nm to 620 nm. The significant increase in PL intensity is attributed to the formation of a surface oxide layer following surface engineering, presenting a recombination route for carriers via localized oxygen surface states. A red shift in peak PL position is well observed in the literature for oxidized SiNCs, which is in agreement with the results shown in Figure 14.6 for both n- and p-type SiNCs. A higher degree of oxidation is indicated by a greater red shift. The red shift observed for n-type SiNCs is greater than that of p-type SiNCs, indicative of a higher degree of oxidation. Following surface engineering, p-type SiNCs tend to exhibit OH- and R–O-terminations, while n-type SiNCs exhibit Si–O–Si oxide network, which was observed in Fourier transform infrared spectroscopy measurements.

A comparison of the PL stability is shown in Figure 14.7. The PL of surface-engineered n-type SiNCs is far less stable than the PL of p-type SiNCs. After storage for 7 days in ethanol, the peak PL of n-type SiNCs is blue shifted by approximately 70 nm and the PL intensity is significantly lower. For p-type SiNCs, the PL is blue shifted by approximately 20 nm. For n-type SiNCs there is a significant reduction in PL intensity after 7 days, while for p-type SiNCs there was a slight increase in PL intensity. The reduction in PL intensity for n-type SiNCs after 7 days is due to Si–O–Si back bond inward oxidation, which is not sufficiently prevented in n-type SiNCs, leading to strained bonds and a reduction in the size of the SiNC core, and thus the observed blue shift in the PL. The high stability observed in p-type SiNCs is due to a well-passivated surface of Si–O–R terminations which prevent back bond inward oxidation and provide good steric stability.

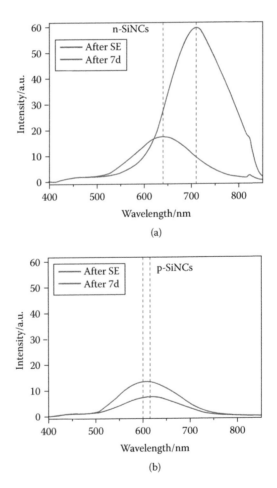

Figure 14.7 Photoluminescence (PL) spectra for SE silicon nanocrystals (SiNCs) following microplasma treatment in ethanol (red line) and after 7 days (blue line) for (a) n-type SiNCs SE for 60 minutes and (b) p-type SiNCs SE for 30 minutes, both in ethanol (the PL maxima is denoted by dotted lines in each spectra). (From Velusamy, T., et al., *ACS Appl. Mater. Interfaces* 7, 28207–28214, 2015.)

The effect of surface engineering on the Fermi level for n- and p-type SiNCs is demonstrated.[52] The position of the Fermi level was determined by scanning Kelvin probe, a technique which scans across a thin film and measures the contact potential difference between a conductive tip and sample surface, which can be translated into the Fermi level. The Fermi level is an important parameter for the design of photovoltaic devices, as this determines the energy band alignment and thus the barriers faced by carriers to travel between materials. The change in the Fermi level is attributed to changes in the surface terminations following surface engineering. Generally speaking, the formation of Si–O surface bonds leads to a change in the Fermi level due to a greater electronegativity of O compared to Si, leading to a shift in electrons from the Si to the O atoms and creating a surface dipole. These results confirm that the plasma–liquid surface engineering technique is not a "one size fits all" method. Different techniques must be developed for intrinsic, p-type, and n-type which are capable of achieving stable SiNCs with controllable and desirable properties.

14.4 SURFACE-ENGINEERED SiNC IN PHOTOVOLTAIC DEVICES

Organic PV have now surpassed 12% power conversion efficiency (PCE),[60] and present a potentially low-cost alternative to the conventional inorganic photovoltaic cells. The best cells are typically based on a bulk-heterojunction architecture, where the active layer consists of a mixture of conjugated donor and fullerene acceptor polymers. Organic solar cells are fabricated using polymers which can be deposited from solution via low-cost printing or coating techniques. These solution-phase coating methods can significantly reduce the cost of manufacture as they can be carried out without the need for high-temperature or high-vacuum atmospheres. There are two types of organic solar cells: (1) the organic bulk-heterojunction solar cell[61–63] based on an electron acceptor and a hole conductor, and (2) the dye-sensitized solar cell[64] based on a photo-anode sensitized by a dye, and an electrolyte for charge transfer.

The major challenge for organic photovoltaic devices such as the bulk-heterojunction cell surrounds the fullerene acceptor. Fullerenes are weak absorbers of visible light, their energy structure is difficult to tune, and production is both difficult and energy intensive. As well as this, fullerenes are unstable within the polymer matrix: fullerene molecules are prone to aggregate within the active layer of the solar cell device leading to a decrease in performance. The formation of large fullerene domains occurs via polymer diffusion which inhibits efficient electron collection. Domain formation can be overcome to some extent through using cosolvents during the deposition process. Despite this, alternatives to fullerenes remain desirable, fueling research both into alternative organic electron acceptors[65] and the emergence of the new field of hybrid solar cells.[66]

In a hybrid solar cell, the fullerene is replaced by an inorganic nanoparticle, typically a quantum-confined nanocrystal. This presents many opportunities to produce high-efficiency devices via band gap tuning, down conversion, and the much sought-after carrier multiplication, while still retaining the advantages of using low-cost and highly customizable organic donors. Plus, nanocrystals are typically much more absorbent than fullerenes, owing to increased contribution to device performance.

Semiconductor nanocrystals have a far broader absorption spectrum than organic compounds, can be doped p- or n-type, and have high surface-to-volume ratio providing a large interfacial area between QDs and the organic compound. Hybrid solar cells based on SiNCs with organic semiconductors have been reported,[2,67–70] most commonly in a bulk-heterojunction architecture. Liu et al. reported a hybrid solar cell based on SiNCs and poly(3-hexylthiophene) (P3HT) with a remarkable PCE of 1.15%[67] by direct replacement of the fullerene with SiNCs. In this case, the SiNCs were produced by nonthermal RF plasma from an SiH_4 precursor, and collection and handling was carried out in a nitrogen glovebox to reduce oxidation. This method expectedly yields H-terminated SiNCs due to the use of the SiH_4 precursor. Three groups of SiNCs were produced with diameters 3–5 nm, 5–9 nm, and 10–20 nm, with the smallest range of SiNCs producing the most efficient devices, likely due to reduced surface traps. Following this, Ding et al. reported a surface-engineered SiNC hybrid solar cell with PCE up to 2.2%.[68] Here, Cl-terminated SiNCs were etched to obtain H-terminated SiNCs. This work was followed up by the same group, who found that a loosely controlled oxidation process over a period of 12 hours yielded SiNC hybrid devices with 3.6% PCE. The oxidation process replaced dangling bonds with Si–O–Si bridge bonds, reducing the amount of trap sites on the SiNC surface. However, excessive oxidation (>12 hr) led to a reduction in PCE, attributed to a deterioration in electrical properties. This work can be tentatively categorized as surface engineering of SiNCs, however, since the SiNCs were simply exposed to oxygen for a given amount of time, this type of method will most probably present issues in reproducibility.

Surface engineering via plasma-induced liquid chemistry has been shown to enhance the performance of SiNC solar cells in a SiNC/polymer hybrid cell. Švrček et al. reported an improvement in solar cell performance for devices produced with surface-engineered SiNCs when compared with as-prepared SiNCs.[69] SiNCs were synthesized by the electrochemical etching of a p-type Si wafer yielding H-terminated SiNCs. Surface engineering by DC microplasma was carried out on the colloidal solution as a means to passivate the SiNC surface. The Si–H terminations along with surface defects are mostly transformed into Si–O–R terminations. This surface engineering led to an enhancement in the PCE, and can be observed in the external quantum efficiency shown in Figure 14.8. This is due to an enhanced exciton dissociation at the polymer interface due to the SiNC surface. While the PCE of the device is low (0.03%), the surface engineering provided to be a 150% improvement in the PCE when compared with untreated SiNCs.

Velusamy et al. also reported SiNC test devices based on both p-type and n-type surface-engineered SiNCs. These test devices were fabricated to gain an understanding of the effect of the surface engineering on the device properties for p- and n-type SiNCs. There was an enhancement in efficiency for p-type surface-engineered SiNCs from 0.002 to 0.005% PCE, mostly attributed to an enhanced fill factor. However, for n-type SiNCs there was a substantial decrease in PCE from 0.01 to 0.001%, attributed to a fall in J_{SC}. The reduced J_{SC} is likely due to the oxide layer, which when too thick can reduce the electrical properties; here the electron mobility was reduced. It is therefore necessary to improve control over the surface engineering process such that the thickness of the oxide layer can be optimized and stabilized, leading to favorable electrical properties.

Figure 14.9 is a schematic diagram displaying the n- and p-type doped silicon nanocrystal Fermi level engineering for the improvement in band alignment in solar cells. The device architecture is taken from the literature for a typical hybrid silicon nanocrystal solar cell.[2] The control of the degree of oxidation via RF microplasma, and thus the control of the Fermi level and electronic transport properties, can be employed to enhance band alignment and increase solar cell device performance. It is also possible to achieve a surfactant-free type-I or type-II alignment (Figure 14.3) through surface engineering. Controlled modification of the surface oxidation can lead to a band alignment in which carriers are either confined to the core, or transported to surface states where they are either extracted or recombine.

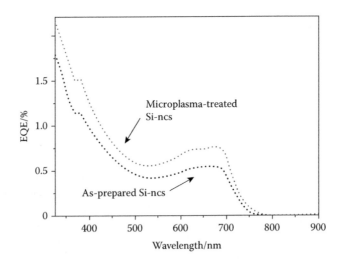

Figure 14.8 External quantum efficiency for a SiNC/PTB7 hybrid solar cell. Surface engineering by microplasma was observed to enhance the external quantum efficiency, and therefore provide an increase in the PCE. (From Švrček, V., et al., *Appl. Phys. Lett.*, 100, 223904, 2012.)

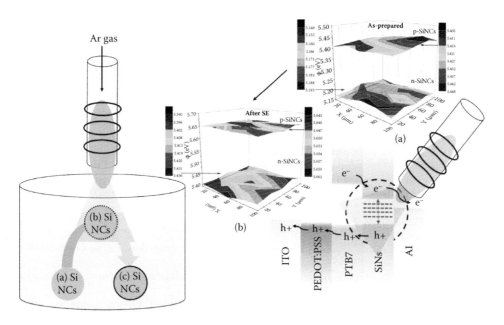

Figure 14.9 Schematic diagram displaying the silicon nanocrystals Fermi level engineering for improvement in band alignment in solar cells. The control of the degree of oxidation via RF microplasma, and thus the control of the Fermi level and electronic transport properties, can be employed to enhance band alignment and increase solar cell device performance. (Adapted from Velusamy, T., et al., *ACS Appl. Mater. Interfaces* 7, 28207–28214, 2015.)

14.5 CONCLUSION

In this chapter, we have overviewed doped SiNC surface engineering by plasma-induced chemistry that can significantly modify absorption and emission properties. The control of SiNC surface properties is highly important for producing useful nanocrystals for electronic and PV devices, while silicon itself possess many desirable features both physically and economically. We demonstrated that the plasma-induced surface chemistry at quantum confinement size is a powerful and flexible tool for the modification of surface chemistries of any nanocrystals, and can be achieved through DC, high-frequency, or laser processes in liquid media. These methods of surface engineering by plasma are particularly interesting given that they can be performed at atmospheric pressures, which is highly favorable for the incorporation into industry.

ACKNOWLEDGMENT

This work was partially supported by a NEDO projects, the Levehulme international network (IN-2012-136) and EPSRC (EP/K022237/1 and EP/M024938/1).

REFERENCES

1. Zhou, Z., Brus, L. & Friesner, R. Electronic structure and luminescence of 1.1- and 1. 4-nm silicon nanocrystals: Oxide shell versus hydrogen passivation. *Nano Lett.* **3**, 163–167 (2003).
2. Ding, Y., Sugaya, M., Liu, Q., Zhou, S. & Nozaki, T. Oxygen passivation of silicon nanocrystals: Influences on trap states, electron mobility, and hybrid solar cell performance. *Nano Energ.* **10**, 322–328 (2014).
3. Kim, K.-H. et al. Enhancement of light extraction from a silicon quantum dot light-emitting diode containing a rugged surface pattern. *Appl. Phys. Lett.* **89**, 191120 (2006).
4. Lien, Y.-C. et al. Fast programming metal-gate Si quantum dot nonvolatile memory using green nanosecond laser spike annealing. *Appl. Phys. Lett.* **100**, 143501 (2012).
5. Conibeer, G., Perez-Wurfl, I., Hao, X., Di, D. & Lin, D. Si solid-state quantum dot-based materials for tandem solar cells. *Nanoscale Res. Lett.* **7**, 193 (2012).
6. Conibeer, G. et al. Silicon quantum dot nanostructures for tandem photovoltaic cells. *Thin Solid Films* **516**, 6748–6756 (2008).

7. Lee, S., Lee, Y., Song, E. B. & Hiramoto, T. Observation of single electron transport via multiple quantum states of a silicon quantum dot at room temperature. *Nano Lett.* **14**, 71–77 (2013).

8. Chan, K. W. et al. Single-electron shuttle based on a silicon quantum dot. *Appl. Phys. Lett.* **98**, 212103 (2011).

9. Rokhinson, L. P., Guo, L. J., Chou, S. Y. & Tsui, D. C. Spintronics with Si quantum dots. *Microelectron. Eng.* **63**, 147–153 (2002).

10. de Boer, W. D. et al. Red spectral shift and enhanced quantum efficiency in phonon-free photoluminescence from silicon nanocrystals. *Nat. Nanotechnol.* **5**, 878–884 (2010).

11. Canham, L. T. Silicon quantum wire array fabrication by electrochemical and chemical dissolution of wafers. *Appl. Phys. Lett.* **57**, 1046–1048 (1990).

12. Iyer, S. S. & Xie, Y. H. Light emission from silicon. *Science* **260**, 40–46 (1993).

13. Hannah, D. C. et al. On the origin of photoluminescence in silicon nanocrystals: Pressure-dependent structural and optical studies. *Nano Lett.* **12**, 4200–4205 (2012).

14. Kanemitsu, Y. Luminescence properties of nanometer-sized Si crystallites: Core and surface states. *Phys. Rev. B* **49**, 16845–16848 (1994).

15. Puzder, A., Williamson, A. J., Grossman, J. C. & Galli, G. Surface chemistry of silicon nanoclusters. *Phys. Rev. Lett.* **88**, 097401 (2002).

16. Reboredo, F. A. & Galli, G. Theory of alkyl-terminated silicon quantum dots. *J. Phys. Chem. B* **109**, 1072–1078 (2005).

17. Godefroo, S. et al. Classification and control of the origin of photoluminescence from Si nanocrystals. *Nat. Nanotechnol.* **3**, 174–178 (2008).

18. Kovalev, D., Diener, J., Heckler, H., Polisski, G., Künzner, N. & Koch, F. Optical absorption cross sections of Si nanocrystals. *Phys. Rev. B* **61**, 4485–4487 (2000).

19. Kovalev, D., Heckler, H., Ben-Chorin, M., Polisski, G., Schwartzkopff, M. & Koch, F. Breakdown of the k-conservation rule in Si nanocrystals. *Phys. Rev. Lett.* **81**, 2803–2806 (1998).

20. Kůsová, K. et al. Brightly luminescent organically capped silicon nanocrystals fabricated at room temperature and atmospheric pressure. *ACS Nano* **4**, 4495–4504 (2010).

21. Groenewegen, V., Kuntermann, V., Haarer, D., Kunz, M. & Kryschi, C. Excited-state relaxation dynamics of 3-vinylthiophene-terminated silicon quantum dots. *J. Phys. Chem. C* **114**, 11693–11698 (2010).

22. Wilson, W. L., Szajowski, P. F. & Brus, L. E. Quantum confinement in size-selected, surface-oxidized silicon nanocrystals. *Science* **262**, 1242–1244 (1993).

23. English, D. S., Pell, L. E., Yu, Z., Barbara, P. F. & Korgel, B. A. Size tunable visible luminescence from individual organic monolayer stabilized silicon nanocrystal quantum dots. *Nano Lett.* **2**, 681–685 (2002).

24. Sykora, M., Mangolini, L., Schaller, R. D., Kortshagen, U., Jurbergs, D. & Klimov, V. I. Size-dependent intrinsic radiative decay rates of silicon nanocrystals at large confinement energies. *Phys. Rev. Lett.* **100**, 067401 (2008).

25. Mastronardi, M. L. et al. Preparation of monodisperse silicon nanocrystals using density gradient ultracentrifugation. *J. Am. Chem. Soc.* **133**, 11928–11931 (2011).

26. Belomoin, G. et al. Observation of a magic discrete family of ultrabright Si nanoparticles. *Appl. Phys. Lett.* **80**, 841 (2002).

27. Mastronardi, M. L. et al. Size-dependent absolute quantum yields for size-separated colloidally-stable silicon nanocrystals. *Nano Lett.* **12**, 337–342 (2012).

28. Shirahata, N. Colloidal Si nanocrystals: A controlled organic–inorganic interface and its implications of color-tuning and chemical design toward sophisticated architectures. *Phys. Chem. Chem. Phys.* **13**, 7284 (2011).

29. Wolkin, M. V., Jorne, J., Fauchet, P. M., Allan, G. & Delerue, C. Electronic states and luminescence in porous silicon quantum dots: The role of oxygen. *Phys. Rev. Lett.* **82**, 197–200 (1999).

30. Žídek, K. et al. Femtosecond luminescence spectroscopy of core states in silicon nanocrystals. *Opt. Express* **18**, 25241–25249 (2010).

31. Dasog, M., Bader, K. & Veinot, J. G. C. Influence of halides on the optical properties of silicon quantum dots. *Chem. Mater.* **27**, 1153–1156 (2015).

32. Yoffe, A. D. Low-dimensional systems: Quantum size effects and electronic properties of semiconductor microcrystallites (zero-dimensional systems) and some quasi-two-dimensional systems. *Adv. Phys.* **42**, 173–262 (1993).

33. Holmes, J. D., Ziegler, K. J., Doty, R. C., Pell, L. E., Johnston, K. P. & Korgel, B. A. Highly luminescent silicon nanocrystals with discrete optical transitions. *J. Am. Chem. Soc.* **123**, 3743–3748 (2001).

34. Jurbergs, D., Rogojina, E., Mangolini, L. & Kortshagen, U. Silicon nanocrystals with ensemble quantum yields exceeding 60%. *Appl. Phys. Lett.* **88**, 233116 (2006).

35. Bulutay, C. Interband, intraband and excited-state direct photon absorption of silicon and germanium nanocrystals embedded in a wide band-gap lattice. *Phys. Rev. B* **76**, 205321 (2007). doi: http://dx.doi.org/10.1103/PhysRevB.76.205321.

36. Öğüt, S., Chelikowsky, J. R. & Louie, S. G. Quantum confinement and optical gaps in Si nanocrystals. *Phys. Rev. Lett.* **79**, 1770–1773 (1997).

37. Vasiliev, I., Öğüt, S. & Chelikowsky, J. R. *Ab initio* absorption spectra and optical gaps in nanocrystalline silicon. *Phys. Rev. Lett.* **86**, 1813–1816 (2001).

38. Furukawa, S. & Miyasato, T. Quantum size effects on the optical band gap of microcrystalline Si:H. *Phys. Rev. B* **38**, 5726–5729 (1988).

39. Lockwood, D., Lu, Z. & Baribeau, J. Quantum confined luminescence in Si/SiO2 superlattices. *Phys. Rev. Lett.* **76**, 539–541 (1996).

40. Zhou, T. et al. Bandgap tuning of silicon quantum dots by surface functionalization with conjugated organic groups. *Nano Lett.* **15**, 3657–3663 (2015).

41. Gupta, A., Swihart, M. T. & Wiggers, H. Plasma nanoparticle synthesis: Luminescent colloidal dispersion of silicon quantum dots from microwave plasma synthesis: Exploring the photoluminescence behavior across the visible spectrum. *Adv. Funct. Mater.* **19**, 696–703 (2009).

42. Cheng, K.-Y., Anthony, R., Kortshagen, U. R. & Holmes, R. J. High-efficiency silicon nanocrystal light-emitting devices. *Nano Lett.* **11**, 1952–1956 (2011).

43. Baldwin, R. K., Pettigrew, K. A., Ratai, E., Augustine, M. P. & Kauzlarich, S. M. Solution reduction synthesis of surface stabilized silicon nanoparticles. *Chem. Commun.* **23**, 1822–1823 (2002).

44. Nolan, B. M., Henneberger, T., Waibel, M., Fässler, T. F. & Kauzlarich, S. M. Silicon nanoparticles by the oxidation of $[Si_4]^{4-}$—And $[Si_9]^{4-}$-containing Zintl phases and their corresponding yield. *Inorg. Chem.* **54**, 396–401 (2015).

45. Huang, S. Y. et al. Customizing electron confinement in plasma-assembled Si/AlN nanodots for solar cell applications. *Phys. Plasmas* **16**, 123504 (2009).

46. Kortshagen, U. Nonthermal plasma synthesis of semiconductor nanocrystals. *J. Phys. D. Appl. Phys.* **42**, 113001 (2009).

47. Mariotti, D. & Sankaran, R. M. Microplasmas for nanomaterials synthesis. *J. Phys. D. Appl. Phys.* **43**, 323001 (2010).

48. Nozaki, T., Sasaki, K., Ogino, T., Asahi, D. & Okazaki, K. Microplasma synthesis of tunable photoluminescent silicon nanocrystals. *Nanotechnology* **18**, 235603 (2007).

49. Sankaran, R. M., Holunga, D., Flagan, R. C. & Giapis, K. P. Synthesis of blue luminescent Si nanoparticles using atmospheric-pressure microdischarges. *Nano Lett.* **5**, 537–541 (2005).

50. Sato, S. & Swihart, M. T. Propionic-acid-terminated silicon nanoparticles: Synthesis and optical characterization. *Chem. Mater.* **18**, 4083–4088 (2006).

51. Švrček, V. Ex situ prepared Si nanocrystals embedded in silica glass: Formation and characterization. *J. Appl. Phys.* **95**, 3158 (2004).

52. Li, H., Wu, Z., Zhou, T., Sellinger, A. & Lusk, M. T. Tailoring the optical gap of silicon quantum dots without changing their size. *Phys. Chem. Chem. Phys.* **16**, 19275–19281 (2014).

53. Wang, G. et al. Type-II core-shell Si-CdS nanocrystals: Synthesis and spectroscopic and electrical properties. *Chem. Commun. (Camb).* **50**, 11922–11925 (2014).

54. Mariotti, D., Mitra, S. & Švrček, V. Surface-engineered silicon nanocrystals. *Nanoscale* **5**, 1385 (2013).

55. Mitra, S., Švrček, V., Mariotti, D., Velusamy, T., Matsubara, K. & Kondo, M. Microplasma-induced liquid chemistry for stabilizing of silicon nanocrystals optical properties in water. *Plasma Process. Polym.* **11**, 158–163 (2014).

56. Mariotti, D., Švrcek, V., Hamilton. W. J., Schmidt, M. & Kondo, M. Silicon nanocrystals in liquid media: Optical properties and surface stabilization by microplasma induced non equilibrium liquid chemistry. *Adv. Funct. Mater.* **22**, 954–961 (2012).

57. Svrcek, V. et al. Dramatic enhancement of photoluminescence quantum yields for surface-engineered Si nanocrystals within the solar spectrum. *Adv. Funct. Mater.* **23**, 6051–6058 (2013).

58. Švrček, V., Mariotti, D. & Kondo, M. Microplasma-induced surface engineering of silicon nanocrystals in colloidal dispersion. *Appl. Phys. Lett.* **97**, 161502/1–161502/3 (2010).

59. Velusamy, T., Mitra, S., Macias-Montero, M., Svrcek, V. & Mariotti, D. Varying surface chemistries for p-doped and n-doped silicon nanocrystals and impact on photovoltaic devices. *ACS Appl. Mater. Interfaces* 7, 28207–28214 (2015).

60. S. S. Li., Y. Li., W. C. Zhao., S. Q. Zhang., S. Mukherjee., H. Ade., and J. H. Hou. Energy-Level Modulation of Small-Molecule Electron Acceptors to Achieve over 12% Efficiency in Polymer Solar Cells. *Adv. Mater.* 28, 9423-9429 (2016)..

61. Yu, G., Gao, J., Hummelen, J. C., Wudl, F. & Heeger, A. J. Polymer photovoltaic cells: Enhanced efficiencies via a network of internal donor-acceptor heterojunctions. *Science* **270**, 1789–1791 (1995).

62. Halls, J. J. M. et al. Efficient photodiodes from interpenetrating polymer networks. *Nature* **376**, 498–500 (1995).

63. Park, S. H. et al. Bulk heterojunction solar cells with internal quantum efficiency approaching 100%. *Nat. Photonics* **3**, 297–302 (2009).
64. O'Regan, B. & Grätzel, M. A low-cost, high-efficiency solar cell based on dye-sensitized colloidal TiO_2 films. *Nature* **353**, 737–740 (1991).
65. Sauvé, G. & Fernando, R. Beyond fullerenes: Designing alternative molecular electron acceptors for solution-processable bulk heterojunction organic photovoltaics. *J. Phys. Chem. Lett.* **6**, 3770–3780 (2015).
66. Milliron, D. J., Gur, I. & Alivisatos, A. P. Hybrid organic–nanocrystal solar cells. *MRS Bull.* **30**, 41–44 (2005).
67. Liu, C.-Y., Holman, Z. C. & Kortshagen, U. R. Hybrid solar cells from P3HT and silicon nanocrystals. *Nano Lett.* **9**, 449–452 (2009).
68. Ding, Y., Gresbacka, R., Liub, Q., Zhoua, S., Pic, X. & Nozakia, T. Silicon nanocrystal conjugated polymer hybrid solar cells with improved performance. *Nano Energ.* **9**, 25–31 (2014).
69. Švrček, V., Yamanari, T., Mariotti, D., Matsubara, K. & Kondo, M. Enhancement of hybrid solar cell performance by polythieno [3,4-b]thiophenebenzodithiophene and microplasma-induced surface engineering of silicon nanocrystals. *Appl. Phys. Lett.* **100**, 223904 (2012).
70. Zhao, S., Pi, X., Mercier, C., Yuan, Z., Sun, B. & Yang, D. Silicon-nanocrystal-incorporated ternary hybrid solar cells. *Nano Energ.* **26**, 305–312 (2016).

Clusters, nanoparticles, and quantum dots

15 Silicon nanocrystals doped with boron and phosphorous

Shu Zhou, Xiaodong Pi, Yi Ding, Firman Bagja Juangsa,
and Tomohiro Nozaki

Contents

15.1 INTRODUCTION

Since the first observation of photoluminescence (PL) of porous silicon (Si) in the visible range,[1] Si-based nanostructures have received increasing attention in the past few decades. As one of the most important Si-based nanostructures, Si nanocrystals (NCs) are promising in various fields ranging from optoelectronics to bioimaging, given the abundance and nontoxicity of Si.[2-6] To fully realize the potential application of SiNCs in these fields, a great deal of effort has been made to tailor the properties of SiNCs. By reducing the size of SiNCs to less than 5 nm, strong quantum confinement occurs, giving rise to a blue shift of the band gap energy of SiNCs, which enables expanding the PL of SiNCs to the whole visible spectrum.[7] Despite the PL energy, the ability to emit light efficiently is also critical. Through tuning the surface of SiNCs via surface functionalization, the PL quantum yields of SiNCs could be significantly improved to be above 60%,[8] on par with direct band gap semiconductor NCs. The surface functionalization may also modify the NC surface to be either hydrophilic or hydrophobic, making them suitable for wet-chemical processing.[9]

In addition to tuning the NC size and surface, doping is a promising way to design the properties of SiNCs. Although SiNCs may be doped by attaching a molecule or electrochemical cell to the

NC surface,[10–12] the doping level of SiNCs is unlikely to be accurately controlled by the remote doping. However, such a problem may be confronted by extrinsically inserting impurities into SiNCs. Up to now, a variety of dopants have been employed for the doping of SiNCs. These dopants can be divided into two categories according to their energy level positions in Si. Metal impurities with deep energy levels have been used to dope SiNCs for combined properties of SiNCs and the metal dopant. For instance, SiNCs doped with erbium (Er) can emit near-infrared PL at around 1.54 μm through the recombination of photogenerated carriers spatially confined in SiNCs, and the subsequent energy transfer to Er^{3+}.[13,14] Manganese (Mn)-doped SiNCs possess a combination of paramagnetic and optical properties that do not exist in undoped SiNCs.[15,16] Iron (Fe)-doped SiNCs with low toxicity show strong and stable PL in water, demonstrating great promise as bimodal agents for optical and magnetic imaging.[17,18]

Compared with metal-doped SiNCs, SiNCs that are doped with B and P have attracted more attention given the fact that they are actually the most widely employed acceptor and donor impurities in Si materials. The B and P doping not only significantly changes the concentration of free carriers that are confined in SiNCs, but also induces localized states (impurity levels) in the band gap of SiNCs, giving rise to added freedom in tuning the optical absorption and emission of SiNCs.[19–21] In this review, we will mainly focus on SiNCs that are doped with B and P; B- and P- doped SiNCs that are either freestanding or embedded in a dielectric matrix will be discussed. We will emphasize recent advances in the synthesis methods of B- and P-doped SiNCs as well as the understanding of their properties.

15.2 SYNTHESIS AND CHARACTERIZATIONS

15.2.1 SYNTHESIS METHODS

The synthesis of intrinsic SiNCs has been well established by using many approaches.[22] The doping of SiNCs is usually adapted from the synthesis of intrinsic SiNCs by including the dopant precursors into the synthesis systems.[23,24] Many research groups have demonstrated the B- and P-doping of SiNCs that are either freestanding, or embedded in a dielectric matrix.

For preparing freestanding B- and P-doped SiNCs, gas dopant precursors are frequently used by a wide range of gas-phase plasma approaches. Kortshagen et al.[25] and Pi et al.[26] employed an Ar–SiH$_4$ nonthermal plasma for synthesizing freestanding intrinsic SiNCs. By introducing dopant precursors (e.g., B_2H_6 and PH_3) into the Ar–SiH$_4$ plasma, freestanding doped SiNCs can be readily obtained.[27–29] Recently, Nozaki et al. have demonstrated that organic gas dopant precursors such as trimethylphosphite (TMP) can also be used to dope SiNCs by a SiCl$_4$-based nonthermal plasma.[30] In addition to nonthermal plasma, microplasma has been employed to produce freestanding P-doped SiNCs by Wiggers et al.[31,32] In this synthesis approach, the Si precursor (SiH$_4$) and dilute gases (Ar and H$_2$) are injected into a microplasma reaction chamber together with a dopant gas (PH$_3$). The microwave energy is used to ignite the plasma and NCs are formed by nucleation and growth processes in the plasma.

Freestanding B- and P-doped SiNCs have also been synthesized via wet-chemical processes. Kauzlarich et al.[33] produce P-doped SiNCs by co-reduction of SiCl$_4$ and PCl$_3$ with magnesium (Mg) in 1,2-dimethoxyethane (glyme) at 50°C. The resulted P-doped SiNCs are capped with octyl groups and soluble in organic solvents such as hexane. Electrochemical etching of bulk Si is a popular method to prepare porous Si in the nanometer regime.[34] Since bulk Si is usually technically doped with B or P, freestanding B- and P-doped SiNCs can thus be obtained by crumbling the porous Si, given the fact that the dopant present in the bulk Si remains in the subsequent NCs in such a top-down approach.[35] Analogous to electrochemical etching, mechanical ball milling of a doped Si wafer in ethanol followed by sedimentation stages also yields freestanding doped SiNCs.[36]

In contrast to freestanding SiNCs, solid dopant precursors are the most employed for doping SiNCs that are embedded in a dielectric matrix. A cosputtering technique has been established to synthesize B- and P-doped SiNCs that are embedded in an SiO$_2$ matrix by various groups.[37–40] Doped SiNCs dispersed in SiO$_2$ films were first prepared by incorporating solid dopant targets (B_2O_3 or P_2O_5) into the sputtering system. After high-temperature annealing in an N$_2$ gas atmosphere, dopants were incorporated into SiNCs as the NCs nucleated and grew in the films. Instead of using the B$_2$O$_3$ target, Sato et al.[41,42] usually synthesized B-doped SiNCs in an SiO$_2$ matrix by a cosputtering of silica and B targets. Perego et al.[43]

presented an alternative approach for the P-doping of SiNCs embedded in an SiO_2 matrix, by sequential deposition of an $SiO/P–SiO_2-SiO_2$ multilayer structure in an electron beam evaporation system. After high-temperature thermal treatment in an N_2 flux, the P atoms contained in the $P–SiO_2$ film segregated into the Si-rich region of the SiO_x films and were incorporated into SiNCs during the growing process. An advantage of this method was the easy control of the dimension and density of doped SiNCs through tuning the thickness and the stoichiometry of the SiO_x films. B- and P-doped SiNCs embedded in an SiO_2 matrix have also been synthesized by ion-implantation.[44–47] Dopants were introduced into the SiO_2 film containing SiNCs uniformly by a series of implantations of dopant ions. Control over the energies and fluencies for each implantation yielded a uniform concentration profile in SiNCs. To release the SiNCs from the oxide matrix, HF acid was usually employed to etch the matrix away.

15.2.2 DOPANT INCORPORATION

The properties of doped SiNCs are greatly dependent on the concentration of dopants that are incorporated into SiNCs. The accurate control of the dopant concentration is actively pursued in doping SiNCs for a variety of applications. A common method to actively tune the dopant concentration in SiNCs is to change the concentration of dopant in the reacting precursors, which may be realized by adjusting the fractional dopant flow rates in the plasma,[29] the sputtering rate of the dopant and Si target via the radio frequency power,[40] the energies and fluencies for implantation of dopant ions,[44] or the concentrations of P in $P–SiO_2$ grains for electron beam evaporation.[43] With the advances in the synthesis of B- and P-doped SiNCs, it has been recently realized that the concentrations of B and P in SiNCs can significantly exceed their solubility limits (0.3% for P and 1% for B) in bulk Si[48] which may enable novel applications. Pi et al.[27] measured the B and P concentrations in plasma-synthesized freestanding doped SiNCs by using inductively coupled plasma atomic emission spectroscopy (ICP-AES). The P concentrations of SiNCs could be widely modulated from 0.06 to 5.6% although the mean size of SiNCs was only 3 nm. By contrast, the B concentrations only varied from 0.1 to 0.3%. In another study of SiNCs with a mean size of 14 nm, Zhou et al.[29] determined the dopant concentration via a chemical titration method and found that the B and P concentrations in SiNCs could be widely modulated from 6 to 31% and from 1 to 18%, respectively. Rowe et al.[28] worked out the P concentrations in ~10 nm SiNCs ranging from 2 to 16% by energy-dispersive X-ray spectroscopy (EDX) measurements. Kauzlarich et al.[33] found the P concentration for SiNCs generated by a solution reduction route was ~6% by using the same method. For B and P codoped SiNCs synthesized by cosputtering, Fujii et al.[49] obtained the B and P concentrations after removing the embedding oxide matrix which were in the ranges of 9–28% and 1.6–7.2%, respectively, by means of ICP-AES measurements.

Imaging dopants in SiNCs is critical to understand if dopants are introduced into SiNCs or not when the dopant concentrations are rather high. Rowe et al.[28] performed an element mapping study for freestanding doped SiNCs by using high-angle annular darkfield (HAADF) scanning transmission electron microscopy (STEM) combined with EDX spectroscopy. The HAADF-STEM and corresponding EDX images for undoped SiNCs and SiNCs heavily doped with P in Figure 15.1a clearly show that P element was associated with the element of Si, indicating that SiNCs were indeed doped with P despite the high dopant concentrations (Figure 15.1b). Instead of EDX spectroscopy, electron energy loss spectroscopy (EELS) has been used for mapping both B and P in heavily doped SiNCs by Zhou et al.[29] and Fujii et al.[50] The results suggested that very large amounts of B and P were incorporated into the SiNCs and/or condensed on the SiNC surfaces. For P-doped SiNCs embedded in an SiO_2 matrix synthesized by ion-implantation, Duguay et al.[47] and Zacharias et al.[51] imaged the dopants at the atomic scale and demonstrated the efficient introduction of P atoms into SiNCs by means of atom probe tomography (APT).

The element mapping images provide necessary information with respect to the dopants in SiNCs, however, it remains unclear whether dopants have been incorporated on the NC surface or to the core of SiNCs. Knowledge of the location of dopants in SiNCs is critical because only dopants residing inside the NC core are regarded as producing free carriers effectively. For freestanding doped SiNCs, the preferential location of dopants may be obtained from the change of dopant concentration after etching the native oxide shell at the surface of SiNCs by HF, as shown in Figure 15.1c.[28] One can also estimate the dopant location by using surface analysis techniques such as X-ray photoelectron spectroscopy (XPS) which enables distinguishing the NC surface from the NC core.[29] To determine the preferential location of dopants in SiNCs that are

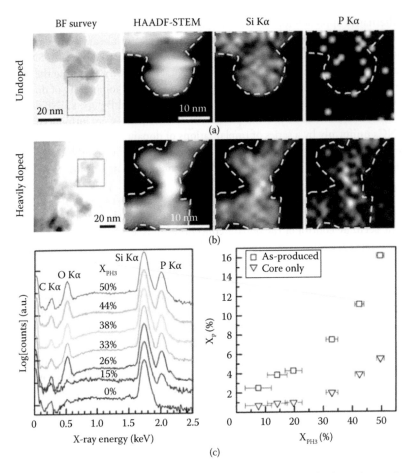

Figure 15.1 P incorporation in SiNCs. (a) Bright field (BF) TEM images for survey high-angle annular dark field (HAADF) STEM images and corresponding EDX maps for Si Kα and P Kα lines for undoped and heavily P-doped SiNCs. The red boxes indicate the analysis area. (b) Semilog plot of SEM-EDX spectra for varying X_{PH3} of "as-produced" SiNCs, with C, O, Si, and P Kα lines identified at 0.277, 0.525, 1.74, and 2.01 keV, respectively. Oxygen and carbon contamination are estimated at less than 3% and are the result of air exposure during sample transfer. Spectra are offset vertically for clarity. (c) Estimated X_P from SEM-EDX spectra for "as produced" samples (squares) and for samples after surface P had been removed to probe the SiNC core (inverted triangles). (Reprinted with permission from Rowe, D.J., et al., 2013, *Nano Lett.*, 13, 1317–1322. Copyright 2013 American Chemical Society.)

embedded in an SiO_2 matrix, depth profiling analysis by using secondary ion mass spectrometry (SIMS) has been proved to be effective by Kim et al.[52] and Perego et al.[43] Study on the location of dopants in SiNCs shows that the preferential location of dopants in SiNCs may greatly depend on the synthesis methods. For freestanding doped SiNCs synthesized by plasma, it is found that B prefers entering the NC core, while P prefers locating at the NC surface.[27–29] By contrast, for SiNCs embedded in an SiO_2 matrix synthesized by cosputtering and electron beam evaporation, P prefers locating at the NC core, while B prefers locating on the NC surface.[43,50,52,53] It looks like there are inconsistencies in the preferential location of dopants which suggests the mechanisms for doping freestanding SiNCs and SiNCs embedded in a matrix may be different.

Zhou et al.[29] have studied the doping mechanism for plasma-synthesized freestanding SiNCs and propose that the growth of doped SiNCs is mainly controlled by kinetics because the thermal equilibrium between SiNCs and background gas is absent in the plasma. The adsorption and desorption of B and P atoms at the NC surface are critical to understand the difference between the doping behavior of B and P. It is regarded that the adsorption of dopants is mainly enabled by the collision of B and P atoms with SiNCs in the plasma. At the same time, desorption of B and P from the NC surface may occur due to the bombardment of electrons and ions. By using orbital motion limited (OML) theory, Mangolini et al.[54] modeled the collision of atomic hydrogen with SiNCs in plasma and illustrated the energy transfer mechanism in the collision. The collisions of B and P atoms with SiNCs were modeled in the same way. After considering the residence time of an SiNC in the plasma, the collision frequency determined dopant concentration could be worked out, which allowed the derivation of the preferential dopant location. It was found that the dopant concentration usually increased from the NC center to the NC surface. This was consistent with the experimental result for P-doped SiNCs, while in contradiction to what was observed in B-doped SiNCs. The authors suggested that this discrepancy could be understood after considering the binding energy of B or P at the NC surface (i.e., 2.85 eV for bond strength of Si–B, 3.64 eV for bond strength of Si–P).[55] It is known that the average bond length of Si–Si bonds at the NC surface decreases with the increase of the NC size, consistent with the change of the surface curvature with the NC size. Because of the shift of the electron cloud of the Si atom shared by the Si–Si bond and Si–B (Si–P) bond toward the Si–Si bond, a shorter Si–Si bond enables a weaker neighboring Si–B (Si–P) bond. The weakening induced by the decrease of Si–Si bond length may be not be serious for P owing to its relatively large binding energy. But less strongly bound B can be affected to be vulnerable enough to dissociate at the NC surface, leading to the observation of the preferential doping in the NC core for B.

For SiNCs that are embedded in a matrix, the effect of a surrounding matrix on the doping of SiNCs should be considered. It is well known that the growth of SiNCs in a matrix requires long-time, high-temperature thermal treatment which enables the initial nucleation of the nanocluster and the following crystallization of the amorphous Si nanostructures in the matrix. In P-doped SiNCs that were embedded in an SiO_2 matrix, Perego et al.[43] suggested that P atoms were expected to segregate and to be trapped in the Si nanocluster region during the initial nucleation of nanocluster. Since the P diffusivity was much lower in SiO_2 than in Si, diffusion of P toward the Si-rich region was strongly favored which increases the probability for the P atom to be trapped in the SiNCs during the crystallization of the amorphous Si nanostructures in the matrix. Kim et al.[52] argued that for B-doped SiNCs embedded in an SiO_2 matrix, the diffusion of B into the Si-rich region comprised two different processes in terms of coordination energetics of the different crystalline sites. B atoms in the Si layers were found to preferentially substitute inactive threefold Si sites in the grain boundaries and then substitute the fourfold Si sites to achieve electrically active doping because the substitution of B in a threefold Si site removing a dangling bond was 0.1 eV more stable than that of the fourfold coordinated active B. In B and P codoped SiNCs that were embedded in an SiO_2 matrix, Fukuda et al.[53] further showed that the preferential location of dopants might be related to the segregation coefficient (k), which was defined by equilibrium concentration of impurity in Si over that in SiO_2. The small k of B (~0.3) and large k of P (>1) indicated that in codoped SiNCs B tended to locate in an outer shell, while P preferred an inner side of SiNCs.

15.2.3 NANOCRYSTAL STRUCTURE AND MORPHOLOGY

Structural characterizations of SiNCs doped with B and P in a wide concentration range have been carried out by means of X-ray diffraction (XRD) and Raman spectroscopy.[28,29,56] The XRD results for both undoped SiNCs and SiNCs doped with B and P indicate that the B and P-doping hardly change the diamond cubic structure of SiNCs. By contrast, significant red shift of vibration of Si–Si bonds together with a broadening toward lower wavenumbers are observed in the Raman spectra of SiNCs with various sizes after B doping, as shown in Figure 15.2a through c. The red shift of the Raman peak for Si–Si vibration bonds suggests the B doping causes significant tensile strain in SiNCs because a B atom is much smaller than an Si atom, similar to what Bustarret et al.[57] have observed for B-hyperdoped Si layers; whereas, the strain is negligible in P-doped SiNCs providing the small size mismatch between a P atom and an Si atom. No shift is thus observed in the Raman spectra of P-doped SiNCs.[29] The tensile strain of Si–Si bonds can be calculated by using $strain(\%) = \dfrac{\Delta\omega}{691.2} \times 100\%$, where $\Delta\omega$ is the change of the Raman shift for Si–Si bonds in SiNCs after B doping. As shown in Figure 15.2d, for ~6.8 nm SiNCs, the B-doping-induced tensile strain of Si–Si bonds increases from 1.2 to 4.2% as the B concentration changes from 17 to 60%. For smaller SiNCs, the B-doping-induced tensile strain of Si–Si bonds is reduced, which means the incorporation of B in smaller SiNCs introduces less significant changes of neighboring Si–Si bonds. The change of full widths at the half maximum (ΔFWHM) of the Raman peaks in Figure 15.2e increases with the B concentration, highlighting the tensile stress-induced structure distortion in SiNCs after doping.

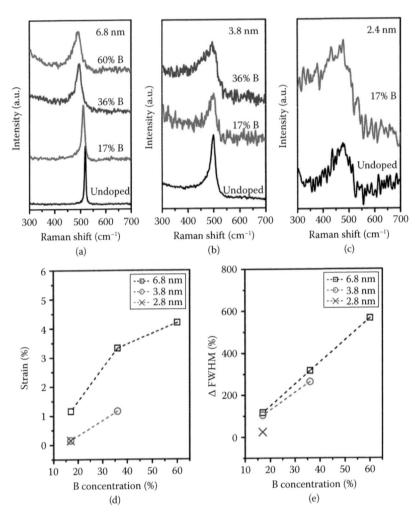

Figure 15.2 Raman spectra of (a) ~6.8, (b) 3.8, and (c) 2.4 nm undoped and B-doped SiNCs. (d) Dependence of the tensile strain of Si–Si bonds on the concentration of B in SiNCs. (e) Dependence of the changes in the FWHM (ΔFWHM) of the Raman peak for Si–Si bonds in B-doped SiNCs on the concentration of B in SiNCs. (Ni, Z., et al., *Adv. Opt. Mater.*, 2016, 4, 700–707. Copyright Wiley-VCH Verlag GmbH & Co. KGaA. Reproduced with permission.)

Despite the strain, the incorporation of dopants into SiNCs may induce variations of the lattice spacing of SiNCs. The lattice spacing of SiNCs doped with various B and P concentrations could be obtained from both the high-resolution transmission electron microscopy (HRTEM) and selected-area electron diffraction (SAED) images.[29] It is seen in Figure 15.3 that all the SiNCs in the transmission electron microscopy (TEM) images are spherical whether they are undoped or doped with B and P. The values of lattice spacing for the Si(111) plane are 0.314, 0.315, 0.314, 0.306, and 0.334 nm for undoped SiNCs, and SiNCs doped with 4% P, 18% P, 7% B, and 31% B, respectively. Clearly, the P-doping essentially does not affect the lattice spacing of SiNCs despite the high P concentration (18%). Again, this should be attributed to the similarity in the atomic size between P and Si and the excellent substitution of P atoms to Si atoms in P-doped SiNCs. While for B-doped SiNCs, the lattice spacing of the Si(111) plane is found to be reduced by 2.5% after SiNCs are doped with 7% B given the fact that the atomic size of B is smaller than Si. The current reduction of 2.5% for the lattice spacing of the Si(111) plane is basically consistent with Vegard's law. However, when the concentration of B increases to 31%, the lattice spacing of the Si(111) plane is enlarged by 6.4%. It is believed that a part of B atoms has been doped into the interstitial sites when the B concentration is high. Besides, it is also observed that twinning occurs to both B- and P-doped SiNCs

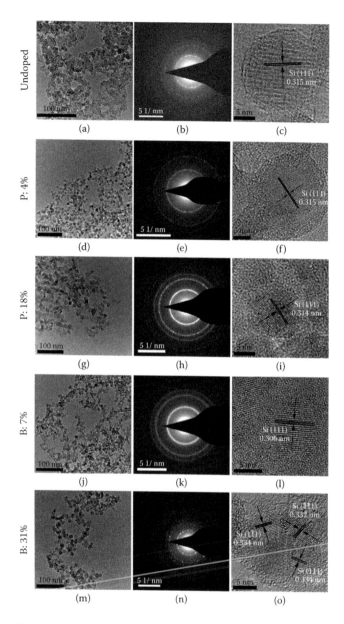

Figure 15.3 Low-magnification TEM images, selected area electron diffraction (SAED) images, and high-magnification TEM images of both undoped and doped SiNCs. (a) through (c) undoped SiNCs; (d) through (f) SiNCs doped with 4% P; (g) through (i) SiNCs doped with 18% P; (j) through (l) SiNCs doped with 7% B, and (m) through (o) SiNCs doped with 31% B. (Zhou, S., et al., Boron- and phosphorus-hyperdoped silicon nanocrystals, *Part. Part. Syst. Charact.*, 2015, 32, 213–221. Copyright Wiley-VCH Verlag GmbH & Co. KGaA. Reproduced with permission.)

when the dopant concentration is high. As the doping concentration increases, the twinning grain boundaries usually increase.[28] The growing twinning grain boundaries can act as scattering interfaces, retarding the movement of carriers in SiNCs.

In addition to changing lattice spacing, Xie et al.[40] indicated that doping might also result in changes of the SiNC morphology. Figure 15.4a through c show the cross-sectional HRTEM images of undoped and B-doped SiNCs embedded in SiO_2 films. It is seen that elliptical-like SiNCs are synthesized in the SiO_2 films for all SiNC samples. Similar to freestanding doped SiNCs, the B-doping-induced narrowing of the lattice spacing is also observed. Besides, change of the shape diversity of the SiNCs is found. The shape

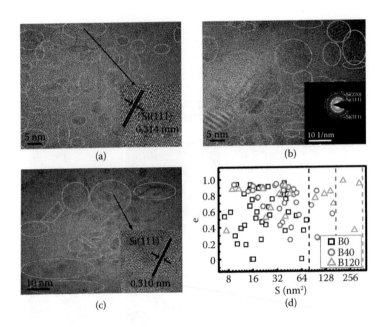

Figure 15.4 Cross-sectional HRTEM of intrinsic and B-doped SRSO films, 1100°C annealed, (a) B0, (b) B40, and (c) B120. The crystallites, detected by the Si lattice fringes, are marked by their borders. The insets in either (a) or (c) show the enlarged HRTEM image for certain regions as marked. The inset in (b) shows the selected area electron diffraction image. (d) Statistical analysis on the shape (eccentricity, e) and size (cross-sectional area, S) of Si–NCs with elliptical shape in (a) through (c). (Reprinted with permission from Xie et al. Appl. Phys. Lett., 102, 123108, 2013. Copyright ©2013 by American Institute of Physics)

diversity of the SiNCs becomes more and more obvious with the increase of B concentration, where trapeziform and trigonal-shaped particles can even be obtained in the image for the sample with the highest B concentration (B120) (Figure 15.4c). The shape of an SiNC may be described by eccentricity of which the value is always between 0 and 1. By calculating the average eccentricity of the SiNCs, one can compare the shape diversity. Figure 15.4d shows that the average eccentricity increases from ~0.57 for undoped SiNCs to ~0.74 for SiNCs with the highest B concentration (B120). To evaluate the size evolution, the authors chose to analysis the cross-sectional area of SiNCs instead. The result in Figure 15.4d indicates that the range of cross-sectional area also increases from 7–80 nm² to 7–330 nm² with the B concentration. This suggests the doping of B enhances the growth of SiNCs through heterogeneous nucleation and decreases the density of SiNCs in the films.

15.3 PROPERTIES

15.3.1 SURFACE CHEMISTRY

The surface chemistry of SiNCs plays a critical role in their applications in various fields. Since part of the doped impurities may inevitably reside at the NC surface, change in the surface chemistry of SiNCs occurs after doping. Figure 15.5 shows the Fourier transform infrared (FTIR) spectra for as-synthesized B- and P-doped SiNCs synthesized by plasma.[58] In Figure 15.5a, a phosphorus-hydride stretch mode at 2276 cm⁻¹ appears and becomes more pronounced as the P concentration increases. This suggests that the surface of the NC contains a significant amount of P. The boron-hydride and boron-oxide stretches at 2500 cm⁻¹, 1800 cm⁻¹, and 1400 cm⁻¹ in the FTIR spectrum of as-synthesized B-doped SiNCs in Figure 15.5b indicate that B also exists at the surface of the SiNCs. Clearly, these dopants residing at the NC surface lead to changes in the surface chemistry of SiNCs.

The variation of surface chemistry induced by dopants may greatly improve the dispersibility and stability of SiNCs in common solvents. Zhou et al.[59] demonstrated that colloidal stability without the use of ligands could be obtained for SiNCs heavily doped with B. The B-doped SiNC colloids were found to be very

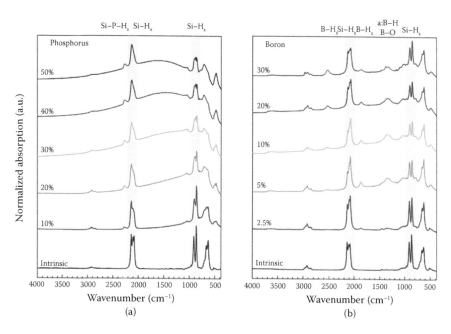

Figure 15.5 FTIR spectra of as-produced (a) P-doped and (b) B-doped SiNCs with increasing fractional doping flow rates. Absorption modes due to silicon-hydride, boron-hydride, and phosphorus-hydride related vibrations are observed, as indicated by the blue and orange regions. (Reprinted with permission from Krammer et al., *Nano Lett.*, 15, 5597–5603, 2015. Copyright 2013 American Chemical Society.)

stable in air, making them suitable for device processing in an ambient atmosphere. Measurement of the zeta potential of B-doped SiNCs indicated that ionized B atoms at the surface/subsurface of doped SiNCs produced a negative potential of −26 mV in the solvent (benzonitrile), approaching ±30 mV, which is usually considered to be sufficient for long-term stability. However, the dispersibility and stability of P-doped SiNCs in benzonitrile was found to be much worse than that of B-doped SiNCs because the P-induced surface potential was not large enough to effectively suppress the van der Waals force–induced agglomeration of SiNCs. Sugimoto et al.[49] and Sasaki et al.[60] additionally showed that SiNCs dispersible in methanol without surface functionalization could be realized by simultaneously doping B and P into SiNCs. Figure 15.6a shows the photos of B and P codoped SiNC colloids prepared after removing the precipitations by centrifugation. The B and P codoped SiNC colloids look transparent despite the dense color. Besides, the concentration of SiNCs remaining in the solution is usually larger for codoped SiNCs with higher B and P concentrations. This means the dispersibility of SiNCs is indeed improved after B and P codoping. The B and P codoped SiNCs were found to be stable in methanol for more than 5 months, while undoped SiNCs prepared in the same procedure formed large agglomerates immediately. The long-term stability suggests that a negative surface potential is likely formed in B and P codoped SiNCs, which is responsible for the colloidal stability of codoped SiNCs in solvent. The results of zeta potential measurements in Figure 15.6a confirm the speculation. The scanning electron microscopy (SEM) image in Figure 15.6b and optical microscopy image in Figure 15.6c representatively show a smooth and continuous film prepared by spin-coating a concentrated solution (~5 mg/ml) of codoped SiNCs on an Si substrate. It is the excellent dispersion of B and P codoped SiNCs in methanol that enables fabricating high-quality film via a low-cost solution process.

15.3.2 ELECTRONIC PROPERTIES

15.3.2.1 Dopant electrical activity

It is well known that only dopants that are electrical activated can produce free carriers in SiNCs. The electrical activity of B in SiNCs can be investigated by means of the Fano effect, which concerns the coupling between discrete optical phonons and the continuum of interband hole excitation in p-type Si.

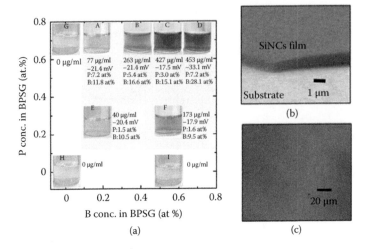

Figure 15.6 (a) Photographs of the solutions after removing precipitates by centrifugation. The abscissa and ordinate are B and P concentrations, respectively, in as-deposited Si-rich BPSG films. The numbers below the pictures are the amounts of SiNCs in the solution, P and B concentrations in Si–NCs, and the zeta-potentials. (b) SEM image and (c) optical microscope image of a spin-coated film from B and P codoped SiNC colloid. (a, reprinted with permission from Sugimoto, H., et al., *J. Phys. Chem. C*, 116, 17969, 2012. Copyright 2012 American Chemical Society; b and c, reprinted with permission from Sasaki, M., et al., *Phys. Chem. C*, 120, 195, 2016. Copyright 2016 American Chemical Society.)

Raman scattering measurements allow the direct observation of Fano broadening of Si optical phonon peaks caused by B doping in SiNCs. Figure 15.7a shows the Raman spectra for both undoped and B-doped SiNCs that are embedded in an SiO_2 matrix.[41] The Si optical phonon peak at ~521 cm^{-1} for undoped SiNCs shows an asymmetric spectral shape with a long tail toward the low wavenumbers due to the phonon confinement effect. After B doping, two peaks at ~618 and 640 cm^{-1}, which can be assigned to the local vibrational modes of ^{11}B and ^{10}B in Si crystal, are identified only for the B-doped SiNCs (inset of Figure 15.7a). The intensity ratio of these two peaks depends on the natural abundance of the two isotopes of ^{11}B (80.2%) and ^{10}B (19.8%), which is usually calculated to be approximately 4:1. In addition, the Si optical phonon peak is observed to be broadened toward high wavenumbers in B-doped SiNCs which is due to the Fano effect. The Fano broadening can also be examined by changing the excitation wavelength employed in the Raman measurements. By increasing the excitation wavelength from 532 to 785 nm, the Si optical phonon peak of B-doped SiNCs is broadened toward high wavenumbers and an antiresonance dip appears at the low wavenumber side of the peak, as shown in Figure 15.7b.[29] They are both characteristics of the Fano effect. The Raman results evidence that B atoms are doped into the substitute site of SiNCs and electrically activated in SiNCs.

The mass of P and Si atoms are very similar, rendering it difficult to detect the local vibrational mode of P atoms in SiNCs by using Raman spectroscopy. However, electron spin resonance (ESR) is a sensitive method to probe single-electron states such as P donors and conduction electrons, in P-doped SiNCs. Stegner et al.[31] performed an ESR study for both undoped and P-doped SiNCs. The spectrum of undoped SiNCs in Figure 15.8a displays a single broad resonance centered at g = 2.006, due to the nonbonding electrons on the three-coordinated Si atoms, named Si dangling bonds at the NC surface. For SiNCs doped with P ([P] = 1.3×10^{18} cm^{-3}), a new resonance centered at g = 1.998 appears. The g = 1.998 resonance is a typical signal in P-doped SiNCs, originating from the exchange-coupled P atoms at the substitutional sites of an SiNC core. Besides, a pair of lines, denoted hf(^{31}P), located symmetrically at the high- and low-field side of the g = 1.998 resonance with a magnetic field splitting of 4.1 mT are observed. These are the typical hyperfine signatures of substitutional P in crystalline Si, where the Zeeman states of the donor electrons are split by interaction with the ^{31}P nucleus (I = 1/2). The P-related electron paramagnetic resonances (EPRs) clearly evidence the existence of substitutional P atoms that are electrically activated in SiNCs.

Clusters, nanoparticles, and quantum dots

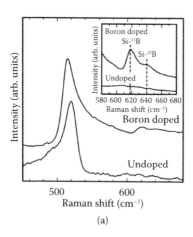
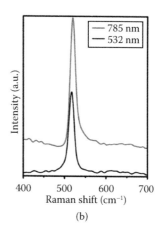

(a)   (b)

Figure 15.7 (a) Raman spectra for undoped and B-doped SiNCs. Inset shows the Raman spectra of local vibrational modes of B in undoped and B-doped SiNCs. (b) Raman spectra of SiNCs hyperdoped with B at the concentration of 7% obtained with the excitation of 532 and 785 nm lasers. (a, reprinted with permission from Sato, K., et al., *Appl. Phys. Lett.*, 94, 161902, 2009. Copyright 2009, American Institute of Physics; b, Zhou, S., et al., *Part. Part. Syst. Charact.*, 2015, 32, 213–221. Copyright Wiley-VCH Verlag GmbH & Co. KGaA. Reproduced with permission.)

As the P concentration increases to 1.2×10^{19} cm^{-3}, the P-induced resonance increases in intensity and prevails over the dangling bonds induced resonance in Figure 15.8a. This suggests P may effectively passivate the dangling bonds at the NC surface. The influence of P-doping on the surface dangling bonds is quantitatively investigated and shown in Figure 15.8b. It is clear that the average Si dangling bond density is reduced by approximately a factor of 2 when the P concentration is changed from 0 to 5×10^{20} cm^{-3}.

The variation of the P-related ESR signal intensity in Figure 15.8c indicates the activation efficiency of P decreases as the size of SiNCs decrease from 11 nm to 4.3 nm. Figure 15.8d shows the P-doping efficiency determined from the spin density of the g = 1.998 peak as a function of the SiNC diameter[32]. For SiNCs with diameters larger than 12 nm, the concentration of electrically active substitutional P is about one order of magnitude lower than the nominal atomic P concentrations of SiNCs. As the NC size decreases to be lower than 12 nm, however, the electrically active P concentration drops to values three orders of magnitude below the nominal atomic concentration. Such a remarkable drop of the P activation efficiency can be attributed to different reasons, such as a size-dependent incorporation probability of P atoms into SiNCs, or a reduced ESR visibility of P for smaller SiNCs due to broadening caused by quantum confinement of the donor electron wave function. Moreover, electron capture by surface Si dangling bonds, which leads to compensation of donors, can be enhanced for smaller SiNCs, where the surface-to-volume ratio is larger.

Zhou et al.[61] obtained the electrical activation efficiency for both B and P by measuring the free carrier absorption in doped SiNCs. The values of the activation efficiency of P in SiNCs with a mean size of ~14 nm were calculated to be 0.3–0.9%, similar to the results obtained by ESR measurements in Figure 15.8d, whereas the calculated values of the activation efficiency of B in SiNCs with the same size were ~2.8–5.2%. Clearly, B produces free carriers much more efficiently than P in SiNCs. However, it should be noted that activation efficiencies for both B and P are currently low. The authors also found that the activation efficiency decreased with the increase of the dopant concentration for both B and P. This indicates that the clustering of B and P may become serious as their concentrations increase, leading to the deactivation of dopants in SiNCs. Future work should be carried out to minimize the number of unionized dopants in SiNCs.

15.3.2.2 Electrical conductivity

The dopants that are electrically activated in SiNCs give rise to extra free carriers which play a critical role in the electronic conductivity of the assembly SiNC films. Understanding the doping effect on the electronic transport of SiNC films is fundamental for various applications. Xie et al.[40] investigated the electrical conductivity of B-doped SiNCs embedded in an SiO$_2$ matrix. The electrical property of the films

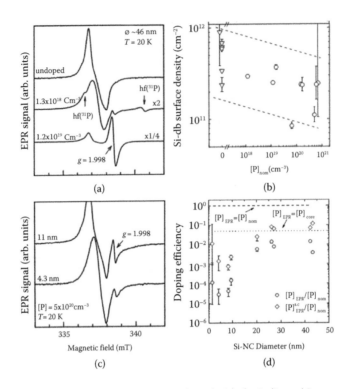

Figure 15.8 (a) EPR spectra of undoped SiNCs and SiNCs doped with the indicated P concentrations. (b) The density of surface Si dangling bonds is plotted as a function of the nominal P concentration. The dashed lines are a guide to the eyes. Triangles and circles represent data from undoped and P-doped Si–NC samples, respectively. To avoid an overlap of the error bars, the three data points at $[P]_{nom} = 10^{20}$ cm^{-3} were slightly displaced on the horizontal axis. (c) EPR spectra for P-doped SiNCs with mean particle diameters of 11 and 4.3 nm. (d) P-doping efficiency determined from the spin density of the g = 1.998 peak as a function of the SiNC diameter. The circles show the measured $[P]_{EPR}$ normalized to $[P]_{nom}$ and the diamonds show the same data after correcting for charge compensation of donors by Si dangling bonds $[P]_{EPR}^{s.c.}$. The dashed and dotted lines show the cases $[P]_{EPR} = [P]_{nom}$ and $[P]_{EPR} = [P]_{core}$, respectively. All the ESR spectra were measured under the same experimental conditions at T = 20 K, and their intensity was normalized to the sample mass. (a and c, reprinted with permission from Stegner, A. R., et al., *Phys. Rev. Lett.*, 100, 026803, 2008. Copyright 2008 by the American Physical Society; b and d, reprinted with permission from Stegner, A.R., et al., *Phys. Rev. B*, 80, 165326, 2009. Copyright 2009 by the American Physical Society.)

was found to be greatly improved after the B doping. A decrease of almost four orders of magnitude in the sheet resistance was achieved, which could be attributed to the significant increase of the carrier density due to the electrically activated doping of B atoms into SiNCs. Another study, on the temperature dependence of the dark conductivity for freestanding SiNCs with various P-doping concentrations, showed that an increase of the P-doping concentration resulted in an increase of the SiNC film dark conductivity and a decrease of the conductivity temperature dependence.[31] The P donors were found to contribute to dark conductivity of the SiNC films via spin-dependent hopping by performing electrically detected magnetic resonance (EDMR) measurement.

Although a great deal of experimental studies have been carried out toward increasing the conductivity of SiNC films by doping,[31,40,42,62] the question still remains whether the metal–insulator transition (MIT) occurs in doped SiNC films at some critical concentrations of free carriers. According to the well-known Mott criterion, the critical electron concentration (n_M) for the MIT in a bulk semiconductor is related to the Bohr radius as

$$n_M a_B^3 \cong 0.02, \tag{15.1}$$

where $a_B = \varepsilon h^2 / m^* e^2$ is the effective Bohr radius (in Gaussian units), ε is the dielectric constant of the semiconductor, m^* is the effective mass of a free carrier, h is the Planck constant, and e is the electronic charge.

By assuming a dense film of semiconductor NCs that touched each other through small facets of radius without any surface ligands that hindered conduction, Han et al.[63] derived the MIT criterion for such NC film as

$$n_M \rho^3 \cong 0.3g, \tag{15.2}$$

where ρ was the radius of the touching facets, and g was the number of equivalent minima in the conduction band of the semiconductor. According to this criterion, MIT occurs in an SiNC film at a critical electron concentration of 5×10^{20} cm^{-3}.

To test this prediction, Chen et al.[64] investigated the electron transport in films of ligand-free P-doped SiNCs with a mean size of 7.5 nm. As expected from previous discussions, the ohmic conductance (G) of P-doped SiNC films increases almost monotonically as the nominal doping concentration increases, as shown in Figure 15.9a. The film conductance over the entire range of doping concentration under investigation is found to follow the Efros-Shklovskii (ES) law as

$$G_f \propto \exp\left[-(T_{ES}/T)^{1/2}\right], \tag{15.3}$$

where

$$T_{ES} = \frac{Ce^2}{\varepsilon_r k_B \xi}, \tag{15.4}$$

ξ is the electron localization length, k_B is the Boltzmann constant, and $C \approx 9.6$ is a numerical coefficient.

From the slope of linear fits for $\ln G$ versus $T^{-1/2}$ the characteristic temperature (T_{ES}) could be readily obtained. The localization length ξ for SiNCs doped with various P concentrations could thus be worked out, as shown in Figure 15.9b. It is seen that the localization length increases with increasing P concentration and exceeds the diameter of an NC when the electron concentration is $> 1.9 \times 10^{20}$ cm^3. This indicates the approach to the MIT in P-doped SiNC films.

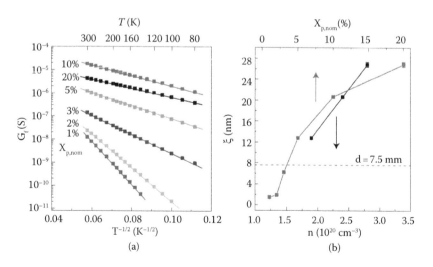

Figure 15.9 Electrical transport in P-doped SiNC films approaching the metal-to-insulator transition. (a) Temperature dependence of the ohmic conductance for films made from SiNCs at different nominal P-doping concentrations. Solid lines are linear fits for each doping concentration. (b) Localization length ξ versus the electron concentration in an NC n and the nominal P doping concentration $X_{P,nom}$. Error bar for each comes from the uncertainty caused by linear fit and is as large as the symbol size. The average diameter of an NC in films is shown by a horizontal dashed line. (Reprinted with permission from Macmillan Publishers Ltd., Chen, T., et al., 2015, Nat. Mater., 15, 299–303, copyright 2015.)

Clusters, nanoparticles, and quantum dots

15.3.3 OPTICAL PROPERTIES

15.3.3.1 Optical emission

Doping of either B or P impurities will quench the PL of SiNCs because of the strong Auger interaction between photoexcited carriers and those induced by doping. Besides, the size mismatch between a dopant and an Si atom causes considerable strain in doped SiNCs, which may further quench the light emission from SiNCs. Figure 15.10 shows the influence of B and P-doping on the PL of freestanding SiNCs with diameters of about 3.6 nm.[27] The PL intensity of SiNCs in Figure 15.10a decreases with increasing B concentration, suggesting strong Auger recombination and strain-induced defect states in B-doped SiNCs. Different from B doping, the PL intensity of SiNCs in Figure 15.10b initially increases when the concentration of P is small, say 0.06%. As we have mentioned before, in contrast to B, strain-induced defect states may be negligible in P-doped SiNCs because the atomic size difference between P and Si is small. The enhancement of the PL intensity may suggest a P atom would first passivate the dangling bond at the surface of an SiNC, which is believed to be the nonradiative recombination center of the SiNC. When the concentration of P exceeds 0.4%, the PL intensity starts to decrease. This means more P atoms have been incorporated into SiNCs and are likely electrically active. It is the P-doping-induced Auger effect that gives rise to strong recombination of photoexcited carriers in SiNCs.

Compared with single B- or P-doped SiNCs, the Auger quenching of the PL may be impaired when B and P impurities are doped simultaneously and compensated.[38] Figure 15.11a shows the PL spectra of B and P codoped SiNCs that are embedded in an SiO_2 matrix. It can be seen that the PL intensity remarkably increases as the P concentration increases. This indicates the B-doping-induced quenching of PL is largely recovered by P-doping. In addition, the PL peak continuously red shifts from 1.4 eV to 1.0 eV with the increase of the P concentration. The observation of PL below the band gap energy of bulk Si crystal indicates that the PL arises from transitions between donor and acceptor states in SiNCs. The below-band-gap PL caused by B and P codoping provides an alternative route to tailor the PL energy of SiNCs without losing the intensity much. By combining the impurity control with the quantum confinement effect—the size control—the energy range of PL can be widely extended from infrared to visible range, as shown in Figure 15.11b.[65]

15.3.3.2 Optical absorption

As an indirect band gap semiconductor, bulk Si is characteristic of an indirect Γ–X transition (T_0) and a direct Γ–Γ transition (T_1), as shown in the inset of Figure 15.12a. It has been established in bulk Si that the B and P-doping could seriously influence excitons associated with the interband transition, leading

Clusters, nanoparticles, and quantum dots

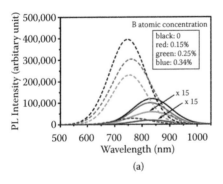
(a)

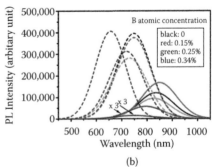
(b)

Figure 15.10 Optical emission of B- and P-doped SiNCs. (a) PL spectra from as-synthesized intrinsic and B-doped SiNCs (solid lines) and the same SiNCs after 5-day exposure to air at room temperature (dashed lines). The B-doped SiNCs are labeled according to B atomic concentrations obtained from ICP-AES measurements. The intensity of PL from as-synthesized SiNCs with B concentrations of 0.25 and 0.34% is magnified by a factor of 15. (b) PL spectra from as-synthesized intrinsic and P-doped SiNCs (solid lines) and the same SiNCs after five-day exposure at air at room temperature (dashed lines). The P-doped SiNCs are labeled according to P atomic concentrations obtained from ICP-AES measurements. The intensity of PL from as-synthesized SiNCs with P concentrations of 1.9 and 5.6% is magnified by a factor of 3. (Reprinted with permission from Pi, X., et al., *Appl. Phys. Lett.*, 92, 123102, 2008. Copyright 2008, American Institute of Physics.)

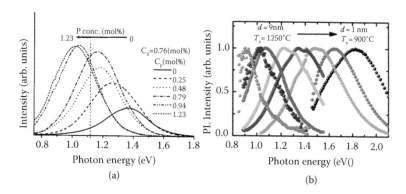

Figure 15.11 Optical emission of B and P codoped SiNCs. (a) P concentration dependence of PL spectra for B and P codoped SiNCs. The B concentration is fixed at 0.76 mol% and P concentration is increased up to 1.26 mol%. Dashed line indicates the band gap energy of bulk Si. (b) PL spectra of B and P codoped SiNCs with different sizes, that is, grown at different temperatures (a, reprinted with permission from Fujii, M., et al., Below bulk-band-gap photoluminescence at room temperature from heavily P- and B-doped Si nanocrystals, *J. Appl. Phys.*, 94, 1990–1995, 2003. Copyright 2003, American Institute of Physics; b, reprinted with permission from Hori, Y., et al., *Nano Lett.*, 16, 2615–2620, 2016. Copyright 2016 American Chemical Society.)

to considerable narrowing of the optical band gap.[66–68] But it remains unclear if B and P-doping affects the interband transition in SiNCs. A focused study on the effect of B doping on the optical absorption of SiNCs has recently been carried out by using UV-Vis absorption measurements because of the excellent dispersion of B-doped SiNCs in polar solvents.[59] Figure 15.12a shows the optical absorption spectra of B-doped SiNCs measured by UV-Vis absorption spectroscopy. The optical absorption spectrum of Cl-passivated undoped SiNCs is included for comparison because the dispersibility of undoped SiNCs that are Cl-passivated is basically as good as that of B-doped SiNCs. After analyzing the optical absorption spectra, it is found that the B doping significantly induces changes to both the T_0 and T_1 transitions in SiNCs. Specifically, the T_0 transition monotonically red shifts from 1.5 to 0.9 eV at the increase of B concentration shown in Figure 15.12b, while the T_1 transition monotonically red shifts from 3.55 to 3.06 eV in Figure 15.12c.

The band gap narrowing associated with the T_0 and T_1 transitions obtained in B-doped SiNCs are summarized in Figure 15.12d. It is seen that the narrowing of T_1 is basically consistent with that of T_0 as the ionized B concentration changes. This indicates the conduction band at both the Γ and X points moves toward the band gap after B doping, leading to the result that the T_0 and T_1 transitions are nearly equally affected. The band gap narrowing (ΔE_g) in bulk Si is given as a function of the concentration (N) of ionized impurity[69]:

$$\Delta E_g = C_1 \left\{ \ln\left(\frac{N}{N_1}\right) + \sqrt{\left[\ln\left(\frac{N}{N_1}\right)\right]^2 + C_2} \right\}, \tag{15.5}$$

where N_1 is a critical parameter, indicating the onset of the band gap narrowing, C_1 and C_2 are constants. After fitting the band gap narrowing of B-doped SiNCs with Equation 15.5, it is observed that the band gap of SiNCs starts decreasing when the ionized B concentration is larger than 1.8×10^{20} cm^{-3}. This value is more than three orders of magnitude higher than that obtained for heavily B-doped bulk Si (1.3×10^{17} cm^{-3}). It is known that the acceptor level of B is deeper in the band gap for SiNCs than bulk Si. Therefore, a much larger B concentration is needed to form an impurity band extending to the valence band and narrow the band gap in SiNCs, than in bulk Si. Additionally, it is noted that the band gap shrinks more significantly in SiNCs than in bulk Si. This may be due to the fact that a number of unionized B atoms give rise to increased disorder in SiNCs, causing the dispersion of impurity energy levels to increase. Therefore, the shrinking of the band gap is enhanced for SiNCs.

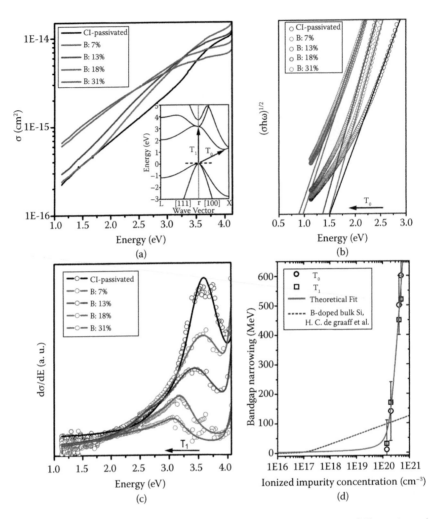

Figure 15.12 Optical absorption of B-doped SiNCs. (a) Per-NC absorption spectra of Cl-passivated undoped SiNCs and B-doped SiNCs in benzonitrile. The concentrations of B are ~7, 13, 18, and 31%,. The inset shows the band structure of bulk Si in which the indirect phonon-assisted Γ–X transition (T_0) and direct Γ–Γ (T_1) transition are indicated. (b) Absorption data for Cl-passivated SiNCs and heavily B-doped SiNCs in benzonitrile plotted as $(\sigma\hbar\omega)^{1/2}$. (c) Derivative absorption spectra of Cl-passivated SiNCs and B-doped SiNCs in benzonitrile. (d) Band gap narrowing associated with the indirect transition (T_0) and direct transition (T_1) obtained in B-doped SiNCs. Graaff et al.'s results for band gap narrowing in heavily B-doped bulk Si are shown for comparison. (Adapted from Zhou, S., et al., *ACS Photonics*, 3, 415–422, 2016. Copyright 2016 American Chemical Society.)

15.3.3.3 Localized surface plasmon resonance

The collective oscillation of free charge carriers that are dielectrically confined in a nanoparticle excited by external electromagnetic field, gives rise to strong light scattering in the appearance of intense absorption bands and enhanced local electromagnetic field which is known as localized surface plasmon resonance (LSPR). LSPR enables tuning the optical response of the nanoparticle through changing the free carrier concentration and the surrounding environment medium according to

$$\omega_{sp} \approx \sqrt{\frac{ne^2}{\varepsilon_0 m^* \left(\varepsilon_\infty + 2\varepsilon_m\right)}}, \tag{15.6}$$

where ω_{sp} is the LSPR frequency, n is the free carrier concentration, ε_0 is the free space permittivity, ε_∞ is the high-frequency dielectric constant of the nanoparticle, and ε_m is the dielectric constant of the surrounding medium.

The free carrier concentration of semiconductor NCs can be significantly modulated by doping. Rowe et al.[28] demonstrated the first observation of LSPR in SiNCs via P-doping by means of plasma. The P-doped SiNCs exhibited tunable mid-infrared LSPR with the energy ranges from 0.07 to 0.3 eV immediately after they were synthesized, as shown in Figure 15.13a. However, LSPR was not obtained in as-synthesized B-doped SiNCs. Zhou et al.[61] indicated that LSPR in the mid-infrared region also occurred to B-doped SiNCs after the NCs were oxidized in air for a long time, as shown in Figure 15.13b. As the P and B concentrations increase, the broad peak induced by LSPR monotonically blue shifts as shown in Figure 15.13a and b, respectively, highlighting the doping-enabled-tunability of LSPR in semiconductor NCs. The sensitivity of the LSPR of P- and B-doped SiNCs to the change of the dielectric constant of the surrounding medium has also been verified by measuring the LSPR frequency in different solvents. This experimentally demonstrates the potential of the LSPR of doped SiNCs in applications such as chemical sensing in the mid-infrared range. Kramer et al.[58] investigated the effect of surface conditions on the plasmonic behavior of both P- and B-doped SiNCs. It was found that P-doped SiNCs exhibited a plasmon resonance immediately after their synthesis, but might lose their plasmonic response with oxidation. In contrast, B-doped SiNCs initially did not exhibit plasmonic response but became plasmonically active through postsynthesis oxidation or annealing. Significantly, B-doped SiNCs show environmentally stable LSPRs after complete oxidation. This makes B-doped SiNCs better positioned for practical implementation than P-doped SiNCs.

Delerue et al.[70] carried out tight-binding calculations on the optical response of ideal P-doped SiNCs and indicated the LSPR energy varied not only with doping concentration, but also with NC size due to a size-dependent screening by valence electrons. A less efficient screening by the valence electrons in small NCs-induced blue shift of the LSPR energy. It also predicted that SiNCs containing a large number of deep defects such as dangling bonds did not give rise to LSPR. The blue shift of the LSPR peak was experimentally observed in B-doped SiNCs when the NC size decreased from 6.8 nm to 3.8 nm.[56] Nevertheless, when the size of SiNCs was smaller than the mean free path, the effect of surface scattering of free carriers could not be neglected. The increase of free carrier scattering at the NC surface might prevail, resulting in the damping of the LSPR when the NC size was further reduced to ~2.4 nm. This trend was similar to that observed in small gold NCs[71] and copper sulfide NCs.[72] The increase of structural disorder and twinning defect may also contribute to the damping of LSPR in small SiNCs.

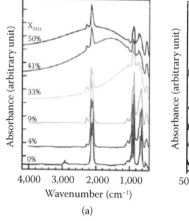
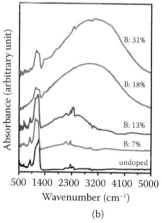

(a)   (b)

Figure 15.13 LSPR of B- and P-doped SiNCs. (a) FTIR spectra for undoped and P-doped SiNCs. The fractional PH$_3$ flow rates, defined as X_{PH3} = [PH$_3$]/([PH$_3$]+[SiH$_4$]) ×100% for the synthesis of P-doped SiNCs are 4, 9, 33, 41, and 50%, respectively. (b) FTIR spectra for undoped and B-doped SiNCs. The concentrations of B are 7, 13, 18, and 31%. (a, adapted from Rowe et al., *Nano Lett.*, 13, 1317–1322, 2013. Copyright 2013 American Chemical Society; b, adapted from Zhou, S., et al., *ACS Nano*, 9, 378–386, 2015. Copyright 2015 American Chemical Society.)

Clusters, nanoparticles, and quantum dots

It is found that the LSPR-induced absorption peaks in P- and B-doped SiNCs can be fitted by using the Drude model in which the Mie absorption of SiNCs including the Drude contribution is expressed by[59,61]

$$\sigma_A(\omega) = \frac{8\pi^2 \sqrt{\varepsilon_m} r^3 \omega}{c} \text{Im} \left\{ \frac{\varepsilon(\omega) - \varepsilon_m}{\varepsilon(\omega) + 2\varepsilon_m} \right\}, \tag{15.7}$$

where $\sigma_A(\omega)$ is the absorption cross section of an SiNC at the frequency of ω, c is the speed of light, r is the NC radius, and $\varepsilon(\omega)$ is the frequency-dependent dielectric constant of the NC. $\varepsilon(\omega)$ is given by the well-known Drude equation:

$$\varepsilon(\omega) = \varepsilon_\infty - \frac{\omega_p^2}{\omega^2 + i\omega\Gamma}, \tag{15.8}$$

where ω_p is the bulk plasma frequency, and Γ is the carrier damping constant. Figure 15.14a representatively shows the good fitting for SiNCs doped with 31% B and 18% P.

Both Γ and ω_p can be derived from the fitting of LSPR-induced absorption. It is known that Γ is related to the carrier mobility (μ) as $\mu = he/(m^*\Gamma)$ (Γ is deemed as energy), while ω_p is related to the free carrier concentration (n) via $\omega_p^2 = ne^2/(m^*\varepsilon_0)$. Since n depends on the activation efficiency (η) of B or P in SiNCs with a concentration of N_d ($n = N_d\eta$), the value of η can be obtained by fitting the ω_p–N_d data according to

$$\omega_p = \sqrt{\frac{N_d \eta e^2}{m^* \varepsilon_0}}. \tag{15.9}$$

The calculated carrier mobility and dopant activation efficiency for each doped SiNC sample are shown in Figure 15.14b and c, respectively. It is thus concluded that LSPR is an effective means to determine the electronic properties of doped SiNCs.

15.3.4 BAND STRUCTURE

The B and P-doping can not only change the concentration of free carriers that are confined in SiNCs, but also induce localized impurity levels in the band gap of SiNCs. These impurity levels aggregate into an impurity band with the increase of dopant concentrations, leading to significant change of the band structure of SiNCs. Zhou et al.[59] investigated the band structure of B-doped SiNCs by using UV-Vis absorption spectroscopy and ultraviolet photoelectron spectroscopy (UPS). The evolution of the SiNC band structure with increasing B doping is shown in Figure 15.15. As the B concentration increases, the Fermi level moves toward the valence band and enters into the original valence band of SiNCs, leading to the degeneracy of the valence band. In addition to the shift of the Fermi level into the original valence band, both the conduction and valence band move toward the band gap. With the further increase of the B concentration, the B-induced impurity band becomes more extended to occupy a larger part of the original band gap. In the meantime, the conduction band more significantly moves toward the band gap, while the Fermi level basically remains within the top region of the original valence band. This picture is consistent with the observed continuous red shift of the optical absorption in B-doped SiNCs in Figure 15.12. Because the B-induced impurity band is merged with the original valence band after heavy B doping, free holes above the Fermi level are largely from the B-induced impurity band. The doping-induced change of the band structure helps understand the origin of fundamental physical behavior (e.g., LSPR) that occurs to heavily doped SiNCs.

The band structure of doped SiNCs changes not only with dopant concentration, but also with the NC size. Analogous to the conduction and valence band edge, both the acceptor and donor levels of B and P exhibit a size dependence in SiNCs. Hori et al.[65] measured the B acceptor and P donor levels in B and P codoped SiNCs with the diameter ranging from 1 to 9 nm by using photoemission yield spectroscopy (PYS) and PL spectroscopy. The highest occupied molecular orbital (HOMO), lowest unoccupied molecular orbital (LUMO), and Fermi level energies determined from the vacuum level are shown

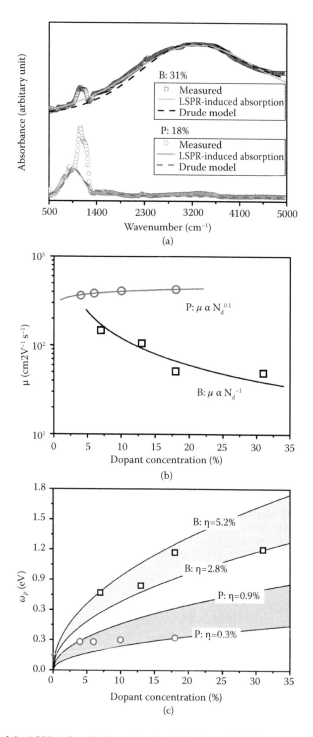

Figure 15.14 (a) Fitting of the LSPR-induced absorption for SiNCs doped with 31% B and that for SiNCs doped with 18% P by using the Drude model. (b) Carrier mobility (μ) of each doped Si–NC sample obtained by fitting the LSPR-induced absorption peak. The black line gives $\mu \propto N_d^{-1}$, while the red line gives $\mu \propto N_d^{0.1}$, where N_d is the dopant concentration. (c) Bulk plasma frequency (ω_p) versus dopant concentration (N_d). The solid lines represent the fitting by using $\omega_p = \sqrt{N_d \eta e^2 / (m^* \varepsilon_0)}$. The values of dopant activation efficiency (η) obtained from the fitting are indicated. (Reprinted with permission from Zhou, S., et al., *ACS Nano*, 9, 378–386, 2015. Copyright 2015 American Chemical Society.)

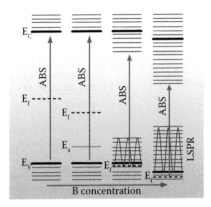

Figure 15.15 Evolution of the SiNC band structure with the increase of the doping level of B. The B concentration increases from left to right. ABS, absorption onset; E_c, conduction band edge; E_v, valence band edge; E_f, Fermi energy level; E_a, impurity energy level. (Reprinted with permission from Zhou, S., et al., *ACS Photonics*, 3, 415–422, 2016. Copyright 2016 American Chemical Society.)

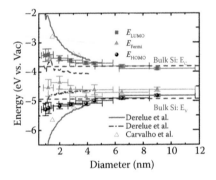

Figure 15.16 HOMO, LUMO, and Fermi level energies of B and P codoped SiNCs measured from the vacuum level. The horizontal dashed lines are the conduction and valence band edges of bulk Si crystal. Valence band and conduction band edges of undoped SiNCs calculated by tight binding approximation (solid curves) and first principle calculations (open triangle) are also shown. Broken curves are acceptor and donor levels calculated by tight binding approximations. (Reprinted with permission from Hori, Y., et al., *Nano Lett.*, 16, 2615–2620, 2016. Copyright 2016 American Chemical Society.)

in Figure 15.16. It is seen that the HOMO–LUMO gap of B and P codoped SiNCs increases as the NC decreases. However, compared with the valence band and conduction band edges of undoped SiNCs, the shift of HOMO and LUMO levels of B and P codoped SiNCs is much smaller, which means B acceptor and P donor states are the HOMO and LUMO levels in codoped SiNCs, respectively. It is noted that both the HOMO and LUMO levels of codoped SiNCs start to change at a critical size of ~5 nm, similar to what is observed in the conduction and valence band edges of undoped SiNCs which is known to be due to the quantum confinement effect. This indicates quantum confinement effect influences the acceptor and donor levels of B and P in codoped SiNCs. In B and P codoped SiNCs, the Fermi level is very close to the HOMO level when the size is >5 nm. As the NC size decreases, it approaches the middle of the band gap. This is mainly because in small codoped SiNCs, donors, and acceptors are more substantially compensated.

15.3.5 OXIDATION PROPERTY

SiNCs are prone to oxidation after exposure to air. Previously, Liptak et al.[73] and Gresback et al.[74] found that the oxidation of SiNCs was dependent on the surface chemistry of SiNCs in that partially fluorine- or chlorine-terminated SiNCs oxidized rapidly compared to purely hydrogen-terminated Si. The authors interpreted that the oxidation of SiNCs might be described by using the Cabrera–Mott model in which the electron tunneling from Si to the oxide interface is critical.

Since the electron concentration of SiNCs can be significantly modulated by B and P-doping, a difference in the oxidation of B- and P-doped SiNCs would be expected if the oxidation of SiNCs indeed follows the Cabrera–Mott mechanism. The oxidation of B- and P-doped SiNCs can be visually characterized by using XPS. Figure 15.17 shows the XPS spectra for undoped SiNCs and SiNCs doped with B (13–31%) and P (6–18%) with a mean size of 14 nm after oxidation for 1 hour, 2 months, and 11 months.[29] Different from undoped SiNCs, the XPS peak related to the oxidation states of Si (~103 eV) emerges for P-doped SiNCs only after 1 h exposure in air. This peak becomes more pronounced in addition to shifting to a higher energy as the oxidation time increases. Clearly P-doping leads to faster oxidation and oxides with higher oxidation states of Si (Si^{4+}), in good agreement with the Cabrera–Mott mechanism.

In this context, oxidation of SiNCs is expected to be slowed after B doping. However, the more pronounced peak related to the oxidation state of Si in the comparison with undoping suggests the B doping actually enhances the oxidation of SiNCs slightly. This discrepancy indicates that electron concentration is not the only factor that impacts the oxidation of SiNCs. It is well known that oxidation of SiNCs is self-limited because of the compressive stress formed at the interface between Si and silicon oxide. This compressive stress could be compensated to some extent by the tensile stress induced by the B doping. It turns out that the oxidation of B-doped SiNCs is slightly enhanced if the stress effect prevails over the B-doping-induced weakening of oxidation in the framework of the Cambrera–Mott mechanism. However, if the

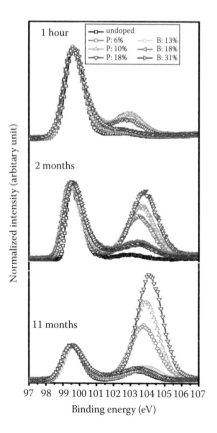

Figure 15.17 XPS spectra of undoped SiNCs and SiNCs doped with B and P. XPS measurements are carried out after SiNCs are stored in air for 1 h, 2 months, and 11 months. The concentrations of P are 6, 10, and 18%, while those of B are 13, 1, and 31%. (Zhou, S., et al., *Part. Part. Syst. Charact.*, 32, 213–221, 2015. Copyright Wiley-VCH Verlag GmbH & Co. KGaA. Reproduced with permission.)

concentration of B is not high enough to render significant tensile stress, the oxidation of SiNCs should be retarded. This was actually demonstrated by the study on the oxidation induced change of PL for SiNCs that are lightly doped with B (0.1–0.3%).[27] Therefore, both the free carriers and tensile stress induced by doping should be accounted for in the investigation on the oxidation of SiNCs.

15.4 CONCLUSIONS AND REMARKS

We have reviewed recent progress in the synthesis of B- and P- doped SiNCs. The advance in the synthesis of doped SiNCs enables modulating the B and P concentrations of SiNCs in a very wide range. It is found that the concentrations of both B and P in SiNCs can significantly exceed their solubility limits in bulk Si. Study on the location of dopants in SiNCs shows the preferential location of dopants in SiNCs may greatly depend on the synthesis methods. For plasma-synthesized freestanding doped SiNCs, B prefers entering into the NC core, while P prefers locating at the NC surface. However, for doped SiNCs that are embedded in an SiO_2 matrix, P prefers locating at the NC core, while B prefers locating at the NC surface. Such an inconsistency in the preferential location of dopants is mainly related to the different mechanisms for doping freestanding SiNCs and SiNCs embedded in a matrix. We would like to point out direct observation of the exact location of these dopants in SiNCs is still lacking. As research on doping SiNCs moves forward, it becomes increasing important to image the location of dopant in doped SiNCs directly.[75]

Although the diamond structure of SiNCs remains after B and P doping, changes of the lattice spacing and twin- and strain-induced defects may appear, especially when the doping concentration is rather high. They may significantly affect the activation of dopants and the behavior of free carriers in SiNCs. Effort to avoid such defects during the synthesis of doped SiNCs is needed in future work.

Doping SiNCs with B and P has enabled novel properties that undoped SiNCs do not possess. The surface chemistry induced by B and P in SiNCs makes it possible to create ligand-free Si–NC colloids which is particularly demanded in low-cost solution-processed devices. Although doping of B or P quenches the light emission of SiNCs, the quenching of light emission will be impaired in B and P codoped SiNCs. Besides, the optical absorption of SiNCs can also be modulated by doping. It is believed that the energies for both optical emission and absorption can be widely extended from infrared to visible range in SiNCs by combining the doping with the quantum confinement effect.

LSPRs have been demonstrated in both B- and P-doped SiNCs. In contrast to noble metal and metal oxide NCs, the LSPR energies for doped SiNCs are all located in the mid-infrared region. Although integrated Si photonic devices for near-IR applications are well known and are attracting great interest, their potential in the mid-IR region has so far remained unexplored.[76] Applications of plasmonic SiNCs in these areas are worth exploring in the future due to their ready compatibility with modern Si technology. In addition, effort to move the LSPR energy of doped SiNCs toward the visible region is needed. It is known that the activation efficiencies of B and P in SiNCs are currently low. This means a great deal of dopants in SiNCs are deactivated and unable to produce free carriers. The mechanism that deactivates the dopant in SiNCs is complex and currently not so clear. If this part of dopants can be activated, the free carrier concentration should be significantly improved, giving rise to shift of LSPR energy toward near-infrared or visible region.

MIT has been proved to occur to P-doped SiNC films at a critical concentration higher than 10^{20} cm^{-3}. Because B can more efficiently produce free carriers than P in SiNCs, it is reasonable that MIT also occurs to B-doped SiNC films. Additionally, the excellent dispersity of B-doped SiNCs in polar solvents facilitates the fabrication of high-quality NC films which is critical for carrier transport. Therefore, it is expected that MIT occurs to B-doped SiNCs much easier than P-doped SiNCs. Further study needs to be carried out to clarify this point.

ACKNOWLEDGMENTS

This work was supported by Grant-in-Aid for Scientific Research (B) of Japan (No.26289045), the National Basic Research Program of China (Grant No. 2013CB632101) and the Natural Science Foundation of China (Grant No. 61222404).

REFERENCES

1. Cullis, A. G., Canham, L. T. Visible Light Emission Due to Quantum Size Effects in Highly Porous Crystalline Silicon. *Nature* **1991**, *353*, 335–338.
2. Pavesi, L., Dal Negro, L., Mazzoleni, C., Franzo, G., Priolo, F. Optical Gain in Silicon Nanocrystals. *Nature* **2000**, *408*, 440–444.
3. Liu, C. Y., Holman, Z. C., Kortshagen, U. R. Hybrid Solar Cells from P3HT and Silicon Nanocrystals. *Nano Letters* **2009**, *9*, 449–452.
4. Cheng, K. Y., Anthony, R., Kortshagen, U. R., Holmes, R. J. High-Efficiency Silicon Nanocrystal Light-Emitting Devices. *Nano Letters* **2011**, *11*, 1952–1956.
5. McVey, B. F. P., Tilley, R. D. Solution Synthesis, Optical Properties, and Bioimaging Applications of Silicon Nanocrystals. *Accounts of Chemical Research* **2014**, *47*, 3045–3051.
6. Ding, Y., Gresback, R., Liu, Q., Zhou, S., Pi, X., Nozaki, T. Silicon Nanocrystal Conjugated Polymer Hybrid Solar Cells with Improved Performance. *Nano Energy* **2014**, *9*, 25–31.
7. Pi, X. D., Liptak, R. W., Nowak, J. D., Pwells, N., Carter, C. B., Campbell, S. A., Kortshagen, U. Air-Stable Full-Visible-Spectrum Emission from Silicon Nanocrystals Synthesized by An All-Gas-Phase Plasma Approach. *Nanotechnology* **2008**, *19*, 245603.
8. Jurbergs, D., Rogojina, E., Mangolini, L., Kortshagen, U. Silicon Nanocrystals with Ensemble Quantum Yields Exceeding 60%. *Applied Physics Letters* **2006**, *88*, 233116.
9. Veinot, J. G. C. Synthesis, Surface Functionalization, and Properties of Freestanding Silicon Nanocrystals. *Chemical Communications* **2006**, 4160–4168.
10. Wolf, O., Dasog, M., Yang, Z., Balberg, I., Veinot, J. G. C., Millo, O. Doping and Quantum Confinement Effects in Single Si Nanocrystals Observed by Scanning Tunneling Spectroscopy. *Nano Letters* **2013**, *13*, 2516–2521.
11. Wheeler, L. M., Neale, N. R., Chen, T., Kortshagen, U. R. Hypervalent Surface Interactions for Colloidal Stability and Doping of Silicon Nanocrystals. *Nature Communications* **2013**, *4*, 2197–2206.
12. Garrone, E., Geobaldo, F., Rivolo, P., Amato, G., Boarino, L., Chiesa, M., Giamello, E., Gobetto, R., Ugliengo, P., Viale, A. A Nanostructured Porous Silicon Near Insulator Becomes Either a P- or an N-Type Semiconductor upon Gas Adsorption. *Advanced Materials* **2005**, *17*, 528–531.
13. Priolo, F., Franzò, G., Pacifici, D., Vinciguerra, V., Iacona, F., Irrera, A. Role of the Energy Transfer in the Optical Properties of Undoped and Er-Doped Interacting Si Nanocrystals. *Journal of Applied Physics* **2001**, *89*, 264–272.
14. Fujii, M., Yoshida, M., Kanzawa, Y., Hayashi, S., Yamamoto, K. 1.54 µm Photoluminescence of Er^{3+} Doped into SiO_2 Films Containing Si Nanocrystals: Evidence for Energy Transfer from Si Nanocrystals to Er^{3+}. *Applied Physics Letters* **1997**, *71*, 1198–1200.
15. Zhang, X., Brynda, M., Britt, R. D., Carroll, E. C., Larsen, D. S., Louie, A. Y., Kauzlarich, S. M. Synthesis and Characterization of Manganese-Doped Silicon Nanoparticles: Bifunctional Paramagnetic-Optical Nanomaterial. *Journal of the American Chemical Society* **2007**, *129*, 10668–10669.
16. McVey, B. F. P., Butkus, J., Halpert, J. E., Hodgkiss, J. M., Tilley, R. D. Solution Synthesis and Optical Properties of Transition-Metal-Doped Silicon Nanocrystals. *The Journal of Physical Chemistry Letters* **2015**, *6*, 1573–1576.
17. Singh, M. P., Atkins, T. M., Muthuswamy, E., Kamali, S., Tu, C., Louie, A. Y., Kauzlarich, S. M. Development of Iron-Doped Silicon Nanoparticles as Bimodal Imaging Agents. *ACS Nano* **2012**, *6*, 5596–5604.
18. Romero, J. J., Wegmann, M., Rodríguez, H. B., Lillo, C., Rubert, A., Klein, S., Kotler, M. L., Kryschi, C., Gonzalez, M. C. Impact of Iron Incorporation on 2–4 nm Size Silicon Nanoparticles Properties. *The Journal of Physical Chemistry C* **2015**, *119*, 5739–5746.
19. Gresback, R., Kramer, N. J., Ding, Y., Chen, T., Kortshagen, U. R., and Nozaki, T. Controlled Doping of Silicon Nanocrystals Investigated by Solution-Processed Field Effect Transistors. *ACS Nano* **2014**, 5650–5656.
20. Mocatta, D., Cohen, G., Schattner, J., Millo, O., Rabani, E., Banin, U. Heavily Doped Semiconductor Nanocrystal Quantum Dots. *Science* **2011**, *332*, 77–81.
21. Norris, D. J., Efros, A. L., Erwin, S. C. Doped Nanocrystals. *Science* **2008**, *319*, 1776–1779.
22. Mangolini, L. Synthesis, Properties, and Applications of Silicon Nanocrystals. *Journal of Vacuum Science & Technology B* **2013**, *31*, 020801.
23. Ni, Z. Y., Pi, X. D., Muhammad, A., Zhou, S., Nozaki, T., Yang, D. Freestanding Doped Silicon Nanocrystals Synthesized by Plasma. *Journal of Physics D: Applied Physics* **2015**, *48*, 314006.
24. Oliva-Chatelain, B. L., Ticich, T. M., Barron, A. R. Doping Silicon Nanocrystals and Quantum Dots. *Nanoscale* **2016**, *8*, 1733–1745.
25. Mangolini, L., Thimsen, E., Kortshagen, U. High-Yield Plasma Synthesis of Luminescent Silicon Nanocrystals. *Nano Letters* **2005**, *5*, 655–659.

26. Pi, X. D., Yu, T., Yang, D. Water-Dispersible Silicon-Quantum-Dot-Containing Micelles Self-Assembled from an Amphiphilic Polymer. *Particle and Particle Systems Characterization* **2014**, *31*, 6, 751–756.

27. Pi, X. D., Gresback, R., Liptak, R. W., Campbell, S. A., Kortshagen, U. Doping Efficiency, Dopant Location, and Oxidation of Si Nanocrystals. *Applied Physics Letters* **2008**, *92*, 123102.

28. Rowe, D. J., Jeong, J. S., Mkhoyan, K. A., Kortshagen, U. R. Phosphorus-Doped Silicon Nanocrystals Exhibiting Mid-Infrared Localized Surface Plasmon Resonance. *Nano Letters* **2013**, *13*, 1317–1322.

29. Zhou, S., Pi, X., Ni, Z., Luan, Q., Jiang, Y., Jin, C., Nozaki, T., Yang, D. Boron- and Phosphorus-Hyperdoped Silicon Nanocrystals. *Particle & Particle Systems Characterization* **2015**, *32*, 213–221.

30. Zhou, S., Ding, Y., Pi, X., Nozaki, T. Doped Silicon Nanocrystals from Organic Dopant Precursor by A SiCl$_4$-Based High Frequency Nonthermal Plasma. *Applied Physics Letters* **2014**, *105*, 183110.

31. Stegner, A. R., Pereira, R. N., Klein, K., Lechner, R., Dietmueller, R., Brandt, M. S., Stutzmann, M., Wiggers, H. Electronic Transport in Phosphorus-Doped Silicon Nanocrystal Networks. *Physical Review Letters* **2008**, *100*, 026803.

32. Stegner, A. R., Pereira, R. N., Lechner, R., Klein, K., Wiggers, H., Stutzmann, M., Brandt, M. S. Doping Efficiency in Freestanding Silicon Nanocrystals from the Gas Phase: Phosphorus Incorporation and Defect-Induced Compensation. *Physical Review B* **2009**, *80*, 165326.

33. Baldwin, R. K., Zou, J., Pettigrew, K. A., Yeagle, G. J., Britt, R. D., Kauzlarich, S. M. The Preparation of a Phosphorus Doped Silicon Film from Phosphorus Containing Silicon Nanoparticles. *Chemical Communications* **2006**, 658–660.

34. Wolkin, M. V., Jorne, J., Fauchet, P. M., Allan, G., Delerue, C. Electronic States and Luminescence in Porous Silicon Quantum Dots: The Role of Oxygen. *Physical Review Letters* **1999**, *82*, 197–200.

35. Velusamy, T., Mitra, S., Macias-Montero, M., Svrcek, V., Mariotti, D. Varying Surface Chemistries for P-Doped and N-Doped Silicon Nanocrystals and Impact on Photovoltaic Devices. *ACS Applied Materials & Interfaces* **2015**, *7*, 28207–28214.

36. Pawlak, B. J., Gregorkiewicz, T., Ammerlaan, C. A. J., Takkenberg, W., Tichelaar, F. D., Alkemade, P. F. A. Experimental Investigation of Band Structure Modification in Silicon Nanocrystals. *Physical Review B* **2001**, *64*, 115308.

37. Mimura, A., Fujii, M., Hayashi, S., Kovalev, D., Koch, F. Photoluminescence and Free-Electron Absorption in Heavily Phosphorus-Doped Si Nanocrystals. *Physical Review B* **2000**, *62*, 12625–12627.

38. Fujii, M., Toshikiyo, K., Takase, Y., Yamaguchi, Y., Hayashi, S. Below Bulk-Band-Gap Photoluminescence at Room Temperature from Heavily P- and B-Doped Si Nanocrystals. *Journal of Applied Physics* **2003**, *94*, 1990–1995.

39. Hao, X. J., Cho, E. C., Scardera, G., Bellet-Amalric, E., Bellet, D., Shen, Y. S., Huang, S., Huang, Y. D., Conibeer, G., Green, M. A. Effects of Phosphorus Doping on Structural and Optical Properties of Silicon Nanocrystals in A SiO$_2$ Matrix. *Thin Solid Films* **2009**, *517*, 5646–5652.

40. Xie, M., Li, D., Chen, L., Wang, F., Zhu, X., Yang, D. The Location and Doping Effect of Boron in Si Nanocrystals Embedded Silicon Oxide Film. *Applied Physics Letters* **2013**, *102*, 123108.

41. Sato, K., Fukata, N., Hirakuri, K. Doping and Characterization of Boron Atoms in Nanocrystalline Silicon Particles. *Applied Physics Letters* **2009**, *94*, 161902.

42. Keisuke, S., Kazuki, N., Naoki, F., Kenji, H., Yusuke, Y. The Synthesis and Structural Characterization of Boron-Doped Silicon-Nanocrystals with Enhanced Electroconductivity. *Nanotechnology* **2009**, *20*, 365207.

43. Perego, M., Bonafos, C., Fanciulli, M. Phosphorus Doping of Ultra-Small Silicon Nanocrystals. *Nanotechnology* **2010**, *21*, 025602.

44. Nakamura, T., Adachi, S., Fujii, M., Miura, K., Yamamoto, S. Phosphorus and Boron Codoping of Silicon Nanocrystals by Ion Implantation: Photoluminescence Properties. *Physical Review B* **2012**, *85*, 045441.

45. Nakamura, T., Adachi, S., Fujii, M., Sugimoto, H., Miura, K., Yamamoto, S. Size and Dopant-Concentration Dependence of Photoluminescence Properties of Ion-Implanted Phosphorus- and Boron-Codoped Si Nanocrystals. *Physical Review B* **2015**, *91*, 165424.

46. Frégnaux, M., Khelifi, R., Muller, D., Mathiot, D. Optical Characterizations of Doped Silicon Nanocrystals Grown by Co-Implantation of Si and Dopants in SiO$_2$. *Journal of Applied Physics* **2014**, *116*, 143505.

47. Khelifi, R., Mathiot, D., Gupta, R., Muller, D., Roussel, M., Duguay, S. Efficient N-Type Doping of Si Nanocrystals Embedded in SiO$_2$ by Ion Beam Synthesis. *Applied Physics Letters* **2013**, *102*, 013116.

48. Kodera, H. Constitutional Supercooling during the Crystal Growth of Germanium and Silicon. *Japanese Journal of Applied Physics* **1963**, *2*, 527–534.

49. Minoru, F., Hiroshi, S., Kenji, I. All-Inorganic Colloidal Silicon Nanocrystals—Surface Modification by Boron and Phosphorus Co-Doping. *Nanotechnology* **2016**, *27*, 262001.

50. Sugimoto, H., Fujii, M., Imakita, K., Hayashi, S., Akamatsu, K. All-Inorganic Near-Infrared Luminescent Colloidal Silicon Nanocrystals: High Dispersibility in Polar Liquid by Phosphorus and Boron Codoping. *Journal of Physical Chemistry C* **2012**, *116*, 17969–17974.

51. Gnaser, H., Gutsch, S., Wahl, M., Schiller, R., Kopnarski, M., Hiller, D., Zacharias, M. Phosphorus Doping of Si Nanocrystals Embedded in Silicon Oxynitride Determined by Atom Probe Tomography. *Journal of Applied Physics* **2014**, *115*, 034304.

52. Hong, S. H., Kim, Y. S., Lee, W., Kim, Y. H., Song, J. Y., Jang, J. S., Park, J. H., Choi, S. H., Kim, K. J. Active Doping of B in Silicon Nanostructures and Development of A Si Quantum Dot Solar Cell. *Nanotechnology* **2011**, *22*, 425203.

53. Fukuda, M., Fujii, M., Sugimoto, H., Imakita, K., Hayashi, S. Surfactant-Free Solution-Dispersible Si Nanocrystals Surface Modification by Impurity Control. *Optics Letters* **2011**, *36*, 4026–4028.

54. Mangolini, L., Kortshagen, U. Selective Nanoparticle Heating: Another Form of Nonequilibrium in Dusty Plasmas. *Physical Review E* **2009**, *79*, 026405.

55. Luo, Y. R. *Comprehensive Handbook of Chemical Bond Energies.* CRC Press, Boca Raton, FL, 2007.

56. Ni, Z., Pi, X., Zhou, S., Nozaki, T., Grandidier, B., Yang, D. Size-Dependent Structures and Optical Absorption of Boron-Hyperdoped Silicon Nanocrystals. *Advanced Optical Materials* **2016**, *4*, 700–707.

57. Bustarret, E., Marcenat, C., Achatz, P., Kacmarcik, J., Levy, F., Huxley, A., Ortega, L., et al. Superconductivity in Doped Cubic Silicon. *Nature* **2006**, *444*, 465–468.

58. Kramer, N. J., Schramke, K. S., Kortshagen, U. R. Plasmonic Properties of Silicon Nanocrystals Doped with Boron and Phosphorus. *Nano Letters* **2015**, *15*, 5597–5603.

59. Zhou, S., Ni, Z., Ding, Y., Sugaya, M., Pi, X., Nozaki, T. Ligand-Free, Colloidal, and Plasmonic Silicon Nanocrystals Heavily Doped with Boron. *ACS Photonics* **2016**, *3*, 415–422.

60. Sasaki, M., Kano, S., Sugimoto, H., Imakita, K., Fujii, M. Surface Structure and Current Transport Property of Boron and Phosphorus Co-Doped Silicon Nanocrystals. *The Journal of Physical Chemistry C* **2016**, *120*, 195–200.

61. Zhou, S., Pi, X., Ni, Z., Ding, Y., Jiang, Y., Jin, C., Delerue, C., Yang, D., Nozaki, T. Comparative Study on the Localized Surface Plasmon Resonance of Boron- and Phosphorus-Doped Silicon Nanocrystals. *ACS Nano* **2015**, *9*, 378–386.

62. Lechner, R., Stegner, A. R., Pereira, R. N., Dietmueller, R., Brandt, M. S., Ebbers, A., Trocha, M., Wiggers, H., Stutzmann, M. Electronic Properties of Doped Silicon Nanocrystal Films. *Journal of Applied Physics* **2008**, *104*, 053701.

63. Fu, H., Reich, K. V., Shklovskii, B. I. Hopping Conductivity and Insulator-Metal Transition in Films of Touching Semiconductor Nanocrystals. *Physical Review B* **2016**, *93*, 125430.

64. Chen, T., Reich, K. V., Kramer, N. J., Fu, H., Kortshagen, U. R., Shklovskii, B. I. Metal-Insulator Transition in Films of Doped Semiconductor Nanocrystals. *Nature Materials* **2016**, *15*, 299–303.

65. Hori, Y., Kano, S., Sugimoto, H., Imakita, K., Fujii, M. Size-Dependence of Acceptor and Donor Levels of Boron and Phosphorus Codoped Colloidal Silicon Nanocrystals. *Nano Letters* **2016**, *16*, 2615–2620.

66. Viña, L., Cardona, M. Effect of Heavy Doping on the Optical Properties and the Band Structure of Silicon. *Physical Review B* **1984**, *29*, 6739–6751.

67. Wagner, J., del Alamo, J. A. Band-gap Narrowing in Heavily Doped Silicon: A Comparison of Optical and Electrical Data. *Journal of Applied Physics* **1988**, *63*, 425–429.

68. Klaassen, D. B. M., Slotboom, J. W., Degraaff, H. C. Unified Apparent Band-gap Narrowing in N-Type and P-Type Silicon. *Solid-State Electronics* **1992**, *35*, 125–129.

69. Slotboom, J. W., Degraaff, H. C. Measurements of Band-gap Narrowing in Si Bipolar-Transistors. *Solid-State Electronics* **1976**, *19*, 857–862.

70. Pi, X. D., Delerue, C. Tight-Binding Calculations of the Optical Response of Optimally P-Doped Si Nanocrystals: A Model for Localized Surface Plasmon Resonance. *Physical Review Letters* **2013**, *111*, 177402.

71. Alvarez, M. M., Khoury, J. T., Schaaff, T. G., Shafigullin, M. N., Vezmar, I., Whetten, R. L. Optical Absorption Spectra of Nanocrystal Gold Molecules. *The Journal of Physical Chemistry B* **1997**, *101*, 3706–3712.

72. Luther, J. M., Jain, P. K., Ewers, T., Alivisatos, A. P. Localized Surface Plasmon Resonances Arising from Free Carriers in Doped Quantum Dots. *Nature Materials* **2011**, *10*, 361–366.

73. Liptak, R. W., Kortshagen, U., Campbell, S. A. Surface Chemistry Dependence of Native Oxidation Formation on Silicon Nanocrystals. *Journal of Applied Physics* **2009**, *106*, 064313.

74. Gresback, R., Nozaki, T., Okazaki, K. Synthesis and Oxidation of Luminescent Silicon Nanocrystals from Silicon Tetrachloride by Very High Frequency Nonthermal Plasma. *Nanotechnology* **2011**, *22*, 305605.

75. Holmberg, V. C., Helps, J. R., Mkhoyan, K. A., Norris, D. J. Imaging Impurities in Semiconductor Nanostructures. *Chemistry of Materials* **2013**, *25*, 1332–1350.

76. Soref, R. Mid-Infrared Photonics in Silicon and Germanium. *Nature Photonics* **2010**, *4*, 495–497.

16 Organically capped silicon nanocrystals

Kateřina Kůsová and Kateřina Dohnalová

Contents

16.1 INTRODUCTION

The interest in silicon nanocrystals (SiNCs) was spurred in the 1990s by Canham's discovery of the naked-eye-visible, room-temperature light emission of porous silicon [21]. This finding stimulated scientific interest in SiNCs capped with silicon oxide, which can form on the NC surface naturally at ambient conditions. Despite intense research and application prospects in both optoelectronics and nanomedicine, oxide-passivated SiNCs were shown to have some shortcomings, mainly the tendency to agglomerate, poor spectral tunability, and poor stability in the case of porous silicon [155]. To remedy these problems, surface passivation (or functionalization*) via direct covalent attachment of preferably organic monolayers

---

* Here, the terminology is not yet clear. Sometimes, these two terms are used interchangeably, whereas sometimes, "passivation" refers solely to the stabilization of the surface, while "functionalization" is used for surface groups providing an additional functionality.

is an excellent alternative to simple oxidation, as known preparation methods do not allow for the lattice-matched core-shell structures common in other semiconductor NCs. The Si–C bond in particular is strong and only weakly polarized [191]. Organic passivation is stable, and, moreover, the attachment of organic groups provides wider versatility of the final product, including tailoring solubility in various solvents, and wider tunability of emission wavelength.

Comparing different surfaces, a "model" SiNC is that terminated with hydrogen (H–SiNC) as hydrogen termination influences the physical properties of the silicon core the least. H–SiNCs provide full spectral tunability [212], following the quantum confinement model well (see Section 16.1.1). However, the Si–H bond on the highly curved surface of an SiNC is much weaker than that on a planar Si, being oxidatively unstable and very sensitive to water [31,41,59,191]. Halogenated, for example, chlorine-terminated surfaces, on the other hand, exhibit only weak photoluminescence (PL) in addition to being unstable in ambient air [64]. Therefore, hydrogenated and halogenated SiNCs are regarded as reactive platforms or sometimes convenient study models under controlled atmosphere [64,212], whereas oxide-covered and organically terminated SiNCs are stable entities, with deeper understanding of physical properties reached in the former, but much wider versatility in the latter. Naturally, other types of attachment, such as Si–S linkage [221], have also been studied.

16.1.1 LEVELS AND BANDS

NCs are usually treated as "artificial atoms," or zero-dimensional objects with discrete energy levels E_N given, in the simplest approximation, by the particle-in-a-box model

$$E_N \sim \frac{\pi^2 \hbar^2 N^2}{2\, m_{\text{eff}} d^2} \tag{16.1}$$

where $\hbar$ is the reduced Planck's constant, m_{eff} the effective mass of the (quasi)particle (exciton), and d the NC diameter. It is this relation that gives NCs their size-dependent optical properties. The concept of NCs as quantum-confined structures (wells) is treated in more detail for example in [157].

However, in reality, NCs consist of several hundreds of core atoms arranged in a periodical lattice, which is why a different point of view, that of solid-state physics, should also have some validity.

In solid-state physics, ideal infinite crystals are described using the so-called band structure. The band structure theory is a single-electron approximation, or, in other words, it describes the behavior of a single selected electron in a solid, and the interaction with the lattice and other electrons is included in a periodical potential. Due to the periodicity of the problem, as stated by the Bloch theorem, the eigenstate wavefunctions $\psi_{n,\vec{k}}^{\text{solid}}$ of such an electron are given by a superposition of plane waves

$$\psi_{n,\vec{k}}^{\text{solid}}(\vec{r}) = u_{n,\vec{k}}(\vec{r}) \exp\left\{ i\vec{k} \cdot \vec{r} \right\}, \tag{16.2}$$

where $u_{n,\vec{k}}$ stands for a periodic mother function (a combination of atomic orbitals) and $\vec{r}$ is the real-space vector. The corresponding energies vary with wave vector $\vec{k}$, which is given as a Fourier transform of the original crystalline lattice. It represents a quantum-mechanical analogue to momentum, but its most important physical property is that a conservation law applies to it.

In reality, the energies corresponding to the states of electrons (and holes) in a solid form allowed and forbidden bands of band structure $E_n(\vec{k})$, characteristic of a particular crystalline lattice and chemical composition. Deep-lying (corresponding to lower orbitals) and high-energy states are usually disregarded, because they do not participate in processes under typical conditions. In semiconductor materials, the highest band occupied at a given temperature with electrons, the valence band (VB), is separated from the higher-lying conduction band (CB), to which electrons can be excited if given extra energy, by a band of forbidden energies, the so-called band gap E_g. Although the band structure of a three-dimensional object is in general a four-dimensional function (energy as a function of the three dimensions of the reciprocal space), it is usually plotted only along selected directions (e.g. $\Gamma \to X$), which contain CB minima (VB maxima) and are therefore most easily accessible to electrons and holes. Examples of band structure are given in Figure 16.1a and b.

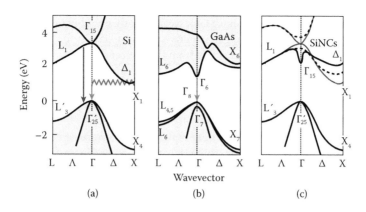

Figure 16.1 (a, b) Bandstructure of indirect-bandgap bulk silicon (a) in comparison with an archetypal direct-bandgap semiconductor GaAs (b). Fundamental (lowest energy) transitions are denoted in dark gray, the zig-zag curve stands for the participation of a phonon quasiparticle. Note that direct transitions are also possible in silicon at higher energies (for non-relaxed carriers). (c) Sketch of bandstructure engineering of SiNCs, which can lead to fundamental direct bandgap through the cooperative effect of quantum confinement (the dotted curve) and tensile strain (the solid black curve). (A physically more correct version is shown in Figure 16.7.)

This concept has been combined with the level-based approach using the effective mass approximation (EMA) [17,62] by considering the CB and VB parabolic near the extremes

$$E_{CB}\left(\vec{k}\right) - E_{VB}\left(\vec{k}\right) = E_g + \left(\frac{\hbar^2 k^2}{2}\right)\left(\frac{1}{m_e} + \frac{1}{m_h}\right) = E_g + \frac{\hbar^2 k^2}{2m_{exc}}, \tag{16.3}$$

where $m_{e/h/exc}$ are the effective masses of electron (hole, exciton). The energy of the corresponding lowest excited state E_{e_1,h_1} then bears close resemblance to the particle-in-a-box model from Equation 16.1

$$E_{e_1,h_1} = E_g^{NC} = E_g^{bulk} + \frac{\hbar^2 \pi^2}{2m_{exc}d^2}, \tag{16.4}$$

leading to band gap opening $E_g^{NC} > E_g^{bulk}$ in NCs. In a less crude approximation, other terms (e.g., electron–hole Coulomb interaction) can be included [36,62,157] and the d^{-2} size dependence can differ from the ideal exponent of 2. For example, in H–SiNCs, the relation

$$E_g^{SiNC}(eV) = 1.17 + 3.73\, d\,(nm)^{-1.39} \tag{16.5}$$

was proposed based on theoretical calculations [36]. This dependence was later applied to both porous silicon [107] and C–SiNCs (prepared by annealing of an SiO$_x$-based precursor) [123,141,208].

A more elaborate treatment of band structure of SiNCs is given in Section 16.5.

16.1.2 TRANSITIONS

The band structure concept is widely utilized in solid-state physics and it allows for the predictions of behavior of a material under various conditions (see also Section 16.5). Perhaps, the most important aspect of band structure is the fact if its VB maximum is located at the same wave vector as the CB minimum, giving rise to either direct or indirect band gap. As shown in Figure 16.1a and b, if an electron is to undergo a CB → VB transition and emit a photon in an indirect band gap material, the difference in wave vector needs to be compensated for by another quasiparticle (due to the wave vector conservation law)[*], which makes the process much less probable than in the case of a direct band gap semiconductor. This is why

---

[*] The wave vector of a photon is very small and thus negligible.

indirect band gap materials, such as silicon (or Ge), are inherently poor light emitters with long radiative lifetimes τ_{rad} in comparison to direct band gap materials (GaAs, CdSe)*.

Importantly, when determined experimentally from luminescence decays, the measured lifetimes τ_{meas} are determined by both radiative and nonradiative $\tau_{rad/nonrad}$ lifetimes (which can be much faster)

$$\frac{1}{\tau_{meas}} = \frac{1}{\tau_{rad}} + \frac{1}{\tau_{nonrad}}; \quad \tau_{rad} = 1\,\mu s, \tau_{nonrad} = 1\,ns \rightarrow \tau_{meas} = 1\,ns. \tag{16.6}$$

Therefore, in order to experimentally determine a material characteristic τ_{rad} and the corresponding radiative rates R_{rad}, either special techniques, using the general dependence of the Fermi's golden rule on the optical density of states (such as the Drexhage experiment [47]), need to be applied, or both decay and luminescence quantum yield (QY) (or internal quantum efficiency) need to be taken into account. When QY is known, a simple conversion can be applied

$$QY = \frac{R_{rad}}{R_{overall}} = \frac{R_{rad}}{R_{rad} + R_{nonrad}} = \frac{\frac{1}{\tau_{rad}}}{\frac{1}{\tau_{rad}} + \frac{1}{\tau_{nonrad}}} = \frac{\frac{1}{\tau_{rad}}}{\frac{1}{\tau_{meas}}} = \frac{\tau_{meas}}{\tau_{rad}}. \tag{16.7}$$

Being more precise, the quantity from Equation 16.7 is usually referred to as internal quantum efficiency, whereas QY also takes into account the dark, non-emitting NCs in an ensemble. For the sake of simplicity, this distinction is not used in this chapter.

16.1.3 PROS AND CONS

Silicon is an interesting material because of its abundance, the mature technology used to fabricate microelectronic circuits, and its low toxicity. Therefore, silicon-based materials are preferred in applications to materials containing, for example, toxic heavy metals. In the form of C–SiNCs, silicon offers very broad spectral tunability, ranging from UV to near-IR [40]. Nevertheless, C–SiNCs have not yet been used commercially, mainly due to the lack of industrial-scale fabrication and functionalization techniques of sufficient amounts of well-defined particles. When compared to other semiconductor NCs, the main challenges yet to be tackled in a C–SiNCs system are the tendency of Si to form stable amorphous structures and the consequent relatively high temperatures for crystallization, its propensity to oxidize, and capping ligand chemistry that is significantly different from that of the well-studied metals and metal chalcogenides. Unlike these materials [62], optical properties of C–SiNCs are, unfortunately, still only superficially understood, partly also because they are determined by a complex interplay of the properties of the core (size, crystallinity), surface (ligands, level of oxidation), and interface (type of linkage, defects). Also, researchers studying C–SiNCs can be viewed as two separate communities, those with a physics and those with a chemistry educational background, each of whom puts emphasis on different aspects with a very small overlap. In order to successfully tackle the problems connected with C–SiNCs, these two approaches need to be fused to a single "interdisciplinary" effort.

The exceptional properties of C–SiNCs hold great promise for future applications, some of which are depicted in Figure 16.2. Those envisioned for C–SiNCs are closely connected to those previously suggested for SiNCs with oxide or other terminations. Specifically, the first force driving the research was the possibility of using them as a material for optoelectronic light-emitting devices [40]. Even though the elusive goal of a silicon laser has not yet been reached, positive optical gain in oxide-capped SiNCs was confirmed many times [156]. Moreover, a band alignment favorable for efficient exciton dissociation between SiNCs and a conjugated polymer P3HT was identified, indicating the feasibility of using a P3HT/SiNC blend as the photoactive layer in photovoltaic applications [120]. Also, SiNCs were studied as a high-capacity material for Li-ion batteries [31]. Later on, semiconductor NCs in general found their way into bio-applications

---

* It is important to realize that, despite the claims in various publications, direct and indirect transitions are not synonyms for dipole-allowed and dipole-forbidden transitions. The radiative recombination rate is a product of the square of (dipole-allowed or -forbidden) matrix transfer element and a factor related to the occupancy of the states participating in the transition, which is the one determined by the direct/indirect nature of the band gap. Thus, both direct and indirect transitions can be dipole-allowed and -forbidden, resulting in the different exponents in the absorption versus photon energy Tauc plot.

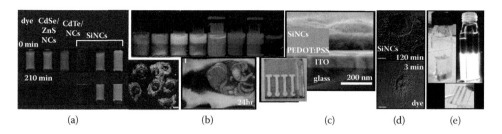

Figure 16.2 Examples of use of C-SiNCs. (a) Photostability of colloidal SiNCs under continuous Xe lamp illumination (0 and 210 min) in comparison with other materials. Bottom: a combined fluorescence image of microtubules of Hela cells labeled by SiNCs and cellular nuclei imaged by Hoechs. (After Zhong Y, et al., *ACS Nano*, 9, 5958–5967, 2015. With permission.) (b) Top: photograph of luminescence of SiNCs. Bottom: overlaid autofluorescence (green) and SiNC (red) fluorescence image of SiNC injected in mice. Uptake in spleen and liver was observed. (After Erogbogbo F, et al., *ACS Nano*, 5, 413–423, 2011. With permission.) (c) TEM image and photo of a SiNC photodiode. (After Lin T, et al., *Adv. Funct. Mater.*, 24, 6016–6022, 2014. With permission.) (d) HeLa cells labeled by SiNCs (top) and conventional dye (botom). SiNCs show excellent photostability. (After He Y, et al., *J. Am. Chem. Soc.*, 133, 14192–14195, 2011. With permission.) (e) Top: photos of SiNCs under ambient and UV ligth. Bottom: SiNC-based hybrid LED. (After Luppi E, et al., *Opt. Mater.*, 27, 1008–1013, 2005. With permission.)

when their photostability superior to dyes had been exploited for fluorescent biological probes [16]. At first, studies employed efficiently emitting direct band gap group II–VI NCs (CdSe), and the scope of possible applications widened [204]. Unfortunately, it was shown that toxic heavy metals forming the core of these NCs such as cadmium can be released despite encapsulation [37,218]. Therefore, research interest shifted to heavy-metal-free materials including nanosilicon, which is biocompatible in many forms [93,159] and after degrading into orthosilicic acid can be excreted by the urinary system [153,218]. Thus, apart from fluorescent markers for cell imaging, SiNCs were also tested as drug delivery and diagnostic tools for nanomedicine [93,158], going as far as commercial use in Prof. Canham's company pSivida. A special chapter could be the use of SiNCs as nanocatalysts [93,158].

Thus, the research of C–SiNCs is experiencing a boom with important recent stimuli being novel, simple high-yield synthesis techniques (see Section 16.2.2) and the confirmation of the possibility of reaching fundamental direct band gap (see Section 16.6.2). This boom can be illustrated in a number of recent reviews both comparing SiNCs with other NCs [56,100] and those specializing solely in SiNCs. These reviews put emphasis on preparation [26,40,59,129,139,183] and surface modification [26,31,59,183] techniques, optical properties [40,59,183] and characterization techniques [26,183], or biological [31,56,93,139,159] and other [31,40,129] applications.

This chapter is organized as follows. First, Section 16.2 gives a detailed overview of various techniques for fabrication and surface modification of SiNCs, which are a key to their physical properties. The following Section 16.3 outlines the basic problems and results of theoretical simulations involving C–SiNCs. Next, Section 16.4 summarizes basic characterization methods, Section 16.5 describes the band structure approach to SiNCs and Section 16.6 focuses on light emission and mentions several surface-induced effects. The last Section 16.7 gives examples of laboratory-tested applications of C–SiNCs.

16.2 FABRICATION TECHNIQUES

The preparation of C–SiNCs is usually a process involving several steps, see Figure 16.3: first, an SiNC core is formed, then the SiNC surface is stabilized or functionalized by organic ligands; other postprocessing steps might be necessary. The resulting sample is typically a colloidal dispersion of freestanding NCs.

16.2.1 FORMATION OF THE CORE

Historically the first SiNCs were porous-silicon-derived ones, which resulted from electrochemical etching [21], a method already known from the 1950s [197]. Soon after that, the very first wet-chemical syntheses [38,74] and gas-phase syntheses [119] (again inspired by earlier approaches yielding larger structures [144]) appeared. A somewhat newer addition to the list lies in the high-temperature anneal of

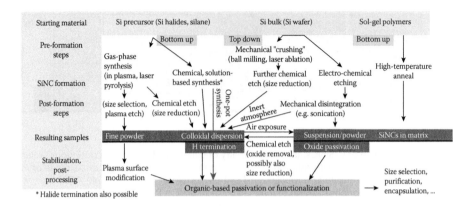

Figure 16.3 Overview of preparation techniques. (Modified after Dohnalová K, et al., *J. Phys. Condens. Matter.*, 26, 173201, 2014. With permission.)

Table 16.1 **Comparison of methods used for the preparation of SiNCs**

METHOD	YIELD	SIZE	EQUIPMENT	CONDITIONS	PROS	CONS
Electrochemical etching	0.5 mg/cm²h	2.5–5 nm	Electrochemical system	Ambient	Easy, excellent crystallinity	Also very large (>100 nm) particles, HF handling
Chemical synthesis	Grams: 10⁻⁵ [194], 0.01 [160], 0.85 [176], 10 [227]	1–10 nm	Glove box, varies based on approach (microwave, UV, …)	Higher T, often inert atmosphere, some at ambient conditions	Good size/shape control	Broad size range requires different chemical routes
Gas-phase synthesis	2–200 mg/h	2–10 nm	Plasma reactor	Controlled T, pressure, atmosphere	High yield, up-scalable, *in-situ* passivation	Expensive equipment, time-consuming to set up
SiOₓ anneal	250 mg/ 3 hod	1–12 nm	Furnace with controlled atmosphere	High T, controlled atmosphere	Monodispersity, size tunability	HF handling

SiOₓ-based precursors [78], inspired by the high-temperature annealing of silicon-rich oxides [76], commonly used to prepare oxide-passivated matrix-embedded SiNCs and SiNC superlattices. These techniques are compared in Table 16.1.

Out of these methods, electrochemical etching, is relatively simple and can be carried out at ambient conditions. It lies in the anodization of a single-crystalline doped Si wafer under constant current in an HF/EtOH-based electrolyte [42,86,133,151,188]. However, it is known to lead to irregular shapes and polydisperse samples with particles and agglomerates ranging from nanometers to micrometers in size [28]. It has also a relatively small yield, considering that the starting material is a high-quality Si wafer. An interesting modification is the polyoxometalate-assisted electrochemical etching followed by controlled H_2O_2/EtOH oxidation, which produces color- and size-tunable water-soluble SiNCs [94].

Various chemical syntheses, on the other hand, generally allow for a much better control over the size distribution and possibly also the shape, however, often only relatively small sizes (diameter less than 2 nm) are reported. Also, due to the sensitivity of both the precursors and uncapped SiNCs to oxygen and moisture, they typically need to be carried out under inert conditions [139] and often under elevated temperatures and pressures. A wide variety of chemical syntheses have been investigated and new types of syntheses are still being reported. Examples of chemically synthetized SiNCs include those prepared by various reduction routes, such as by the reduction of halide salts (typically $SiCl_4$) using the sodium metal (Na) at higher temperatures and pressures (385°C, >100 atm.) [74] or using other reducing agents [232]. Reduction of halide salts can also be carried out in a nonpolar solvent in the presence of a surfactant (e.g., tetraoctylammonium bromide—TOAB) encapsulating the forming nanoparticle and thus defining its size (the so-called inverse-micelle approach or microemulsion synthesis) [176,194,209]. This method yields near-gram scale amounts of NCs [176], but usually only with relatively small sizes. The oxidation route, for example, involving the oxidation of magnesium silicide with bromide [160], has also been explored. Moreover, metathesis reactions of alkali silicides (MSi, M = Na, K, or Mg) with $SiCl_4$, Br_2 or ammonium bromide (NH_4Br) have been very successful [13,122,146,216]. Yet another synthesis technique is the preparation from supercritical fluids, such as the thermal decomposition of diphenylsilane in octanol [52,83]. To avoid high temperatures and pressures, microwaves can be applied to increase reaction efficiency of the normal process [6]. Last but not least, chemical syntheses from dopant-rich (e.g., Mn) precursors can give rise to paramagnetic SiNCs [195].

Next, gas-phase approaches are usually based on the thermolysis of silane [119], require both high temperatures (850°C and more) and inert atmosphere. In contrast to chemical syntheses, they tend to produce somewhat larger SiNCs as the forming SiNCs in a reactor under high temperature are prone to agglomeration and coalescence. On the plus side, this technique is upscalable to macroscopic yield, which is interesting for future industrial applications. Pyrolysis of silane can also be driven by a CO_2 laser as was done in [51,113], where the reported sizes were 3–7 and 5–20 nm, respectively, and where high yields were reached (20–200 mg/h) [113]. A special approach is that of nonthermal plasma, where the free electrons have a much higher temperature than heavier species, which leads to charging of the forming SiNCs and consequently significantly reduces their coalescence. Thus, much narrower size distribution and smaller sizes can be reached [99,131].

Thermal processing of SiO_x-based precursors such as HSQ relies on a high-temperature anneal (>1100°C) in a controlled atmosphere [78,81,123,134] and reaches reasonable yields. These NCs are produced as SiO_2-matrix-embedded and therefore require further processing, which might be, on the one hand, challenging for potential industrial use; on the other hand, they feature excellent monodispersity and size tunability.

Other, less frequently exploited techniques, include disintegration of Si targets, such as laser ablation in a gaseous [148,207] or liquid [184] environment, and high-energy ball milling [67,75]. These methods can produce several-nm-large SiNCs. A more exotic method is then the production of SiNCs from rice husks, where SiO_2 nanoparticles are extracted and subsequently annealed, with the yield of about 1 mg SiNCs with sizes around 20 nm per 20 mg of rice husks [121]. Although still unconventional, this method of extracting SiNCs from a wide variety of agricultural-waste "silicon accumulator plants" including rice husks, bamboo, or sugarcane, is of immense interest for commercial use due to its cost-effectiveness [9].

When the formation of the silicon core is finished, other postprocessing treatments might be necessary. Porous-silicon-based SiNCs are liberated from the porous Si structure mechanically [42] or using sonication [87]. In matrix-embedded SiNCs, the sample is ground to powder and the matrix is chemically etched away using HF [78,81]. Whereas HF etches away only silicon oxide and therefore can be used to remove undesirable oxide passivation [79], cycles of HF etching and subsequent oxidation are needed to reduce the size of the SiNC core if necessary. Such cycles may proceed via repeated HF etching and exposure to air [162,207], or via a (photoinitiated) combined HF/HNO_3 etch [28], where HNO_3 is responsible for the oxidation step. This further chemical etch is most commonly used with some of the SiNCs prepared by gas-phase syntheses [66,85,113] or ablation techniques [207] to reach sizes small enough to allow for light emission. Size reduction through etching can also be achieved in plasma, for example, by using CF_4- or SF_6-based plasma [118,161].

Clusters, nanoparticles, and quantum dots

16.2.2 SURFACE PASSIVATION OR FUNCTIONALIZATION

Typically, SiNCs freshly prepared by etching, gas-phase or most chemical syntheses possess hydrogenated surfaces [200]; in chemical reactions involving halogens (Cl, Br), the surface is halogenated.* These surfaces are unstable and reactive; therefore, they need to be passivated. Early organic passivations of SiNCs were carried out in the late 1990s [96,216]. The passivation can be carried out either chemically by a solution route or in plasma; theoretical aspects of passivation are described further in Section 16.3.1.

Numerous chemical methods have been established to generate Si–C or other linkages on silicon [19]. However, inhomogeneous surfaces of SiNCs are much more complex and their chemistry can differ from that of flat bulk Si [11,191]. By far the most important wet-chemistry surface passivation method is hydrosilylation, a transformation of an Si–H to an Si–C bond. It was pioneered in the early 1990s on flat bulk Si [11,117] and later also on porous silicon [96,190]. Hydrosilylation results in grafting of a variety of ligands ranging from alkyl [66,81,85,86,89,92,112,178,194] to terminal carboxyl–COOH [24,174,220] or amine NH_2 [6] groups. The reaction is often oxygen-sensitive with reaction times typically reaching tens of hours. It can be facilitated by heat, light, or a catalyst. Thermal hydrosilylation requires higher temperatures (~150°C) and therefore only high-boiling chemicals (long alkyl chains) can be grafted onto SiNC surfaces. Recently, it has been demonstrated to lead to ligand oligomerization [217], in contrast to the long-assumed monolayer coverage, which can be a complication for optoelectronic applications. A possible photoinduced alternative is then limited by size-dependent reactivity [95]. Recently, it was shown that hydrosilylation can occur even at room temperature without any catalysts provided it involves certain terminal groups (e.g., carboxyl COOH) [220]. Room-temperature hydrosilylation of aliphatic ketones can also proceed under microwave irradiation, resulting into Si–O–C alkoxy-terminated SiNCs as a reactive platform for further functionalization [165]. Apart from hydrosilylation, alkyls, as well as other ligands, can be also attached to halogenated surfaces which are reactive as well and have strong electrophilic activity [67,109,216]. Methanol washing of chloride-terminated surfaces then leads to alkoxy linkage [232]. The list of possible surface treatments can be complemented with surface polymerization on Si–H surfaces [49,112], or direct transformation between Si–O and Si–C surfaces by laser irradiation in aromatic hydrocarbons [104,145] and vice versa [44].

Plasma-based surface modifications were pioneered later than the solution-route passivations, in the 2000s [115,130]. Various alkyls, alkenes, and long alkyl chains can be grafted onto the SiNC surface and reaction times are very short [99]. In addition to aerosol-based plasma, surface modifications can also be achieved in the liquid plasma phase (microplasma), yielding Si–O–C linkage [133,192].

Several approaches were proposed to bypass the complicated procedure of forming the SiNC core as the first step and passivating it as the second step. Such one-pot approaches include some of the chemical synthesis techniques, which can be designed to result directly into ligand-passivated SiNCs. Reports on one-pot chemical syntheses often utilize $C_6H_{17}NO_3Si$ (APTMS) as the source of silicon and are usually of a relatively recent date. Examples of such reactions are given in Table 16.2. Moreover, ball milling [75] or ablation in liquids [184] can also lead directly to ligand-passivated SiNCs. As regards gas-phase syntheses, the two steps of the formation of the Si core and its surface passivation can conveniently be combined into a single reactor [99,130]. This all-gas-phase approach eliminates the need of oxygen- and moisture-free transfer of the forming SiNCs; it also features very fast reaction times (~1 s) and allows for plasma-induced grafting of a wider range of ligands, including shorter alkyl chains [99].

In situations where the desired surface moieties have more than one functional group, a multistep approach for the passivation itself might be preferred. An early multistep approach was adopted in alkoxy-capped SiNC, where further treatment generates much more stable siloxane-alkyl Si–O–Si–C surfaces [231]. Very recently, alkoxy-functionalized SiNCs were also used as reactive platform for further functionalization via ligand exchange [165]. Moreover, thiol-ene "click" chemistry was proposed, where the terminal alkene group can further react with thiols (HS–R) with a wide variety of distal groups, allowing for the grafting of various terminal (bio)functional groups [177]. Multistep approaches can also be utilized to combine SiNCs with various polymers (e.g., a SiNC/polymer composites moldable into various shapes [180]) or to conjugate them with antibodies [196].

---

* Plasma-etched SiNCs possess fluorine-terminated surfaces.

Table 16.2 **Examples of one-pot syntheses**

SYNTHESIS	SURFACE	SIZE (nm)	ADVANTAGES	REF.
Microwave reduction of APTMS	Amine groups	2.2	QY 20–25%, no harsh conditions in synthesis, yield 0.1 g/10 min	[226]
UV reduction of APTMS	Naphthalimide ($C_{12}H_7NO_2$)	2–3	<40 min, 10 g, QY 25%, ambient air conditions, photostability under strong UV	[227]
Hydrothermal reduction of APTMS	Ethylamine	2.1	QY 22%, simple	[8]
NaSi + NH_4Br in neat glutaric acid	Si–O–C_4H_7COOH	4	Medium is the surface agent, versatililty (also for other passivants)	[132]
Microwave break-up of Si nanowires with glutaric acid	Glutaric acid	3.1	Water dispersible, photostability	[73]

16.2.3 FURTHER POSTPROCESSING

Produced SiNCs, either bare or ligand-passivated, are often quite polydisperse, and therefore various size-selection strategies are exploited. Obviously the simplest technique is filtration [85,104,114,121,227]. More elaborate approaches then include density-gradient ultracentrifugation [24,135,141], which utilizes the different effective densities of smaller and larger SiNCs due to a smaller weight fraction of the heavier core. This method was successfully applied to separate 1.3–2.2 nm alkyl-capped SiNCs using the centrifugation of ~10^5 G [135]. Another method is size-selective precipitation [123,137,210], based on the different packing densities of the surface ligands, which give rise to size-dependent solubility. Therefore, larger SiNCs can aggregate and be separated with the addition of a small amount of anti-solvent. As an example, application of this method allows the separation of 12 size fractions of undecenoic-acid-capped SiNCs sized from approx. 4.5 to 2.5 nm [24]. Moreover, size-exclusion (column) chromatography can also be carried out [85,209]. A size-selection process is possible also in a gas-phase environment, such as using a differential mobility analyzer after ablation [148].

Further purification might be necessary to remove unreacted surfactants or side products. Since SiNCs are larger than chemicals used for their production, the above mentioned size-exclusion chromatography tools [112,177,182] are often employed for this purpose. Subsequently, further encapsulation of the produced and suitably terminated SiNCs can be desirable for specific applications, typically to impart solubility in water for biological applications. The shell material can be phospholipid micelles [53,61], hydrophilic polymer [72], lipid nanoparticles [77], amphiphilic polymer micelles [80], polymer nanoparticles [71], or silica [151]. An added benefit of the silica coating (silanization) is that it allows for further functionalization with amines, carboxylic acids, and thiols, in turn facilitating the linkage of other biomolecules; this process has already been developed for II–VI NCs [16]. The encapsulation often results in more NCs in a single nanosphere with a significantly larger diameter (~100 nm) than the original NCs [53,72].

16.3 THEORETICAL CALCULATIONS

The electronic and optical properties of C–SiNCs are in general more interesting and complex than those of uncapped SiNCs. Purely due to the large surface-to-volume ratio [85], the surface atoms contribute considerably to the electronic states of the whole C–SiNCs system [168,181,216,229,231]. The truncated bulk surface of the SiNC leads to complex reconstructions [11] and capping of the surface with ligands is, unlike in metals, site-specific and directional, requiring rigorous molecular-like simulations [11,111,171,173,203,222,230]. Further complexity arises with broken symmetry, additional states, or charge shifts due to electronegative capping.

The difficulty with rigorous and robust simulations of C–SiNCs arises with the lack of powerful computational techniques. Current simulations rely to a large degree on an initial by-hand design of the system by the theoretician, while at the same time it has been established that the computed electronic and optical properties of C–SiNCs are often very sensitive to these parameters. The most rigorous *ab initio* approaches, such as the density-functional theory (DFT), have been successfully used to simulate C–SiNCs up to 2 nm [167,171,202]. However, it is unsuitable for the larger, 2–4 nm C–SiNCs that are often of great interest to experimentalists. For such larger SiNCs, the computationally efficient bulk-like techniques such as $\vec{k} \cdot \vec{p}$, pseudopotential, or tight-binding (TB) can be used with some success, but the role of the organic ligands cannot be rigorously implemented. Currently, the most accurate and computationally viable technique is the density-functional TB (DFTB) method [22,90,97,110,111,147]—TB parametrized using DFT. For the indirect band structure of silicon, additional TB [44,163] and DFT–TB [70,106] $\vec{k}$-space tools were developed (see Section 16.5).

16.3.1 SURFACE RECONSTRUCTION AND COVERAGE WITH LIGANDS

First, to simulate the NC core, a spherical shape is assumed. However, the truncated surface of bulk silicon leads to unsatisfied (dangling) bonds, which can act as nonradiative recombination centers [15] and cause the reactivity of SiNCs [166,167]. Subsequently, the surface is reconstructed to minimize the surface energy by decreasing the number of dangling bonds [11,46,143,169,171,198], which leads to various nonbulk Si–Si conformations and, in the extreme case of very small Si clusters (~1 nm), to the total reconstruction of the whole core to amorphous [45,171] or icosahedral shapes [171,224]. For larger particles, the resulting surface has bulk-like facets. Each of these facets has different densities of atoms and dangling bonds [one for (111) and (110) and two for (100)], which gives them different reactivity, leading to different reconstructions, inhomogeneous adsorption of ligands, and different charge transfer [11]. For example, the unreconstructed (100) facet cannot be completely passivated even with small H atoms [169,173,230], but the large density of single dangling bonds on the (111) facet might lead to larger steric effects [171]. One can, by a careful design, achieve a system with only single dangling bonds per surface Si atoms [171] for all facets, so they become terminated with the same type of chemical group [230]. After the ligand attachment, on the one hand some of the weak reconstructed Si–Si bonds might relax [11] when replaced with stable Si–C bonds [30]; on the other hand, using ligands with considerable electronegativity (such as S, N or O), the surface outer-layer Si–Si bonds can show additional elongation [110]. As regards the core's symmetry, methyl groups have only a small influence, but larger passivants are more difficult to place at the surface, are more reactive [202], and break the original symmetry. The important parameter of capping density is in general limited by steric hindrance, evaluated as the difference in energy between different configurations of the passivants on the surface. It has been demonstrated that full surface coverage of SiNCs with unreconstructed surface is not realistic [111,171] and that maximum surface coverage can be as low as ~30–40% [89,202]. In reality, the coverage will also vary depending on the used capping chemistry [7,52,83,160], but in general, high coverage by organic ligand is desirable, to prevent oxidation and improve stability. Surface-related effects (see also Section 16.6.2 for experimental results) induced by forced full coverage include the reduction in optical gap [230] as well as the rise of considerable tensile strain [70,106] (increased Si–Si average bond length), but also reduced chemical activity [111] and eventually the destabilization of the C–SiNCs system showing sooner for larger clusters and longer ligands [171,186].

16.3.2 ENERGY LEVELS AND TRANSITIONS

For C–SiNCs, a large drop in both highest occupied molecular orbital (HOMO) and lowest unoccupied molecular orbital (LUMO) energies is typically observed (with respect to H–SiNCs) [111,171,230]. This shift is larger for larger NCs and only weakly depends on the number of ligands [110,111,203]. It is caused by the different electronegativity [154] between Si and ligand elements*, which causes the formation of a dipole on the surface [230]. In particular, alkyls will appear slightly positively charged with respect to Si [111,230], resulting in charge transfer between the core and the ligand [111]. For alkyl ligands, however, charge transfer is very weak [111], leaving the iso-surface of frontier orbitals similar to those of H–SiNCs [110] and depending only weakly on the size of the NC and the number of ligands [203]. More pronounced effects then occur for more electronegative elements such as O [98,110,168,171,229], F, Cl [98,125,126], or N [98],

---

* Electronegativity values: 2.55 for C, 2.1 for H, N 3.04 and O 3.44, and 1.9 for Si.

leading ultimately to size-insensitive surface-dominated behavior. It was pointed out that for SiNCs, LUMO and HOMO energies behave in a way qualitatively similar to the electron affinity (energy given by adding an electron) and ionization potential (energy needed to remove an electron), respectively [140], which has been used to explain some of the modified properties of C–SiNCs. For example, the lowered ionization potential for C–SiNCs implies higher reactivity of C–SiNCs with respect to H–SiNCs [111], reflected in easier photoinduced hydrosylilation processes [173]. Also, it might explain the erratic blinking patterns observed for C–SiNCs [102] (see Section 16.6), as C–SiNCs are much more easily ionized than H–SiNCs [171]; size dependence of this effect further implies that larger NCs could possibly be charged under conditions which leave the small ones neutral [171]. While the HOMO and LUMO energies shift dramatically, the band gap does not seem to depend on the length of the alkyl ligand [58,171] and the band gap size follows the size quantum confinement effect (see Section 16.1.1 and Figure 16.4a) [58,110,111,171,203]. Band gap is only slightly smaller, compared to H–SiNCs, as the confinement is effectively lowered by the presence of the dense alkyl layer [171] (see Figure 16.4a and Figure 16.7c and d). More changes occur to higher states in the bands, where at higher coverages, the absorption spectrum exhibits new peaks above the band gap [58], corresponding to transitions involving states localized on the surface of Si–C bonds. This close relation between electronegativity and the HOMO–LUMO states [1] has also allowed for the implementation of electronegative methyl-like ligands in TB [44,163], leading to surprisingly good agreement with more rigorous methods in terms of band gap. At the same time, drastically enhanced radiative rates were found for a large range of electronegative elements [163].

Optical properties are typically defined by the density of states, the corresponding radiative rates R_{rad}, as shown in Figure 16.4b, and the optical band gap E_g, which is evaluated as the difference between the HOMO–LUMO states.* Whereas, according to Kasha's rule, optical emission always occurs from the lowest (LUMO) state, it is important to realize that in molecules and small nanostructures, optical excitation and de-excitation can lead to considerable structural changes [57] and hence absorption and emission might differ in energy by a "Stokes shift" [110,111,203]. This structural change cannot be ignored [124], as it can be as high as 0.1–1 eV, reaching values known for ionic solids. It differs for various ligands and surface coverage [110,124,168,203] and is less pronounced in larger NCs due to increased structural rigidity [203]. Most of the distortions happen in the core (elongation of Si–Si bonds) or the outer Si layer [110] rather than in the surface Si–ligand bonds [35,111,203]. In larger NCs, Stokes shift is dominantly caused by a change in the overall shape of the NC upon electron–hole excitation [57].

The presence of more complex ligands involving conjugated alkenes or various functional groups in close proximity to the SiNC surface has a more profound influence on the optical properties [166,202,230].

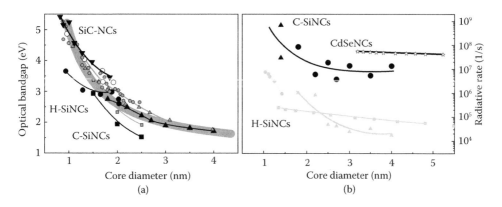

Figure 16.4 Theoretical calculations of (a) optical bandgap and (b) radiative rates (b) as a function of core size of C-SiNCs (solid black symbols) taken from different references [70,98,163,171,202] (different symbols). Data are complemented with results for H-SiNCs (gray symbols) [65,70,88,108,150,163] and other (open symbols) NCs [48,171]. The thick light-gray curve in panel (a) corresponds to Equation 16.5.

---

* In the case of a quasicontinuous character of the bands, the HOMO and LUMO states are identified as the states with considerable electronic density and oscillator strength [199].

For example, $-H_2C=CH-COOH$ was shown to lead to enhanced radiative rate [202], $-H_2C=CH-C_6H_5$ to lowered rates [202], conjugated alkenes to reduced rate and smaller gap [202], and double bonded $=CH_2$ to a reduction in band gap energy even for a single contaminant [166,230].

Last but not least, since C-containing compounds are used as reactants during many types of syntheses, one might consider the possibility that C is incorporated into the Si lattice, forming a cluster with the SiC structure [171,172]. At fixed core sizes, the width of band gap of alkyl-terminated SiNCs is similar to that of Si-rich SiC dots, see Figure 16.4a, suggesting that, unless radiative and nonradiative lifetimes are very different, PL measurements cannot be used to differentiate SiNCs from Si-rich SiC NCs below 2 nm. Luckily, the LUMO energies of alkyl-terminated SiNCs change compared with Si-rich SiC NCs, thus, ionization energies or electron affinities may be used to physically distinguish between them.

16.4 STRUCTURAL AND CHEMICAL PROPERTIES

16.4.1 SIZE AND CRYSTALLINITY

One of the most important parameters to be determined in a nanoparticle is clearly its size, or more precisely speaking the distribution of sizes. Despite its crucial role, it is not easy to reliably quantify the distribution of sizes in a few nm size regime. One of the most widespread direct (imaging) techniques is high-resolution transmission electron microscopy (HRTEM) (or TEM), giving directly, and very precisely, the measure of the size of the core and also its shape. However, SiNCs are more difficult to image than other NCs due to the low electron density contrast between Si and carbon [95,152], utilized as a support for HRTEM measurements. Also, HRTEM results often suffer from poor statistics and bias toward larger NCs. Other microscopy techniques, such as AFM [52,185], can be applied too, this time in principle evaluating the size for both the core and the passivating layer.[*] Although experimentally easier, AFM measurements cannot resolve as much detail as HRTEM and the decisive parameter to look at is the NC height with respect to a surface on which it is deposited. The procedure of determining a height difference between a surface and the highest point of a very small highly curved NC can bring about large uncertainty in the determined value. Other indirect methods include dynamic light scattering (DLS) [104,226,227] and small-angle X-ray scattering (SAXS) [60,81,95,220], including the grazing-incidence SAXS [18], which is even more precise for smaller NCs. DLS is based on measuring temporal correlations of a coherent laser beam scattered by NCs dispersed in a liquid undergoing a Brownian motion. These temporal correlations can then be recalculated into a diffusion coefficient characterizing the particles, which is then transformed into the so-called hydrodynamic diameter (diameter of a spherical particle with the same diffusion coefficient). In SAXS, the X-ray diffraction (XRD) pattern of the studied particles recorded under small angles allows for the determination of the distribution of sizes (and possibly also other longer range spatial arrangements) [81]. In both DLS and SAXS, the distribution of sizes is a result of a complicated fitting analysis and therefore extreme care needs to be taken especially in highly polydisperse samples, since the shape of the size distribution, for example, needs to be assumed in (and confirmed by) the fit.

Thus, on the one hand, the results of the direct imaging methods can be relatively easily and reliably analyzed, but they can be distorted by the experimenter who can preselect which particles they choose to image, while on the other hand, indirect DLS and SAXS inherently probe all the particles at once, but problems might arise due to automatic uneducated analysis, sample polydispersity, or low concentration. DLS in particular is also highly sensitive to the presence of larger particles,[†] where the signal of small NCs can be screened off by just a few larger ones, and also to the precise determination of the colloid's viscosity, which can be altered by the addition of NCs or production by-products. Therefore, the safe way is obviously always to combine results of different methods [81].

Apart from size, the fact whether the studied nanoparticles are crystalline is also of high importance. This fact cannot be automatically presumed and might depend on the preparation technique. Here, HRTEM (and the corresponding selected area electron diffraction pattern) can be clearly applied

---

[*] However, the apparent thickness of the surface layer might vary as the surface groups are much softer than the Si core and might behave differently under different conditions [152].

[†] For NCs smaller than 100 nm, the scattered signal rises as the 6th power of size.

to image lattice fringes, see [152]. Furthermore, XRD is commonly exploited to confirm the diamond Si lattice.* The next method is Raman scattering experiments, in which the signal of the transverse optical (TO) phonon of crystalline Si lattice appears around 520 cm^{-1} [4,79]. Both the XRD and Raman methods actually combine the information about crystallinity and size. In XRD, the width of the diffraction lines is inversely proportional to the NC size [33,81] (the so-called Scherrer formula). In Raman measurements, the situation is more complicated since the TO phonon mode shifts with size, NC deformation and due to possible (laser-induced) heating. Moreover, it is still unclear which model quantifying the relationship between the size and Raman shift is the correct one to apply [79,179] and therefore Raman spectroscopy should not be used as a technique for the determination of size without the comparison with other techniques.

Two noteworthy examples of excellent size and structural analyses were both performed on SiNCs prepared from HSQ. In the former set of experiments, several complementary techniques determining the distribution of sizes were applied [81], whereas the latter was a thorough HRTEM analysis, where the generally low contrast between Si and amorphous carbon supports was circumvented by utilizing graphene as a support [152]. In these experiments, it was even possible to visualize lattice defects and organic capping ligands. These and some other characterization techniques are shown in Figure 16.5.

16.4.2 SURFACE CHEMISTRY, COMPOSITION, AND DISPERSIBILITY

Surface chemistry is assessed using common techniques such as Fourier transform infrared spectroscopy (FT-IR) (often in the attenuated total reflection mode, which allows the detection of lower concentrations), X-ray photoelectron spectroscopy (XPS), nuclear magnetic resonance (NMR), and energy-dispersive X-ray spectroscopy for elemental composition. The emphasis is usually put on verifying whether the ligands are attached via a direct Si–C (or possibly Si–N, Si–S) or alkoxy Si–O–C bond, and on assessing the oxidation (ligand coverage) of the surface; see also Section 16.3.1. In FT-IR, the Si–C stretching mode is located around 680 cm^{-1}, but it may be observable only with surface methyl–CH$_3$ groups rather than longer alkyl chains [20,104]. Rocking modes of Si–alkyl groups are usually found at slightly higher wavenumbers around 770–850 cm^{-1} accompanied by the alkyl deformation mode at 1250 cm^{-1} [20,75,104,216]. On the other hand, vibrations corresponding to Si–O–Si bonds, generally strong in FT-IR, appear between 1000 and 1100 cm^{-1} [10,83,104,138,216] (Si–O–C tend to be situated roughly in the same region). These vibrations can then be compared with the 2100 cm^{-1} Si–H stretches [10,20]. In XPS, the Si2p peak is located at 99.5 eV (although it can be situated at higher binding energies due to charging [23,49,83,177]), the Si–C peak at ~102 eV and silicon oxide peak at 103–104 eV [29,220]; integrated intensities of these peaks can be utilized to estimate surface coverage (elemental composition) [83,176,177]. As for NMR, typically ^{1}H or ^{13}C NMR spectra are collected, however, to confirm the origin of the observed lines, it is possible to employ more elaborate techniques, such as diffusion-coefficient resolved NMR [104,177]. For examples of surface chemistry analyses, see Figure 16.5.

An important task in colloidal SiNCs is to impart dispersibility in a desired medium. SiNCs capped with nonpolar groups, such as alkyl or alkenyl, are highly soluble in nonpolar solvents such as toluene. However, the aim is often to use water or a water-based biological environment as the dispersing medium, which makes it necessary to either attach polar terminal groups to the NC surface, or to use some of the encapsulation techniques (see Section 16.2.3). Popular polar groups include carboxyl–COOH groups [24,174,220], such as glutaric COOH(CH$_2$)$_3$COOH or propionic C$_2$H$_5$COOH acids, and amine NH$_2$ [6] groups.

In aqueous dispersions, stability can be assessed using zeta potential (see Figure 16.5j), a potential difference between the dispersing medium and the stationary layer of fluid attached to the dispersed NC. If the absolute value of zeta potential is greater than 30 mV, the dispersion is stable. A zeta potential titration curve for varying pH, which influences stability of the colloid, can also be evaluated [24,132]. An oxidized surface [133], or a surface containing carboxyl groups [132] (changing into COO− in a stable colloid), is characterized by a negative surface charge, whereas amine groups take on the positively charged form of NH$_3^+$.

Apart from dispersibility in an aqueous medium, the production of various hybrid nonagglomerating SiNC/polymer nanocomposites can be of tremendous interest in applications [49,228].

---

* Unlike II–VI NCs, in Si the diamond lattice is the only stable configuration up to pressures of at least 10 GPa [69].

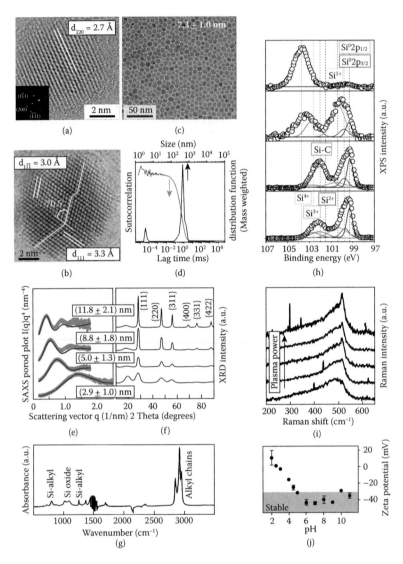

Figure 16.5 Examples of structural and chemical analyses of SiNCs. (a, b) Bright-field scanning TEM images of a (a) defect-free and (b) twinned SiNC. (After Panthani MG, et al., *J. Phys. Chem. C*, 116, 22463–22468, 2012. With permission.) (c) TEM image of SiNCs yielding a distribution of sizes. (After Hessel CM, et al., *Chem. Mater.*, 24, 393–401, 2012. With permission.) (d) DLS-derived size distribution of colloidal SiNCs (black, the top axis) and the corresponding DLS autocorrelation curve (gray, the bottom axis). The two different size ranges contribute to the two rising slopes at different timescales of the correlation curve. Note the much smaller contribution on the smaller size range. (Modified after Kůsová K, et al., *ACS Nano*, 4, 4495–4504, 2010. With permission.) (e) SAXS measurement of SiNCs (gray) and a fit assuming the spherical particle model with Gaussian distribution (black curves), yielding distributions of sizes. (f) The corresponding XRD characterizations for the SiNCs with different sizes from panel (e); note peak broadening with decreasing size. [Panels (e, f) were modified after Hessel CM, et al., *Chem. Mater.*, 24, 393–401, 2012. With permission.] (g) FTIR spectra of SiNCs. (After Heintz AS, et al., *Adv. Mater.*, 19, 3984. With permission.) (h) XPS analysis of SiNCs: the circles are data, and the curves represent the individual components of the fit. From the top downward: oxide-embedded SiNCs, SiNCs produced by etching off the oxide matrix and exposure to air, dodecene-capped SiNCs, and dodecene-capped SiNCs after 3-week exposure to air. (After Yu, Y., et al., *Langmuir*, 29, 1533–1540, 2013. With permission.) (i) Raman spectra of SiNCs: the line near 520 cm⁻¹ corresponding to the crystalline TO phonon emerges with increasing plasma power, indicating increasing crystallinity. (After Anthony, R., and Kortshagen, U., *Phys. Rev. B*, 80, 115407, 2009. With permission.) (j) Zeta potential of SiNCs: the gray area marks the stable pH range. (After Chen, K.K., et al., *Part. Part. Syst. Charact.*, 32, 301–306. With permission.)

16.5 BAND STRUCTURE

Despite the prevalence of the band structure concept in solid-state physics, its applicability to NCs has long been disputed by theoreticians, since in its pure form (see Section 16.1.1) it presumes an infinite periodical lattice. Although the band structure of an SiNC was addressed as early as the 1980s [101], it has been treated in detail theoretically, including its indirect nature, only recently [70,163]. A general method [70] allowing the reconstruction of band structure from computed real-space molecular orbitals was introduced. It is based on restricting the infinite Bloch plane wave wavefunction from Equation 16.2 in real space using a window function $w(\vec{r})$

$$\psi_{n,\vec{k}}^{NC}(\vec{r}) = w(\vec{r})\psi_{n,\vec{k}}^{solid}(\vec{r}) = w(\vec{r})u_{n,\vec{k}}(\vec{r})\exp\left\{i\vec{k}\cdot\vec{r}\right\}. \tag{16.8}$$

this wavefunction is then projected into reciprocal space using Fourier transform, resulting in a convolution of the Fourier transforms of three terms

$$\widetilde{\psi_{n,\vec{k}}^{NC}}(\vec{k}') = \tilde{w}(\vec{k}') * \widetilde{\tilde{u}_{n,\vec{k}}(\vec{k}')} * \delta(\vec{k}' - \vec{k}). \tag{16.9}$$

whereas the trivial delta-function term $\delta(\vec{k}' - \vec{k})$ coming from the Fourier transform of the plane wave $exp\left\{i\vec{k}\cdot\vec{r}\right\}$ "shifts" the projection to wave vector $\vec{k}$, the Fourier-transformed mother function $\tilde{u}_{n,\vec{k}}$ can be, due to its periodicity, expanded into a discrete Fourier series with coefficients c_m. In reality, one of these coefficients usually dominates. The Fourier-transformed window function term $\tilde{w}$ then introduces blurring of the wavefunction; in the simplest case when the window function has a step-like shape, the blurring is characterized by a sinc function.* For a schematic representation of Equation 16.9, the determining factor of what the band structure of NCs looks like, see Figure 16.6.

The application of the above approach implies that, unlike in a "bulk" band structure such as those in Figure 16.1, (i) the wavevector of an NC becomes discretized, assuming only certain values, and (ii) the real-space localization imparts delocalization of the wavevector. Both these implications can be easily understood qualitatively, since (i) the introduction of surface (finite size) also causes discretization of the wavevector and (ii) Heisenberg's uncertainty relations connect real- and reciprocal-space vectors, implying that localization in real space must cause delocalization (blurring) in $\vec{k}$-space.

Using this approach, the band structure of an SiNC can be reconstructed. In practice, the quantity which is plotted is the real-valued momentum density $\widetilde{\rho_N^{NC}}$

$$\widetilde{\rho_N^{NC}}(\vec{k}') = \left| \left\langle \sum_{a=bands} d_a \psi_{n_a,\vec{k}_a}^{NC}(\vec{r})|\exp\left\{i\vec{k}\cdot\vec{r}\right\}(\vec{k}') \right\rangle \right|^2 \tag{16.10}$$

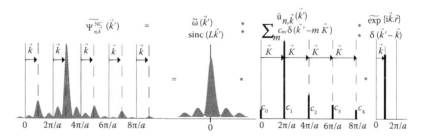

Figure 16.6 Schematics of the reciprocal-space representation of an electron in an NC for a simple one-dimensional case from Equation 16.9 (a is a lattice constant of a one-dimensional atomic chain, L is its length, $\vec{k}$ is a reciprocal vector, and m indexes reciprocal unit cells). (Adapted from Hapala, P., et al., *Phys. Rev. B*, 87, 195420, 2013.)

---

* This approach yields the unfolded band structure.

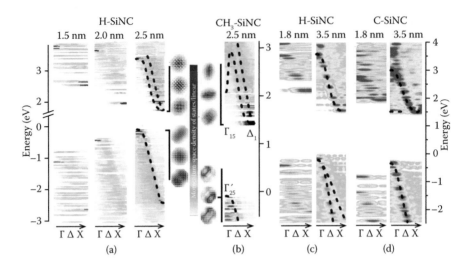

Figure 16.7 Calculated bandstructure for (a,c) H-SiNCs and (b,d) methyl-capped SiNCs. The red and blue areas represent real-space densities of states close to the band edges. Panels (a, b) represent DFT calculations, and panel (c) corresponds to TB approximation. In some of the panels, the calculated bandstructure is overlaid with the bandstructure of bulk silicon for comparison; panels (a) and (b) are to be compared with Figure 16.1a and c, respectively. (Panel (a): after Hapala, P., et al., *Phys. Rev. B*, 87, 195420, 2013; panel (b): after Kůsová, K., et al., *Adv. Mater. Interfaces*, 1, 1300042, 2014; panels (c,d): after Poddubny, A.N., and Dohnalová, K., *Phys. Rev. B*, 90, 245439, 2014.)

corresponding to an energy level E_N. Results of this projection for the case of an H–SiNC, as shown in Figure 16.7, confirm that the band structure concept is applicable to SiNCs starting from a certain critical size. The band structure for NCs is "fuzzy," delocalized in reciprocal space, but the maxima of momentum density follow the trends of bulk silicon very well. Apart from the delocalization, minigaps of forbidden energy intervals between certain energies appear, in particular close to band edges. These minigaps are an important feature, because single-NC measurements (see also Section 16.6) of various semiconductors confirm that the states in an NC (in at least some of the NCs) are very sharp energetic levels [62]. This observation would have been in complete contradiction to the existence of energy bands in NCs if it were not for the minigaps. (Nevertheless, the precise structure of the fine features in the theoretical calculation is not fixed for a NC of a certain size as it can vary for small changes in surface reconstruction or ligand attachment, making every NC unique.)

It is difficult to find an exact size under which the band structure concept is no longer valid, but it certainly lies, for silicon, somewhere between 1.5 and 2.0 nm.* Experiments probing the validity of band structure–related behavior are relatively scarce; however, in optical absorption data, bulk-like transitions were observed to persist down to the smallest studied SiNCs of 1.8 nm [209].

Another question is then the influence of surface passivation on the band structure. In Figure 16.7a and c, the band structure of SiNCs with "model" hydrogen capping is presented. Figure 16.7b and d then confirm that the band structure concept does not lose its sense in SiNCs with C–linked organic capping, represented by –CH₃ methyl groups in the calculations. One of the reasons for the applicability of band structure in SiNCs with C–linked capping is the fact that the capping does not induce strict real-space localization, for example, on surface states, as is evident from Figure 16.7b. On the other hand, the surface-states-related localization of wave functions corresponding to the states close to the band edges occurs for the case of oxide capping, where the band-edge states are much more delocalized in reciprocal space. The most important consequence of the above band structure–based description of SiNCs is that it directly connects various aspects of their behavior to the well-understood behavior of bulk

---

* We note in passing that this size roughly corresponds to an SiNC comprising $3 \times 3 \times 3$ basic cells—to a size limit above which at least some level of periodicity is present. Thus, it can be expected that in other semiconductors the size limit for the validity of band structure will be slightly different, depending on their lattice constant.

silicon (see also Section 16.6). For example, both the indirect Γ–X and the direct Γ–Γ transitions (see Figure 16.1a) appear in the optical absorption spectra of SiNCs with various surfaces (oxidized, Cl, dodecene) [64,209] provided that the size dispersion is narrow enough. Interestingly, the Γ_{15} CB minimum was predicted to split due to quantum confinement, with the lower branch downshifting and the upper branch upshifting in energy with decreasing size (see Figure 16.1c) [164] due their different effective masses (convex/concave curvatures of the bands). Whereas the lower branch downshift was experimentally observed in PL experiments on oxidized SiNCs [14], the above absorption experiments [64,209] reported an opposite trend—a blue shift both when compared to bulk Si and with decreasing NC size, suggesting a connection with the upper split-off branch. These parallels signify that the methods of band structure engineering are transferable from bulk silicon to NCs, as detailed in Section 16.6.2.

Apart from optical experiments, scanning tunneling spectroscopy also allows for the detection of band-edge states. This technique was successfully applied to the study of band gap size dependence in C–SiNCs with varied surfaces [211] as well as to the identification of in-gap states. The results were in accord with the corresponding PL behavior [3,211]. Last but not least, electronic states can be studied directly using X-ray spectroscopy techniques [187].

16.6 LIGHT EMISSION AND SURFACE-INDUCED EFFECTS

16.6.1 RED VERSUS BLUE EMISSION

The effectiveness of light emission of SiNCs is a property that sets them apart from bulk silicon the most. Based on PL properties and in contradiction to the EMA theory from Section 16.1.1, C–SiNCs can be divided into two distinct classes: those with orange–red, and blue–green emission,* with very few exceptions in between (possessing non-alkyl-based, e.g., allylamine, surface ligands). The distinction is obvious from Figure 16.8a, plotting experimental results of PL maximum versus NC diameter.† Some studies even reported an abrupt switch between these two spectral regions [29,33,34,44,84]. The stark difference

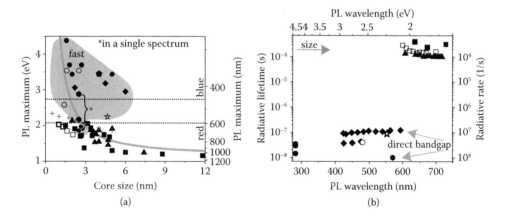

Figure 16.8 Experimentally obtained (a) PL maximum as a function of SiNC size and (b) radiative lifetimes as a function of PL wavelength for alkyl-capped (solid symbols) and alkyl-linked (open symbols) C-SiNCs prepared by different methods: annealing of SiO$_x$-based precursors (squares) [33,60,81,123,127,137], gas-phase synthesis (triangles) [84,85,112,130,141,208], chemical synthesis from SiCl$_4$ (circles) [33,109,176,182,194,205,216], and alkali silicides (diamonds) [33,43,44,160,185], electrochemical etching followed by (re)passivation (hexagon) [39,104,145], ball milling (pentagon) [75], and arrested precipitation (star) [52]. Panel (a) is complemented by data for H-SiNCs (in gray): experimental values for etched SiNCs [28] (crosses) and theoretical model from Equation 16.5 (the curve). Compare with Figure 16.4.

Clusters, nanoparticles, and quantum dots

---

* Interestingly, a similar scenario also applies to SiNCs with an oxidized surface.
† In order to simplify, this figure neglects other possible factors, such as the influence of size dispersion, excitation wavelength, solvent, or the surface-passivating species.

between these two classes becomes even more obvious when one realizes the order-of-magnitude different radiative lifetimes associated with the red and blue PL, as depicted in Figure 16.8b: red-emitting C–SiNCs are connected with long τ_{rad} (>10 μs), whereas in the blue PL τ_{rad} is short (~ns). The preparation method seems to determine PL properties—NCs prepared by low-temperature* chemical syntheses emit in blue while those fabricated by high-temperature techniques, such as anneal of SiO_x-based precursors, gas-phase synthesis, and etching from prefabricated crystalline Si (wafer) are characterized by slow red–orange PL. Another factor might be surface passivation, as the orange slowly emitting SiNCs tend to be terminated with long alkyl chains.

The described behavior is in contradiction to the predictions of theoretical calculations based on the transitions between quantum-confined core-related states—compare Figures 16.4 and 16.8—and therefore, unfortunately, the mechanisms of PL in C–SiNCs are still poorly understood. The orange–red PL is usually reported to be due to quantum confinement and to be connected with the SiNC core [69], although the effect of surface oxidation was also suggested [29,44,127]. As for the blue PL, on the one hand, it was suggested to be connected to surface states and arise from (partial) surface oxidation [34,84], the presence of nitrogen and oxygen [33], or residual chlorine impurities [31]; on the other hand, in blue-green emitting butyl-capped C–SiNCs prepared by (chlorine- and nitrogen-free) low-temperature chemical synthesis, the PL was confirmed to arise from the NC core [43]. Moreover, unlike in defect PL, the blue PL is often reported to be excitation wavelength dependent [33,43,75,182,185]. Therefore, it is highly probable that in reality a combination of all these effects has to be taken into consideration to properly explain the origin of PL in these NCs, and the mechanisms might vary based on preparation and passivation method.

The QY of PL is influenced by many factors. Naturally, surface passivation and its quality play an important role. QYs as high as 62% were reported for C–SiNCs with slow PL [92], although this value decreased to 40% after several days in air. The QY values of 30–40% were also reached in other slowly emitting C–SiNCs [4,137,141]. Interestingly, the QY of the slow PL was observed to decrease at shorter wavelength emission (at smaller sizes), which was attributed to a larger amount of nonradiative surface traps in smaller SiNCs [137]. As regards the fast emission (associated usually with smaller NCs), a size-selective study has not yet been performed, but QY values between 10 and 30% were confirmed [43,104,109,176]. High QY of ~50 and 30–75% were observed in thiol-capped SiNCs [196] and N–linked organically capped SiNCs [109,201]. Also, much higher QYs were reported for crystalline than for amorphous particles [4]. Furthermore, brightening of PL under illumination was observed in ordered ensembles of (monodispersed) SiNCs [141], being attributed to the electronic interaction of the surface trap states.

Similarly to other semiconductor NCs [62], SiNCs also exhibit differences in PL when studied on a single-NC basis. Single-NC studies of C–SiNCs are relatively scarce so far, but narrowing of spectra and spectral replicas on a single-NC level were confirmed for C–SiNCs with fast PL [43,104], whereas ultranarrow linewidths were observed in slowly emitting C–SiNCs [193]. Blinking studies revealed, in parallel to other NCs, the presence of power-law scaling of on- and off-times in fast [102] and slowly emitting [142] SiNCs; however, a nonblinking regime at short timescales (<10 ms) [102] and suppressed off-times in aggregates [142] were confirmed. Also, low-temperature single-NC spectroscopy measurement results can be explained by the observation of the radiative recombination of a trion (a quasiparticle containing one electron and two holes) [103], which is a novel radiative recombination channel.

Other aspects of PL have so far been studied only in connection with the slowly emitting SiNCs. A combined pressure-dependent XRD and PL investigations of alkane-capped SiNCs revealed that compression induces changes in PL analogous to those of the indirect Γ–X transition in bulk Si [69], connecting the observed PL to this transition. Temperature-dependent PL studies also confirmed band gap evolution similar to bulk silicon, with larger deviation for smaller NC sizes [127].[†] Furthermore, hot-carrier

---

* Here, the temperature should be compared to the temperature needed for crystallization of SiNCs, which is usually around 1,000 K, although lower values of ~800 K were reported [81,82,129].

[†] Interestingly, both of these trends also appeared in oxidized SiNCs [105].

relaxation has very recently been assessed using THz measurements, yielding size-dependent values of hot-carrier lifetimes between 0.5 and 0.9 ps with slowed-down cooling at smaller sizes [12]. Although the majority of hot-carrier relaxation occurred in <1 ps, a small fraction of hot carriers also remained on the order of 10 ps [12].

16.6.2 SURFACE-INDUCED FUNDAMENTAL DIRECT BAND GAP AND OTHER EFFECTS

As indicated by the long radiative lifetimes, the slow red PL is associated with the indirect Γ–X recombination [69]. Relatively recently, the formation of direct band gap with fundamental direct Γ–Γ transitions has been confirmed by both experimental results and theoretical calculations. This was found using electrochemically etched and repassivated methyl-capped C–SiNCs [106] and SiNCs chemically synthetized from alkali silicides and hydrosilylated with butyl capping [44,163]. These NCs possibly constitute a third class of C–SiNCs, in addition to the red- and blue-emitting ones. The formation of the direct bandgap was confirmed both theoretically using the "fuzzy" band structure approach outlined in Section 16.5 by enhanced radiative rates (see Figure 16.7b and d), as well as experimentally by τ_{rad} measurements and other experiments. It was attributed to the cooperative effect of quantum confinement and tensile strain induced by the surface capping, both shifting the Γ and X band structure extrema in the "desired" direction, see Figure 16.1c [106], and to the effect of the suitable value of the electronegativity of the ligands with weak charge transfer, leading to the appearance of direct band gap states at the bottom of the CB [163], respectively. (Interestingly, the variations in electronegativity showed direct correlation with enhanced radiative rates, see also Section 16.3.2). These results are important, because they confirm the possibility of direct band gap emission with high rates in silicon, disproving the popular claim of silicon's indirect band gap.

Other candidates for possible direct band gap emission in SiNCs with short radiative lifetimes were demonstrated both experimentally [52,109,185,201] and theoretically [202,203], but the lack of $\vec{k}$-space information in those reports could not lead to the direct band gap interpretation.

As surface ligands influence the band gap, the PL wavelength can naturally be tuned without changing the NC size if the range of surface ligands of SiNCs is broadened beyond the simple alkyl termination, as was experimentally demonstrated [32,201].

The problem of charge transport in a layer with SiNCs can be alleviated by carefully choosing suitable surface groups. In general, long alkyl ligands may block interparticle charge carrier transport. Shorter chains are less insulating, but may be difficult to prepare in sufficient quality [116]. Furthermore, noncovalent interactions between aromatic rings (π–π stacking) can also contribute to increased packing density and electronic coupling between NCs. Examples of C–SiNCs with properties tuned for charge transport include those capped with triphenylamine-based surface ligands, which were suggested as a tool providing macroscopic charge transport properties [228]. Moreover, allylbenzene was the passivant of choice in C–SiNCs LEDs [136]. Alternatively, the allyl disulfide ligand allowed the formation of stable high-concentration dispersions (inks) with enhanced conductivity after spin-coating utilized for the fabrication of a photodiode [116].

Although direct band gap in emission has been demonstrated, the corresponding absorption transitions might remain inefficient due to low density of direct band gap states. This drawback can be overcome by attaching a "light-antenna" fluorophore molecule to the surface of SiNCs, resulting in efficient excitation through energy transfer. Efficient (55% [175] or even 95% [189]), and fast [189] energy transfer with the use of various surface molecules has been confirmed.

Finally, the attachment of suitable surface ligands can also induce a surface doping-like effect by shifting the Fermi energy toward CB or VB, as confirmed by scanning tunneling spectroscopy [211].

16.7 APPLICATIONS

Since many of the fabrication and functionalization methods for C–SiNCs were reported much more recently than those for oxide-capped SiNCs, C–SiNCs somewhat fall behind their oxide-capped counterpart in terms of the broadness of the scope of possible applications; on the other hand, when designed for a certain application, C–SiNCs surpass oxide-capped ones since the surface properties can be fitted for the desired purpose, see also Figure 16.2.

Table 16.3 **Overview of C-SiNCs-Based laboratory LEDs**

SINC	LED COLOR	LED CHARACTERISTICS	REF.
Dodecene, 5 nm	Infrared	Peak EQE of 8.6%, V_{on} ~1.5 V @ 853 nm, peak optical power output 0.13 mW/cm²	[25]
Decyl, 3–5 nm	Red	Peak EQE of 0.7% at V ~10 V	[170]
Allylbenzene, 1.3–1.8 nm	Orange–red	EQE of 1.1%, V_{on} ~2 V @ 670 nm, device lifetime up to at least 40 h	[128]
Allylbenzene, 2.5–5 nm	Orange–red	Peak EQE of 0.17%, V_{on} ~4.6 V @ 700 nm, maximum luminance of 22.7 cd/m²	[136]
Alkyl and oxide, 0.8 and 2.4 nm	White–blue	EQE 4.2×10^{-4}% at the peak optical power density 700 nW/cm², V ~6 V	[214]
Octadecene, 2.1 nm	White	Peak EQE of 0.3%, V_{on} ~4 V, active region is a hybrid-bilayer of SiNCs and a luminescent polymer	[60]

16.7.1 LIGHT SOURCES

In order to realize an electrically pumped light-emitting device, both optical and electrical properties need to be excellent. C–SiNCs exhibit enhanced PL properties when compared to oxide capping, however, especially long alkyl chains are insulating (see also Section 16.6.2). Therefore, SiNC film with high packing densities combined with carefully optimized cutting-edge polymer transport layers are an imperative to reach reasonable external quantum efficiencies (EQE) and turn-on voltages V_{on}. In accord with the trend reported in radiative lifetimes and PL QYs, mostly red–infrared light emission of C–SiNCs has been exploited, see Table 16.3, with the highest EQE of 8.6% @ 853 nm [25]. Alternatively, SiNCs can be used just as a passive phosphor and spectral shaper absorbing UV/blue excitation from a different material in a hybrid device [60].

16.7.2 BIOIMAGING AND DRUG DELIVERY

Since alkyl-capped SiNCs are hydrophobic, dispersibility in aqueous environment under biological conditions* has to be ensured (see Section 16.4.2). In addition to dispersibility for *in vitro* studies, *in vivo* imaging requires NCs that emit within the transparency window of biological tissue (650–900 nm) [206] and are of a suitable size; NC with hydrodynamic diameter smaller than 5.5 nm are rapidly filtered by the kidneys [27], whereas particles larger than 100 nm are rapidly removed from blood by the liver and spleen [149]. Following on the application suggested for porous Si nanoparticles, C–SiNCs are now also being tested as, for example, bioimaging fluorescent markers for long-term cellular *in vitro* as well as *in vivo* imaging or as drug delivery nanocarriers, as summarized in Table 16.4. The added benefit of the organic capping is usually the bright tunable PL, as can be illustrated in the example of concurrent tumor imaging and chemotherapy delivery [91]. Interestingly, C–SiNCs exhibit photostability superior not only to conventional organic dyes, but also to group II–VI NCs [205,227], see Figure 16.2a. C–SiNCs can even be used for fluorescence imaging under two-photon excitation [195], exploiting excitation wavelength which falls into the tissue transparency window.

16.7.3 OTHER APPLICATIONS

An emerging field is the use of SiNCs as contrast agents for magnetic resonance imaging (MRI). As silicon is not paramagnetic, paramagnetism can be achieved by combining SiNCs with a paramagnetic species, such as the incorporation of (doping with) Mn [195] or the coencapsulation with paramagnetic nanoparticles [54]. Alternatively, hyperpolarization can be exploited [5]. Such SiNCs

---

* i.e., pH of ~7.4 and ionic strength of ~150 mM [206].

Table 16.4 Overview of various bio-tagging and nanomedicine laboratory applications

SURFACE	COATING	SIZE (nm)	PL (nm)	ADVANTAGE, USE	REF.
Alkyl	DMSO (lyophilic sol)	5	430, 650	Cellular imaging, lack of *in vitro* toxicity, different accumulation in human malignant and normal cells	[2,39]
Amine groups	None	2.2	470	Biocompatibility, photostability and pH stability, cellular imaging	[226]
Allylamine	None	1.4	450	Photostability, cellular imaging	[205]
Polyacrylic acid	None	<10	620	Photostability, cellular imaging	[114]
Polyacrylic acid		60, more	580	Real-time and long-term cell imaging	[72,225]
Styrene, ethyl-undecylenate	Phospolipid micelles (PEG)	50–120	450–900	Photostability with pH, temperature, *in vitro* and *in vitro* imaging in mice, *in vivo* toxicity assessed	[53,55]
Glutaric acid	Possibly proteins	3	660	Photostability, microtubules labeled with SiNCs for real-time and long-term cellular imaging	[73,225]
Alkyl (dodecene)	Amphiphilic polymer	18	620	Dispersion pH stability, small size, tissue imaging	[80]
Poly(methacrylic acid)		50–100	550	pH-regulated drug delivery, high-anticancer activity (Hela cells), blood compatibility, PL tracking	[215]
Amine groups	Chemotherapy drug	4	500	Concurrent tumor imaging and chemotherapy	[91]
Propylamine	Dextransulfate	8–40	440	Cellular imaging, two-photon PL excitation, nontoxic to mammalian cells, accumulation in macrophages (diagnostics)	[195]
Thiol	Bovine serum albumin	130	600	Exceptional biocompatibility, QY 5–10% in 0.1 M potassium phosphate buffer, immunostaining on live cancer cells, specific binding similar to conventional dyes	[196]
Alminoprofen (pain killer)	None	6–9	blue	Functional drug, maintaining biological effect of the drug with reduced cytotoxicity	[68]

possess long spin-relaxation times favorable for MRI and can then exhibit dual modalities, allowing for example for concurrent MR and fluorescence imaging, (see Table 16.5).

Also, as regards optoelectronic applications, a nanocomposite of polystyrene and polystyrene-chainscapped SiNCs has been suggested as a charge trapping material for memory devices [49]. C–SiNCs have been employed in photodiodes, both as the active material [116] and to increase performance of a hybrid SiNC/graphene/Si photodetector [219]. As for photovoltaic applications, a hybrid solar cell with power conversion efficiency of 12.9% was demonstrated [50].

Last but not least, as outlined in Table 16.5, quenching of PL in the presence of certain species can be utilized for the fabrication of sensors of explosives, biological, or toxic species both in the environment and in tissue.

Clusters, nanoparticles, and quantum dots

Table 16.5 **Other applications of C-SiNCs**

SURFACE	MECHANISM	APPLICATION	REF.
Alkyl or aromatic + oxide	Long spin-relaxation time	*in vivo* hyperpolarized MRI imaging agent	[5]
Propylamine, dextran sulfate	Mn incorporation	Concurrent MRI and fluorescence imaging	[195]
Ethyl undecylenate	Co-encapsulation with Fe_3O_4 nanoparticles	A magnetofluorescent probe, enhanced cellular uptake in magnetic field, *in vivo* tumor fluorescence imaging	[54]
Octadecene	Energy transfer	Improved efficiency of a hybrid Si nanowire-SiNC solar cell to 12.9%	[50]
Dodecene	Scattering, absorption, and charge transfer	Improved performance of a hybrid SiNC/graphene/Si photodetector (higher built-in voltage, reduced reflection)	[219]
Allyl disulfide	Absorption	Photodiode with a peak photoresponse of 0.02 AW^{-1} to UV light, fabricated simply by spin coating from a C-SiNCs ink	[116]
Dodecyl	PL quenching	Paper-based sensor of nitroaromatic explosives	[63]
Ethylamine	PL quenching	Water-soluble probe for TNT detection (1–500 nM)	[8]
Amine groups	PL quenching	Selective detection of dopamine (0.3 nM–10 µM)	[223]
Undecenoic acid	PL quenching	Detection of Cu^{2+} in HeLa cells	[213]

16.8 SUMMARY

C–SiNCs started to be studied about 25 years ago. Since then, a wide variety of preparation techniques have been developed, allowing for wide spectral tunability from UV to infrared featuring reasonable QYs and excellent photostability. Similarly, going hand-in-hand with preparation techniques, a broad range of surface functionalizations were realized, expanding the possibilities of utilizing C–SiNCs in more technology-relevant contexts.

In contrast to the more well-known group III–V and II–VI NCs, C–SiNCs are available, environmentally friendly, and nontoxic. Therefore, they offer exciting opportunities for both fundamental and applied research, as well as for commercial use in applications ranging from light sources and detectors, photovoltaic cells, fluorescent imaging tools, diagnostic, and therapeutic agents, to environmental sensors. Although proof-of-concept devices have been amply demonstrated, there are still issues to be addressed. Probably the most pressing one is that of a suitable preparation method, as the current methods often require special expertise to reach the desired outcome. Also, thorough control of size/shape and optical properties using a single preparation method is still not possible. Furthermore, the optical properties are determined by a complex interplay of the properties of the core, the surface, and the interface, which makes it more difficult for the characterization to provide sound feedback to technology. Last but not least, despite positive outcomes for now, toxicity studies are still underway, determining the impact of C–SiNCs in live tissue.

With further advancements in technology, C–SiNCs promise to have huge impact, despite their tiny size.

ACKNOWLEDGMENT

Dr. Štěpán Stehlík is gratefully acknowledged for going through parts of the manuscript. K.D. acknowledges the MacGilllavry Fellowship of the University of Amsterdam and Dutch funding Fundamenteel Onderzoek der Materie (FOM) Projectruimte 15PR3230.

REFERENCES

1. Allen LC. (1989). Electronegativity is the Average One-Electron Energy of the Valence-Shell Electrons in Ground-State Free Atoms. *J. Am. Chem. Soc.*, 111(25):9003–9014.
2. Alsharif NH, Berger CEM, Varanasi SS, Chao Y, Horrocks BR and Datta HK. (2009). Alkyl-Capped Silicon Nanocrystals Lack Cytotoxicity and have Enhanced Intracellular Accumulation in Malignant Cells via Cholesterol-Dependent Endocytosis. *Small*, 5(2):221–228.
3. Angı A, et al. (2016). Photoluminescence through In-Gap States in Phenylacetylene Functionalized Silicon Nanocrystals. *Nanoscale*, 8:7849–7853.
4. Anthony R and Kortshagen U. (2009). Photoluminescence Quantum Yields of Amorphous and Crystalline Silicon Nanoparticles. *Phys. Rev. B*, 80:115407.
5. Atkins TM, Cassidy MC, Lee M, Ganguly S, Marcus CM and Kauzlarich SM. (2013). Synthesis of Long T_1 Silicon Nanoparticles for Hyperpolarized ^{29}Si Magnetic Resonance Imaging. *ACS Nano*, 7(2):1609–1617.
6. Atkins TM, Louie AY and Kauzlarich SM. (2012). An Efficient Microwave-Assisted Synthesis Method for the Production of Water Soluble Amine-Terminated Si Nanoparticles. *Nanotechnology*, 23(29):294006.
7. Baldwin RK, et al. (2002). Room Temperature Solution Synthesis of Alkyl-Capped Tetrahedral Shaped Silicon Nanocrystals. *J. Am. Chem. Soc.*, 124(7):1150–1151.
8. Ban R, Zheng F and Zhang J. (2015). A Highly Sensitive Fluorescence Assay for 2,4,6-Trinitrotoluene using Amine-Capped Silicon Quantum Dots as a Probe. *Anal. Methods*, 7:1732–1737.
9. Batchelor L, Loni A, Canham LT, Hasan M and Coffer JL. (2012). Manufacture of Mesoporous Silicon from Living Plants and Agricultural Waste: An Environmentally Friendly and Scalable Process. *Silicon*, 4(4):259–266.
10. Bateman JE, Eagling RD, Horrock BR and Houlton A. (2000). A Deuterium Labeling, FTIR, and *Ab Initio* Investigation of the Solution-Phase Thermal Reactions of Alcohols and Alkenes with Hydrogen-Terminated Silicon Surfaces. *J. Phys. Chem. B*, 104(23):5557–5565.
11. Bent SF. (2002). Organic Functionalization of Group IV Semiconductor Surfaces: Principles, Examples, Applications, and Prospects. *Surf. Sci.*, 500(1–3):879–903.
12. Bergren MR, Palomaki PKB, Neale NR, Furtak TE and Beard MC. (2016). Size-Dependent Exciton Formation Dynamics in Colloidal Silicon Quantum Dots. *ACS Nano*, 10(2):2316–2323.
13. Bley RA and Kauzlarich SM. (1996). A Low-Temperature Solution Phase Route for the Synthesis of Silicon Nanoclusters. *J. Am. Chem. Soc.*, 118(49):12461–12462.
14. de Boer WDAM, et al. (2010). Red Spectral Shift and Enhanced Quantum Efficiency in Phonon-Free Photoluminescence from Silicon Nanocrystals. *Nature Nanotech.*, 5(12):878–884.
15. Brawand NP, Voros M and Galli G. (2015). Surface Dangling Bonds Are a Cause of B-Type Blinking in Si Nanoparticles. *Nanoscale*, 7:3737–3744.
16. Bruchez M, Jr, Moronne M, Gin P, Weiss S and Alivisatos AP. (1998). Semiconductor Nanocrystals as Fluorescent Biological Labels. *Science*, 281(5385):2013–2016.
17. Brus LE. (1984). Electron–Electron and Electron–Hole Interactions in Small Semiconductor Crystallites: The Size Dependence of the Lowest Excited Electronic State. *J. Chem. Phys.*, 80(9):4403–4409.
18. Buljan M, et al. (2015). Self-Assembly of Ge Quantum Dots on Periodically Corrugated Si Surfaces. *Appl. Phys. Lett.*, 107(20):203101.
19. Buriak JM. (2002). Organometallic Chemistry on Silicon and Germanium Surfaces. *Chem. Rev.*, 102(5):1271–1308.
20. Canaria CA, Lees IN, Wun AW, Miskelly GM and Sailor MJ. (2002). Characterization of the Carbon-Silicon Stretch in Methylated Porous Silicon—Observation of an Anomalous Isotope Shift in the FTIR Spectrum. *Inorg. Chem. Commun.*, 5(8):560–564.
21. Canham LT. (1990). Silicon Quantum Wire Array Fabrication by Electrochemical and Chemical Dissolution of Wafers. *Appl. Phys. Lett.*, 57(10):1046–1048.
22. Casida ME and Salahub DR. (2000). Asymptotic Correction Approach to Improving Approximate Exchange Correlation Potentials: Time-Dependent Density-Functional Theory Calculations of Molecular Excitation Spectra. *J. Chem. Phys.*, 113(20):8918–8935.
23. Chao Y, et al. (2005). Reactions and Luminescence in Passivated Si Nanocrystallites Induced by Vacuum Ultraviolet and Soft-X-Ray Photons. *J. Appl. Phys.*, 98(4):044316.
24. Chen KK, Mastronardi ML, Kübel C and Ozin GA. (2015). Size-Selective Separation and Puriftcation of "Water-Soluble" Organically Capped Brightly Photoluminescent Silicon Nanocrystals. *Part. Part. Syst. Charact.*, 32(3):301–306.
25. Cheng KY, Anthony R, Kortshagen UR and Holmes RJ. (2011). High-Efficiency Silicon Nanocrystal Light-Emitting Devices. *Nano Lett.*, 11(5):1952–1956.
26. Cheng X, Lowe SB, Reece PJ and Gooding JJ. (2014). Colloidal Silicon Quantum Dots: From Preparation to the Modification of Self-Assembled Monolayers (SAMs) for Bio-Applications. *Chem. Soc. Rev.*, 43:2680–2700.

27. Choi HS, et al. (2007). Renal Clearance of Quantum Dots. *Nat. Biotechnol.*, 25(10):1165–1170.

28. Choi J, Wang NS and Reipa V. (2007). Photoassisted Tuning of Silicon Nanocrystal Photoluminescence. *Langmuir*, 23(6):3388–3394.

29. Coxon PR, Wang Q and Chao Y. (2011). An Abrupt Switch between the Two Photoluminescence Bands within Alkylated Silicon Nanocrystals. *J. Phys. D Appl. Phys.*, 44(49):495301.

30. Cucinotta C, Bonferroni B, Ferretti A, Ruini A, Caldas M and Molinari E. (2006). First-Principles Investigation of Functionalization-Defects on Silicon Surfaces. *Surf. Sci.*, 600(18):3892–3897.

31. Dasog M, Kehrle J, Rieger B and Veinot JGC. (2016). Silicon Nanocrystals and Silicon-Polymer Hybrids: Synthesis, Surface Engineering, and Applications. *Angew. Chem. Int. Ed.*, 55(7):2322–2339.

32. Dasog M, los Reyes GBD, Titova LV, Hegmann FA and Veinot JGC. (2014). Size vs Surface: Tuning the Photoluminescence of Freestanding Silicon Nanocrystals across the Visible Spectrum via Surface Groups. *ACS Nano*, 8(9):9636–9648.

33. Dasog M, et al. (2013). Chemical Insight into the Origin of Red and Blue Photoluminescence Arising from Freestanding Silicon Nanocrystals. *ACS Nano*, 7(3):2676–2685.

34. DeBenedetti WJI, et al. (2015). Conversion from Red to Blue Photoluminescence in Alcohol Dispersions of Alkyl-Capped Silicon Nanoparticles: Insight into the Origins of Visible Photoluminescence in Colloidal Nanocrystalline Silicon. *J. Phys. Chem. C*, 119(17):9595–9608.

35. Degoli E, et al. (2004). *Ab Initio* Structural and Electronic Properties of Hydrogenated Silicon Nanoclusters in the Ground and Excited State. *Phys. Rev. B*, 69:155411.

36. Delerue C, Allan G and Lannoo M. (1993). Theoretical Aspects of the Luminescence of Porous Silicon. *Phys. Rev. B*, 48:11024–11036.

37. Derfus AM, Chan WCW and Bhatia SN. (2004). Probing the Cytotoxicity of Semiconductor Quantum Dots. *Nano Lett.*, 4(1):11–18.

38. Dhas NA, Raj CP and Gedanken A. (1998). Preparation of Luminescent Silicon Nanoparticles: A Novel Sonochemical Approach. *Chem. Mater.*, 10(11):3278–3281.

39. Dickinson FM, et al. (2008). Dispersions of Alkyl-Capped Silicon Nanocrystals in Aqueous Media: Photoluminescence and Ageing. *Analyst*, 133:1573–1580.

40. Dohnalová K, Gregorkiewicz T and Kůsová K. (2014). Silicon Quantum Dots: Surface Matters. *J. Phys. Condens. Matter.*, 26(17):173201.

41. Dohnalová K, Kůsová K and Pelant I. (2009). Time-Resolved Photoluminescence Spectroscopy of the Initial Oxidation Stage of Small Silicon Nanocrystals. *Appl. Phys. Lett.*, 94(21):211903.

42. Dohnalová K, Ondic L, Kůsová K, Pelant I, Rehspringer JL and Mafouana RR. (2010). White-Emitting Oxidized Silicon Nanocrystals: Discontinuity in Spectral Development with Reducing Size. *J. Appl. Phys.*, 107(5):053102.

43. Dohnalová K, et al. (2012). Microscopic Origin of the Fast Blue-Green Luminescence of Chemically Synthesized Non-Oxidized Silicon Quantum Dots. *Small*, 8(20):3185–3191.

44. Dohnalová K, et al. (2013). Surface Brightens Up Si Quantum Dots: Direct Bandgap-Like Size-Tunable Emission. *Light Sci. Appl.*, 2(1):e47.

45. Draeger EW, Grossman JC, Williamson AJ and Galli G. (2003). Influence of Synthesis Conditions on the Structural and Optical Properties of Passivated Silicon Nanoclusters. *Phys. Rev. Lett.*, 90:167402.

46. Draeger EW, Grossman JC, Williamson AJ and Galli G. (2004). Optical Properties of Passivated Silicon Nanoclusters: The Role of Synthesis. *J. Chem. Phys.*, 120(22):10807–10814.

47. Drexhage K. (1970). Influence of a Dielectric Interface on Fluorescence Decay Time. *J. Lumin.*, 1:693–701.

48. van Driel AF, Allan G, Delerue C, Lodahl P, Vos WL and Vanmaekelbergh D. (2005). Frequency-Dependent Spontaneous Emission Rate from CdSe and CdTe Nanocrystals: Influence of Dark States. *Phys. Rev. Lett.*, 95:236804.

49. Dung MX, Choi JK and Jeong HD. (2013). Newly Synthesized Silicon Quantum Dot-Polystyrene Nanocomposite Having Thermally Robust Positive Charge Trapping. *ACS Appl. Mater. Interfaces*, 5(7):2400–2409.

50. Dutta M, Thirugnanam L, Trinh PV and Fukata N. (2015). High Efficiency Hybrid Solar Cells Using Nanocrystalline Si Quantum Dots and Si Nanowires. *ACS Nano*, 9(7):6891–6899.

51. Ehbrecht M, Kohn B, Huisken F, Laguna MA and Paillard V. (1997). Photoluminescence and Resonant Raman Spectra of Silicon Films Produced by Size-Selected Cluster Beam Deposition. *Phys. Rev. B*, 56(11):6958–6964.

52. English DS, Pell LE, Yu Z, Barbara PF and Korgel BA. (2002). Size Tunable Visible Luminescence from Individual Organic Monolayer Stabilized Silicon Nanocrystal Quantum Dots. *Nano Lett.*, 2(7):681–685.

53. Erogbogbo F, Yong KT, Roy I, Xu G, Prasad PN and Swihart MT. (2008). Biocompatible Luminescent Silicon Quantum Dots for Imaging of Cancer Cells. *ACS Nano*, 2(5):873–878.

54. Erogbogbo F, et al. (2010). Biocompatible Magnetofluorescent Probes: Luminescent Silicon Quantum Dots Coupled with Superparamagnetic Iron(III) Oxide. *ACS Nano*, 4(9):5131–5138.

55. Erogbogbo F, et al. (2011). *In Vivo* Targeted Cancer Imaging, Sentinel Lymph Node Mapping and Multi-Channel Imaging with Biocompatible Silicon Nanocrystals. *ACS Nano*, 5(1):413–423.

56. Fan J and Chu PK. (2010). Group IV Nanoparticles: Synthesis, Properties, and Biological Applications. *Small*, 6(19):2080–2098.

57. Franceschetti A and Pantelides ST. (2003). Excited-State Relaxations and Franck-Condon Shift in Si Quantum Dots. *Phys. Rev. B*, 68:033313.

58. Gali A, Vörös M, Rocca D, Zimanyi GT and Galli G. (2009). High-Energy Excitations in Silicon Nanoparticles. *Nano Lett.*, 9(11):3780–3785.

59. Ghosh B and Shirahata N. (2014). Colloidal Silicon Quantum Dots: Synthesis and Luminescence Tuning from the Near-UV to the Near-IR Range. *Sci. Tech. Adv. Mater.*, 15(1):014207.

60. Ghosh B, et al. (2014). Hybrid White Light Emitting Diode Based on Silicon Nanocrystals. *Adv. Funct. Mater.*, 24(45):7151–7160.

61. Goller B, Polisski S, Wiggers H and Kovalev D. (2010). Silicon Nanocrystals Dispersed in Water: Photosensitizers for Molecular Oxygen. *Appl. Phys. Lett.*, 96(21):211901.

62. Gómez DE, Califano M and Mulvaney P. (2006). Optical Properties of Single Semiconductor Nanocrystals. *Phys. Chem. Chem. Phys.*, 8:4989–5011.

63. Gonzalez CM, et al. (2014). Detection of High-Energy Compounds Using Photoluminescent Silicon Nanocrystal Paper Based Sensors. *Nanoscale*, 6:2608–2612.

64. Gresback R, Murakami Y, Ding Y, Yamada R, Okazaki K and Nozaki T. (2013). Optical Extinction Spectra of Silicon Nanocrystals: Size Dependence upon the Lowest Direct Transition. *Langmuir*, 29(6):1802–1807.

65. Guerra R and Ossicini S. (2010). High Luminescence in Small Si/SiO_2 Nanocrystals: A Theoretical Study. *Phys. Rev. B*, 81(24):245307.

66. Gupta A, Swihart MT and Wiggers H. (2009). Luminescent Colloidal Dispersion of Silicon Quantum Dots from Microwave Plasma Synthesis: Exploring the Photoluminescence Behavior Across the Visible Spectrum. *Adv. Funct. Mater.*, 19(5):696–703.

67. Hallmann S, Fink MJ and Mitchell BS. (2015). Williamson Ether Synthesis: An Efficient One-Step Route for Surface Modifications of Silicon Nanoparticles. *J. Exp. Nanosci.*, 10(8):588–598.

68. Hanada S, Fujioka K, Futamura Y, Manabe N, Hoshino A and Yamamoto K. (2013). Evaluation of Anti-Inflammatory Drug-Conjugated Silicon Quantum Dots: Their Cytotoxicity and Biological Effect. *Int. J. Mol. Sci.*, 14(1):1323–1334.

69. Hannah DC, et al. (2012). On the Origin of Photoluminescence in Silicon Nanocrystals: Pressure-Dependent Structural and Optical Studies. *Nano Lett.*, 12(8):4200–4205.

70. Hapala P, Kůsová K, Pelant I and Jelínek P. (2013). Theoretical Analysis of Electronic Band Structure of 2- to 3-nm Si Nanocrystals. *Phys. Rev. B*, 87(19):195420.

71. Harun NA, Horrocks BR and Fulton DA. (2011). A Miniemulsion Polymerization Technique for Encapsulation of Silicon Quantum Dots in Polymer Nanoparticles. *Nanoscale*, 3:4733–4741.

72. He Y, Kang ZH, Li QS, Tsang C, Fan CH and Lee ST. (2009). Ultrastable, Highly Fluorescent, and Water-Dispersed Silicon-Based Nanospheres as Cellular Probes. *Angew. Chem. Int. Ed.*, 48(1):128–132.

73. He Y, et al. (2011). One-Pot Microwave Synthesis of Water-Dispersible, Ultraphoto- and pH-Stable, and Highly Fluorescent Silicon Quantum Dots. *J. Am. Chem. Soc.*, 133(36):14192–14195.

74. Heath JR. (1992). A Liquid-Solution-Phase Synthesis of Crystalline Silicon. *Science*, 258(5085):1131–1133.

75. Heintz AS, Fink MJ and Mitchell BS. (2007). Mechanochemical Synthesis of Blue Luminescent Alkyl/Alkenyl-Passivated Silicon Nanoparticles. *Adv. Mater.*, 19:3984.

76. Heitmann J, Müller F, Zacharias M and Gösele U. (2005). Silicon nanocrystals: Size matters. *Adv. Mat.*, 17(7):795–803.

77. Henderson EJ, et al. (2011). Colloidally Stable Silicon Nanocrystals with Near-Infrared Photoluminescence for Biological Fluorescence Imaging. *Small*, 7(17):2507–2516.

78. Hessel CM, Henderson EJ and Veinot JGC. (2006). Hydrogen Silsesquioxane: A Molecular Precursor for Nanocrystalline $Si\text{-}SiO_2$ Composites and Freestanding Hydride-Surface-Terminated Silicon Nanoparticles. *Chem. Mater.*, 18:6139–6146.

79. Hessel CM, Wei J, Reid D, Fujii H, Downer MC and Korgel BA. (2012). Raman Spectroscopy of Oxide-Embedded and Ligand-Stabilized Silicon Nanocrystals. *J. Phys. Chem. Lett.*, 3(9):1089–1093.

80. Hessel CM, et al. (2010). Alkyl Passivation and Amphiphilic Polymer Coating of Silicon Nanocrystals for Diagnostic Imaging. *Small*, 6(18):2026–2034.

81. Hessel CM, et al. (2012). Synthesis of Ligand-Stabilized Silicon Nanocrystals with Size-Dependent Photoluminescence Spanning Visible to Near-Infrared Wavelengths. *Chem. Mater.*, 24(2):393–401.

82. Hirasawa M, Orii T and Seto T. (2006). Size-Dependent Crystallization of Si Nanoparticles. *Appl. Phys. Lett.*, 88(9):093119.

83. Holmes JD, Ziegler KJ, Doty RC, Pell LE, Johnsto KP and Korgel BA. (2001). Highly Luminescent Silicon Nanocrystals with Discrete Optical Transitions. *J. Am. Chem. Soc.*, 123:3743–3748.

84. Hua F, Erogbogbo F, Swihart M and Ruckenstein E. (2006). Organically Capped Silicon Nanoparticles with Blue Photoluminescence Prepared by Hydrosilylation Followed by Oxidation. *Langmuir*, 22:4363.

85. Hua F, Swihart MT and Ruckenstein E. (2005). Efficient Surface Grafting of Luminescent Silicon Quantum Dots by Photoinitiated Hydrosilylation. *Langmuir*, 21(13):6054–6062.

86. Hwang J, et al. (2015). Simple Preparation of Fluorescent Silicon Nanoparticles from Used Si Wafers. *Ind. Eng. Chem. Res.*, 54(22):5982–5989.

87. Hwang TH, Lee YM, Kong BS, Seo JS and Choi JW. (2012). Electrospun Core-Shell Fibers for Robust Silicon Nanoparticle-Based Lithium Ion Battery Anodes. *Nano Lett.*, 12:802–807.

88. Hybertsen MS. (1994). Absorption and Emission of Light in Nanoscale Silicon Structures. *Phys. Rev. Lett.*, 72(10):1514–1517.

89. Jariwala BN, Dewey OS, Stradins P, Ciobanu CV and Agarwal S. (2011). In Situ Gas-Phase Hydrosilylation of Plasma-Synthesized Silicon Nanocrystals. *ACS Appl. Mater. Interfaces*, 3(8):3033–3041.

90. Jelinek P, Wang H, Lewis J, Sankey OF and Ortega J. (2005). Multicenter Approach to the Exchange-Correlation Interactions in *Ab Initio* Tight-Binding Methods. *Phys. Rev. B*, 71:235101.

91. Ji X, et al. (2015). Highly Fluorescent, Photostable, and Ultrasmall Silicon Drug Nanocarriers for Long-Term Tumor Cell Tracking and *In-Vivo* Cancer Therapy. *Adv. Mater.*, 27(6):1029–1034.

92. Jurbergs D, Rogojina E, Mangolini L and Kortshagen U. (2006). Silicon Nanocrystals with Ensemble Quantum Yields Exceeding 60%. *Appl. Phys. Lett.*, 88(23):233116–233118.

93. Kafshgari MH, Voelcker NH and Harding FJ. (2015). Applications of Zero-Valent Silicon Nanostructures in Biomedicine. *Nanomedicine*, 10(16):2553–2571.

94. Kang Z, et al. (2007). A Polyoxometalate-Assisted Electrochemical Method for Silicon Nanostructures Preparation: From Quantum Dots to Nanowires. *J. Am. Chem. Soc.*, 129(17):5326–5327.

95. Kelly JA, Shukaliak AM, Fleischauer MD and Veinot JGC. (2011). Size-Dependent Reactivity in Hydrosilylation of Silicon Nanocrystals. *J. Am. Chem. Soc.*, 133(24):9564–9571.

96. Kim NY and Laibinis PE. (1998). Derivatization of Porous Silicon by Grignard Reagents at Room Temperature. *J. Am. Chem. Soc.*, 120(18):4516–4517.

97. Köhler C, Hajnal Z, Deák P, Frauenheim T and Suhai S. (2001). Theoretical Investigation of Carbon Defects and Diffusion in α-Quartz. *Phys. Rev. B*, 64:085333.

98. König D, Rudd J, Green MA and Conibeer G. (2008). Role of the Interface for the Electronic Structure of Si Quantum Dots. *Phys. Rev. B*, 78(3):035339.

99. Kortshagen U. (2009). Nonthermal Plasma Synthesis of Semiconductor Nanocrystals. *J. Phys. D: Appl. Phys.*, 42(11):113001.

100. Kovalenko MV, et al. (2015). Prospects of Nanoscience with Nanocrystals. *ACS Nano*, 9(2):1012–1057.

101. Künne L, Skála L and Bílek O. (1983). Tight-Binding Cluster Model for C, Si, and Ge. II. Size and Shape Dependence of Some Cluster Properties. *Phys. Stat. Sol. B*, 118(1):173–178.

102. Kůsová K, Pelant I, Humpolíčková J and Hof M. (2016). Comprehensive Description of Blinking-Dynamics Regimes in Single Direct-Band-Gap Silicon Nanocrystals. *Phys. Rev. B*, 93(3):035412.

103. Kůsová K, Pelant I and Valenta J. (2015). Bright Trions in Direct-Bandgap Silicon Nanocrystals Revealed by Low-Temperature Single-Nanocrystal Spectroscopy. *Light Sci. Appl.*, 4:e336.

104. Kůsová K, et al. (2010). Brightly Luminescent Organically Capped Silicon Nanocrystals Fabricated at Room Temperature and Atmospheric Pressure. *ACS Nano*, 4(8):4495–4504.

105. Kůsová K, et al. (2012). Luminescence of Free-Standing versus Matrix-Embedded Oxide-Passivated Silicon Nanocrystals: The Role of Matrix-Induced Strain. *Appl. Phys. Lett.*, 101(14):143101.

106. Kůsová K, et al. (2014). Direct Bandgap Silicon: Tensile-Strained Silicon Nanocrystals. *Adv. Mater. Interfaces*, 1(2):1300042.

107. Ledoux G, et al. (2000). Photoluminescence Properties of Silicon Nanocrystals as a Function of their Size. *Phys. Rev. B*, 62:15942–15951.

108. Lehtonen O, Sundholm D and Vanska T. (2008). Computational Studies of Semiconductor Quantum Dots. *Phys. Chem. Chem. Phys.*, 10:4535–4550.

109. Li Q, et al. (2013). Surface-Modified Silicon Nanoparticles with Ultrabright Photoluminescence and Single-Exponential Decay for Nanoscale Fluorescence Lifetime Imaging of Temperature. *J. Am. Chem. Soc.*, 135(40):14924–14927.

110. Li QS, Zhang RQ, Lee ST, Niehaus TA and Frauenheim T. (2008). Optimal Surface Functionalization of Silicon Quantum Dots. *J. Chem. Phys.*, 128(24):244714.

111. Li QS, Zhang RQ, Niehaus TA, Frauenheim T and Lee ST. (2007). Theoretical Studies on Optical and Electronic Properties of Propionic-Acid-Terminated Silicon Quantum Dots. *J. Chem. Theor. Comput.*, 3(4):1518–1526.

112. Li X, He Y and Swihart MT. (2004). Surface Functionalization of Silicon Nanoparticles Produced by Laser-Driven Pyrolysis of Silane followed by HF/HNO$_3$ Etching. *Langmuir*, 20(11):4720–4727.

113. Li X, He Y, Talukdar S and Swihart M. (2003). Process for Preparing Macroscopic Quantities of Brightly Photoluminescent Silicon Nanoparticles with Emission Spanning the Visible Spectrum. *Langmuir*, 19:8490–8496.

114. Li Z and Ruckenstein E. (2004). Water-Soluble Poly(acrylic acid) Grafted Luminescent Silicon Nanoparticles and Their Use as Fluorescent Biological Staining Labels. *Nano Lett.*, 4:1463.

115. Liao YC and Roberts JT. (2006). Self-Assembly of Organic Monolayers on Aerosolized Silicon Nanoparticles. *J. Am. Chem. Soc.*, 128(28):9061–9065.

116. Lin T, Liu X, Zhou B, Zhan Z, Cartwright AN and Swihart MT. (2014). A Solution-Processed UV-Sensitive Photodiode Produced Using a New Silicon Nanocrystal Ink. *Adv. Funct. Mater.*, 24(38):6016–6022.

117. Linford M and Chidsey C. (1993). Alkyl Monolayers Covalently Bonded to Silicon Surfaces. *J. Am. Chem. Soc.*, 115:12631–12632.

118. Liptak RW, Devetter B, Thomas JH 3rd, Kortshagen U and Campbell SA. (2009). SF$_6$ Plasma Etching of Silicon Nanocrystals. *Nanotechnology*, 20(3):035603.

119. Littau KA, Szajowski PJ, Muller AJ, Kortan AR and Brus LE. (1993). A Luminescent Silicon Nanocrystal Colloid via a High-Temperature Aerosol Reaction. *J. Phys. Chem.*, 97:1224–1230.

120. Liu CY, Holman ZC and Kortshagen UR. (2009). Hybrid Solar Cells from P3HT and Silicon Nanocrystals. *Nano Lett.*, 9(1):449–452.

121. Liu N, Huo K, McDowell MT, Zhao J and Cui Y. (2013). Rice Husks as a Sustainable Source of Nanostructured Silicon for High Performance Li-Ion Battery Anodes. *Sci. Rep.*, 3:1919.

122. Liu Q and Kauzlarich SM. (2002). A New Synthetic Route for the Synthesis of Hydrogen Terminated Silicon Nanoparticles. *Mater. Sci. Eng. B*, 96(2):72–75.

123. Liu SM, Yang Y, Sato S and Kimura K. (2006). Enhanced Photoluminescence from Si Nano-Organosols by Functionalization with Alkenes and Their Size Evolution. *Chem. Mater.*, 18(3):637–642.

124. Luppi E, et al. (2005). The Electronic and Optical Properties of Silicon Nanoclusters: Absorption and Emission. *Opt. Mater.*, 27(5):1008–1013.

125. Ma Y, Chen X, Pi X and Yang D. (2011). Theoretical Study of Chlorine for Silicon Nanocrystals. *J. Phys. Chem. C*, 115(26):12822–12825.

126. Ma Y, Pi X and Yang D. (2012). Fluorine-Passivated Silicon Nanocrystals: Surface Chemistry versus Quantum Confinement. *J. Phys. Chem. C*, 116(9):5401–5406.

127. Maier-Flaig F, et al. (2012). Photophysics of Organically-Capped Silicon Nanocrystals—A Closer Look into Silicon Nanocrystal Luminescence Using Low Temperature Transient Spectroscopy. *Chem. Phys.*, 405:175–180.

128. Maier-Flaig F, et al. (2013). Multicolor Silicon Light-Emitting Diodes (SiLEDs). *Nano Lett.*, 13(2):475–480.

129. Mangolini L. (2013). Synthesis, Properties, and Applications of Silicon Nanocrystals. *J. Vac. Sci. Technol.*, 31(2):020801.

130. Mangolini L and Kortshagen U. (2007). Plasma-Assisted Synthesis of Silicon Nanocrystal Inks. *Adv. Mater.*, 19(18):2513–2519.

131. Mangolini L, Thimsen E and Kortshagen U. (2005). High-Yield Plasma Synthesis of Luminescent Silicon Nanocrystals. *Nano Lett.*, 5(4):655–659.

132. Manhat BA, et al. (2011). One-Step Melt Synthesis of Water-Soluble, Photoluminescent, Surface-Oxidized Silicon Nanoparticles for Cellular Imaging Applications. *Chem. Mater.*, 23(9):2407–2418.

133. Mariotti D, Švrček V, Hamilton JWJ, Schmidt M and Kondo M. (2012). Silicon Nanocrystals in Liquid Media: Optical Properties and Surface Stabilization by Microplasma-Induced Non-Equilibrium Liquid Chemistry. *Adv. Funct. Mater.*, 22(5):954–964.

134. Mastronardi ML, Henderson EJ, Puzzo DP and Ozin GA. (2012). Small Silicon, Big Opportunities: The Development and Future of Colloidally-Stable Monodisperse Silicon Nanocrystals. *Adv. Mater.*, 24(43):5890–5898.

135. Mastronardi ML, et al. (2011). Preparation of Monodisperse Silicon Nanocrystals Using Density Gradient Ultracentrifugation. *J. Am. Chem. Soc.*, 133(31):11928–11931.

136. Mastronardi ML, et al. (2012). Silicon Nanocrystal OLEDs: Effect of Organic Capping Group on Performance. *Small*, 8(23):3647–3654.

137. Mastronardi ML, et al. (2012). Size-Dependent Absolute Quantum Yields for Size-Separated Colloidally-Stable Silicon Nanocrystals. *Nano Lett.*, 12(1):337–342.

138. Mawhinney DB, Glass JA Jr. and Yates JT. (1997). FTIR Study of the Oxidation of Porous Silicon. *J. Phys. Chem. B*, 101(7):1202–1206.

139. McVey BFP and Tilley RD. (2014). Solution Synthesis, Optical Properties, and Bioimaging Applications of Silicon Nanocrystals. *Acc. Chem. Res.*, 47(10):3045–3051.

140. Melnikov DV and Chelikowsky JR. (2004). Electron Affinities and Ionization Energies in Si and Ge Nanocrystals. *Phys. Rev. B*, 69:113305.

141. Miller JB, Sickle ARV, Anthony RJ, Kroll DM, Kortshagen UR and Hobbie EK. (2012). Ensemble Brightening and Enhanced Quantum Yield in Size-Purified Silicon Nanocrystals. *ACS Nano*, 6(8):7389–7396.

142. Miller JB, et al. (2015). Enhanced Luminescent Stability through Particle Interactions in Silicon Nanocrystal Aggregates. *ACS Nano*, 9(10):9772–9782.

143. Mitas L, Therrien J, Twesten R, Belomoin G and Nayfeh MH. (2001). Effect of Surface Reconstruction on the Structural Prototypes of Ultrasmall Ultrabright Si_{29} Nanoparticles. *Appl. Phys. Lett.*, 78(13):1918–1920.

144. Murthy T, Miyamoto N, Shimbo M and Nishizawa J. (1976). Gas-Phase Nucleation during the Thermal Decomposition of Silane in Hydrogen. *J. Cryst. Growth*, 33(1):1–7.

145. Nakamura T, Yuan Z, Watanabe K and Adachi S. (2016). Bright and Multicolor Luminescent Colloidal Si Nanocrystals Prepared by Pulsed Laser Irradiation in Liquid. *Appl. Phys. Lett.*, 108(2):023105.

146. Neiner D, Chiu HW and Kauzlarich SM. (2006). Low-Temperature Solution Route to Macroscopic Amounts of Hydrogen Terminated Silicon Nanoparticles. *J. Am. Chem. Soc.*, 128(34):11016–11017.

147. Niehaus TA, et al. (2001). Tight-Binding Approach to Time-Dependent Density-Functional Response Theory. *Phys. Rev. B*, 63:085108.

148. Orii T, Hirasawa M and Seto T. (2003). Tunable, Narrow-Band Light Emission from Size-Selected Si Nanoparticles Produced by Pulsed-Laser Ablation. *Appl. Phys. Lett.*, 83(16):3395–3397.

149. Osaki F, Kanamori T, Sando S, Sera T and Aoyama Y. (2004). A Quantum Dot Conjugated Sugar Ball and Its Cellular Uptake. On the Size Effects of Endocytosis in the Subviral Region. *J. Am. Chem. Soc.*, 126(21):6520–6521.

150. Ossicini S, Pavesi L and Priolo F, editors. (2003). *Light Emitting Silicon for Microelectronics, Volume 194 of Springer Tracts in Modern Physics*. Springer, Berlin.

151. Pan GH, Barras A, Boussekey L, Addad A and Boukherroub R. (2013). Alkyl Passivation and SiO_2 Encapsulation of Silicon Nanoparticles: Preparation, Surface Modification and Luminescence Properties. *J. Mater. Chem. C*, 1:5261–5271.

152. Panthani MG, Hessel CM, Reid D, Casillas G, Jos-Yacamn M and Korgel BA. (2012). Graphene-Supported High-Resolution TEM and STEM Imaging of Silicon Nanocrystals and their Capping Ligands. *J. Phys. Chem. C*, 116(42):22463–22468.

153. Park JH, Gu L, von Maltzahn G, Ruoslahti E, Bhatia SN and Sailor MJ. (2009). Biodegradable Luminescent Porous Silicon Nanoparticles for *In Vivo* Applications. *Nat. Mat.*, 8(4):331–336.

154. Pauling L. (1960). *The Nature of the Chemical Bond and the Structure of Molecules and Crystals*. Cornell University Press, New York, NY.

155. Pavesi L and Turan R. (2010). *Silicon Nanocrystals: Fundamentals, Synthesis and Applications*. Wiley-VCH Verlag, Weinheim.

156. Pelant I. (2011). Optical Gain in Silicon Nanocrystals: Current Status and Perspectives. *Phys. Stat. Sol. (a)*, 208(3):625–630.

157. Pelant I and Valenta J. (2012). *Luminescence Spectroscopy of Semiconductors*. Oxford University Press, Oxford.

158. Peng F, Cao Z, Ji X, Chu B, Su Y and He Y. (2015). Silicon Nanostructures for Cancer Diagnosis and Therapy. *Nanomedicine*, 10(13):2109–2123.

159. Peng F, Su Y, Zhong Y, Fan C, Lee ST and He Y. (2014). Silicon Nanomaterials Platform for Bioimaging, Biosensing, and Cancer Therapy. *Acc. Chem. Res.*, 47(2):612–623.

160. Pettigrew KA, Liu Q, Power PP and Kauzlarich SM. (2003). Solution Synthesis of Alkyl- and Alkyl/Alkoxy-Capped Silicon Nanoparticles via Oxidation of Mg_2Si. *Chem. Mater.*, 15(21):4005–4011.

161. Pi XD, Liptak RW, Campbell SA and Kortshagen U. (2007). In-Flight Dry Etching of Plasma-Synthesized Silicon Nanocrystals. *Appl. Phys. Lett.*, 91(8):083112.

162. Pi XD, Mangolini L, Campbell SA and Kortshagen U. (2007). Room-Temperature Atmospheric Oxidation of Si Nanocrystals after HF Etching. *Phys. Rev. B*, 75:085423.

163. Poddubny AN and Dohnalová K. (2014). Direct Band Gap Silicon Quantum Dots Achieved via Electronegative Capping. *Phys. Rev. B*, 90:245439.

164. Prokofiev AA, et al. (2009). Direct Bandgap Optical Transitions in Si Nanocrystals. *JETP Lett.*, 90(12):758–762.

165. Purkait TK, et al. (2016). Alkoxy-Terminated Si Surfaces: A New Reactive Platform for the Functionalization and Derivatization of Silicon Quantum Dots. *J. Am. Chem. Soc.*, 138(22):7114–7120.

166. Puzder A, Williamson AJ, Grossman JC and Galli G. (2002). Surface Chemistry of Silicon Nanoclusters. *Phys. Rev. Lett.*, 88:097401.

167. Puzder A, Williamson AJ, Grossman JC and Galli G. (2002). Surface Control of Optical Properties in Silicon Nanoclusters. *J. Chem. Phys.*, 117(14):6721–6729.

168. Puzder A, Williamson AJ, Grossman JC and Galli G. (2003). Computational Studies of the Optical Emission of Silicon Nanocrystals. *J. Am. Chem. Soc.*, 125(9):2786–2791.

169. Puzder A, Williamson AJ, Reboredo FA and Galli G. (2003). Structural Stability and Optical Properties of Nanomaterials with Reconstructed Surfaces. *Phys. Rev. Lett.*, 91:157405.

170. Puzzo DP, Henderson EJ, Helander MG, Wang Z, Ozin GA and Lu Z. (2011). Visible Colloidal Nanocrystal Silicon Light-Emitting Diode. *Nano Lett.*, 11(4):1585–1590.

171. Reboredo FA and Galli G. (2005). Theory of Alkyl-Terminated Silicon Quantum Dots. *J. Phys. Chem. B*, 109(3):1072–1078.

172. Reboredo FA, Pizzagalli L and Galli G. (2004). Computational Engineering of the Stability and Optical Gaps of SiC Quantum Dots. *Nano Lett*, 4(5):801–804.

173. Reboredo FA, Schwegler E and Galli G. (2003). Optically Activated Functionalization Reactions in Si Quantum Dots. *J. Am. Chem. Soc.*, 125(49):15243–15249.

174. Rogozhina E, Eckhoff D, Gratton E and Braun P. (2006). Carboxyl Functionalization of Ultrasmall Luminescent Silicon Nanoparticles through Thermal Hydrosilylation. *J. Mater. Chem.*, 16:1421–1430.

175. Rosso-Vasic M, De Cola L and Zuilhof H. (2009). Efficient Energy Transfer between Silicon Nanoparticles and a Ru-Polypyridine Complex. *J. Phys. Chem. C*, 113(6):2235–2240.

176. Rosso-Vasic M, Spruijt E, van Lagen B, De Cola L and Zuilhof H. (2008). Alkyl-Functionalized Oxide-Free Silicon Nanoparticles: Synthesis and Optical Properties. *Small*, 4(10):1835–1841.

177. Ruizendaal L, Pujari SP, Gevaerts V, Paulusse JMJ and Zuilhof H. (2011). Biofunctional Silicon Nanoparticles by Means of Thiol-Ene Click Chemistry. *Chem. Asian J.*, 6(10):2776–2786.

178. Ryabchikov YV, Alekseev S, Lysenko V, Bremond G and Bluet JM. (2013). Photoluminescence of Silicon Nanoparticles Chemically Modified by Alkyl Groups and Dispersed in Low-Polar Liquids. *J. Nanopart. Res.*, 15(4):1535.

179. Sagar DM, et al. (2015). Quantum Confined Electron–Phonon Interaction in Silicon Nanocrystals. *Nano Lett.*, 15(3):1511–1516.

180. Sato K, Fukata N, Hirakuri K, Murakami M, Shimizu T and Yamauchi Y. (2010). Flexible and Transparent Silicon Nanoparticle/Polymer Composites with Stable Luminescence. *Chem. Asian J.*, 5(1):50–55.

181. Sato S and Swihart MT. (2006). Propionic-Acid-Terminated Silicon Nanoparticles: Synthesis and Optical Characterization. *Chem. Mater.*, 18(17):4083–4088.

182. Shiohara A, et al. (2011). Sized Controlled Synthesis, Purification, and Cell Studies with Silicon Quantum Dots. *Nanoscale*, 3:3364–3370.

183. Shirahata N. (2011). Colloidal Si Nanocrystals: A Controlled Organic-Inorganic Interface and Its Implications of Color-Tuning and Chemical Design toward Sophisticated Architectures. *Phys. Chem. Chem. Phys.*, 13(16):7284–7294.

184. Shirahata N, et al. (2009). Laser-Derived One-Pot Synthesis of Silicon Nanocrystals Terminated with Organic Monolayers. *Chem. Commun.* (31):4684–4686.

185. Siekierzycka JR, Rosso-Vasic M, Zuilhof H and Brouwer AM. (2011). Photophysics of n-Butyl-Capped Silicon Nanoparticles. *J. Phys. Chem. C*, 115(43):20888–20895.

186. Sieval AB, van den Hout B, Zuilhof H and Sudhölter EJR. (2001). Molecular Modeling of Covalently Attached Alkyl Monolayers on the Hydrogen-Terminated Si(111) Surface. *Langmuir*, 17(7):2172–2181.

187. Šiller L, et al. (2009). Core and Valence Exciton Formation in X-Ray Absorption, X-Ray Emission and X-Ray Excited Optical Luminescence from Passivated Si Nanocrystals at the Si $L_{2,3}$ Edge. *J. Phys. Condens. Matter*, 21(9):095005.

188. Smith R and Collins S. (1992). Porous Silicon Formation Mechanisms. *J. Appl. Phys.*, 71(8):R1–R22.

189. Sommer A, Cimpean C, Kunz M, Oelsner C, Kupka HJ and Kryschi C. (2011). Ultrafast Excitation Energy Transfer in Vinylpyridine Terminated Silicon Quantum Dots. *J. Phys. Chem. C*, 115(46):22781–22788.

190. Stewart MP and Buriak JM. (2001). Exciton-Mediated Hydrosilylation on Photoluminescent Nanocrystalline Silicon. *J. Am. Chem. Soc.*, 123(32):7821–7830.

191. Stewart MP and Buriak JM. (2002). New Approaches toward the Formation of Silicon-Carbon Bonds on Porous Silicon. *Comm. Inorg. Chem.*, 23(3):179–203.

192. Švrček V, Mariotti D and Kondo M. (2010). Microplasma-Induced Surface Engineering of Silicon Nanocrystals in Colloidal Dispersion. *Appl. Phys. Lett.*, 97(16):161502.

193. Sychugov I, Fucikova A, Pevere F, Yang Z, Veinot JGC and Linnros J. (2014). Ultranarrow Luminescence Linewidth of Silicon Nanocrystals and Influence of Matrix. *ACS Photonics*, 1(10):998–1005.

194. Tilley R, Warner J, Yamamoto K, Matsui I and Fujimori H. (2005). Micro-Emulsion Synthesis of Monodisperse Surface Stabilized Silicon Nanocrystals. *Chem. Commun.*, 14:1833–1835.

195. Tu C, Ma X, Pantazis P, Kauzlarich SM and Louie AY. (2010). Paramagnetic, Silicon Quantum Dots for Magnetic Resonance and Two-Photon Imaging of Macrophages. *J. Am. Chem. Soc.*, 132(6):2016–2023.

196. Tu CC, Chen KP, Yang TA, Chou MY, Lin LY and Li YK. (2016). Silicon Quantum Dot Nanoparticles with Antifouling Coatings for Immunostaining on Live Cancer Cells. *ACS Appl. Mater. Interfaces*, 8(22):13714–13723.

197. Uhlir A, Jr. (1956). Electrolytic Shaping of Germanium and Silicon. *Bell Syst. Tech. J.*, 35:333.

198. Vasiliev I and Martin R. (2002). Optical Properties of Hydrogenated Silicon Clusters with Reconstructed Surfaces. *Phys. Stat. Sol. (B)*, 233(1):5–9.

199. Vasiliev I, Oğüt S and Chelikowsky JR. (2001). *Ab Initio* Absorption Spectra and Optical Gaps in Nanocrystalline Silicon. *Phys. Rev. Lett.*, 86:1813–1816.

200. Veinot J. (2006). Synthesis, Surface Functionalization, and Properties of Freestanding Silicon Nanocrystals. *Chem. Commun.*, 40:4160–4168.

201. Wang L, et al. (2015). Ultrafast Optical Spectroscopy of Surface-Modified Silicon Quantum Dots: Unraveling the Underlying Mechanism of the Ultrabright and Color-Tunable Photoluminescence. *Light Sci. Appl.*, 4(1):e245.

202. Wang R, Pi X and Yang D. (2012). First-Principles Study on the Surface Chemistry of 1.4 nm Silicon Nanocrystals: Case of Hydrosilylation. *J. Phys. Chem. C*, 116(36):19434–19443.

203. Wang X, Zhang R, Niehaus T and Frauenheim T. (2007). Excited State Properties of Allylamine-Capped Silicon Quantum Dots. *J. Phys. Chem. C*, 111(6):2394–2400.

204. Wang Y, Hu R, Lin G, Roy I and Yong KT. (2013). Functionalized Quantum Dots for Biosensing and Bioimaging and Concerns on Toxicity. *ACS Appl. Mater. Interfaces*, 5(8):2786–2799.

205. Warner JH, Hoshino A, Yamamoto K and Tilley RD. (2005). Water-Soluble Photoluminescent Silicon Quantum Dots. *Angew. Chem. Int. Ed.*, 44(29):4550–4554.

206. Weissleder R. (2001). A Clearer Vision for *In Vivo* Imaging. *Nat. Biotechnol.*, 19(4):316–317.

207. Werwa E, Seraphin AA, Chiu LA, Zhou C and Kolenbrander KD. (1994). Synthesis and Processing of Silicon Nanocrystallites Using a Pulsed Laser Ablation Supersonic Expansion Method. *Appl. Phys. Lett.*, 64(14):1821–1823.

208. Wheeler LM, Anderson NC, Palomaki PKB, Blackburn JL, Johnson JC and Neale NR. (2015). Silyl Radical Abstraction in the Functionalization of Plasma-Synthesized Silicon Nanocrystals. *Chem. Mater.*, 27(19):6869–6878.

209. Wilcoxon JP, Samara GA and Provencio PN. (1999). Optical and Electronic Properties of Si Nanoclusters Synthesized in Inverse Micelles. *Phys. Rev. B*, 60:2704–2714.

210. Wilson WL, Szajowski PF and Brus L. (1993). Quantum Confinement in Size-Selected, Surface-Oxidized Silicon Nanocrystals. *Science*, 262(5137):1242–1244.

211. Wolf O, Dasog M, Yang Z, Balberg I, Veinot JGC and Millo O. (2013). Doping and Quantum Confinement Effects in Single Si Nanocrystals Observed by Scanning Tunneling Spectroscopy. *Nano Lett.*, 13(6):2516–2521.

212. Wolkin MV, Jorne J, Fauchet PM, Allan G and Delerue C. (1999). Electronic States and Luminescence in Porous Silicon Quantum Dots: The Role of Oxygen. *Phys. Rev. Lett.*, 82(1):197–200.

213. Xia B, Zhang W, Shi J and Xiao S. (2013). Fluorescence Quenching in Luminescent Porous Silicon Nanoparticles for the Detection of Intracellular Cu^{2+}. *Analyst*, 138:3629–3632.

214. Xin Y, Nishio K and Saitow K. (2015). White-Blue Electroluminescence from a Si Quantum Dot Hybrid Light-Emitting Diode. *Appl. Phys. Lett.*, 106(20):201102.

215. Xu Z, et al. (2012). Photoluminescent Silicon Nanocrystal-Based Multifunctional Carrier for pH-Regulated Drug Delivery. *ACS Appl. Mater. Interfaces*, 4(7):3424–3431.

216. Yang CS, Bley RA, Kauzlarich SM, Lee HWH and Delgado GR. (1999). Synthesis of Alkyl-Terminated Silicon Nanoclusters by a Solution Route. *J. Am. Chem. Soc.*, 121(22):5191–5195.

217. Yang Z, Iqbal M, Dobbie AR and Veinot JGC. (2013). Surface-Induced Alkene Oligomerization: Does Thermal Hydrosilylation Really Lead to Monolayer Protected Silicon Nanocrystals? *J. Am. Chem. Soc.*, 135(46):17595–17601.

218. Yong KT, et al. (2013). Nanotoxicity Assessment of Quantum Dots: From Cellular to Primate Studies. *Chem. Soc. Rev.*, 42:1236–1250.

219. Yu T, Wang F, Xu Y, Ma L, Pi X and Yang D. (2016). Graphene Coupled with Silicon Quantum Dots for High-Performance Bulk-Silicon-Based Schottky-Junction Photodetectors. *Adv. Mater.*, 28(24):4912–4919.

220. Yu Y, Hessel CM, Bogart TD, Panthani MG, Rasch MR and Korgel BA. (2013). Room Temperature Hydrosilylation of Silicon Nanocrystals with Bifunctional Terminal Alkenes. *Langmuir*, 29(5):1533–1540.

221. Yu Y, Rowland CE, Schaller RD and Korgel BA. (2015). Synthesis and Ligand Exchange of Thiol-Capped Silicon Nanocrystals. *Langmuir*, 31(24):6886–6893.

222. Zhang RQ, Costa J and Bertran E. (1996). Role of Structural Saturation and Geometry in the Luminescence of Silicon-Based Nanostructured Materials. *Phys. Rev. B*, 53:7847–7850.

223. Zhang X, et al. (2015). Highly Sensitive and Selective Detection of Dopamine Using One-Pot Synthesized Highly Photoluminescent Silicon Nanoparticles. *Anal. Chem.*, 87(6):3360–3365.

224. Zhao Y, Kim YH, Du MH and Zhang SB. (2004). First-Principles Prediction of Icosahedral Quantum Dots for Tetravalent Semiconductors. *Phys. Rev. Lett.*, 93:015502.

225. Zhong Y, et al. (2012). Microwave-Assisted Synthesis of Biofunctional and Fluorescent Silicon Nanoparticles Using Proteins as Hydrophilic Ligands. *Ang. Chem. Int. Ed.*, 51(34):8485–8489.

226. Zhong Y, et al. (2013). Large-Scale Aqueous Synthesis of Fluorescent and Biocompatible Silicon Nanoparticles and Their Use as Highly Photostable Biological Probes. *J. Am. Chem. Soc.*, 135(22):8350–8356.

227. Zhong Y, et al. (2015). Facile, Large-Quantity Synthesis of Stable, Tunable-Color Silicon Nanoparticles and Their Application for Long-Term Cellular Imaging. *ACS Nano*, 9(6):5958–5967.

228. Zhou T, et al. (2015). Bandgap Tuning of Silicon Quantum Dots by Surface Functionalization with Conjugated Organic Groups. *Nano Lett.*, 15(6):3657–3663.

229. Zhou Z, Brus L and Friesner R. (2003). Electronic Structure and Luminescence of 1.1- and 1.4-nm Silicon Nanocrystals: Oxide Shell versus Hydrogen Passivation. *Nano Lett.*, 3(2):163–167.

230. Zhou Z, Friesner RA, and Brus L. (2003). Electronic Structure of 1 to 2 nm Diameter Silicon Core/Shell Nanocrystals: Surface Chemistry, Optical Spectra, Charge Transfer, and Doping. *J. Am. Chem. Soc.*, 125(50):15599–15607.

231. Zou J, Baldwin RK, Pettigrew KA and Kauzlarich SM. (2004). Solution Synthesis of Ultrastable Luminescent Siloxane-Coated Silicon Nanoparticles. *Nano Lett.*, 4(7):1181–1186.

232. Zou J, Sanelle P, Pettigrew KA and Kauzlarich SM. (2006). Size and Spectroscopy of Silicon Nanoparticles Prepared via Reduction of $SiCl_4$. *J. Cluster Sci.*, 17(4):565–578.

17

Near-infrared luminescent colloidal silicon nanocrystals

Hiroshi Sugimoto and Minoru Fujii

Contents

17.1 INTRODUCTION

Semiconductor nanocrystals, or quantum dots, with diameters comparable to, or smaller than the exciton Bohr radius, exhibit unique optoelectronic properties due to the quantum size effects. The energy state structures of nanocrystals can be controlled by size, shape, composition (alloying), and surface chemistry (Talapin et al. 2010). Among different types of nanocrystals, "colloidal" dispersion of nanocrystals is promising as fluorescent nanoprobes in biomedical applications (Michalet et al. 2005; Yao et al. 2014) and precursors for printed optoelectronic devices (Talapin 2005; Tang et al. 2011; Kwak et al. 2012).

From the material point of view, cadmium (Cd) and lead (Pb) chalcogenide nanocrystals lead the research on colloidal semiconductor nanocrystals. However, the toxic heavy metal elements raise concern for the applications, especially in biomedical fields. Silicon (Si) nanocrystals are one of the promising alternatives due to the high environmental friendliness and the high compatibility with biological substances. There have been a numerous researches on the luminescence properties of Si nanocrystals since the discovery of room-temperature visible photoluminescence (PL) from porous Si (Canham 1990). However, the majority of them are on Si nanocrystals embedded in solid matrices, and the development of "colloidal" Si nanocrystals had been a challenging task for many years. The covalent bond nature of Si crystal, which requires high-crystallization energy exceeding the limitation of a typical liquid system, makes solution synthesis of highly crystalline Si nanocrystals difficult. Fortunately, several alternative approaches have been developed in the past decade and colloidally stable Si nanocrystals have been realized (Veinot 2006; Mangolini and Kortshagen 2007; Sugimoto et al. 2012). The quality of colloidal Si nanocrystals (i.e., crystallinity, size distribution, stability in solution, and PL quantum yield [QY]) has been rapidly catching up with that of Cd and Pb chalcogenide nanocrystals.

Among large number of researches on colloidal Si nanocrystals made in the past decade, in this chapter, we focus on the development of near-infrared (NIR) luminescent Si nanocrystals for targeting the applications in biomedical fields. Figure 17.1 shows the attenuation coefficient of bloods, skin, and fatty tissues

Clusters, nanoparticles, and quantum dots

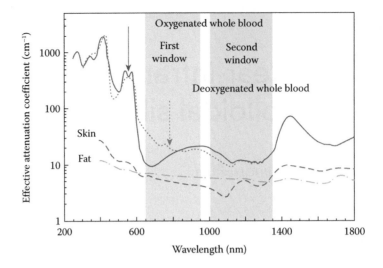

Figure 17.1 Effective attenuation coefficient of oxygenated blood, deoxygenated blood, skin, and fatty tissue. The first and second near-IR biological window is shown. (Reprinted by permission from Macmillan Publishers Ltd., *Nature Nanotechnology*, 2009, 4, 710–711, copyright 2009.)

of the human body as a function of light wavelength. There is an optimum range where the attenuation coefficient is small, the so-called biological window spanning from 700 to 1350 nm. Therefore, biocompatible near-IR luminescent Si nanocrystals possess an advantage as nanoprobes in deep-tissue imaging and in biosensing devices used in biological substances.

17.2 GROWTH OF SIZE-CONTROLLED COLLOIDAL Si NANOCRYSTALS

During the past decade, various processes have been developed for the growth of colloidal Si nanocrystals. Among them, a widely used process for the development of biomedical application of Si nanocrystals is a low-temperature, solution-based precursor reduction process. However, Si nanocrystals prepared by this process always have the luminescence in the UV and blue range and no near-IR luminescence have been reported. We therefore do not refer to the process in this chapter.

A very successful approach for the development of near-IR luminescent colloidal Si nanocrystals is the growth of nanocrystals in silicate matrices followed by hydrofluoric acid (HF) etching to liberate nanocrystals in solution. Si nanocrystals in SiO_2 are prepared by high-temperature annealing of various SiO_x ($x < 2$) materials such as commercial SiO_x powder (Liu et al. 2006; Sato and Swihart 2006), sputter-deposited SiO_x (Sato et al. 2010; Sugimoto et al. 2012), and solid sol-gel polymers $(HSiO_{1.5})_n$ (Hessel et al. 2006, 2012; Mastronardi et al. 2012; Sugimoto et al. 2014). J. G C. Veinot of the Veinot Group and B. A. Korgel of the Korgel Group contributed greatly to the improvement of the productivity and quality of Si nanocrystals by using hydrogen silsesquioxane (HSQ) as a starting polymer. Figure 17.2 shows the schematic illustration of the preparation of colloidal Si nanocrystals from HSQ (Veinot 2006; Hessel et al. 2012; Dasog et al. 2014). HSQ is annealed in a slightly reducing atmosphere and is decomposed into Si nanocrystals and SiO_2 matrices. By HF etching, the Si nanocrystals are liberated from matrices, resulted in freestanding hydrogen-terminated Si nanocrystals in solution. The surface functionalization by long alkene chains such as 1-dodecene and 1-octadecene makes Si nanocrystals dispersible in nonpolar solvents by steric barriers (Hessel et al. 2012). As will be discussed later, the choice of the molecules for the surface functionalization is particularly important to achieve efficient PL in the near-IR range.

In these methods, the size of Si nanocrystals can be controlled by the growth temperature and the stoichiometry of the starting material, x in SiO_x. Figure 17.3 shows the transmission electron microscopy (TEM) images of colloidal Si nanocrystals prepared by the method in Figure 17.2 with

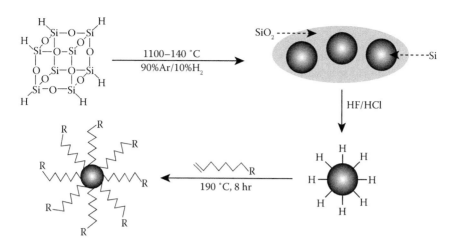

Figure 17.2 Schematic illustration of synthetic pathway from HSQ to alkyl passivated Si nanocrystals. (Reprinted with permission from Hessel, C.M., et al., 2012, 24, 393–401. Copyright 2012 American Chemical Society.)

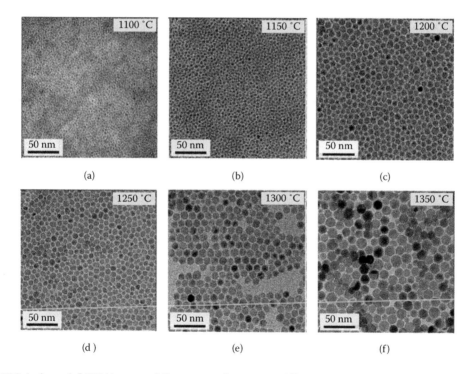

Figure 17.3 (a through f) TEM images of Si nanocrystals grown at different temperatures. (Reprinted with permission from Hessel, C.M., et al., 2012, 24, 393–401. Copyright 2012 American Chemical Society.)

different growth temperatures (1100–1350°C). The average diameter of Si nanocrystals can be controlled from 2.7 to 11.8 nm. The size is rather uniform and the standard deviation is 10–15% of the average diameter.

Another successful approach for the preparation of near-IR luminescence colloidal Si nanocrystals is a plasma synthesis method. Kortshagen and coworkers have pioneered the preparation of freestanding Si nanocrystals by nonthermal plasma synthesis using silane (Mangolini and Kortshagen 2007).

Plasma-synthesized Si nanocrystals are also functionalized with 1-dodecene and dispersed in nonpolar solvents. In this case, the size of nanocrystals is controlled by the reactant flow rates.

In all the Si nanocrystal growth methods, size distribution as large as ~10% is not avoidable. Therefore, in addition to the efforts to reduce the size distribution during the growth process, postgrowth size purification methods have been developed. By using a density-gradient ultracentrifugation method, colloidal Si nanocrystals with very small size distribution are successfully developed in Mastronardi et al. 2011 (average diameter: 2.2 nm; polydispersity index (PDI): 1.04) and in Miller et al. 2012 (average diameter: 3.79 nm; PDI: 1.005).

In the conventional synthesis approach discussed so far, a stable colloidal solution is produced only by properly functionalizing the surface of nanocrystals with organic ligands. Recently, a new type of colloidal Si nanocrystals that do not have organic ligands on the surface have been developed. The preparation procedure is summarized in Figure 17.4. Si-rich borophosphosilicate glasses (BPSG) are first prepared by cosputtering of Si, SiO_2, B_2O_3, and P_2O_5 (Sugimoto et al. 2012), or HSQ with dopant acids (H_3BO_3 and H_3PO_4) (Sugimoto et al. 2014). By annealing the Si-rich BPSG, boron (B) and phosphorus (P) codoped Si nanocrystals are grown in BPSG, and they are liberated in solution by the HF etching. The B and P codoped Si nanocrystals are dispersible in alcohol or water, which is particularly important for the bio-applications, without organic ligands (Sugimoto et al. 2012). The high solution dispersibility arises from the negative surface potential induced by codoping. The size of codoped colloidal Si nanocrystals can be controlled in a wide range by the growth temperature (Sugimoto et al. 2013).

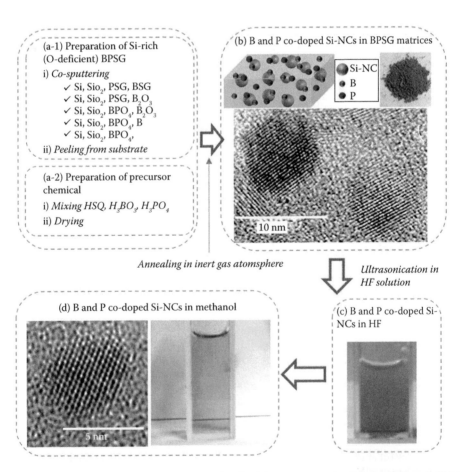

Figure 17.4 Preparation procedure of B and P codoped colloidal Si nanocrystals. (Reprinted from Fujii, M., et al., *Nanotechnology*, 27, 262001, 2016.)

17.3 PHOTOLUMINESCENCE PROPERTIES

The PL wavelength, spectral shape, quantum efficiency, and lifetime of Si nanocrystals depend strongly on the kinds of surface termination as well as the size. Therefore, the PL of Si nanocrystals in different surface terminations have to be discussed differently. In Si nanocrystals embedded in silica matrices, the PL energy can be controlled from bulk Si bandgap to ~2 eV, and the PL is considered to originate from the recombination of quantum-confined excitons in Si nanocrystals (exciton Bohr radius of ~4.5 nm) (Takeoka et al. 2000; Ledoux et al. 2002; Barbagiovanni et al. 2014). On the other hand, in colloidal Si nanocrystals, due to the complexity of the surface termination and the effect of solvent, the PL property is rather complicated. Colloidal Si nanocrystals can be classified into two groups from the point of view of PL characteristics: one shows PL in the UV to blue region (300–500 nm), and the other shows size-dependent PL in the yellow to near-IR wavelength region. As described above, in this article we focus on the latter case.

Here, we summarize recent reports on the size-tunable near-IR luminescence in colloidal Si nanocrystals. Figure 17.5 shows room-temperature PL and PL excitation spectra of octadecene/dodecene-stabilized colloidal Si nanocrystals in toluene (Hessel et al. 2012). The PL peak wavelength changes from 720 nm (3 nm in diameter) to 1060 nm (12 nm in diameter) depending on the size. For the largest Si nanocrystals, the PL spectrum is a nonGaussian shape due to the fundamental long-wavelength limit for PL of Si nanocrystals (Takeoka et al. 2000).

As shown in Figure 17.6a, plasma-synthesized Si nanocrystals capped with dodecene also exhibit size-tunable PL (Hannah et al. 2012). The longest PL wavelength is also limited to around 1000 nm. Figure 17.6b shows the PL decay curve detected at 680 nm of Si nanocrystals 2.6 nm in diameter. The decay curve is well fitted by a single-exponential function with a 75.9 μsec lifetime. The PL lifetimes of several tens of μsec have commonly been observed in near-IR luminescent colloidal Si nanocrystals irrespective of the preparation methods

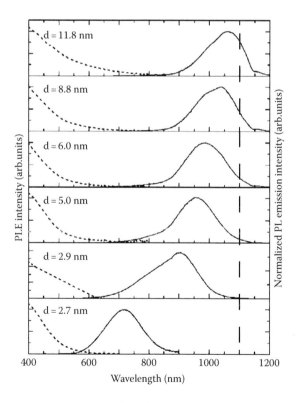

Figure 17.5 Room-temperature PL (excitation wavelength = 420 nm) and photoluminescence emission (PLE) (measured at PL peak wavelength) spectra of alkane-stabilized Si nanocrystals dispersed in toluene. The vertical dashed line indicates the position of the bulk Si band gap (1100 nm). (Reprinted with permission from Hessel, C.M., et al., 2012, 24, 393–401. Copyright 2012 American Chemical Society.)

Clusters, nanoparticles, and quantum dots

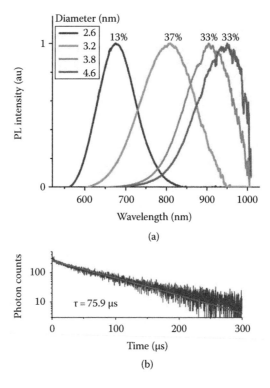

Figure 17.6 (a) Normalized PL spectra of Si nanocrystals with indicated average nanocrystal diameters and PL QYs. (b) PL decay curve of 2.6 nm Si nanocrystal sample measured at 680 nm with a fitting curve of a single-exponential decay function. (Reprinted from Hannah, D.C., et al., *Nano Lett.*, 12, 4200–4205, 2012.)

(Mastronardi et al. 2012; Sangghaleh et al. 2015). The lifetimes are similar to those of Si nanocrystals embedded in silica matrices with comparable sizes (Linnros et al. 1999; Takeoka et al. 2000; Vinciguerra et al. 2000).

The PL energy of Si nanocrystals can be extended to longer wavelengths by introducing impurity states in the band gap. This has been successfully achieved by B- and P-doping in Si nanocrystals (Fujii et al. 2003). Figure 17.7 shows the normalized PL spectra of B and P codoped Si nanocrystals dispersed in methanol (Sugimoto et al. 2013). Unlike conventional Si nanocrystals in Figures 17.5 and 17.6, the PL peak shift is not limited to the bulk Si band gap; it reaches nearly 1500 nm. Several theoretical calculations have been performed on the energy state structures of impurity doped Si nanocrystals (Iori et al. 2007).

In Figure 17.8a, we summarize the PL peak energy of colloidal Si nanocrystals as a function of diameter. The red curve shows the relations obtained from effective mass approximations (Trwoga et al. 1998; Barbagiovanni et al. 2012; Hessel et al. 2012):

$$E(d) = E_g + \frac{\hbar^2}{2d^2}\left(\frac{1}{m_e^*} + \frac{1}{m_h^*}\right) - \frac{1.786e^2}{\varepsilon_r d},$$

where d is the nanocrystal diameter, e is the elementary charge, ε_r is the relative permittivity of Si (11.7), and m_e^* and m_h^* are the effective masses of the electron and hole, which are $0.19m_0$ and $0.286m_0$, respectively. Apparently, the data can be classified into three groups. The first one is PL in UV to blue region (enclosed by green line), the second one is that red to near-IR region and following the theoretical prediction (enclosed by blue line), and the third is the PL below the theoretical prediction (magenta plots). Colloidal Si nanocrystals in the first group, which exhibit the PL in UV to blue region (390–500 nm), are synthesized by low-temperature solution processes (typically lower than 160°C) using halogenated silane precursors (SiX$_4$, X = Cl, Br) (Tilley et al. 2005; Shirahata et al. 2010; Cheng et al. 2012) or oxidation of Zintl salts (Neiner et al. 2006). The surface of such nanocrystals are usually terminated by oxygen,

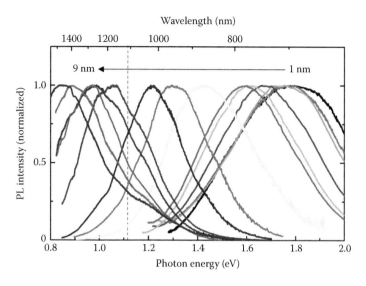

Figure 17.7 Normalized PL spectra of B and P codoped colloidal Si nanocrystals with different diameters. The vertical red line indicates the bulk Si bandgap. (Reprinted with permission from Sugimoto, H., et al., 2013, 117, 11850–11857, 2013. Copyright 2013 American Chemical Society.)

nitrogen and thiols (Warner et al. 2005; Lin and Chen 2009; Yu et al. 2015). Similar UV to blue PL is observed for Si nanocrystals prepared by thermal decomposition of HSQ after some specific chemical treatments (Figure 17.8b) (Dasog et al. 2013, 2014). By treating hydrogen-terminated Si nanocrystals with tetraoctylammonium bromide (TOAB) or ammonium bromide (NH$_4$Br), the luminescence color immediately changes from red to blue. The characteristic features of the UV to blue PL are the nsec lifetimes and the 10–20% of PL QYs. Moreover, the PL peak energy depends on the excitation energy rather than the size (Dasog et al. 2013). The most probable origin of the blue to UV PL is defects on the surface of Si nanocrystals.

The PL peak energy of Si nanocrystals in the second group follows the theoretical prediction when the diameter is larger than ~2 nm, despite the different preparation methods. This is a clear evidence that the PL arises from quantum-confined excitons. The nanocrystals are functionalized with alkene and the PL lifetimes are several tens of μsec.

The third group is B and P codoped Si nanocrystals, which show a similar trend to undoped Si nanocrystals. However, the PL energy is 300–400 meV lower than that of undoped Si nanocrystals (Sugimoto et al. 2013). The low-energy PL is considered to arise from electron transitions between donor and acceptor states in the band gap of Si nanocrystals. The size dependence of the donor and acceptor states of codoped Si nanocrystals is studied by photoemission yield spectroscopy (Hori et al. 2016).

As can be seen in Figures 17.5 through 17.7, the PL of colloidal Si nanocrystals are very broad. The PL full width at half maximum (FWHM) of colloidal Si nanocrystals reaches 300–400 meV around 1.6 eV. This is mainly to the inhomogeneous broadening caused by the size distribution. Mastronardi et al. succeeded in reducing the FWHM to 230 meV through size purification by density-gradient ultracentrifugation (Mastronardi et al. 2011). It should be stressed here that the FWHM of the PL of a single Si nanocrystal at room temperature is much broader than that of a single II–VI and IV–VI semiconductor nanocrystal and is 100–150 meV. This is due to the coupling of the optical transitions with momentum conservation phonons (Valenta et al. 2002).

Figure 17.9 shows the external PL QY of colloidal Si nanocrystals luminescing in the NIR range. Interestingly, the external QY always has a maximum around 800 nm irrespective of the preparation processes. Therefore, the trend seems to be universal in the exciton luminescence of Si nanocrystals. Although the reason of the trend is not very clear, a plausible explanation is that the QY is determined by the competition between the enhanced radiative recombination rate by the quantum confinement, and increased surface defects due to larger surface curvature in smaller nanocrystals. As a result, there is an optimum size

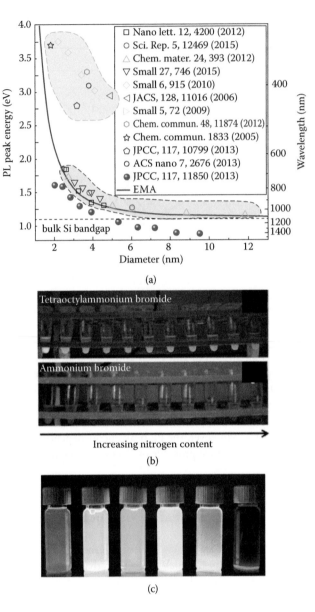

(a)

(b)

(c)

Figure 17.8 (a) PL peak energies of colloidal Si nanocrystals as a function of diameter. The data are taken from the literatures indicated in the figure. (b) Photograph of H-terminated Si nanocrystals under UV illumination that are reacted with TOAB and NH₄Br. (Reprinted with permission from Dasog, M., et al., 2013, 7, 2676–2685. Copyright 2013 American Chemical Society.)

range to achieve the highest PL QY, which is 3.0–3.5 nm in diameter. The external QY 65% at the wavelength of 700–900 nm in nonpolar solvents (Sangghaleh et al. 2015).

17.3.1 COLLOIDAL GERMANIUM NANOCRYSTALS AND SILICON GERMANIUM ALLOY NANOCRYSTALS

In this section, we briefly summarize the research on colloidal germanium (Ge) nanocrystals, which have similar properties to Si nanocrystals. The advantage of Ge nanocrystals is the larger absorption cross section than Si nanocrystals in the visible range due to the smaller direct band gap (0.88 eV). However, the synthesis of NIR luminescent Ge nanocrystals had been challenging for many years and the researches are still limited

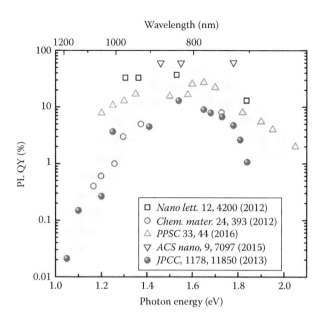

Figure 17.9 External PL QY of colloidal Si nanocrystals. The data are taken from literature indicated in the figure.

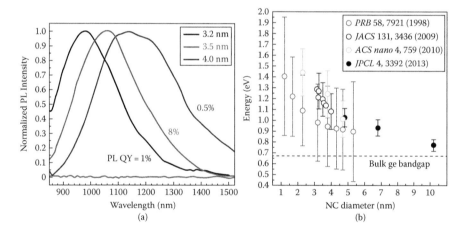

Figure 17.10 (a) Normalized PL spectra (excited at 808 nm) of ODE-capped Ge nanocrystals at room temperature. Green curve shows PL (not normalized) of TOP-capped Ge nanocrystals (3.8 nm). (Reprinted with permission from Lee, D.C., et al., 2009, 131, 3436–3437. Copyright 2009 American Chemical Society.) (b) PL peaks of Ge nanocrystals synthesized by colloidal methods and embedded in SiO₂. (Reprinted with permission from Wheeler, L.M., et al., 2013, 4, 3392–3396. Copyright 2013 American Chemical Society.)

(Lee et al. 2009; Ruddy et al. 2010; Wheeler et al. 2013). Klimov and coworkers succeeded in producing crystalline Ge nanocrystals dispersible in solution from GeI₂ at low-temperature (<300°C) (Lee et al. 2009). Figure 17.10 shows the PL spectra of colloidal Ge nanocrystals with different diameters. The PL QY of 8% around 1000–1100 nm is achieved. The right panel of Figure 17.10 shows the PL peak energy of colloidal and matrix-embedded Ge nanocrystals taken from different literatures. Wheeler et al. synthesized colloidal Ge nanocrystals capped with Grignard reagents by a nonthermal plasma reactor by decomposing GeCl₄. The 10.2 nm Ge nanocrystals exhibit the PL at 0.77 eV, which is very close to the bulk band gap.

It is well known that Si and Ge form solid solutions in the whole composition range ($Si_{1-x}Ge_x$, $x = 0 - 1$). To take advantages of both Si and Ge nanocrystals, $Si_{1-x}Ge_x$ alloy nanocrystals have been developed

(Barry et al. 2011; Erogbogbo et al. 2011a; Kanno et al. 2014). Although composition dependence of the absorption and PL properties is reported, the quality of the material is still not high enough compared to Si nanocrystals.

17.4 APPLICATION IN BIOIMAGING

One of the promising applications of near-IR luminescent colloidal Si nanocrystals is the luminescent nanoprobes for bioimaging, due to the high biocompatibility of Si. An essential requirement for the use of Si nanocrystals in bioimaging is that nanocrystals should be dispersible and stable in aqueous solution. Furthermore, the PL wavelength is preferably within the transparent window of tissues.

Uptake of Si nanocrystals by cell via endocytosis has been studied by fluorescence imaging *in vitro* (Henderson et al. 2011; Erogbogbo et al. 2013; Das and Jana 2014; Chandra et al. 2016). Erogbogbo et al. synthesized water-dispersible Si nanocrystal clusters (50–150 nm in diameter) using phospholipid micelles (Erogbogbo et al. 2008). The bright red PL are observed in living pancreatic cancer cells (Erogbogbo et al. 2008). Kalbacova and coworkers have studied the impact of B and P codoped Si nanocrystals on human osteoblast cells (Ostrovska et al. 2016). Thanks to the highly stable luminescence of codoped Si nanocrystals in a biological environment, cocultivation for more than 24 hours was possible. Figure 17.11 shows the fluorescence microscope image with a green layer of cell autofluorescence and red layer of the signal above 785 nm. As shown in the spectra in Figure 17.11b, a large wavelength separation of the cell autofluorescence and PL of Si nanocrystals enable high-contrast imaging.

Erogbogbo et al. demonstrated for the first time *in vivo* tumor targeting using micelle encapsulated Si nanocrystals (Erogbogbo et al. 2011b). Figure 17.12 shows the PL properties of bioconjugated Si nanocrystals in a mouse, which shows the detectable PL from developed nanocrystals. They have assessed *in vivo* cytotoxicity of the micelle encapsulated Si nanocrystals in mice, along with *in vitro* cytotoxicity for Si nanocrystals with different size, surface functionalization, and encapsulation. Their work demonstrates that Si nanocrystals can be used as biocompatible fluorescent probes for both *in vitro* and *in vivo* imaging, including multicolor imaging at near-IR wavelengths.

An important advantage of near-IR luminescent Si nanocrystals in bioimaging is the long PL lifetime. Joo et al. demonstrated the time-gated imaging, where the image is acquired at a certain time after an excitation pulse, allowing discrimination of a probe signal from other endogenous signals (Gu et al. 2013; Joo et al. 2015). Figure 17.13 shows the luminescence imaging and PL decay curves of Si nanocrystals

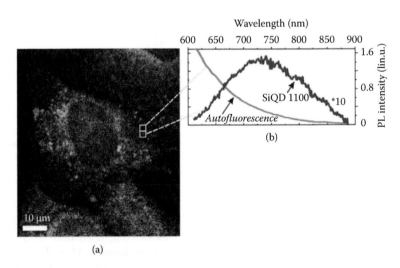

Figure 17.11 (a) Combined fluorescence image with the green layer showing the full signal (dominated by cell autofluorescence) and the orange layer showing signal above 785 nm (dominated by Si nanocrystal luminescence). (b) PL spectra of cell autofluorescence (blue) and Si nanocrystals (red) from area indicated by a rectangle in the panel (a). (Reprinted with permission from Ostrovska, L., et al., 2016, The impact of doped silicon quantum dots on human osteoblasts, *RSC Adv.*, 6(68), 63403–63413. Reproduced by permission of The Royal Society of Chemistry.)

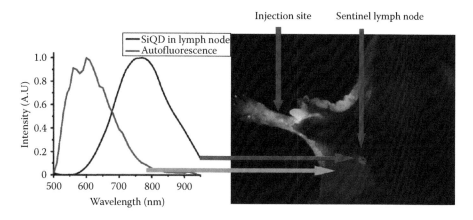

Figure 17.12 Sentinel lymph node imaging following localization of Si nanocrystals in an axillary position. Autofluorescence is coded in green, and the unmixed Si nanocrystal signal is coded in red. (Reprinted with permission from Erogbogbo, F., et al., 2011b, 5, 413–423. Copyright 2011 American Chemical Society.)

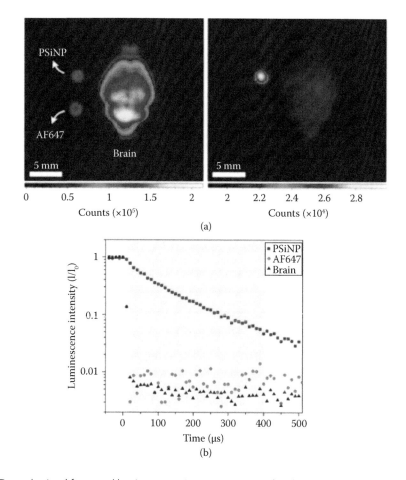

Figure 17.13 Data obtained for gated luminescence imaging compared with steady-state imaging. (a) Continuous-wave and gated luminescence images of the same brain under UV excitation (λ_{ex} = 365 nm, gate width, 400 µs, gate delay for gated luminescence =5 µs). (b) Normalized PL decay curves of Si nanoparticles (PSiNP), AF647, and brain tissue. (Reprinted with permission from Joo, J., et al., 2015, 9, 6233–6241. Copyright 2015 American Chemical Society.)

Clusters, nanoparticles, and quantum dots

obtained by electrochemical etching of Si wafers, the molecular dye Alexa Fluor 647 (AF647) and brain tissue autofluorescence. Note that the signals from the AF647 sample and the brain tissue are readily visible at steady state, almost disappear in the gated luminescence image, whereas the longer lived PL from Si nanocrystals is clearly seen in the gated image (Joo et al. 2015).

17.5 SUMMARY

For the past decade, great advancements have been made in the development of near-IR luminescent colloidal Si nanocrystals in terms of colloidal stability, tunability of PL wavelength, quantum efficiency, etc. In this chapter, most of the recent important researches on the synthesis process, the luminescence properties, and the bioimaging applications are summarized. The excellent results suggest that near-IR luminescent Si nanocrystals are now ready for a wide range of application researches, especially in the biomedical fields.

ACKNOWLEDGMENT

This work was partly supported by 2015 JST Visegrad Group (V4)-Japan Joint Research Project on Advanced Materials and JSPS KAKENHI Grant Number 16H03828. H.S. acknowledges the support from the Grant-in-Aid for JSPS Fellows (No. 26-3120).

REFERENCES

Barbagiovanni, E.G., Lockwood, D.J., Simpson, P.J., and Goncharova, L.V., 2012. Quantum confinement in Si and Ge nanostructures. *Journal of Applied Physics*, 111 (3), 034307.

Barbagiovanni, E.G., Lockwood, D.J., Simpson, P.J., and Goncharova, L.V., 2014. Quantum confinement in Si and Ge nanostructures: Theory and experiment. *Applied Physics Reviews*, 1 (1), 011302.

Barry, S.D., Yang, Z., Kelly, J.A., Henderson, E.J., and Veinot, J.G.C., 2011. Synthesis of Si x Ge 1–x nanocrystals using hydrogen silsesquioxane and soluble germanium diiodide complexes. *Chemistry of Materials*, 23 (22), 5096–5103.

Canham, L.T., 1990. Silicon quantum wire array fabrication by electrochemical and chemical dissolution of wafers. *Applied Physics Letters*, 57 (10), 1046.

Chandra, S., Ghosh, B., Beaune, G., Nagarajan, U., Yasui, T., Nakamura, J., Tsuruoka, T., Baba, Y., Shirahata, N., and Winnik, F.M., 2016. Functional double-shelled silicon nanocrystals for two-photon fluorescence cell imaging: Spectral evolution and tuning. *Nanoscale*, 8 (16), 9009–9019.

Cheng, X., Gondosiswanto, R., Ciampi, S., Reece, P.J., and Gooding, J.J., 2012. One-pot synthesis of colloidal silicon quantum dots and surface functionalization via thiol-ene click chemistry. *Chemical Communications*, 48 (97), 11874.

Das, P., and Jana, N.R., 2014. Highly colloidally stable hyperbranched polyglycerol grafted red fluorescent silicon nanoparticle as bioimaging probe. *ACS Applied Materials & Interfaces*, 6 (6), 4301–4309.

Dasog, M., De los Reyes, G.B., Titova, L.V., Hegmann, F.A., and Veinot, J.G.C., 2014. Size vs surface: Tuning the photoluminescence of freestanding silicon nanocrystals across the visible spectrum via surface groups. *ACS Nano*, 8 (9), 9636–9648.

Dasog, M., Yang, Z., Regli, S., Atkins, T.M., Faramus, A., Singh, M.P., Muthuswamy, E., Kauzlarich, S.M., Tilley, R.D., and Veinot, J.G.C., 2013. Chemical insight into the origin of red and blue photoluminescence arising from freestanding silicon nanocrystals. *ACS Nano*, 7 (3), 2676–2685.

Erogbogbo, F., Liu, T., Ramadurai, N., Tuccarione, P., Lai, L., Swihart, M.T., and Prasad, P.N., 2011a. Creating ligand-free silicon germanium alloy nanocrystal inks. *ACS Nano*, 5 (10), 7950–7959.

Erogbogbo, F., Liu, X., May, J.L., Narain, A., Gladding, P., Swihart, M.T., and Prasad, P.N., 2013. Plasmonic gold and luminescent silicon nanoplatforms for multimode imaging of cancer cells. *Integrative Biology*, 5 (1), 144–150.

Erogbogbo, F., Yong, K., Roy, I., Hu, R., Law, W., Zhao, W., Ding, H., Wu, F., Kumar, R., Swihart, M.T., and Prasad, P.N., 2011b. In vivo targeted cancer imaging, sentinel lymph node mapping and multi-channel imaging with biocompatible silicon nanocrystals. *ACS Nano*, 5 (1), 413–423.

Erogbogbo, F., Yong, K., Roy, I., Xu, G., Prasad, P.N., and Swihart, M.T., 2008. Biocompatible luminescent silicon quantum dots for imaging of cancer cells. *ACS Nano*, 2 (5), 873–878.

Fujii, M., Sugimoto, H., and Imakita, K., 2016. All-inorganic colloidal silicon nanocrystals—Surface modification by boron and phosphorus co-doping. *Nanotechnology*, 27 (26), 262001.

Fujii, M., Toshikiyo, K., Takase, Y., Yamaguchi, Y., and Hayashi, S., 2003. Below bulk-band-gap photoluminescence at room temperature from heavily P- and B-doped Si nanocrystals. *Journal of Applied Physics*, 94 (3), 1990–1995.

Gu, L., Hall, D.J., Qin, Z., Anglin, E., Joo, J., Mooney, D.J., Howell, S.B., and Sailor, M.J., 2013. In vivo time-gated fluorescence imaging with biodegradable luminescent porous silicon nanoparticles. *Nature Communications*, 4, 2326.

Hannah, D.C., Yang, J., Podsiadlo, P., Chan, M.K.Y., Demortière, A., Gosztola, D.J., Prakapenka, V.B., Schatz, G.C., Kortshagen, U., and Schaller, R.D., 2012. On the origin of photoluminescence in silicon nanocrystals: Pressure-dependent structural and optical studies. *Nano Letters*, 12 (8), 4200–4205.

Henderson, E.J., Shuhendler, A.J., Prasad, P., Baumann, V., Maier-Flaig, F., Faulkner, D.O., Lemmer, U., Wu, X.Y., and Ozin, G.A., 2011. Colloidally stable silicon nanocrystals with near-infrared photoluminescence for biological fluorescence imaging. *Small*, 7 (17), 2507–2516.

Hessel, C.M., Henderson, E.J., Veinot, J.G.C., Uni, V., February, R.V., Re, V., Recei, M., and August, V., 2006. Hydrogen silsesquioxane: A molecular precursor for nanocrystalline Si–SiO$_2$ composites and freestanding hydride-surface-terminated silicon nanoparticles. *Chemistry of Materials*, 18 (1), 6139–6146.

Hessel, C.M., Reid, D., Panthani, M.G., Rasch, M.R., Goodfellow, B.W., Wei, J., Fujii, H., Akhavan, V., and Korgel, B.A., 2012. Synthesis of ligand-stabilized silicon nanocrystals with size-dependent photoluminescence spanning visible to near-infrared wavelengths. *Chemistry of Materials*, 24 (2), 393–401.

Hori, Y., Kano, S., Sugimoto, H., Imakita, K., and Fujii, M., 2016. Size-dependence of acceptor and donor levels of boron and phosphorus codoped colloidal silicon nanocrystals. *Nano Letters*, 16 (4), 2615–2620.

Iori, F., Degoli, E., Magri, R., Marri, I., Cantele, G., Ninno, D., Trani, F., Pulci, O., and Ossicini, S., 2007. Engineering silicon nanocrystals: Theoretical study of the effect of codoping with boron and phosphorus. *Physical Review B—Condensed Matter and Materials Physics*, 76 (8), 1–14.

Joo, J., Liu, X., Kotamraju, V.R., Ruoslahti, E., Nam, Y., and Sailor, M.J., 2015. Gated luminescence imaging of silicon nanoparticles. *ACS Nano*, 9 (6), 6233–6241.

Kanno, T., Fujii, M., Sugimoto, H., and Imakita, K., 2014. Colloidal hydrophilic silicon germanium alloy nanocrystals with a high boron and phosphorus concentration shell. *Journal of Materials Chemistry C*, 2 (28), 5644–5650.

Kwak, J., Bae, W.K., Lee, D., Park, I., Lim, J., Park, M., Cho, H., et al., 2012. Bright and efficient full-color colloidal quantum dot light-emitting diodes using an inverted device structure. *Nano Letters*, 12 (5), 2362–2366.

Ledoux, G., Gong, J., Huisken, F., Guillois, O., and Reynaud, C., 2002. Photoluminescence of size-separated silicon nanocrystals: Confirmation of quantum confinement. *Applied Physics Letters*, 80 (25), 4834.

Lee, D.C., Pietryga, J.M., Robel, I., Werder, D.J., Schaller, R.D., and Klimov, V.I., 2009. Colloidal synthesis of infrared-emitting germanium nanocrystals. *Journal of the American Chemical Society*, 131, 3436–3437.

Lin, S.W. and Chen, D.H., 2009. Synthesis of water-soluble blue photoluminescent silicon nanocrystals with oxide surface passivation. *Small*, 5 (1), 72–76.

Linnros, J., Lalic, N., Galeckas, A., and Grivickas, V., 1999. Analysis of the stretched exponential photoluminescence decay from nanometer-sized silicon crystals in SiO$_2$. *Journal of Applied Physics*, 86 (11), 6128.

Liu, S.-M., Yang, Y., Sato, S., and Kimura, K., 2006. Enhanced photoluminescence from Si nano-organosols by functionalization with alkenes and their size evolution. *Chemistry of Materials*, 18 (3), 637–642.

Mangolini, L. and Kortshagen, U., 2007. Plasma-assisted synthesis of silicon nanocrystal inks. *Advanced Materials*, 19 (18), 2513–2519.

Mastronardi, M.L., Hennrich, F., Henderson, E.J., Maier-Flaig, F., Blum, C., Reichenbach, J., Lemmer, U., Kübel, C., Wang, D., Kappes, M.M., and Ozin, G.A., 2011. Preparation of monodisperse silicon nanocrystals using density gradient ultracentrifugation. *Journal of the American Chemical Society*, 133 (31), 11928–11931.

Mastronardi, M.L., Maier-Flaig, F., Faulkner, D., Henderson, E.J., Kübel, C., Lemmer, U., and Ozin, G.A., 2012. Size-dependent absolute quantum yields for size-separated colloidally-stable silicon nanocrystals. *Nano Letters*, 12 (1), 337–342.

Michalet, X., Pinaud, F.F., Bentolila, L.A., Tsay, J.M., Doose, S., Li, J.J., Sundaresan, G., Wu, A.M., Gambhir, S.S., and Weiss, S., 2005. Quantum dots for live cells, in vivo imaging, and diagnostics. *Science*, 307, 538–544.

Miller, J.B., Van Sickle, A.R., Anthony, R.J., Kroll, D.M., Kortshagen, U.R., and Hobbie, E.K., 2012. Ensemble brightening and enhanced quantum yield in size-purified silicon nanocrystals. *ACS Nano*, 6 (8), 7389–7396.

Neiner, D., Chiu, H.W., and Kauzlarich, S.M., 2006. Low-temperature solution route to macroscopic amounts of hydrogen terminated silicon nanoparticles. *Journal of the American Chemical Society*, 128 (34), 11016–11017.

Ostrovska, L., Broz, A., Fucikova, A., Belinova, T., Sugimoto, H., Kanno, T., Fujii, M., Valenta, J., and Kalbacova, M.H., 2016. The impact of doped silicon quantum dots on human osteoblasts. *RSC Advances*, 6 (68), 63403–63413.

Ruddy, D.A., Johnson, J.C., Smith, E.R., and Neale, N.R., 2010. Size and bandgap control in the solution-phase synthesis of near-infrared-emitting germanium anocrystals. *ACS Nano*, 4 (12), 7459–7466.

Sangghaleh, F., Sychugov, I., Yang, Z., Veinot, J.G.C., and Linnros, J., 2015. Near-unity internal quantum efficiency of luminescent silicon nanocrystals with ligand passivation. *ACS Nano*, 9 (7), 7097–7104.

Sato, K., Fukata, N., Hirakuri, K., Murakami, M., Shimizu, T., and Yamauchi, Y., 2010. Flexible and transparent silicon nanoparticle/polymer composites with stable luminescence. *Chemistry—An Asian Journal*, 5 (1), 50–55.

Sato, S. and Swihart, M.T., 2006. Propionic-acid-terminated silicon nanoparticles: Synthesis and optical characterization. *Chemistry of Materials*, 18 (17), 4083–4088.

Shirahata, N., Hasegawa, T., Sakka, Y., and Tsuruoka, T., 2010. Size-tunable UV-luminescent silicon nanocrystals. *Small*, 6 (8), 915–921.

Sugimoto, H., Fujii, M., and Imakita, K., 2014. Synthesis of boron and phosphorus codoped all-inorganic colloidal silicon nanocrystals from hydrogen silsesquioxane. *Nanoscale*, 6 (21), 12354–12359.

Sugimoto, H., Fujii, M., Imakita, K., Hayashi, S., and Akamatsu, K., 2012. All-inorganic near-infrared luminescent colloidal silicon nanocrystals: High dispersibility in polar liquid by phosphorus and boron codoping. *The Journal of Physical Chemistry C*, 116 (33), 17969–17974.

Sugimoto, H., Fujii, M., Imakita, K., Hayashi, S., and Akamatsu, K., 2013. Codoping n- and p-type impurities in colloidal silicon nanocrystals: Controlling luminescence energy from below bulk band gap to visible range. *The Journal of Physical Chemistry C*, 117 (22), 11850–11857.

Takeoka, S., Fujii, M., and Hayashi, S., 2000. Size-dependent photoluminescence from surface-oxidized Si nanocrystals in a weak confinement regime. *Physical Review B—Condensed Matter and Materials Physics*, 62 (24), 16820–16825.

Talapin, D.V., 2005. PbSe nanocrystal solids for n- and p-channel thin film field-effect transistors. *Science*, 310 (5745), 86–89.

Talapin, D.V., Lee, J.-S., Kovalenko, M.V., and Shevchenko, E.V., 2010. Prospects of colloidal nanocrystals for electronic and optoelectronic applications. *Chemical Reviews*, 110 (1), 389–458.

Tang, J., Kemp, K.W., Hoogland, S., Jeong, K.S., Liu, H., Levina, L., Furukawa, M., et al., 2011. Colloidal-quantum-dot photovoltaics using atomic-ligand passivation. *Nature Materials*, 10 (10), 765–771.

Tilley, R.D., Warner, J.H., Yamamoto, K., Matsui, I., and Fujimori, H., 2005. Micro-emulsion synthesis of monodisperse surface stabilized silicon nanocrystals. *Chemical Communications*, (14), 1833.

Trwoga, P.F., Kenyon, A.J., and Pitt, C.W., 1998. Modeling the contribution of quantum confinement to luminescence from silicon nanoclusters. *Journal of Applied Physics*, 83 (7), 3789.

Valenta, J., Juhasz, R., and Linnros, J., 2002. Photoluminescence spectroscopy of single silicon quantum dots. *Applied Physics Letters*, 80 (6), 1070.

Veinot, J.G.C., 2006. Synthesis, surface functionalization, and properties of freestanding silicon nanocrystals. *Chemical Communications*, (40), 4160–4168.

Vinciguerra, V., Franzò, G., Priolo, F., Iacona, F., and Spinella, C., 2000. Quantum confinement and recombination dynamics in silicon nanocrystals embedded in Si/SiO_2 superlattices. *Journal of Applied Physics*, 87 (11), 8165.

Warner, J.H., Hoshino, A., Yamamoto, K., and Tilley, R.D., 2005. Water-soluble photoluminescent silicon quantum dots. *Angewandte Chemie—International Edition*, 44 (29), 4550–4554.

Wheeler, L.M., Levij, L.M., and Kortshagen, U.R., 2013. Tunable band gap emission and surface passivation of germanium nanocrystals synthesized in the gas phase. *The Journal of Physical Chemistry Letters*, 4 (20), 3392–3396.

Yao, J., Yang, M., and Duan, Y., 2014. Chemistry, biology, and medicine of fluorescent nanomaterials and related systems: New insights into biosensing, bioimaging, genomics, diagnostics, and therapy. *Chemical Reviews*, 114, 6130–6178.

Yu, Y., Rowland, C.E., Schaller, R.D., and Korgel, B.A., 2015. Synthesis and ligand exchange of thiol-capped silicon nanocrystals. *Langmuir*, 31 (24), 6886–6893.

18 Hydrogen-terminated silicon quantum dots

Rui-Qin Zhang and Yanoar Pribadi Sarwono

Contents

18.1 INTRODUCTION

After oxygen, silicon is the second most abundant element on Earth. It has many useful properties that facilitate its many applications such as electronics and optoelectronic devices. Recent trends of miniaturizing electronic devices have uncovered many novel properties. First, silicon can gradually transform into a direct semiconductor if the size of the silicon is less than 5 nm.[1] Second, the ratio of the surface to volume increases, and lastly the energy band gap is wider. The measurement of 5 nm is known as Bohr radius of the exciton or the electron–hole pair of silicon. Photogenerated electrons and holes show more obvious quantum confinement effect when their distance is reduced to, or below, the exciton Bohr radius of bulk silicon. Silicon nanostructures with such a quantum confinement effect can be applied in the optoelectronics devices due to their efficient photoluminescence properties.

Silicon quantum dots have been intensively studied dating back to the 1980s. They are silicon-based nanocrystals and have interesting quantum-mechanical properties which lie between those of bulk silicon and those of discrete molecules. Research on silicon quantum dots is very active; it promises both more sophisticated applications, and interesting fundamental physical properties. Since the silicon quantum dot is very sensitive to its size, shape, and surface composition, it is an excellent model to study its property–structure relationships.

Stability of silicon nanostructures is important because a stable nanosized silicon cluster maintains its properties and functionality. Unsaturated silicon clusters cause the structural instability and the lack of crystalline tetrahedral, symmetry. Hydrogen, the simplest element in the periodic table, can passivate the dangling bonds of unsaturated silicon nanostructures. This process is known as hydrogenation. Due to the importance of the hydrogenation, this chapter will begin with a detailed discussion of it, its mechanism, and its effect on the structure and stability of silicon nanoclusters. The last two sections will address the ground state and excited-state properties of silicon nanoparticles.

18.2 HYDROGENATION AND ITS MECHANISM

18.2.1 DANGLING BONDS, SURFACE RECONSTRUCTION, AND SURFACE PASSIVATION

Low-dimensional silicon nanostructures have attracted the scientific community because they possess potential applications in the field of nanoscience and nanotechnology. Moreover, their properties can be tuned by varying their size and morphology with the help of microprocessing fabrication.[2–8] One-dimensional silicon nanowires and zero-dimensional quantum dots are potentially applied to nanoscale electronics and optical devices such as field-effect transistors,[9] light-emitting diodes,[10] molecular electronics,[4] and nanoscale sensors[5,11,12] because they are compatible to the conventional silicon technology.

In nanoscale materials such as in silicon quantum dots or silicon nanowires, the surface-to-volume ratio increases and the effect of quantum confinement becomes more pronounced. Because of the increased specific surface area and the reduced particle size, more silicon atoms of reduced coordination numbers appear on the surface of quantum dots. The reduced coordination of surface silicon atoms results in unsaturated dangling bonds which make the surface of the nanoparticles very reactive and unstable. The surface effect affects the nanoparticles' structure, the electronic structure, the charge transport, and the formation of surface defects.

Let us consider a spherical unsaturated Si_5 cluster with a tetrahedral T_d symmetry of the silicon bulk which is not the global energy minimum as shown in Table 18.1.[13] In other words, the $T_d Si_5$ cluster is not the most stable structure because there exist unpaired electrons in a nonbonding orbital namely the dangling bonds on the surface, which are almost isolated and form a weak π bond with high symmetry around the surface. This weak and highly symmetric π bond keeps the cluster in a substable structure with T_d symmetry. Based on our first-principles calculations, several dangling bond states exist inside the gap. Those states are localized and originally come from the dangling bonds of the surface atoms. The computed density of states (DOSs) confirmed this. The most stable structure of the Si_5 cluster has a D_{3h} symmetry. But in this structure, no silicon atoms have fourfold coordination therefore the atoms cannot be marked as bulk or surface. Besides that, the structure is no longer a crystalline domain and shows no gap state, and each atom displays similar local DOS.

Table 18.1 **Summary of some calculated energy gaps**

CLUSTER	SYMMETRY	DF[a]	PM3	HARTREE—FOCK		MP2 6-31G*	CIS 6-31G*
				6-31G*	6-311++G**		
Si_5[b]	T_d		7.14	12.64			
Si_5	L_{3h}		4.82	7.74			
Si_5H_{12}	C_{2v}	5.5 (5.8)	6.83	12.25	11.02	12.35	10.83
Si_5H_{12}	T_d	6.3	7.28	13.44	11.57	13.47	
$Si_{17}H_{36}$	T_d	5.2	6.09	11.32			
$Si_{29}H_{36}$	T_d	4.9	5.59	10.58			
$Si_{35}H_{36}$	T_d	4.8	5.41				
$Si_{41}H_{60}$	T_d		5.29				

Source: Reprinted with permission from Zhang, R.Q., et al., 1996, *Phys. Rev. B*, 53, 7847. Copyright 1996 by the American Physics Society.

[a] Data of density-functional approach with self-energy corrections were estimated from Ref. 20, and the datum in the parentheses is experimental; also cited in Ref. 20.

[b] A substable structure as described in the text.

A stable spherical cluster Si_5H_{12} with T_d symmetry will be obtained if the T_d Si_5 cluster is saturated with 12 hydrogen atoms. After hydrogenation, all the localized dangling bond states have been moved to the valence and the conduction bands. Thus the Si_5H_{12} cluster can be considered as a crystallite. The energy gap of Si_5H_{12} with T_d symmetry is larger than that of T_d Si_5, since hydrogenation on the surface gets rid of the dangling bonds' influence on the bulk silicon atoms. For a larger unsaturated silicon cluster, the dangling bond affects several atomic layers underneath and leads to deviation of the bonds from the bulk ones. The deviated bonds develop the tail states which reduce the energy gap.

Dangling bonds of silicon atoms on bulk-terminated surfaces or impurities on the surface cause a clean silicon surface to be unstable and chemically reactive. Surface reconstruction minimizes such effects. The reconstruction takes place in the bulk-terminated silicon surfaces and makes the clean silicon surface more stable. However, surface reconstruction gives the silicon nanomaterial different properties than the bulk-like properties.

Most reconstructed silicon surfaces or amorphous silicon surfaces contain threefold coordinated silicon atoms. Due to the surface reconstruction, the silicon surface atoms do not have the original tetrahedral symmetry in the bulk. However, the threefold coordinated silicon atoms still keep the sp^3 hybridization on the reconstructed surface. In such a site, the silicon structure has a defect because the site includes an unsaturated dangling bond. The electronic states of the dangling bond are deep in the band gap and around the mid gap. Consequently, the surface has low conductivity. Due to the strong bond of silicon–hydrogen, if hydrogen is adsorbed on the silicon crystal surface such as Si(100), Si(111), or amorphous silicon surfaces, it will easily passivate the undercoordinated silicon atoms by saturating the dangling bonds of silicon atoms. Silicon surface passivation removes the deep states in the band gap, and returns the system back to bulk-like properties. Besides that, the electronic states of the induced hydrogen are passive as they lie in very deep levels and far away from the band gap. This method is used widely to eliminate the deep gap states in silicon-based semiconductor materials. Therefore, silicon surface passivation is needed since it efficiently prevents the surface reconstruction and preserves the bulk-like structure.

18.2.2 PROPERTIES OF HYDROGENATION OF SEMICONDUCTOR SURFACES

Hydrogenation of semiconductor surfaces changes the physical and chemical properties of semiconductors. For example, a clean diamond surface is an insulator while a hydrogen-terminated diamonds surface is a p-type semiconductor.[14] The reactivity of semiconductors becomes low after hydrogenation which removes the surface dangling bonds. There are many applications of hydrogen-terminated silicon surfaces in the field of nanolithography and molecular electronic devices. Thus, an understanding of the properties of the hydrogen-terminated semiconductor is very useful and obtaining a controllable hydrogen-terminated semiconductor is an important issue. In the following section, the structure and stability of hydrogen-terminated silicon nanostructures are discussed.

18.2.3 HYDROGEN-TERMINATED SILICON NANOSTRUCTURES

Hydrogen termination and oxygen termination are the two main ways applied for silicon surface passivation. They effectively terminate silicon surface dangling bonds. In silicon microprocessing fabrication, one uses gas-phase exposure (e.g., plasma treatment and oven anneals) and wet-chemical etching for the silicon surface passivation. A perfectly saturated surface is important in order to get perfect electrical and optical properties of silicon nanowires and silicon quantum dots.

Here we discuss our previous theoretical studies[15] of fully and partially hydrogen-terminated silicon nanostructures. Our models were fully hydrogen-terminated silicon and partially hydrogen-terminated silicon for simulating hydrogenated silicon nanocrystals and partially hydrogenated silicon nanoclusters, respectively. We made use of the tight-binding method for molecular dynamics, geometry optimization, and lattice parameters. As for structural formation energy, we made use of the density-functional theory. It turned out that our models could characterize the structural properties of silicon nanostructures and reveal the surface effect on material stability. Our model of structural formation energies also could predict structural stability and is a good reference for more complex silicon nanostructures.

18.2.3.1 Fully hydrogen-terminated silicon nanocrystals

The fully hydrogenated silicon surfaces are commonly believed to be inert or passive to other materials. There are at least three reasons. First, all silicon atoms on the crystal surface are fourfold coordinated. Second, all of the sp³ orbitals are saturated. Third, the bond between silicon and hydrogen is strong. Nonetheless there is some reaction in a fully hydrogen-terminated silicon system. Lin et al.[16] reported that the reaction of disilane (Si_2H_6) has been found on the monohydride Si (100) surface. Thus, we need to understand further the structural stability and properties of such fully hydrogen-terminated silicon nanomaterials.

Motivated by the finding and the success of previous production in experiments, we considered the silicon nanocrystals with unreconstructed monohydride and trihydride Si(111)-like facets, symmetric dihydride Si(100)-like facets, and monohydride Si(110)-like facets. With the referencing of several experimental works,[8,17–19] we used the tight-binding potential model for the silicon–hydrogen system to geometrically optimize all the hydrogenated silicon nanocrystals. The distance of silicon–hydrogen is 1.5 Å in the unreconstructed surfaces of hydrogenated silicon crystals. Therefore, we may assume that they interact weakly and neglect the interactions between themselves.

As a theoretical model, we used ball-and-stick diagrams shown in Figure 18.1.[15] Figure 18.1a is a spherical hydrogen-terminated silicon particle with a diameter of $D = 3.0$ nm. The sphere was centered on one of the silicon atoms so there is little difference from those centered on tetrahedral interstitials.[20–22] The other is a cubic slab-like nanocrystal of 1.5 nm edge (Figure 18.1b). To support our aim which was to study various unreconstructed facets of nanoparticles by fully hydrogen termination, we covered the particle with various facets. For example, the silicon nanoparticle of diameter $D = 3.0$ nm was covered by some unreconstructed facets such as symmetric dihidryde structures on Si(100)-like facets, monohydride structures on Si(111)-like facets, and monohydride structures on Si(110)-like facets. The other smaller particles were also covered by the various facets above but with smaller area. Therefore we had optimized the hydrogen-terminated spherical silicon nanoparticles of diameters 3.0, 2.5, 2.4, 2.2, 2.1, 2.0, 1.8, 1.7, 1.4, and 1.2 nm. We calculated the effect of particle diameters of the models above on the electronic structures. The band gap is inversely proportional to the diameter. This result agrees with the previous ones of tight-binding and density-functional calculations.[20,21,23,24] The relationship between the particle gap and its diameter is $E_g = 0.45 + 0.33/D$ where E_g is the band gap in eV, and D is the diameter in nm units. Based on our calculation, the silicon particles of diameter 2.0 nm and 3.0 nm have the band gap of 2.12 eV and 1.56 eV, respectively, which are supported by previous tight-binding results.[23,24]

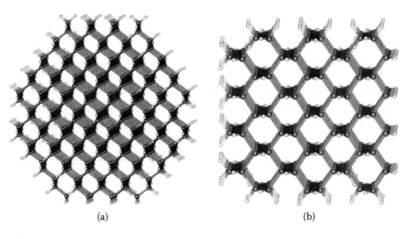

(a) (b)

Figure 18.1 Ball-and-stick diagrams of two hydrogenated silicon nanocrystals fully relaxed using the tight-binding method. The root mean square force is 0.005 eV/Å, and the maximum force is 0.01 eV/Å. The black dots represent Si atoms, and the gray dots represent H atoms. (a) $Si_{705}H_{300}$, $D = 3.0$ nm particle; and (b) $Si_{251}H_{172}$, 1.5×1.5×1.5 nm³, denoted as slab 1. (Reprinted with permission from Yu, D.K., et al., 2002, *J. Appl. Phys.*, 92, 7453. Copyright 2002, American Institute of Physics.)

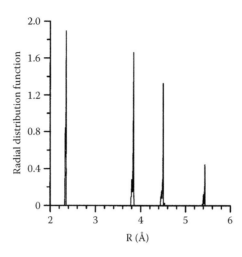

Figure 18.2 Radial distribution function of hydrogenated silicon nanocrystal $Si_{705}H_{300}$ [Figure 2]. (Reprinted with permission from Yu, D.K., et al., 2002, *J. Appl. Phys.*, 92, 7453. Copyright 2002, American Institute of Physics.)

After examining the electronic structure of silicon nanoparticles, let us focus on the structural properties of silicon nanoparticles. For that purpose, we need to introduce a radial distribution function which is known as RDF and is defined as $g(r) = (1/N) \sum_{i=1}^{N} \sum_{j \neq i}^{N} \delta(r - r_{ij})$. The RDF tells us the probability of finding two atoms at a distance r. Figure 18.2 shows the RDF of a 3 nm diameter of an $Si_{705}H_{300}$ hydrogenated silicon nanocrystal.[15] Figure 18.2 displays narrow peaks at 2.35 Å, 3.84 Å, 4.50 Å, and 5.43 Å. Because the positions of those peaks are the same as the positions of the peaks of the bulk diamond structure, the result implied that the hydrogen-terminated silicon nanoparticles retain the bulk diamond structure. The relative height of the peaks is not necessarily the same as that of the bulk structure since the number of the clusters is finite. In Figure 18.2, we also found subpeaks which are located at the left side of the main peaks, indicating that some of the silicon atoms are slightly closer to each other than those in bulk silicon. The location of these displaced atoms is at the surface layers of the clusters. It turns out that the surface structural relaxation causes the distance between silicon atoms to become shorter. For a 1.5 nm diameter core inside the particle of 2.3521 Å lattice constant, the lattice expansion is $\frac{\Delta a}{a} = 8.9 \times 10^{-4}$.

For the particles of diameter 2.5 nm, 2.2 nm, and 2.0 nm, the lattice expansion of the core is 9.2×10^{-4} 9.2×10^{-4} 9.0×10^{-4}, respectively. This result is consistent with the result of X-ray diffraction measurements. Bellet et al.[25] reported that the hydrogenated nanocrystals show a small lattice expansion of the order of 10^{-4} to 10^{-3}. Based on our calculations and the experimental results, we could not conclude that the lattice expansion depends on the particle diameter because their relationships are not noticeable. Our calculations simply confirmed the experimental results that the hydrogenated silicon nanocrystals exhibit a very small lattice expansion. Besides, the lattice contraction exists only on the surface layers for all calculated hydrogenated silicon nanocrystals.

We summarized the surface relaxation of several nanocrystals in Table 18.2.[15] For the monohydride Si(111)-like facets of the 3.0 nm diameter nanocrystals in Figure 18.1a, the distance from the first layer to the second layer D_{12} is 2.338 Å. The value is contracted by 0.012 Å from the bulk value of 2.35 Å. Meanwhile D_{23}, D_{34}, D_{45} are 2.346 Å, 2.350 Å, and 2.351 Å, respectively, which can be considered as having no significant contraction. What could we deduce from the results? The surface relaxation of a silicon nanocrystal is mainly due to the local bonding environment. Consequently, the relaxation length of the bond at the edge of the facet is longer than the relaxation length of the bond at the center of the facet. Also, the length of the bond beneath a certain facet exhibits either negligible or no contraction. Structural relaxation takes place in the two outermost layers only and the effect is small from the third layers to the next layers beneath it.

Table 18.2 **Surface relaxation of the hydrogenated silicon nanocrystals. $D_{i,\,i+1}$ indicates the average atomic distance between the ith and (i + 1)th Si layer. The unit is in Å. The data in the parentheses show the dispersion of the bond lengths**

Si NANOCRYSTALS SATURATED BY H		Si(111)-(1X1) MONOHYDRIDE	Si(111)-(1X1) TRIHYDRIDE	Si(100)-(1X1) DIHYDRIDE	Si(110)-(1X1) MONOHYDRIDE
(a) D = 3.0 nm	D_{12}	2.338(0.001)		2.331(0.001)	2.330 2.338(0.001)
	D_{23}	2.346(0.001)		2.342(0.001)	2.345(0.001) 2.347(0.001)
	D_{34}	2.350(0.001)		2.349(0.001)	2.350
	D_{45}	2.351(0.001)		2.351	2.352
(b) D = 2.5 nm	D_{12}	2.337(0.001)	2.337(0.001)		
	D_{23}	2.346(0.001)	2.343(0.001)		
	D_{34}	2.352(0.001)	2.350(0.001)		
	D_{45}	2.352	2.352		
(c) slab 1	D_{12}			2.333(0.001)	2.330(0.001) 2.338(0.001)
	D_{23}			2.343(0.001)	2.346(0.001) 2.348(0.001)
	D_{34}			2.350(0.001)	2.351(0.001) 2.351(0.001)
	D_{45}			2.350	2.351
(d) slab 2	D_{12}	2.338(0.001)	2.337(0.001)		
	D_{23}	2.344(0.001)	2.343(0.001)		
	D_{34}	2.351(0.001)	2.350(0.001)		
	D_{45}	2.351	2.351		

Source: Reprinted with permission from Yu, D.K., et al., 2002, *J. Appl. Phys.*, 92(12). Copyright 2002, American Institute of Physics.

The trend of the surface relaxation in smaller particles is the same. It shows no evident dependency on the particle diameter. So, the structural relaxation in the fully hydrogen-terminated silicon takes place within several of the outermost layers. Based on our calculation, it occurs from the outermost to the third layer.

We also studied the structural relaxation in slab-shape nanocrystals. As stated in Table 18.2, the structural relaxation of the slabs is similar to that of spherical nanoparticles. Thus the overall size and shape of the hydrogenated nanocrystals have little effect on the structural relaxation.

As a summary, fully hydrogenated silicon nanocrystals have a small surface relaxation. The contraction is from 0.001 Å until 0.002 Å which happens in the first layer to the third layer. Beneath the third layer, we found either negligible contraction or no contraction at all. The local bonding environment mainly causes the relaxation. The size and the shape of nanocrystals do not affect the structural relaxation. Inside the hydrogenated silicon nanocrystals, we found small lattice expansion which agrees with the X-ray diffraction experiments. Both the lattice expansions and the surface relaxation do not clearly depend on the size of the nanoscrystals. However, in the next two sections, we will see that the size of nanocrystals affects their electronic properties.

18.2.3.2 Partially hydrogen-terminated silicon nanocrystals

Having studied the structural properties of fully hydrogen-terminated silicon nanocrystals, we focused on the study of the structural properties of the partially hydrogen-terminated silicon nanocrystals. The presence of the surface dangling bonds on silicon nanoclusters may affect the structural properties differently than that of the fully hydrogen-terminated silicon nanocrystals.

Where can we find partially hydrogen-terminated silicon nanoclusters? Rechtsteiner et al.[26] reported that partially hydrogen-terminated silicon nanoclusters may exist in chemical vapor deposition (CVD) films[26] because the wide range of [H]/[Si] ratios can be observed. Also, partially hydrogen-terminated silicon nanocrystals can be produced from the fully hydrogen-terminated silicon nanocrystals, when the surface hydrogen is desorbed under high temperatures.

We[15] carried out the tight-binding molecular dynamics simulation that was combined with a simulated annealing technique to obtain the energetically favorable structures of Si_mH_x clusters with $m \leq 151$. Figure 18.3a is a diagram of a fully hydrogenated $Si_{100}H_{86}$ cluster.[15] The cluster keeps the ideal bulk structure after the equilibration and cooling. Thus, the fully hydrogen-terminated silicon nanocrystals are stable structures. The $Si_{100}H_{86}$ shows the same behavior as that of fully hydrogenated silicon nanocrystals, for example, its lattice expansion is about 10^{-3} and the contraction takes place in the outermost three layers. Figure 18.3b and c shows the structures of $Si_{100}H_{60}$ and $Si_{100}H_{40}$, respectively. If we remove 50% of the hydrogen atoms, the dangling bonds will be on the surface of nanocrystals. The remaining hydrogen atoms saturate the dangling bond and prevent any drastic change of the structures. As a result, there are only lattice distortions in the clusters and the tetrahedral structures are still maintained. In Figure 18.3d, if we remove 70–80% of the hydrogen atoms of $Si_{100}H_{20}$, more dangling bonds will be on the surface of the nanocrystal. More distortions are found in the clusters. The structure loses the tetrahedral structure and becomes more compact, showing small cage-like structures which look like the structures of intermediate size pure silicon clusters.

Miyazaki[27] introduced the structural formation energy defined mathematically as $E_{form}(x) = E_{tot}(x) - E_{tot}(x=0) - xE_{tot}(H_2)$. $E_{tot}(x)$ is the Si_mH_x clusters total energy, and $E_{tot}(H_2)$ is the total energy per hydrogen atom of an H_2 molecule. We made use of the structural formation energy to explore the relative stability of the Si_mH_x clusters with a different number for x. Here the formation energy means a relative energy over the pure Si_m clusters, and estimates the total silicon–hydrogen bond energies. In Figure 18.4, if the number of hydrogen atoms of $Si_{60}H_x$, $Si_{100}H_x$, and $Si_{151}H_x$ clusters are increased,

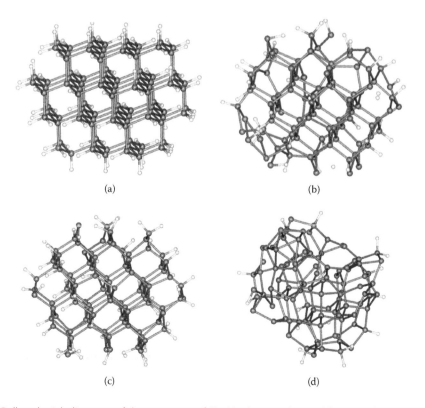

(a)   (b)

(c)   (d)

Figure 18.3 Ball-and-stick diagrams of the structures of $Si_{100}H_x$ clusters obtained from simulated annealings. (a) Fully H saturated $Si_{100}H_{86}$; (b) $Si_{100}H_{60}$; (c) $Si_{100}H_{40}$; and (d) $Si_{100}H_{20}$. (Reprinted with permission from Yu, D.K., et al., 2002, *J. Appl. Phys.*, 92, 7453. Copyright 2002, American Institute of Physics.)

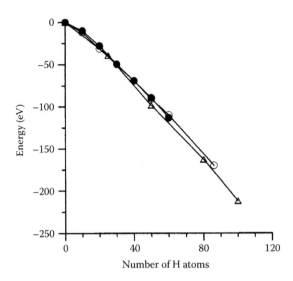

Figure 18.4 Formation energy of the Si_mH_x clusters. Closed circle: m = 60; open circle: m = 100; triangle: m = 151. (Reprinted with permission from Yu, D.K., et al., 2002, *J. Appl. Phys.*, 92, 7453. Copyright 2002, American Institute of Physics.)

the formation energies will decrease monotonically.[15] The end of the lines in Figure 18.4 has the lowest energy, for example, the points of $Si_{60}H_{60}$, $Si_{100}H_{80}$, and $Si_{150}H_{100}$ clusters. Such clusters are fully hydrogenated silicon nanocrystals. Thus, the fully hydrogenated silicon nanocrystals are the most stable structures. Miyazaki[27] found the same fact, meaning that in Si_6H_x clusters the "bulk" structure is the most stable. Rechtsteiner et al.[26] showed that the entropy is very important to determine small silicon–hydrogen cluster structures, for example, $Si_{23}H_x$ at high temperature. The effect of entropy in larger nanosized Si_mH_x clusters was not clear at that time. However, it was already clear that at low temperatures, the total energy is more dominant than the entropy effect.

18.2.4 STABILITY OF HYDROGEN-TERMINATED SILICON NANOSTRUCTURES

Chabal,[2] we,[13] and Ma et al.[8] found that hydrogen saturation efficiently prevents surface reconstruction and preserves a bulk-like structure. Zhao et al.[3] and Ma et al.[8] showed that the surface saturation of silicon nanowires causes the energy band gap to depend on the size of the particle. Since the surface structure affects the electrical and the optical properties of the material, it is crucial to accurately predict the structures of silicon nanomaterials. We searched for a well-defined quantity that can predict the material stability precisely. Kagimura et al.[28] and Aradi et al.[29] tried to use the diameter to explain the stability of silicon nanoclusters. Others[30,31] described the stability of silicon nanoclusters from a surface area point of view. We,[31] and Justo et al.[32] worked on the surface-to-volume ratio and perimeter respectively to reveal the stability of silicon nanoclusters. Unfortunately, those quantities are not appropriate to explain the material stability.

We found that the ratio of hydrogen to silicon (H/Si ratio) depends linearly on the cohesive energy. This quantity serves to be the predictor of structural stability. It is the reason behind a series of magic numbers in the silicon quantum dots. We[33] proved that the H/Si ratio characterizes the structural properties of silicon nanoclusters excellently and even explains the role of the surface effect in material stability. After analyzing the relationship between the surface energy and the nature of bonding, we also explained the physics behind it.

Let us find the general trend of the stability of silicon nanostructures to start with. The structure of eight representative silicon nanowires enclosed by low-index facets is shown in Figure 18.5. Then their surfaces are saturated with hydrogen atoms. By doing so, any dangling bonds will be removed and their perfect tetrahedral sp³ hybridization is maintained. So the silicon–silicon bonds and the silicon–hydrogen

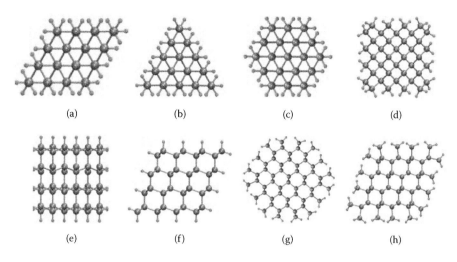

Figure 18.5 Top views of the relaxed structures of the SiNWs. (a) a <111> SiNW with a rhombic cross section, (b) a <111> SiNW with a triangular cross section, (c) a <111> SiNW with a hexagonal cross section, (d) a <100> SiNW with a rectangular cross section, (e) a <112> SiNW with a rectangular cross section, (f) a <110> SiNW with a rhombic cross section, (g) a <110> SiNW with a hexagonal cross section and its (111) facets passivated with SiH₃ radicals, and (h) a <110> SiNW with a rhombic cross section and its (111) facets passivated with SiH₃ radicals. (Reprinted with permission from Xu, H., et al., 2009, *Appl. Phys. Lett.*, 95, 253106. Copyright 2009, American Institute of Physics.)

bonds are the same as those in bulk silicon and silane. After counting the number of bonds we found an equation for the cohesive energy:

$$E_{coh} = 2E_{Si-Si} - (0.5E_{Si-Si} - E_{Si-H})\alpha \qquad (18.1)$$

where E_{Si-Si} and E_{Si-H} are the silicon–silicon and silicon–hydrogen bond energies, respectively, α is the H/Si ratio, a ratio of n_H to n_{Si} where n_H and n_{Si} are the numbers of hydrogen atoms and silicon atoms, respectively. From Equation 18.1, it is straightforward to see that the cohesive energies depend linearly on the H/Si ratio and that the slope is the difference between $0.5E_{Si-Si}$ and E_{Si-H}.

We applied the model above to find the energetic stability of various silicon nanowires in Figure 18.5 and of silicon quantum dots by using density-functional theory[34–35] within local density approximation.[36] Figure 18.6 shows a linear relationship between the cohesive energies and the H/Si ratio.[33] The straight line corresponds to the equation of E_{coh}=5.230 + 2.399α. The slope of 2.399 agrees very well with 2.412 of Equation 18.1 if we evaluate the E_{Si-Si} and the E_{Si-H} with the bulk silicon and silane. When the H/Si ratio is close to zero, the cohesive energy approaches the value of the bulk material.

The total energy expression of the hydrogen-terminated silicon nanostructures is given by

$$E_{tot} = n_{Si}E_{Si} + n_H E_H - [2E_{Si-Si} - (0.5E_{Si-Si} - E_{Si-H})\alpha]n_{Si}. \qquad (18.2)$$

In low-dimensional systems, the surface energy is very difficult to calculate because it depends on both the surface index and the surface atomic configuration. In Equation 18.2 of our system, the surface energy of silicon nanostructures is composed of the energy of the silicon–silicon bond and the energy of the silicon–hydrogen bond. Unlike the low-dimensional system, we can immediately see and calculate the system energy by using our model. Based on Equation 18.2 showing that the stability of different structures depends on the silicon–hydrogen bond density, the (111) structure has more surface stability than the (110) structure, which in turn has more surface stability than the (100) structure. Our work confirmed this result[31] and explained why the disilane reaction have been found on the monohydride Si(100) surface in Subsection 18.2.3.1.

Chan et al.[37] and Lu et al.[38] studied the magic numbers of <110> and <112> silicon nanowires using a genetic algorithm and density-functional theory calculations. However, our model can describe the total

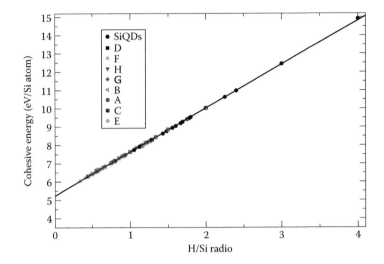

Figure 18.6 Cohesive energies of <100>, <110>, <111>, and <112> SiNWs, and SiQDs as a function of the H/Si ratio. (Reprinted with permission from Xu, H., et al., 2009, *Appl. Phys. Lett.*, 95, 253106. Copyright 2009, American Institute of Physics.)

energy of a system more efficiently than other methods such as first-principles or empirical calculations. Moreover, our results agree with those obtained from tight-binding models[31] and first-principles calculations[37,38] and thus is reliable. Our model explains the origin of the surface energy and predicts the structural stability of a nanostructure. Furthermore, it can be considered as a practical, straightforward, accurate, and efficient expression to calculate the total energy of the system.

Williamson et al.[39] and Puzder et al.[40] discussed that a stable structure has to be identified by its symmetry and the evaluation depends on the experience of the researchers. To calculate the total energy accurately, high-computational resources are needed. Equation 18.2 of our model solves the problem easily. For silicon quantum dots with a given number of atoms, the stable structures are the ones involving fewer hydrogen atoms. Figure 18.7 shows stable silicon quantum dots with various numbers of silicon atoms.[33] We found $Si_{10}H_{16}$, $Si_{14}H_{20}$, $Si_{18}H_{24}$, $Si_{22}H_{28}$, $Si_{26}H_{30}$, $Si_{30}H_{34}$, $Si_{35}H_{36}$, $Si_{39}H_{40}$, $Si_{44}H_{42}$, and $Si_{48}H_{46}$ to

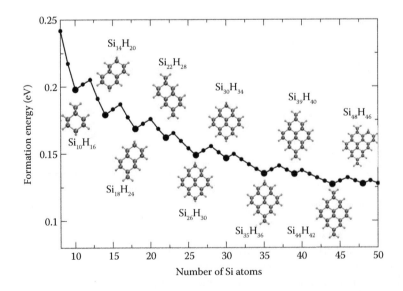

Figure 18.7 Formation energies as a function of the number (n) of Si atoms for SiQDs. (Reprinted with permission from Xu, H., et al., 2009, *Appl. Phys. Lett.*, 95, 253106. Copyright 2009, American Institute of Physics.)

be the local minima of the formation energies. All of them are stable structures and possess a perfect sp^3 hybridization. Another important feature is that our model can be applied to other group IV nanostructures. We calculated and found that the cohesive energy shows a straight-line relationship to the H/M (M=C, Si, Ge, and Sn) ratios. Therefore, other group IV nanostructures are supposed to have some magic numbers. Landt et al.[41] confirmed that $C_{10}H_{16}$, $C_{14}H_{20}$, $C_{18}H_{24}$, $C_{22}H_{28}$, and $C_{26}H_{30}$ are stable C nanoclusters. Our theoretical results are useful to understand other stable quantum dots.

18.3 GROUND-STATE PROPERTIES: SIZE-DEPENDENT GAP

The light absorption is a natural property of the ground, equilibrium state of the system. The light absorption energy is the energy of optically allowed electron transition crossing the energy gap between the bottom of the conduction band and the top of the valence band.[42] What is the behavior of the absorption energy of the hydrogenated silicon nanoparticles? Many groups of researchers have put their efforts into understanding this. For example, Dow and Ren used the tight-binding method[23] to study the DOS and the energy gap of the hydrogenated silicon nanoparticles. When the size of the hydrogenated silicon nanoparticles is increased to 4.9 nm diameter, the particle shows the same DOS and band gap as those of the silicon nanoparticles in the bulk phase. Steigmeier and Delley[20,21] studied silicon nanocrystals using the density-functional approach. They found two interesting results: First, the energy difference or the band gap depends linearly on the inverse of the particle diameters. Second, the luminescence intensity decreases strongly, if the size of the particles increases. Delerue et al.[43] studied theoretically the luminescence aspect of silicon nanoparticles and nanowires. They found that the energy gap is proportional to $d^{-1.39}$, where d is the diameter, and the absorption energies of the nanosized structures agree with the observed photon energies in the luminescence experiment. Their work verified that the quantum confinement model in the photoluminescence of silicon nanostructures is valid. Wang and Zunger[22] used a pseudopotential method to study the energy gap and the radiative recombination rates and tried to relate them to the size, the shape, and the orientation of silicon nanocrystals. It turned out that the band gap does not depend on both the surface orientation and the shape of the particles.

Zhang et al.[13] studied small silicon particles focusing on their surface saturation. We related it to the electronic structure and discussed the photoluminescence behavior. We assumed that the bulk phase has an ideal structure and the surface dangling bonds are terminated with hydrogen atoms. In our work[15] diameters of 3.0, 2.5, 2.4, 2.2, 2.1, 2.0, 1.8, 1.7, 1.4, and 1.2 nm hydrogen-terminated spherical silicon nanoparticles were optimized by an empirical tight-binding approach. In the previous section, Figure 18.5a is a hydrogen-terminated silicon nanoparticle of diameter $D = 3.0$ nm centered on one of the silicon atoms. We found that the particle diameters affect the electronic structure and the band gap is inversely proportional to the diameter. The results agree with those of previous studies which used either tight-binding or density-functional calculations. The straight-line equation of the energy gap versus diameter is $E_g = 0.45 + 3.33/D$, where E_g and D are the energy gap in eV and diameter in nm, respectively. In the case of silicon particles with 2.0 and 3.0 nm diameter, the energy gap is 2.12 eV and 1.56 eV, respectively, in accordance with the previous tight-binding calculations.[23,24] We found that the energy gap decreases when the cluster size is increased, in agreement with the quantum confinement effect. The energy gap approaches the bulk value for a large cluster size of 200 silicon atoms. The absorption energy (E_{abs}, in eV) versus the diameter of silicon nanoparticles (d_0, in nanometers) fits the formula $E_{abs} = 7.156 \exp(-d_0/1.032) + 1.773$ (eV).[44] To obtain the equation, we used a self-consistent-charge density-functional tight-binding (SCC-DFTB) method.[45,46]

18.4 EXCITED-STATE PROPERTIES

18.4.1 EMISSION ENERGY

The emission is the main characteristic of the excited, non-equilibrium state of the silicon dots. It turns out that the emission is more complex than the absorption. The relaxation dynamics of the dot at its excited state plays an important role for the emission. The aim of studying the emission energy is to reveal the Stoke's shift. By doing so, we could point out techniques for maximizing the quantum yield and increasing the efficiency of photoluminescence.

Photoexcitation destabilizes the dot. It causes the system to to be excited to a state of a higher energy than the stable ground state. Simultaneously, an exciton (a virtual particle or a quasiparticle) or an electron–hole pair is created. Another way to understand this is that an electron is promoted from HOMO to LUMO. HOMO stands for the highest occupied molecular orbital and is a bonding state lower than LUMO. LUMO stands for the lowest unoccupied molecular orbital and is either a nonbonding state of higher energy, or an antibonding state of higher energy. The destabilization of the dot after photoexcitation appears in either shape deformation or in the form of the weakening or the stretching of some bonds. We know from Section 18.2 that the energy gap will be reduced if there is any shape deformation from the tetrahedral structure. Because symmetry brings degeneracy in electronic states, the broken tetrahedral symmetry due to the shape deformation at the excited state will split the degeneracy at the LUMO ground state, so there is a state that decouples out and appears between the HOMO and LUMO ground state. This means the gap at the excited state becomes narrower.

We may use bond weakening to explain the reduction in the energy gap. When the length of the bond between two atoms increases, the bond is weakened, and there is a shift in the electronic cloud in the bonding region. The magnitude of the electronic cloud shift is proportional to the two atoms in the bond. This induces some degree of charge localization on these two atoms. Due to the charge localization, there is a state that appears in the energy gap; therefore, the gap at the excited state is reduced. When the charge localization at the excited state or the magnitude of the structural deformation is greater, the LUMO at the excited state will be displaced more downward with respect to the LUMO at the ground state. Therefore, the emission energy decreases, and the Stoke's shift increases. From our calculation, the hydrogen electronic structure in the hydrogen–silicon quantum dots is not affected after photoabsorption or photoexcitation. In the excited state, the electronic cloud around the silicon atoms of the cluster is reorganized. This electronic structure of silicon atoms in tetrahedral symmetry causes the energetic distribution of the electronic cloud of the dot. The role of hydrogen is to passivate the dot. It simply saturates the dangling bonds of silicon and maintains the T_d symmetry of the silicon core.

Unfortunately there were few studies on the emission gap while there had been many groups studying the absorption energy or the energy gap theoretically.[40,47–54] The main reason is that the method of excited-state geometric optimization is inefficient. Nonetheless, understanding the excited state and the dynamics of it is very important. By doing so, we will understand the silicon nanoparticle light emission.

18.4.2 CLUSTER SIZE EFFECT ON EXCITED STATES

How does the particle size affect the emission energy? Puzder et al.[40] and Williamson et al.[39] studied the light emission of ideal hydrogenated particles using the density-functional theory with local density approximation and a quantum Monte Carlo method and found that Stoke's shift depends strongly on the particle size. Pantelides and Franceschetti[55] and Luppi et al.[56] observed the same dependency when studying Stoke's shift. However Luppi et al.[56] reported that the emission energies are different than the absorption energies because the emission energies of nanoparticles with less than 35 silicon atoms show a slow and non-monotonic change between 2.6 eV and 0.5 eV. When reducing the size of nanoparticles, the band gaps of silicon nanostructures experience an increase in frequency, from infrared wavelengths into visible light. Nevertheless, the theoretical and experimental results disagree with each other, mainly due to three reasons: the uncertainty of the particle size, the impurities on the surface of the hydrogenated silicon particles, and the approximations methods to study the excited-state properties.

We studied the ground-state properties of silicon nanoparticles using the self-consistent-charge density-functional tight-binding (SCC-DFTB) method[45,46] for structural optimization. The numerical basis sets that describe the silicon atomic orbital s, p, and d, and the hydrogen atomic orbital s were used in the calculations. Then, a coupling matrix was built. The matrix describes the response of the potential to a change in the electron density and explains the emission energies. The use of the γ approximation brings more efficiency to the time-dependent density-functional tight-binding (TD-DFTB) formula. Therefore, we could study a nanoscale system of several hundred atoms via this approach. After the excited-state geometric optimization by finding the energy gradient, the emission spectrum could be calculated.

We compared the results of the TD-DFTB method with those of the time-dependent density-functional theory (TD-DFT) method at the B3LYP/6-311G* level[57,58] using the GAUSSIAN 03 package to check the accuracy of the TD-DFTB. In Table 18.3 we see that in general, the energy gaps from TD-DFTB are lower than that from TD-DFT.[44] For the Si_5H_{12} particle, the TD-DFT (B3LYP) energy gap is 6.68 eV, larger than the experimental value of 6.5 eV.[59] The TD-DFTB method produces an energy gap of 6.40 eV which is lower but closer to the experimental value. For larger particles such as $Si_{29}H_{36}$ and $Si_{35}H_{36}$, TD-DFTB gives 4.42 eV and 4.37 eV, respectively, while TD-DFT produces 4.66 eV and 4.48 eV, respectively. Both methods agree with the results of the second order perturbation theory which gives 4.45 eV for $Si_{29}H_{36}$ and 4.33 eV for $Si_{35}H_{36}$.[60,61] The results of TD-DFTB are much better than the results of other methods such as Quantum Monte Carlo, the gradient-corrected Perdew-Burke-Ernzerhof functional, known as GGA-PBE. Such a comparison indicated that the results of the TD-DFTB method are close to those of the first-principles TD-DFT calculation. Another benefit of TD-DFTB is the computational efficiency for studying the optical properties of silicon nanoparticles.

In Table 18.4, we see the absorption energies and the emission energies of silicon nanoparticles with different diameters.[44] When the diameter increases, the absorption energies decrease from 6.40 eV to 2.81 eV. The decrease is slower when the diameter is bigger than 1.5 nm or the number of silicon atoms is more than 87. We studied the dependency of the optical properties on the size by using spherical hydrogenated silicon clusters that have a 5–199 silicon atoms. Figure 18.8 shows that the emission energy decreases monotonically when the size of the cluster increases, especially when the clusters have a diameter larger than 1.5 nm or the number of silicon atoms is greater than 87.[62] This is in accordance with the quantum size effect. But when the diameter of the clusters is less than 1.5 nm, the emission energies do not agree with the quantum confinement effect model because they decrease non-monotonically. The reason is that small clusters are weakly bound by their constituent atoms or that small clusters have low cohesive energy.

Table 18.3 Calculated optical gaps (in eV) for several small hydrogenated silicon nanoparticles. Data in parentheses are the corresponding emission energies

	Si_5H_{12}[a]	$Si_{17}H_{36}$	$Si_{29}H_{36}$	$Si_{35}H_{36}$
TD-DFTB[b]	6.40 (2.29)	4.47 (2.40)	4.42 (2.57)	4.37 (2.89)
TD-DFT/B3LYP[b]	6.68 (2.96)	5.04 (2.54)	4.66	4.48
TD-DFT/B3LYP[c]	6.66	5.03	4.53	4.42

Source: Reprinted with permission from Wang, X., et al., 2007, Appl. Phys. Lett., 90, 123116. Copyright 2007, American Institute of Physics.
[a] The experimental absorption value is 6.5 eV.[59]
[b] Our work.[44]
[c] Reference.[61]

Table 18.4 Calculated absorption energies (E_{abs}) and emission energies (E_{emi}) of silicon nanoparticles with different diameters (d_0)

PARTICLE	d_0 (nm)	E_{ABS} (eV)	E_{EMI} (eV)
Si_5H_{12}	0.45	6.40	2.29
$Si_{17}H_{36}$	0.98	4.47	2.40
$Si_{29}H_{36}$	1.03	4.42	2.57
$Si_{35}H_{36}$	1.09	4.37	2.89
$Si_{59}H_{60}$	1.36	3.72	3.18
$Si_{73}H_{78}$	1.41	3.51	3.12
$Si_{87}H_{76}$	1.48	3.47	3.25
$Si_{123}H_{100}$	1.74	3.11	3.04
$Si_{147}H_{100}$	1.76	3.08	2.94
$Si_{199}H_{140}$	2.00	2.81	2.76

Source: Reprinted with permission from Wang, X., et al., 2007, Appl. Phys. Lett., 90, 123116. Copyright 2007, American Institute of Physics.

Clusters, nanoparticles, and quantum dots

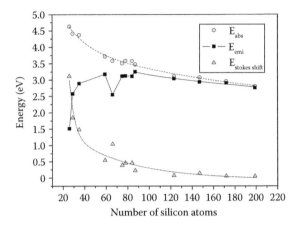

Figure 18.8 Absorption and emission energies and Stokes shift of silicon particles (ranging from $Si_{26}H_{32}$ to $Si_{199}H_{140}$) versus the number of silicon atoms in them. (Reprinted with permission from Wang, X., et al., 2007, 111, 12588–12593. Copyright 2007 American Chemistry Society.)

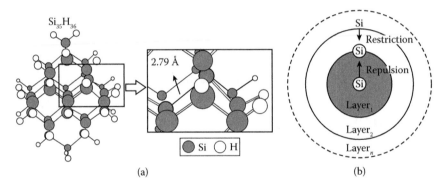

Figure 18.9 Schematic diagrams representing structural relaxation: (a) structure of $Si_{35}H_{36}$ in its excited state, (b) structure relaxation diagram of general spherical silicon nanoparticles. (Reprinted with permission from Wang, X., et al., 2007, *Appl. Phys. Lett.*, 90, 123116. Copyright 2007, American Institute of Physics.)

The cohesive energy increases when the size of the clusters increases; thus small clusters are less resistant to photoexcitation than larger clusters.

The inner silicon–silicon bonds are longer than the surface silicon–silicon bonds. Due to longer bonds, the inner silicon–silicon bonds are weaker if photoexcitation takes place. The weak bonds are considered as defects in the nanoparticle and are affected much by photoexcitation. So the weak or stretched bonds are the exciton traps or centers for optical activity. The silicon atom that is connected to the central silicon atom moves away from its original position because of excited-state relaxation shown in Figures 18.9a and 18.10b.[44] In the Si_5H_{12} cluster, the bonds are stretched more or weakened further because it has smaller size and lower cohesion. Consequently, the electronic cloud localization takes place strongly and the gap between the LUMO level at the excited state and the LUMO level at the ground state is significantly lowered. Figure 18.11 shows the situation while Figure 18.9b is the diagram of the structure relaxation of general silicon nanoparticles.[44] In larger clusters, the stretching of the silicon–silicon bonds is prevented due to the rigidity of the outer layer structures. Figure 18.11 shows that as the cluster size increases, the drop of the LUMO level at the excited state is less. Stoke's shift is also decreased when the cluster size increases. Therefore, if we prevent shape deformation or minimize charge localization, Stoke's shift will be minimized.

Frontier orbitals at both ground and excited state are shown in Figure 18.12.[62] If two orbitals resemble each other, then the energy levels of the two orbitals are very close. The HOMO orbitals at ground state and excited state look like being the same. This means that there is only a little shift of the HOMO level in the excited state

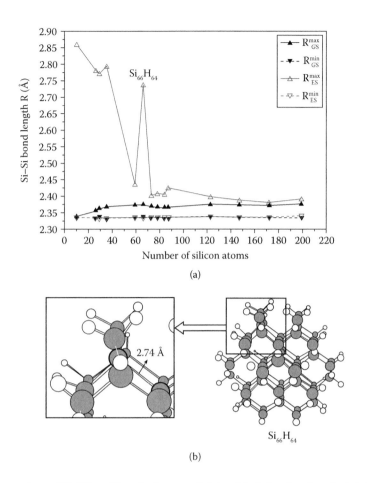

(a)

(b)

Figure 18.10 (a) Comparison of Si–Si bond lengths in ground states (black lines) and excited states (red lines). The solid lines correspond to the maximal bond lengths, while the dashed lines correspond to the minimal bond lengths. (b) Schematic geometrical relaxation diagram of the $Si_{66}H_{64}$ in excited state. (Reprinted with permission from Wang, X., et al., 2007, 111, 12588–12593. Copyright 2007 American Chemistry Society.)

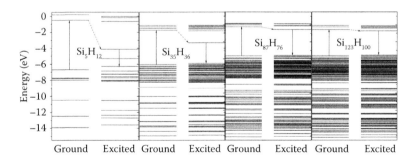

Figure 18.11 Calculated energy levels of valence orbitals of several silicon nanoparticles in their ground and excited states. (Reprinted with permission from Wang, X., et al., 2007, *Appl. Phys. Lett.*, 90, 123116. Copyright 2007, American Institute of Physics.)

to the HOMO level in the ground state. The situation is not the same for the LUMO case. In larger clusters, LUMO in the excited state looks the same as LUMO in the ground state because the energy levels are quasicontinuous if we increase the cluster size and approach the bulk behavior. However, in smaller clusters, LUMO in the excited state looks very different from LUMO in the ground state. The reason is that the excited state moves down considerably.

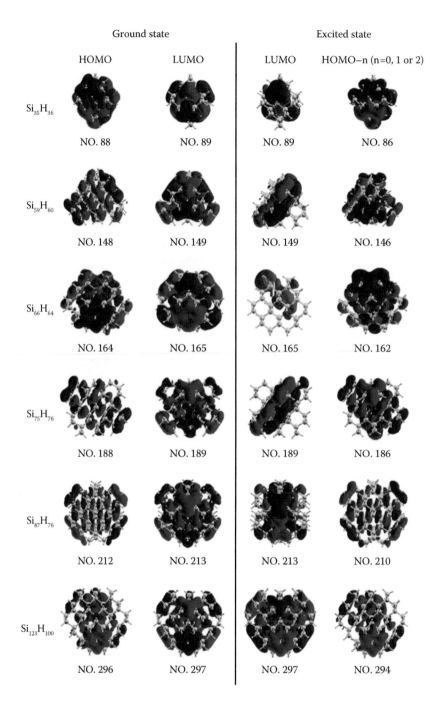

Figure 18.12 Isosurfaces of wave functions of silicon nanoparticles in ground and excited states plotted at the same isovalue. (Reprinted with permission from Wang, X., et al., 2007, 111, 12588–12593. Copyright 2007, American Chemistry Society.)

18.5 SUMMARY

In this chapter, we reviewed the studies of fully hydrogenated silicon nanocrystals and partially hydrogenated silicon nanoclusters. In the fully hydrogenated silicon nanocrystals, the surface relaxation is small. The contraction, which takes place in the first three outermost layers, is from 0.001 Å to 0.002 Å. The lattice expansion of our model agrees with X-ray diffraction measurements. The size and the shape of nanocrystals do not affect either the surface relaxation or the lattice expansion. Based on the study of the Si_mH_x (m≤151) silicon nanoclusters, the fully hydrogenated silicon nanocrystals are the most stable structures compared to those of partially hydrogenated ones. Removing up to 50% of the total terminating hydrogen atoms causes lattice distortions to the crystal structure although the tetrahedral symmetry still remains. But, removing more than 70% of the total hydrogen-terminating atoms causes the clusters to evolve to more compact structures. It turns out that the H/Si ratio is an ideal and superior quantity to describe the stability of hydrogen-terminated silicon nanostructures.

In the ground state, the particle diameter affects the electronic structure. Also, the band gap is found to be inversely proportional to the diameter of the silicon particle. When the number of silicon atoms is more than 87 or when the diameter of the clusters is larger than 1.5 nm, the emission energies decrease. If their size is less than the size above, the emission energies will decrease non-monotonically. The reason is that small clusters have less cohesive energy than larger clusters. In larger clusters, the drop of the LUMO at the excited state is less and therefore Stoke's shift will also decrease. The HOMO orbitals at ground state and at the excited state look the same. In larger clusters, LUMO in the excited state looks the same as LUMO in the ground state. But in smaller clusters, they look very different, meaning there is a large energy shift. So the key to minimize the Stoke's shift is to prevent the shape deformation or to reduce the charge localization.

ACKNOWLEDGMENT

Yanoar thanks Sven Ahrens of Beijing Computational Science Research Center and William Walker of PENABUR Secondary Kelapa Gading for proofreading the draft of the chapter.

REFERENCES

1. Q. Wu, X. Wang, Q. S. Li and R. Q. Zhang, Excited State Relaxation and Stabilization of Hydrogen Terminated Silicon Quantum Dots, *Journal of Cluster Science 24*, 381–397, 2013.
2. Y. J. Chabal, Hydride formation on the Si (100): H_2O surface, *Physical Review B 29*, 3677, 1984.
3. X. Zhao, C. M. Wei, L. Yang and M. Y. Chou, Quantum confinement and electronic properties of silicon nanowires, *Physical Review Letters 92*, 236805, 2004.
4. J. T. Yates, A new opportunity in silicon-based microelectronics, *Science 279*, 335–336, 1998.
5. Y. Cui, Q. Wei, H. Park and C. M. Lieber, Nanowire nanosensors for highly sensitive and selective detection of biological and chemical species, *Science 293*, 1289–1292, 2001.
6. K. Q. Peng, A. J. Lu, R. Q. Zhang and S. T. Lee, Motility of metal nanoparticles in silicon and induced anisotropic silicon etching, *Advanced Functional Materials 18*, 3026–3035, 2008.
7. M. F. Ng and R. Q. Zhang, Dimensionality dependence of optical properties and quantum confinement effects of hydrogenated silicon nanostructures, *The Journal of Physical Chemistry B 110*, 21528–21535, 2006.
8. D. D. D. Ma, C. S. Lee, F. C. K. Au, S. Y. Tong and S. T. Lee, Small-diameter silicon nanowire surfaces, *Science 299*, 1874–1877, 2003.
9. Y. Cui, Z. Zhong, D. Wang, W. U. Wang and C. M. Lieber, High performance silicon nanowire field effect transistors, *Nano Letters 3*, 149–152, 2003.
10. L. Pavesi, L. Dal Negro, C. Mazzoleni, G. Franzo and F. Priolo, Optical gain in silicon nanocrystals, *Nature 408*, 440–444, 2000.
11. T. Strother, W. Cai, X. Zhao, R. J. Hamers and L. M. Smith, Synthesis and characterization of DNA-modified silicon (111) surfaces, *Journal of the American Chemical Society 122*, 1205–1209, 2000.
12. J. I. Hahm and C. M. Lieber, Direct ultrasensitive electrical detection of DNA and DNA sequence variations using nanowire nanosensors, *Nano Letters 4*, 51–54, 2003.
13. R. Q. Zhang, J. Costa and E. Bertan, Role of structural saturation and geometry in the luminescence of silicon-based nanostructured materials, *Physical Review B 53*(12), 7847–7850, 1996.
14. K. Hayashi, S. Yamanaka, H. Okushi and K. Kajimura, Study of the effect of hydrogen on transport properties in chemical vapor deposited diamond films by Hall measurements, *Applied Physics Letters 68*, 376–378, 1996.
15. D. K. Yu, R. Q. Zhang and S. T. Lee, Structural properties of hydrogenated silicon nanocrystals and nanoclusters, *Journal of Applied Physics 92*(12), 7453–7458, 2002.
16. D. S. Lin, T. Miller, T. C. Chiang, R. Tsu and J. E. Greene, Thermal reactions of disilane on Si (100) studied by synchrotron-radiation photoemission, *Physical Review B 48*, 11846, 1993.
17. P. Dumas, Y. J. Chabal and G. S. Higahi, Coupling of an adsorbate vibration to a substrate surface phonon: H on Si (111), *Physical Review Letters 65*, 1124, 1990.

18. Y. Morita and H. Tokumoto, Ideal hydrogen termination of Si (001) surface by wet-chemical preparation, *Applied Physics Letters 67*, 2654–2656, 1995.
19. P. Jakob, Y. J. Chabal, K. Kuhnke and S. B. Christman, Monohydride structures on chemically prepared silicon surfaces, *Surface Science 302*, 49–56, 1994.
20. B. Delley and E. F. Steigmeier, Quantum confinement in Si nanocrystals, *Physical Review B 47*, 1397, 1993.
21. B. Delley and E. F. Steigmeier, Size dependence of band gaps in silicon nanostructures, *Applied Physics Letters 67*, 2370–2372, 1995.
22. L. W. Wang and A. Zunger, Electronic structure pseudopotential calculations of large ([approximately] 1000 atoms) Si quantum dots, *The Journal of Physical Chemistry 98*, 2158–2165, 1994.
23. S. Y. Ren and J. D. Dow, Hydrogenated Si clusters: band formation with increasing size, *Physical Review B 45*, 6492, 1992.
24. M. Hirao and T. Uda, Electronic structure and optical properties of hydrogenated silicon clusters, *Surface Science 306*, 87–92, 1994.
25. D. Bellet, G. Dolino, M. Ligeon, P. Blanc and M. Krisch, Studies of coherent and diffuse x-ray scattering by porous silicon, *Journal of Applied Physics 71*, 145–149, 1992.
26. G. A. Rechtsteiner, O. Hampe and M. F. Jarrold, Synthesis and temperature-dependence of hydrogen-terminated silicon clusters, *The Journal of Physical Chemistry B 105*, 4188–4194, 2001.
27. T. Miyazaki, T. Uda, I. Stich and K. Terakura, Theoretical study of the structural evolution of small hydrogenated silicon clusters: Si6Hx, *Chemical Physics Letters 261*, 346–352, 1996.
28. R. Kagimura, R. W. Nunes and H. Chacham, Structures of Si and Ge nanowires in the subnanometer range *Physical Review Letters 95*, 115502, 2005.
29. B. Aradi, L. E. Ramos, P. Deák, T. Köhler, F. Bechstedt, R. Q. Zhang and Th. Frauenheim, Theoretical study of the chemical gap tuning in silicon nanowires, *Physical Review B 76*, 035305, 2007.
30. P. B. Sorokin, P. V. Avramov, A. G. Kvashnin, D. G. Kvashnin, S. G. Ovchinnikov and A. S. Fedorov, Density functional study of <110>-oriented thin silicon nanowires, *Physical Review B 77*, 235417, 2008.
31. R. Q. Zhang, Y. Lifshitz, D. D. D. Ma, Y. L. Zhao, Th. Frauenheim, S. T. Lee and S. Y. Tong, Structures and energetics of hydrogen-terminated silicon nanowire surfaces, *The Journal of Chemical Physics 123*, 144703, 2005.
32. J. F. Justo, R. D. Menezes and L. V. C. Assali, Stability and plasticity of silicon nanowires: The role of wire perimeter, *Physical Review B 75*, 045303, 2007.
33. H. Xu, X. B. Yang, C. S. Guo and R. Q. Zhang, An energetic stability predictor of hydrogen-terminated Si nanostructures, *Applied Physics Letters 95*, 253106, 2009.
34. P. Hohenberg and W. Kohn, Inhomogeneous electron gas, *Physical Review 136*, B864, 1964.
35. W. Kohn and L. J. Sham, Self-consistent equations including exchange and correlation effects, *Physical Review 140*, A1133, 1965.
36. M. C. Payne, M. P. Teter, D. C. Allan, T. A. Arias and J. D. Joannopoulos, Iterative minimization techniques for ab initio total-energy calculations: molecular dynamics and conjugate gradients, *Reviews of Modern Physics 64*, 1045, 1992.
37. T. L. Chan, C. V. Ciobanu, F. C. Chuang, N. Lu, C. Z. Wang and K. M. Ho, Magic Structures of H-Passivated<110> Silicon Nanowires, *Nano Letters 6*, 277–281, 2006.
38. N. Lu, C. V. Ciobanu, T. L. Chan, F. C. Chuang, C. Z. Wang and K. M. Ho, The structure of ultrathin H-passivated [112] silicon nanowires, *The Journal of Physical Chemistry C 111*, 7933–7937, 2007.
39. A. J. Williamson, J. C. Grossman, R. Q. Hood, A. Puzder and G. Galli, Quantum Monte Carlo calculations of nanostructure optical gaps: Application to silicon quantum dots, *Physical Review Letters 89*, 196803, 2002.
40. A. Puzder, A. J. Williamson, J. C. Grossman and G. Galli, Computational studies of the optical emission of silicon nanocrystals, *Journal of the American Chemical Society 125*, 2786–2791, 2003.
41. L. Landt, K. Klunder, J. E. Dahl, R. M. K. Carlson, T. Moller and C. Bostedt, Optical response of diamond nanocrystals as a function of particle size, shape, and symmetry, *Physical Review Letters 103*, 047402, 2009.
42. M. Hirao and T. Uda, First principles calculation of the optical properties and stability of hydrogenated silicon clusters, *International Journal of Quantum Chemistry 52*, 1113–1119, 1994.
43. C. Delerue, G. Allan and M. Lannoo, Theoretical aspects of the luminescence of porous silicon, *Physical Review B 48*, 11024, 1993.
44. X. Wang, R. Q. Zhang, S. T. Lee, T. A. Niehaus and Th. Frauenheim, Unusual size dependence of the optical emission gap in small hydrogenated silicon nanoparticles, *Applied Physics Letters 90*, 123116, 2007.
45. D. Porezag, Th. Frauenheim, T. Koehler, G. Seifert and R. Kaschner, Construction of tight-binding-like potentials on the basis of density-functional theory: Application to carbon, *Physical Review B 51*, 12947, 1995.
46. M. Elstner, D. Porezag, G. Jungnickel, J. Elsner, M. Haugk, Th. Frauenheim, S. Suhai and G. Seifert, Self-consistent-charge density-functional tight-binding method for simulations of complex materials properties, *Physical Review B 58*, 7260, 1998.

47. J. R. Chelikowsky, L. Kronik and I. Vasiliev, Time-dependent density-functional calculations for the optical spectra of molecules, clusters, and nanocrystals, *Journal of Physics: Condensed Matter 15,* R1517, 2003.
48. D. Prendergast, J. C. Grossman, A. J. Williamson, J. L. Fattebert and G. Galli, Optical properties of silicon clusters in the presence of water: a first principles theoretical analysis, *Journal of the American Chemical Society 126,* 13827–13837, 2004.
49. A. R. Porter, M. D. Towler and R. J. Needs, Excitons in small hydrogenated Si clusters, *Physical Review B 64,* 035320, 2001.
50. H. C. Weissker, J. Furthmuller and F. Bechstedt, Optical properties of Ge and Si nanocrystallites from ab initio calculations. II. Hydrogenated nanocrystallites, *Physical Review B 65,* 155328, 2002.
51. G. Onida, L. Reining and A. Rubio, Electronic excitations: density-functional versus many-body Green's-function approaches, *Reviews of Modern Physics 74,* 601, 2002.
52. P. H. Hahn, W. G. Schmidt and F. Bechstedt, Molecular electronic excitations calculated from a solid-state approach: Methodology and numerics, *Physical Review B 72,* 245425, 2005.
53. L. X. Benedict, A. Puzder, A. J. Williamson, J. C. Grossman, G. Galli, J. E. Klepeis, J. Y. Raty and O. Pankratov, Calculation of optical absorption spectra of hydrogenated Si clusters: Bethe-Salpeter equation versus time-dependent local-density approximation, *Physical Review B 68,* 085310, 2003.
54. I. Vasiliev, Optical excitations in small hydrogenated silicon clusters: comparison of theory and experiment, *Physica Status Solidi 239,* 19, 2003.
55. A. Franceschetti and S. T. Pantelides, Excited-state relaxations and Franck-Condon shift in Si quantum dots, *Physical Review B 68,* 033313, 2003.
56. E. Luppi, E. Degoli, G. Cantele , S. Ossicini, R. Magri, D. Ninno and O. Bisi, The electronic and optical properties of silicon nanoclusters: absorption and emission, *Optical Materials 27,* 1008–1013, 2005.
57. A. D. Becke, Density-functional thermochemistry. III. The role of exact exchange, *The Journal of Chemical Physics 98,* 5648–5652, 1993.
58. C. Lee, W. Yang and R. G. Parr, Development of the Colle-Salvetti correlation-energy formula into a functional of the electron density, *Physical Review B 37,* 785, 1988.
59. F. Fehler, *Forschungsberichte des Landes Nordrhein-Westfalen,* Westdeutscher, Köln, 1977.
60. A. D. Zdetsis, Optical and electronic properties of small size semiconductor nanocrystals and nanoclusters, *Reviews on Advanced Materials Science, 11,* 56–78, 2006.
61. C. S. Garoufalis, A. D. Zdetsis and S. Grimme, High level ab initio calculations of the optical gap of small silicon quantum dots, *Physical Review Letters 87,* 276402, 2001.
62. X. Wang, R. Q. Zhang, T. A. Niehaus, Th. Frauenheim and S. T. Lee, Hydrogenated silicon nanoparticles relaxed in excited states, *The Journal of Physical Chemistry C 111,* 12588–12593, 2007.

Nanowires and nanotubes

19

Silicon nanowires as electron field emitters

Javier Palomino, Deepak Varshney, Brad R. Weiner, and Gerardo Morell

Contents

19.1 INTRODUCTION

Over the years, silicon has proved to be the standard material in the electronics market, dominating the microelectronics industry with approximately 90% of all semiconductor devices sold worldwide. Silicon is the second most abundant element in the earth's crust and has demonstrated repeatedly to be a practical and versatile material for a wide range of applications in electronic, photovoltaic devices, and electron field emitters. Silicon nanomaterials are also popular since they can be synthesized on a large scale and inexpensively by several methods, and in various morphologies including silicon nanowires (SiNWs), nanorods, nanobelts, and nanoparticles. Quasi-one-dimensional (1D) SiNW semiconductors have attracted much attention because of their distinctive optical, electronic, and mechanical properties (Lee et al. 2016; Schwartz et al. 2016), suitable for modern applications in electron field emission due to their high electron mobility (Liu and Fan 2005; Chen et al. 2006; Chan et al. 2008; Shao et al. 2010; Fan et al. 2011). Recently, there has been an increase in applications of quasi-1D SiNWs due to their compatibility with existing semiconductor technology, evidenced by the successful incorporation of SiNWs in field-effect transistors and in UHD full-color flat panel displays (Chen et al. 2011).

19.1.1 OVERVIEW OF SiNW FIELD EMITTERS

The subject of 1D nanostructures, such as nanotubes and nanowires, has been extensively studied because of their novel physical and chemical properties, and their prospective applications in device development (Morales and Lieber 1998; Feng et al. 2000; Liu et al. 2000). They were first fabricated via a vapor–liquid–solid (VLS) mechanism (Wagner and Ellis 1964), followed by other approaches, such as laser ablation (Morales and Lieber 1998), thermal evaporation (Feng et al. 2000), and chemical etching (Peng et al. 2004). Fabrication of electron-emitting nanomaterials (Iijima 1991; Rinzler et al. 1995; Chen et al. 2010) and their application to flat panel displays (Lee et al. 2001; Biaggi-Labiosa et al. 2008) have attracted much attention to the study of 1D materials having high aspect ratios, stable structures, and enhanced electron field emission (EFE) properties (Chen et al. 2002; Wu et al. 2002; Zhu et al. 2003; Xiang et al. 2005; Valentín et al. 2013). Since silicon plays a significant role in the microelectronics field, SiNW-based emitters have been widely studied (Kulkarni et al. 2005; Huang et al. 2007). In order to improve the EFE properties, various kinds of modifications have been made to SiNWs, such as H_2 plasma surface treatment (Au et al. 1999), Mo-modification (Ha et al. 2002), Ni-implantation (Ok et al. 2006), IrO_2 coating (Chen et al. 2009), gold decoration (Zhao et al. 2011), and diamond coating (Liu et al. 1994; She et al. 1999; Tzeng et al. 2007; Thomas et al. 2012).

Diamond films have also been widely studied for use as electron field emitters in vacuum microelectronic devices because of their negative electron affinity (NEA) and low effective work function (Himpsel et al. 1979). Numerous approaches have been developed to enhance the EFE properties of diamond films, including the modification of grain shape, reduction of grain size (Lu et al. 2006), increasing the conductivity by doping with boron and nitrogen species (Okano et al. 1996), and fabricating diamond tips through the utilization of high aspect ratio templates (Wang et al. 2006) such as aligned silicon tip arrays, which were prepared by employing conventional chemical vapor deposition (CVD) methods, electron beam lithography, and chemical etching techniques (Liu 1995; She et al. 1999; Kiselev et al. 2005; Tzeng et al. 2008). This chapter shows that bare SiNWs possess good EFE properties, nevertheless they can be significantly enhanced by growing ultrananocrystalline diamond (UNCD) on the SiNWs.

19.1.2 PROPERTIES OF SiNWs

The sp^3-bonded SiNW is a quasi-1D material (Figure 19.1), where the atoms in crystalline silicon are arranged in a diamond lattice structure with a lattice constant of 5.430 Å (Tang et al. 2012). Silicon is a solid at room temperature, with a high melting point of 1414°C, and a relatively high thermal conductivity of 149 $W \cdot m^{-1} \cdot K^{-1}$. In its crystalline form, pure silicon has a gray color with a metallic luster, and is hard and brittle. Silicon is a semiconductor material with a band gap of ~1.12 eV, and four valence electrons; its four bonding electrons provide the opportunity to combine with many other elements to form a wide range of compounds.

In a quasi-1D SiNW (an NW is a one degree of freedom structure and is often called a one-dimensional nanostructure if its diameter is less than 100 nm), the momentum of an electron is mostly confined to one direction. This reduction in dimensionality results in dramatic quantum effects dependent on wire material, axis orientation, length, and diameter. These quantum effects in NWs change the electrical, chemical, and mechanical properties. SiNWs exhibit properties and applications that are quite different from their bulk form. In the last decade, they have been intensively studied by experimentalists and theorists because of their applicable properties and compatibility with conventional silicon microtechnology, already

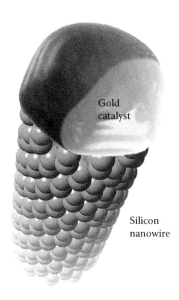

Gold
catalyst

Silicon
nanowire

Figure 19.1 Representation of a silicon nanowire with the gold catalyst at the tip.

implemented in the industry (Shao et al. 2010; Won 2010). To accomplish broad application of SiNWs, control and manipulation of their properties, including size and geometry, are crucial (Shao et al. 2010).

Due to their high surface-to-volume ratio and a quasi-1D structure, SiNWs possess properties that can outperform the current state of the art of silicon devices in modern applications such as sensors with high sensitivity, efficient solar cells, and enhanced lithium-ion (Li-ion) batteries. The integration of diamond with SiNW leads to a composite material, having the right combination of functional properties of the constituent materials, which may have potential applications as cold field emitters, UHD flat panel displays, and enhanced Li-ion batteries.

19.1.3 APPLICATIONS

Silicon nanowire-based sensors: A sensor is a device that transforms a physical or chemical response of the target into an electrical signal. The sensor is composed of an active sensing part, which translates the input into a temporary signal, and a transducer that decodes the temporary signal into an electrical response. The quasi-1D properties of SiNWs can be utilized in transducers and in some cases as the active sensing part. Applications, such as chemical, biochemical, and biological sensors, utilize the small geometry and the electrical and mechanical properties of nanowires, as shown by Zheng et al. (2005), who demonstrated the capabilities of SiNWs to detect cancer markers electrically. Mechanical sensors can benefit from the high piezo-resistance effect in SiNWs (Bhaskar et al. 2013).

Silicon nanowire-based solar cells and anodes for Li-ion batteries: Energy harvesting from renewable sources and the storage of electrical energy are among the most demanding challenges of our society (Armaroli and Balzani 2007). Light harvesting can benefit from SiNWs, because they have the capability to increase the optical absorption and collection efficiency in solar cells, where light trapping is enhanced by forming radial structures, such that the p-n junction (Tian et al. 2007) can be placed much closer to the carrier generating region.

Li-ion batteries are currently one of the best technologies to store electrical energy, with graphite as the standard anode material. Theoretical calculations predict that silicon can improve the capacity by one order of magnitude, due to the accommodation of up to 4.4 Li atoms per silicon atom. Nevertheless, the huge lattice expansion upon lithiation/delithiation leads to pulverization of the anode. Researchers have reported that SiNWs can expand laterally and still maintain the current transport in the vertical direction. These research results have been shown in half-cells (Wang et al. 2011), but the enhanced cycle ability in full cells remains to be shown. Furthermore, it has been reported that silicon below a critical size (150 nm) does not crack or fracture during lithiation/delithiation (Liu et al. 2012).

Nanowires and nanotubes

19.2 METHOD OF PREPARING SiNWs AND UNCD/SiNWs

Silicon-based nanostructures can be synthesized from silicon nanoparticles (SiNPs) to avoid the pyrophoric hazards associated to silane (LaDou 1983). A novel route to synthesize diamond films using aliphatic polymers as a seeding source, has been shown to have the best possible nucleation density, high growth rates, and excellent morphology (Varshney et al. 2013; Palomino et al. 2014). This approach enables the fabrication of ultrananocrystalline diamond on silicon nanowires (UNCD/SiNWs) for cold electron field emitters.

19.2.1 ROLE OF SILICON NANOPARTICLES IN THE GROWTH OF SILICON NANOSTRUCTURED FILMS BY CVD

Silane gas (SiH_4) is a commonly used precursor to grow silicon nanostructures, but it is hazardous because it is toxic and pyrophoric with a wide flammable range and high reactivity. In order to overcome the potential hazards of silane, SiNPs can be used as a source material to fabricate silicon nanostructures, such as nanowires, nanoposts, nanorods, and hybrid silicon-carbon nanostructures, which present new capabilities for modern applications. SiNPs are cheap, nontoxic, and nonpyrophoric. SiNPs have been used successfully to grow SiNWs and SiCNTs at temperatures below the melting point of bulk silicon (Palomino et al. 2014, 2015).

As the size of a particle decreases, the surface-to-volume ratio increases, promoting the surface atoms to detach from their positions and diffuse on the surface, which is followed by a partial melting of the thin layer close to the surface of the nanoparticle. The surface melting mostly occurs at temperatures much lower than the bulk melting, depending on particle size, morphology, defects, and reactor conditions. As the melted surface of the nanoparticle is exposed to high temperatures, energetic molecules can decompose by an exothermic process (Kumar and Ando 2010) providing enough energy to form nanostructures from the melted surface or even to evaporate it to facilitate pure material transport toward the substrate, where the growth of the nanostructure is desired. The decrease in melting point with decreasing particle size is a well-known phenomenon that has been described by thermodynamic models and simulated using molecular dynamics (Dick et al. 2002). Experimental evidence has been demonstrated by several research groups using different techniques (Qi and Wang 2004).

19.2.2 ROLE OF POLYMERS IN THE GROWTH OF CARBON-BASED FILMS BY CVD

Isolated polymer molecules such as paraffin wax or polyethylene can be converted into single crystals by heating them up to their melting point under controlled conditions inside a reactor. Such crystals are known as single-chain crystals and represent potential seeds to enhance the growth of carbon allotropes. Weber et al. (2007) have prepared such polyethylene nanocrystals with hexagonal structures comprising an inner crystalline layer, attributed to compact folding of the polymer chains. They fabricated large polyethylene diamond-shaped crystals, and proposed an explanation on how the hexagonal nanocrystals grow into diamonds, asserting that the hexagonal structure grows quickly when the crystals are small, but transform into a thermodynamically more stable diamond-shaped structure when they reach a critical size (Cheng 2007).

Recently, Varshney et al. have demonstrated that the use of aliphatic polymers, such as polyethylene, paraffin wax, n-tetracosane, and n-octacosane as seeding sources, significantly enhances the synthesis of CNTs and diamond films and their composites/hybrids (Varshney et al. 2011, 2013; Varshney 2012) by providing a simple, cost-effective, and non-abrasive method of diamond nucleation.

19.2.3 CHEMICAL VAPOR DEPOSITION (CVD)

CVD is a widely used technique for solid thin film deposition on a substrate from vapor species through chemical reactions inside the reactor. In a CVD process, the reactant gases pass through energetic zones—hot filament—to be decomposed into reactive species, then transported toward the substrate surface by forced flow or diffusion and/or convection, either to be deposited directly or transformed into condensable species via chemical reactions at the substrate surface forming nucleation centers. The process continues as long as the reactants are available, resulting in the fabrication of the desired material in the form of a continuous thin film. The CVD technique is accompanied by the production of chemical residual products that are swept out of the chamber together with unreacted precursor gases.

In the semiconductor industry, CVD is one of the most important methods for the film deposition, due to its high production, high purity, and low cost of operation. CVD has many advantages as compared to physical vapor deposition (PVD) methods such as sputtering. One of the main advantages of CVD is that films can be grown uniformly all around and inside complex shapes, such as high aspect ratio holes (carbon nanotubes). Due to the high precursor flow rates, the CVD deposition rates are several times higher than PVD methods. CVD stoichiometry is easy to control by verifying the flow rates of precursors and does not require ultrahigh vacuum. Other advantages of CVD include growth of high purity films, the ability to fabricate abrupt junctions, and the deposition of a wide variety of materials. However, CVD presents some disadvantages. CVD requires gas or volatile precursors at near-room temperature, limiting the use of some elements in the periodic table. Although this problem has been alleviated by using metal-organic precursors, these compounds are relatively expensive. Additionally, CVD precursors are often hazardous or toxic and the residual products may also be toxic, requiring extra steps in handling the reactants and in the treatment of reactor exhaust. Another CVD disadvantage is that high deposition temperatures for some CVD processes (often greater than 600°C) are often harmful for structures that have already been fabricated on the substrates, although the use of plasma-enhanced CVD may reduce the deposition temperatures.

19.2.4 STAGES OF CVD

The stages of the CVD process comprise complex flow dynamics, where the gases flow into the chamber, react, deposit material, and then residual products are removed from the chamber. Figure 19.2 shows the schematic representation of the sequence of events during a CVD reaction, which is described as follows:

1. Input of reactant gases into the chamber by pressurized gas lines forming a main gas flow region
2. Gas phase reactions from main flow region
3. Mass transport (product of gas phase reactions) to the substrate
4. Adsorption and diffusion of products on the substrate (normally heated)
5. Chemical reaction on the surface in order to dissociate the molecules
6. Diffusion of atoms on the surface toward the nucleation sites and growth of islands
7. Desorption of residual products of the reactions and mass transport (residual products) to the main flow region

19.2.5 THERMAL CVD

In thermal CVD, temperatures can reach as high as 2000°C by use of resistive heating, radio frequency induction heating, or radiant heating. Hot wall thermal CVD reactors are typically large furnaces into

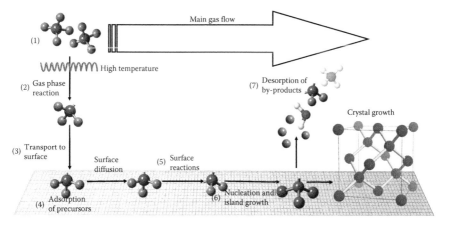

Figure 19.2 Sequence of events during the CVD process.

which the substrates are placed and deposition is done on a large area substrate or several substrates at once, as shown in Figure 19.3.

In a representative thermal CVD process, the source material is heated to an elevated temperature, which allows the precursors (gas, liquid, or solid phase) to evaporate, and/or decompose. The substrate is exposed to one or more precursors (gas phase), which react, decompose, and form nucleation centers on the substrate surface producing the desired nanostructured material film. Volatile by-products are also formed and are removed by carrier gas flow (typically argon gas) through the reaction chamber. The advantages of thermal CVD include: (i) diversity in hydrocarbon and other source materials (gas, liquid, solid), (ii) synthesis of high quality and large area materials, (iii) variety in nanostructured products, and (iv) simplicity of the system.

19.2.6 SYNTHESIS OF SiNWs BY THERMAL CVD

A schematic of a custom-built hot wall thermal CVD reactor is shown in Figure 19.4. It includes a resistive heater at the wall of the electric tube furnace, a silicon source material consisting of a mixture of Si nanoparticles and graphite (1:1) placed at the center of a quartz tube, and a gold (Au) coated copper (Cu) substrate that is placed 7–10 cm away from the source material on the downstream side of the Ar flow. The thin gold layer (100–200 nm) coated on the Cu substrate by radio frequency sputtering acts as a catalyst to grow SiNWs because of its significant silicon solubility. Before each deposition, the thermal CVD chamber is evacuated to ~5 m Torr, then the reactor is filled by a continuous flow (100 sccm) of Ar gas reaching a constant pressure of ~100 mTorr during the whole growth process; Ar is introduced from one closed end of the quartz tube and acts as a carrier gas. The temperature of the source material is first raised from room temperature to 1050°C, and then kept constant for 3 hr (until the end of the growth process). In contact with graphite at high temperature, the native SiO_x contained in the source material is reduced to form Si and CO_x vapors, where Si vapor is transported by the carrier gas toward the substrate (950°C) to form nucleation centers and subsequently the growth of SiNWs. The by-products, including CO_x, are exhausted out of the chamber together with unreacted precursor gases, as explained in the stages of CVD section.

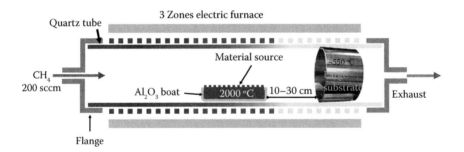

Figure 19.3 Schematic representation of the hot wall thermal CVD system.

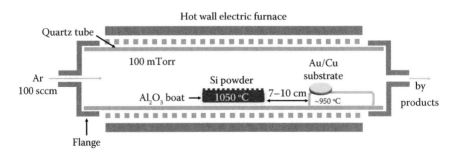

Figure 19.4 Schematic representation of the thermal CVD system showing the configuration of raw materials and the corresponding temperatures. (Adapted from Palomino, J., et al., *ACS Appl. Mater. Interfaces*, 6, 13815–13822, 2014.)

19.2.7 GROWTH MECHANISM OF Si NANOWIRES

The synthesis of SiNWs follows the VLS (Wagner and Ellis 1964; Schmidt et al. 2009) growth mechanism, with Au as the catalyst in the present case (Al can be also used as catalyst). Figure 19.5 shows a schematic representation of the VLS mechanism, where the metal (Au) catalyst thin film forms liquid Au-Si alloy droplets at high temperatures (950°C). Au drops adsorb vapors of the source material (Si) due to its high solubility, achieving a super saturation stage, and the subsequent precipitation of the source material (Si) at the liquid–solid interface occurs to reach a minimum free energy of the alloy system. The 1D crystal growth begins and continues as long as the vapor components are supplied. Since vapor (carrying Si nanoparticles), liquid (gold catalyst), and solid (precipitated 1D structures) phases are involved, it is known as the VLS mechanism (Wagner and Ellis 1964; Schmidt et al. 2009; Klimovskaya et al. 2011).

19.2.8 CVD DIAMOND NUCLEATION

Diamond nucleation during the CVD process is the result of various reactions, which lead to the formation of diamond nuclei (sp^3), stabilization of diamond nuclei (sp^3) with respect to graphite nuclei and preferential surface etching of sp^2-bonded carbon (Messier 1990). In order to accomplish these conditions, most current CVD diamond growth environments employ large amounts of molecular hydrogen as compared to carbon source gas, which allows supersaturation of atomic hydrogen resulting from gas activation such as hot filament. Atomic hydrogen plays an important role in CVD diamond growth: (i) it etches sp^2 carbon faster than sp^3, so graphite and other nondiamond phases can be removed preferentially from the substrate and only clusters with diamond structure remain and continue to grow; (ii) it stabilizes the diamond surface and preserves the sp^3 hybridization configuration; (iii) it dissociates hydrocarbons into radicals, a necessary precursor for diamond formation; and (iv) it dissociates hydrogen from the radicals adsorbed on the surface creating active sites for further adsorption of the diamond precursors.

In 1987, Mitsuda et al. (1987) reported that scratching the substrate surface with diamond powder significantly enhances the nucleation density, thus substrate surface scratching is the most common and powerful method for achieving high nucleation diamond density and fine uniform grain size. Diamond fragments of small size get locked into scratches and provide nucleation sites and carbon species diffuse into the existing nucleation sites. Diffusion and bonding of carbon species result in the formation of critical nuclei and the diamond growth process continues as long as the precursors are supplied. However, the scratching seeding process can be detrimental to the substrates, devices and nanostructures (microprocessors, nanowires, nanotubes, etc.) on which diamond is required to grow. The use of aliphatic saturated hydrocarbon polymers provides a non-abrasive, simple, and cost-effective process for diamond nucleation, leading to the fabrication of high-quality diamond films and their composites.

19.2.9 SYNTHESIS OF UNCD/SiNWs

The SiNWs were coated with UNCD by using a hot filament chemical vapor deposition (HF-CVD) process. A schematic of the HF-CVD reactor is shown in Figure 19.6. The filament can be various materials, such as W, Ta, and Re, but rhenium (Re) wire is preferred as it does not react with carbon and is

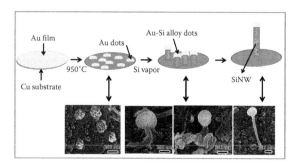

Figure 19.5 Schematic representation of the VLS growth mechanism of SiNWs, supported by FESEM images. (Adapted from Palomino, J., et al., *ACS Appl. Mater. Interfaces*, 6, 13815–13822, 2014.)

Nanowires and nanotubes

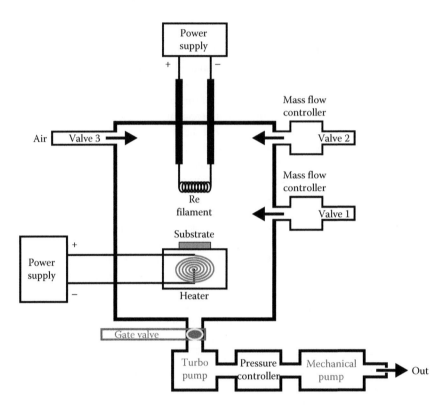

Figure 19.6 Schematic of the HF-CVD System.

not consumed during the growth process. Typically, 8 cm of Re wire of 0.5 mm diameter was rolled as a helical spring (or other configuration), and positioned at 8 mm above the substrate. A DC power supply in current controlled mode was used to heat the filament resistively to ~2400°C, and the substrate holder made of Mo was heated by a graphite resistance heating element positioned right below the filament, as shown in Figure 19.6.

The SiNWs were first coated with melted aliphatic saturated hydrocarbon polymer (~10 um thick film of paraffin wax) that was used as a diamond seeding source, and then introduced into the HF-CVD reactor. Under the HF-CVD conditions, the paraffin wax decomposes, leaving behind abundant sp³-C crystallites (Varshney et al. 2014) on the Si nanowire surface that result in enhanced nucleation of UNCD. Paraffin wax is more efficient in the creation of diamond nuclei than traditional detrimental methods, such as polishing and ultrasonication, which produce substantial surface damage.

Before each deposition, the HF-CVD chamber was evacuated to 5×10^{-7} Torr and then filled with a gas mixture consisting of 0.3% CH_4 and 99.7% H_2, maintaining the combined flow of gases at 100 sccm, and total pressure at 20 Torr. These parameters can be tailored depending on the desired material to grow. The gas mixture was introduced in the reaction chamber using two gas valves and two mass flow controllers as shown Figure 19.6. The growth process was carried out at relatively low temperatures (~400–500°C) (Auciello and Sumant 2010). The deposition time can be optimized in order to obtain a uniform coverage of the substrate.

19.2.10 GROWTH MECHANISM OF ULTRANANOCRYSTALLINE DIAMOND

Figure 19.7 shows the schematic representation of the UNCD growth on SiNW seeded by paraffin wax, which consists of a long straight chain of n-alkanes ($CH_3[CH_2]nCH_3$) with melting point of ~50°C and flash point starting at ~150°C. At the substrate temperature of ~400°C, the polymer decomposes into hydrocarbon radicals (CH_x) leaving behind large amounts of sp³-C rich nano fragments that result in enhanced nucleation centers for UNCD. In the HF-CVD environment, the precursor gases and radicals

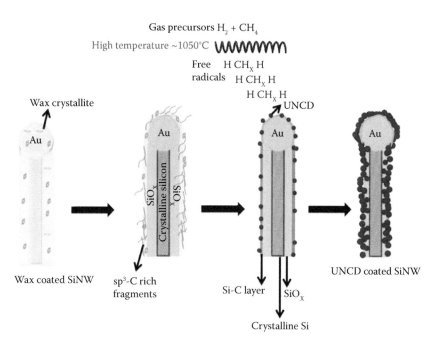

Figure 19.7 Schematic representation of the growth mechanism of UNCD on SiNWs. (Adapted from Palomino, J., et al., *ACS Appl. Mater. Interfaces*, 6, 13815–13822, 2014.)

decompose into H and C; the atomic H etches the SiO_x shell layer of SiNW, and the C atoms form an SiC interface layer and react with the active nuclei to produce UNCD (Varshney et al. 2013). Two factors limit the ability of the nanodiamond grains to grow beyond 5–10 nm: the three-dimensional network structure of the SiNWs, and the presence of silicon oxide remnants. On flat Cu substrates, similar parameters can yield nanodiamond grains of 10–15 nm (Varshney et al. 2014).

The thin SiO_x layer on the wires together with the presence of an sp^3-C rich environment may also play an important role. Active Si sites are generated from the reduction of the SiO_2 layer with the assistance of hydrogen atoms (Chen et al. 2007). This leads to the formation of an SiC interfacial layer that also acts as a nucleation site for UNCD growth (Liu 1995; Kiselev et al. 2005; Chen et al. 2008).

19.3 CHARACTERIZATION OF SiNWs AND UNCD/SiNWs

The morphology and the chemical composition of the fabricated films were examined using a field emission scanning electron microscope (FE-SEM) and high-resolution transmission electron microscopy (HR-TEM). The films were also characterized by Raman spectroscopy performed at room temperature, using a 514 nm Ar-ion laser as an excitation source. The EFE characteristics of the films were measured with an electrometer. All the measurements were recorded at an anode–cathode distance, $d_{CA} = 100 \pm 2$ μm, and at a pressure of $\sim 5 \times 10^{-7}$ Torr (6.7×10^{-5} Pa). Currents lower than 1×10^{-12} A were assumed as background noise.

19.3.1 FIELD EMISSION SCANNING ELECTRON MICROSCOPY (FE-SEM)

FE-SEM images (Figure 19.8) reveal the morphology of the bare SiNWs. Figure 19.8a shows a straight nanowire consisting of a core-shell structure of the SiNW, where the SiOx shell thickness is around 10–20 nm. The low magnification image (Figure 19.8b) exhibits uniform and high-density growth of bare SiNWs, where the diameter of the nanowires estimated from the image is in the range of 100–180 nm and the tip of each wire is bulbous with a diameter of about 200–250 nm due to the presence of (Au) catalyst, and is consistent with the VLS growth mechanism.

Figure 19.9a shows a low magnification image of high-density growth of UNCD-coated SiNWs. The uniform UNCD grains coated on SiNWs, obtained using paraffin wax as diamond seed, indicate that

Figure 19.8 SEM images of the as-grown (a) core/shell structure SiNW/SiO$_x$ showing (Au) catalyst at its tip. (b) High density of SiNWs at low magnification.

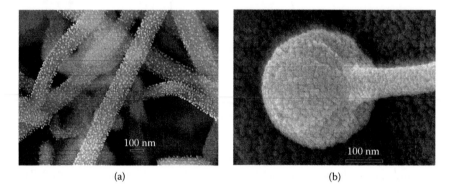

Figure 19.9 SEM images of the as-grown (a) UNCD uniformly coated SiNWs and (b) apex at high magnification coated with UNCD.

the paraffin wax on as-grown SiNW acts as diamond nucleation sites in the HF-CVD environment. The UNCD coating is uniform throughout the wire length, as compared with published reports (Liu et al. 1994; Kiselev et al. 2005; Tzeng et al. 2007). The magnified image of a single nanowire also shows a uniformly UNCD-coated bulbous tip (shown in Figure 19.9b).

19.3.2 RAMAN SPECTROSCOPY

The room temperature Raman spectrum of the UNCD-coated SiNWs is shown in Figure 19.10. It was taken in the range of 100–2000 cm^{-1}, revealing a strong first-order Si transverse optical (TO) phonon mode at 514 cm^{-1}. The symmetric and high intensity peak point to the high crystallinity of the NWs (Niu et al. 2004). Compared to bulk Si (521 cm^{-1}), a downshift is observed for the samples, which is not caused by the phonon confinement effect (Meier et al. 2006), since the diameter of the SiNWs is relatively large compared to the excitonic Bohr radius of Si. The redshift can be attributed to the tensile strain experienced by the NWs (Dhara and Giri 2011) due to the coating (shown in Figure 19.9). There are two small and broad peaks at around 925 and 290 cm^{-1} corresponding to the second-order transverse optical (2TO) and transverse acoustic (TA) phonon modes of Si, respectively (Liu et al. 2001).

The inset of Figure 19.2 (magnified by 25x) shows a typical visible Raman spectrum of UNCD (Wang et al. 2012) coating over SiNWs (Tzeng et al. 2007) in the range from 1000 to 2000 cm^{-1}, revealing a broad D band around 1347 cm^{-1} arising from a disordered sp^2 carbon (Wang et al. 2012) present at the UNCD grain boundaries. The Raman cross-section of Si is much higher than that of diamond, accounting for the relative intensities. The broadening of the D peak can be attributed to the phonon confinement effect due to the small grain size of UNCD (~5 nm) (Maillard-Schaller et al. 1999). The highly optically absorbing sp^2 bonded carbon precludes observation of the sp^3 Raman signal, since the visible Raman is about 50–250 times

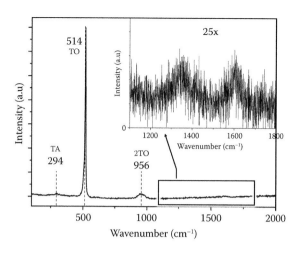

Figure 19.10 Raman spectrum of UNCD-coated SiNWs. Inset shows the UNCD spectrum in the range of 1000–2000 cm⁻¹. (Adapted from Palomino, J., et al., *ACS Appl. Mater. Interfaces*, 6, 13815–13822, 2014.)

more sensitive to sp^2-bonded carbon than to the sp^3-bonded carbon (Xiao et al. 2004). The band around 1600 cm⁻¹ corresponds to sp^2-hybridized carbon, indicating the presence of graphitic carbon accumulated at the grain boundaries (Varshney et al. 2012).

19.3.3 HIGH-RESOLUTION TRANSMISSION ELECTRON MICROSCOPY

Examination of nanoparticles using TEM and electron energy loss spectroscopy (EELS) is necessary to unambiguously identify the nature of the material. Figure 19.11 shows TEM images of the UNCD-decorated SiNWs revealing that they are uniformly coated with dense nanoparticles (UNCD crystals). Figure 19.11a shows a constant diameter wire, which consists of a core-shell structure with numerous nano-sized particles of sizes ranging from 3 to 10 nm, dispersed uniformly along the whole wire. Figure 19.11b shows a magnified image of the bulbous wire tip revealing that it is also covered with UNCD nanoparticles. Figure 19.11c is a magnified image of the linear portion of the UNCD/SiNW that reveals the core-shell structure of the SiNW having a UNCD shell thickness of around 10–20 nm.

Figure 19.12a depicts the HR-TEM micrograph of UNCD grown on SiNWs, suggesting that the UNCD is comprised of small grains (~5 nm). On further magnification (Figure 19.12b), the crystalline core-shell structure of the wire with uniform coating of UNCD crystals is observed. Figure 19.12c shows the lattice fringes of the nanowire crystalline core, revealing an interplanar spacing of about 0.31 nm, which matches with the (111) orientation of Si and is consistent with published data (Suzuki et al. 1991). The HR-TEM image of the wires coated with UNCD particles is shown in Figure 19.12d, indicating that the lattice spacing is about 0.205 nm, which is a typical lattice parameter for diamond (Tzeng et al. 2008), consistent with the diamond (111) planes. The data were obtained by HR-TEM measurements and the subsequent calculations were performed using DigitalMicrograph™ software.

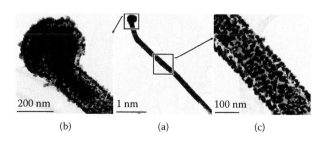

Figure 19.11 TEM images of (a) UNCD-coated Si NW (b) bulbous tip of the wire (c) linear part of the wire. (Reprinted from Palomino, J., et al., *ACS Appl. Mater. Interfaces*, 6, 13815, 2014. With permission.)

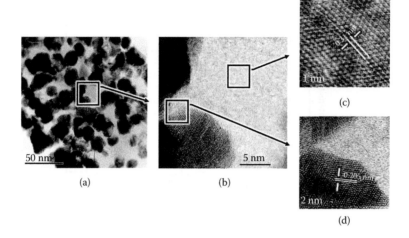

Figure 19.12 High-resolution transmission electron microscopic scanning mode (STEM) images of (a) UNCD-coated SiNW (b) magnified image of the wire showing (c) interplanar spacing of the Si crystal and (d) diamond crystals. (Reprinted from Palomino, J., et al., *ACS Appl. Mater. Interfaces*, 6, 13815, 2014. With permission.)

19.3.4 ELECTRON ENERGY LOSS SPECTROSCOPY

The UNCD growth was further explored by EELS, which is sensitive to the local chemical bonding of carbon, and was carried out (using a beam size of 20 nm) on different regions (see Figure 19.13). The EELS spectrum was recorded for the nanowire core (shown in Figure 19.13a) and it exhibits distinct and well-established spectral features of Si in the range of 100–160 eV, as shown in Figure 19.13b. The small peak at 107 eV is attributed to the presence of SiO_x (Shakerzadeh et al. 2011) coming from atmospheric contamination. Figure 19.13c was recorded for the UNCD coating showing the presence of a weak band feature at 285.23 eV, corresponding to the C 1s → π^* transition and indicating the presence of sp^2-C, in agreement with the Raman spectrum discussed above. It shows the C 1s→ σ^* transition around 292.5 eV which is consistent with sp^3-bonded carbon (Chen et al. 2013). Figure 19.13c reveals a dip at ~302.4 eV corresponding to the second absolute band gap of diamond, which is a fingerprint that confirms the diamond nature of the dense nanograins (Chen et al. 2008).

19.3.5 ENERGY DISPERSIVE SPECTROSCOPY

Energy dispersive spectroscopic (EDS) elemental mapping (Figure 19.14) was used to study the spatial distribution of Si, Au, O, and C on the UNCD-coated SiNWs. Figure 19.14a is the TEM morphological image of a representative wire, and Figures 19.14b through 19.14e show its elemental mapping analysis. Figure 19.14b reveals a strong C signal indicating a high density, uniform coating of UNCD throughout the wire as well as the tip. The Si elemental distribution profile points out that the silicon is located

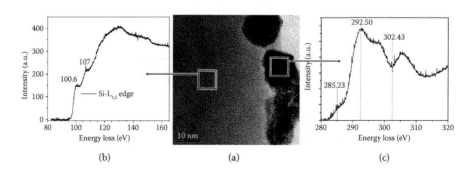

Figure 19.13 Electron energy loss spectra (EELS) of (b) SiNWs (c) UNCD crystals recorded at the region shown in Figure 19.3.6a. (Reprinted from Palomino, J., et al., *ACS Appl. Mater. Interfaces*, 6, 13815, 2014. With permission.)

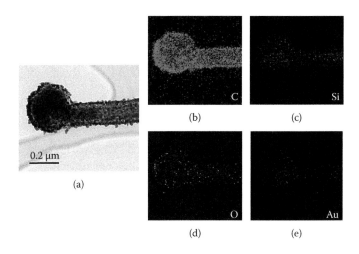

Figure 19.14 EDS color mapping of (a) UNCD-coated SiNW, with the elemental composition for (b) carbon, (c) silicon, (d) oxygen, and (e) gold. (Reprinted from Palomino, J., et al., *ACS Appl. Mater. Interfaces*, 6, 13815, 2014. With permission.)

mostly in the middle region (core) as compared to the edge area (shell), as shown in Figure 19.14c. The low intensity signals detected in the core region are due to the cylindrical geometry of the shell that covers the SiNWs, dissipating the signal from the cores. Oxygen is mainly distributed in the shell region of the wires (Figure 19.14d), which is ascribed to the SiO_x shell. Figure 19.14e shows that elemental gold is mainly distributed at the tip of the NW, confirming the VLS (Wagner and Ellis 1964) mechanism of the Si nanowire growth. The elemental analysis indicates that the SiNWs are coated with SiO_x and UNCD, forming core/shell heterostructures. This result is consistent with the observed bright/dark contrast in the TEM images discussed earlier.

19.4 ELECTRON FIELD EMISSION (EFE) BACKGROUND

19.4.1 FOWLER–NORDHEIM (F-N) THEORY

At low temperatures, the metal can be considered as a potential box filled with electrons up to the Fermi level, which lies below the vacuum level by some amount of energy (Figure 19.15), and the difference in energy between Fermi and vacuum levels is known as the work function, ϕ. In this context, the vacuum level represents the potential energy of a stationary electron outside the metal in the absence of an external field. When a strong external field is applied, the potential outside the metal is deformed along the line *V–eEx*, forming a triangular barrier, through which electrons can tunnel, and the emission occurs mostly from the Fermi level proximities, as shown in Figure 19.15.

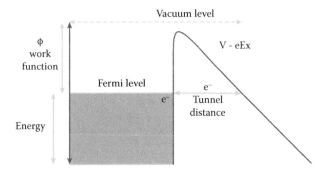

Figure 19.15 Diagram of the energy-level scheme for EFE.

Nanowires and nanotubes

In the context of the F-N plot, the $\ln\left[\dfrac{J}{E^2}\right] - \dfrac{1}{E}$ plot describes the exponential dependence relationship between the applied electric field and emission current. The straight line is indicative that the EFE from the material under analysis follows F-N behavior, and is represented by the simplified F-N equation.

$$J = \frac{A\beta^2 E^2}{\phi} \cdot \exp\left[\frac{-B\phi^{3/2}}{\beta E}\right] \tag{19.1}$$

where J is the current density, E is the applied field strength, β is the field enhancement factor, ϕ is the work function of the emitter material, and A and B are constants, corresponding to A = 1.56 x 10^{-10} AV^{2}eV and B = -6.83 x 10^3 V(eV)$^{3/2}$ (μm)$^{-1}$ respectively (Kulkarni et al. 2005). The local electric field E_L can be related to the macroscopic field E_M by $E_L = \beta E_M$, which indicate that high values of β signify high local electric field due to the presence of nanostructures such as SiNWs, CNTs, and UNCD coating. Therefore, β is a parameter dependent on the geometry of the nanostructure, crystal structure, and density of emitting points, and is used to determine the degree of field emission enhancement. The work function of a particular nanomaterial, β, can be calculated from the slope of the F-N plot using the expression

$$\beta = \frac{-B\phi^{3/2}}{slope} \tag{19.2}$$

The F-N plot can provide information related to the number of materials participating in the electron emission. For instance, the F-N plot of bare SiNWs is fitted with a single slope which corresponds to a specific β value. The UNCD-coated SiNWs give an F-N plot where the enhancement factor β can be calculated from two slopes corresponding to UNCD and SiNWs.

19.4.2 EFE MEASUREMENT SYSTEM

EFE measurement systems are typically comprised of the following components: (i) a vacuum chamber, capable of minimum pressures of ~5 x 10^{-7} Torr, (ii) a power supply to provide the potential difference between anode and cathode, and (iii) an electrometer to measure the field emission currents. Figure 19.16 shows a schematic of the EFE system where the measuring configuration

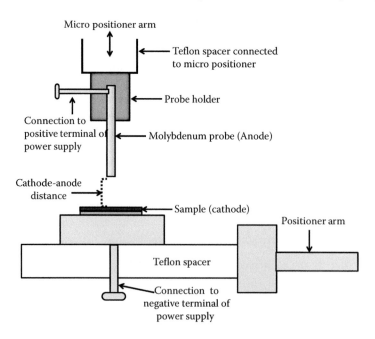

Figure 19.16 Schematic of EFE system.

consists of a cylindrical anode configuration. In this configuration, the macroscopic electric field (E_M) on the sample (i.e., cathode) is calculated accurately by $E_M = \dfrac{V}{d_{AC}}$, where V is the voltage applied to the anode and d_{AC} is the distance between the anode and cathode. The anode is a solid Mo cylindrical rod of 3 mm diameter and two inches long, which is mounted on a micropositioner with an accuracy of ±2 μm. The sample (cathode) is fixed on an aluminum support by using silver paint to provide good electrical conduction. A power supply is used to provide the voltage across anode and cathode, and the field emitted current is measured using an electrometer with a detection limit of 10^{-14} A. However, currents lower than 1×10^{-12} A were considered as background noise level. The turn-on electric field (E_t) is defined as the electric field necessary to emit a current of 1 nA.

19.5 SiNWs AND UNCD/SiNWs AS ELECTRON FIELD EMITTERS

Silicon- and carbon-based nanomaterials have attracted attention as electron field emitters. Most of the carbon materials containing both sp^2 and sp^3 bonds have been studied, including carbon nanotubes (CNTs) (Bonard et al. 1998) and diamond (Lu et al. 2006), which demonstrated excellent field emission characteristics. SiNWs and nanorods (Au et al. 1999), have also shown strong field emission properties. Thus carbon- and silicon-based composite nanomaterials are potential candidates for enhanced cold emitters.

Typical cold cathode materials like SiNWs possess a high turn-on field compared to CNTs. CNTs have some disadvantages, including poor stability and short lifetime. Some researchers have observed the modification of surface morphology due to the evaporation of CNTs in EFE experiments. Diamond can overcome these problems due to its good mechanical and chemical stability, high thermal conductivity, and high resilience against radiation (Varshney et al. 2013). Furthermore, diamond possesses NEA (Yamaguchi et al. 2009), which facilitates the electron emission at low electric fields.

19.5.1 CURRENT DENSITY MEASUREMENT OF SiNWs AND UNCD/SiNWs FIELD EMITTERS

Field emission measurements corresponding to a cathode–anode distance of ~100 μm, and at room temperature after conditioning the films at ~1 μA at their respective field for about 2 h are shown below. The measured current density as a function of the macroscopic electric field is shown in Figure 19.17, indicating that bare SiNWs can be turned on at $E_0 = 4.3 ± 0.1$ V/μm, while the UNCD/SiNW heterostructures exhibit a lower threshold field of $E_0 = 3.7 ± 0.1$ V/μm. Emission current density of bare SiNWs goes as high as ~0.1 mA/cm² at 25 V/μm, whereas the current density for UNCD/SiNW is one order of magnitude higher, reaching ~2 mA/cm² at 25 V/μm, and saturation was not observed at this field.

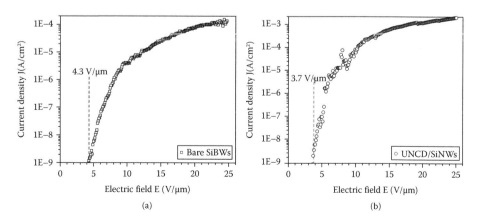

Figure 19.17 Field emission J vs. E plots for (a) as-grown bare SiNWs and (b) UNCD/SiNWs film. (Adapted from Palomino, J., et al., *ACS Appl. Mater. Interfaces*, 6, 13815–13822, 2014.)

Table 19.1 Comparison of field emission properties for CNTs, SiNWs, and UNCD

NANOMATERIAL	E_o (V/m)[a]	JMAX (mA/cm²)[b]	EJMAX (V/μm)[c]	REFERENCES
CNTs	2.3	0.3	7.5	(Katar et al. 2008, 2009)
SiNWs	4.3	0.1	25	present work
UNCD	5.6	0.1	18	(Varshney et al. 2014)
UNCD/CNTs	1.9	2.0	20	(Varshney et al. 2013)
UNCD/SiNWs	3.7	2.0	25	present work

Source: Palomino, J., et al., *ACS Appl. Mater. Interfaces,* 6, 13815–13822, 2014.
[a] E_o: turn-on field
[b] J_{max}: maximum EFE current density
[c] E_{Jmax}: field required for J_{max}

Table 19.1 shows the enhanced EFE properties of UNCD/SiNWs as compared to those of UNCD, bare SiNWs, CNTs, and UNCD/CNTs. All of these materials alone exhibit relatively low current densities of 0.1, 0.3, and 0.1 mA/cm², respectively. However, by coating the SiNWs and CNTs with UNCD, a synergistic effect comes into action and the current densities are enhanced by about one order of magnitude. The enhancement in EFE properties of UNCD/SiNWs can be attributed to the good electrical contact between the sp²-terminated UNCD and SiNWs, which facilitates the transfer of electrons from Si to UNCD, thus circumventing the silicon oxide layer. Moreover, the presence of sp² hybridized carbon around the diamond crystallites and the geometrical enhancement factor (diameters ~5–10 nm) also play an important role in facilitating the electron emission process itself. The high density of UNCD crystallites is a key parameter influencing the EFE properties. The increment of current density can be attributed to the combination of high density of UNCD particles coating the SiNWs and the high density of the SiNWs network.

19.5.2 FOWLER-NORDHEIM PLOT ANALYSIS FOR SiNWs AND UNCD/SiNWs FIELD EMITTERS

The F-N plots are shown in Figure 19.18, where linear regions are observed indicating that the field emission from SiNWs and UNCD/SiNWs follows the F-N behavior. From the slope of linear regions and assuming that the work function of SiNWs is close to 4.7 eV (Minami et al. 1998), β can be calculated using Equation 19.2. High values of β indicate high local electric field due to the presence of one-dimensional SiNWs, enhanced by UNCD coating.

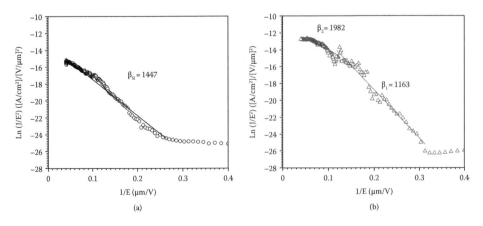

Figure 19.18 F-N Ln[J/E²] vs [1/E] plots for (a) as-grown base SiNWs and (b) UNCD/SiNWs films. (Adapted from Palomino, J., et al., *ACS Appl. Mater. Interfaces,* 6, 13815–13822, 2014.)

Nanowires and nanotubes

For bare SiNWs, the F-N plot is fitted with a single slope which corresponds to a β value of ~1447. The enhancement factor calculated for the two slopes for UNCD/SiNWs correspond to the β values of ~1163 at low field and ~1982 at high field. These β values are superior to those reported by Thomas et al. (2012), and much higher than values reported by Tzeng et al. (2007, 2008). The two emission regimes distinguished for UNCD correspond to the low field regime <7 V/m, where electrons are emitted mainly from the sp² carbon around the UNCD crystallites that are conformally distributed along the SiNWs. The direct contact between UNCD and Si avoids the oxide layer barrier and takes advantage of the NEA of diamond (Liao et al. 1998). The second slope at fields >7 V/m corresponds to higher local electric fields that are required to obtain emission directly from the SiNWs.

19.6 SUMMARY

Silicon nanowires can be produced by different growth methods and using a variety of catalyst materials, with Au as the most popular material because of its good silicon solubility. Thermal CVD is a good method to grow SiNWs which follows the tip growth VLS mechanism. Ultrananocrystalline diamond coating on SiNWs can be fabricated by HF-CVD using paraffin wax (polymer) as a non-abrasive seeding method. The HR-TEM and EELS analyses of the UNCD coating on SiNWs revealed a grain size of 5–10 nm. The UNCD/SiNW nanostructures show enhanced electron field emission properties as compared to the EFE performance of other field emitters, including bare SiNWs. The enhanced EFE properties of UNCD/SiNW are ascribed to the synergistic effects of UNCD crystallites and SiNWs, where the presence of sp² hybridized carbon around the diamond crystallites and the geometrical enhancement factor play an important role in facilitating the electron emission process.

ACKNOWLEDGMENTS

This research was carried out under the auspices of the Institute for Functional Nanomaterials (NSF Grant 1002410), PR NASA EPSCoR (NASA Cooperative Agreement NNX07AO30A) and PR DOE EPSCoR (DOE Grant DE-FG02-08ER46526, J.P.). We gratefully acknowledge the use of research facilities of various researchers at the University of Puerto Rico: Sputtering (Dr. Luis F. Fonseca); Micro-Raman Spectroscopy (Dr. Ram S. Katiyar and Mr. William Pérez).

REFERENCES

Armaroli N, Balzani V. (2007). The future of energy supply: Challenges and opportunities. *Angew Chem Int Ed Engl* 46: 52–66.

Au FCK, Wong KW, Tang YH, Zhang YF, Bello I, Lee ST. (1999). Electron field emission from silicon nanowires. *Appl Phys Lett* 75: 1700–1702.

Auciello O, Sumant AV. (2010). Status review of the science and technology of ultrananocrystalline diamond (UNCD™) films and application to multifunctional devices. *Diam Relat Mater* 19: 699–718.

Bhaskar K, Pardoen T, Passi V, Raskin J-P. (2013). Piezoresistance of nano-scale silicon up to 2 GPa in tension. *Appl Phys Lett* 102: 31911.

Biaggi-Labiosa A, Solá F, Resto O, Fonseca LF, González-Berríos A, De Jesús J, Morell G. (2008). Nanocrystalline silicon as the light emitting material of a field emission display device. *Nanotechnology* 19: 225202.

Bonard J-M, Salvetat J-P, Stöckli T, de Heer WA, Forró L, Châtelain A. (1998). Field emission from single-wall carbon nanotube films. *Appl Phys Lett* 73: 918–920.

Chan CK, Peng H, Liu G, McIlwrath K, Zhang XF, Huggins RA, Cui Y. (2008). High-performance lithium battery anodes using silicon nanowires. *Nat Nanotechnol* 3: 31–35.

Chen H-C, Chen S-S, Wang W-C, Lee C-Y, Guo J, Lin I-N, Chang C-L. (2013). The potential application of ultrananocrystalline diamond films for heavy ion irradiation detection. *AIP Adv* 3: 62113.

Chen J, Deng SZ, Xu NS, Wang S, Wen X, Yang S, Yang C, Wang J, Ge W. (2002). Field emission from crystalline copper sulphide nanowire arrays. *Appl Phys Lett* 80: 3620–3622.

Chen K, Deng J, Zhao F, Cheng G, Zheng R. (2010). Influence of Zn ion implantation on structures and field emission properties of multi-walled carbon nanotube arrays. *Sci China Technol Sci* 53: 776–781.

Chen KI, Li BR, Chen YT. (2011). Silicon nanowire field-effect transistor-based biosensors for biomedical diagnosis and cellular recording investigation. *Nano Today* 6: 131–154.

Chen L-J, Tai N-H, Lee C-Y, Lin I-N. (2007). Effects of pretreatment processes on improving the formation of ultra-nanocrystalline diamond. *J Appl Phys* 101: 64308.

Chen T-M, Hung J-Y, Pan F-M, Chang L, Wu S-C, Tien T-C. (2009). Pulse electrodeposition of iridium oxide on silicon nanotips for field emission study. *J Nanosci Nanotechnol* 9: 3264–3268.

Chen W, Yao H, Tzang CH, Zhu J, Yang M, Lee S-T. (2006). Silicon nanowires for high-sensitivity glucose detection. *Appl Phys Lett* 88: 213104.

Chen YC, Zhong XY, Konicek AR, Grierson DS, Tai NH, Lin IN, Kabius B, et al. (2008). Synthesis and characterization of smooth ultrananocrystalline diamond films via low pressure bias-enhanced nucleation and growth. *Appl Phys Lett* 92: 133113.

Cheng SZD. (2007). Materials science: Polymer crystals downsized. *Nature* 448: 1006–1007.

Dhara S, Giri PK. (2011). Effect of growth temperature on the catalyst-free growth of long silicon nanowires using radio frequency magnetron sputtering. *Int J Nanosci* 10: 13–17.

Dick K, Dhanasekaran T, Zhang Z, Meisel D. (2002). Size-dependent melting of silica-encapsulated gold nanoparticles. *J Am Chem Soc* 124: 2312–2317.

Fan G, Zhu H, Wang K, Wei J, Li X, Shu Q, Guo N, Wu D. (2011). Graphene/silicon nanowire Schottky junction for enhanced light harvesting. *ACS Appl Mater Interfaces* 3: 721–725.

Feng SQ, Yu DP, Zhang HZ, Bai ZG, Ding Y. (2000). The growth mechanism of silicon nanowires and their quantum confinement effect. *J Cryst Growth* 209: 513–517.

Ha JK, Chung BH, Han SY, Choi JO. (2002). Drastic changes in the field emission characteristics of a Mo-tip field emitter array having PH[sub 3]-doped a-Si:H as a resistive layer material throughout vacuum packaging processes in a field emission display. *J Vac Sci Technol B Microelectron Nanometer Struct* 20: 2080–2084.

Himpsel F, Knapp J, VanVechten J, Eastman D. (1979). Quantum photoyield of diamond(111)—A stable negative-affinity emitter. *Phys Rev B* 20: 624–627.

Huang CT, Hsin CL, Huang KW, Lee CY, Yeh PH, Chen US, Chen LJ. (2007). Er-doped silicon nanowires with 1.54 μm light-emitting and enhanced electrical and field emission properties. *Appl Phys Lett* 91: 93133.

Iijima S. (1991). Helical microtubules of graphitic carbon. *Nature* 354: 56–58.

Katar S, Labiosa A, Plaud AE, Mosquera-Vargas E, Fonseca L, Weiner BR, Morell G. (2009). Silicon encapsulated carbon nanotubes. *Nanoscale Res Lett* 5: 74–80.

Katar SL, González-Berríos A, De Jesus J, Weiner B, Morell G. (2008). Direct deposition of bamboo-like carbon nanotubes on copper substrates by sulfur-assisted HFCVD. *J Nanomater* 2008: 1–7.

Kiselev NA, Hutchison JL, Roddatis VV, Stepanova AN, Aksenova LL, Rakova EV, Mashkova ES, Molchanov VA, Givargizov EI. (2005). TEM and HREM of diamond crystals grown on Si tips: Structure and results of ion-beam-treatment. *Micron* 36: 81–88.

Klimovskaya A, Sarikov A, Pedchenko Y, Voroshchenko A, Lytvyn O, Stadnik A. (2011). Study of the formation processes of gold droplet arrays on Si substrates by high temperature anneals. *Nanoscale Res Lett* 6: 151.

Kulkarni NN, Bae J, Shih C-K, Stanley SK, Coffee SS, Ekerdt JG. (2005). Low-threshold field emission from cesiated silicon nanowires. *Appl Phys Lett* 87: 213115.

Kumar M, Ando Y. (2010). Chemical vapor deposition of carbon nanotubes: A review on growth mechanism and mass production. *J Nanosci Nanotechnol* 10: 3739–3758.

LaDou J. (1983). Potential occupational health hazards in the microelectronics industry. *Scand J Work Environ Health* 9: 42–46.

Lee NS, Chung DS, Han IT, Kang JH, Choi YS, Kim HY, Park SH, et al. (2001). Application of carbon nanotubes to field emission displays. *Diam Relat Mater* 10: 265–270.

Lee SH, Kim JW, Lee TI, Myoung JM. (2016). Inorganic nano light-emitting transistor: P-type porous silicon nanowire/n-type ZnO nanofilm. *Small* 12: 4222–4228.

Liao M, Zhang Z, Wang W, Liao K. (1998). Field-emission current from diamond film deposited on molybdenum. *J Appl Phys* 84: 1081–1084.

Liu J. (1995). Field emission characteristics of diamond coated silicon field emitters. *J Vac Sci Technol B Microelectron Nanom Struct* 13: 422–426.

Liu J, Zhirnov VV., Wojak GJ, Myers AF, Choi WB, Hren JJ, Wolter SD, McClure MT, Stoner BR, Glass JT. (1994). Electron emission from diamond coated silicon field emitters. *Appl Phys Lett* 65: 2842–2844.

Liu XH, Zhong L, Huang S, Mao SX, Zhu T, Huang JY. (2012). Size-dependent fracture of silicon nanoparticles during lithiation. *ACS Nano* 6: 1522–1531.

Liu Y, Fan S. (2005). Field emission properties of carbon nanotubes grown on silicon nanowire arrays. *Solid State Commun* 133: 131–134.

Liu Z., Xie S., Zhou W., Sun L., Li Y., Tang D., Zou X., Wang C., Wang G. (2001). Catalytic synthesis of straight silicon nanowires over Fe containing silica gel substrates by chemical vapor deposition. *J Cryst Growth* 224: 230–234.

Liu Z, Pan Z, Sun L, Tang D, Zhou W., Wang G, Qian L., Xie S. (2000). Synthesis of silicon nanowires using AuPd nanoparticles catalyst on silicon substrate. *J Phys Chem Solids* 61: 1171–1174.

Lu X, Yang Q, Chen W, Xiao C, Hirose A. (2006). Field electron emission characteristics of diamond films with different grain morphologies. *J Vac Sci Technol B Microelectron Nanometer Struct* 24: 2575–2580.

Maillard-Schaller E, Kuettel OM, Diederich L, Schlapbach L, Zhirnov VV, Belobrov PI. (1999). Surface properties of nanodiamond films deposited by electrophoresis on Si(100). *Diam Relat Mater* 8: 805–808.

Meier C, Lüttjohann S, Kravets VG, Nienhaus H, Lorke A, Wiggers H. (2006). Raman properties of silicon nanoparticles. *Phys E Low Dimens Syst Nanostruct* 32: 155–158.

Messier R, Yarbrough WA. (1990). Current issues and problems in the chemical vapor deposition of diamond. *Science* 247: 688–696.

Minami T, Miyata T, Yamamoto T. (1998). Work function of transparent conducting multicomponent oxide thin films prepared by magnetron sputtering. *Surf Coatings Technol* 108–109: 583–587.

Mitsuda Y, Kojima Y, Yoshida T, Akashi K. (1987). The growth of diamond in microwave plasma under low pressure. *J Mater Sci* 22: 1557–1562.

Morales A, Lieber C. (1998). A laser ablation method for the synthesis of crystalline semiconductor nanowires. *Science* 279: 208–211.

Niu J, Sha J, Yang Q, Yang D. (2004). Crystallization and Raman shift of array-orderly silicon nanowires after annealing at high temperature. *Jpn J Appl Phys* 43: 4460–4461.

Ok Y-W, Seong T-Y, Choi C-J, Tu KN. (2006). Field emission from Ni-disilicide nanorods formed by using implantation of Ni in Si coupled with laser annealing. *Appl Phys Lett* 88: 0431061–0431063.

Okano K, Koizumi S, Silva SRP, Amaratunga GAJ. (1996). Low-threshold cold cathodes made of nitrogen-doped chemical-vapour-deposited diamond. *Nature* 381: 140–141.

Palomino J, Varshney D, Resto O, Weiner BR, Morell G. (2014). Ultrananocrystalline diamond-decorated silicon nanowire field emitters. *ACS Appl Mater Interfaces* 6: 13815–13822.

Palomino J, Varshney D, Weiner BR, Morell G. (2015). Study of the structural changes undergone by hybrid nanostructured Si-CNTs employed as an anode material in a rechargeable lithium-ion battery. *J Phys Chem C* 119: 21125–21134.

Peng K, Huang Z, Zhu J. (2004). Fabrication of large-area silicon nanowire p–n junction diode arrays. *Adv Mater* 16: 73–76.

Qi WH, Wang MP. (2004). Size and shape dependent melting temperature of metallic nanoparticles. *Mater Chem Phys* 88: 280–284.

Rinzler AG, Hafner JH, Nikolaev P, Nordlander P, Colbert DT, Smalley RE, Lou L, Kim SG, Tománek D. (1995). Unraveling nanotubes: Field emission from an atomic wire. *Science* 269: 1550–1553.

Schmidt V, Wittemann JV, Senz S, Gösele U. (2009). Silicon nanowires: A review on aspects of their growth and their electrical properties. *Adv Mater* 21: 2681–2702.

Schwartz M, Nguyen TC, Vu XT, Weil M, Wilhelm J, Wagner P, Thoelen R, Ingebrandt S. (2016). DNA detection with top-down fabricated silicon nanowire transistor arrays in linear operation regime. *Phys Status Solidi Appl Mater Sci* 1519: 1510–1519.

Shakerzadeh M, Teo EHT, Sorkin A, Bosman M, Tay BK, Su H. (2011). Plasma density induced formation of nanocrystals in physical vapor deposited carbon films. *Carbon* 49: 1733–1744.

Shao M, Ma DDD, Lee S-T. (2010). Silicon nanowires—Synthesis, properties, and applications. *Eur J Inorg Chem* 2010: 4264–4278.

She JC, Huq SE, Chen J, Deng SZ, Xu NS. (1999). Comparative study of electron emission characteristics of silicon tip arrays with and without amorphous diamond coating. *J Vac Sci Technol B Microelectron Nanometer Struct* 17: 592–595.

Suzuki M, Kudoh Y, Homma Y, Kaneko R. (1991). Monoatomic step observation on Si(111) surfaces by force microscopy in air. *Appl Phys Lett* 58: 2225–2227.

Tang W, Dayeh SA, Picraux ST, Huang JY, Tu KN. (2012). Ultrashort channel silicon nanowire transistors with nickel silicide source/drain contacts. *Nano Lett* 12: 3979–3985.

Thomas JP, Chen H-C, Tseng S-H, Wu H-C, Lee C-Y, Cheng HF, Tai N-H, Lin I-N. (2012). Preferentially grown ultranano c-diamond and n-diamond grains on silicon nanoneedles from energetic species with enhanced field-emission properties. *ACS Appl Mater Interfaces* 4: 5103–5108.

Tian B, Zheng X, Kempa TJ, Fang Y, Yu N, Yu G, Huang J, Lieber CM. (2007). Coaxial silicon nanowires as solar cells and nanoelectronic power sources. *Nature* 449: 885–889.

Tzeng Y-F, Lee C-Y, Chiu H-T, Tai N-H, Lin I-N. (2008). Electron field emission properties on ultra-nano-crystalline diamond coated silicon nanowires. *Diam Relat Mater* 17: 1817–1820.

Tzeng Y-F, Liu K-H, Lee Y-C, Lin S-J, Lin I-N, Lee C-Y, Chiu H-T. (2007). Fabrication of an ultra-nanocrystalline diamond-coated silicon wire array with enhanced field-emission performance. *Nanotechnology* 18: 435703.

Valentín LA, Carpena-Nuñez J, Yang D, Fonseca LF. (2013). Field emission properties of single crystal chromium disilicide nanowires. *J Appl Phys* 113: 14308.

Varshney D. (2012). Electron emission of graphene-diamond hybrid films using paraffin wax as diamond seeding source. *World J Nano Sci Eng* 2: 126–133.

Varshney D, Kumar A, Guinel MJ-F, Weiner BR, Morell G. (2012). Spontaneously detaching self-standing diamond films. *Diam Relat Mater* 21: 99–102.

Varshney D, Makarov VI, Saxena P, Guinel MJF, Kumar A, Scott JF, Weiner BR, Morell G. (2011). Electron emission from diamond films seeded using kitchen-wrap polyethylene. *J Phys D Appl Phys* 44: 85502.

Varshney D, Palomino J, Gil J, Resto O, Weiner BR, Morell G. (2014). New route to the fabrication of nanocrystalline diamond films. *J Appl Phys* 115: 54304.

Varshney D, Sumant AV, Resto O, Mendoza F, Quintero KP, Ahmadi M, Weiner BR, Morell G. (2013). Single-step route to hierarchical flower-like carbon nanotube clusters decorated with ultrananocrystalline diamond. *Carbon* 63: 253–262.

Wagner RS, Ellis WC. (1964). Vapor-liquid-solid mechanism of single crystal growth. *Appl Phys Lett* 4: 89–90.

Wang W, Epur R, Kumta PN. (2011). Vertically aligned silicon/carbon nanotube (VASCNT) arrays: Hierarchical anodes for lithium-ion battery. *Electrochem Commun* 13: 429–432.

Wang X, Ocola LE, Divan RS, Sumant AV. (2012). Nanopatterning of ultrananocrystalline diamond nanowires. *Nanotechnology* 23: 75301.

Wang ZL, Luo Q, Li JJ, Wang Q, Xu P, Cui Z, Gu CZ. (2006). The high aspect ratio conical diamond tips arrays and their field emission properties. *Diam Relat Mater* 15: 631–634.

Weber CHM, Chiche A, Krausch G, Rosenfeldt S, Ballauff M, Harnau L, Göttker-Schnetmann I, Tong Q, Mecking S. (2007). Single lamella nanoparticles of polyethylene. *Nano Lett* 7: 2024–2029.

Won R. (2010). Photovoltaics: Graphene–silicon solar cells. *Nat Photonics* 4: 411.

Wu ZS, Deng SZ, Xu NS, Chen J, Zhou J, Chen J. (2002). Needle-shaped silicon carbide nanowires: Synthesis and field electron emission properties. *Appl Phys Lett* 80: 3829–3831.

Xiang B, Zhang Y, Wang Z, Luo XH, Zhu YW, Zhang HZ, Yu DP. (2005). Field-emission properties of TiO_2 nanowire arrays. *J Phys D Appl Phys* 38: 1152–1155.

Xiao X, Birrell J, Gerbi JE, Auciello O, Carlisle JA. (2004). Low temperature growth of ultrananocrystalline diamond. *J Appl Phys* 96: 2232–2239.

Yamaguchi H, Masuzawa T, Nozue S, Kudo Y, Saito I, Koe J, Kudo M, Yamada T, Takakuwa Y, Okano K. (2009). Electron emission from conduction band of diamond with negative electron affinity. *Phys Rev B* 80: 165321.

Zhao F, Cheng G-A, Zheng R-T, Zhao D-D, Wu S-L, Deng J-H. (2011). Field emission enhancement of Au-Si nanoparticle-decorated silicon nanowires. *Nanoscale Res Lett* 6: 176.

Zheng G, Patolsky F, Cui Y, Wang WU, Lieber CM. (2005). Multiplexed electrical detection of cancer markers with nanowire sensor arrays. *Nat Biotechnol* 23: 1294–1301.

Zhu YW, Zhang HZ, Sun XC, Feng SQ, Xu J, Zhao Q, Xiang B, Wang RM, Yu DP. (2003). Efficient field emission from ZnO nanoneedle arrays. *Appl Phys Lett* 83: 144–146.

20 Silicon nanowires for Li-based battery anode applications

Didier Pribat

Contents

20.1 INTRODUCTION

In our large cities, we are getting slowly asphyxiated and poisoned by the exhaust emissions from internal combustion engines (ICE) of automobiles, trucks, and motorcycles (Lagally et al. 2012), to the point that anthropogenic carbon nanotubes have been found recently in the respiratory system of children in Paris (Kolosnjaj-Tabi et al. 2015). Even though most modern automotive vehicles have been equipped with catalytic converters and an electronic regulation of fuel injection for quite a long time (Pribat and Velasco 1988), they still emit pollutants, particularly in developing countries, where regulations are not so tight and controls are more difficult to organize. The electrification of transportation, which until recently could be viewed as a trendy (albeit expensive) bourgeois-bohemian option, is becoming a public health issue as more and more people die from air pollution, particularly in India and China (see e.g., http://www.bbc.com/news/science-environment-35568249). Although fuel cells burning hydrogen are probably the best option for the long term (Schlapbach 2009; Tollefson 2010) battery-powered vehicles are today cheaper and seem to be within reach. Actually, one must realize that batteries and fuel cells both power the same electric motors, so that even for hydrogen fuel-cell vehicles, car companies are considering using auxiliary batteries to take advantage of regenerative braking, which would extend the driving range further. In other words, whether used as the main power source or as an auxiliary one, batteries will be needed for electric vehicles in the future. At this point, we would like to point out that even if electricity originates from coal-burning plants, so that electric vehicles generate in the end more CO_2 than their gasoline-powered counterparts (Larcher and Tarascon 2015), CO_2 sequestration techniques (which are not realistic for individual ICE vehicles) can be used in large power plants/factories using fossil fuels to produce electricity (Ciferno et al. 2009) or hydrogen.

When it comes to choosing a battery for transportation, although several chemistries/technologies are available, the lithium-ion option is the most realistic since it offers the highest energy density, more than five times higher than lead-acid systems and more than twice higher than nickel-metal hydride (Ni-MH) systems (Armand and Tarascon 2008). In lithium-ion batteries (LIBs), lithium ions are shuttled back and forth between a graphite anode and an oxide cathode ($LiMO_2$, where M is a transition metal such as Ni, Co, or Mn; all three metals can be used together yielding the so-called NCM cathodes) or a phosphate cathode (e.g., LiFePO$_4$, abbreviated as LFP) as the battery is charged and discharged (the so-called rocking chair operation, see below). For the anode, graphite is an interesting insertion material because it exhibits only a ~10% increase of the interlayer graphite spacing (along the c-axis) and less than 1% expansion in the basal graphite plane at maximum lithium (Li) insertion. Hence graphite anodes can withstand many charge–discharge cycles without degradations. However, graphite can only host one lithium atom for six carbon atoms, which corresponds to a modest capacity of 370 mAh/g[*].

Actually, LIBs, which have steadily progressed over the past 25 years, seem to have reached their asymptotic capacity value with the present combination of graphite at the anode and spinel, layered, or olivine oxides at the cathode. Progress is now marginal, while higher battery capacities are in demand for applications to electric vehicles as well as storage systems backing up renewable intermittent energy sources (solar, wind). Recently, silicon (Si) has made a remarkable breakout in the field of energy storage, aiming at replacing carbon in the anodes of lithium-based rechargeable batteries (Kasavajjula et al. 2007). The major interest of Si lies in the fact that it can store about 10 times more Li than graphite at room temperature (RT). However, this large Li uptake induces a strong volume expansion as well as phase changes of the host material, which result in cracking and rapid pulverization of the electrode upon charge–discharge cycling. The recent use of nanostructured Si (particularly Si nanowires) has enabled significant reliability/lifetime progress, as anodes based on such "nano Si" materials can now routinely withstand more than 1000 charge–discharge cycles. Because of their high surface-to-volume ratio, and because surface atoms have more freedom to relax, nanostructures can better accommodate volume changes without fracturing. An add-on advantage of anodes making use of nanomaterials is their improved charge–discharge rates, due to the much shorter diffusion distances for the Li species inside the nanostructures.

With an abundance of 28.2% by weight in the Earth's crust, Si ranks second, only overtaken by oxygen. Si is a so-called rock-forming element, which is always found in nature in oxidized compounds, mostly in association with other elements such as aluminum, magnesium, calcium etc. Even though most extracted and purified Si[†] is used in alloys with iron, aluminum, and copper, for example, Si is better known for its use in electron devices, electronic circuits, electronic systems, and brown products[‡]. Actually, Si (whether crystalline, polycrystalline or amorphous) is really the cornerstone of the electronics industry, supporting all possible sectors, from power devices and microprocessors to solar cells, including flat-panel displays, all sorts of mobile equipment, and medical X-ray detectors for digital radiography. It is therefore rather remarkable that Si can also extend its stranglehold to batteries, which are intimately linked to electronic devices, particularly mobile devices and systems.

This book chapter is devoted to the use of Si nanowires (SiNWs) for the fabrication of high-capacity Li-based battery anodes. SiNWs are such a realistic option for future anodes of Li-based batteries, that they have already been included as a milestone in some timeline history of batteries (http://www.upsbattery-center.com/blog/history-batteries-timeline/). Figure 20.1 shows their interest in battery systems used for electric vehicles.

After this short introduction, the next paragraph (Section 20.2) presents the basic functioning principle of LIBs. Section 20.3 is devoted to the synthesis of SiNWs, whether by growth techniques or by anisotropic etching of a crystalline structure. Section 20.4 highlights the electrochemical interactions of Li with Si, with emphasis on SiNWs. Various NW-based structures (essentially composites) are presented in Section 20.5 and their advantages and drawbacks are discussed. Finally, this chapter is concluded by an overview of industrial activities in the field of Si-based anodes/batteries, as well as by a few comments on future research and development directions.

---

[*] The performance of battery electrodes are usually measured by their capacities, expressed in Ah (1 Ah = 3600 coulombs) per unit weight or unit volume, corresponding, respectively, to the specific or volumetric capacity; the capacity represents the actual electrical charge (i.e., the number of Li ions) that can be stored in the host material, whether at the anode or the cathode.

[†] Metallurgical grade Si is obtained by the carboreduction of SiO$_2$ at ~2000°C. It contains about 2% impurities and is not suited for electronic applications.

[‡] For electronic applications, metallurgical Si needs to be refined first and then regrown as monocrystalline ingots (e.g., by using the Czochralski process). Those ingots are machined and sliced into wafers before being purchased by companies manufacturing semiconductor devices and systems.

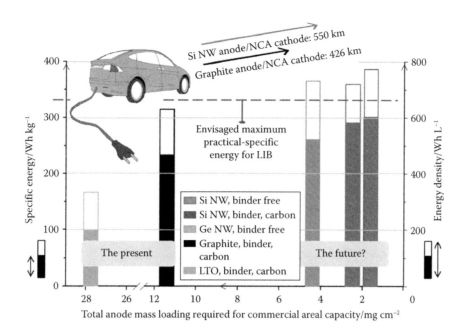

Figure 20.1 Estimated specific energy and energy density values evaluated for LIBs containing various anode materials. The values are plotted against the anode mass (active material plus conductive carbon and binder if applicable) required to achieve a commercially relevant loading of 4 mA h/cm². In each case, it is assumed that the cathode is LiNiCoAlO₂ (NCA), delivering a capacity of 200 mAh/g. The calculated electric vehicle ranges are based on the performance of the 85 kW h battery pack in a 373 hp Tesla Model S 85. (Kennedy, T., et al.: Advances in the application of silicon and germanium nanowires for high-performance lithium-ion batteries. *Advanced Materials*. 2016. 27. 5696–5704.Copyright with permission from Wiley-VCH. Reproduced with permission.)

20.2 HOW DO LIBs OPERATE?

There are many text books as well as review papers explaining the principle of batteries and LIBs (e.g., Winter and Brodd 2004; Nazri and Pistoia 2008; Reddy 2010; Ozawa 2012; Zhang and Zhang 2015), so we will be succinct here and just highlight the basics. Figure 20.2 shows schematic illustrations of the working principle of an LIB with a graphite anode and an LiCoO₂ cathode (Thackeray et al. 2012).

As the battery is operating, lithium ions are extracted from the graphite anode ($Li_xC_6 \rightarrow xLi^+ + xe^- + 6C$), they travel through the electrolyte, and are inserted into the layered cathode ($xLi^+ + xe^- + Li_{1-x}CoO_2 \rightarrow LiCoO_2$). Electrons released by the oxidation of Li at the anode are used in the external circuit to produce some work, before being reinjected at the cathode. When the battery is being recharged, lithium ions are extracted from the cathode ($LiCoO_2 \rightarrow yLi^+ + ye^- + Li_{1-y}CoO_2$; y is typically limited to ~0.5 for stability reasons) (Melot and Tarascon 2013) and the electrons injected in the external circuit reduce Li^+ ions at the anode ($yLi^+ + ye^- + 6C \rightarrow Li_yC_6$). Figure 20.3a shows the energy diagram of a charged LIB in open-circuit conditions (Goodenough and Kim 2010). Note however that electrons are not directly available at the Fermi levels of the electrodes, since they have to be released or absorbed through electrode reactions involving Li ions. This represents a fundamental difference between a battery and a capacitor.

Even if there are many variations concerning the cathode material (Thackeray et al. 2012; Melot and Tarascon 2013), most commercial LIBs use a graphite anode. However, the number of Li ions that can be accommodated by graphite (one Li for six carbons) corresponds to a capacity of 372 mAh/g*. This is a rather small value, all the more as the graphite load in an anode is only ~85% since it is necessary to add a binder,

---

* The energy stored in a battery is the product of capacity and voltage.

Nanowires and nanotubes

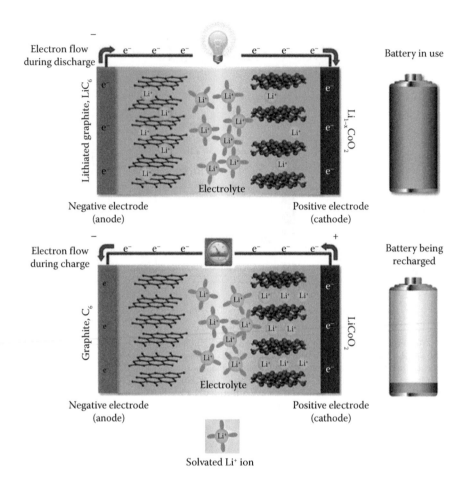

Figure 20.2 Schematic illustrations of the working principle of an LIB with a graphite anode and an $LiCoO_2$ cathode. Top: the battery is in use and Li ions are extracted from the anode, travel through the electrolyte (usually solvated, see Xu 2004) and are inserted into the $Li_{1-x}CoO_2$-layered cathode. Note that cations (Co^{4+}) have to change oxidation number and transform to Co^{3+} as Li ions are inserted into the cathode. Bottom: the battery is being recharged and Li ions are extracted from the cathode (Co^{3+} ions change to Co^{4+}), travel through the electrolyte, and are inserted into the graphite anode. At full charge, the anode composition is LiC_6, which corresponds to one Li entity for six carbon atoms. (Thackeray MM, et al., 2012, Electrical energy storage for transportation-approaching the limits of, and going beyond, lithium-ion batteries, *Energy Environ. Sci.* 5, 7854–7863, 2012. Reproduced by permission of The Royal Society of Chemistry.)

as well as a conductive powder for a proper functioning[*] (see a schematic representation, Figure 20.3b). Actually, metallic lithium (instead of graphite) should be the best choice for anodes of Li-based batteries, since (i) Li is the lightest metal (density ~0.53 g/cm³) and (ii) Li exhibits the lowest redox potential of the periodic table (–3.045 V against the standard hydrogen potential). Moreover, the capacity of a pure lithium anode is ~3860 mAh/g, more than 10 times that of graphite. Unfortunately, upon battery charging, Li metal tends to grow in the form of dendrites (Brissot et al. 1998; Xu 2004), which after several charge–discharge cycles can short-circuit the two electrodes, inducing risks of thermal runaway and fire (Wang et al. 2012). Because of this security problem, Li metal has been replaced by an insertion-type anode, resulting in the concept of Li-ion technology, where no Li metal appears anymore (Tarascon and Armand 2001).

---

[*] In a traditional battery electrode fabrication process, the Li-storing material (graphite for an anode) is incorporated into a slurry containing a binder (polymer) as well as a conductive additive (a powder based on amorphous carbon that does not react with Li). This slurry is subsequently deposited on the current collector of the battery electrode (usually a copper foil), typically by tape casting and dried in an oven before being used, in order to remove the solvents from the slurry. The anode is usually not lithiated when it is assembled in the battery cell and Li is provided by the cathode content.

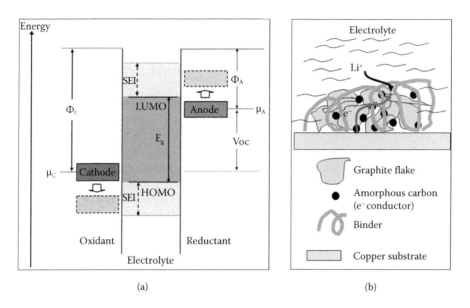

Figure 20.3 Energy levels of the different elements of an LIB and schematic representation of the microstructure of a commercial anode. (a) Energy levels (Fermi levels) of the anode and cathode relative to a vacuum reference (ϕ_A and ϕ_C represent, respectively, the work functions of the anode and cathode). V_{oc} is the open-circuit voltage. E_g (in green) is the energy window in which the electrolyte is thermodynamically stable (the HOMO and LUMO energy levels correspond, respectively, to the highest occupied and lowest unoccupied molecular orbitals of the less stable electrolyte component). However, the stability domain of the electrolyte can be kinetically extended (and the V_{oc} increased) by the formation of a passivating film (the so-called solid electrolyte interphase—SEI) on each electrode; this is represented by the pastel colors (light blue and pink) inside the dotted boxes. (Reprinted with permission from Goodenough, J.B., and Kim, Y., 2010, 22, 587–603. Copyright 2010 American Chemical Society.) (b) A typical anode is constituted by a copper current collector on which graphite flakes (one to several microns in diameter) are held together by an organic binder (e.g., polyvinylidene fluoride (PVDF)); an amorphous carbon powder (with nanometer-size grains) is usually added, in order to improve electron transport (red arrow) between the graphite flakes and the current collector as well as between the graphite flakes themselves. The anode has to be porous, in order to facilitate electrolyte penetration.

When considering the use of a battery electrode, several parameters have to be evaluated and taken into account (Pribat 2015). Although there is no hierarchical order, maybe the first parameter to consider is the capacity of the electrode, which has already been defined above (First footnote on page 454). Actually, a more relevant parameter is often the electrode capacity per unit area of the current collector, expressed in mAh/cm². This number takes into account the mass loading of active material per unit area of the current collector, a quantity which is always limited in practice. For instance, in a commercial Li-ion battery, the thickness of the graphite-based deposit on the current collector of the anode is usually limited to ~80 μm, because of delamination/adhesion problems at larger thickness values. Taking a density of 2 g/cm³ for graphite (whose capacity is 372 mAh/g), as well as a 85–90% proportion of active material in the slurry and assuming a ~30% porosity of the graphite-based deposit (the porosity is necessary for electrolyte permeation), this yields an anode capacity of 3.5 to 4 mAh/cm², depending on the exact porosity and exact proportion of active graphite in the slurry. Note that 4 mAh/cm² is the commercially relevant value chosen in Figure 20.1.

The next parameter concerns the open circuit voltage (V_{oc}) delivered by the battery hosting this electrode; for an anode, the idea is to maximize the V_{oc} for a given cathode, that is, exhibit (once the battery charged) a chemical potential as close as possible to that of pure Li. The third important parameter is the ability of the electrode to generate the so-called solid electrolyte interphase (SEI) film, which deserves a few words of explanation. As far as anodes are concerned, the SEI film usually forms naturally upon the first battery charge, as the imposed voltage induces a partial decomposition of the electrolyte. This decomposition occurs because the imposed voltage goes outside the thermodynamic stability window of the electrolyte. The SEI is an insulating film which blocks electron transfer, but is highly permeable to Li ions. If the SEI is mechanically and chemically stable, there is no more electrolyte decomposition after the first battery charge and the anode is stabilized, even though its chemical potential lies outside the electrolyte stability window (as shown in Figure 20.3a). The SEI composition depends on the particular electrolyte (solvent and lithium salt) used for battery fabrication and to a lesser extent on the anode material itself. For a graphite-based anode and the common $LiFP_6$/ethylene carbonate (EC)/diethyl carbonate (DEC) electrolyte, the SEI is typically composed of lithium compounds (Li_2CO_3, LiF, Li_2O), polyolefins, semicarbonates, etc.). Note that the stability of the SEI on graphite-based anodes is another reason for their large commercial success.

The fourth and extremely important parameter, determining the practical usefulness of an anode, is its cycle lifetime—the number of battery charge–discharge cycles it can withstand without significant degradation and capacity loss. Intuitively, one understands that this parameter will depend both on the magnitude of the capacity (the larger the capacity, the more Li can be packed, but the larger the swelling/straining/deformation of the host structure) as well as on the depth of charge/discharge of the electrode; for instance, if an alloying reaction is concerned, a small charge/discharge might avoid some phase transformations, the latter always being detrimental to the integrity of the electrode. The Coulombic efficiency (CE) is a closely related parameter, corresponding to the fraction of the prior charge that can be delivered during the following discharge. Of course, the CE has to be as close to 100% as possible. For a 99.9% value, corresponding to a 0.1% only loss per charge–discharge cycle (and which could look like a good performance), the capacity will be reduced to ~82% of its original value after 200 cycles and down to ~37% after 1000 cycles. For a ~90% capacity retention after 1000 cycles, one needs a 99.99% CE.

A fifth parameter to be considered is the charge/discharge rate of the anode, the so-called C-rate. The C-rate is determined by the magnitude of the current used to charge/discharge the battery, assuming the nominal capacity. For instance, for a pure graphite anode (capacity = 372 mAh/g), applying a 372 mA/g current will charge or discharge the electrode in 1 hour; this corresponds to cycling the electrode at 1C (or C/1) rate. In order to cycle the anode at a C/10 rate, the experimenter will apply a 37.2 mA/g charge/discharge current (in this case, 10 hours are needed to charge or discharge the electrode); cycling the electrode with a current of 744 mA/g will correspond to a 2C rate and an half an hour charge/discharge time. Note that high cycling rates necessarily decrease the capacity of the electrode, since kinetic problems (slow electrode reactions, Li^+ diffusion in the electrolyte, Li^+ desolvatation before insertion, etc.) always limit electron transfer.

Finally, since electrolytes used in advanced Li-ion batteries are highly flammable, security is also a parameter of paramount importance. This is the reason why Li metal is not used.

As already stated in the introduction, new battery applications, particularly for the electrification of transportation require higher energy densities, which can be achieved in practice by increasing the open-circuit voltage or increasing the capacity of the electrodes, or both. Concerning cathodes, although insertion-type compounds are still being marginally improved (Melot and Tarascon 2013), many studies are devoted to Li-air and Li-S systems, which would increase the electrode capacity by a factor of at least 5 (Bruce et al. 2012; Christensen et al. 2012).

Concerning anodes, many materials have been studied and tested (Obrovac and Chevrier 2014; Pribat 2015), but Si is the most promising because its capacity can be roughly similar to that of pure Li, which is more than 10 times larger than graphite. Unfortunately, because of a ~300% volume increase upon full lithiation, Si films, or powders do not resist charge–discharge cycling, so that Si-based anodes rapidly degrade and cannot be commercialized. However, when used in nanostructured form (such as nanoparticles, nanowires, nanotubes), Si can better withstand the huge volume variations accompanying Li uptake/extraction without cracking, and Si-based anodes become viable. For instance, crystalline Si (c-Si) nanoparticles can withstand full lithiation operation when their diameter is kept below ~150 nm (Liu et al. 2012b). Particles with larger diameters (up to ~900 nm) escape cracking if amorphous Si (a-Si), instead of c-Si, is the starting material (McDowell et al. 2013). Similarly, depending on lithiation rate, c-Si nanopillars/nanowires keep their physical integrity upon full lithiation, unless their diameter is increased above ~300–360 nm (Ryu et al. 2011; Lee et al. 2012).

Among all possible Si nanostructures, NWs are probably the most attractive, because they can be grown directly (and hence individually contacted) on the current collector of the anode, so that no binder, nor conductive additives need to be employed; this is a definitive advantage over other nanostructures such as nanoparticles. Moreover, the dense NW arrays usually obtained by direct growth (using CVD-VLS, see later, Section 20.3) provide a highly porous medium in which electrolyte penetration is easy; this is illustrated in Figure 20.4, which shows a typical random array of SiNWs grown on a stainless steel foil (Laïk et al. 2008).

Figure 20.4 Scanning electron microscope viewgraphs of a typical SiNW array grown on a stainless steel foil and showing its high porosity. The scale bar is 10 μm on the left-hand picture and 3 μm on the right-hand picture. (Reproduced from *Electrochimica Acta*, 53, Laïk, B., et al., Silicon nanowires as negative electrode for lithium-ion microbatteries, 5528–5532, Copyright 2008, with permission from Elsevier.)

Furthermore, because of the large void around each NW (Figure 20.4), swelling upon Li uptake is not hampered, which reduces the mechanical stress imposed to the electrode and therefore increases its lifespan.

The first paper concerning the RT electrochemical lithiation of SiNWs dates from 1999 (Zhou et al. 1999). However, this first study was not aiming at the development of a new anode material for Li-ion batteries, but rather at studying the doping of SiNWs. The same group realized the interest of SiNWs for Li-ion batteries 1 year later (Li et al. 2000). Another study concerning the lithiation of SiNWs (and nanoparticles) appeared in 2001 (Gao et al. 2001). It is quite remarkable that those three papers went quite unnoticed by the battery community and it took 7 more years for the subject to reappear (Chan et al. 2008; Laïk et al. 2008). Since then, the huge interest in SiNWs has been clearly assessed and a huge number of papers have been published on the lithiation mechanism as well as on the effects of lithiation, including the size-dependent cracking phenomenon, the stress effect, the stabilization of the SEI, and so on. In order to give the reader a global overview, the next section provides a summary of the various methods used to synthesize SiNWs, either by growth/deposition or by etching techniques.

20.3 SYNTHESIS OF Si NANOWIRES

Various approaches have been used to synthesize SiNW arrays, which can be categorized as growth (Schmidt et al. 2010) or deposition methods on the one hand, and etching methods on the other hand (Huang et al. 2011). Those are briefly presented below, with some comments on their potential advantages and drawbacks.

Concerning growth methods, the most popular one is based on the so-called vapor–liquid–solid (VLS) mechanism, discovered more than 50 years ago (Wagner and Ellis 1964). VLS growth is usually performed in a chemical vapor deposition (CVD) reactor under a flowing silicon-bearing gas (e.g., silane—SiH_4— or silicon tetrachloride—$SiCl_4$) and at temperatures ranging from ~300 up to above 1000°C, depending on the gas precursor and the type of eutectic-forming metal employed (Schmidt et al. 2010). In the VLS growth mechanism, a divided metal first forms a liquid solution (usually droplets of eutectic composition) with Si atoms, the latter originating from the thermal decomposition of the Si-bearing gas. As more Si atoms are added, the liquid droplets become supersaturated with Si, thus forcing precipitation to occur at the solid–liquid interface. A dynamic equilibrium is rapidly established, where the flux of Si atoms incorporated in the liquid eutectic drop is balanced by the flux of Si atoms precipitating at the liquid–solid interface. Since the surface of the liquid eutectic alloy is ideally rough, growth is highly anisotropic, resulting in long, whisker-like crystals with a spherical eutectic cap at their tips (see Figure 20.4). The VLS growth mechanism is summarized in Figure 20.5, with Au as the eutectic-forming metal (Laïk et al. 2008). The eutectic-forming metal is often called a catalyst, but this is not necessarily justified. For instance, there is no clear evidence that Au catalyzes the decomposition of SiH_4 at low temperature.

As far as Si is concerned, the VLS mechanism operates across a wide range of wire diameters, from several nm to 100s of microns. The NW diameter depends on the diameter of the metal particles. Although we have

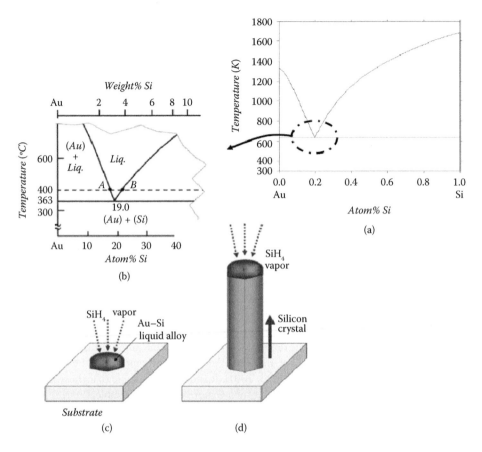

Figure 20.5 Schematic mechanism of the VLS growth process: (a) the Au–Si phase diagram; (b) enlargement of the Au-rich side of the phase diagram, corresponding to the eutectic region; (c) formation of the Au–Si liquid drop close to eutectic composition; (d) anisotropic growth of the Si nanowire. (Reproduced from *Electrochimica Acta*, 53, Laïk, B., et al., Silicon nanowires as negative electrode for lithium-ion microbatteries, 5528–5532, Copyright 2008, with permission from Elsevier.)

represented the growth mechanism with Au in Figure 20.5, SiNWs can be grown from a number of different catalysts, which can be schematically classified according to three types (Schmidt et al. 2010). Type A catalysts exhibit a phase diagram with Si comprising a simple eutectic (e.g., Au, Cu, Ag); type B catalysts also show a simple eutectic, but at much lower Si concentrations, typically below 1% (and sometimes well below); In, Ga, Zn are such examples. Type C catalysts form silicides and examples are Ni, Pt, or Ti. For the latter elements, the lowest eutectic temperature is higher than 800°C (Schmidt et al. 2010).

The VLS-grown SiNWs are generally monocrystalline, even if they can be highly kinked and twinned, particularly when type B catalysts are employed (He et al. 2015). Three main growth directions are usually observed depending on the NW diameter: <111>, <110>, and <112>. For NWs with a diameter below 20 nm, the <110> growth direction is mostly encountered, whereas above 50 nm, the <111> direction is favored. For diameters between 20 and 50 nm, the <112> direction is often observed (Schmidt et al. 2010). Twinning and kinking in NWs seem to depend critically on the mechanical stability of the triple boundary line, where the liquid droplet, the solid NW, and the gas phase coexist. A dynamic model has been developed for the VLS process (Schwarz and Tersoff 2009, 2011), where growth is controlled by the balance of capillary forces acting on the triple boundary line, combined with the difference in chemical potential between liquid and solid phases. This model reproduces most of the experimental observations concerning NW growth, in particular kinking.

Although the growth of NWs by the VLS method seems to be well understood, there are a number of subtleties in the process, which can yield very different NW structures and morphologies. For instance, Au-catalyzed VLS growth in a hot wall CVD reactor will yield long and straight NWs with low defect densities (Baïk 2008), whereas highly kinked and curved NWs are obtained after Au-catalyzed growth in a hot plate (cold wall) reactor (Nguyen et al. 2011). This noteworthy difference in NW morphology is related to the peculiarities of the VLS growth mechanism, where the liquid eutectic droplet at the tip of the NW tends to migrate toward high temperature regions (Wagner and Lewitt 1975); the liquid droplet is said to climb the liquidus of the phase diagram. Consequently, since in a hot wall reactor the tube wall tends to be at a higher temperature than the substrate at the center of the tube, a radial growth of long NWs is observed. On the other hand, when a planar furnace is used, the temperature in the gas phase above the furnace decreases rapidly so that after a short period of vertical growth in the stagnant layer, the wires will tend to kink and curve their growth direction toward the substrate where the temperature is higher. Because of this change in growth direction, the NWs become highly entangled and they can then be interconnected by modifying the deposition conditions during the CVD growth sequence, thus yielding a highly porous and rigid block, from which individual specimens cannot detach (Nguyen et al. 2011). The net result is an improvement in the SiNW anode stability as will be discussed in Section 20.5. The growth situations explained above are summarized in Figure 20.6.

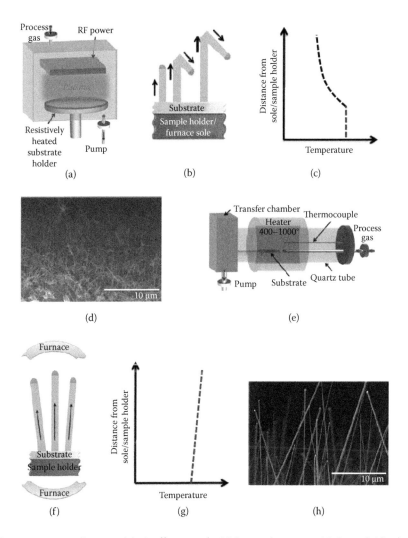

Figure 20.6 Temperature gradients and their effects on the VLS growth process: (a) through (d) a hot plate furnace (cold wall reactor) induces bending of the SiNWs during growth because of a negative thermal gradient above the substrate; (e) through (h) in a hot wall reactor, SiNWs grow straight because the thermal gradient above the substrate is positive. (Zamfir, M.R., et al., 2013, Silicon nanowires for Li-based battery anodes: A review, *J. Mater. Chem. A.*, 1, 9566–9586., Reproduced by permission of The Royal Society of Chemistry.)

Nanowires and nanotubes

At the end of the 1990s, a laser ablation method was developed for the synthesis of Si (as well as Ge) NWs (Morales and Lieber 1998; Zhang et al. 1998). This method, highly inspired from the synthesis of single-wall carbon nanotubes developed by the Smalley group in Rice University (Guo et al. 1995; Thess et al. 1996), makes use of an Si target which is doped with the metal catalyst (or rather alloyed, since the metal concentration can be up to 10 atoms%). The target is situated in a quartz-tube furnace, so that the environmental conditions (pressure, temperature, neutral gas flow) can be controlled easily. A pulsed laser ablates the target, vaporizing both catalyst and semiconductor atoms. Atoms in the vapor phase rapidly condense into Si-rich liquid metal-Si clusters from which Si precipitation begins as the supersaturation of the liquid phase increases. Altogether, this laser ablation method also exploits the VLS process (Morales and Lieber 1998). SiNWs with very small diameters (6–20 nm) can be obtained using this synthesis method. However, it has rapidly been replaced by CVD-type methods, even for producing NWs with small diameters.

Another SiNW synthesis approach, with a mechanism similar to the VLS process, takes place directly in solution (Holmes et al. 2000). Typically, metal nanoparticles (e.g., Au colloids) are mixed in a solvent (e.g., hexane, toluene, benzene) with a silane-based reactant (e.g., alkylsilane, arylsilane, polysilane) and the mixture is heated above the metal-Si eutectic temperature (363°C for Au) inside a high-pressure vessel (the pressure can reach 100–200 atm, depending on the solvent). In such conditions, the silane reactant decomposes, releasing Si atoms which alloy with the metal, inducing eutectic formation, followed by Si precipitation and nanowire growth from the Si-saturated nanoparticles. With this solution-based, VLS-like method (called supercritical fluid–liquid–solid, SFLS, because the solvent becomes supercritical under the temperature and pressure conditions) bulk quantities of SiNWs can be synthesized during each run. By using high boiling temperature solvents, SiNW growth can be performed at atmospheric pressure, transforming the process into a simpler solution–liquid–solid (SLS) mechanism (Heitsch et al. 2008). Battery anodes have been fabricated with SFLS-grown SiNWs (see Section 20.4 later); however, NWs grown this way are not directly connected to any current collector and they usually need to be further engineered according to a traditional battery electrode process, and be incorporated into a slurry containing a binder (polymer) as well as a conductive additive (carbon-based powder).

Before concluding this section on growth/deposition, let us mention electro-deposition as another technique of NW synthesis. For instance, amorphous SiNWs have been synthesized by electrodeposition inside nanoporous membranes (Mallet et al. 2008), using an electrolytic bath composed of $SiCl_4$ dissolved in an ionic liquid (1-Butyl-1-methylpyrrolidinium bis(trifluoromethanesulfonyl) imide (P1,4)). Si was electrodeposited inside porous ion track-etched polycarbonate membranes, with pore diameters of 15, 110, and 400 nm, resulting in NWs with, respectively, identical diameters. Those nanowires (deposited at RT) can be crystallized by a moderate temperature annealing, although for battery anode application, there is no need to use crystalline Si as we shall see later in Section 20.4.

As mentioned already, SiNWs can also be obtained by etching standard crystalline Si substrates. Various approaches can be used, such as deep reactive ion etching (D-RIE) using a Cl_2– or SF_6-based chemistry in a plasma reactor (Hsu et al. 2008; Garnett and Yang 2010). However, RIE-type techniques are usually not cheap to implement (expensive vacuum vessels + expensive gases + eventually photolithography), even though submicron lithography operations can be avoided to synthesize NWs with diameters in the 100 nm range; for instance, in the above-cited work, the authors have used a monolayer of submicron diameter silica beads deposited on the Si wafer surface as a hard mask.

Alternatively, SiNWs with high aspect ratios can be obtained by metal-assisted wet-chemical etching (MACE), which can simply be performed in a beaker. In the MACE process, the SiNWs are fabricated by localized catalytic etching of silicon substrates in aqueous acid solutions (Peng et al. 2002). The catalysts are metal nanoparticles obtained by a galvanic displacement reaction (usually called electroless deposition) on the surface of the substrate. For instance, starting with an aqueous solution containing HF and $AgNO_3$, the galvanic displacement reaction between Ag^+ ions and Si first produces local deposition of metallic Ag clusters as well as a concomitant local oxidation of Si (i.e., local SiO_2 formation) under the Ag cluster. The SiO_2 produced is readily dissolved by HF, locally producing a nanocup, right under the Ag cluster. As more Ag is deposited from the solution on top of the original cluster, the formation and subsequent dissolution of SiO_2 proceeds, thus "digging" the Si substrate underneath the Ag particle. The result is an array of SiNWs, corresponding to unetched Si between buried Ag nanoparticles. A faster, two-step MACE process has also been developed (Peng et al. 2006; Sivakov et al. 2010), where the nucleation and deposition of Ag nanoparticles on the Si surface takes place in an initial bath (as explained above) and Si etching is subsequently performed in separate HF/Fe(NO$_3$)$_3$ or HF/H$_2$O$_2$ solutions (Sivakov et al. 2010; Huang et al. 2011). Recent improvements have been published, particularly concerning the precise patterning of the metal layer (Brodoceanu et al. 2016).

Figure 20.7 shows an example of an array of MACE SiNWs synthesized in our laboratory with a two-step process, first using an AgNO$_3$/HF solution to deposit Ag nanoparticles, followed by etching in a second HF/H$_2$O$_2$ solution. Note that if etching is stopped and started again, the NWs are laterally etched at the interface between the two etching operations, which eases their fracture and detachment from the substrate, using, for example, a very brief ultrasonic treatment (Weisse et al. 2011). Interestingly, starting with highly doped Si substrates, the MACE process can produce highly porous NWs, which is an advantage for battery applications (see Section 20.5.1). Again, NWs synthesized using a MACE process need to be harvested and further suspended into some kind of slurry before being spread on the anode current collector by tape casting.

Etching processes are certainly useful at laboratory scale for feasibility experiments. However, their use for the industrial fabrication of SiNWs for battery anodes looks hardly viable from an economical point of view, even though the cost of crystalline Si has significantly decreased over the past few years, because of economies of scale due to increased manufactured volumes driven by the development of wafer-type Si solar

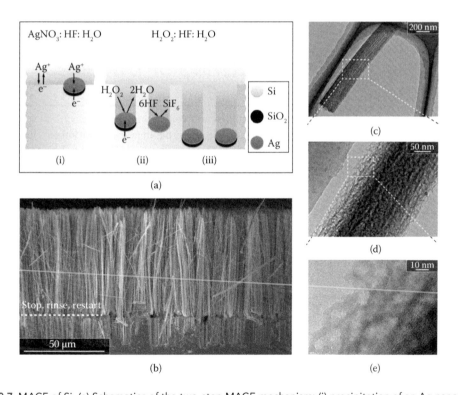

Figure 20.7 MACE of Si. (a) Schematics of the two-step MACE mechanism; (i) precipitation of an Ag nanoparticle and Si oxidation under it in a first AgNO$_3$-based bath, (ii) details of the Ag-assisted oxidation and etching of Si in a second H$_2$O$_2$ + HF bath, (iii) continuous etching leaves pillars (NWs) in between nanoholes. (b) scanning electron microscope (SEM) cross-sectional view of a SiNW array synthesized by a two-step MACE of a (100)-oriented, p-doped Si wafer. The second etching step in H$_2$O$_2$ + HF was interrupted and started again after rinsing in order to induce an easy fracture plane, which is highlighted by the white dotted line. (c) through (e) Transmission electron microscopy (TEM) pictures showing the structure of one p-doped SiNW obtained by the MACE process. Note the high porosity of the NW, particularly in the enlarged view of (e). (Zamfir, M.R., et al., 2013, Silicon nanowires for Li-based battery anodes: A review, *J. Mater. Chem. A.*, 1, 9566–9586., Reproduced by permission of The Royal Society of Chemistry.)

cells (Green 2016). Moreover, assuming the use of Si wafers, ingot machining, and slicing operations can consume up to 40% of the original ingot (Wang et al. 2008) and a MACE-type process would probably waste another 50% of a wafer, resulting in a mere 20% of the original ingot being used. Even if Si powders instead of wafers are used (Bang et al. 2011) the fabrication of crystalline Si is still an energy-hungry process, involving melting of Si above 1410°C (e.g., the Czochralski crystal pulling method). Maybe the use of metallurgical grade Si should be explored in order to decrease costs.

In comparison, VLS SiNWs can be grown at moderate temperature, around 500°C (or even lower if plasma-enhanced CVD—PECVD—is used, see Toan et al. 2016), and on very large-area substrates, which increases productivity. For instance, in the active-matrix liquid-crystal display industry, silicon thin films are routinely deposited by PECVD on the so-called Generation 10 glass substrates measuring 2.88 x 3.13 m² (Samukawa et al. 2012). Finally, the VLS-CVD synthesis of SiNWs essentially consumes the Si which is needed for growing the wires (plus admittedly some stray deposition on the walls of the reactor, which can be minimized), resulting in much less Si waste. SFLS or SLS growth approaches also seem more economically viable than etching techniques.

Finally, we would like to point out that when it comes to using nanostructured Si in a traditional battery electrode process (see Footnote on page 456), NWs are probably not the best choice compared to nanoparticles/nanopowders, which are easier to fabricate and cheaper. The real advantage of SiNWs is that they can be used binder-free, at least in the situation where they are grown directly on the anode's current collector by a CVD-VLS-type process. However this novel processing technique is not necessarily well accepted by battery manufacturers, since they would have to profoundly modify their fabrication tools/habits, not to mention cost issues.

20.4 THE ROOM-TEMPERATURE ELECTROCHEMICAL INTERACTION AND ALLOYING OF LITHIUM WITH SILICON

Since it is performed far from thermodynamic equilibrium conditions, the RT electrochemical alloying of Li with Si does not follow the Li–Si phase diagram. However, detailed investigations of the RT lithiation of Si have been performed over the years, using *in situ* X-ray diffraction (XRD) analysis (Limthongkul et al. 2003; Obrovac and Christensen 2004; Hatchard and Dahn 2004; Li and Dahn 2007; Obrovac and Krause 2007), *in situ* transmission electron microscopy observations (Liu et al. 2011, 2012a) *ex situ* pair distribution function analysis (Key et al. 2011) to name a few (see also Obrovac and Chevrier 2014). All those characterization methods shown that crystalline Si (c-Si) converts to an Li-rich amorphous phase (a-Li_xSi) during the first lithiation, as first reported by Li et al. in 2000 and confirmed by Limthongkul et al. in 2003. This a-Li_xSi amorphous phase (where x ~3.5 is the most accepted value) usually crystallizes into the metastable c-$Li_{15}Si_4$ compound toward the end of lithiation, when the voltage (against Li^+/Li^0) goes below ~50 mV. At this point, the capacity reaches ~3580 mAh/g. However, this c-$Li_{15}Si_4$ polycrystalline phase exists over a range of compositions, so Dahn and coworkers have labeled it $Li_{15\pm\delta}Si_4$ (Hatchard and Dahn 2004). The maximum δ value seems to be ~0.9, corresponding to a maximum capacity of ~3800 mAh/g at RT.

The first delithiation starts with an equilibrium between c-$Li_{15}Si_4$ and an amorphous Li_2Si phase; once all the Li is extracted from c-$Li_{15}Si_4$, further delithiation of the Li_2Si phase yields amorphous Si (a-Si). The atomic structure of the Li_2Si phase is very similar to that of a-Si, since the delithiation of the former resembles that of a solid solution of Li in a-Si, with the voltage rising continuously on a galvanostatic curve (Zamfir et al. 2013; Obrovac and Chevrier 2014). In brief, the first total lithiation–delithiation cycle transforms crystalline Si into amorphous Si.

For the second and following lithiation cycles, there is a first equilibrium between a-Si and another amorphous phase of composition $Li_{2.5}$Si (Wang et al. 2013a; McDowell et al. 2013). Once all the a-Si consumed, a third amorphous phase appears (with Li/Si > 2.5), in equilibrium with $Li_{2.5}$Si. When the lithiation voltage approaches zero, the c-$Li_{15}Si_4$ phase crystallizes again. The two-phase lithiation mechanisms of c-Si and a-Si are very similar, with a sharp phase boundary between c-Si and a-$Li_{3.5}$Si or between a-Si and a-$Li_{2.5}$Si

(Liu et al. 2012a; Wang et al. 2013a). In summary, the only crystalline phase that can appear after the first lithiation–delithiation cycle is c-$Li_{15}Si_4$ (assuming a deep charging, close to zero volts versus Li^+/Li^0).

The second and following delithiation steps are similar to the first one. At this point, note that if a-Si instead of c-Si is the starting material, the first lithiation step of a-Si will correspond to the second one of c-Si and will eventually end up by the crystallization of the c-$Li_{15}Si_4$ phase if the voltage goes close to zero. The first delithiation step will be the same as that described for c-Si since in both cases the delithiation of c-$Li_{15}Si_4$ is of concern. The major phases and their compositions are summarized in Table 20.1 (Zamfir et al. 2013).

While most of the above observations have been performed on Si powders or thin films, the behavior of nanostructured Si is not so clear, particularly concerning the crystallization of the c-$Li_{15}Si_4$ phase at the end of lithiation. This is the reason why we have indicated c-$Li_{15}Si_4$ or a-$Li_{3.75}Si$ at the end of lithiation in Table 20.1. Actually, because of stress effects in SiNWs (Liu et al. 2011; Yang et al. 2015), the voltage curve of Si can be depressed during lithiation (Sethuraman et al. 2010a, 2010b; Ichitsubo et al. 2011). Consequently, once the electrode is brought to the cut-off charging voltage, close to zero volts (vs. Li^+/Li^0), the Li loading corresponding to the $Li_{15}Si_4$ stoichiometry is not yet reached, so the crystallization of the c-$Li_{15}Si_4$ phase cannot take place. Of course, the capacity of the corresponding electrode is reduced compared to the nominal value. Incomplete volume lithiation is another stress-related effect in c-Si NWs (Liu et al. 2013). Finally, doping also influences the formation of the c-$Li_{15}Si_4$ phase in NWs, as we have only observed it in highly doped specimens (Zamfir et al. 2013).

20.5 SOME RECENT RESULTS CONCERNING THE USE OF SiNWs IN LIBs

The major problem to be sorted out when building an anode based on SiNWs is the lifespan, which essentially translates into three subproblems of (i) NW cracking, (ii) NW detachment, and (iii) SEI mechanical instability. Some possible solutions to those problems are presented below. Maybe another problem is the charge–discharge rate, which is usually quite low; as far as NW are concerned, there seems to be a compromise between the areal mass loading of the anode and its charge–discharge rate (Zamfir et al. 2013).

20.5.1 NANOWIRE CRACKING AND DETACHMENT FROM THE SUBSTRATE

We have seen already (end of Section 20.2) that there was a critical size below which NW cracking was avoided both for crystalline and amorphous Si. However, even when cracking is avoided, the swelling of NWs upon Li uptake is still taking place (see e.g., Liu et al. 2011). This swelling induces NW detachment from the substrate, due to the stress developed at their roots on volume change (Nguyen et al. 2011; Cho and Picraux 2013). In other words, while the wires can expand longitudinally and laterally on Li insertion, they are constrained at their roots by the Li-neutral current collector, whose volume does not change.

Table 20.1 Summary of the major phases and their composition during the two first lithiation–delithiation cycles of c-Si. The first and second delithiation steps are identical. Actually, after the first full lithiation–delithiation cycle, the following ones are similar (except for some capacity loss). If a-Si is the starting material (instead of c-Si) the reaction paths for any cycle are those of the second lithiation–delithiation cycle in the table.

First full lithiation (of c-Si)	c-Si $\rightarrow$ c-Si + a-$Li_{3.5}Si$ $\rightarrow$ $\begin{cases} \text{a-}Li_{3.75}Si \\ \text{c-}Li_{15}Si_4 \end{cases}$
First delithiation	$\left. \begin{array}{c} \text{a-}Li_{3.75}Si \\ \text{c-}Li_{15}Si_4 \end{array} \right\} \longrightarrow \begin{cases} \text{a-}Li_{3.75}Si \\ \text{c-}Li_{15}Si_4 \end{cases} + \text{a-}Li_2Si \longrightarrow \text{a-}Li_2Si \longrightarrow \text{a-Si}$
Second full lithiation	a-Si $\longrightarrow$ a-Si + a-$Li_{2.5}Si$ $\longrightarrow$ a-$Li_{2.5}Si$ + a-$Li_{w(?)}Si$ $\underset{w > 2.5}{\longrightarrow}$ $\begin{cases} \text{a-}Li_{3.75}Si \\ \text{c-}Li_{15}Si_4 \end{cases}$
Second delithiation	$\left. \begin{array}{c} \text{a-}Li_{3.75}Si \\ \text{c-}Li_{15}Si_4 \end{array} \right\} \longrightarrow \begin{cases} \text{a-}Li_{3.75}Si \\ \text{c-}Li_{15}Si_4 \end{cases} + \text{a-}Li_2Si \longrightarrow \text{a-}Li_2Si \longrightarrow \text{a-Si}$

Source: Adopted from Zamfir, M.R., et al., 2013, Silicon nanowires for Li-based battery anodes: A review, *J. Mater. Chem. A.*, 1, 9566–9586., Reproduced by permission of The Royal Society of Chemistry.

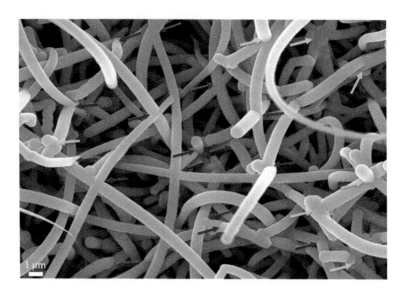

Figure 20.8 Scanning electron microscope view of a highly interconnected SiNW array, synthesized by plasma-enhanced CVD at 420°C. The red arrows indicate simple interconnections between two wires; the blue arrows highlight interconnections between more than two wires. The result of the interconnection is a compact, highly porous medium where individual NWs cannot detach from the current collector on which they have been grown. (Nguyen, H.T., et al.: Highly interconnected Si nanowires for improved stability Li-ion battery anodes. *Advanced Energy Materials*. 2011. 1. 1154–1161. Copyright Wiley-VCH. Reproduced with permission.)

As the wires detach from their substrate, the circulation of electrons is stopped and Li remains trapped in the now inactive Si material. One solution is to provide some kind of physical link between the NWs, which can hold them together, even if some detach from the current collector.

This has been achieved by interconnecting the NWs during their growth in a PECVD-type reactor (Nguyen et al. 2011), as already explained in Section 20.3 previously. Figure 20.8 shows a typical scanning electron microscope viewgraph of such an interconnected SiNW network; note the difference with "straight-type" NWs shown in Figure 20.4 earlier. The other advantage of those interconnected wires is that they provide a high areal mass loading of Si, up to 1.2 mg/cm² (Nguyen et al. 2011), which is one of the largest reported for nanostructured Si.

20.5.2 STABILITY OF THE SEI

The next problem to be solved is related to the instability of the SEI layer. We have already explained in Section 20.2 that the SEI was a passivating film which was generated on the surface of the anode during the first battery charge, due to partial decomposition of the electrolyte. Because of the large volume/area change of SiNWs during lithiation, the SEI tends to crack and delaminate from the Si surface, rendering the passivation inefficient and repeatedly exposing parts of the bare Si surface to the electrolyte (Wu et al. 2012; Tokranov et al. 2014; Yoon et al. 2016). As a consequence, the Si surface keeps reacting with the electrolyte, trapping more Li into SEI products, thus inducing capacity fading (remember that the amount of Li in the battery is limited by the cathode content). Moreover, the SEI tends to thicken, which slows down Li exchanges, affecting the charge–discharge rate of the electrode. The problem is amplified by the use of nanostructures which exhibit large surface-to-volume ratios. Finally, because of incomplete surface passivation, side reactions between the electrolyte and the Si material can be anticipated (Nguyen et al. 2011; Lux et al. 2012), which also consume some of the electrode, thus contributing to the capacity fading phenomenon.

The best way to avoid trouble would be to suppress the SEI formation by controlling the charging voltage of the battery, in order to prevent crossing the stability boundary of the electrolyte. Unfortunately, for most organic electrolytes, decomposition starts around 0.7 V (against Li⁺/Li⁰), well before the onset of the Li reaction with Si. The formulation of new electrolytes, including ionic liquids, is indeed a research axis (Li et al. 2016), particularly concerning additives (Haregewoina et al. 2016). Several other strategies have been developed to improve the stability of the SEI layer on SiNWs; schematically, there are two main approaches, namely (i) limiting the volume increase of the NWs on lithiation, whether mechanically or by appropriate voltage cuts (without sacrificing too much capacity), and (ii) finding some kind of coating, sufficiently elastic to withstand the NW deformation, which would replace the SEI. Most techniques related to those approaches (such as porous NWs, core-shell NWs, coated NWs, and embedded NW) have already been extensively reviewed (Zamfir et al. 2013; Su et al. 2013; Mai et al. 2014; Pribat 2015; Du et al. 2016) and we will only focus here on recent developments, particularly those concerning graphene–SiNW composites, which we believe carry the most promise for the future.

20.5.3 GRAPHENE–SiNW COMPOSITES

In these composites, the SiNWs are "wrapped" inside flexible graphene foils that can be stacked on top of each other. Graphene is a one-atom thick, two-dimensional sheet, where carbon atoms form a "chicken wire" pattern (Geim and Novoselov 2007). One intuitively feels that graphene can provide a highly deformable electrical contact that can follow the expansion and contraction of the Si nanostructures as they are lithiated and delithiated (see Figure 20.9, reproduced from Ren et al. 2014). Moreover, even if a NW breaks, the broken bits and fragments stay in electrical contact with the surrounding graphene sheets, so there is no capacity loss. The only constraint is that the graphene layers have to be uniformly porous, in order to let the Li ions go through.

Graphene is the material of choice for this kind of encapsulation, since it is not only flexible, but also chemically inert and it exhibits a low electrical resistivity (a few $\mu\Omega$.cm, depending on doping, substrate, etc.). Although CVD methods have been developed for the controlled fabrication of large-area graphene sheets (Bartelt and McCarty 2012; Zhang et al. 2013), the simplest route to obtain graphene flakes/films for use in batteries relies on solution-based processes involving the chemical exfoliation of graphite or graphite derivatives such as graphite oxide (Park and Ruoff 2009). One such process is the so-called Hummers method (Hummers and Offeman 1958) which uses pure graphite powder as starting material. Note that according to the schematics shown in Figure 20.9, the graphene–SiNW composite can be stacked on the current collector without any binder.

Although the structure obtained after step 3 in Figure 20.9 seems to be an obvious solution to accommodate Si volume variations while keeping electrical contact, making uniform composites of porous graphene sheets (e.g., reduced graphite oxide, rGO) and Si nanostructures is by no means an easy task. The simplest approach would be to prepare a liquid suspension (preferably water-based) with the SiNWs and porous graphite oxide (GO) sheets, spread/spray the suspension on the current collector of the anode (Cu foil) and

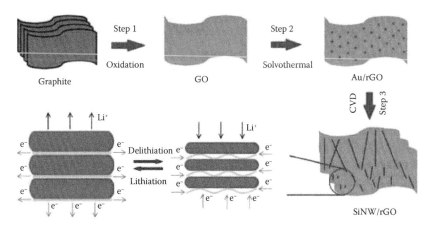

Figure 20.9 Schematic illustration of the NW-graphene composite. Here, rGO stands for "reduced graphite oxide" (graphene). Note that the rGO sheets can easily follow the SiNW volume variations upon lithiation–delithiation as there are no rigid links. (Ren, J.-G., et al., 2013, Silicon-graphene composite anodes for high-energy lithium batteries, *Energy Technol.*, 1, 77–84. Reproduced by permission of The Royal Society of Chemistry.)

Nanowires and nanotubes

then reduce the GO back to graphene, which can be done either thermally (Becerril et al. 2008) or by a chemical treatment using hydrazine hydrate (Stankovich et al. 2007). However, SiNWs are always oxidized, so their OH-terminated surface tends to be negatively charged in solution. On the other hand, GO is also negatively charged because of the ionization of carboxylic acid and phenolic hydroxyl groups (Li et al. 2008). Consequently, the GO flakes tend to repel the Si nanostructures, which yields highly nonuniform composites. Several authors have used electrostatic self-assembly by modifying the surface charges of the Si nanostructures (particularly nanoparticles) and graphene sheets using various polymers (Zhou et al. 2012; Ye et al. 2014). Once the electrostatic assembly has been performed in a water-based solution, the composites are dried and fired at high temperature ($\geq$500°C) in a controlled reducing atmosphere.

The direct growth of graphene around SiNWs has also been attempted, but the process temperature is too high (1200°C) and not compatible with VLS-grown SiNWs on a Cu substrate (Yang et al. 2013). More sophisticated (and expensive) processes have also been explored for packing the SiNWs in graphene-like sheets, where they are first "wrapped" within a CVD graphene-like deposit and then further embedded within rGO sheets (Cho et al. 2011; Wang et al. 2013b).

Because the above methods are all quite complex (and expensive), the direct growth of SiNWs inside catalyst-seeded graphene sheets (as shown in Figure 20.9) has been developed and appears as a more realistic option (Ren et al. 2014; Hassan et al. 2014). This trend follows the pioneering work of Ren and coworkers who, to the best of our knowledge, were the first to grow Si nanoparticles directly inside graphene sheets and without any seeding metal (Ren et al. 2013). Note that embedding Si nanoparticles (instead of NWs) into porous graphene is also a highly explored route to improve the lifetime of Si-based anodes (see e.g., Zhao et al. 2011; Zhou et al. 2012; Ye et al. 2014).

As a final remark on these composites, let us point out that graphene is an ideal substrate for a stable SEI, which obviously is another huge advantage. However, if the electrolyte penetrates between the graphene sheets, then the SEI will form on SiNWs on the first battery charge, electronically isolating them from the graphene. This will lead to rapid capacity fading. So, the porosity of the graphene must be adapted to let only desolvated Li ions go through, which is not an easy task. In particular, although rGO is naturally porous, its porosity is not necessarily adapted for this kind of task. We believe is the main reason why graphene–Si composites have not been more investigated in the literature so far. Controlling the porosity of graphene can be performed by various means (Russo et al. 2013; Jiang and Fan 2014; Sun et al. 2014) but at a certain cost.

20.6 CONCLUDING REMARKS

Silicon is the way to go in order to increase the capacity of Li-ion batteries in the near future. Although Panasonic has announced the use of Si in their 18650-type batteries it seems to be limited to small amounts only (2 to 3 mass% of the active material in the anode). This is because Si is difficult to tame, since anodes based on Si powders or Si thin films rapidly degrade and fail because of repeated volume changes on lithiation–delithiation. The use of nanostructured Si, particularly nanowires, is a game-changing progress, since anodes using such nanostructures can now withstand more than 1000 charge–discharge cycles with capacity values over 1250 mAh/g—3 to 4 times that of graphite-based anodes. Further improvements will be possible with graphene–SiNW composite which can in principle be used without binder, an add-on advantage over Si nanoparticles caged inside carbon shells (Li et al. 2012).

However, as with all industrial products and because we live in a market-based economy, cost is an essential parameter. Growing NWs is expensive, all the more as cheap grinding and tape-casting processes have to be replaced by CVD-type methods involving vacuum operations and handling of toxic gases. Moreover, unless large companies are concerned (e.g., Samsung or LG), battery manufacturers do not have the "technological culture" (inherited from the semiconductor industry) required to grow SiNWs. A cost objective for the electric cars to be economically viable, compared to their gasoline-powered counterparts, is a battery pack at ~190 $/kWh in 2020 and below 100 $/kWh in 2030 (it is now ~250 $/kWh …). Even if economies of scale will be possible at Elon Musk's Gigafactory, because of the large volumes anticipated (as with solar cells, see Green 2016), there is at present no certainty concerning the above cost target, even with the traditional graphite-based anodes.

In the meantime, several start-up or small companies are developing anodes based on SiNWs or nanostructured Si. For instance, Nexeon Ltd. in the UK has patented a nanostructured Si pillars technology (US patents no. 7402829, 7683359, and 7842535) which is supposed to overcome the problem of cycling, as well as provide larger Li storage capacities. In France, EnWireS exploits patents from the CEA and commercializes cheap SiNWs, 10 nm in diameter, and available in powder or in solution; such wires are obviously sold for further incorporation into a slurry, according to the traditional electrode fabrication process. In the USA, SiNode, founded in 2012 through Northwestern University's NUvention program, is commercializing a patented composite of silicon nanoparticles and rGO (Zhao et al. 2011). They claim to achieve capacities between 1000 mAh/g and over 2500 mAh/g. California Lithium Battery (CalBattery) uses a process developed at Argone National Laboratory; again it is a composite of graphene and Si, where the Si is directly deposited on/inside graphene sheets by CVD from a chlorosilane compound (Ren et al. 2013). Enovix is developing their 3D silicon Li-ion battery. Although they do not say much about the technology, it seems that they are using some kind of MACE process to etch SiNWs in Si wafers. The company was started in 2007 and is backed by Cypress Semiconductor, Intel, and Qualcomm. Envia Systems is another Silicon Valley company developing Si-based anodes. Finally, Amprius which is a spin-off from Yi Cui's laboratory in Stanford also develops Si-based anodes. The company owns an R&D lab and their corporate headquarters are in Sunnyvale; they also run a pilot production line in Nanjing, China, as well as a manufacturing facility in Wuxi, China.

REFERENCES

Armand M, Tarascon J-M. (2008). Building better batteries. *Nature.* 451:652–657.
Bang BM, Kim H, Song H-K, Cho J, Park S. (2011). Scalable approach to multi-dimensional bulk Si anodes *via* metal-assisted chemical etching. *Energ. Environ. Sci.* 4:5013–5019.
Bartelt NC, McCarty KF. (2012). Graphene growth on metal surfaces. *MRS Bull.* 37:1158–1165.
Becerril HA, et al. (2008). Evaluation of solution-processed reduced graphene oxide films as transparent conductors. *ACS Nano.* 2:463–470.
Brissot C, Rosso M, Chazalviel J-N, Baudry P, Lascaud S. (1998). In situ study of dendritic growth in lithium/PEO-salt/lithium cells. *Electrochim. Acta* 43:1569–1574.
Brodoceanu D, et al. (2016). Fabrication of silicon nanowire arrays by near-field laser ablation and metal-assisted chemical etching. *Nanotechnology.* 27:075301, 1–8.
Bruce PG, Freunberger SA, Hardwick LJ, Tarascon J-M. (2012). Li–O$_2$ and Li–S batteries with high energy storage. *Nat. Mater.* 11:19–29.
Chan CK, et al. (2008). High-performance lithium battery anodes using silicon nanowires. *Nat. Nanotechnol.* 3:31–35.
Cho J-H, Picraux ST. (2013). Enhanced lithium ion battery cycling of silicon nanowire anodes by template growth to eliminate silicon underlayer islands. *Nano Lett.* 13:5740–5745.
Cho YJ, et al. (2011). Nitrogen-doped graphitic layers deposited on silicon nanowires for efficient lithium-ion battery anodes. *J. Phys. Chem. C.* 115:9451–9457.
Christensen J, et al. (2012). A critical review of Li/air batteries. *J. Electrochem. Soc.* 159:R1–R30.
Ciferno JP, Fout TE, Jones AP, Murphy JT. (2009). Capturing carbon from existing coal-fired power plants. *CEP Magazine*, April 2009, pp. 33–41.
Du F-H, Wang K-X, Chen J-S. (2016). Strategies to succeed in improving the lithium-ion storage properties of silicon nanomaterials. *J. Mater. Chem. A.* 4:32–50.
Gao B, Sinha S, Fleming L, Zhou O. (2001). Alloy formation in nanostructured silicon. *Adv. Mater.* 13:816–819.
Garnett E, Yang P. (2010). Light trapping in silicon nanowire solar cells. *Nano Lett.* 10:1082–1087.
Geim AK, Novoselov KS. (2007). The rise of graphene. *Nat. Mater.* 6:183–191.
Goodenough JB, Kim Y. (2010). Challenges for rechargeable Li batteries. *Chem. Mater.* 22:587–603.
Green MA. (2016). Commercial progress and challenges for photovoltaics. *Nat. Energ.* 1:1–6.
Guo T, Nikolaev P, Thess A, Colbert DT, Smalley RE. (1995). Catalytic growth of single-walled nanotubes by laser vaporization. *Chem. Phys. Lett.* 243:49–54.
Haregewoina AM, Wotangoa AS, Hwang B-J. (2016). Electrolyte additives for lithium ion battery electrodes: Progress and perspectives. *Energ. Environ. Sci.* 9:1955–1988.
Hassan FM, et al. (2014). Subeutectic growth of single-crystal silicon nanowires grown on and wrapped with graphene nanosheets: High-performance anode material for lithium-ion battery. *ACS Appl. Mater. Interfaces.* 6:13757–13764.
Hatchard TD, Dahn JR. (2004). In situ XRD and electrochemical study of the reaction of lithium with amorphous silicon. *J. Electrochem. Soc.* 151:A838–A842.

He Z, Nguyen HT, Toan LD, Pribat D. (2015). A detailed study of kinking in indium-catalyzed silicon nanowires. *Cryst. Eng. Comm.* 17:6286–6296.

Heitsch AT, Fanfair DD, Tuan H-Y, Korgel BA. (2008). Solution-liquid-solid (SLS) growth of silicon nanowires. *J. Am. Chem. Soc.* 130:5436–5437.

Holmes JD, Johnston KP, Doty RC, Korgel BA. (2000). Control of thickness and orientation of solution-grown silicon nanowires. *Science.* 287:1471–1473.

Hsu C-M, Connor ST, Tang MX, Cui Y. (2008). Wafer-scale silicon nanopillars and nanocones by Langmuir–Blodgett assembly and etching. *Appl. Phys. Lett.* 93:133109, 1–3.

Huang Z, Geyer N, Werner P, de Boor J, Gösele U. (2011). Metal-assisted chemical etching of silicon: A review. *Adv. Mater.* 23:285–308.

Hummers WS, Offeman RE. (1958). Preparation of graphitic oxide. *J. Am. Chem. Soc.* 80:1339–1340.

Ichitsubo T, et al. (2011). Mechanical-energy influences to electrochemical phenomena in lithium-ion batteries. *J. Mater. Chem.* 21:2701–2708.

Jiang L, Fan Z. (2014). Design of advanced porous graphene materials: From graphene nanomesh to 3D architectures. *Nanoscale.* 6:1922–1945.

Kasavajjula U, Wang C, Appleby AJ. (2007). Nano- and bulk-silicon-based insertion anodes for lithium-ion secondary cells. *J. Power Sour.* 163:1003–1039.

Kennedy T, Brandon M, Ryan KM. (2016). Advances in the application of silicon and germanium nanowires for high-performance lithium-ion batteries. *Adv. Mater.* 27:5696–5704.

Key B, Morcrette M, Tarascon JM, Grey CP. (2011). Pair distribution function analysis and solid state NMR studies of silicon electrodes for lithium ion batteries: Understanding the (de)lithiation mechanisms. *J. Am. Chem. Soc.* 133:503–512.

Kolosnjaj-Tabi J, et al. (2015). Anthropogenic carbon nanotubes found in the airways of Parisian children. *EBio Med.* 2:1697–1704.

Lagally CD, Reynolds CCO, Grieshop AP, Kandlikar M, Rogak SN. (2012). Carbon nanotube and fullerene emissions from spark-ignited engines. *Aerosol Sci. Technol.* 46:156–164.

Laïk B, et al. (2008). Silicon nanowires as negative electrode for lithium-ion microbatteries. *Electrochim. Acta.* 53:5528–5532.

Larcher D, Tarascon J-M. (2015). Towards greener and more sustainable batteries for electrical energy storage. *Nat. Chem.* 7:19–29.

Lee SW, et al. (2012). Fracture of crystalline silicon nanopillars during electrochemical lithium insertion. *Proc. Natl. Acad. Sci. U. S. A.* 109:4080–4085.

Li D, Müller MB, Gilje S, Kaner RB, Wallace GG. (2008). Processable aqueous dispersions of graphene nanosheets. *Nat. Nanotechnol.* 3:101–104.

Li H, et al. (2000). The crystal structural evolution of nano-Si anode caused by lithium insertion and extraction at room temperature. *Solid State Ion.*135:181–191.

Li J, Dahn JR. (2007). An in situ X-ray diffraction study of the reaction of Li with crystalline Si. *J. Electrochem. Soc.* 154:A156–A161.

Li Q, Chen J, Fan L, Kong X, Lu Y. (2016). Progress in electrolytes for rechargeable Li-based batteries and beyond. *Green Energ. Environ.* 1:18–42.

Li X, et al. (2012). Hollow core-shell structured porous Si-C nanocomposites for Li-ion battery anodes. *J. Mater. Chem.* 22:11014–11017.

Limthongkul P, Jang Y-I, Dudney NJ, Chiang Y-M. (2003). Electrochemically-driven solid-state amorphization in lithium-silicon alloys and implications for lithium storage. *Acta Mater.* 51:1103–1113.

Liu XH, Fan F, Yang H, Zhang S, Huang JY, Zhu T. (2013). Self-limiting lithiation in silicon nanowires. *ACS Nano.* 7:1495–1503.

Liu XH, et al. (2012a). In situ atomic-scale imaging of electrochemical lithiation in silicon. *Nat. Nanotechnol.* 7:749–756.

Liu XH, et al. (2011). Ultrafast electrochemical lithiation of individual Si nanowire anodes. *Nano Lett.* 11:2251–2258.

Liu XH, et al. (2012b). Size-dependent fracture of silicon nanoparticles during lithiation. *ACS Nano.* 6:1522–1531.

Lux SF, et al. (2012). The mechanism of HF formation in LiPF$_6$ based organic carbonate electrolytes. *Electrochem. Comm.* 14:47–50.

Mai L, Tian X, Xu X, Chang L, Xu L. (2014). Nanowire electrodes for electrochemical energy storage devices. *Chem. Rev.* 114:11828–11862.

Mallet J, et al. (2008). Growth of silicon nanowires of controlled diameters by electrodeposition in ionic liquid at room temperature. *Nano Lett.* 8:3468–3474.

McDowell MT, et al. (2013). In situ TEM of two-phase lithiation of amorphous silicon nanospheres. *Nano Lett.* 13:758–764.

Melot BC, Tarascon JM. (2013). Design and preparation of materials for advanced electrochemical storage. *Acc. Chem. Res.* 45:1226–1238.

Morales AM, Lieber CM. (1998). A laser ablation method for the synthesis of crystalline semiconductor nanowires. *Science*. 279:208–211.

Nazri GA, Pistoia G. (2008). *Lithium batteries: Science and technology*. Springer, New York.

Nguyen HT, et al. (2011). Highly interconnected Si nanowires for improved stability Li-ion battery anodes. *Adv. Energ. Mater*. 1:1154–1161.

Nguyen HT, et al. (2012). Alumina-coated silicon-based nanowire arrays for high quality Li-ion battery anodes. *J. Mater. Chem*. 22:24618–24626.

Obrovac MN, Chevrier VL. (2014). Alloy negative electrodes for Li-ion batteries. *Chem. Rev*. 114:11444–11502.

Obrovac MN, Christensen L. (2004). Structural changes in silicon anodes during lithium insertion/extraction. *Electrochem. Solid-State Lett*. 7:A93–A96.

Obrovac MN, Krause LJ. (2007). Reversible cycling of crystalline silicon powder batteries and energy storage. *J. Electrochem. Soc*. 154:A103–A108.

Ozawa K. (2012). *Lithium ion rechargeable batteries: Materials, technology and new applications*, Wiley, Weinheim.

Park S, Ruoff RS. (2009). Chemical methods for the production of graphenes. *Nat. Nanotechnol*. 4:217–224.

Peng K-Q, et al. (2006). Fabrication of single-crystalline silicon nanowires by scratching a silicon surface with catalytic metal particles. *Adv. Funct. Mater*. 16:387–394.

Peng K-Q, Yan Y-J, Gao S-P, Zhu J. (2002). Synthesis of large-area silicon nanowire arrays via self-assembling nanoelectrochemistry. *Adv. Mater*. 14:1164–1167.

Pribat D. (2015). Alloy-based anode materials. In: *Rechargeable batteries, materials, technologies and new trends* (Zhang Z, Zhang SS, eds.), pp. 189–229. Springer International Publishing, Cham, Switzerland.

Pribat D, Velasco G. (1988). Microionic gas sensors for pollution and energy control in the consumer market. *Sensors Actuators*. 13:173–194.

Reddy T, ed. (2010). *Linden's book on batteries*, 4th edition. McGraw Hill, New York.

Ren J-G, et al. (2013). Silicon-graphene composite anodes for high-energy lithium batteries. *Energ. Technol*. 1:77–84.

Ren J-G, et al. (2014). Silicon nanowires-reduced graphene oxide composite as a high-performance lithium-ion battery anode material. *Nanoscale*. 6:3353–3360.

Russo P, Hu A, Compagnini G. (2013). Synthesis, properties and potential applications of porous graphene: A review. *Nano-Micro Lett*. 5:260–273.

Ryu I, Choi JW, Cui Y, Nix WD. (2011). Size-dependent fracture of Si nanowire battery anodes. *J. Mech. Phys. Solids*. 59:1717–1730.

Samukawa S, et al. (2012). The 2012 Plasma Roadmap. *J. Phys. D: Appl. Phys*. 45:253001, 1–37.

Schlapbach L. (2009). Hydrogen-fuelled vehicles. *Nature*. 460:809–811.

Schmidt V, Wittemann JV, Gösele U. (2010). Growth, thermodynamics, and electrical properties of silicon nanowires. *Chem. Rev*. 110:361–388.

Schwarz KW, Tersoff J. (2009). From droplets to nanowires: Dynamics of vapor-liquid-solid growth. *Phys. Rev. Lett*. 102:206101, 1–4.

Schwarz KW, Tersoff J. (2011). Elementary processes in nanowire growth. *Nano Lett*. 11:316–320.

Sethuraman VA, Chon MJ, Shimshak M, Srinivasan V, Guduru PR. (2010a). In situ measurements of stress evolution in silicon thin films during electrochemical lithiation and delithiation. *J. Power Sour*. 195:5062–5066.

Sethuraman VA, Srinivasan V, Bower AF, Guduru PR. (2010b). In situ measurements of stress-potential coupling in lithiated silicon. *J. Electrochem. Soc*. 157:A1253–A1261.

Sivakov VA, et al. (2010). Realization of vertical and zigzag single crystalline silicon nanowire architectures. *J. Phys. Chem. C*. 114:3798–3803.

Stankovich S, et al. (2007). Synthesis of graphene-based nanosheets via chemical reduction of exfoliated graphite oxide. *Carbon*. 45:1558–1565.

Su X, et al. (2013). Silicon-based nanomaterials for lithium-ion batteries: A review. *Adv. Energ. Mater*. 4:1300882, 1–23. DOI: 10.1002/aenm.201300882.

Sun B, Huang X, Chen S, Munroe P, Wang G. (2014). Porous graphene nanoarchitectures—An efficient catalyst for low charge-overpotential, long life and high capacity lithium-oxygen batteries. *Nano Lett*. 14:3145–3152.

Tarascon J-M, Armand M. (2001). Issues and challenges facing rechargeable lithium batteries. *Nature*. 414:359–367.

Thackeray MM, Wolverton C, Isaacs ED. (2012). Electrical energy storage for transportation-approaching the limits of, and going beyond, lithium-ion batteries. *Energ. Environ. Sci*. 5:7854–7863.

Thess A, et al. (1996). Crystalline ropes of metallic carbon nanotubes. *Science*. 273:483–487.

Toan LD, et al. (2016). Si nanowires grown by Al-catalyzed plasma-enhanced chemical vapor deposition: Synthesis conditions, electrical properties and application to lithium battery anodes. *Mater. Res. Express*. 3:015003, 1–10.

Tokranov A, Sheldon BW, Li C, Minne S, Xiao X. (2014). In situ atomic force microscopy study of initial solid electrolyte interphase formation on silicon electrodes for Li-ion batteries. *Appl. Mater. Interfaces*. 6:6672–6686.

Tollefson J. (2010). Fuel of the future? *Nature*. 464:1262–1264.

Wagner RS, Ellis WC. (1964). Vapor-solid-liquid mechanism of single crystal growth. *Appl. Phys. Lett*. 4:89–91.

Wagner RS, Levitt AP. (1975). *Whisker technology*. Wiley Interscience, New York.

Wang JW, et al. (2013a). Two-phase electrochemical lithiation in amorphous silicon. *Nano Lett.* 13:709–715.

Wang Q, et al. (2012). Thermal runaway caused fire and explosion of lithium ion battery. *J. Power Sour.* 208:210–224.

Wang TY, et al. (2008). A novel approach for recycling of kerf loss silicon from cutting slurry waste for solar cell applications. *J. Cryst. Growth.* 310:3403–3406.

Wang B, et al. (2013b). Adaptable silicon-carbon nanocables sandwiched between reduced graphene oxide sheets as lithium ion battery anodes. *ACS Nano.* 7:1437–1445.

Weisse JM, Kim DR, Lee CH, Zheng X. (2011). Vertical transfer of uniform silicon nanowire arrays via crack formation. *Nano Lett.* 11:1300–1305.

Winter M, Brodd JR. (2004). What are batteries, fuel cells and supercapacitors? *Chem. Rev.* 104:4245–4269.

Wu H, et al. (2012). Stable cycling of double-walled silicon nanotube battery anodes through solid–electrolyte interphase control. *Nat. Nanotechnol.* 7:310–315.

Xu K. (2004). Electrolytes and interphases in Li-ion batteries and beyond. *Chem. Rev.* 114:11503–11618.

Yang H, Liang W, Guo X , Wang C-M, Zhang S. (2015). Strong kinetics-stress coupling in lithiation of Si and Ge anodes. *Extreme Mech. Lett.* 2:1–6.

Yang Y, et al. (2013). Graphene encapsulated and SiC reinforced silicon nanowires as an anode material for lithium-ion batteries. *Nanoscale.* 5:8689–8694.

Ye Y-S, et al. (2014). Improved anode materials for lithium-ion batteries comprise non-covalently bonded graphene and silicon nanoparticles. *J. Power Sour.* 247:991–998.

Yoon I, Abraham DP, Lucht BL, Bower AF, Guduru PR. (2016). In situ measurement of solid electrolyte interphase evolution on silicon anodes using atomic force microscopy. *Adv. Energ. Mater.* 6:1600099. DOI: 10.1002/aenm.201600099.

Zamfir MR, Nguyen HT, Moyen E, Lee YH, Pribat D. (2013). Silicon nanowires for Li-based battery anodes: A review. *J. Mater. Chem. A.* 1:9566–9586.

Zhang Y, Zhang L, Zhou C. (2013). Review of chemical vapor deposition of graphene and related applications. *Acc. Chem. Res.* 46:2329–2339.

Zhang YF, et al. (1998). Silicon nanowires prepared by laser ablation at high temperature. *Appl. Phys. Lett.* 72:1835–1837.

Zhang Z, Zhang SS, eds. (2015). *Rechargeable batteries, materials, technologies and new trends.* Springer International Publishing, Cham, Switzerland.

Zhao X, Hayner CM, Kung MC, Kung HH. (2011). In-plane vacancy-enabled high-power Si-graphene composite electrode for lithium-ion batteries. *Adv. Energ. Mater.* 1:1079–1084.

Zhou GW, et al. (1999). Controlled Li doping of Si nanowires by electrochemical insertion method. *Appl. Phys. Lett.* 75:2447–2449.

Zhou X, Yin Y-X, Wan L-J, Guo Y-G. (2012). Self-assembled nanocomposite of silicon nanoparticles encapsulated in graphene through electrostatic attraction for lithium-ion batteries. *Adv. Energ. Mater.* 2:1086–1090.

21 Coated silicon nanowires for battery applications

Alexandru Vlad and Rico Rupp

Contents

21.1 INTRODUCTION

Silicon has been the most important material for the semiconductor industry since its emergence and Si is still the governing element in this field due to its superior carrier mobility, advanced standards of fabrication, as well as high abundance and limited toxicity (Liu et al. 2010). Nanostructured Si enables the extremely high transistor density in processors of modern electronic devices, and is frequently used for bio- or chemical sensors, and optoelectronics. One of the most popular fields of application is energy conversion, since about 90% of all photovoltaic cells are currently based on silicon (Peng et al. 2013). Further improvement of photovoltaic cells is, for example, promised by the use of Si nanowires (SiNWs). These SiNWs are not only employed in energy conversion, but also in energy storage applications, such as supercapacitors. In the following, however, we will discuss the use of coated SiNWs in lithium-ion batteries (LIBs), another type of electrical energy storage. Being the element with the highest known capacity for the storage of lithium ions, silicon shows the ability to improve energy storage dramatically and could therefore have a huge impact in future developments, since energy storage becomes more and more important on the global market (Gupta and Hawtin 2015).

Batteries can generally be divided into primary (disposable) and secondary (rechargeable). Obviously, secondary batteries are the first choice for many applications, especially if they are used on a daily basis and require high amounts of energy. The principle of an LIB, which is a secondary battery, was first proposed by Whittingham in the 1970s (Whittingham 1976) and is based on lithium ions. Li^+ can shuttle between two electrodes, separated by a porous, electrically insulating separator sheet and an ion-conducting liquid electrolyte. During charging of the battery, electrons are pumped into the anode by an externally applied electric potential. Li ions travel from the cathode through the electrolyte to the anode and are used to compensate the depleted charge generated from the reduction of the anode material. This builds up a potential difference between the two electrodes. If the external current source is replaced by an electric device, discharging occurs and the potential difference leads to the opposite effect: Li ions travel back from the anode to the cathode, while electrons flow through the external circuit, where they power the device. Since the first commercialization by Sony in 1991, the branch of secondary batteries is heavily dominated by LIBs batteries. But what are the advantages of LIBs that lead to this trend? Since the electrochemical decomposition of water by electrolysis takes place at potential differences of about 1.23 V or more, aqueous electrolyte batteries are restricted to operate at low voltages, typically below 1.4V. LIBs, however, utilize nonaqueous electrolytes that have a much wider electrochemical stability window and thus allow the use of more reactive (electropositive–electronegative) electrode materials. Therefore, the first commercial LIB was made with an $LiPF_6$–carbonates liquid electrolyte, an $LiCoO_2$ cathode (positive electrode), and a graphite anode (negative electrode), and was able to achieve an electromotive force of about 4 V, defined by this specific anode–cathode combination (Yoshino 2012). Although several changes have been applied to the electrode composition and cell engineering, this is still the most common version of secondary batteries (Park 2013; Pistoia 2014).

Other factors that have favored LIB intrusion and persistence into commercial use are their high specific energy (stored energy per mass) and energy density (stored energy per volume), if compared to other battery types. This again originates from high operating voltage and high specific charge storage capacity of the applied materials. Nevertheless, safety remains a concern with LIBs, mainly due to the flammability of the organic electrolytes and high energy content. The safety aspect is important for the application of LIBs in user devices, which was first tested by Yoshino in 1986 (Yoshino 2012). While safety was ameliorated by replacing the Li metal with graphite anode materials, it still represents a major technological challenge and recent accidents (such as several occurring battery problems in the Boeing 787 Dreamliner in 2013) have shown to what extent this is critical.

Among the main challenges of next-generation LIBs systems, energy and power remain the most important. Even if at the moment we are satisfied by the energy content provided by LIBs, it still remains 10 times lower than that of gasoline. For electrification of vehicular applications, this represent a major bottleneck in further development. Finding new materials and chemistries able to considerably overcome this barrier is thus of utmost importance and silicon is one possible candidate, given the huge amount of electrochemical energy it can store. In the following, we will detail the use of Si nanostructures for Li-ion technology, in particular surface-coated nanostructures—an approach found to be highly beneficial for reversible and high-efficiency cycling of silicon.

21.2 SILICON FOR ENERGY STORAGE

21.2.1 WHY SILICON?

Within the last decade, silicon has increasingly been in the focus of research as a very promising anode material for next-generation LIBs. To understand the reason for this development, the specific capacity C_{spec} can be used. This is a tool for determining the amount of stored charge for a certain electroactive electrode material. More specifically, the gravimetric specific capacity (mAh/g) will be used in this section and is defined as

$$C_{spec} = \frac{x\,F}{n\,M} \qquad (21.1)$$

where x is the number of transferred electrons (for example, $x = 1$ for Li$^+$), $F = 96485.33$ C/mol is the Faraday constant, n is the number of moles of electroactive component involved in the electrochemical reaction, and M is the molecular mass of the active material. Another tool is the volumetric specific capacity (mAh/cm^3) that gives the amount of stored charge per unit volume of electroactive material.

The specific capacity versus electrochemical potential for some of the most used battery materials is depicted in Figure 21.1. Taking a look at the (gravimetric) specific capacity, it becomes clear why silicon is promising as an anode material. If we consider the relevant lithiated phase Li$_{15}$Si$_4$ ($n = 3.75$), which is closest to the random dispersion of lithium in silicon and therefore kinetically favorable over the fully lithiated phase Li$_{22}$Si$_5$, Equation 21.1 leads to a specific capacity $C_{spec}(Li_{15}Si_4) = 3589$ mAh/g. This is about 10 times higher than for commercial graphite anodes with only $C_{spec}(LiC_6) = 372$ mAh/g and close to that of metallic lithium. Another important advantage of silicon over many other anode materials is its low lithiation potential of about 0.2 V (measured vs. Li/Li$^+$). This leads to a high nominal battery voltage, since the cell voltage is equal to the difference between the anode and cathode redox potentials.

Although lithium has a slightly higher specific capacity and a lower potential, metallic lithium as an anode material is not in commercial use anymore, due to safety reasons: unstable lithium plating (i.e., dendrite formation during lithium deposition) and electrolyte consumption ultimately causes perforation of the separator and therefore a short circuit with the cathode. Lithium dendrite growth and the associated combustion of the electrolyte is not characteristic of the alloying–dealloying process of silicon, making this material potentially safer. Besides the electrochemical properties of silicon, the nearly unlimited availability of Si, its nontoxic nature, and high safety in processing speak in favor of this material. Despite this, there are still major issues to be solved with respect to stable and reversible cycling of silicon electrodes and electrolyte consumption during cycling, and only after providing a solid solution to these we can dream of high-capacity silicon inside batteries.

21.2.2 WHY NANO?

Aside from the previously mentioned advantages, the use of silicon for lithium storage also brings along several problems. These problems are mainly caused by the volume expansion of silicon accompanying lithium uptake during electrochemical reaction. The lithiation of silicon takes place in form of alloying, meaning that the Si crystal structure is completely changed and an (amorphous) alloy is formed. In commercial graphite anodes, in contrast, lithiation occurs in form of intercalation of lithium into

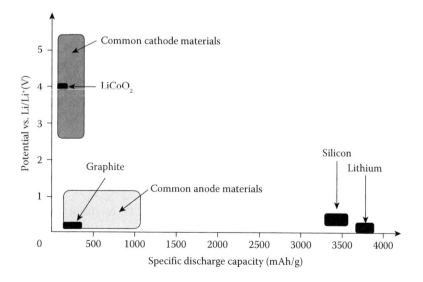

Figure 21.1 Electrochemical redox potential against lithium and specific capacities mapped for commonly used cathode and anode materials. The use of additional inactive materials, for example, binders, additives, metal foils, cell encasing, and wiring, reduces the capacity of the final cells to less than 30% of the respective material's-specific capacity.

existing vacancies. The stoichiometry of the lithiated compounds further highlights the different mechanisms: while only one Li can be accommodated per six carbons in graphite (LiC_6 stoichiometry), around four Li can be accommodated by Si ($Li_{3.75}Si$) (Chan et al. 2012). The volume expansion of silicon is thus over 300%, much higher than that of graphite with only up to about 10%. Furthermore, lithiation does not happen homogeneously within the whole electrode material, since Li^+ has to diffuse into the anode and the silicon is lithiated starting from the interphase with the electrolyte toward the bulk of the material. Lithiated and nonlithiated volumes of the silicon are separated by an atomically sharp boundary, which has been observed by *in situ* transmission electron microscopy (TEM) (McDowell et al. 2013). This leads, together with the expanding "outer" part of the silicon, to high stresses, which increase with the extent of lithiation. The stresses within the active material can reach critical levels, enough to cause fracture of the anode material. With the buildup of stress and fractures during the repeated lithiation and delithiation, progressive amounts of silicon detach and thus lose electrical contact with the current collector. This leads to electrical isolation and stops the possibility of electron exchange and therefore further electrochemical reaction, finally leading to a rapid decrease of capacity retention.

The above-described process is schematically illustrated in Figure 21.2 for different silicon electrode configurations. It is also called the electrode pulverization process and prohibits the use of bulk silicon (but also other materials like tin or aluminum) as anode material. The fracture was shown to take place above a critical size value of 150 nm, independently of applied configuration—thin film, particle, or wire. Silicon films, even if processed at a thickness below 100 nm, still fracture, given the anisotropic stress evolution due to adhesion to the current collector.

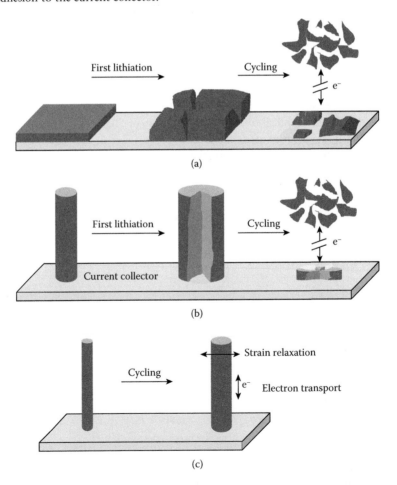

Figure 21.2 (a) Pulverization of a bulk film or (b) micrometer-sized wires during electrochemical cycling and (c) possibility of strain relaxation with the use of silicon nanowires.

One strategy to handle stress within the active material due to the lithiation-induced volume expansion, is the use of nanostructured materials. It is shown by experiments and modeling that the volume change during (de-)lithiation can be compensated by introducing enough free volume for the expansion (Zhao et al. 2012). While bulk materials result in inhomogeneous stress along the interface with the substrate and the lithiation front, nanostructures are not strongly constrained by the current collector and they show fast strain relaxation. It can be shown that a sufficiently small dimension leads to resistance against fracture. Different values for the critical size of features in nanostructured materials are reported, but generally this threshold is stated to be in the range of <200 nm for the diameter of particles or nanowires, depending on the system. Another effect of nanostructures is the reduced diffusion length of the Li-ions by an increased surface-to-volume ratio of the active material. With a shorter diffusion length, an increased rate of lithiation can be achieved, improving times for charge/discharge of the battery and vice versa increasing the capacity at higher rates.

Various nanostructures have been proposed and are currently in the focus of research. The strategies used include nanoparticles (NPs), 1D nanowires (NWs), and 2D thin films. Furthermore, porous materials or more complex structures can be used. Nanowires, which are in the focus of this section, have the special property that they can be also grown or assembled to be in direct contact with the current collector. This enables electron transport along the axis of the 1D structures. For the use of NPs in a binding matrix, for instance, a conducting agent—typically carbon black—is required for electrical conductivity. This conducting agent can be avoided by using NW arrays. Since silicon is a semiconductor, its resistivity at room temperature, however, is high with respect to metals. Doped silicon NWs with a resistivity in the range of about 1–10 Ωcm are therefore often used.

21.2.3 WHY COATED?

The contact between electrolyte and anode in typical LIBs leads to the reduction of nonaqueous electrolyte components at low potentials, and therefore to the decomposition of a part of the electrolyte and the formation of a solid electrolyte interphase (SEI). The electrolyte is chosen so the SEI can still allow ion conduction and avoid further contact between electrode and electrolyte. Furthermore, the electrolyte has to be able to dissolve a lithium salt (e.g., $LiPF_6$) and have an appropriate viscosity to ensure sufficient ionic conductivity. Typically, mixtures of ethylene carbonate (EC), diethyl carbonate (DEC), propylene carbonate (PC), dimethyl carbonate (DMC), vinylene carbonate (VC), or ethyl methyl carbonate (EMC) are used. Many possible reactions, including one- and multielectron reactions, are possible when these are subjected to cathodic polarization of typically below 0.8 V versus Li/Li^+. Some are leading to a stable and some to a less stable SEI, depending on the respective reaction products. The SEI formation can also proceed in several steps at different potentials. Zhang et al. showed, for example, the formation of SEI with low conductivity at potentials of the anode versus Li/Li^+ above 0.15 V, and further formation of SEI with higher conductivity at lower potentials (Zhang et al. 2006). The stability of the SEI can be influenced by the nature of the electrolyte, the anode, and additive components. VC, for example, is especially beneficial for Si-based anodes and increases the stability of the SEI by polymerization before lithiation. The formation of SEI typically takes place during the first few cycles and stops after a certain thickness is reached. If the SEI is stable and has a strong adhesion to the electrode, it decreases concentration polarization and overvoltage, leading to an improvement in cycling of the LIB.

In LIBs with anodes based on the alloying–dealloying mechanism, as in the case of silicon, the formation of SEI is, in contrast to conventional LIBs, a problem and can lead to the failure of the battery. This is caused by the change in volume between the delithiated and lithiated states. The SEI is formed at low potentials of the lithiated state. Since the volume decreases during delithiation, the previously formed SEI fractures and electrolyte can again reach the surface of the active material, resulting in the formation of more SEI during lithiation in the following cycle (Figure 21.3). This leads to a thick, loosely bound layer of SEI and therefore to an increased impedance/polarization of the electrode, and the Li^+ diffusion is decreased excessively. The electrochemical reactivity and therefore the cycle life of the battery is thereby reduced.

Besides the use of nanostructures, the second basic principle for the use of Si-based anodes is the coating of the active material, or the use of a buffer matrix with the included active nanostructures. Shielding of

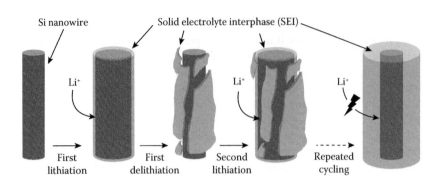

Figure 21.3 Repeated SEI formation and fracturing at the surface of silicon nanowires due to volume changes during cycling of the anode. Charge collection is eventually inhibited after repeated cycling, because Li⁺ can only efficiently diffuse through a thin SEI.

the silicon from the electrolyte can prevent the formation of a thick SEI layer, if materials and fabrication techniques are chosen wisely. Furthermore, a conducting coating can improve the electron transfer to/from the current collector, which is poor in bulk silicon. If the coating or matrix offer the required space for the volume expansion during lithiation, pulverization of the anode can be avoided. Various methods and materials are employed for this purpose. Commonly used are amorphous carbon/graphite/graphene coatings, metal coatings, metaloxide/–silicate/–nitrate coatings, a ductile buffer matrix—which often consists of a polymeric binder and added carbon for electrical conductivity—or a conductive polymer. These materials need to allow Li⁺ diffusion for the lithiation of the active material, while preventing electrolyte components from passing through. A low electrical resistivity of the coating enhances electron transport between active material and current collector, and therefore the cyclability of the anode. In the case of coated NPs, this electrical conductivity is required, while electrical conducting nanowires do not necessarily need a conductive coating, since electrons can travel along the wire itself between the active material and the current collector. As mentioned before, the resistivity of doped Si is typically in the range of 1–10 Ωcm, therefore coatings with a resistivity lower than this can have a strongly beneficial effect on the cycling of SiNWs.

Not only electrical and ionic conductivity of the NW coating are to be considered, but also its mechanical properties. Since SiNWs expand during lithiation, stress is not only caused by the volume difference between lithiated and nonlithiated material, but arises from all other constraining effects. Coatings have been recorded to slow down lithiation kinetics of SiNWs (Zhao et al. 2012). The induced mechanical stress thereby counteracts the electrochemical driving force. Depending on the coating material and thickness, as well as the size of the SiNWs, this driving force can be compensated by compressive stresses within the NWs and hoop stresses within the coating. Under certain circumstances, lithiation can be stopped even after reaching only a fraction of the nominal specific capacity. The effect of mechanical constraint onto the lithiation process of silicon can be expressed in terms of Gibbs free energy ΔG of the lithiation reaction (Equation 21.2).

$$\Delta G = \Delta G_r - eE - \Omega \sigma_m \tag{21.2}$$

ΔG_r is the Gibbs free energy of the lithiation reaction of silicon without applied potential or occurring stress. These two contributing effects are expressed by the work eE that is induced by the applied potential E, with the elementary charge e, and the work $-\Omega \sigma_m$ done by the mechanical stress if one atom is inserted. Ω is the change of volume due to insertion of one Li atom and σ_m is the mean stress within the particle. Strictly speaking, σ_m is a sum of the stress within the lithiated and nonlithiated volume along the reaction front. It can be shown by numerical calculations that the constraint increases with progressing lithiation. If more lithium atoms are inserted, the stress increases and therefore also the work $-\Omega \sigma_m$. For amorphous $Li_{2.1}Si$, for example, ΔG_r is equal to −0.18 eV and can be completely compensated by the mechanical work, which increases with state of charge in an exponential manner. It can be shown that the swelling, and therefore the induced stress, in an SiNW depend strongly on the crystallographic orientation

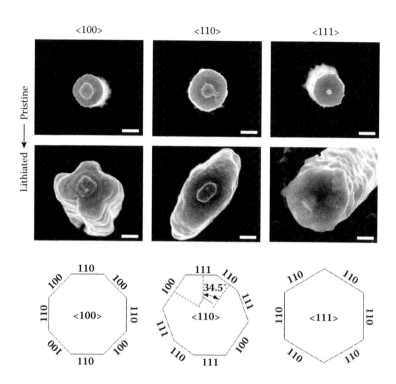

Figure 21.4 Silicon NWs before and after lithiation for different crystalline orientations. The reaction/lithiation rate is faster along the <110> direction and the swelling of the nanowires thus depends strongly on the crystalline orientation with respect to the NW axis. The <111> orientation leads to homogeneous swelling and is therefore often preferred. (Adapted with permission from Lee, S. W., et al., 2011, 11, 3034–3039. Copyright 2011 American Chemical Society.)

due to differences in reaction/lithiation rate at the lithiation front. Thus, swelling happens faster along the <110> direction, while it is considerably slower along the <111> direction, and the electrochemical performance of Si anodes is altered by the choice of crystallographic orientation. This is illustrated in Figure 21.4, where scanning electron microscope (SEM) images of SiNWs of three different orientations are shown before and after lithiation. Homogeneous lithiation appears only if the axis is oriented along the <111> direction of silicon.

The use of hollow NWs, or in other words nanotubes (NTs), instead of plain structures has been proposed and showed that stress within the active material and the coating can be reduced (Zhao et al. 2012). The possibility of hollow structures to expand toward the inside reduces stress within the particles and the surrounding coating in comparison to plain structures. Furthermore, it has been shown, among others by Zhao et al., that the size reduction of the active NWs can reduce the constraining effect of the coating. This increases the acceptable state of charge (or extent of lithiation) before fracture occurs in the coating, or before the lithiation driving force is compensated.

Figure 21.5 summarizes four possible configurations for coated silicon nanostructures, which can all increase the cycle lifetime of Si-based anodes in comparison to pure SiNWs. The last three of these approaches are more promising, since mechanical constraint and SEI formation are minimized:

1. A stiff coating on SiNWs can reduce lithiation due to a mechanical constraint. If the coating is not strong enough it can fracture and therefore lose its ability to stop progressing SEI formation.
2. If the coating is flexible, it can expand in a reversible or irreversible manner together with the active NW. The first case, however, brings the possibility that the SEI, which forms on the surface of the coating, cracks due to expansion of the NW and repeated cycling forms a thick SEI, as described for the noncoated case. Irreversible expansion can theoretically lead to stable cycling.
3. A stiff cage can allow volume expansion and protect the NW from contact with the SEI, which is only formed during the first cycles.

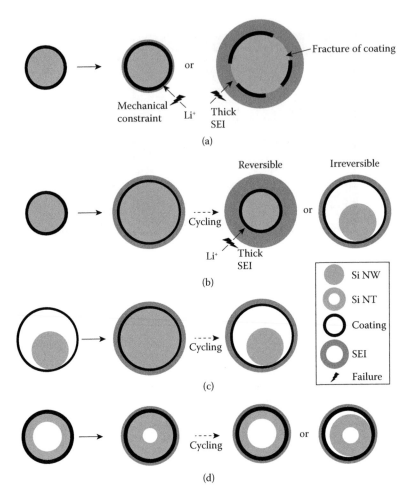

Figure 21.5 Overview of possible lithiation and degradation mechanisms in coated silicon structures. Lithiation of silicon nanowires with a (a) stiff or (b) flexible coating, (c) a stiff cage, as well as silicon nanotubes with (d) a stiff coating.

4. The use of NTs with a stiff coating, instead of NWs, allows expansion within the active material itself, while the formed SEI is stable. After cycling, the NTs have been shown to either adhere to the coating (Wu et al. 2012) or to retract and only contact the stiff shell at some points (Hertzberg et al. 2010).

For case b, carbon- or polymer-based coatings show promising properties, since they allow for more expansion than metal coatings, before they fracture and expose the active material to the electrolyte. However, these properties can change after lithiation and carbon, for example, can become brittle with cycling (Liu et al. 2011). Furthermore, even if the coating is extremely flexible, it is likely that the state of charge is limited to some extent, since the forming amorphous Li_xSi shell around the SiNW core can already induce self-limiting stresses, as Liu et al. showed (Liu et al. 2012). McDowell et al. showed further that even the native oxide layer of SiNWs can suppress the volume expansion for diameters lower than about 50 nm (McDowell et al. 2011).

For the case of a stiff coating, certain metals can be used. They typically show a lower diffusion coefficient of Li^+ than graphite or polymers. The relatively low diffusion coefficient of Li^+, for example, about 10^{-9}–10^{-11} cm²/s in Ni, however, is not necessarily a restricting factor since the diffusion through the lithiated Li_xSi shell is even much slower, at about 10^{-12}–10^{-14} cm²/s. Lang et al. showed by *ab initio* molecular dynamics simulations that the amorphous Li_xSi shell around an SiNW can exert attractive forces on the lithium ions and block lithiation at a certain thickness (Lang et al. 2013). Therefore, the Li_xSi shell seems to be the limiting factor for diffusion.

21.3 FABRICATION TECHNIQUES

In this section, we will discuss techniques for the fabrication of coated SiNWs and nanotubes. Theoretically, the techniques are not limited to the methods presented here and other possibilities for the production of SiNWs/NTs and their coating are conceivable. Rather than to give a complete list, the goal of this section is to give a short and comprehensive review and evaluation of techniques that are used in recent research for the production of coated Si-based anodes in LIBs.

21.3.1 Si NANOWIRE/NANOTUBE FABRICATION

As described previously, the use of SiNWs in LIB anodes has many advantages over bulk material. The small radial dimensions prevent fracture during repeated lithiation and delithiation, reduce diffusion lengths of Li$^+$ within the active material, and facilitate charge transfer between silicon and the current collector. All this increases capacity retention during cycling, as well as the achievable charging/discharging rate. Nanotubes can further reduce stress within the active material and the surroundings by allowing expansion during lithiation toward the inside of the NTs.

21.3.1.1 Chemical vapor deposition

Chemical vapor deposition (CVD) is a widely adopted method for the production of solid materials, most often in form of thin films. By adapting the process through the use of nanostructured catalysts or templates, not only thin films, but also NWs and NTs can be grown. CVD is a bottom-up approach and is based on the adsorption of a precursor gas onto a heated substrate, where the precursor is decomposed. The most commonly used precursor for CVD of silicon is silane (SiH$_4$), which decomposes following Equation 21.3 and results in silicon deposition. Thermal silane decomposition takes place at temperatures above 420°C. Higher temperatures are chosen for the growth of crystalline silicon, while temperatures just above 420°C result in amorphous silicon due to the slow kinetics of the deposited silicon.

$$SiH_4 \xrightarrow{\Delta T} Si + 2H_2 \tag{21.3}$$

Several other precursors can theoretically be used for silicon CVD, for example, silicon tetrachloride (SiCl$_4$) (Peng et al. 2013). However, these other options for the growth of SiNWs are not favorable for the applications under review, since silane allows the growth of NWs with a smaller diameter and at a lower temperature.

21.3.1.1.1 Template-assisted CVD

Template-assisted CVD is a frequently used method for the production of silicon NTs or core-shell NWs. The principle is based on the deposition of silicon by CVD onto a template. For NTs, a porous template membrane with well-defined pores (Figure 21.6a), or sacrificial NWs (Figure 21.6b) can be used. The length l and inner diameter d$_{in}$, as well as outer diameter d$_{out}$ are controlled by template geometry and the thickness of the coatings. A silicon layer is deposited onto the inside of the template or onto the outside of the sacrificial NWs. To release the silicon NTs, the template is selectively removed. Examples are the use of an alumina (Al$_2$O$_3$) template, which can be removed by etching in hydrogen fluoride (HF) solution, or carbon fibers, which can be removed by oxidation in air. A protective and conducting coating can be added in both cases by an additional CVD step. Core-shell structures with a conductive core are produced by using a suitable nanowire template as a core, which is not removed at the end.

The path length for electron transport through silicon, which has a relatively low conductivity, is decreased by these core-shell structures and coated nanotubes. This has a beneficial effect on the cycling rate of the anodes with respect to pure silicon NWs or NTs. Another advantage is the reduced capacity loss due to mechanical stabilization of silicon by the inactive core or coating. Coated NTs are furthermore protected from contact with the electrolyte and resulting excessive SEI formation, as described in Section 21.2.3.

Template-assisted CVD of silicon nanostructures for LIB anodes has been conducted in the past by atmospheric-pressure CVD (APCVD) (Hertzberg et al. 2010; Wu et al. 2012; Liu et al. 2015), by

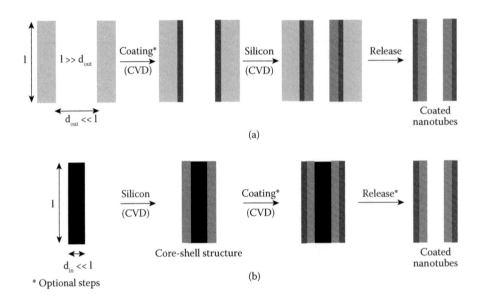

Figure 21.6 Schematic representation of the template-assisted CVD for silicon nanotubes with (a) porous membrane and (b) nanowire sacrificial templates. A coating can be applied with an additional CVD step. Core-shell structures are achieved for nanowire templates if the optional steps are avoided.

low-pressure CVD (LPCVD) (Karki et al. 2013), or by plasma-enhanced CVD (PECVD) (Nguyen et al. 2012; Qu et al. 2012). In the case of LPCVD, the pressure within the reaction chamber is kept below atmospheric pressure. The advantage of this technique over APCVD is an improved film uniformity and purity. However, if compared to APCVD, the lower pressure is not only advantageous; it also reduces the throughput of the precursor gas proportionally with the chamber pressure itself. Therefore, the deposition rate is reduced for LPCVD. The typically used temperature, which also influences the deposition rate, can be increased to about 600°C for LPCVD to counteract the effect of the lower gas throughput. The third possibility, PECVD, is more complex than APCVD or LPCVD, since it utilizes a plasma within the deposition chamber. High-energetic electrons in this plasma are used in addition to temperature to decompose the precursor gas on the substrate surface. The reaction rate is thereby increased, while the temperature can be reduced to a value of typically 250°C.

21.3.1.1.2 Vapor–liquid–solid growth

Vapor–liquid–solid (VLS) growth is a bottom-up growth method for the production of silicon NWs and currently the most often used approach for application in LIBs (Chen et al. 2011; McDowell et al. 2011, 2012; Cho et al. 2012; Memarzadeh et al. 2012; Wu et al. 2012; Yao et al. 2012; Cho and Picraux 2013; Kohandehghan et al. 2013). Similar to CVD, this technique is also based on the decomposition of a precursor gas, typically silane, as a source of silicon. The main difference is the use of a catalyst, which forms a liquid phase at elevated temperatures. The most common catalyst for the growth of SiNWs is gold due to its ability to dissolve and then reprecipitate silicon atoms. The first step for the growth of SiNWs by VLS is the deposition of a thin gold film or Au NPs onto a substrate. This can be done, for example, by sputtering, electrodeposition, or electroless plating. After gold is deposited, CVD at temperatures above 363°C leads to the formation of liquid droplets of a gold–silicon mixture. The temperature limit is 363°C, since this is the eutectic temperature of the system at a Au:Si composition of 4:1. The gold droplets will then supersaturate with silicon by decomposition of more silane. Since the melting temperature of silicon is 1414°C, much higher than the VLS temperature, silicon will nucleate at the liquid–solid interphase. This leads to the growth of silicon NWs underneath the gold catalyst (Figure 21.7).

The diameter of these NWs is determined by the size of the droplets. The smallest possible diameter D_{min} of a droplet, in turn, is given by Equation 21.4, where V_m is the molar volume of the liquid, R the

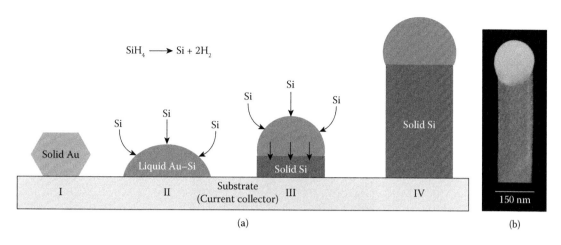

SiH$_4$ ⟶ Si + 2H$_2$

Solid Au

Liquid Au–Si

Solid Si

Solid Si

Si   Si   Si

Si   Si   Si

Substrate
(Current collector)

I   II   III   IV

(a)

Solid Si

150 nm

(b)

Figure 21.7 (a) Schematic illustration of the four steps in vapor–liquid–solid growth of silicon nanowires. I: catalyst deposition; II: formation of liquid eutectic phase; III: nucleation of supersaturated silicon; IV: nanowire growth, and (b) SEM image of a SiNW with the Au catalyst at the top. (Memarzadeh, E. L., et al., 2012, Silicon Nanowire Core Aluminum Shell Coaxial Nanocomposites for Lithium Ion Battery Anodes Grown with and without a TiN Interlayer. *Journal of Materials Chemistry*, 22, 6655–6668. Reproduced by permission of The Royal Society of Chemistry.)

ideal gas constant, T the temperature, s the degree of supersaturation of Si in Au, and σ_{lv} the liquid–vapor interfacial energy. This value is typically in the range of several nanometers.

$$D_{min} = \frac{4V_m}{RT\ln(s)}\sigma_{lv} \tag{21.4}$$

Temperature control during VLS is of high importance. Not only does it affect the nanowire diameter (Equation 21.4), but it also affects the formation of unwanted species. Silicides can form on the support substrate (an element that might be used subsequently as a current collector when assembling the lithium cell), especially at high temperatures, and reduce the specific capacity of the anode, and the possible cycling rate. The appearance of silicides can be minimized by the right choice of temperature, which depends on the deposition parameters and the nature of the support substrate, but is rather low. Another reason for the importance of temperature control is the formation of a continuous silicon film underneath the NWs, which reduces the cycle lifetime of the anode due to fracturing and therefore loss of capacity. Silicon islands have been reported to be eliminated through the use of a porous template for the VLS growth of silicon NWs (Cho and Picraux 2013). The template avoids formation of the silicon islands and is removed after the growth process, similar to the template-assisted CVD method that was described in Section 21.3.1.1.1.

21.3.1.1.3 Metal-assisted chemical etching

The most commonly used top-down approach for the production of SiNWs for LIB anodes is the metal-assisted chemical etching (MACE) of silicon wafers (Huang et al. 2009; Vlad et al. 2012; Lee et al. 2013; Sandu et al. 2014). This technique is based on the etching of silicon, which is partially covered by a noble metal film—typically gold or platinum. Generally, an aqueous HF/H$_2$O$_2$ mixture can etch silicon, since H$_2$O$_2$ is introducing holes in the silicon valence band. However, the etching rate is only a few nanometers per hour for pure silicon. This is where the noble metal comes into play. It facilitates the reduction of H$_2$O$_2$ drastically and therefore increases the injection of holes into the silicon substrate, which is thus preferentially etched at the interface with the noble metal. As a consequence, the metal sinks into the substrate (Figure 21.8) and creates, depending on the coverage of silicon with the catalyst, a porous structure or SiNWs.

MACE is a very versatile, simple, and inexpensive method. By controlling the noble metal deposition, for example, by lithography, the size and density of nanowires can be controlled, down to about 5 nm. An advantage of MACE is the possibility to grow not only round nanowires, but also silicon structures with

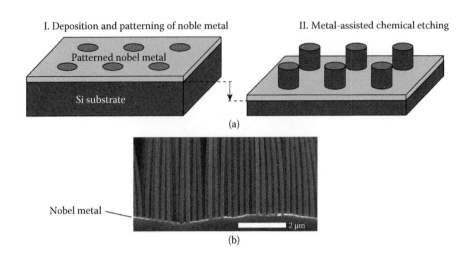

I. Deposition and patterning of noble metal

Patterned nobel metal

Si substrate

II. Metal-assisted chemical etching

(a)

Nobel metal

2 μm

(b)

Figure 21.8 (a) Schematic of the two steps for MACE and b) an SEM image of SiNWs produced by this technique. The NWs in (b) are bundled up and the inlay shows the single wires, which have a uniform diameter.

different cross sections, which is not the case for the previously described VLS growth. Furthermore, the crystal quality and crystallographic orientation of the SiNWs can be easily controlled by the choice of the silicon substrate. In contrast to methods like reactive ion etching, MACE does not introduce crystallographic defects into the silicon.

21.3.1.2 Inductively coupled plasma etching

Inductively coupled plasma (ICP) etching is, similar to MACE, a top-down process for the fabrication of SiNWs (Lee et al. 2014; Ye et al. 2014). In this process, etching of a silicon substrate occurs physically and chemically. To control shape, size, and areal density of the final nanowires, a photomask is first applied and patterned on top of the Si substrate. For the etching process, sulfur hexafluoride (SF_6) is typically used as the etchant gas. Fluorine ions are generated under plasma excitation, which is sustained by RF electromagnetic induction. These ions are then accelerated toward the silicon, where they eject Si ions (mechanical etching if sufficient energy is acquired), or react with silicon to form volatile SiF_4 (chemical etching). Since the ions are accelerated isotropically, the etching process appears mainly in the direction of the acceleration. However, lateral etching can still occur to a certain extent. To avoid this and to ensure anisotropic profile etching, the walls of the etched structures can be passivated by use of octafluorocyclobutane (C_4F_8) in the plasma gas, which forms a polymeric passivation layer on the walls. Etching and passivation can either be conducted simultaneously or in alternative steps, known as the Bosch process, which is schematically illustrated in Figure 21.9a. If gas flow, applied power, and other variables are chosen carefully, the passivation layer is thick enough to prevent lateral etching, but still allows etching in direction of ion acceleration. Steps III and IV in Figure 21.9a can be repeated until the desired depth of etching is reached. This enables the fabrication of high aspect ratio NWs (Vlad et al. 2013). After the process is finished, the passivation layer can be removed by heating the sample. An SEM image of the final NWs is shown in Figure 21.9b.

In addition to these advances, similar to MACE, ICP etching allows adaption of the shape of the NW cross section by changing the mask. Figure 21.10 shows TEM images of SiNWs with round and square cross sections, produced by ICP etching and consecutively coated by atomic layer deposition (ALD) (see Section 21.3.2.4) with TiO_2. The square (Ye et al.) shows that the geometry design can affect stress relaxation and therefore cycle lifetime.

21.3.2 COATING TECHNIQUES

In Section 21.2.3, we discussed why SiNWs as anode materials are often coated. In this section, we will take a look at how these coatings can actually be applied. Similar to the previously discussed techniques for the fabrication of NWs, this is a nonexhaustive list and other methods could in principle be used. However,

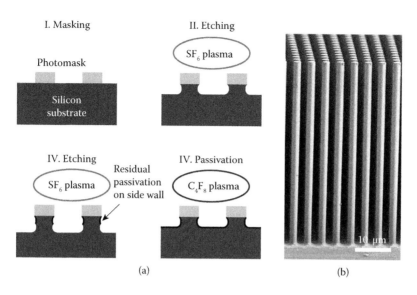

Figure 21.9 (a) Schematic representation of the four steps for the Bosch process (ICP etching), where steps III and IV are repeated to increase the length of the NWs, and (b) SEM image of SiNWs produced by ICP etching. (In part from *Journal of Power Sources*, 248, Ye, J. C., et al., Enhanced Lithiation and Fracture Behavior of Silicon Mesoscale Pillars via Atomic Layer Coatings and Geometry Design, 447–456, Copyright 2014, with permission from Elsevier.)

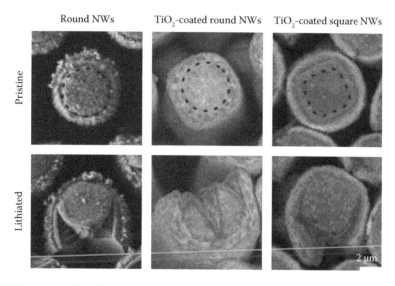

Figure 21.10 SiNWs produced by ICP etching with round or square cross sections, before and after lithiation. The square cross section leads here to improved stress relaxation, and thus less fracture, and the TiO_2 coating supports the formation of a thin SEI, as well as ion transport into the bulk. The outside layer is not the coating, which is too thin to be observed here, but the SEI, which is smoother for the NWs with coating. (Adapted from *Journal of Power Sources*, 248, Ye, J. C., et al., Enhanced Lithiation and Fracture Behavior of Silicon Mesoscale Pillars via Atomic Layer Coatings and Geometry Design, 447–456, Copyright 2014, with permission from Elsevier.)

the approaches presented here give a summary of methods that have been used for this specific application in the past.

21.3.2.1 Magnetron sputtering

Magnetron sputtering is a physical vapor deposition (PVD) method and is frequently used to apply metallic or nonmetallic coatings to SiNWs for LIB anodes. Copper (Chen et al. 2011; McDowell et al. 2012; Lee et al. 2014), aluminum (Memarzadeh et al. 2012), nickel (Karki et al. 2013), gold (Lee et al. 2014), titanium nitride

(Kohandehghan et al. 2013), and others have been applied to silicon NWs or NTs in the past. All these materials can provide good electrical conductivity and protection against continuous SEI formation, if the coating layer is uniform and if cracks can be avoided.

In magnetron sputtering, argon plasma is sustained by a magnetron, which contains an applied magnetic field to increase the path length of electrons in the plasma. Ar ions are then accelerated onto the target, where they act as a mechanical ablation source and eject part of the substrate material. For conductive targets the acceleration can be done by a DC bias, while nonconductive targets require a radio frequency potential to avoid charging. The sputtered material reaches the substrate mainly in form of neutral atoms and condenses to form the desired coating. To sputter ceramic coatings by the use of metallic targets, or to control stoichiometry of the coating, reactive magnetron sputtering can be utilized. It involves the use of a reactive process gas, which is present in the chamber in addition to argon as the main constituent of the plasma. For reactive sputtering of TiN, for example, a mixture of Ar and N_2 can be used and nitrogen atoms are incorporated into the coating together with the sputtered titanium.

Since magnetron sputtering does not require heating of the target, it allows coating with materials that show a very high evaporation temperature, which is not the case for other PVD methods, like thermal evaporation techniques. However, conformal coating of high aspect ratio NWs is problematic. Since sputtered atoms reach the substrate in a rather directed way, parts of the nanowires can be shadowed and therefore be unevenly or insufficiently coated. If the material is to be used as an anode for lithium batteries, uneven coatings can still provide some mechanical stability to the silicon NWs and therefore increase cycling stability. However, it can lead to bending of the NWs due to uneven constraint of the volume expansion during lithiation, result in fracture, or even lack complete protection against the electrolyte in the first place. It is thus important to monitor the uniformity if this technique is to be employed.

21.3.2.2 Evaporation

As its name states, coating of nanowires by evaporation is based on introduction of energy into the respective source material, which then evaporates. The vapor can travel through the vacuum chamber and condense on the NWs. This process is mainly used for the deposition of metals, since they are relatively easy to evaporate. An example is the thermal evaporation of copper onto SiNWs (McDowell et al. 2012). But, coating with other materials like carbon is also possible using this technique (Chen et al. 2011).

Several variations of the evaporation method exist. Their main difference is the applied energy source. This can be the thermal evaporation of material in a directly heated crucible, evaporation by an electron beam, heating by applying a large electric current to a wire-shaped material source, and others.

The main disadvantage of evaporation techniques for coatings in battery applications is the high directionality, being even less uniform than for magnetron sputtering. Conformal coating, even of low aspect ratio nanostructures, is nearly impossible. Bending during lithiation of an SiNW, which was directionally coated with Cu by thermal evaporation, can be seen in the TEM image in Figure 21.11. Evaporation is therefore not often recorded as a technique for the coating of SiNWs.

21.3.2.3 Chemical vapor deposition

The working principle of CVD of coatings on silicon NWs or NTs is the same as for the production of these nanostructures, which was described in Section 21.3.1.1. The advantage of CVD over magnetron

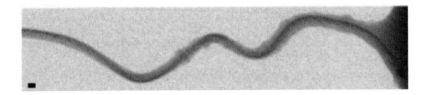

Figure 21.11 TEM image of a lithiated SiNW, which was coated with Cu by thermal evaporation. The copper coating and therefore its constraint on the NW during expansion is uneven due to the directional nature of the process and bending occurs as a result. (Reprinted from *Nano Energy*, 1(3), McDowell, M. T., et al., The Effect of Metallic Coatings and Crystallinity on the Volume Expansion of Silicon during Electrochemical Lithiation/Delithiation, 401–410, Copyright 2012, with permission from Elsevier.)

sputtering is the high uniformity of the coating, since deposition of the respective material is conducted by decomposition from the gas phase. A disadvantage, however, is the need for relatively high temperatures to decompose the precursor gas. Depending on the materials used, this high temperature can lead to the formation of silicides or oxides, which reduce the specific capacity of the anode.

An example for the use of CVD is the coating step in template-assisted CVD fabrication of carbon-coated silicon nanotubes (SiNTs) (Hertzberg et al. 2010), which is schematically illustrated in Figure 21.6a. An SEM image before, and a TEM image after lithiation of an SiNT@C are shown in Figure 21.12. Carbon was deposited by the decomposition of C_3H_6 at 700°C. Although the C coating seems to remain intact after the lithiation, the TEM image shows the separation between coating and NTs after cycling (one possibility in Figure 21.5d). Since defects are typically present in coatings, this can potentially lead to electrolyte exposure and repeated SEI generation in the created voids, meaning on the surface of the retracting and expanding NTs.

21.3.2.4 Atomic layer deposition

ALD offers the possibility to deposit a variety of materials with high uniformity and an extremely well-controlled layer thickness with a precision down to single atomic layers. Examples of the use of ALD in battery applications are the deposition of Al_2O_3 (Nguyen et al. 2012; Ye et al. 2014) or TiO_2 (Ye et al. 2014) on SiNWs.

The principle of ALD is the alternated use of two different precursor gases. The first precursor is introduced into the reaction chamber together with a carrier gas. It will then chemisorb onto the surface of the SiNWs. When the surface is saturated, the residual precursor is removed from the chamber and a second precursor is introduced, which reacts with the chemisorbed layer and forms the first atomic layer of the coating. This process is repeated until the final coating thickness is reached. In the case of an alumina coating, trimethyl aluminum and water vapor can be used as precursors with N_2 as the carrier gas.

Besides its well-defined coating, another advantage of ALD is that this process can often be conducted at low temperatures, even down to room temperature. However, since the deposition of each atomic layer requires the flushing and reaction of two different precursors, ALD is a rather slow process. The typical deposition rate is only a few hundred nanometers per hour and restricts the application for high-through-put fabrication. Another disadvantage is the need of reactive surfaces to promote the reaction with the precursor, but in the case of Si this is ensured by the native surface oxide layer. An example of TiO_2-coated SiNWs by ALD was given earlier in Figure 21.10.

21.3.2.5 Electroless plating

Electroless plating is a technique that provides another possibility for the coating of silicon NWs with metals. It has mainly been used for copper (Vlad et al. 2012) and nickel (Sandu et al. 2014) coatings. An example of nickel coatings with varying thickness on SiNWs is shown in Figure 21.13.

In contrast to conventional electroplating, the reduction of metal ions in electroless plating does not depend on an external electron source. This independence from current flow makes this process also

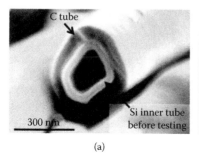
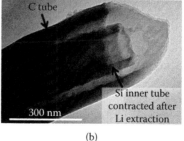

(a) (b)

Figure 21.12 Carbon-coated Si NTs (a) before cycling and (b) after 10 cycles. Carbon and silicon were consecutively coated by CVD on a porous alumina membrane. (Adapted with permission from Hertzberg, B., et al., 2010, 132(25), 8548–8549. Copyright 2010 American Chemical Society.)

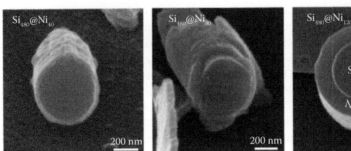

Figure 21.13 Electroless plated nickel coatings with thicknesses of 40, 80, and 120 nm on SiNWs with a diameter of 480 nm. The NWs were fabricated by MACE. (Adapted with permission from Sandu, G., et al., 2014, 8, 9427–9436. Copyright 2014 American Chemical Society.)

suitable for substrates with low electrical conductivity. It also allows uniform plating of substrates with more complex surface geometries (e.g., nanowires and nanotubes), which would locally vary the current density and therefore coating properties of electroplated metals. These restrictions are avoided by electroless plating since a reducing agent is added to the aqueous medium, which contains the metal source. Reduction of the metal source, which results in the metal deposition, and oxidation of the reducing agent form a redox process without the need of additional electrons. However, further reactions can occur during the plating process, for example, hydrogen evolution, and also the process is very sensitive to the surface state of the substrate material.

Besides metal source and reducing agent, other components can be added to the medium. These can be complexing agents or stabilizing/inhibiting agents. Furthermore, moderate heating to typically about 65°C is employed as an energy source.

21.3.2.6 Carbonization

For the coating of SiNWs by carbon, carbonization can be employed. This method belongs to the group of thermolysis, which is the chemical decomposition at elevated temperatures. More specifically, pyrolysis is the decomposition of organic material if no oxygen is present. Carbonization, in turn, is a form of pyrolysis, which results mainly in carbon as a residue, ideally without any other solid or liquid components. Most people use pyrolysis on a regular basis without even knowing, for example, by roasting their breakfast toast or other food. The golden-brown or, less ideally, black color of roasted food is a result of pyrolysis.

Carbon coatings on SiNWs by carbonization can be achieved by the use of many different organic materials, which are applied to the substrate as precursors. Examples are the carbonization of furfuryl alcohol at 500°C (Liu et al. 2015), shown in Figure 21.14, or of a resorcinol/formaldehyde mixture at 650°C (Huang et al. 2009). The carbonization step is typically done in an argon atmosphere to avoid a reaction with oxygen during heating. Furthermore, an additional heating step at lower temperatures is used before carbonization, in order to form a polymerized precursor layer on top of the NWs.

21.3.2.7 Surface oxides: native and controlled

Besides the coating of SiNWs with an externally applied material, oxidation of SiNWs also leads to a layer, namely an SiO_x around the Si core, which can be treated as a coating and has been shown to have similar mechanically constraining effects on the lithiation process (McDowell et al. 2011). The native oxide layer is about 2–5 nm thick and can significantly limit lithiation if the diameter of the SiNWs is too small, especially in the range of 50 nm or less. It can also lead to the appearance of Li_2O and Li_4SiO_4 during the first cycle, which are stiffer than silicon and increase impedance of the anode. An artificially grown and well-defined SiO_2 layer can be created by heating the SiNWs or NTs in air (Wu et al. 2012). This does not only change the mechanical properties, but also surface chemistry, especially with respect to Si–OH groups. Interaction with the SEI or an additional coating/matrix can be influenced by this chemical factor.

The effect of a well-defined SiO_2 layer, especially in combination with the use of NTs, is shown in Figure 21.15. The reduction of SEI formation and good conservation of the NT shape is clearly visible in

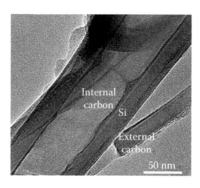

Figure 21.14 Si NT with an inner and outer carbon coating. The Si NTs were produced with a ZnO nanowire sacrificial template (see Figure 21.6b) with a carbonization step before and after the Si CVD deposition. (Adapted with permission from Liu, J., et al., 2015, 9, 1985–1994. Copyright 2015 American Chemical Society.)

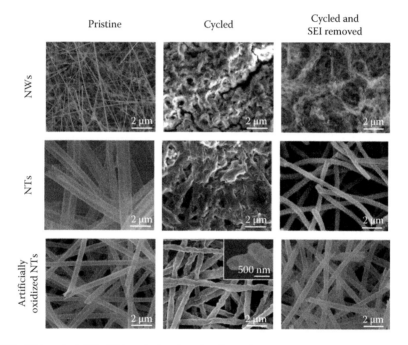

Figure 21.15 SiWs (200 cycles), NTs (200 cycles) and artificially oxidized NTs (2000 cycles) before and after cycling. The SEI was selectively etched in the last column to observe the nanostructures after cycling. (Adapted by permission from Macmillan Publishers Ltd: *Nature Nanotechnology*, Wu, H., et al., 2012, Stable Cycling of Double-Walled Silicon Nanotube Battery Anodes through Solid–electrolyte Interphase Control, 7(5), 310–315, copyright 2012.)

this case. Note further, that the artificially oxidized tubes in this image were cycled 2,000 times, while the other structures were cycled for only 200 times and nonetheless show more degradation and SEI formation.

21.3.2.8 Others

In addition to the techniques already mentioned, there are many other feasible approaches for the coating of SiNWs. Examples, which will not be covered in detail here, are the electrochemical polymerization and simultaneous deposition of conductive PEDOT polymer coatings (Yao et al. 2012), or the coating of NWs with graphene flakes (Lee et al. 2013). The first example is represented in Figure 21.16, but the possible material combinations, variations (e.g., porous coatings) and coating techniques are manifold. In fact, many of these possibilities are still unexplored and especially the SEI formation on the coating, lithium diffusion, and other mechanisms still require more research in this area.

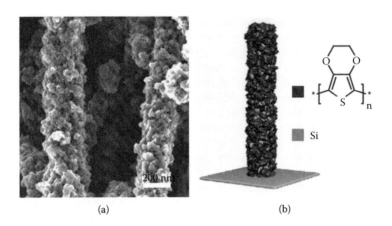

Figure 21.16 (a) SEM image and (b) schematic representation of PEDOT-coated SiNWs. (Yao, Y., et al., 2012, Improving the Cycling Stability of Silicon Nanowire Anodes with Conducting Polymer Coatings. *Energy & Environmental Science*, 5, 7927–7930. Reproduced by permission of The Royal Society of Chemistry.)

21.4 DISCUSSION OF FUTURE DEVELOPMENTS

After the basics of LIBs, as well as the advantages of coated SiNWs/nanotubes and techniques for their production have been discussed, one question is still unanswered—are these coated nanostructures the final solution? In order to answer this, the ideal case for anodes has to be considered. Looking only at specific capacity and voltage, this would be pure lithium. As mentioned before, however, dendrite growth is a big problem for the use of pure lithium and makes this anode type currently unsafe to use, which is why storage materials are used to contain the lithium. Research is currently being done to create anode structures or solid electrolytes that prevent dendrite growth of lithium anodes. This would increase the specific capacity of the anodes, but many problems have still to be overcome. Until these approaches are successful, silicon is one of the most promising options for anodes in next-generation LIBs, especially in form of the coated nanostructures discussed here.

Regarding Si-based anodes, a tendency of research toward hollow or porous structures can be observed, since free volume can make up for expansion of the active material, as a trade-off for a decreased volumetric energy density. Another target is the use of microscopic instead of nanoscopic structures, which would reduce the costs drastically, but are heavily affected by stress, induced by volume changes during cycling. Thus, contact loss to the current collector has to be avoided and, furthermore, diffusion paths of lithium are longer in larger structures and could lead to lower cycling rates due to diffusion limitations.

REFERENCES

Chan, Maria K. Y., Christopher Wolverton, and Jeffrey P. Greeley. 2012. First Principles Simulations of the Electrochemical Lithiation and Delithiation of Faceted Crystalline Silicon. *Journal of the American Chemical Society* 134(35): 14362–74. doi:10.1021/ja301766z.

Chen, Huixin, Ying Xiao, Lin Wang, and Yong Yang. 2011. Silicon Nanowires Coated with Copper Layer as Anode Materials for Lithium-Ion Batteries. *Journal of Power Sources* 196: 6657–62. doi:10.1016/j.jpowsour.2010.12.075.

Cho, Jeong-Hyun, Xianglong Li, and Samuel Thomas Picraux. 2012. The Effect of Metal Silicide Formation on Silicon Nanowire-Based Lithium-Ion Battery Anode Capacity. *Journal of Power Sources* 205: 467–73. doi:10.1016/j.jpowsour.2012.01.037.

Cho, Jeong Hyun, and S. Tom Picraux. 2013. Enhanced Lithium Ion Battery Cycling of Silicon Nanowire Anodes by Template Growth to Eliminate Silicon Underlayer Islands. *Nano Letters* 13 (11): 5740–7. doi:10.1021/nl4036498.

Gupta, Sujata, and Nigel Hawtin. 2015. From Gadgets to the Smart Grid. *Nature* 526: 90–1. doi:10.1038/526S90a.

Hertzberg, Benjamin, Alexander Alexeev, and Gleb Yushin. 2010. Deformations in Si-Li Anodes upon Electrochemical Alloying in Nano-Confined Space. *Journal of the American Chemical Society* 132 (25): 8548–9. doi:10.1021/ja1031997.

Huang, Rui, Xing Fan, Wanci Shen, and Jing Zhu. 2009. Carbon-Coated Silicon Nanowire Array Films for High-Performance Lithium-Ion Battery Anodes. *Applied Physics Letters* 95 (13): 2009–11. doi:10.1063/1.3238572.

Karki, Khim, Yujie Zhu, Yihang Liu, Chuan Fu Sun, Liangbing Hu, Yuhuang Wang, Chunsheng Wang, and John Cumings. 2013. Hoop-Strong Nanotubes for Battery Electrodes. *ACS Nano* 7 (9): 8295–302. doi:10.1021/nn403895h.

Kohandehghan, Alireza, Peter Kalisvaart, Kai Cui, Martin Kupsta, Elmira Memarzadeh, and David Mitlin. 2013. Silicon Nanowire Lithium-Ion Battery Anodes with ALD Deposited TiN Coatings Demonstrate a Major Improvement in Cycling Performance. *Journal of Materials Chemistry A* 1: 12850–61. doi:10.1039/c3ta12964k.

Lang, Li, Chuanding Dong, Guohong Chen, Jihui Yang, Xiao Gu, Hongjun Xiang, Ruqian Wu, and Xingao Gong. 2013. Self-Stopping Effects of Lithium Penetration into Silicon Nanowires. *Nanoscale* 5 (24): 12394–8. doi:10.1039/c3nr03301e.

Lee, Gibaek, Stefan L. Schweizer, and Ralf B. Wehrspohn. 2014. CMOS-Compatible Metal-Stabilized Nanostructured Si as Anodes for Lithium-Ion Microbatteries. *Nanoscale Research Letters* 9 (1): 613. doi:10.1186/1556-276X-9-613.

Lee, Sang Eon, Han-Jung Kim, Hwanjin Kim, Jong Hyeok Park, and Dae-Geun Choi. 2013. Highly Robust Silicon Nanowire/Graphene Core-Shell Electrodes without Polymeric Binders. *Nanoscale* 5 (19): 8986–91. doi:10.1039/c3nr00852e.

Lee, Seok Woo, Matthew T. McDowell, Jang Wook Choi, and Yi Cui. 2011. Anomalous Shape Changes of Silicon Nanopillars by. *Nano Letters* 11: 3034–9. doi:10.1021/nl201787r.

Liu, Hua, Zaiping Guo, Jiazhao Wang, and Konstantin K. Konstantinov. 2010. Si-Based Anode Materials for Lithium Rechargeable Batteries. *Journal of Materials Chemistry* 20: 10055–7. doi:10.1039/c0jm01702g.

Liu, Jinyun, Nan Li, Matthew D. Goodman, Hui Gang Zhang, Eric S. Epstein, Bo Huang, Zeng Pan, et al. 2015. Mechanically and Chemically Robust Sandwich-Structured C@Si@C Nanotube Array Li-Ion Battery Anodes. *ACS Nano* 9 (2): 1985–94. doi:10.1021/nn507003z.

Liu, Xiao Hua, Jiang Wei Wang, Shan Huang, Feifei Fan, Xu Huang, Yang Liu, Sergiy Krylyuk, et al. 2012. Self-Limiting Lithiation in Silicon Nanowires. *ACS Nano* 7: 749–56. doi:10.1038/nnano.2012.170.

Liu, Yang, He Zheng, Xiao Hua Liu, Shan Huang, Ting Zhu, Jiangwei Wang, Akihiro Kushima, et al. 2011. Lithiation-Induced Embrittlement of Multiwalled Carbon Nanotubes. *ACS Nano* 5 (9): 7245–53. doi:10.1021/nn202071y.

McDowell, Matthew T., Seok Woo Lee, Justin T. Harris, Brian A. Korgel, Chongmin Wang, William D. Nix, and Yi Cui. 2013. In Situ TEM of Two-Phase Lithiation of Amorphous Silicon Nanospheres. *Nano Letters* 13 (2): 758–64. doi:10.1021/nl3044508.

McDowell, Matthew T., Seok Woo Lee, Ill Ryu, Hui Wu, William D. Nix, Jang Wook Choi, and Yi Cui. 2011. Novel Size and Surface Oxide Effects in Silicon Nanowires as Lithium Battery Anodes. *Nano Letters* 11 (9): 4018–25. doi:10.1021/nl202630n.

McDowell, Matthew T., Seok Woo Lee, Chongmin Wang, and Yi Cui. 2012. The Effect of Metallic Coatings and Crystallinity on the Volume Expansion of Silicon during Electrochemical Lithiation/Delithiation. *Nano Energy* 1 (3): 401–10. doi:10.1016/j.nanoen.2012.03.004.

Memarzadeh, Elmira L., W. Peter Kalisvaart, Alireza Kohandehghan, Beniamin Zahiri, Chris M. B. Holt, and David Mitlin. 2012. Silicon Nanowire Core Aluminum Shell Coaxial Nanocomposites for Lithium Ion Battery Anodes Grown with and without a TiN Interlayer. *Journal of Materials Chemistry* 22: 6655–68. doi:10.1039/c2jm16167b.

Nguyen, Hung Tran, Mihai Robert Zamfir, Loc Dinh Duong, Young Hee Lee, Paolo Bondavalli, and Didier Pribat. 2012. Alumina-Coated Silicon-Based Nanowire Arrays for High Quality Li-Ion Battery Anodes. *Journal of Materials Chemistry* 22 (47): 24618–26. doi:10.1039/c2jm35125k.

Park, Jung-Ki. 2013. *Principles and Applications of Lithium Secondary Batteries*. Weinheim, Germany: Wiley.

Peng, Kui-Qing, Xin Wang, Li Li, Ya Hu, and Shuit-Tong Lee. 2013. Silicon Nanowires for Advanced Energy Conversion and Storage. *Nano Today* 8 (1): 75–97. doi:10.1016/j.nantod.2012.12.009.

Pistoia, Gianfranco. 2014. *Lithium-Ion Batteries: Advances and Applications*. Amsterdam, Oxford: Elsevier.

Qu, Jun, Huaqing Li, John J. Henry, Surendra K. Martha, Nancy J. Dudney, Hanbing Xu, Miaofang Chi, et al. 2012. Self-Aligned Cu-Si Core-Shell Nanowire Array as a High-Performance Anode for Li-Ion Batteries. *Journal of Power Sources* 198: 312–17. doi:10.1016/j.jpowsour.2011.10.004.

Sandu, Georgiana, Laurence Brassart, Jean François Gohy, Thomas Pardoen, Sorin Melinte, and Alexandru Vlad. 2014. Surface Coating Mediated Swelling and Fracture of Silicon Nanowires during Lithiation. *ACS Nano* 8 (9): 9427–36. doi:10.1021/nn503564r.

Vlad, Alexandru, Andreas Frölich, Thomas Zebrowski, Constantin Augustin Dutu, Kurt Busch, Sorin Melinte, Martin Wegener, and Isabelle Huynen. 2013. Direct Transcription of Two-Dimensional Colloidal Crystal Arrays into Three-Dimensional Photonic Crystals. *Advanced Functional Materials* 23 (9): 1164–71. doi:10.1002/adfm.201201138.

Vlad, Alexandru, Arava Leela Mohana Reddy, Anakha Ajayan, Neelam Singh, Jean-François Gohy, Sorin Melinte, and Pulickel M. Ajayan. 2012. Roll up Nanowire Battery from Silicon Chips. *Proceedings of the National Academy of Sciences of the United States of America* 109 (38): 15168–73. doi:10.1073/pnas.1208638109.

Whittingham, M. Stanley. 1976. Electrical Energy Storage and Intercalation Chemistry. *American Association for the Advancement of Science* 192 (4244): 1126–7. doi:10.1126/science.192.4244.1126.

Wu, Hui, Gerentt Chan, Jang Wook Choi, Ill Ryu, Yan Yao, Matthew T. McDowell, Seok Woo Lee, et al. 2012. Stable Cycling of Double-Walled Silicon Nanotube Battery Anodes through Solid–electrolyte Interphase Control. *Nature Nanotechnology* 7 (5): 310–15. doi:10.1038/nnano.2012.35.

Yao, Yan, Nian Liu, Matthew T. McDowell, Mauro Pasta, and Yi Cui. 2012. Improving the Cycling Stability of Silicon Nanowire Anodes with Conducting Polymer Coatings. *Energy & Environmental Science* 5: 7927–30. doi:10.1039/c2ee21437g.

Ye, Jianchao , Yonghao An, Taewook Heo, Monika M. Biener, Rebecca J. Nikolić, Meijie Tang, Hanqing Jiang, Yinmin Wang. 2014. Enhanced Lithiation and Fracture Behavior of Silicon Mesoscale Pillars via Atomic Layer Coatings and Geometry Design. *Journal of Power Sources* 248: 447–56. doi:10.1016/j.jpowsour.2013.09.097.

Yoshino, Akira. 2012. The Birth of the Lithium-Ion Battery. *Angewandte Chemie, International Edition* 51: 5798–800. doi:10.1002/anie.201105006.

Zhang, Sheng S., Kang Xu, Richard Jow. 2006. EIS Study on the Formation of Solid Electrolyte Interface in Li-Ion Battery. *Electrochimica Acta* 51: 1636–40. doi:10.1016/j.electacta.2005.02.137.

Zhao, Kejie, Matt Pharr, Lauren Hartle, Joost J. Vlassak, and Zhigang Suo. 2012. Fracture and Debonding in Lithium-Ion Batteries with Electrodes of Hollow Core-Shell Nanostructures. *Journal of Power Sources* 218: 6–14. doi:10.1016/j.jpowsour.2012.06.074.

Ion-implanted silicon nanowires

Bennett E. Smith and Peter J. Pauzauskie

Contents

22.1 INTRODUCTION

Silicon is the second most abundant element found in the earth's crust [1] and has been identified as a component of interplanetary dust resulting from nucleosynthesis in supernovae [2]. Both its electrical properties and terrestrial abundance have propelled silicon to its foundational status in modern integrated circuit technology. Low-cost, high-volume production of high purity single crystals through the Czochralski crystal-growth process have made silicon the material of choice for microprocessors that are nearly ubiquitous in electronic devices. In terms of renewable energy, silicon is used heavily in photovoltaic solar panels. With a global market share of about 90%, crystalline silicon is by far the most important photovoltaic technology today [3]. In terms of biomedical applications, silicon has been shown to be both nonimmunogenic [4] and also biodegradable with the reported *in vivo* half-life of silicic acid being approximately one week [5].

Along with elemental composition, silicon materials research has been continuously exploring the properties of micro- and nanoscaled devices and structures. One such structure is the nanowire (NW), which provides many novel characteristics due to its high aspect ratio, one-dimensional morphology. Nanowires of silicon have been synthesized and doped using several different methods and have been demonstrated to be advantageous in many applications. The intent of the chapter is to provide the reader with an outline of synthetic routes for the production of silicon nanowires (SiNWs) and their electronic doping through ion implantation, followed by a discussion of how doped SiNWs are influencing research in a range of fields including microelectronics [6] and biomedicine [7].

Current research interests directed toward size control and ion implantation are looking to understand how the combination of morphology-dependent resonances and doping within SiNWs will improve their optical absorption capabilities for targeted photothermal and photodynamic therapy. For example, certain properties of silicon (such as its indirect band gap) are less than ideal in applications that require a high optical absorption coefficient. Ion implantation provides a unique path toward engineering SiNWs with precise carrier concentration profiles and also different degrees of crystallinity that have a major impact on the optical absorption coefficient of the final materials.

Following this introduction this chapter is organized into three primary sections focused on the (1) formation, (2) properties, and (3) applications of ion-implanted SiNWs. The primary focus of the formation section is to describe the main processing and fabrication routes to prepare ion-implanted SiNWs. The formation section is followed by a discussion of the optical, electronic, and thermal properties of ion-implanted SiNWs. Finally, a range of current applications is discussed with particular focus on microelectronic and biomedical arenas.

22.2 FORMATION OF SILICON NANOWIRES

Silicon nanowires must first be made before ion implantation is possible. The production of SiNWs is accomplished through one of two methods: (1) nanowires can be grown through a bottom-up approach based on molecular precursors in either a gas or aqueous phase, or (2) nanowires can be chemically etched in a top-down fashion from a silicon substrate. Historically, the former bottom-up process was developed first, and several methods based on using molecular precursors will be highlighted at the beginning of this section, to be followed by a discussion of top-down methods.

22.2.1 BOTTOM-UP NANOWIRE GROWTH VIA MOLECULAR PRECURSORS

A wide variety of bottom-up methods have been developed to synthesize SiNWs with a range of diameters, lengths, and high aspect ratios (L/d 1000). The section below provides an overview of the most common of these bottom-up methods, including the vapor–liquid–solid method, solution-based growth, and also supercritical fluid–liquid–solid synthesis.

22.2.1.1 Vapor–liquid–solid synthesis

The vapor–liquid–solid (VLS) approach to nanowire synthesis was first reported in 1964 by Wager and Ellis while working at the Bell Telephone Laboratories (Figure 22.1) [8]. A metal particle is used as a catalyst to absorb an alloy with a molecular semiconductor precursor (i.e., silicon tetrachloride, $SiCl_4$) to form a eutectic. Once the metal catalyst become saturated with the silicon, a semiconductor nanowire precipitates from the droplet. A steady state is reached where new silicon atoms adsorb onto the metal droplet, while a solid silicon wire continues to grow. The diameter of the resulting nanowire is controlled through the size of the corresponding metallic catalyst particle. Many different methods have been used to synthesize semiconductor nanowires using the VLS approach, including laser ablation followed by growth in the gas phase [9], epitaxial growth from single-crystal substrates [10,11], and also growth from solid metallic particles in high vacuum that can be observed with *in situ* transmission electron microscopy (TEM) microscopy [12].

One constraint of the VLS process discussed above is that it is difficult to scale these reactions to commercially relevant volumes. Although VLS nanowires are extremely useful for the purposes of research and development, it is necessary to use other processing strategies if larger volumes of nanowire materials are to be made available for a given application. Additional processing pathways are discussed below which offer advantages in terms of the scalability of nanowire production.

22.2.1.2 Solution phase growth

Solution-based synthetic routes to SiNW production offer the advantage of generating large quantities of crystalline nanowires with extremely small diameters. In a typical solution-based process, a silicon

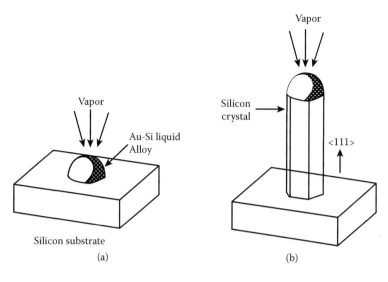

Figure 22.1 Schematic of vapor–liquid–solid nanowire growth process. (a) Initial condition with liquid metal droplet on substrate; the droplet could also be aerosolized in a gas phase environment. (b) Growing silicon nanowire crystal with liquid droplet at the tip. (Reprinted with permission from Wagner, R. S., and Ellis, W. C., *Appl. Phys. Lett.* 4, 89–90, 1964.)

precursor molecule is mixed into a solvent (either liquid or supercritical fluid) along with a metallic catalyst particle. In comparison with the VLS process discussed above, solution-based processes typically are conducted with metallic catalyst particles freely suspended in the fluid phase, rather than fixed to a substrate surface. In this way, a three-dimensional volume becomes the space for nanowire growth, rather than being confined to a two-dimensional surface.

Analogous to the VLS process, nanowires are induced to grow when the molecular precursor reacts at the catalysis surface, yielding a silicon nanowire. For example, Holmes et al. showed that a supercritical (500°C and 270 bar) solution of diphenylsilane and alkanethiol-capped gold nanocrystals in hexane will produce high aspect ratio (>1000) SiNWs with diameters in the 40–50Å range with less than a ±10% standard deviation [13]. Additionally, they demonstrated that pressure can be altered to influence the growth direction of the SiNWs.

Much of the literature regarding solution growth of SiNWs refers to high temperatures and pressures to achieve the supercritical fluid necessary for silane degradation. Recently, Heitsch et al. demonstrated that the use of a silane with lower stability and a high boiling point solvent could eliminate the need for supercritical conditions [14]. Specifically, they showed that the use of trisilane (Si_3H_8) in either octacosane ($C_{28}H_{58}$), or squalane ($C_{30}H_{52}$) was sufficient to generate SiNWs with an average diameter of 25.8 ± 5.3 nm and length of 2.0 ± 0.9 μm.

One constraint of VLS and solution-based growth is that the nanowires prepared with these methods are single crystalline. As such, it is difficult to prepare nanowire samples with porosity using VLS or solution-based growth. Porosity can be desirable for SiNWs in that an increase in internal surface area can lead to internal pores for the adsorption of molecular pharmaceuticals, discussed in more detail in Section 22.4.1.1. One alternative to bottom-up VLS and solution-based growth that can produce large quantities of either single crystalline or porous SiNWs is discussed in the following subsection.

22.2.2 TOP-DOWN METAL-ASSISTED CHEMICAL ETCHING

Another popular and reliable technique to produce SiNWs from the crystalline wafer is metal-assisted chemical etching (MACE) [15]. In contrast to the additive VLS and solution-based synthesis methods already discussed, MACE is based on top-down subtractive etching of single-crystalline silicon wafers [15]. The selection and preparation of a single-crystal silicon wafer (crystallographic orientation, thickness, carrier type, carrier concentration, etc.) is a critical first step in producing SiNWs through subtractive, top-down etching. High purity, single-crystal boules of silicon with n- or p-type conductivity usually are produced using the well-known Czochralski method. The boule is then sliced along the desired crystallographic axis and polished to a low surface roughness (<10 Å) to produce single-crystal wafers which can be used for a variety of applications, including microprocessor fabrication, micro- and nanoscale silicon device etching, and as a substrate for growth and analysis of other materials.

MACE nanowire synthesis requires that the wafer then be partially covered by a noble metal (e.g., silver) through lithographic patterning, and immersed in a solution containing HF and an oxidizing reagent. The metal/semiconductor interface forms a Schottky junction where the semiconductor is etched downward in a continuous fashion to produce high aspect ratio nanowires, as seen in Figure 22.2a, over a wide area (inset) [16].

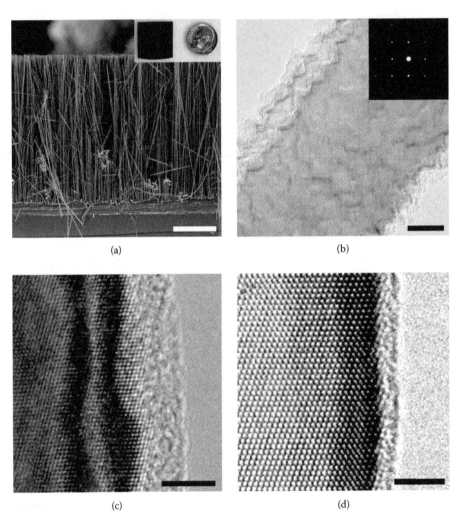

(a) (b)

(c) (d)

Figure 22.2 (a) Cross-sectional SEM of an MACE Si nanowire array. Dendritic Ag growth can be seen within the array as a product of Ag$^+$ reduction onto the wafer during reaction. The Ag is etched in nitric acid after the synthesis, and elemental analysis confirms it is dissolved completely. Inset, a MACE Si nanowire array Si wafer chip of the typical size used for the syntheses. Similar results are obtained on entire four-inch wafers. The chip is dark and nonreflective owing to light scattering by, and absorbing into, the array. (b) Bright-field TEM image of a segment of an MACE Si nanowire. The roughness is clearly seen at the surface of the wire. The selected area electron diffraction pattern (inset) indicates that the wire is single crystalline all along its length. (c) High-resolution TEM image of an MACE Si nanowire. The roughness is evident at the interface between the crystalline Si core and the amorphous native oxide at the surface, and by undulations of the alternating light/dark thickness fringes near the edge. (d) High-resolution TEM of a VLS-grown Si nanowire. Scale bars for a through d are 10 μm, 20 nm, 4 nm, and 3 nm, respectively. (Reprinted with permission from Hochbaum, A. I., et al., *Nature*, 451, 163–167, 2008.)

It is also possible to use similar etching methodology to prepare silicon nanomaterial with a disc-like, oblate-spheroid morphology [17] with dimensions comparable to blood cells (~10μm). There is a strong demand for techniques that allow the fabrication of biocompatible porous nanoparticles (pSi) for drug delivery applications. This disc fabrication process relies on a combination of colloidal lithography and MACE. The height and diameter of the pSi nanodiscs can be easily adjusted through the choice of the colloidal template. Furthermore, the nanodiscs are degradable in a physiological milieu and are nontoxic to mammalian cells. In order to highlight the potential of the pSi nanodiscs in drug delivery, an *in vitro* investigation was reported that involved loading nanodiscs with the anticancer agent camptothecin and functionalization of the nanodisc periphery with an antibody that targets receptors on the surface of neuro-blastoma cells [18]. These disc-like nanocarriers were shown to selectively attach to, and kill, cancer cells.

22.2.2.1 Non-patterned MACE growth

Although lithography can be used to control morphologies much more accurately, the techniques employed for lithographic patterning can be time consuming and costly. For easier preparation of SiNWs where control of the diameter is not required, an aqueous solution of metal ions can be used to create a film on the wafer's surface. In this approach, the oxidizing solution is prepared with the addition of solvated metal cations. The metal ions deposit onto the silicon surface randomly and lead to the selective oxidation of the underlying Si wafer, generating a dense arrays of nanowires. TEM images shown in Figure 22.2b and 22.2c indicate that the nanowires grown through the MACE process can have significant surface texture and porosity, which is not simple to achieve through additive nanowire syntheses. However, the simplicity of this MACE approach involving random metal deposition also requires one to sacrifice control over the diameter of the resulting silicon nanowire array.

Another significant factor in MACE synthesis is that the metal used to grow the nanowires is often found attached to the surface of the nanowires, as shown in Figure 22.3a. These small metallic nano-crystals must be removed for many biomedical applications given the toxicity of metal ions *in vivo*. It is possible to remove these surface-bound metallic nanocrystals through a second liquid-phase etching step, as shown with TEM (Figure 22.3b), neutron activation analysis (Figure 22.3c), and atom probe tomography (Figure 22.3d) [19].

Despite the vast number of synthetic pathways that are available for the production of SiNWs, there are still great challenges to producing nanowires with well-defined, arbitrary doping profiles. Both additive and subtractive silicon nanowire production methods can leave behind metal atoms in the resulting nanowires. Also, it is challenging to achieve complex p- or n-type doping profiles, or to produce structures with a combination of both crystalline and amorphous microstructure. In the following section, ion implantation will be presented as a means to preparing large amounts of silicon nanowire materials with complex doping profiles and also different degrees of crystallinity with the aim of opening up new properties and applica-tions of silicon nanowire materials.

22.2.3 ION IMPLANTATION

Ion implantation has been used for many decades to create specific p- or n-type carrier concentrations in semiconductor materials [20]. An ion accelerator is used to adjust the kinetic energy of positive or negative ions to achieve a precise implantation depth within the semiconductor host crystal Figure 22.4. For the case of silicon, boron is used to create p-type doping, and either phosphorus (P) or arsenic (As) is used for n-type doping. The depth of implantation depends of several factors including the ion's kinetic energy and also the crystallographic orientation of the semiconductor wafer. Increasing the kinetic energy of the ion leads to deeper implantation within the 2D semiconductor substrate. The final concentration of doped ions is controlled through regulating the flux of ions introduced into the carrier lattice with the dosage quanti-fied as a fluence with units of ions/cm^2.

The primary deceleration mechanism for an ion propagating through a crystal is based on collisions with other atoms in the crystal lattice [22]. Each collision reduces the kinetic energy of the implanted ion, leading to the breaking of ionic or covalent bonds within the crystal, and also the formation of point defects (vacancies, interstitals, antisites, etc.). The deceleration trajectory is random with many potential pathways being possible as the ion gradually comes to a stop within the lattice. For a fixed acceleration

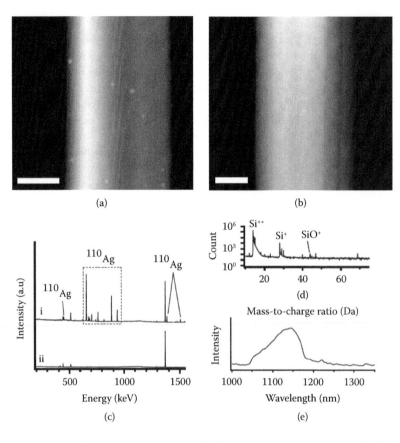

(a) (b)

(c) (d) (e)

Figure 22.3 Compositional analysis and microstructure of silicon nanowires prepared via MACE. (a) High-angle annular dark-field image of a single silicon nanowire without silver etching, demonstrating the presence of silver deposits. Scale bar = 50 nm. (b) High-angle annular dark-field image of a silver-etched silicon nanowire with no detectable silver. Scale bar = 50 nm. (c) Neutron activation analysis of silicon nanowire array before (i) and after (ii) the silver etching process. (d) Atom probe tomography (APT) mass spectrum from a single SiNW, demonstrating no detectable silver signal (Ag$^+$: 107 Da, Ag^{2+}: 53.5 Da). (e) Photoluminescence of SiNWs excited by a 975 nm laser source. (Reprinted with permission from Smith, B. E., et al. *ACS Photonics* 2, 559–564, 2015).

Figure 22.4 Schematic of gold ion implantation. Gold ions are accelerated to an energy of 500 keV which leads to an average implantation depth of approximately 50 nm. (Reprinted with permission from Roder, P. B., et al. *Adv. Opt. Mater.*, 3, 1362–1367, 2015.)

energy, crystallographic orientations with high atomic packing densities (e.g., the cubic <111> direction) lead to shorter stopping distances relative to orientations with lower packing densities (e.g., the cubic <001> direction) [23]. Predicting the depth profile of implanted ions is possible through Monte Carlo simulations. Accurate predictions for stopping ranges are possible through the use of freely available software, including the stopping range of ions in matter, or SRIM package [24].

Ion implantation not only introduces heteroatoms within the semiconductor crystal lattice, but also a large number of Frenkel (interstitial/vacancy-pairs) and Schottky (neutral vacancies) defects that must be thermally annealed in order to activate carriers from the implanted ions. Inert atoms such as gold or silicon may also be used for implantation with the primary intention being to introduce a specific amount of damage to the crystal lattice. Crystalline silicon has an indirect band gap, and therefore does not have a high absorption coefficient. Amorphous silicon, however, has a significantly larger absorption coefficient than crystalline silicon, leading to its widespread application in solar cells.

Recently, ion implantation of SiNWs with silicon or gold atoms (see Figure 22.5) has been shown to introduce a large amount of crystallographic point defects within silicon nanomaterials [21], and can be used to increase the optical absorption coefficient of SiNWs. This has potential advantages for future

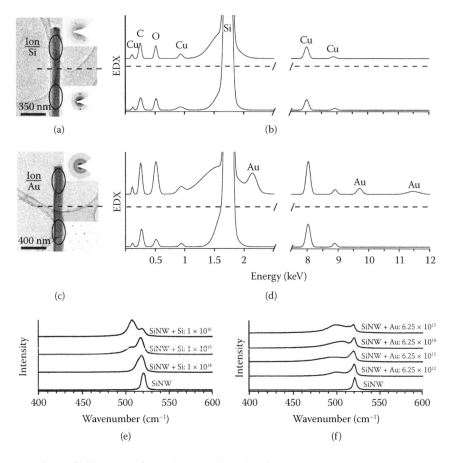

Figure 22.5 TEM bright-field images of a) a silicon-implanted and c) gold-implanted SiNW. Comparing the select area electron diffraction (SAED) of the tip of the silicon-implanted SiNW ((a), top-right inset) with the base of the same SiNW ((a), bottom-right inset) shows partial amorphization throughout the SiNW as predicted by SRIM calculations. Energy-dispersive X-ray (EDX) spectra of the same regions (b) also suggest no change in elemental composition. Comparing the SAED of the tip of the gold-implanted SiNW ((c), top-right inset) with the base of the same SiNW ((c), bottom-right inset) suggests that amorphization from gold implantation is contained at one end of the SiNW. EDX spectra of the same regions (d) further suggest that the gold is constrained to the SiNW tip. Increased amorphization and lattice damage from increasing (e) silicon and (f) gold implantation dosage is also evidenced by the presence of an increasing shoulder off the silicon 520 cm⁻¹ Raman peak. (Reprinted with permission from Roder, P. B., et al. *Adv. Opt. Mater.*, 3, 1362–1367, 2015.)

applications in photothermal therapy, where chemically inert material with high optical absorption coefficients can be used to photoablate tumor cancerous tissues. The physical properties, including optical absorption, of silicon are discussed in more detail below.

22.3 PROPERTIES

22.3.1 OPTICAL ABSORPTION COEFFICIENT

The optical properties of silicon are determined by its indirect band gap which results in a low optical absorption coefficient in both visible and near-infrared (NIR) spectral regions. However, the introduction of dopant atoms through ion implantation can increase silicon's optical absorption either through the activation of free carriers, or through the formation of amorphous silicon which has a higher optical absorption coefficient [25].

Several other approaches have been used to increase the optical absorption coefficient of silicon. One such method has involved the use of nanoscale clusters of gold as a means to further increase the photothermal absorption coefficient of SiNWs for enhanced photothermal therapy [26]. During the rapid VLS growth of SiNWs in a low-oxygen environment, it is possible to harvest nanowires that contain a high density of gold nanoclusters (Au NCs) with a uniform coverage over the entire length of the nanowire sidewalls. The presence of gold nanocrystals leads to significant plasmon-induced optical absorption, which is larger than would occur for SiNWs alone.

Another surprising property of SiNWs during photothermal heating is related to their propensity for combustion. In particular, SiNWs synthesized by the thermal evaporation of a silicon monoxide powder or a mixture of silicon and silica powders were observed to burn fiercely in air and exhibited a large photoacoustic effect when exposed to a conventional photographic flash; see Figure 22.6. The energy required to ignite the Si nanowires was 0.1–0.2 J/cm². The remaining material obtained after burning Si nanowires consisted of various forms of nanostructures (e.g., nanoparticles, amorphous wires, and nanotubes) [27]. The as-prepared Si nanowire samples were stable in air. No further oxidization was observed on the Si nanowires stored in air

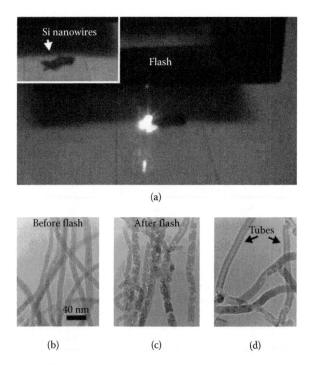

Figure 22.6 (a–d) Optical ignition of silicon nanowire combustion using a camera flash with a power of 0.1–0.2 J/cm². (Reprinted with permission from Wang, N., et al., *Nano Lett.* 3, 475–477, 2003.)

for years. A gas flame could not ignite the Si nanowires. The authors observed that the SiNWs were slowly oxidized in the flame, forming silica nanowires. However, when exposed to a camera flash at short range (~3 cm), Si nanowires ignited and burned in air. The pulse duration was about 5 ms. When the optical flash ignition experiment was performed in inert gases, no ignition was observed.

The SiNWs showed a strong ability to confine energy from visible light which is one hypothesis for what enabled their ability to ignite under a brief flash of light. The following section discusses this light-confining effect in great depth in the context of the well-developed theory of morphology-dependent cavity (Mie) resonances.

22.3.1.1 Morphology-dependent resonances

The large range of accessible diameters for SiNWs leads to the potential to form internal electromagnetic standing waves within individual nanowire cavities. Analogous to acoustic "whispering-gallery" modes, these optical standing waves are often referred to as morphology-dependent resonances (MDRs) that can lead to large increases in the optical absorption coefficient of SiNWs [28]. Analytical (Mie) solutions to Maxwell's equations have been derived for the case of infinite cylinders [29], and can be used to predict what diameters of SiNWs will lead to internal MDRs/standing waves within the nanowire cavity.

Heat transfer in a material can be modeled with the following time-dependent partial differential equation (PDE) that is based on energy conservation within a differential volume element:

$$C_p \rho \frac{\partial T}{\partial t} = \kappa \nabla^2 T + S, \qquad (22.1)$$

In this equation, the term on the left-hand side represents the time-dependent rate of thermal energy increase or decrease within a material based on its physical properties including heat capacity (C_p) and density (ρ). The first term on the right-hand side of the energy PDE accounts for heat diffusion within a material and is based on the material's thermal conductivity (κ), and also the local thermal environment surrounding a given point. The Laplacian term ($\nabla^2 T$) can be thought of as representing the difference between the temperature at a differential point in space relative to the average temperature surrounding it. Therefore, the local temperature at a point will rise or fall depending on whether the average surrounding temperature is hotter or cooler than at the particular point. The thermal conductivity of a material is a proportionality constant that has a large impact on the corresponding rate of change of the local temperature, given a fixed temperature gradient.

The final term on the right-hand side of the energy PDE (Equation 22.1) represents a local source (or sink) of thermal energy within the material. Dimensional analysis can be used to show that S has units of $\left[\frac{W}{m^3} \right]$. In the context of ion-implanted SiNWs, this term is a source of thermal energy that depends the local absorption of electromagnetic energy, and generation of heat within the material. The source term is highly dependent on the carrier type, concentration, and also the degree of crystallinity within the silicon nanowire material.

The final coordinate system used in the energy equation depends heavily on the morphology of a given structure, where spherical coordinates are used for objects with spherical symmetry, Cartesian coordinates are used for objects with rectangular symmetry, etc. In the case of SiNWs, the most appropriate choice for the Laplacian operator is the cylindrical coordinate system. Analytical solutions have been reported for the heat transport equation using an infinite cylindrical coordinate system [28] shown in Figure 22.7, and also for the finite cylinder [30].

When solving the energy equation it is convenient to rewrite it in a nondimensional form. For example, in the case of the finite cylinder [30], the energy equation can be rewritten in the following nondimensional way:

$$\frac{\partial \Theta}{\partial \tau} = \frac{1}{\xi} \frac{\partial}{\partial \xi} \left(\xi \frac{\partial \Theta}{\partial \xi} \right) + \frac{1}{\xi^2} \frac{\partial^2 \Theta}{\partial \phi^2} + \left(\frac{R}{L} \right)^2 \frac{\partial^2 \Theta}{\partial \zeta^2} + S^*(\xi, \zeta, \tau) \qquad (22.2)$$

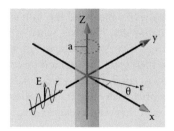

Figure 22.7 Coordinates used for silicon nanowire photothermal heating simulations. (Reprinted with permission from Roder, P. B., et al., *Langmuir*, 28, 16177–16185, 2012.)

in which the dimensionless variables are defined by

$$\Theta = \frac{T_{NW} - T_\infty}{T_\infty}, \xi = \frac{r}{R}, \zeta = \frac{z}{L}, \tau = \frac{\kappa_{NW}}{\rho_{NW} C_{p,NW}} \frac{t}{R} \tag{22.3}$$

Here T_{NW} is the NW's temperature, T_∞ is the fluid temperature far from the NW, κ_{NW} is the thermal conductivity of the NW, ρ_{NW} is the NW's density, L is the NW's length, R is the NW's radius, and $C_{p,NW}$ is the NW's specific heat capacity at constant pressure per unit mass.

The dimensionless source function is

$$S^*(\xi, \phi, \zeta, \tau) = \frac{R^2}{\kappa_{NW} T_\infty} S(\xi, \phi, \zeta, \tau) \tag{22.4}$$

which represents the intrinsic electromagnetic heating of the NW. The time-averaged heating point-source function is given by [31]:

$$S = \frac{1}{2} \sigma \mathbf{E} \cdot \mathbf{E}^* \tag{22.5}$$

where E is the complex internal electric field and the variable σ is the conductivity of the NW at optical frequencies given by

$$\sigma = \frac{4\pi \mathrm{Re}\{N_{NW}\} \mathrm{Im}\{N_{NW}\}}{\lambda_{inc} \mu c} \tag{22.6}$$

in which N_{NW} is the complex index of refraction of the NW, λ_{inc} is the wavelength of the incident wave, μ is the magnetic permeability of the NW, and c is the velocity of light in vacuum. The electric field amplitude of the incident planewave, E_{inc}, is related to the laser irradiance by:

$$E_i^2 = \frac{2}{N_B c \epsilon_0} I_{inc} \tag{22.7}$$

where N_B is the complex index of refraction of the bath (water), ϵ_0 is the permittivity of free space, and I_{inc} is the laser irradiance in the fluid. Clearly, this source function $S(\xi, \phi, \zeta, \tau)$ for electromagnetic heating depends critically on the internal electromagnetic field.

In general, the internal electromagnetic heating source is a function of radial coordinate, angle ϕ, axial coordinate ζ, and time τ as well as the optical properties of the material and the characteristics of the laser light source. For the high frequencies associated with continuous wave (CW) laser irradiation we can use a time-averaged source. Consequently, the dimensionless heat source reduces to the constant three-dimensional source $S^*(\xi, \phi, \zeta)$. If the internal electromagnetic field does not vary across the NW's diameter, the heat source reduces further to a function of axial position only, that is, to $S^*(\zeta)$.

Homogeneous internal electromagnetic fields have recently been shown to be a good approximation for SiNWs with diameters in the order of 10s of nanometers [28]. If necessary, it is possible to extend the analysis presented above to heating by pulsed lasers by using a time-dependent source function.

The solution of the inhomogeneous energy equation can be written in terms of these orthonormal functions as

$$\Theta(\xi,\phi,\zeta,\tau) = \sum_{\ell=1}^{\infty}\sum_{m=1}^{\infty}\sum_{n=0}^{\infty} A_{\ell mn}(\tau)u_{\ell n}(\xi)v_n(\phi)w_m(\zeta) \tag{22.8}$$

where the eigenfunctions are

$$u_{\ell n} = \frac{X_{\ell n}(\xi)}{\|X_{\ell n}\|} = \frac{J_n(\gamma_{\ell n}\xi)}{\|X_{\ell n}\|},$$

$$v_n = \frac{Y_n(\phi)}{\|Y_n\|} = \frac{\cos(n\phi)}{\|Y_n\|}, \tag{22.9}$$

$$w_m = \frac{Z_m(\zeta)}{\|Z_m\|} = \frac{\cos(\delta_m\zeta) + (Bi_1/\delta_m)\sin(\delta_m\zeta)}{\|Z_m\|},$$

Here $J_n(\gamma_{\ell n}\xi)$ is an nth order Bessel function, $\|X_{\ell n}\|$, $\|Y_n\|$, and $\|Z_m\|$ are the norms of the eigenfunctions, and $\gamma_{\ell n}$, n, and δ_m are their respective eigenvalues. Substituting Equation 22.8 into Equation 22.2 and applying the principle of orthogonality for each of the eigenfunctions, the time-dependent coefficients $A_{\ell mn}(\tau)$ are found to be

$$A_{\ell mn}(\tau) = \int_0^1\int_0^\pi\int_0^1\int_0^\tau S^*(\xi',\phi',\zeta',\tau')\exp[-\lambda_{\ell mn}^2(\tau-\tau')]\xi'u_{\ell n}(\xi')v_n(\phi')w_m(\zeta')d\xi'd\phi'd\zeta'd\tau' \tag{22.10}$$

where the primes indicate dummy variables of integration, and

$$\lambda_{\ell mn}^2 = \gamma_{\ell n}^2 + \left(\frac{R}{L}\right)^2\delta_m^2 \tag{22.11}$$

If the source function is not a function of time, that is, if the time-averaged electromagnetic source function is used, Equation 22.10 can be integrated over time to yield

$$A_{\ell mn}(\tau) = \frac{1-\exp[-\lambda_{\ell mn}^2\tau]}{\lambda_{\ell mn}^2}\int_0^1\int_0^\pi\int_0^1 S^*(\xi',\phi',\zeta')\xi'u_{\ell n}(\xi')v_n(\phi')w_m(\zeta')d\xi'd\phi'd\zeta' \tag{22.12}$$

At steady state (as $\tau\to\infty$) this result reduces to constants given by

$$A_{\ell mn} = \frac{1}{\lambda_{\ell mn}^2}\int_0^1\int_0^\pi\int_0^1 S^*(\xi',\phi'\zeta')\xi'u_{\ell n}(\xi')v_n(\phi')w_m(\zeta')d\xi'd\phi'd\zeta' \tag{22.13}$$

and the steady state temperature distribution reduces to

$$\Theta(\xi,\phi,\zeta) = \sum_{\ell=1}^{\infty}\sum_{m=1}^{\infty}\sum_{n=0}^{\infty}\frac{u_{\ell n}(\xi)v_n(\phi)w_m(\zeta)}{\lambda_{\ell mn}^2}\int_0^1\int_0^\pi\int_0^1 S^*(\xi',\phi',\zeta')\xi'u_{\ell n}(\xi')v_n(\phi')w_m(\zeta')d\xi'd\phi'd\zeta' \tag{22.14}$$

One of the most useful features of the theory presented above is the ability to predict MDRs in silicon nanowire materials. MDRs can be thought of as internal electromagnetic standing waves based on multiple total internal reflections of photons within a silicon nanowire cavity. As photons reflect multiple times within a cavity, there is an increase in the probability that they are absorbed to generate heat within the silicon nanowire. Figure 22.8 shows how an incident plane wave can lead to an internal whispering-gallery mode resonance within a silicon nanowire. It is this internal electromagnetic standing wave profile that is the source of photothermal heating in SiNWs materials.

Figure 22.9a shows how photothermal heating is predicted to change for a fixed laser excitation wavelength as the diameter of a given cylinder increases. At specific diameters, large internal cavity resonances are established that lead to large increases in optical absorption and subsequent heating of silicon nanowire materials. These particular resonant diameters also exhibit large light scattering cross sections that are calculated using Mie theory; these are shown in Figure 22.9b.

Beyond inducing local heating, the absorption of light within SiNWs can also lead to other significant chemical effects when energy is transferred from the silicon crystal to surrounding oxygen molecules. This surprising change in the chemical reactivity of oxygen is discussed in more detail in the next section.

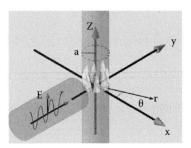

Figure 22.8 Schematic of an internal electromagnetic cavity resonance within a silicon nanowire that can be used to increase the amount of photothermal heating.

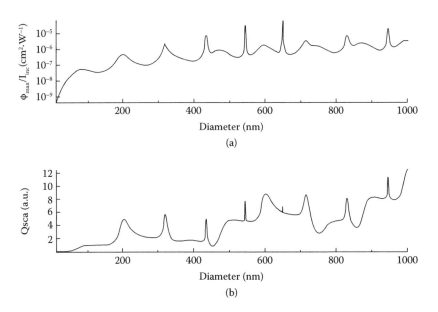

Figure 22.9 Comparison of calculated values for the maximum reduced temperatures of laser-heated silicon nanowires in water scaled by the incident irradiance with the calculated diameter-dependent light scattering efficiency. (a) Diameter-dependent maximum reduced temperature scaled by the incident irradiance of a single-crystalline silicon nanowire irradiated in water at a free-space wavelength of $\lambda = 980$ nm. (b) Calculated scattering cross section for a single-crystalline silicon nanowire in water irradiated at a free-space wavelength of $\lambda = 980$ nm as a function of nanowire diameter. (Reprinted with permission from Roder, P. B., et al., *Langmuir*, 28, 16177–16185, 2012.)

22.3.1.2 Singlet oxygen generation

Although it is well known that the indirect band structure of silicon leads to low optical absorption coefficients [25], there are additional optoelectronic features related to the band structure of silicon that are less well known. First, excitons within silicon single crystals have been shown to have long (millisecond) excited-state lifetimes. Second, three-quarters of the excitons in silicon have a triplet configuration of electron and hole spins [32]. The combination of large excited-state lifetimes with triplet spin states leads to the possibility of significant energy exchange (Figure 22.10) between triplet oxygen molecules, to form singlet oxygen excited states (Figure 22.11). The band gap of silicon (1.1eV) is close to that of the triplet-singlet excitation energy of diatomic oxygen (0.98ev). Singlet oxygen generation has been reported to occur through a triplet-triplet annihilation (TTA) process [32].

One of the most useful features of the TTA process is that highly reactive oxygen molecules can react with cancerous tissue as a novel way to treat malignant tumors. This processes is known as photodynamic therapy, or PDT. Historically, PDT has been pursued with molecular photosensitizers that absorb light, and then transfer energy to triplet diatomic oxygen to produce singlet oxygen molecules [33].

Although PDT with molecular photosensitizers is a very useful approach for cancer treatment, it has a few short-term and long-term side effects arising from reactive oxygen species (ROS) generation, and the long half-live of molecular photosensitizers *in vivo*. Recently, a new PDT based not on the ROS generation capability of photosensitizers but on porous silicon (pSi) can also be utilized as a therapeutic agent that generates sufficient heat to kill cancer cells without toxicity [34]. As discussed in more detail later in this chapter, porous silicon is biodegradable and also has a shorter half-life than molecular photosensitizers [5].

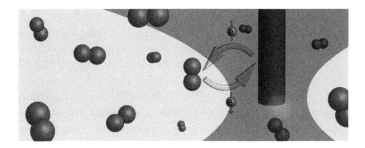

Figure 22.10 Schematic of singlet oxygen generation. (Reprinted with permission from Smith, B. E., et al., *ACS Photonics*, 2, 559–564, 2015.)

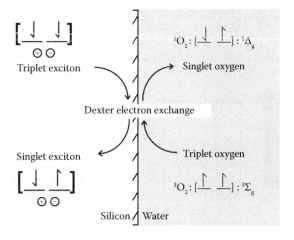

Figure 22.11 Silicon–water interface diagram showing the transfer of electrons from surface excitons in silicon to dissolved oxygen molecules. (Reprinted with permission from Smith, B. E., et al., *ACS Photonics*, 2, 559–564, 2015.)

22.3.2 ELECTRICAL PROPERTIES

Ion implantation of nanowires has also been used to modify their electrical and optical properties. For instance, the implantation of hydrogen in ZnO nanowires has been shown to generate shallow donors, while the implantation of transition metal and rare earth elements has been shown to lead to optical emission that is difficult to generate with other synthetic methods [35].

In the case of silicon, early work was focused on the implantation of bistable shallow donors in monocrystalline quantum wires [36]. Electrical and optical properties of bistable shallow donors in monocrystalline silicon, which are introduced by proton implantation followed by annealing at ~450°C, have been studied. The temperature dependencies of equilibrium and nonequilibrium carrier concentration and relaxation kinetics were investigated and infrared absorption lines of bistable shallow donor electronic excitations were detected. The obtained experimental data demonstrate that the bistable shallow donors can be identified as quantum wire defect nanoclusters.

More recently, ion implantation has been used to form axial doping profiles (Figure 22.12) in that they have enabled the formation of p-n junctions [37]. Ion implantation has also been used to create radial doping profiles in SiNWs with a crystalline silicon core that is surrounded by an oxide shell [6].

Recrystallization of SiNWs after ion implantation strongly depends on the ion doses and species. Full amorphization by high-dose implantation induces polycrystal structures in SiNWs even after high-temperature annealing, with this tendency more pronounced for heavy ions. Hot-implantation techniques dramatically suppress polycrystallization in SiNWs, resulting in reversion to the original single-crystal structures and consequently high reactivation rate of dopant atoms. The chemical bonding states and electrical activities of implanted boron and phosphorus atoms have recently been evaluated by Raman scattering and electron spin resonance, demonstrating the formation of p- and n-type SiNWs [38].

Ion implantation can be a very useful technique to dope SiNWs heavily to improve their electrical properties. However, heavy implantation can amorphize the nanowires completely. Subsequently, a complete recovery of their crystallinity, which is of utmost importance to ensure their improved electrical properties, becomes nontrivial. Das and coworkers [39] performed a controlled study of nanowire recrystallization using vertical Si(111) nanowires that were amorphized during doping by arsenic ion implantation. Upon a single-step thermal anneal by furnace (500–650°C) or by rapid thermal annealing (800–1200°C), the nanowires turned partly single-crystalline from the bottom and partly polycrystalline from the top, owing to a competition between solid phase epitaxial regrowth from the substrate, and random nucleation and growth. A complete recrystallization of the amorphized nanowires was achieved only after the furnace-annealed nanowires were annealed for a second time at a higher temperature (950–1200°C). The polycrystalline grains formed during the first anneal were successfully aligned to the <111> direction, leading to a recovery of the single-crystalline structure of the nanowires [39].

Vertical epitaxial short (200–300 nm long) SiNWs grown by molecular beam epitaxy on Si(111) substrates were separately doped *p*- or *n*-type *ex situ* by implanting with B or P and As ions respectively at room temperature. Multienergy implantations were used for each case, with fluences of the order of 10^{13}–10^{14} cm^{-2}, and the NWs were subsequently annealed by rapid thermal annealing (RTA).

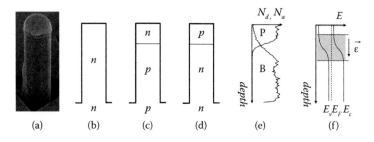

Figure 22.12 (a) SEM image of a silicon nanowire with the gold catalyst still on top. (b) through (d) Doping scheme for samples 1–3. (e) Simulated doping profile of sample 2. (f) Corresponding band structure; the depletion region is gray shaded and the direction of the electric field indicated. (Reprinted with permission from Hoffmann, S., et al., *Nano Letters* 9, 1341–1344, 2009.)

TEM showed no residual defect in the volume of the NWs. Electrical measurements of single NWs with a Pt/Ir tip inside a scanning electron microscope (SEM) showed significant increase of electrical conductivity of the implanted NWs compared to that of a nominally undoped NW. The p-type (B-implanted) NWs showed the conductivity expected from the intended doping level. However, the n-type NWs (P- and As-implanted), showed one to two orders of magnitude lower conductivity. The authors suggested that a stronger surface depletion is mainly responsible for this behavior of the n-type NWs [40].

22.4 APPLICATIONS

The wide range of optical and electrical property modification enabled by ion implantation has led to a large range of promising applications for these materials. This final section discusses several kinds of applications including those in (1) biomedicine/photothermal heating, and (2) optoelectronics.

22.4.1 PHOTOTHERMAL HEATING

One of the main advantages of silicon in biomedical applications is its biocompatibility [41] and biodegradability [42]. Studies continuously show that silicon is not inherently cytotoxic [41] and is naturally decomposed *in vivo* into silicic acid [42]. Popplewell and colleagues [5] reported using ^{32}Si as a tracer in a human uptake experiment to determine a gastrointestinal uptake factor for silicic acid, and to elucidate the kinetics of renal elimination. Urine collections were made for extending intervals from 2 to 12 h over two days following ingestion by a single human subject of a neutral silicic acid solution containing tracer levels of ^{32}Si ($t_{1/2}$ ~150 y). Silicon was isolated as SiO_2 and the ^{32}Si content determined by accelerator mass spectrometry, using a gas-filled magnet technique to eliminate a prolific isobaric interference from ^{32}S. Silicon uptake appears to have been essentially complete within 2 h of ingestion. Elimination occurred by two simultaneous first-order processes with half-lives of 2.7 and 11.3 h, representing around 90% and 10%, respectively, of the total output. The rapidly eliminated ^{32}Si was probably retained in the extracellular fluid volume, while the slower component may represent intracellular uptake and release. Elimination of absorbed ^{32}Si was essentially complete after 48 h and was equivalent to 36% of the ingested dose. This establishes only a lower limit for gastrointestinal absorption as, although there was no evidence for longer term retention of additional ^{32}Si, the possibility could not be excluded by their results [5].

Additionally, membranes based on porous silicon have been demonstrated to be biocompatible and applied to support human ocular cells *in vitro* and *in vivo* within the rat eye [43]. A colorimetric assay for silicic acid showed that membranes with pore sizes of 40–60 nm slowly dissolved, but the material could be maintained in a tissue culture medium *in vitro* for at least two weeks without visible degradation. When implanted under the rat conjunctiva, the material did not erode the underlying or overlying tissue. The implant underwent slow dissolution, but remained visible at the operating microscope for over eight weeks. End-stage histology indicated the presence of a thin fibrous capsule surrounding the implant, but little evidence of any local accumulation of acute inflammatory cells or vascularization. Human lens epithelial cells and primary human corneal explants adhered to the porous silicon membranes, where they remained viable and underwent division. Primary corneal epithelial cells supported on membranes were labeled with a cell tracker dye and implanted under the rat conjunctiva. Seven days later, labeled cells had moved from the membrane into the ocular tissue spaces. A porous silicon membrane may have value as a biomaterial that can support the delivery of cells to the ocular surface and improve existing therapeutic options in patients with corneal epithelial stem cell dysfunction, and ocular surface disease [43].

Photothermal therapy based on porous silicon (pSi) has also been investigated in combination with NIR laser [44]. *In vivo* animal test results showed that the murine colon carcinoma tumors were completely resorbed without giving damage to surrounding healthy tissue within five days of pSi and NIR laser treatment. Tumors were not observed to have recurred at all in the pSi/NIR treatment groups thereafter. Both the *in vitro* cell test and *in vivo* animal test results suggest that thermotherapy based on pSi in combination with NIR laser irradiation is an efficient technique to selectively destroy cancer cells without damaging the surrounding healthy cells [44].

As mentioned in the formation section above, the implantation process introduces damage to the lattice and can generate active dopants for increased absorption. Recent experiments with gold-implanted

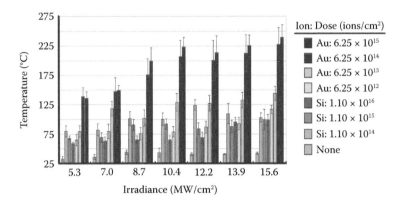

Figure 22.13 Photothermal superheating of water using optically trapped, ion-implanted SiNWs. Increasing the ion implantation fluence is observed to increase the observed temperature through analysis of the nanowires' Brownian dynamics. (Reprinted with permission from Roder, P. B., et al., *Adv. Opt. Mater.*, 3, 1362–1367, 2015.)

SiNWs have shown that it is possible to superheat water to as much as 225°C using a highly focused NIR laser trap [21]. Figure 22.13 shows experimental temperature measurements of ion-implanted SiNWs with a range of ion implantation fluences. Increasing the fluence of gold implantation is observed to increase the observed temperatures of optically trapped SiNWs.

The Au-NC coated SiNWs with an antibody-coated surface obtain the unique capability to capture breast cancer cells at twice the highest efficiency currently achievable (~88% at 40 min cell incubation time) from a nanostructured substrate. It was also shown that the irradiation of breast cancer cells captured on Au-NC coated SiNWs with a NIR light resulted in a high mortality rate of these cancer cells, raising the prospect for simultaneous capture and plasmonic photothermal therapy for circulating tumor cells [26].

Silicon nanocrystals (SiNCs) have been the subject of intense research interest since the discovery of room-temperature photoluminescence (PL) from porous silicon [45]. Nanostructured Si materials exhibit size-dependent PL arising from the influences of quantum confinement can have a small hydrodynamic radius and are biocompatible and biodegradable. In this context, SiNCs have received significant attention as *in vivo* imaging agents and for use in a variety of optoelectronic applications. Si nanocrystals (SiNCs) exhibiting relatively high near-IR photoluminescent quantum yields also exhibit a notable photothermal (PT) response. The PT effect has been quantified as a function of NC size, defect concentration, and irradiating energy, suggesting that the origin of the PT response is a combination of carrier thermalization and defect-mediated heating. The PT effect observed under NIR irradiation suggests that SiNCs could find use in combined *in vivo* PL imaging and PT therapy [45].

22.4.2 OPTOELECTRONICS

22.4.2.1 Solar panel materials

Typical solar cell devices utilize a planar p-n junction which can result in low extraction efficiency due to the relatively low carrier diffusion lengths in inexpensive, candidate photovoltaic materials. Recent research has explored the benefits of one-dimensional materials in solar cells [46, 47]. The model used in these devices is based on a radial p-n junction (Figure 22.14) where the collection path is significantly reduced thereby providing superior efficiency.

22.4.2.2 Light-emitting diodes (LEDs)

Silicon's indirect band gap not only results in low absorption and, consequently, low efficiency for photovoltaics devices, it also means that applications of intrinsic silicon in LED technology are not optimal due to low radiative recombination. Although silicon has been shown to produce visible emission [48], direct band gap materials are typically selected for their more efficient light emission.

One recent demonstration of visible emission from silicon found a way of increasing emission by approximately three orders of magnitude. Fabbri et al. show that a chemical–vapor–deposition synthesis of

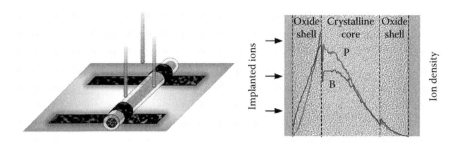

Figure 22.14 (Left) Schematic of NW ion implantion. (Right) High-resolution TEM micrograph of a SiNW superimposed with the simulated implantation profiles for a corresponding $SiO_2/Si/SiO_2$ layered structure. (Reprinted with permission from Colli, A., et al., *Nano Lett.*, 8, 2188–2193, 2008.)

hexagonal-structure SiNWs doped with boron produces efficient luminescence in the visible and NIR [49]. Ion-implanted SiNWs are promising materials for energy-efficient light-emitting diode devices as well.

22.5 FUTURE OUTLOOK

The wide variety of processing routes to silicon nanowire materials has enabled a great deal of control over the size, porosity, and also morphology of silicon nanowire materials, ranging from single crystals with ultrathin, high aspect ratios to flat, disc-like structures with significant amounts of porosity. However, regardless of the synthetic processing route, controlling the composition and extent of amorphization of SiNWs has remained a challenge.

Ion implantation continues to offer additional degrees of freedom in designing silicon nanowire materials with a variety of different electronic doping profiles, as well as control over local amorphization, that can lead to significant changes in the NW optoelectronic properties. There is still a great amount of space to explore in terms of fine tuning the properties of ion-implanted SiNWs for applications in energy harvesting, photothermal/photodynamic tumor therapies, photocatalysis, and also for controlled drug release. The material presented in this chapter is intended to present a broad overview of what has been reported to date for ion-implanted SiNWs with the hope of stimulating new ideas for future students and current scientific and engineering investigators.

REFERENCES

1. Lutgens, F. K. *Essentials of geology.* 11th ed. (Prentice Hall, Boston, MA, 2012).
2. Davis, A. M. Stardust in meteorites. *Proceedings of the National Academy of Sciences* **108**, 19142–19146 (2011).
3. Battaglia, C., Cuevas, A. & Wolf, S. D. High-efficiency crystalline silicon solar cells: Status and perspectives. *Energy & Environmental Science* **9**, 1552–1576 (2016).
4. Godin, B. et al. Discoidal porous silicon particles: Fabrication and biodistribution in breast cancer bearing mice. *Advanced Functional Materials* **22**, 4225–4235 (2012).
5. Popplewell, J. F. et al. Kinetics of uptake and elimination of silicic acid by a human subject: A novel application of 32Si and accelerator mass spectrometry. *Journal of Inorganic Biochemistry* **69**, 177–180 (1998).
6. Colli, A. et al. Ion beam doping of silicon nanowires. *Nano Letters* **8**, 2188–2193 (2008).
7. Hong, C., Kang, J., Kim, H. & Lee, C. Photothermal properties of inorganic nanomaterials as therapeutic agents for cancer thermotherapy. *Journal of Nanoscience and Nanotechnology* **12**, 4352–4355 (2012).
8. Wagner, R. S. & Ellis, W. C. Vapor-liquid-solid mechanism of single crystal growth. *Applied Physics Letters* **4**, 89–90 (1964).
9. Morales, A. M. & Lieber, C. M. A laser ablation method for the synthesis of crystalline semiconductor nanowires. *Science* **279**, 208–211 (1998).
10. Kuykendall, T. et al. Crystallographic alignment of high-density gallium nitride nanowire arrays. *Nature Materials* **3**, 524–528 (2004).
11. Pauzauskie, P. J. & Yang, P. Nanowire photonics. *Materials Today* **9**, 36–45 (2006).

12. Boston, R., Schnepp, Z., Nemoto, Y., Sakka, Y. & Hall, S. R. In situ TEM observation of a microcrucible mechanism of nanowire growth. *Science* **344**, 623–626 (2014).

13. Holmes, J. D., Johnston, K. P., Doty, R. C. & Korgel, B. A. Control of thickness and orientation of solution-grown silicon nanowires. *Science* **287**, 1471–1473 (2000).

14. Heitsch, A. T., Fanfair, D. D., Tuan, H.-Y. & Korgel, B. A. Solution-liquid-solid (SLS) growth of silicon nanowires. *Journal of the American Chemical Society* **130**, 5436–5437 (2008).

15. Huang, Z., Geyer, N., Werner, P., de Boor, J. & Gösele, U. Metal-assisted chemical etching of silicon: A review. *Advanced Materials* **23**, 285–308 (2011).

16. Hochbaum, A. I. *et al.* Enhanced thermoelectric performance of rough silicon nanowires. *Nature* **451**, 163–167 (2008).

17. Wolfram, J., Shen, H. & Ferrari, M. Multistage vector (MSV) therapeutics. *Journal of Controlled Release* **219**, 406–415 (2015).

18. Alhmoud, H. et al. Porous silicon nanodiscs for targeted drug delivery. *Advanced Functional Materials* **25**, 1137–1145 (2015).

19. Smith, B. E. *et al.* Singlet-Oxygen Generation from Individual Semiconducting and Metallic Nanostructures during Near-Infrared Laser Trapping. *ACS Photonics* **2**, 559–564 (2015).

20. Schmidt, B. & Wetzig, K. *Ion beams in materials processing and analysis* (Springer Vienna, Vienna, 2013).

21. Roder, P. B. *et al.* Photothermal Superheating of Water with Ion-Implanted Silicon Nanowires. *Advanced Optical Materials* **3**, 1362–1367 (2015).

22. Biersack, J. P. & Haggmark, L. G. A Monte Carlo computer program for the transport of energetic ions in amorphous targets. *Nuclear Instruments and Methods* **174**, 257–269 (1980).

23. Robinson, M. T. & Oen, O. S. The channeling of energetic atoms in crystal lattices. *Applied Physics Letters* **2**, 30–32 (1963).

24. Ziegler, J. F., Biersack, J. P. & Ziegler, M. D. *SRIM—The stopping and range of ions in matter* (Pergamon, New York, 2008).

25. Yu, P. Y. & Cardona, M. *Fundamentals of semiconductors* (Springer, Berlin, 2010).

26. Park, G.-S. et al. Full surface embedding of gold clusters on silicon nanowires for efficient capture and photothermal therapy of circulating tumor cells. *Nano Letters* **12**, 1638–1642 (2012).

27. Wang, N., Yao, B. D., Chan, Y. F. & Zhang, X. Y. Enhanced photothermal effect in Si nanowires. *Nano Letters* **3**, 475–477 (2003).

28. Roder, P. B., Pauzauskie, P. J. & Davis, E. J. Nanowire heating by optical electromagnetic irradiation. *Langmuir* **28**, 16177–16185 (2012).

29. Bohren, C. F. & Huffman, D. R. *Absorption and scattering of light by small particles* (Wiley-VCH, New York, 1998).

30. Roder, P. B., Smith, B. E., Davis, E. J. & Pauzauskie, P. J. Photothermal heating of nanowires. *The Journal of Physical Chemistry C* **118**, 1407–1416 (2014).

31. Jackson, J. *Classical electrodynamics*. 2nd ed. (Wiley, New York, 1975).

32. Kovalev, D. & Fujii, M. Silicon nanocrystals: Photosensitizers for oxygen molecules. *Advanced Materials* **17**, 2531–2544 (2005).

33. Jarvi, M. T., Patterson, M. S. & Wilson, B. C. Insights into photodynamic therapy dosimetry: Simultaneous singlet oxygen luminescence and photosensitizer photobleaching measurements. *Biophysical Journal* **102**, 661–671 (2012).

34. Lee, C., Kim, H., Cho, Y. & Lee, W. I. The properties of porous silicon as a therapeutic agent via the new photodynamic therapy. *Journal of Materials Chemistry* **17**, 2648–2653 (2007).

35. Ronning, C. et al. Tailoring the properties of semiconductor nanowires using ion beams. *Physica Status Solidi (b)* **247**, 2329–2337 (2010).

36. Abdullin, K. A. *et al.* Shallow hydrogen-induced donor in monocrystalline silicon and quantum wires. *Materials Science in Semiconductor Processing.* Papers presented at the E-MRS 2004 Spring Meeting Symposium C: New Materials in Future Silicon Technology **7**, 447–451 (2004).

37. Hoffmann, S. et al. Axial p-n junctions realized in silicon nanowires by ion implantation. *Nano Letters* **9**, 1341–1344 (2009).

38. Fukata, N. et al. Recrystallization and reactivation of dopant atoms in ion-implanted silicon nanowires. *ACS Nano* **6**, 3278–3283 (2012).

39. Das Kanungo, P. et al. Characterization of structural changes associated with doping silicon nanowires by ion implantation. *Crystal Growth & Design* **11**, 2690–2694 (2011).

40. Kanungo, P. D. et al. Ex situ n and p doping of vertical epitaxial short silicon nanowires by ion implantation. *Nanotechnology* **20**, 165706 (2009).

41. Bayliss, S. C., Heald, R., Fletcher, D. I. & Buckberry, L. D. The culture of mammalian cells on nanostructured silicon. *Advanced Materials* **11**, 318–321 (1999).

42. Park, J.-H. et al. Biodegradable luminescent porous silicon nanoparticles for in vivo applications. *Nature Materials* **8**, 331–336 (2009).

43. Low, S. P., Voelcker, N. H., Canham, L. T. & Williams, K. A. The biocompatibility of porous silicon in tissues of the eye. *Biomaterials* **30**, 2873–2880 (2009).

44. Hong, C., Lee, J., Zheng, H., Hong, S.-S. & Lee, C. Porous silicon nanoparticles for cancer photothermotherapy. *Nanoscale Research Letters* **6**, 321 (2011).

45. Regli, S., Kelly, J. A., Shukaliak, A. M. & Veinot, J. G. C. Photothermal response of photoluminescent silicon nanocrystals. *The Journal of Physical Chemistry Letters* **3**, 1793–1797 (2012).

46. Kayes, B. M., Atwater, H. A. & Lewis, N. S. Comparison of the device physics principles of planar and radial p-n junction nanorod solar cells. *Journal of Applied Physics* **97**, 114302 (2005).

47. Ko, M.-D., Rim, T., Kim, K., Meyyappan, M. & Baek, C.-K. High efficiency silicon solar cell based on asymmetric nanowire. *Scientific Reports* **5**, 11646 (2015).

48. Aharoni, H. & du Plessis, M. The spatial distribution of light from silicon LEDs. *Sensors and Actuators A: Physical* **57**, 233–237 (1996).

49. Fabbri, F., Rotunno, E., Lazzarini, L., Fukata, N. & Salviati, G. Visible and infra-red light emission in boron-doped wurtzite silicon nanowires. *Scientific Reports* **4**, (2014).

23 Si nanowires for evolutionary nanotechnology

Ming Hu, Hua Bao, and Yaping Dan

Contents

23.1 INTRODUCTION

One-dimensional (1D) nanostructures, such as nanowires (NWs), rods, belts, and tubes, have been studied extensively for more than two decades, due to their novel and remarkable mechanical, electrical, optical, thermal, and chemical properties. Among these, Si nanowires (SiNWs) receive exceptional attention, primarily due to their semiconductor nature and compatibility with the current Si-based semiconductor technology. A tremendous amount of progress has been made in the field of 1D nanostructures in the past decades. It is generally accepted that NWs provide a good system to investigate the dependence of mechanical, electrical, or thermal transport properties on dimensionality, size reduction, and quantum confinement. SiNWs also play a critical role in both interconnects and functional units in fabricating electronic, electromechanical, optoelectronic, and electrochemical devices with small dimensions. As compared with 0D quantum dots and quantum wells, the significant advancement of 1D SiNWs is benefited from the relatively easy, reasonably low cost, and robust synthesis and fabrication with well-controlled dimensions, surface morphology, phase purity, and chemical composition.

With the state of the art, this chapter aims to summarize major advances in the field of SiNWs in the past decades. Particular emphasis will be put on the synthesis, physical and chemical characterization and properties, and application of SiNWs. We will also present brief concluding remarks at the end of the chapter and provide some prospects for new directions of SiNWs for future evolutionary nanotechnology.

23.2 EXPERIMENTAL FABRICATION AND SYNTHESIS OF SiNWs

23.2.1 BOTTOM-UP APPROACH

Significant progress has been made in the development of facile and controlled methods for SiNW fabrication in recent years. Generally speaking, the bottom-up and top-down approaches are two basic approaches for fabricating SiNWs. In the bottom-up approach one has to figure out various ways (usually by chemical reaction) to assemble or join the Si atoms together to form SiNWs. There are numerous routes to the bottom-up fabrication of SiNWs. Among them, the vapor–liquid–solid (VLS) growth mechanism, and its analogues, is the most commonly used approach for experimental synthesis of SiNWs (Wagner and Ellis 1964; Holmes et al. 2000; Barth et al. 2010). This growth concept was first proposed by Wagner and Ellis (1964), where the material growth takes place via phase changes that are mediated through a catalyst particle. This requires the desired material to be provided in the gas phase as a material source. Usually this can be realized either in a molecular form that already has individual atoms, or in the form of a gas compound that can be decomposed into individual atoms later on. For synthesis of SiNWs, the commonly used Si precursor gas is SiH_4, $SiHCl_3$, or higher order silanes, and gold (Au) is often selected as the catalyst particles. It should be noted that a good choice of an appropriate seed material will facilitate the robust control over the diameter of the produced nanowires. In the meantime, the seed material can also significantly affect the quality of crystallinity of SiNWs. From this point of view, one should be very careful in selecting an appropriate precursor material. For example, the reasons for choosing SiH_4 as precursor material in fabricating SiNWs are (1) the Si–H bond is sufficiently labile under NW synthesis conditions (e.g., usually elevated temperature) to directly liberate reactive Si species for further NW growth; (2) The -H group liberated on precursor decomposition is ideally preferable to form a gas phase (H_2), which not only prevents the NW product from being contaminated by liquid or solid phase by-products, but also serves to inhibit the undesirable oxidization of the formed SiNWs.

Essentially, the VLS mechanism works via speeding up the adsorption of a gas phase onto a solid surface by introducing a catalytic liquid alloy phase. The process can be briefly described as follows: first, the Si–Au system (in the case of Au as catalyst) becomes the liquid phase above the eutectic temperature (about 640 K). Second, when there is a sufficient and continuous Si source to flow in, more and more Si atoms will diffuse into the Si–Au liquid. Because this Si–Au system is thermodynamically unstable, partial supersaturated or excess silicon atoms will crystalize into solids. In this way, nucleation of Si solids gradually takes place in the track of the Si–Au catalyst. As the Si–Au catalyst contact area/line goes further, the solidification of Si atoms will continue. As a result, a pure silicon block or plate is formed. By adjusting the size (diameter) of the catalyst particles, the shape and diameter of the as-formed SiNWs can be well controlled and manipulated. Other typical analogues of the VLS mechanism, such as supercritical fluid–liquid–solid (SFLS), supercritical fluid–solid–solid (SFSS), solution–liquid–solid (SLS), vapor–solid–solid (VSS), and oxide-assisted growth mechanisms, are also widely adopted in literature to synthesize SiNWs.

23.2.2 TOP-DOWN APPROACH

In addition to the well-known bottom-up approach, top-down fabrication of SiNWs is also a well-established technology. To form horizontal nanowires that are electrically isolated from the substrate, two approaches are commonly used. The simplest approach is to use a silicon-on-insulator (SOI) substrate and etch the SiNW into the thin active Si layer (Yang et al. 2004). The other approach is to use bulk Si and a deep reactive ion etch (RIE) process together to construct a stack of SiNWs (Sacchetto et al. 2009). In Figure 23.1 we illustrate these two approaches schematically. The second method is usually thought to have precise control of a small footprint and there will be product of several parallel nanowires that will be beneficial for some specific nanoelectronic device applications.

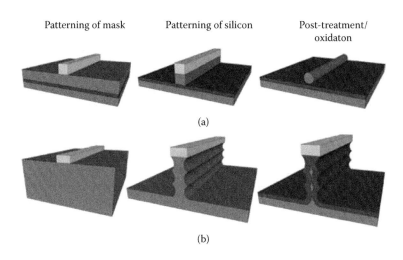

Patterning of mask   Patterning of silicon   Post-treatment/
                                                           oxidaton

(a)

(b)

Figure 23.1 Schematic illustration of fabricating the horizontal SiNWs by top-down approach. (a) The simplest approach (for details see text), and (b) involving deep reactive ion etching process. (Reproduced from Mikolajick, T., et al., *Phys. Status Solidi-Rapid Res. Lett.*, 7, 793–799. With permission.)

Obviously, according to the above-described procedures the top-down approach will lead to a horizontal arrangement of nanowires. This gives us the benefit of being able to predefine various patterns of SiNWs and then realize the efficient one-time fabrication without further postprocessing the alignment and/or re-arrangement of SiNWs. This is normally trivial as NWs are bundles in the bottom-up approach, and are hard to split into individuals. However, one benefit of bottom-up nanowire growth over the top-down approach is that SiNWs can be easily doped or alloyed *in situ* during crystal growth by incorporating dopant or other alloying precursors during the bottom-up NW synthesis. This is highly advantageous for targeting enhanced charge or electronic transport properties of SiNWs, since it does not require further process (usually the techniques available such as ion implanting are destructive and may destroy the atomic structure of the formed NWs) to generate additional charge carriers. Actually, both methods have been widely used to fabricate SiNW arrays for vast amounts of applications (Dupré et al. 2013; Calaza et al. 2015; Fonseca et al. 2016; Rey et al. 2016).

23.3 PHYSICAL AND CHEMICAL PROPERTIES OF SiNWs

23.3.1 MECHANICAL PROPERTIES

The mechanical properties of SiNWs have received much less attention, particularly at the beginning of the extensive application of SiNWs as electronic devices. However, they are of paramount importance in device processing as the atomic structures of these building blocks undergo changes in both synthesis and later realistic applications due to strain and stress induced by temperature and environmental variation. For instance, dislocations can be generated inside SiNWs and the resultant electrical transport properties may be quite different. This greatly motivated researchers to investigate the mechanical properties of SiNWs in the early 21st century.

Of the many mechanical properties, Young's modulus (denoted as E) is the basic mechanical parameter involved in almost all device-level applications. According to the elastic theory, Young's modulus can be reflected by average binding forces between the atoms. Therefore, in the case of one-dimensionality, Young's modulus is expected to be highly dependent on the surface-to-volume ratio and the surface conditions such as tension effects. From the experimental aspect, Young's modulus is often measured by a simple tension test incorporated within transmission electron microscopy (TEM) or atomic force microscope (AFM), or home-built instruments, or via noncontact tapping, or a nanoindentation process conducted by AFM. The AFM-based technique holds the advantage of robust measurement (can go back-and-forth by multiple times) and nondestructive detection. In the AFM contact mode, the AFM tip scans across the vertically aligned NW forest. The Young's modulus can be extracted by simultaneously recording the surface topography and lateral force exerted on the AFM tip. This technique was first developed by Song et al. (2005) who measured the elastic modulus of individual [0001] ZnO NWs/nanorods grown on a sapphire surface with an average diameter of 45 nm. The measured elastic modulus of 29 GPa for ZnO NWs is much lower compared to the bulk counterpart. Although this instrument can measure Young's modulus without destructively manipulating the sample, the behavior of the NW sample is unknown. To overcome this problem, Hoffmann et al. (2006) improved the experimental setup by incorporating an AFM in the scanning electron microscope (SEM). They used the AFM tip to bend the SiNWs to a large extent, while the SEM was used to obtain the image such that the mechanical deformation of SiNWs can be visualized *in situ*. They found that the average strength before fracture is around 12 GPa, which is 6% of the Young's modulus of Si along the NW direction. A correlation between strength and length was also generated, meaning that shorter SiNWs will have larger fracture strength. The AFM technique was further advanced by Wu et al. (2005) who presented a reproducible method to measure broad mechanical properties including Young's modulus, yield strength, and fracture stress. They applied their technique to metallic (Au) NWs. It was found that the Young's modulus of Au NWs is essentially independent of the NW diameter, while the smallest diameter NWs are shown to have the largest yield strength, with a value up to 100 times that of bulk materials, and substantially larger than that for bulk nanocrystalline metals.

Generally speaking, both experiments (San Paulo et al. 2005; Tabib-Azar et al. 2005; Heidelberg et al. 2006) and theoretical simulations (Menon et al. 2004; Kang and Cai 2007) predict that the Young's modulus of SiNWs (normally depending on the crystal orientation) is in the range of 94–210 GPa, which is more or less the same as that for the bulk counterpart (in the range of 135–190 GPa). In contrast, the fracture strength of SiNWs ranges from 12 to 23 GPa (Hoffmann et al. 2006; Kang and Cai 2007) with the upper limit close to the value for the bulk counterpart, meaning that NWs possess a higher ability of containing cracks to resist fracture, because they have far fewer slip planes that allow the dislocation and cracks to move. It should be noted that a diameter-dependent character of Young's modulus for SiNWs was found recently in ultrathin Si nanotubes (SiNTs) (Wingert et al. 2015), due to the strong surface effect and local surface strains, leading to a significant softening of Young's modulus.

23.3.2 THERMAL TRANSPORT PROPERTIES

23.3.2.1 Intrinsic lattice thermal conductivity of smooth SiNWs

Thermal transport property is another important physical parameter for SiNWs, in particular for the realistic application of energy conversion (thermoelectrics) and thermal management for nanoelectronics. The first experimental measurement of thermal conductivity of individual single-crystalline SiNWs was performed by Li et al. (2003) with a suspended heater method, where SiNW bridges the two heater pads and thermal conductivity is directly calculated by Fourier's law. The measured SiNWs have a diameter ranging from 22 to 115 nm, with thermal conductivity falling in the range of 10–50 W/mK, which is much lower than the bulk value. They attributed the mechanism to the phonon-boundary scattering. From the aspect of theoretical modeling, Mingo and Yang (2003) theoretically predicted the temperature-dependent thermal conductivity of Si and Ge NWs using the full-phonon dispersion relations. They compared their results with experiments and also two other traditional theoretical methods: the Callaway and Holland approaches. Their theoretical modeling is in good agreement with experimental results for the case of SiNWs, while the traditional Callaway and Holland approaches produce poor predictions, indicating the necessity of using modelistic phonon dispersion relations and the inadequate form of the anharmonic scattering rates. From atomistic simulation, Wang et al. (2009) predicted thermal conductivity of SiNWs using the nonequilibrium molecular dynamics (NEMD) method with the Stillinger–Weber (SW) potential and the Nose–Hoover thermostat. The thermal conductivity of 2.2, 4.3, and 6.5 nm diameter SiNW was found to be 2.35, 6.86, and 10.00 W/mK, respectively. The slightly higher thermal conductivity found in NEMD simulation as compared with experiments is understandable, considering that the atomic structure used in NEMD is perfect and the interatomic potential may be inaccurate. Nevertheless, both atomistic simulation and theoretical modeling reproduce the experimental trend and highlight the importance of phonon-boundary scattering and the phonon-confinement effect.

23.3.2.2 Manipulation of intrinsic lattice thermal conductivity of SiNWs

23.3.2.2.1 Surface morphology modification

Since the experimental instruments (in particular the direct bridge method) were established, tremendous research efforts have been dedicated to measuring thermal transport properties of SiNWs under various conditions. First of all, many researches focused on studying the effect of surface morphology control, in particular surface roughness, which is motivated by the inevitable random structure on the NW surface and more importantly, the demand for ultralow thermal conductivity driven by waste heat energy conversion (thermoelectric effect). Hochbaum et al. (2008) proposed the concept of rough SiNWs and measured the thermal conductivity to be as low as 1–2 W/mK for NW diameter of 50–75 nm. Such low thermal conductivity was verified by independent experimental work from Boukai et al. (2008), who reported that the 10-nm-wide NWs exhibited a thermal conductivity value of 0.76 W/mK. In addition, atomistic equilibrium molecular dynamics (EMD) along with Boltzmann transport equation (BTE) calculations also confirm this (Donadio and Galli 2009). The mechanism underlying the rough SiNWs was identified as two combined effects: the presence of extended nonpropagating modes induced by the disordered surfaces, and the significantly decreased phonon lifetimes of propagating modes. Based on a full-phonon dispersion relation coupled with BTE simulation, Martin et al. (2009) introduced a frequency-dependent model of boundary scattering which shows excellent agreement with the experimental work by Hochbaum et al. (2008), where a remarkably strong effect of surface roughness on the thermal conductivity is demonstrated. A new scaling law of thermal conductivity proportional to $(D/\Delta)^2$ at small NW diameters is also predicted by their model, where D and Δ are the NW diameter and root mean square (rms) surface roughness of the NW, respectively. These studies highlight the importance of realizing low thermal conductivity by enhancing the phonon-boundary scattering via introducing roughness/disordered structure on NW surfaces.

23.3.2.2.2 Radial and longitudinal structure modification

Instead of making the NW surface rough, Hu et al. (2011a, 2011b) proposed radial and longitudinal structure modification. By these approaches core-shell NWs (Hu et al. 2011a) and superlattice nanowires (SLNWs) (Hu and Poulikakos 2012) can be formed, respectively. By performing NEMD simulations, Hu et al. (2011a) studied the effect of germanium (Ge) coatings on the thermal transport properties of SiNWs. It has been shown that simply depositing Ge shell with a thickness of only 1 to 2 unit cells on single crystalline SiNW leads to a dramatic decrease of up to 75% in the thermal conductivity at room temperature, compared with pristine SiNW. They further analyzed the mode level vibrational density states (VDS) of phonons and the participation ratio (PR), which are the two fundamental methods to characterize phonon behavior. The remarkable reduction in thermal conductivity of Si-core/Ge-shell NWs is attributed to the strong depression and localization of long-wavelength phonon modes at the Si–Ge interface and of high-frequency nonpropagating diffusive modes. The simulation results clearly pinpoint the importance of phonon interference at the crystalline interface. Interestingly enough, Yang and Chen (2005) presented BTE modeling results on the thermal transport of core-shell and tubular NWs. They quantified how the surface conditions and the core-shell geometry affect the thermal conductivity by taking a representative case of Si-core/Ge-shell at room temperature. It was found that the effective thermal conductivity changes not only with the composition of the constituents but also with the radius of the nanowires. It is worth pointing out that the phonon scattering at the core-shell interface is assumed to be diffusive in their BTE model, while in Hu et al.'s direct NEMD simulation, all interfacial phonon scatterings along with the possible phonon interference effect are naturally included in the MD simulation (Hu et al. 2011a). This probably is responsible for the even lower thermal conductivity found in NEMD as compared with BTE modeling.

By longitudinal structure modification to pristine SiNW, Hu and Poulikakos (2012) proposed another concept of Si/Ge SLNWs to further reduce the thermal conductivity. Thermal transport in Si/Ge SLNWs is drastically suppressed by taking advantage of the inherent one-dimensionality (nanowire itself) and the combined effect of surface and interfacial phonon scattering. More interestingly, different from the previous concept of Si-core/Ge-shell NWs, the thermal conductivity of an Si/Ge SLNWs changes non-monotonically with both the Si/Ge periodic length and the NW diameter. Such abnormal nonmonotonic dependence is attributed to the two competing mechanisms: (1) for large periodic lengths the longitudinal interface modulation significantly depresses the phonon group velocities and induces strong phonon scattering at the interface, and thus the heat conduction is hindered; (2) when periodic lengths become extremely short, coherent phonons occur in the structure, which facilitate heat conduction in the SLNW. Such an effect counteracts the interface effect and the two competing mechanisms finally yield the nonmonotonic thermal conductivity dependence on the Si/Ge periodic length. Similar coherent phonon transport has already been found in both previous theoretical modeling (Yang and Chen 2003; Chen et al. 2005; Murphy and Moore 2007) and experiments (Wang et al. 2010; Luckyanova et al. 2012) on thin superlattice films.

23.3.2.2.3 Interior/central region control

Chen et al. (2010) proposed a method to reduce the thermal conductivity of SiNWs by consistently drilling a small hole at the center of pristine SiNW, to construct an SiNT structure. By taking advantage of strong phonon scattering on both sides (inner and outer) of SiNTs, they demonstrate that only 1% reduction in the cross-sectional area (equivalent to diameter) induces a 35% reduction to the lattice thermal conductivity. They further explained the simulation results by analyzing the spatial distribution of vibrational energy, and found that the strong localization modes concentrated on the inner and outer surfaces of SiNTs (the percentage of delocalized modes decreases with the enhanced surface-to-volume ratio in SiNTs) are responsible for the remarkable reduction in thermal conductivity. A similar idea was recently realized by Wingert et al. (2015) who experimentally fabricated ultrathin crystalline SiNTs and measured the thermal conductivity. The thickness of the crystalline SiNTs wall ranges from 5 to 10 nm and the diameter of the SiNTs is larger than 40 nm. The measured thermal conductivity is in the range of 1.1–1.7 W/mK, which is below the apparent boundary scattering limit and is even about 30% lower compared to the amorphous SiNTs with similar geometries. This study suggests that the amorphous materials do not always hold the lower limit of thermal transport. They attribute these findings to the strong elastic softening effect observed in the ultrathin crystalline SiNTs, by measuring Young's modulus and comparing the results with bulk Si and other values for pristine SiNWs reported in literature. The ultralow thermal conductivity and the softening effect in elastic modulus are further supported by the full atomistic simulation. The studies on SiNTs show that significant reduction in the thermal conductivity of SiNWs does not necessarily require disordered structures.

23.3.2.2.4 Other structure manipulation

There are also some other approaches proposed in terms of reducing the thermal conductivity of pristine SiNWs. The first approach is isotope doping. Yang et al. (2008) reported that the thermal conductivity of SiNWs is reduced exponentially with isotopic concentration by performing NEMD simulation. The minimum thermal conductivity (about 27% of thermal conductivity of pristine ^{28}SiNW) is reached, when the NW is doped with 50% isotope ^{42}Si. They also studied the thermal conductivity of superlattice SiNWs composed of isotopic-regular lattices (^{29}Si/^{28}Si and ^{42}Si/^{28}Si) and found that at the lattice period length of 1.09 nm the thermal conductivity is only 25% of that in pristine ^{28}SiNW. The same enhancement of thermal conductivity is observed at extremely short periodic lengths due to the well-known coherent phonons. They explained the ultralow thermal conductivity of isotope SLNWs with phonon spectrum theory. The second approach is alloying. By alloying the pristine SiNW with Ge and using molecular dynamics simulation, Chen et al. (2009) demonstrated that the thermal conductivity of Si–Ge alloyed NWs ($Si_{1-x}Ge_x$) is remarkably dependent on the Ge composition. Similar to what they found for isotope NWs, the thermal conductivity of Si–Ge alloyed NWs reaches the minimum (about 18% of that of pristine SiNW), when the Ge concentration is 50%. With only 5% Ge adatoms, thermal conductivity of alloyed $Si_{0.95}Ge_{0.05}$ NWs is reduced by half. The reduction of thermal conductivity is attributed to the localization of phonon modes induced by the random site phonon scattering. The third approach is nanoparticle inclusions: Zhang et al. (2012) found that creating small numbers of Ge nanoparticles, with a volumetric concentration less than 4%, inside the pristine SiNWs as inclusions, results in a large scattering of propagating phonons. The thermal conductivity of the as-formed nanoparticle-inclusion SiNWs is decreased by up to 70%. They further found that this method is robust considering that the reduction magnitude has weak dependence on the difference in atom mass or the interfacial interaction strength between the nanoparticle inclusion and the nanowire matrix.

23.3.2.3 Lattice thermal conductivity of strained SiNWs

Stress or strain is inevitable not only in experimental growth of pure crystalline and the various forms of SiNWs, but also in the realistic integration into the nanodevices, such as being fabricated as metal–oxide–semiconductor field-effect transistors. Therefore, an intuitive question is how the thermal transport property is affected by the local stress or strain, or equivalently how can the thermal conductivity be manipulated by mechanical strain. Li et al. (2010) examined the thermal conductivity of smooth SiNWs with diameter of 2.2 and 4.4 nm by performing EMD with Tersoff potential. It is found that the thermal conductivity monotonically decreases with uniaxial tensile strain to +12% and increases with uniaxial compressive strain to -12%. However, the magnitude of thermal conductivity change is not that much: only about 30% for the entire +/-12% strain range, as compared with the previously mentioned structural modulations. This is understandable considering that the mechanical strain (either compression or tension) can only effectively tune the elastic or harmonic part of the interatomic interaction in NW (such as phonon dispersion) and can hardly affect the anharmonic part (phonon relaxation time), which is mainly governed by phonon-boundary scattering. Unless severe plastic deformation occurs, which induces phonon-grain boundaries or phonon-dislocations interactions, phonon scattering will be enhanced and thus the thermal conductivity is expected to change dramatically. The same trend is found by recent experiments conducted by Murphy et al. (2014) who reported the first experimental measurements of the effect of spatially uniform tensile strain on the thermal conductivity of an individual suspended SiNW using *in situ* Raman piezothermography. No noticeable change was observed in thermal conductivity for the engineering strain less than 0.8%. One should note that the SiNW measured in their experiments has a quite large diameter (171–177 nm) and is <111> oriented, while the SiNW used in MD simulation is aligned along the <100> direction. Recently, the thermal conductivity of pristine SiNWs has been found to be strongly dependent on the surface orientation (Zhou et al. 2016). Therefore, care should be taken when comparing the absolute value of thermal conductivity of SiNWs with different diameters and surface orientations. Nevertheless, these studies indicate that mechanical strain is not as effective as the isotope and structural modulation in reducing the thermal conductivity.

23.3.3 OPTICAL AND OPTOELECTRONIC PROPERTIES

23.3.3.1 Optical properties of an individual nanowire

The nanowire has a finite size in the radial direction, which has two major effects on the optical properties of nanowires. The first one is the quantum confinement effect: when the nanowire size is below or comparable to the Bohr radius of a material (about 4.9 nm for silicon (Cao et al. 2010b)), the electronic structure of the nanowire can be strongly modified by the finite dimension. The quantum confinement effect can be understood through the particle-in-a-box model, which is discussed in standard quantum mechanics textbooks (Griffiths 2004). Calculations based on density functional theory and the tight-binding model show that ultrathin SiNWs have direct band gaps and the band gap value decreases with the diameter (Nolan et al. 2007). The same conclusion has also been reached by GW calculations (Bruno et al. 2007) and experiments (Ma et al. 2003). In addition, for thin SiNWs with a diameter smaller than 10 nm, due to the large surface-to-volume ratio, surface states become dominant to the electronic properties (Nolan et al. 2007). The electronic property is thus also strongly dependent on the detailed surface morphology and the surface passivation of the nanowires. The optical properties of the ultrathin nanowires are related to the electric properties. The investigations on these ultrathin nanowires are mostly theoretical, based on first-principles calculations and tight-binding theory, and experimental investigations are rare. Due to the difficulty to control the growth of such nanowires, the practical application is still challenging. Therefore, in this chapter we will not be focusing on such ultrathin SiNWs. Interested readers are recommended to read some review articles, such as Rurali (2010), and Hasan et al. (2013).

The second effect is the optical confinement effect: when the nanowire diameter is comparable to the wavelength of the external electromagnetic wave, the finite dimension of the nanowire only allows certain optical modes to propagate (Yeh and Shimabukuro 2008). These modes are usually resonant modes that can strongly interact (including both absorption and scattering) with the external electric field (Bohren and Huffman 1983). Since the optical wavelength is generally the most important regime, the diameter of the nanowire discussed here is at least tens of a nanometer, much larger than the Bohr radius. Therefore the surface effect and quantum confinement are not important to the optical properties of the SiNWs. The dielectric function of the nanowires can be assumed to be that of bulk silicon. The SiNWs can thus be regarded as a dielectric cylinder.

If the diameter of the nanowire is comparable to the wavelength of electromagnetic wave, an individual nanowire can be viewed as a dielectric waveguide or optical antenna. Certain optical modes can be excited by the external electromagnetic wave, which makes the spectral response of the cylinders different from their bulk counterparts. Such effects are purely classical, and therefore can be well described by Maxwell's electromagnetic theory. The most common case of discussion involves a plane wave incident on a dielectric cylinder. The optical response is obtained by solving the Maxwell equations, which is also known as the Mie-scattering theory (Bohren and Huffman 1983). In the framework of the Mie-scattering theory, the incident plane wave and scattered waves are expanded by the cylindrical harmonic functions, then the expansion coefficients are obtained by matching the boundary conditions. The details can be found in many books (e.g., Bohren and Huffman 1983) and thus will not be presented here. The optical property of the cylinder can be quantified by two parameters, including the frequency-dependent scattering cross section and absorption cross section. The former quantifies the ability of the cylinder to redirect the incident plane wave, and the latter gives the ability to absorb incident waves.

Although the Mie-scattering theory for optical scattering by a cylinder has been developed for many years, such an analysis was not applied to investigate the optical property of a nanowire until Cao et al. (2009) published their work in 2009. In their work, the authors measured the photocurrent of a germanium nanowire lying horizontally on a gold substrate, as shown in Figure 23.2a and b.

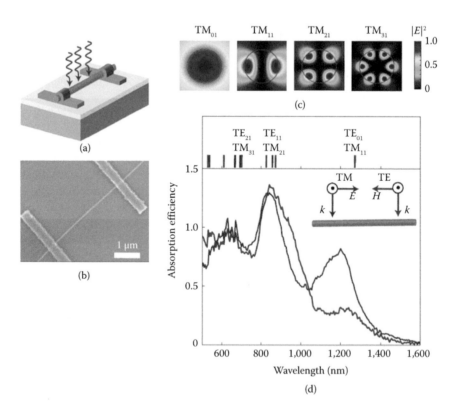

Figure 23.2 (a) Schematic illustration of the germanium nanowire device used for photocurrent measurements. (b) Scanning electron microscopy image of a 25 nm radius germanium nanowire device. (c) The configuration of the electric field intensity for typical transverse-magnetic leaky modes. The blue circle refers to the nanowire–air interface. (d) Absorption efficiency spectra of a 110 nm radius germanium nanowire taken using linearly polarized transverse-electric (TE; red) or transverse-magnetic (TM; blue) light. (Reproduced from Cao, L., et al., *Nature Mater.*, 8, 643–647. With permission.)

The authors noticed evident spectral absorption peaks in the photocurrent, and these peaks are observed at the wavelength corresponding to those optical resonant modes, which are shown in Figure 23.2c and d. The resonant modes can be regarded as multiple total internal reflections of light within the cylindrical geometry, and thus results in light confinement and field enhancement with the nanowires. Since the optical absorption is proportional to the field intensity, these resonant modes will result in enhanced light absorption and thus a large photocurrent at certain excitation wavelengths. Later, the same group of authors demonstrated leaky mode resonance-enhanced optical absorption for individual SiNWs (Cao et al. 2010a). The spectral features of scattering due to the resonance have also been demonstrated (Broenstrup et al. 2010; Cao et al. 2010b). Since the resonance mode is strongly related to the diameter of the nanowires, it is possible to further tailor the spectral absorption and scattering properties of individual nanowires by tuning the diameter of nanowires. For example, the SiNWs with different diameters appear to be different colors under white light illumination (Cao et al. 2010b).

23.3.3.2 Optical properties of silicon nanowire arrays and mats

23.3.3.2.1 Vertically aligned array (simulation and experiment)

The emerging interest on the optical properties of vertically aligned nanowire arrays originated from a design of nanowire-based solar cell architecture proposed by Kayes et al. (2005), as shown in Figure 23.3a. Such a design was initially proposed to reduce the carrier collection length and enhance the carrier collection efficiency of thin film solar cells. Later Hu and Chen (2007) employed a transfer-matrix method to study the optical properties of an ordered SiNW square array and found that such a structure has an additional benefit: the array structure can help to reduce light reflection at the top surface and thus reduces the reflection loss. The antireflection effect was also demonstrated by experiments, such as Tsakalakos et al. (2007), and Garnett and Yang (2008). By bottom-up or top-down approaches, the core nanowires are first grown or etched on a substrate, and then shell is deposited by a chemical vapor deposition approach. The SEM image of the SiNW solar cell in Tsakalakos et al. (2007) is

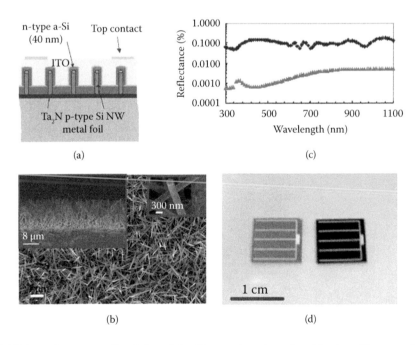

Figure 23.3 (a) Schematic cross-sectional view of the Si nanowire solar cell architecture. The nanowire array is coated with a conformal a-Si:H thin film layer. (b) Scanning electron micrograph (plan view) of a typical Si nanowire solar cell on stainless steel foil, including a-Si and ITO layers with insets showing a cross-sectional view of the device and a higher magnification of an individual Si nanowire coded with a-Si and ITO. (c) Specular reflectance of silicon solar cell. (d) Picture of a planar a-Si:H solar cell fabricated on a degenerately doped Si substrate (left) and an SiNW solar cell fabricated on the same type of substrate with a 1 cm² area. (Reproduced from Tsakalakos, L., et al., *Appl. Phys. Lett.*, 91, 233117, 2007. With permission.)

shown in Figure 23.3b. The diameter of the nanowires is 109±30 nm, and they are not exactly verti-cally aligned. Nevertheless, it experimentally demonstrated the extremely low reflectivity over the entire solar spectrum for such SiNW array structures (Tsakalakos et al. 2007). It can be seen from Figure 23.3c and d that the SiNW array solar cell is much darker compared to a planar solar cell, showing that the reflectivity is much smaller.

It should be noted that reduced reflection does not necessarily result in enhanced energy conversion efficiency. For solar application, high optical absorption in the active region is more important. In order to provide a design guideline for enhancing the ultimate solar absorption efficiency, Lin and Povinelli (2009) performed further numerical electromagnetic simulations based on a square lattice arrangement of nanow-ires, and investigated how the structure could be related to the optical absorption. For such geometry, there are only three tunable structural parameters, including the height, diameter, and interwire spacing (or lattice constant), as shown in Figure 23.4a. The optimizations are carried out by fixing the length

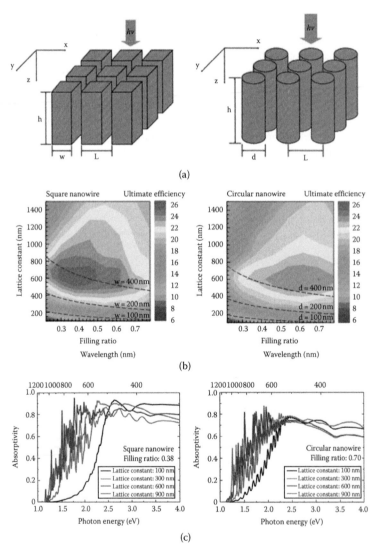

Figure 23.4 (a) Schematic of the simulation structure of vertically aligned silicon nanowire array with a square lattice. The tunable parameters in such geometry are height h, lattice spacing L, and diameter d. (b) The inte-grated optical absorption efficiency for nanowire arrays with fixed height (2.33 um) but different lattice constant and filling ratio. (c) The absorption spectra of square and circular nanowire arrays with different lattice constant. (Reproduced from Fang, X., et al., *J. Quant. Spectrosc. Radiat. Transf.*, 133, 579–588. With permission.)

and tuning the diameter of the nanowire and the lattice constants of the array (Lin and Povinelli 2009). Investigations of optical properties of periodically arranged arrays have also been extended to different shaped nanowire (such as a square shape [Fang et al. 2014]), as well as different lattices (such as hexagonal [Li et al. 2012]). Interestingly, in all these simulation results, the optimal design for solar absorption purposes usually have a lattice constant of approximately 600 nm, as shown in Figure 23.4b. This is because when the lattice constant is around 600 nm, significant absorption enhancement can be induced at a similar wavelength. For silicon, the indirect band gap makes the intrinsic absorption in this spectral regime very weak. As a result, the absorption enhancement around 600 nm is important to enhance the ultimate efficiency.

Due to the difficulty to fabricate the periodic nanowire arrays with uniform diameter and spacing, a question that naturally arises is whether the periodic arrays can be representative enough to the properties of the experimentally fabricated nanowire arrays. Bao and Ruan (2010) first investigated the optical properties of vertically aligned SiNW arrays with 100 nm diameter using the finite–different time domain (FDTD) simulations. They built pseudorandom nanowire arrays with aperiodic arrangements, different diameters, and random heights. It was found that all the aperiodic nanowire arrays on average enhance the optical absorption of the nanowire array, and that the array with an uneven nanowire top surface can also reduce optical reflection. Such a result is explained by the enhanced multiple scattering among nanowires. Later Du et al. (2011) checked the randomly positioned nanowire array with the diameter and wire spacing that are closer to the optimal values, and further confirm the possibility of enhanced optical absorption using aperiodicity. Lin et al. (2012) further compared the broadband antireflection effect of aperiodic arrays, and used a random walk approach to optimize the position of nanowires. They also found that the antireflection effect of aperiodic structures can be better than the ordered counterparts.

23.3.3.2.2 Inclined or completely random nanowire mats

Although the vertically aligned nanowire arrays have extraordinary performance in optical absorption, they are more difficult and costly to fabricate. The nanowires grown on a substrate are often completely randomly oriented, like mat structures. The completely random SiNW mats were observed to have a light yellow–brown color (Street et al. 2009) and are therefore not a good candidate material for photovoltaic application. Further measurements show that they have a single reflection peak in the visible range. However, it is fundamentally an interesting question to understand why these SiNW mats have such unique properties.

The reflection spectra of the optical reflectance can vary across studies, depending on how the NW mats are fabricated. For example, Convertino et al. (2010) have grown tapered silicon NW mats using Au films as the catalyst. Depending on the mean length and diameters of NWs, the reflectance spectra can have a Fabri–Perot-like oscillation with respect to wavelength, or have a single maximum. Street et al. (2009) measured the total reflectance of the two NW mat samples grown using a gold thin film as catalyst, which contain NWs with diameters from 50 to 200 nm. Their NW mats have a yellow or brown color and each shows a single reflection maximum within a 600–800 nm wavelength range. They explained the observed reflection spectra by the different optical characteristics in distinct spectral regimes, including strong absorption in the short wavelength regime, medium absorption and strong scattering regime in the middle, and low absorption and significant transmission in the long wavelength regime. The different maxima locations in the samples are explained by various contributions of the three effects. Brönstrup et al. (2011) proposed a statistical model based on a random walk approach and Mie-scattering theory which can fit the experimental optical property in the visible range with only three fitting parameters. Bao et al. (2012) fabricated SiNW mats using the chemical vapor deposition method with gold nanoparticles as the catalyst. The nanowires grown by this method often have a diameter similar to the size of the gold nanoparticles. Therefore the diameter of nanowires can be controlled using this method. Figure 23.5a shows the random nanowire mats with an 80 nm average diameter. It can be seen that the nanowires all have quite similar diameters. The samples with average diameters of 40, 60, and 80 nm were fabricated, and the optical reflectance spectra were measured using a spectrometer and integrating sphere. As shown in Figure 23.5b, a single maximum was observed for all three samples and the maximum frequency shifts to lower value

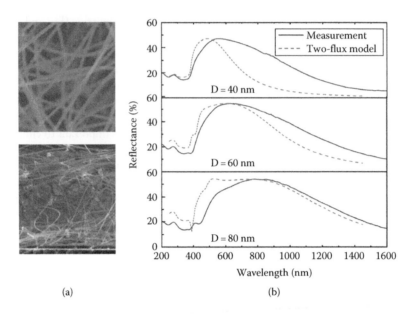

Figure 23.5 (a) SEM image of the random nanowire mats grown using the chemical vapor deposition (CVD) method. The diameter is controlled by the Au nanoparticle catalyst. (b) The measured reflectivity of silicon nanowire mats with different diameters. The results agree relatively well with the prediction from the Mie-scattering theory and the two-flux model. (Reproduced from Bao, H., et al., *J. Appl. Phys.* 112, 124301, 2012. With permission.)

as the average diameter becomes larger (Bao et al. 2012). By a comparison between the measurement and theoretical modeling based on Mie-scattering and a two-flux model, it was demonstrated that reflection maximum is located at a similar frequency at which the Mie-scattering is the strongest for the particular diameter. This shows that such an optical reflection spectrum is originated from the optical scattering of individual SiNWs (Bao et al. 2012).

Note that the optical properties of random SiNW mats are quite complicated, and strongly dependent on not only the average diameter, but also the volume fraction and arrangements. All the investigations discussed above are on the SiNWs arrays with relatively small volume fraction, so the effects such as multiple scattering and localization are less important. It has been reported that for random GaP nanowires, strong scattering and localization was observed (Muskens et al. 2009). The authors perceive that a similar effect could also be observed in SiNWs mats.

23.3.3.4 Surface passivation and carrier transport

Surfaces are where the periodic crystal lattice of semiconductors terminates. The terminated bonds on surfaces, if left unsaturated (so-called dangling bonds), will create localized defect energy levels in the band gap. The defect levels, particularly those deep in the band gap, are efficient recombination centers which increase recombination rate and shorten the lifetime of minority charge carriers, resulting in inferior performance of optoelectronic devices including photodetectors (Adivarahan et al. 2000; Dan et al. 2011), solar cells (Baxter and Aydil 2005; Tsakalakos et al. 2007; LaPierre et al. 2013), and light-emitting devices (Demichel et al. 2010). Surface passivation is to suppress the surface recombination rate of charge carriers. Up to date, two surface passivation techniques have been developed. The first one is to introduce static electric fields near the surfaces so that electrons and holes are separated to reduce their recombination probability. For instance, an Al_2O_3 coating is an excellent surface passivation material thanks to a high density of static charges in the film that separate electrons and holes near the surfaces (Hoex et al. 2008). The second technique is to extend the broken crystal lattice by coating the semiconductor with a wider band gap material. The resultant heterojunction confines the photogenerated electron–hole pairs within the semiconductor which has a reduced density of recombination centers at the interface due to the saturation of dangling bonds by the material coating. For instance, the wide band gap insulators such as SiO_2 and Si_3N_4 are often used as a surface passivation coating for silicon (Adivarahan et al. 2000; Arulkumaran et al. 2004). It is worthwhile to note that hydrogen-rich amorphous silicon (a-SiH) films are sometimes employed as surface passivation material for crystalline silicon due to the fact that a-SiH has a band gap (~1.7eV) wider than silicon (De Wolf and Kondo 2007; Fujiwara and Kondo 2007).

For planar semiconducting devices, surface passivation has been extensively investigated in the 1980s and 1990s (Nannichi et al. 1988; Fenner et al. 1989; Aberle 2001). For nanoscale semiconductor devices, surface recombination has a much bigger impact on the device performance due to their significantly enhanced surface-to-volume ratio and the fact that multiple crystalline-oriented surfaces are exposed. However, surface passivation of nanostructured optoelectronic devices had remained largely unexplored until 2011 when researchers from the Institute of California Technology (Kelzenberg et al. 2008) and Harvard University (Dan et al. 2011) first demonstrated that surface passivation can significantly extend the minority-carrier lifetime of microwire and nanowire devices (Figure 23.6). Since then, surface passivation has been widely applied to nanostructured devices such as nanowire solar cells to improve performances (Kim et al. 2011; LaPierre et al. 2013). It is known that the impact of surface passivation is largely dependent on the surface-to-volume ratio, the device size. To achieve high performance, aggressive passivation techniques are required for ultrascaled nano or quantum devices. For example, optoelectronic devices based on quantum dots often suffer from short carrier lifetime and low carrier mobility (Konstantatos et al. 2006). In recent years, a breakthrough has been made for quantum dot solar cells and light-emitting devices after quantum dots are surface passivated by functionalizing the surfaces with small molecules (Konstantatos et al. 2006; Clifford et al. 2009; Ip et al. 2012; Dai et al. 2014).

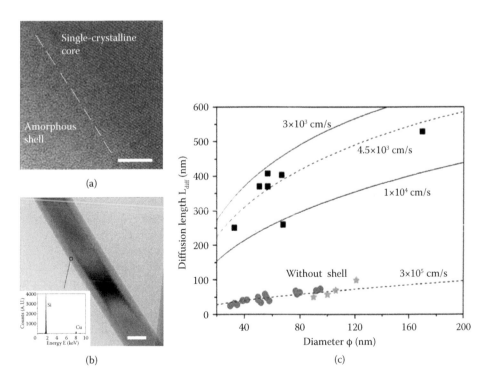

Figure 23.6 (a) TEM image shows that the nanowire has a single-crystalline core and amorphous shell. Scale bar: 3 nm. (b) The shell thickness is approximately 10 nm. Energy dispersive X-ray spectroscopy (EDS) indicates that the amorphous shell is amorphous silicon (inset). Scale bar: 30 nm. (c) Minority-carrier diffusion length in bare and core-shell nanowires as a function of nanowire diameter. (Reproduced from Dan, Y., et al., *Nano Lett.*, 11, 2527–2532, 2011. With permission.)

23.4 APPLICATIONS OF SiNWs

23.4.1 SENSORS

Since a sensor device generally needs to transform a physical or chemical signal of the environment into an electrical signal, which is usually easy to detect, SiNWs are excellent candidates for high-resolution sensors, due to their very sensitive physical and chemical properties to the external environment, including shape, length, temperature, and pressure. In the past few decades, many chemical and biochemical sensors have been intensively involved with SiNWs (Rashid et al. 2013). In addition, SiNWs receive exceptional attention for mechanical sensors, primarily benefiting from their high piezoresistance (He and Yang 2006). Moreover, similar to single-walled carbon nanotubes, freestanding SiNWs can be used as excellent oscillators. Feng et al. (2007) demonstrated a very high frequency nanomechanical resonator based on single-crystalline SiNWs grown by the bottom-up approach. The SiNW resonators operate around 200 MHz and possess a very high quality factor (about 2000–2500). Such an excellent resonance property is gained from the fundamental resonances as high as 215 MHz of pristine SiNWs. The pristine resonators provide the quality factor as high as 13,100 for an 80 MHz device. The SiNW-based resonators exhibit great potential for future applications in resonant sensing, quantum electromechanical systems, and high-frequency signal processing. Nichol et al. (2012) detected the statistical polarization of ^{1}H spins in polystyrene by using a radio frequency SiNW mechanical oscillator as a low-temperature nuclear magnetic resonance force sensor. This is fundamentally a new route to ultrasensitive magnetic resonance force microscopy (MRFM) detection using SiNWs oscillators. This bottom-up approach of using SiNWs oscillators as force detectors offers a new opportunity for greatly improved force sensitivity. In addition, the nanoscale magnetic resonance imaging could also benefit from this new technique considering that large time-dependent field gradients can be generated. Last but not least, the new tools are promising for advanced MRFM toward high-resolution molecular imaging. Recently, Kim et al. (2015) proposed a flexible low-power ring oscillator, which is composed of three CMOS inverters with a gain of 70 at a supply voltage of 1 V. p- and n-channel SiNW field-effect transistors forming the component inverters show on/off current ratios on the order of 10^4 and 10^5, respectively. The oscillator is reported to have generated a sinusoidal wave with frequency of 6.6 MHz. The good mechanical bendability of the ring oscillator shows promise for future flexible electronic device applications.

23.4.2 THERMAL SCIENCE: THERMOELECTRICS

Thermoelectrics offer an attractive pathway for addressing an important niche in the globally growing landscape of energy demand, since they can convert heat, and in particular waste heat, to electricity, the highest form of energy in terms of thermodynamic quality. In the past decades, developing schemes to improve thermoelectric conversion efficiency were guided by the concept of phonon glass-electron crystal—reducing the lattice contribution to the thermal conductivity as closely as possible to an amorphous state, while keeping relatively high electrical conductivity and Seebeck coefficient. Under this guideline, a considerable amount of scientific effort has been dedicated to reducing the lattice thermal conductivity. The phonon glass-electron crystal approach has stimulated a great amount of new research, resulting in significant increases in ZT for several compounds such as skutterudites, clathrates, and half-Heusler intermetallic compounds. Recently, due to the ability gained to create nanostructured materials, the nanostructuring of existing thermoelectric materials of interest and low dimensionality especially 1D nanowire have emerged as promising pathways to greatly reduce the lattice thermal conductivity to low values previously thought impossible, as a result of which the thermoelectric performance is significantly improved. Typical examples in this route include low-dimensional nanostructures such as quantum dots and nanowires, along with subsequent lateral and longitudinal structure modulation (Hochbaum et al. 2008; Hu et al. 2011a; Hu and Poulikakos 2012), nanocomposites, superlattices, and bulk nanostructured materials. By exploiting nanoscale effects, such as strong boundary or interfacial phonon scattering, and by taking advantage of the nanoconfinement effect, the nanostructured materials are able to achieve decent *ZT* values at room temperature and record-high *ZT* values in the range of 1.5–2.0 at medium and high temperatures.

One-dimensional nanowires receive extensive attention in the thermoelectric research due to their intrinsic high carrier mobility (good for electronic transport) and considerably low lattice thermal conductivity. After the inspiring study of Hochbaum et al. (2008), continuous research has been dedicated to further reduce the lattice thermal conductivity while in the meantime maintain a decent electronic transport property. Those typical approaches are already mentioned in the previous subsections. Here we would like to mention a few more strategies that have not been touched yet. The nanoporous concept was first proposed by Lee et al. (2008) who calculated the thermal conductivity of crystalline bulk Si with periodically arranged nanometer-sized pores by classical MD simulations. The thermal conductivity is found to be as low as 0.6 W/mK with porous size of 1.0 nm, regardless of the porous shape (circular or square-like). The substantial reduction in thermal conductivity of porous Si as compared with bulk Si was confirmed by recent MD simulations (Yang et al. 2014). This porous strategy is expected to combine with the low-dimensionality nature of SiNWs, which is anticipated to further reduce the thermal conductivity and thus enhance the *ZT* coefficient.

Another strategy to reduce the intrinsic thermal conductivity of SiNWs is to use polycrystalline material. Bulk polycrystalline Si has been proved to have subamorphous thermal conductivity as compared with the bulk amorphous counterpart (Ju and Liang 2012). Such low thermal conductivity can be explained in terms of effectively suppressing the propagative phonons (propagons) by many random grain boundaries existing in the material, reaching a similar phonon state to amorphous silicon. If combined with a 1D structure, the propagons will be further suppressed and thus the polycrystalline SiNW is expect to reach a record low thermal conductivity, which should be well below the amorphous limit. Since the typical grain size involved in experimental synthesis is much larger than the mean free path of electrons, introducing grain boundaries will not substantially affect the electronic transport properties. Therefore, the polycrystalline SiNW is very promising for realizing relatively high *ZT* coefficients for thermoelectrics.

23.4.3 ELECTRONICS: OPTOELECTRONICS (PHOTOVOLTAICS AND PHOTODETECTOR)

One of the most important applications for SiNWs is in the optoelectronics area, especially for SiNW-based solar cells. To achieve high energy conversion efficiency, solar cells require efficient optical absorption and exciton separation and collection. To reduce the cost of solar cells, the base material should be abundant and low cost, such as silicon. However, due to the indirect bandgap of silicon, the intrinsic absorption in the long wavelength regime is low. It often requires a few hundred micron thickness to efficiently absorb sunlight. On the other hand, to achieve good carrier collection efficiency, the solar cell should not be thicker than a few times of the minority-carrier diffusion length. As such, high-quality silicon is needed to enhance the minority-carrier diffusion length, which consequently increases the material cost. To resolve this dilemma, a vertically aligned SiNW array with a radial p–n junction was first proposed by Kayes et al. (2005). The radial junction structure allows a radial carrier collection, while the photon absorption occurs in the nanowire axial direction, which inspires a number of investigations on the SiNW-based solar cells. In 2007, Tian et al. (2007) fabricated the first radial junction p-i-n solar cells with an individual nanowire; the efficiency of the single wire was estimated to be 3.4%. Since then, large area solar cells based on SiNWs have been fabricated. For example, Tsakalakos et al. (2007) fabricated SiNW solar cells with a bottom-up approach. The p-type SiNWs with average diameter of ~109 nm were grown by the vapor–liquid–solid (VLS) mechanism and an n-type amorphous silicon of ~40 nm thickness was coated to form a p–n junction. However, the energy conversion efficiency is extremely low (estimated to be 0.1%). Later, Garnett and Yang (2010) demonstrated an ordered SiNW array solar cell using the top-down approach, and the efficiency was found to be around 5–6%. These nanowire array-based silicon solar cells failed to achieve high efficiency, due to several reasons. First, the nanowires are too thin which is not optimal for carrier collection. Theoretical analysis and numerical simulations show that to achieve the optimal carrier collection efficiency, the wires should have a comparable to the minority-carrier diffusion length, which is typically a few microns.

Second, the surface-to-volume ratio of a nanowire is large. Surface states due to the dangling bonds can induce additional carrier recombination, which lowers the carrier collection efficiency. Effective surface passivation of these nanowires is still technically quite challenging. Inspired by these works on silicon solar cells, better photovoltaic performances were later achieved by nano-patterned silicon surface (Jeong et al. 2013), radial junction microwires array with effective surface passivation (Gharghi et al. 2012), and nanowire array with III–V materials (Krogstrup et al. 2013; Wallentin et al. 2013). These are beyond the scope of this chapter so will not be presented in more detail. For more information about nanowire-based photovoltaics, readers are recommended to look up these excellent review articles: Yu et al. (2012); Dasgupta and Yang (2014); Li et al. (2014); Brongersma et al. (2014).

Another important application of SiNWs is for photo detection. The usage of nanowires as photodetectors is motivated by increasing computing speed and lowering power consumption for silicon photonic-integrated circuits. However, the ultrascaled volume of nanowires significantly reduces their light absorption efficiency, resulting in low sensitivity and signal-to-noise ratio. High gain is therefore an important parameter for nanowire photodetectors. P. D. Yang's group first observed extraordinarily high gain (up to five orders of magnitude) in ZnO nanowire photoconductors in 2002 (Soci et al. 2007). Since then, SiNW photoconductors formed by bottom-up approach, such as the VLS method, and top-down approach, such as a RIE were reported to be highly sensitive to photoillumination (Soci et al. 2010; Zhang et al. 2010).

In fact, semiconducting photoconductors have already been reported to have high gain even for bulk devices in the past few decades (Matsuo et al. 1984). For nanowire photoconductors, the gain is even higher (Lee et al. 2012). A large number of works have focused on decoding the underlying high-gain mechanism of nanowire photoconductors (Soci et al. 2010). Two mechanisms have been proposed and widely accepted. The first mechanism is that the high gain is due to the charge circulation, a model that has already been used to explain the photogain observed in traditional bulk semiconducting photoconductors (Neamen 2011). In this model, one type of charge carriers in the photogenerated electron–hole pairs is trapped by nanowire surface trap states. The counterpart charge carriers driven by electric field will circulate in the circuit many times before recombination, equivalent to generating many times more charge carriers. The longer the trap lifetime, the higher the gain will be. This mechanism seems to be consistent with the experimental observation: the high photogain often comes along with slow response time (long trap lifetime). But quantitatively the theory cannot predict the experimental results both for bulk and nanowire photoconductors (Kim et al. 2011). This inconsistency motivates researchers to explore other possible mechanisms. The gating effect is widely believed to play another important role in the photogain of nanowire photoconductors. It works in the way described as follows. Once the surface trap states capture the photogenerated charge carriers, electron–hole pairs will be separated, forming a local electric field perpendicular to the surfaces. This local electric field will bend the energy band and modulate the majority charge carriers, acting like a gate on the nanowire surfaces. This gating effect will significantly increase the nanowire conductivity, resulting in the seemingly high photogain. The high surface-to-volume ratio of nanowires simply enhances this effect.

Although the nanowire photoconductors have high photogain, these devices have a major drawback—the response time is slow (microseconds to seconds). Consequently, these photoconductive devices are not suitable for high-speed application in that the gain will be reduced to below unity if operating at the GHz band (Soci et al. 2007). To develop photodetectors with high gain at high speed, Hayden et al. (2006) demonstrated SiNW avalanche photodiodes in 2006 that can operate at tens of GHz with a gain of hundreds. The common issue with avalanche photodiodes is that the operating voltage needs to be as high as tens of volts, which is incompatible with integrated circuits. Recently, Tan et al. (2014, 2016) showed that SiNW bipolar phototransistors can operate at low bias voltage (~1V) but with high gain (~5500). The potential 3dB bandwidth of these nanowire bipolar phototransistors is as high as tens of GHz. Nevertheless, for applications that do not require high speed, SiNW photoconductors still have great potential. For example, Lee et al. (2012) demonstrated SiNW photoconductors could be used as photosensitive retina in human eyes. The successful integration of these SiNW photonconductors with neural nerves might allow the blind to regain their vision (Figure 23.7).

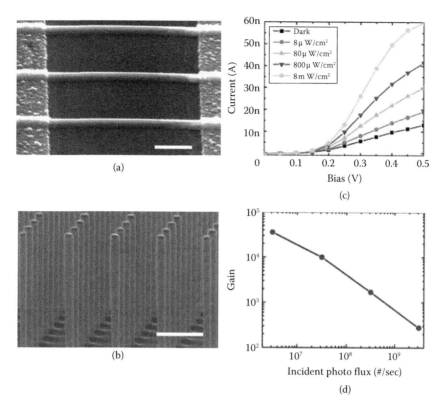

Figure 23.7 (a) SEM image of horizontal silicon nanowires. Scale bar: 2 μm. (b) SEM image of vertically standing silicon nanowires. Scale bar: 1 μm. (c) I–V curve of the nanowire under illumination of different light intensity. (d) Gain versus incident light flux. (Reproduced from Zhang, A., et al., *Appl. Phys. Lett.*, 93, 121110, 2008. With permission.)

23.5 CONCLUDING REMARKS AND PERSPECTIVE

Silicon nanowires are usually regarded as quasi-1D structures with a diameter in the nanometer range. In the past decades, SiNWs have exhibited critical device function and thus can be exploited as device elements in future architectures for micro- or nanosystems. SiNWs are also expected to play a leading role in serving as building blocks for novel micro-or nanoscale electronics assembled without the need for complex and costly fabrication facilities. This chapter reviews some major advances in the synthesis, physical and chemical characterization and properties, and application of SiNWs. Some representative properties covered include mechanical, thermal transport, optical, and optoelectronic properties, with emerging applications to the relevant fields. Despite of the significant development gained in characterizing and applying single or individual SiNW in the past decades, we anticipate that the future perspective will be focused more on the integration and construction of SiNW-based devices and systems, where both bottom-up and top-down approaches require the rational design and synergetic optimization of physical and chemical properties of SiNWs over different length scales.

REFERENCES

Aberle AG. (2001). Overview on SiN surface passivation of crystalline silicon solar cells. *Solar Energy Materials and Solar Cells.* 65: 239–248.

Adivarahan V, et al. (2000). SiO$_2$-passivated lateral-geometry GaN transparent Schottky-barrier detectors. *Applied Physics Letters.* 77: 863–865.

Arulkumaran S, Egawa T, Ishikawa H, Jimbo T, Sano Y. (2004). Surface passivation effects on AlGaN/GaN high-electron-mobility transistors with SiO$_2$, Si$_3$N$_4$, and silicon oxynitride. *Applied Physics Letters.* 84: 613–615.

Bao H, Ruan X. (2010). Optical absorption enhancement in disordered vertical silicon nanowire arrays for photovoltaic applications. *Optics Letters.* 35: 3378–3380.

Bao H, Zhang W, Chen L, Huang H, Yang C, Ruan X. (2012). An investigation of the optical properties of disordered silicon nanowire mats. *Journal of Applied Physics.* 112: 124301.

Barth S, Hernandez-Ramirez F, Holmes JD, Romano-Rodriguez A. (2010). Synthesis and applications of one-dimensional semiconductors. *Progress in Materials Science.* 55: 563–627.

Baxter JB, Aydil ES. (2005). Nanowire-based dye-sensitized solar cells. *Applied Physics Letters.* 86: 053114.

Bohren CF, Huffman DR. (1983). *Absorption and scattering of light by small particles.* New York, FL: Wiley.

Boukai AI, Bunimovich Y, Tahir-Kheli J, Yu J-K, Goddard WA, Heath JR. (2008). Silicon nanowires as efficient thermo-electric materials. *Nature.* 451: 168–171.

Broenstrup G, Jahr N, Leiterer C, Csaki A, Fritzsche W, Christiansen S. (2010). Optical properties of individual silicon nanowires for photonic devices. *ACS Nano.* 4: 7113–7122.

Brongersma ML, Cui Y, Fan S. (2014). Light management for photovoltaics using high-index nanostructures. *Nature Materials.* 13: 451–460.

Brönstrup G, Garwe F, Csáki A, Fritzsche W, Steinbrück A, Christiansen S. (2011). Statistical model on the optical properties of silicon nanowire mats. *Physical Review B.* 84: 125432.

Bruno M, Palummo M, Ossicini S, Del Sole R. (2007). First-principles optical properties of silicon and germanium nanowires. *Surface Science.* 601: 2707–2711.

Calaza C, et al. (2015). Bottom-up silicon nanowire arrays for thermoelectric harvesting. *Materials Today: Proceedings.* 2: 675–679.

Cao L, et al. (2010a). Semiconductor nanowire optical antenna solar absorbers. *Nano Letters.* 10: 439–445.

Cao L, Fan P, Barnard ES, Brown AM, Brongersma ML. (2010b). Tuning the color of silicon nanostructures. *Nano Letters.* 10: 2649–2654.

Cao L, White JS, Park J-S, Schuller JA, Clemens BM, Brongersma ML. (2009). Engineering light absorption in semi-conductor nanowire devices. *Nature Materials.* 8: 643–647.

Chen J, Zhang G, Li B. (2009). Tunable thermal conductivity of $Si_{1-x}Ge_x$ nanowires. *Applied Physics Letters.* 95: 073117.

Chen J, Zhang G, Li B. (2010). Remarkable reduction of thermal conductivity in silicon nanotubes. *Nano Letters.* 10: 3978–3983.

Chen Y, Li D, Lukes JR, Ni Z, Chen M. (2005). Minimum superlattice thermal conductivity from molecular dynamics. *Physical Review B.* 72: 174302.

Clifford JP, Konstantatos G, Johnston KW, Hoogland S, Levina L, Sargent EH. (2009). Fast, sensitive and spectrally tuneable colloidal quantum-dot photodetectors. *Nature Nanotechnology.* 4: 40–44.

Convertino A, Cuscunà M, Martelli F. (2010). Optical reflectivity from highly disordered Si nanowire films. *Nanotechnology.* 21: 355701.

Dai X, et al. (2014). Solution-processed, high-performance light-emitting diodes based on quantum dots. *Nature.* 515: 96–99.

Dan Y, Seo K, Takei K, Meza JH, Javey A, Crozier KB. (2011). Dramatic reduction of surface recombination by in situ surface passivation of silicon nanowires. *Nano Letters.* 11: 2527–2532.

Dasgupta NP, Yang P. (2014). Semiconductor nanowires for photovoltaic and photoelectrochemical energy conversion. *Frontiers of Physics.* 9: 289–302.

De Wolf S, Kondo M. (2007). Abruptness of a-Si : H/c-Si interface revealed by carrier lifetime measurements. *Applied Physics Letters.* 90: 042111.

Demichel O, Heiss M, Bleuse J, Mariette H, Morral AFI. (2010). Impact of surfaces on the optical properties of GaAs nanowires. *Applied Physics Letters.* 97: 201907.

Donadio D, Galli G. (2009). Atomistic simulations of heat transport in silicon nanowires. *Physical Review Letters.* 102: 195901.

Du QG, Kam CH, Demir HV, Yu HY, Sun XW. (2011). Broadband absorption enhancement in randomly positioned silicon nanowire arrays for solar cell applications. *Optics Letters.* 36: 1884–1886.

Dupré L, et al. (2013). Ultradense and planarized antireflective vertical silicon nanowire array using a bottom-up technique. *Nanoscale Research Letters.* 8: 123.

Fang X, Zhao CY, Bao H. (2014). Radiative behaviors of crystalline silicon nanowire and nanohole arrays for photovoltaic applications. *Journal of Quantitative Spectroscopy and Radiative Transfer.* 133: 579–588.

Feng XL, He R, Yang P, Roukes ML. (2007). Very high frequency silicon nanowire electromechanical resonators. *Nano Letters.* 7: 1953–1959.

Fenner DB, Biegelsen DK, Bringans RD. (1989). Silicon surface passivation by hydrogen termination: A comparative study of preparation methods. *Journal of Applied Physics.* 66: 419–424.

Fonseca L, et al. (2016). Smart integration of silicon nanowire arrays in all-silicon thermoelectric micro-nanogenerators. *Semiconductor Science and Technology.* 31: 084001.

Fujiwara H, Kondo M. (2007). Effects of a-Si : H layer thicknesses on the performance of a-Si : H/c-Si heterojunction solar cells. *Journal of Applied Physics.* 101: 054516.

Garnett E, Yang P. (2010). Light trapping in silicon nanowire solar cells. *Nano Letters.* 10: 1082–1087.

Garnett EC, Yang P. (2008). Silicon nanowire radial p-n junction solar cells. *Journal of the American Chemical Society.* 130: 9224–9225.

Gharghi M, Fathi E, Kante B, Sivoththaman S, Zhang X. (2012). Heterojunction silicon microwire solar cells. *Nano Letters.* 12: 6278–6282.

Griffiths DJ. (2004). *Introduction to quantum mechanics.* Upper Saddle River, NJ: Pearson Prentice Hall.

Hasan M, Huq MF, Mahmood ZH. (2013). A review on electronic and optical properties of silicon nanowire and its different growth techniques. *Springerplus.* 2: 151.

Hayden O, Agarwal R, Lieber CM. (2006). Nanoscale avalanche photodiodes for highly sensitive and spatially resolved photon detection. *Nature Materials.* 5: 352–356.

He R, Yang P. (2006). Giant piezoresistance effect in silicon nanowires. *Nature Nanotechnology.* 1: 42–46.

Heidelberg A, et al. (2006). A generalized description of the elastic properties of nanowires. *Nano Letters.* 6: 1101–1106.

Hochbaum AI, et al. (2008). Enhanced thermoelectric performance of rough silicon nanowires. *Nature.* 451: 163–167.

Hoex B, Gielis JJH, van de Sanden MCM, Kessels WMM. (2008). On the c-Si surface passivation mechanism by the negative-charge-dielectric Al_2O_3. *Journal of Applied Physics.* 104: 113703.

Hoffmann S, et al. (2006). Measurement of the bending strength of vapor-liquid-solid grown silicon nanowires. *Nano Letters.* 6: 622–625.

Holmes JD, Johnston KP, Doty RC, Korgel BA. (2000). Control of thickness and orientation of solution-grown silicon nanowires. *Science.* 287: 1471–1473.

Hu L, Chen G. (2007). Analysis of optical absorption in silicon nanowire arrays for photovoltaic applications. *Nano Letters.* 7: 3249–3252.

Hu M, Giapis KP, Goicochea JV, Zhang X, Poulikakos D. (2011a). Significant reduction of thermal conductivity in Si/Ge core-shell nanowires. *Nano Letters.* 11: 618–623.

Hu M, Poulikakos D. (2012). Si/Ge superlattice nanowires with ultralow thermal conductivity. *Nano Letters.* 12: 5487–5494.

Hu M, Zhang X, Giapis KP, Poulikakos D. (2011b). Thermal conductivity reduction in core-shell nanowires. *Physical Review B.* 84: 085442.

Ip AH, et al. (2012). Hybrid passivated colloidal quantum dot solids. *Nature Nanotechnology.* 7: 577–582.

Jeong S, McGehee MD, Cui Y. (2013). All-back-contact ultra-thin silicon nanocone solar cells with 13.7% power conversion efficiency. *Nature Communications.* 4: 2950.

Ju S, Liang XJ. (2012). Thermal conductivity of nanocrystalline silicon by direct molecular dynamics simulation. *Journal of Applied Physics.* 112: 064305.

Kang K, Cai W. (2007). Brittle and ductile fracture of semiconductor nanowires-molecular dynamics simulations. *Philosophical Magazine.* 87: 2169–2189.

Kayes BM, Atwater HA, Lewis NS. (2005). Comparison of the device physics principles of planar and radial p-n junction nanorod solar cells. *Journal of Applied Physics.* 97: 114302.

Kelzenberg MD, et al. (2008). Photovoltaic measurements in single-nanowire silicon solar cells. *Nano Letters.* 8: 710–714.

Kim DR, Lee CH, Rao PM, Cho IS, Zheng X. (2011). Hybrid Si microwire and planar solar cells: Passivation and characterization. *Nano Letters.* 11: 2704–2708.

Kim Y, Jeon Y, Kim S. (2015). Flexible silicon nanowire low-power ring oscillator featuring one-volt operation. *Microelectronic Engineering.* 145: 120–123.

Konstantatos G, et al. (2006). Ultrasensitive solution-cast quantum dot photodetectors. *Nature.* 442: 180–183.

Krogstrup P, et al. (2013). Single-nanowire solar cells beyond the Shockley-Queisser limit. *Nature Photonics.* 7: 306–310.

LaPierre RR, et al. (2013). III-V nanowire photovoltaics: Review of design for high efficiency. *Physica Status Solidi-Rapid Research Letters.* 7: 815–830.

Lee J-H, Galli GA, Grossman JC. (2008). Nanoporous Si as an efficient thermoelectric material. *Nano Letters.* 8: 3750–3754.

Lee S, et al. (2012). Ultra-high responsivity, silicon nanowire photodetectors for retinal prosthesis. *Paper presented at 25th IEEE International Conference on Micro Electro Mechanical Systems (MEMS)*, Paris, France, January 29–February 2.

Li D, Wu Y, Kim P, Shi L, Yang P, Majumdar A. (2003). Thermal conductivity of individual silicon nanowires. *Applied Physics Letters.* 83: 2934.

Li J, Yu H, Li Y. (2012). Solar energy harnessing in hexagonally arranged Si nanowire arrays and effects of array symmetry on optical characteristics. *Nanotechnology.* 23: 194010.

Li X, Maute K, Dunn M, Yang R. (2010). Strain effects on the thermal conductivity of nanostructures. *Physical Review B.* 81: 245318.

Li Y, Chen Q, He D, Li J. (2014). Radial junction Si micro/nano-wire array photovoltaics: Recent progress from theoretical investigation to experimental realization. *Nano Energy*. 7: 10–24.

Lin C, Huang N, Povinelli ML. (2012). Effect of aperiodicity on the broadband reflection of silicon nanorod structures for photovoltaics. *Optics Express*. 20: A125–A132.

Lin C, Povinelli ML. (2009). Optical absorption enhancement in silicon nanowire arrays with a large lattice constant for photovoltaic applications. *Optics Express*. 17: 19371–19381.

Luckyanova MN, et al. (2012). Coherent phonon heat conduction in superlattices. *Science*. 338: 936–939.

Ma DDD, Lee CS, Au FCK, Tong SY, Lee ST. (2003). Small-diameter silicon nanowire surfaces. *Science*. 299: 1874–1877.

Martin P, Aksamija Z, Pop E, Ravaioli U. (2009). Impact of phonon-surface roughness scattering on thermal conductivity of thin Si nanowires. *Physical Review Letters*. 102: 125503.

Matsuo N, Ohno H, Hasegawa H. (1984). Mechanism of high gain in GaAs photoconductive detectors under low excitation. *Japanese Journal of Applied Physics*. 23: L299–L301.

Menon M, Srivastava D, Ponomareva I, Chernozatonskii LA. (2004). Nanomechanics of silicon nanowires. *Physical Review B*. 70: 125313.

Mikolajick T, et al. (2013). Silicon nanowires-a versatile technology platform. *Physica Status Solidi-Rapid Research Letters*. 7: 793–799.

Mingo N, Yang L. (2003). Predicting the thermal conductivity of Si and Ge nanowires. *Nano Letters*. 3: 1713–1716.

Murphy KF, Piccione B, Zanjani MB, Lukes JR, Gianola DS. (2014). Strain-and defect-mediated thermal conductivity in silicon nanowires. *Nano Letters*. 14: 3785–3792.

Murphy PG, Moore JE. (2007). Coherent phonon scattering effects on thermal transport in thin semiconductor nanowires. *Physical Review B*. 76: 155313.

Muskens OL, et al. (2009). Large photonic strength of highly tunable resonant nanowire materials. *Nano Letters*. 9: 930–934.

Nannichi Y, Jia-Fa F, Oigawa H, Koma A. (1988). A model to explain the effective passivation of the GaAs surface by $(NH_4)_2S_x$ treatment. *Japanese Journal of Applied Physics*. 27: 2367–2369.

Neamen DA. (2011). *Semiconductor physics and devices: Basic principles*. Fourth Edition in Beijing. FL: Publishing House of Electronics Industry, p. 634 SE–634, Beijing, China.

Nichol JM, Hemesath ER, Lauhon LJ, Budakian R. (2012). Nanomechanical detection of nuclear magnetic resonance using a silicon nanowire oscillator. *Physical Review B*. 85: 054414.

Nolan M, O'Callaghan S, Fagas G, Greer JC, Frauenheim T. (2007). Silicon nanowire band gap modification. *Nano Letters*. 7: 34–38.

Rashid JIA, Abdullah J, Yusof NA, Hajian R. (2013). The development of silicon nanowire as sensing material and its applications. *Journal of Nanomaterials*. 3: 1–16.

Rey BM, et al. (2016). Fully tunable silicon nanowire arrays fabricated by soft nanoparticle templating. *Nano Letters*. 16: 157–163.

Rurali R. (2010). Colloquium: Structural, electronic, and transport properties of silicon nanowires. *Reviews of Modern Physics*. 82: 427–449.

Sacchetto D, Ben-Jamaa MH, Micheli GD, Leblebici Y. (2009). Fabrication and characterization of vertically stacked gate-all-around Si nanowire FET arrays. *IEEE Proceedings European Solid State Device Research Conference ESSDERC*, pp. 245–248, Athens, Greece.

San Paulo A, et al. (2005). Mechanical elasticity of single and double clamped silicon nanobeams fabricated by the vapor-liquid-solid method. *Applied Physics Letters*. 87: 053111.

Soci C, Zhang A, Bao XY, Kim H, Lo Y, Wang DL. (2010). Nanowire photodetectors. *Journal of Nanoscience and Nanotechnology*. 10: 1430–1449.

Soci C, et al. (2007). ZnO nanowire UV photodetectors with high internal gain. *Nano Letters*. 7: 1003–1009.

Song J, Wang X, Riedo E, Wang, ZL. (2005). Elastic property of vertically aligned nanowires. *Nano Letters*. 5: 1954–1958.

Street RA, Wong WS, Paulson C. (2009). Analytic model for diffuse reflectivity of silicon nanowire mats. *Nano Letters*. 9: 3494–3497.

Tabib-Azar M, et al. (2005). Mechanical properties of self-welded silicon nanobridges. *Applied Physics Letters*. 87: 113102.

Tan SL, Zhao X, Chen K, Crozier KB, Dan Y. (2016). High performance silicon nanowire bipolar phototransistors. *Applied Physics Letters*. 109: 033505.

Tan SL, Zhao X, Dan Y. (2014). High-sensitivity silicon nanowire phototransistors. *Nanoengineering: Fabrication, Properties, Optics, and Devices XI*. 9170: 917002.

Tian B, et al. (2007). Coaxial silicon nanowires as solar cells and nanoelectronic power sources. *Nature*. 449: 885–889.

Tsakalakos L, Balch J, Fronheiser J, Korevaar BA, Sulima O, Rand J. (2007). Silicon nanowire solar cells. *Applied Physics Letters*. 91: 233117.

Wagner RS, Ellis WC. (1964). Vapor-liquid-solid mechanism of single crystal growth. *Applied Physics Letters*. 4: 89–90.

Wallentin J, et al. (2013). InP nanowire array solar cells achieving 13.8% efficiency by exceeding the ray optics limit. *Science*. 339: 1057–1060.

Wang S, Liang X, Xu X, Ohara T. (2009). Thermal conductivity of silicon nanowire by nonequilibrium molecular dynamics simulations. *Journal of Applied Physics*. 105: 014316.

Wang Y, Liebig C, Xu X, Venkatasubramanian R. (2010). Acoustic phonon scattering in Bi_2Te_3/Sb_2Te_3 Bi_2Te_3/Sb_2Te_3 superlattices. *Applied Physics Letters*. 97: 083103.

Wingert MC, Kwon S, Hu M, Poulikakos D, Xiang J, Chen R. (2015). Sub-amorphous thermal conductivity in ultra-thin crystalline silicon nanotubes. *Nano Letters*. 15: 2605–2611.

Wu B, Heidelberg A, Boland JJ. (2005). Mechanical properties of ultrahigh-strength gold nanowires. *Nature Materials*. 4: 525–529.

Yang B, Chen G. (2003). Partially coherent phonon heat conduction in superlattices. *Physical Review B*. 67, 195311.

Yang FL, et al. (2004). 5 nm-Gate nanowire FinFET. *IEEE Symposium VLSI Technology*. 196–197, Honolulu, HI

Yang L, Yang N, Li B. (2014). Extreme low thermal conductivity in nanoscale 3D Si phononic crystal with spherical pores. *Nano Letters*. 14: 1734–1738.

Yang N, Zhang G, Li B. (2008). Ultralow thermal conductivity of isotope-doped silicon nanowires. *Nano Letters*. 8: 276–280.

Yang R, Chen G. (2005). Thermal conductivity modeling of core-shell and tubular nanowires. *Nano Letters*. 5: 1111–1115.

Yeh C, Shimabukuro FI. (2008). *The essence of dielectric waveguide*. New York, FL: Springer Science + Business Media.

Yu M, Long YZ, Sun B, Fan Z. (2012). Recent advances in solar cells based on one-dimensional nanostructure arrays. *Nanoscale*. 4: 2783–2796.

Zhang A, Kim H, Cheng J, Lo YH. (2010). Ultrahigh responsivity visible and infrared detection using silicon nanowire phototransistors. *Nano Letters*. 10: 2117–2120.

Zhang A, You S, Soci C, Liu Y, Wang D, Lo YH. (2008). Silicon nanowire detectors showing phototransistive gain. *Applied Physics Letters*. 93: 121110.

Zhang X, Hu M, Giapis KP, Poulikakos D. (2012). Schemes for and mechanisms of reduction in thermal conductivity in nanostructured thermoelectrics. *Journal of Heat Transfer*. 134: 102402.

Zhou Y, Chen Y, Hu M. (2016). Strong surface orientation dependent thermal transport in Si nanowires. *Scientific Reports*. 6: 24903.

Fundamentals of silicon nanotubes

Neda Ahmadi

Contents

24.1 INTRODUCTION

After the discovery of carbon nanotubes (CNTs) by Iijima (1991) scientists predicted other kinds of nanotubes, such as Gallium nitride (GaN), Boron nitride (BN), Aluminium nitride (AlN) (Goldberger et al. 2003; Xi et al. 2006; Zhukovskii et al. 2006), and silicon nanotubes (SiNTs) (Fagan et al. 2000) while silicon nanowires with SP^3 hybridization had been reported as a one-dimensional silicon nanostructure for many years (Namastsu et al. 1995; Ono et al. 1997; Leobandung et al. 1997; Tang et al. 2000; Wu et al. 2001; Zhang et al. 2001; Schi et al. 2001).

From a chemical point of view, silicon forms four covalent σ bonds with SP^3 hybridization similar to other kinds of elements in GGup four of the periodic table. Unlike carbon which is in group four and has a strong π bond with SP^2 hybridization, the π bond in a silicon compound is weak (about 25 Kcal/mol) compared with carbon (about 60 Kcal/mol) (Dmitrii et al. 2006).

Scientists investigated the method of fabrication of SiNTs with a diameter of more than 50 nm for the first time in 2002 (Sha et al. 2002; Jeong et al. 2003; Chen et al. 2005; Tang et al. 2005). The experimental evidences show the fabrication of SiNTs with a small diameter of about 2 nm in 2005 (Crescenzi et al. 2005; Castrucci et al. 2006), and with two kinds of hybridization (SP²/SP³) (Castrucci et al. 2012).

Recently, single and multiwalled SiNTs have been discovered and open the door to the possibility that a graphene-like sheet with SP² hybridization can be obtained.

A silicon sheet with SP² hybridization and two kind of atoms, –A and –B type, was named silicene and was reported in experimental evidence (Kara et al. 2009; De Padova et al. 2011). Scanning tunneling microscopy (STM) and angle-resolved photomission spectroscopy (ARPES) confirmed the synthesis of silicon sheet (Vogt et al. 2012).

Theoretical studies (Yang and Ni 2005; Guzmán-Verri et al. 2007) show that the single-walled SiNTs can be classified according to their hybridization into two categories—silicon hexagonal nanotubes (Si h-NTs) and silicon gearlike nanotubes (Si h-NTs). These are formed by rolling up silicene and Si(111) sheet. The Si(111) has a structure similar to silicene except that the B atoms are vertically displaced due to SP³ hybridization.

With respect to the eminence of SiNTs in modern microelectronics, study about these kind of nanostructures is an important discipline. This chapter is organized as follows: after a brief overview of SiNTs we explain about possible structures of SiNTs, then the electronic and optical properties of SiNTs will be defined. In Section 24.4, fabrication methods are detailed, and in the last section the application of SiNTs in lithium-ion batteries (LIBs) will be explained.

24.1.1 POSSIBLE STRUCTURE OF SILICON NANOTUBES

The tubular silicon nanostructure such as SiNTs had been previously investigated theoretically according to the density-functional theory (DFT) (Bai et al. 2003). These previous studies depend on Hartree-Fock (Zhang et al. 2002), Møller-Plesset perturbation (MP2) and electron correlation (Bai et al. 2003), semi-empirical (Zhang et al. 2002), quantum chemical calculation, and molecular dynamics simulations (Bai et al. 2003). These studies show that for one-dimensional silicon nanostructures the SP³ hybridization is more stable than SP² (Guzmán-Verri et al. 2007). Furthermore, the theoretical studies have predicted different structures for SiNTs. Figure 24.1 illustrates some different structures of SiNTs (Dmitrii et al. 2006). Figure 24.1a and b show the tubular structure with SP² and SP³ hybridization, respectively. Figure 24.1c shows a completely different model for tubular structures based on a tetragon of SP³ hybridization silicon atoms. One of the most interesting tubular structures of silicon is ladder-type oligo (cyclotetrasilane)s. This structure, which is shown in Figure 24.1d, is more stable than tubular structures with SP² hybridization (Dmitrii et al. 2006).

Guzmán-Verri et al. have investigated the structure and electronic properties of silicon nanostructures such as silicon graphene-like sheet and SiNTs; they used the tight-binding Hamiltonian (Guzmán-Verri et al. 2007).

Here we briefly explain about the structure of silicene according to the tight-binding model. Silicene is a two-dimensional (2D) silicon nanostructure with a honeycomb lattice, SP² hybridization, and lattice constant a. In this lattice, we can see two kinds of atom: A and B. All are located in the plane. The single-walled Si h-NTs are formed by rolling up this kind of lattice.

Another kind of 2D silicon nanostructure is Si(111) with SP³ hybridization, and honeycomb lattice. In this structure, we have two kinds of atom as in silicene except that the atoms labeled A are in the xy plane and all the B atoms are located below, the plane at $z = -\dfrac{a}{2\sqrt{6}}$. Si h-NTs are formed by rolling up Si(111). The basic vectors of these lattices are $a_1 = \dfrac{a}{2}(\sqrt{3}, -1)$ and $a_2 = \dfrac{a}{2}(\sqrt{3}, 1)$. The basic vectors and unit cell of these lattices are similar to graphene sheet. If the distance between Si–Si, is l then $a = \dfrac{1}{\sqrt{2}}$. The position vectors of Si(111) sheet for the nearest neighbor (1NN) are (Guzmán-Verri et al. 2007):

$$\delta_1^{(1)} = \left(\frac{a}{\sqrt{3}}, 0, \frac{-a}{2\sqrt{6}} \right), \ \delta_2^{(1)} = \left(\frac{-a}{2\sqrt{3}}, \frac{a}{2}, \frac{-a}{2\sqrt{6}} \right), \ \delta_3^{(1)} = \left(\frac{-a}{2\sqrt{3}}, \frac{-a}{2}, \frac{-a}{2\sqrt{6}} \right), \quad (24.1)$$

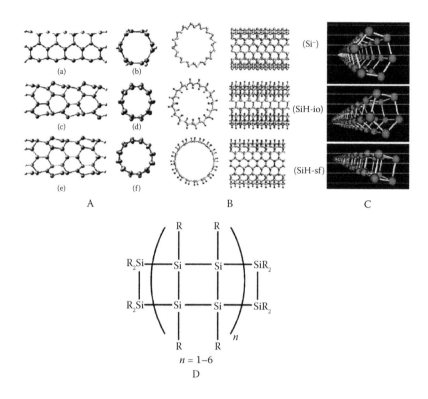

Figure 24.1 (A) (a) An sp² armchare (3,3) SiNT, (Si$_{54}$H$_{12}$) with axis along the plane of the page. (b) The top view of (3,3) nanotube. (c) Zigzag (5,0) SiNT (Si$_{50}$H$_{10}$), with axis along the plane of page. (d) Top view of (5,0). (e) Zigzag (6,0) SiNT (Si$_{60}$H$_{10}$), with axis along the plane of page. (f) Top view of (6,0). (B) sp³ SiNT, hydrogen terminated. (C) sp³ tetragonal SiNT with different diameters. (D) The structure of oligo (cytrotetrasilane)s. (From Dmitrii et al., *Small*, 2:22–25, 2006. With permission.)

and for second-nearest neighbor (2NN)

$$\delta_1^{(2)} = (0, a, 0), \; \delta_2^{(2)} = (0, -a, 0), \; \delta_3^{(2)} = \left(\frac{a\sqrt{3}}{2}, \frac{-a}{2}, 0 \right) \delta_4^{(2)} = \left(\frac{-a\sqrt{3}}{2}, \frac{a}{2}, 0 \right),$$

$$\delta_5^{(2)} = \left(\frac{a\sqrt{3}}{2}, \frac{a}{2}, 0 \right), \; \delta_6^{(2)} = \left(\frac{-a\sqrt{3}}{2}, \frac{-a}{2}, 0 \right). \tag{24.2}$$

The position vector of silicene are similar to Si(111) except that there is no Z component. Thus we can get the position vector of silicene by making the Z component of the above equations equal to zero.

Guzmán-Verri et al. investigated the band structure of silicene and Si(111) using the orthogonal tight-binding model for (1NN) and (2NN) approximation. They used 10 × 10 and 8 × 8 Hamiltonian for silicene and Si(111), respectively.

The analytical equation for the dispersion relation of Si(111) sheet considering second-nearest neighbor at the Γ point is given as (Guzmán-Verri et al. 2007)

$$E_{p+}(\Gamma) = E_P + 3\left[(PP\sigma)_2^{AA} + (PP\pi)_2^{AA} \right] + \frac{1}{3}[4(PP\sigma)_1^{AB} + 5(PP\pi)_1^{AB}],$$

$$E_{p-}(\Gamma) = E_P + 3\left[(PP\sigma)_2^{AA} + (PP\sigma)_2^{AA} \right] - \frac{1}{3}[4(PP\sigma)_1^{AB} + 5(PP\pi)_1^{AB}], \tag{24.3}$$

where $Ep\pm$ is twofold degenerate.

Table 24.1 Silicon two-center parameters obtained from Vogl et al. and Grosso and Piermarocchi blank

PARAMETER	VOGL ET AL.	GROSSO AND PIERMAROCCHI
Es	–4.2000	–4.0497
Ep	1.7150	1.0297
Es*	6.6850	
$(ss\sigma)_1^{AB}$	–2.0750	–2.0662
$(sp\sigma)_1^{AB}$	2.4808	2.0850
$(pp\sigma)_1^{AB}$	2.7163	3.1837
$(pp\pi)_1^{AB}$	–0.7150	–0.9488
$(s^*p\sigma)_{AB}^1$	2.3274	
$(ss\sigma)_2^{AA}$		0.0000
$(sp\sigma)_2^{AA}$		0.0000
$(pp\sigma)_2^{AA}$		0.8900
$(pp\pi)_2^{AA}$		–0.3612

Table 24.1 shows parameters which correspond to Vogl et al. (SP^{3}S*) (Vogl et al. 1983) and to Grosso and Piermarocchi SP^3 (Grosso and Piermarocchi 1995), which can be used in the above equation. One can use the above equation for silicene by making the z component of 1NN equal to zero and substituting the appropriate direct cosines of the 1NN position vectors. In this case, the π and σ band are decoupled and the bands become independent. For the π band we can use Equation 24.4; this equation comes from the well-known 2×2 tight-binding Hamiltonian considering 2NN approximation (Equation 24.11)

$$E(K)=E_P + (PP\pi)_2^{AA}\, g_{25}(K) \pm (PP\pi)_1^{AB}\,\omega(K),\qquad(24.4)$$

where

$$\omega(K)=|g_{12}|=\sqrt{\left(1+4\cos\frac{\sqrt{3}k_x a}{2}\cos\frac{k_y a}{2}+4\cos^2\left(\frac{k_y a}{2}\right)\right)},\qquad(24.5)$$

here $g_{25}(k)$ and $g_{12}(k)$ are given by

$$\begin{cases} g_{25}(K)= e^{ik.\delta_1^{(2)}} + e^{ik.\delta_2^{(2)}} + e^{ik.\delta_3^{(2)}} + e^{ik.\delta_4^{(2)}} + e^{ik.\delta_5^{(2)}} + e^{ik.\delta_6^{(2)}}, \\ f(k)=g_{12}(K)= e^{ik.\delta_1^{(1)}} + e^{ik.\delta_2^{(1)}} + e^{ik.\delta_3^{(1)}}. \end{cases}\qquad(24.6)$$

Here $pp\pi_1^{AB} = \gamma_0 = -0.9488\ eV$ and $(pp\pi)_2^{AA} = \gamma_1 = -0.3612$ are the first and the second-nearest neighbor transfer internal, respectively (Guzmán-Verri et al. 2007). By considering the boundary condition on the K in Equation 24.4 one can obtain the band structure of nanotubes (Saito et al. 1998).

Figure 24.2 shows the band structure of silicene and Si(111) according to the tight-binding models (Guzmán-Verri et al. 2007) and *ab initio* (Yang and Ni 2005). Figure 24.2a and b shows the band structure of silicene considering 1NN (SP^{3}S*) and 2NN (SP^3), respectively. The π and σ band in silicene are decoupled similar to graphene due to the planar and orbital symeteries (Guzmán-Verri et al. 2007). If we compare the band structure of silicene from the 1NN model with the band structure of graphene, we will understand they have a similar form. This occurs due to silicene and graphene have the same lattice except

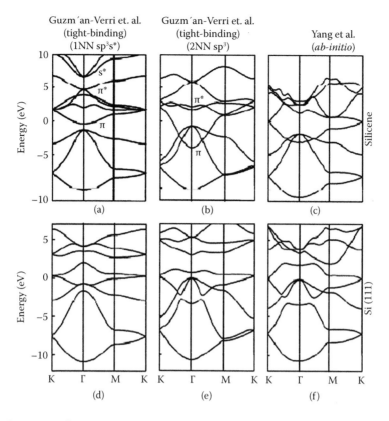

Figure 24.2 Band structure of silicene and Si(111) according to tight-binding model and compared with *ab initio*. (From Guzmn-Verri et al., *Phys. Rev. B*, 76:075131–075140, 2007. With permission.)

that the σ valence bands in silicene have been lowered, therefore the crossing between σ and π bands does not occur in silicene unlike graphene.

Using the SP³ model (2NN) causes the different results in silicene. Figure 24.2b shows the silicene band structure considering 2NN approximation. In contrast, in Figure 24.2a, the π and σ valence bands do cross. Figure 24.2c shows the band structure of silicene which was obtained by Yang and coworker (Yang and Ni 2005). There are clear differences between the 1NN approximation and the *ab initio* model. For example, at the Γ point in Figure 24.2a, conduction and valence bands have the opposite curvature unlike Yang's result in Figure 24.2c.

Figure 24.2d and e show the band structure of Si(111) for SP³S* (1NN) and SP³ (2NN) models, respectively. The SP³ hybridization causes a coupling between the π and σ states while the SP³S* model causes different results especially at the Γ point. When the band approaches this point according to the SP³ (2NN) model, the curvature of some of them changes. The differences might be because of second-nearest neighbor interaction.

The band structure of Si(111) that was computed by Yang et al. is shown in Figure 24.2f. We can see good agreement between this model and the SP³ model especially along KΓM directions. On the other hand, most of the differences occur between SP³ and the *ab initio* model along the MK direction.

24.2 ELECTRONIC PROPERTIES OF SILICON NANOTUBES

Guzmán-Verri and coworkers used the orthogonal tight-binding model and have calculated the band structure of Si h-NTs and Si g-NTs (Guzmán-Verri et al. 2007). According to these calculations, there are 2N bands in the band structure of Si h- NTs under consideration π band. The number of 2 is due to 2 × 2 Hamiltonian and the number of N is for quantization of the wave vector K. These calculations show that

Si h-NTs follow the Hamada's rule (Hamada et al. 1992) similar to CNTs and have semiconducting and metallic behaviors. These results are agreement with Ahmadi and coworkers' results (Ahmadi et al. 2016).

According to Hamada's rule the zigzag nanotube (n,m = 0) is a conductor if n is a multiple of 3 and otherwise are semiconductors. Furthermore, all armchair nanotubes (n = m) are conductors.

Due to using 8 × 8 Hamiltonian for Si g-NTs, the number of bands in this kind of nanotubes is 8N bands. On the other hand, in Si g-NTs for each Si atoms there are four electrons and a total of 2N atoms therefore each unit cell has 8N electrons (N comes from the quantization of wave vector K). For this reason, one can see that there is a proliferation of the number of bands in the Si g-NTs compared with Si h-NTs.

Figure 24.3 shows the band structure of zigzag and armchair nanotubes according to Guzmán-Verri et al. (Guzmán-Verri et al. 2007) and Yang et al. (Yang and Ni 2005) calculations. The dashed line corresponds to the Fermi level. The left-hand panels show the band structure of (4,4) and (6,6) armchair Si h-NT. It is obvious the Si h-NTs (4,4) and (6,6) are conductors. Furthermore it is clear that (8,0) zigzag Si h-NT behaves similar to a semiconducting nanotube because at the Γ point there is a band gap. On the other hand, (12,0) zigzag Si h-NT is conductor because the band gap at the Γ point is zero. Thus the band structure as shown in the furthermost left panel confirms Hamada's rule.

The middle panels show the band structure of armchair and zigzag Si g-NTs. It is clear that (4,4) and (6,6) armchair Si g-NTs are conductors and zigzag Si g-NT (8,0) behaves similarly to a semiconductor due to the band gap at the Γ point. Furthermore, zigzag Si g-NT (12,0) is a conductor according to these calculations.

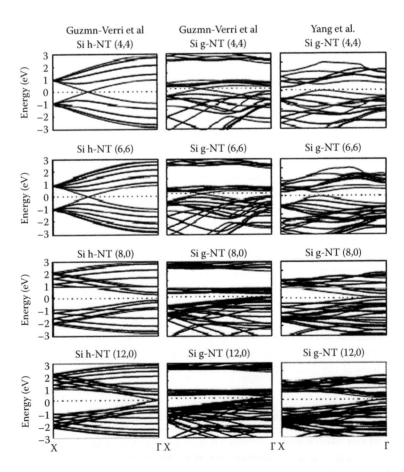

Figure 24.3 Band structure of Si h-NTs and Si g-NTs according to tight-binding and *ab-initio* calculations. (From Guzmn-Verri et al., *Phys. Rev. B*, 76:075131–075140, 2007. With permission.)

Thus these results confirm Hamada's rule and are not in agreement with Yang et. al.'s calculations (Yang and Ni 2005).

Yang et al. reported that Si(n,0) h -NTs with n varying from 5 to 11 are metal, unlike CNTs and Si (n,0) h-NTs with n varying from 12 to 18 which are metal for 12, 15, and 18. They also show that armchair Si (n,n) g-NTs for n = (3–11) are semiconductors whose gap decreases when the diameter increases. Moreover, they showed that zigzag Si g-NTs for n = (12–24) are semiconductors, and for n = (5–9) they are metal. In additional, these results show that Si NTs do not obey Hamada's rule. The right-hand panels of Figure 24.3 show the band structure of armchair and zigzag Si g-NTs according to Yang et al. with calculations based on the first-principle method. It is clear that armchair Si g-NT (4,4) and (6,6) are semiconductors while zigzag Si g-NT (8,0) is a conductor. Moreover, zigzag Si g-NT (12,0) is semiconductor and a gap opens at the Γ point—these are absolutely the opposite of Guzmán-Verri and coworkers' results.

24.2.1 ELECTRONIC PROPERTIES OF HYDROGENATED Si NANOTUBES

Tight-binding calculations have also been used to investigate the band structure of fully hydrogenated silicon nanotubes (Si h-NTs) (Guzmán-Verri et al. 2011). These results represent that for some chiralities, hydrogenated silicon nanotubes show a new type of semiconductor nanotube with existing direct and indirect band gaps.

Figure 24.4 shows the band structure of armchair (n,n) and zigzag (n,0) nanotubes with even and odd *n*. For armchair and zigzag nanotubes with *n* odd in Figure 24.4b and d, it is clear there is a direct gap at the Γ point. This gap equals the hydrogenated Si sheet ($\cong$2.2 eV) (Guzmán-Verri et al. 2011). Furthermore, for armchair and zigzag nanotubes with *n* even (Figure 24.4a and c) it is obvious that there are direct and indirect band gaps at the Γ and X points at the same energy ($\cong$2.2 eV). In addition, Si h-NTs do not follow Hamada's rule. These results are not agreement with the density-functional tight-binding (DFTB) method (Seifert et al. 2001).

The DFTB shows that Si h-NTs are semiconductors whose band gap depends on the chirality and equals 2.5 eV, and it is not clear whether the band gap is direct or indirect.

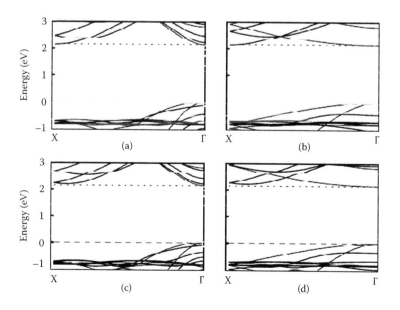

Figure 24.4 Band structure of (a) armchair (6,6), (b) armchair (7,7), (c) zigzag (6,0), (d) zigzag (7,0) hydrogenated Si nanotubes. (From Guzman-Verri GG et al., *J. Phys. Condens. Matter*, 23:145502–145507, 2011. With permission.)

24.2.2 ELECTRONIC PROPERTIES OF NANOTUBE UNDER INTRINSIC CURVATURE EFFECT

Ahmadi et al. (Ahmadi et al. 2016) applied the tight-binding approach to obtain the band structure of zigzag Si h-NTs under intrinsic curvature effect. The band structure and wave function are obtained by using tight-binding Hamiltonian. A one-dimensional wave vector is considered in the first Brillouin zone. The electron energy spectrum of silicene sheet is similar to graphene and is obtained by (Saito et al. 1998)

$$E_l(k_z) = st\sqrt{|f(k_z)|},$$ (24.7)

where

$$f(k_z) = \sum_{i=1}^{3} t_i e^{ik_z \cdot u_i}.$$ (24.8)

Here u_i is the position vectors for the ith nearest neighbors and t_i is the ith transfer integral, and the direction of nanotube axis is z. For the silicene sheet $t_i = t = -0.949$ eV (Guzmán-Verri et al. 2007). $s = \pm 1$ corresponds to the conduction and valence band. We can use Equation 24.7 for obtaining the energy spectrum of zigzag Si h-NTs by using the periodic boundary condition. Note, however, during the rolling up of a silicene sheet u_1 remains constant, but u_2 and u_3 are reduced. The modified position vectors for the nearest neighbor are $\tilde{u}_1 = \left(\dfrac{a}{\sqrt{3}}, 0\right)$, $\tilde{u}_2 = \left(\dfrac{-a}{2\sqrt{3}}, -a_y\right)$ and $\tilde{u}_3 = \left(\dfrac{-a}{2\sqrt{3}}, a_y\right)$. Here $a_y = \sqrt{A_1^2 + A_2^2}$ where

$$|A_1| = R\sin\frac{a}{2R} \text{ and } |A_2| = R\left[1 - \cos(\frac{a}{2R})\right]$$ "R" and "a" indicate the radius of the nanotube and lattice

constant, respectively. The transfer integral modifies due to the curvature. There is a relationship between the original t_i and modified $\tilde{t}_i$ transfer integral via (Harrison 1989)

$$\frac{\tilde{t}_i}{t_i} = \left(\frac{|u_i|}{|\tilde{u}_i|}\right)^2.$$ (24.9)

With substituting Equation 24.9 in Equations 24.7 and 24.8, the modified electron energy spectrum for Si h-NTs is given by (Ahmadi et al. 2016)

$$E_l(k_z) = s|t|\sqrt{\left(1 + 4\frac{\tilde{t}_2}{t}\cos\left(\frac{\sqrt{3}k_z a}{2}\right)\cos\left(k_y a_y\right) + 4\left(\frac{\tilde{t}_2}{t}\right)^2 \cos^2\left(k_y a_y\right)\right)}$$ (24.10)

Using the periodic boundary conditions on K along the chiral direction for zigzag nanotubes indicates $k_y = \dfrac{2\pi l}{na}$ (Saito et al. 1998). l is an integer number which takes value 1,2, … 2n for valence and conduction bands thus k_y is a quantized parameter and "a" is the lattice constant. For all zigzag silicon tubes, without applying the curvature effect the band with index l and $(2n-l)$ are degenerated except for the cases $l = n$ and 2n. Equation 24.10 strongly depends on the value of the lattice constant. For this reason, our numerical calculation shows that the band degeneracy is broken in zigzag Si h-NTs under the intrinsic curvature effect while the band degeneracy is not broken in CNTs. Moreover, our numerical results show that zigzag Si h-NTs follow Hamada's rule and these results are in agreement with Guzmán-Verri and coworkers' results (Guzmán-Verri et al. 2007). Figure 24.5a and c show the band structure of (6,0) and (8,0) as a conductor and semiconductor zigzag nanotubes, respectively, without curvature effect. In Figure 24.5b, applying the curvature effect for (6,0) causes a direct band gap to open at the Γ point, thus we have a metallic-semiconductor phase transition. Figure 24.5d shows in contrast to (6,0), the curvature effect decreases the band gap of (8,0) as a semiconductor tube. These results are agreement with previous works (Popov 2004). Figure 24.5b and d illustrate the breaking in the band degeneracy. This effect is related to symmetry breaking which is resulted from Equation 24.10.

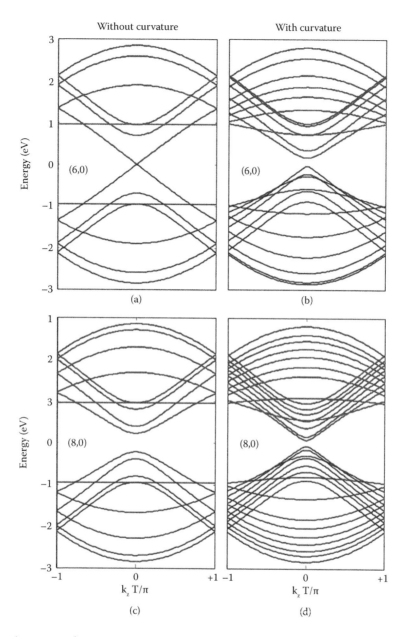

Figure 24.5 Band structure of SiH-NTs (6,0) and (8,0) with and without including the curvature effect in the first Brillouin zone. (Ahmadi et al. 2016).

24.2.3 TIGHT-BINDING DESCRIPTION OF ELECTRONIC PROPERTIES OF SILICON NANOTUBE UP TO SECOND-NEAREST NEIGHBOR

Shokri and coworker (Shokri and Ahmadi 2015) investigated the dispersion relation of SiNTs using the tight-binding method by applying the second-nearest neighbor interaction, 2NN, for the π electrons. They used the well-known 2×2 Hamiltonian for 2NN interaction

$$H(\mathbf{K}) = \begin{bmatrix} E_p + (pp\pi)_2^{AA} g_{25}(\mathbf{k}) & \gamma_0 g_{12}(\mathbf{k}) \\ \gamma_0 g_{12}^*(\mathbf{k}) & E_p + (pp\pi)_2^{AA} g_{25}(\mathbf{k}) \end{bmatrix}. \tag{24.11}$$

Where $g_{25}(K)$ and $g_{12}(K)$ are obtained by Equation 24.6, $\delta_i^{(1)}$ and $\delta_i^{(2)}$ are the position vectors for the first-nearest neighbors (1NN) and second-nearest neighbors (2NN), respectively. By substituting for δ_i (Equations 24.1 and 24.2) and K in $g_{25}(k)$ and $g_{12}(k)$ one can obtain

$$
\begin{cases}
g_{25}(\mathbf{k}) = 2\cos(k_y a) + 4\cos\left(\frac{1}{2}k_y a\right)\cos\left(\frac{\sqrt{3}}{2}k_z a\right), \\
f(\mathbf{K})g_{12}(\mathbf{k}) = \exp\left(i\frac{1}{\sqrt{3}}k_z a\right) + 2\exp\left(-i\frac{k_z a}{2\sqrt{3}}\right)\cos\left(\frac{1}{2}k_y a\right).
\end{cases}
\tag{24.12}
$$

According to the periodic boundary condition for zigzag nanotube $k_y = \frac{2\pi l}{na}$. Thus by substituting all parameters in Equation 24.4, the band structure of Si h-NTs has been obtained considering a 2NN approximation. In this model (Shokri and Ahmadi 2015), the overlapping integral (S parameters) has been ignored for the first- and second-nearest neighbor. Moreover, constant term E_P is subtracted in Equation 24.4. Figure 24.6 shows the band structure of (6,0) and (7,0) considering 1NN and 2NN approximation.

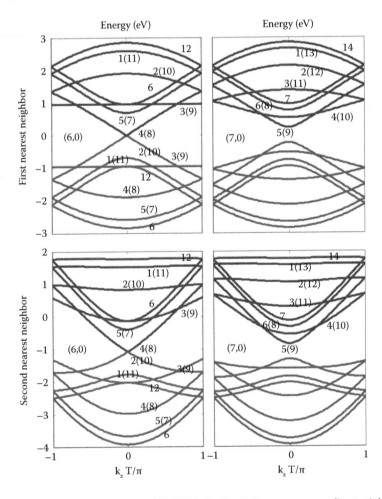

Figure 24.6 (a) Band structure of Si h-NTs (6,0) and (b) (7,0) in the first Brillioun zone according to tight-binding model and nearest neighbor approximation (1NN). (c) and (d) according to tight-binding model up to the second nearst-neighbor approximation (2NN) at $l = \frac{2n}{3}$. (From Shokri and Ahmadi, *Opt. Quant. Electron*, 47:2169–2179, 2015. With permission.)

In Figure 24.6c and d, it is clear that the energy spectrum shifts for 2NN approximation because of 2NN transfer integral $pp\pi_2^{AA}$ and the energy shift strongly depends on subband index l.

24.2.4 THE ELECTRONIC PROPERTIES OF Si h-NTs UNDER AXIAL MAGNETIC FIELD

The tight-binding model can be used to achieve the electronic structure for SiNTs under an axial magnetic field. The effective matrix Hamiltonian under an axial magnetic field is given by

$$H(\mathbf{K}) = \gamma_0 \begin{bmatrix} 0 & f(\mathbf{k}) \\ f(\mathbf{k})^* & 0 \end{bmatrix}. \tag{24.13}$$

Here $\gamma_0 = -0.949\text{eV}$ is the transfer integral for 1NN and $f(\mathbf{k})$ equals $g_{12}(\mathbf{k})$. Note that the periodic boundary conditions for a zigzag Si h-NTs shows $k_y = \dfrac{2\pi l}{na}$ (Saito et al. 1998). k_y is a quantized parameter because l is an integer which indicates the band number of valence and conduction. By applying an axial magnetic field along the nanotube axis, the transfer integral is modified by the phase factor, $\gamma_{ij} = \gamma_0 \exp\left(\dfrac{i2\pi\varphi_{ij}}{\varphi_0}\right)$ which $\varphi_{ij} = \displaystyle\int_{r_i}^{r_j} A.dl$ (Hatami et al. 2011). Here, the vector potential has been used in the Landau gage as $A = (B_y, 0,0)$. $\phi = \pi\rho^2 B$ is the magnetic flux flow through the nanotube corresponding section with a radius ρ and an applied uniform magnetic field B. $\varphi_0 = \dfrac{hc}{e}$ is the magnetic flux quantum which shows the period of energy gap variation. By applying the magnetic field, the angular momentum number changes to $l + F$ which $= \dfrac{\varphi}{\varphi_0}$. By substituting $f(\mathbf{k})$ in Equation 24.13 and diagonalization of it, the modified electron energy spectrum for a zigzag nanotube is given by $E_l(k_z,\varphi) = \pm\gamma_0\sqrt{1 + 4\cos\left(\dfrac{\sqrt{3}k_z a}{2}\right)\cos\left(\dfrac{\pi(1+F)}{n}\right) + 4\cos^2\left(\dfrac{\pi(1+F)}{n}\right)}$,

which $\pm$ denotes conduction and valence band. Figure 24.7 shows the band structure of (6,0) in the absence and presence of a magnetic field as a function of wave vector k_z. As an example, the value of the magnetic field is $\phi = 0.1\phi_0$. It is observed that (6,0) is a conductor nanotube in the absence of a magnetic field and the band gap is zero at the Γ point. By applying the magnetic field, the system tends toward semiconducting behavior and one can see metal-semiconductor phase transition. In the metallic system, by increasing the magnetic field the band gap is increased and this gap reaches the maximum at half flux quantum (Lu 1995). Furthermore,

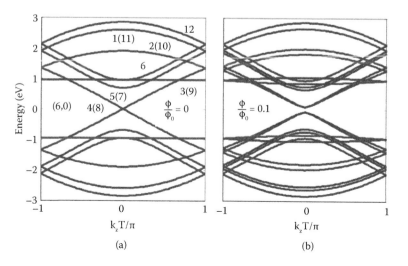

Figure 24.7 (a) Band structure of Si h-NTs (6,0) in the absence of magnetic field and (b) in the presence of magnetic field, $\Phi = 0.1\Phi_0$ in the first Brillouin zone.

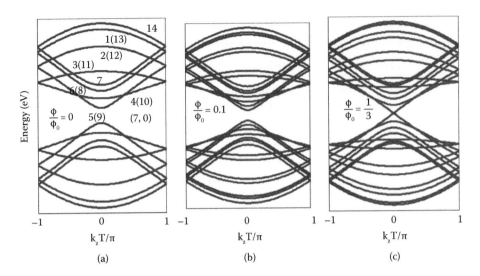

Figure 24.8 (a) Band structure of Sih-NTs (7,0) in the absence of magnetic field (b) in the presence of magnetic field when $\Phi = 0.1\Phi_0$ (c) when $\Phi = \dfrac{1}{3}\Phi_0$ in the first Brillouin zone for light parallel polarization to the tube axis.

it is clear that by applying the magnetic field the band degeneracy is broken. The intrinsic chirality of Bloch electrons and the curvature of nanotube at the K_+ and K_- points cause the band splitting in the magnetic field.

There is difference between the behavior of semiconductor and conductor SiNTs in the presence of a magnetic field. Figure 24.8 shows the band structure of (7,0) as a function of wave vector k_z in the absence and presence of a magnetic field. As an example, the value of the magnetic field is $0.1\phi_0$. In Figure 24.8b it is clear that the band gap decreases by applying the magnetic field. In addition, the band gap of the semiconducting nanotubes decreases by increasing the magnetic field and reaches zero at $\phi = \dfrac{1}{3}\phi_0$ and $\phi = \dfrac{2}{3}\phi_0$ (Ahmadi and Shokri 2016). Figure 24.8c shows the band structure of (7,0) at $\phi = \dfrac{1}{3}\phi_0$. It is obvious that the band gap at the Γ point is zero and the transition $9^v \rightarrow 9^c$ is forbidden. Furthermore, the band degeneracy is broken in the presence of a magnetic field.

24.3 THEORETICAL STUDY ON OPTICAL PROPERTIES OF SILICON NANOTUBES

24.3.1 OPTICAL PROPERTIES OF Si h-NTs UNDER INTRINSIC CURVATURE EFFECT FOR LIGHT PARALLEL POLARIZATION

Tight-binding calculations have been used to obtain the optical properties of nanotubes. For calculating the intraband optical matrix elements for transitions between an initial valence band to a final conduction band, we need the polarization vector $\mathbf{P}$ and the electric dipole vector $\mathbf{D}$. The optical transition matrix elements equal the inner product of $\mathbf{P}$ and $\mathbf{D}$. In order to obtain electric dipole vector $\mathbf{D}$, we should find the wave function of valence and conduction bands which are obtained by solving the Schrodinger equation of the tight-binding Hamiltonian matrix. For incident light polarized parallel to the tube axis (vertical transition), there are just transitions with the same K along the nanotube axis ($\mathbf{k} = \mathbf{k}'$) (Ahmadi et al. 2016).

The wave function of the valence ($s = -1$) and conduction ($s = +1$) is written as a linear combination of Bloch wave functions

$$\Psi_{c,v}(\mathbf{k},\mathbf{r}) = \frac{1}{\sqrt{N}} \sum_{i=1}^{N} \sum_{\alpha=A,B} \sum_{P=1}^{2n} C_{\alpha p}^{c,v}(\mathbf{k}) e^{i\mathbf{k}\cdot\mathbf{R}_{ij}^{\alpha p B q}} \varphi(\mathbf{r} - \mathbf{R}_i^{\alpha p}),$$ where α refers to the A- and B-type

atoms, $\mathbf{R}_i^{\alpha p}$ refers to the position of the Pth A-type atom α at site i and $\varphi(\mathbf{r} - \mathbf{R}_i^{\alpha p})$ is the π- orbital wave function of the Pth A-type atom in site i. Furthermore, $N = \dfrac{L}{T}$ is the number of unit cells of nanotubes, L is the tube length, and T is the length of unit cell. The unit cell has been considered a cylindrical segment.

The corresponding normalized energy eigenvectors are found as: $C_\alpha^{cv} = \dfrac{1}{\sqrt{2}}\left(\pm \dfrac{s\,f(k_z)}{\sqrt{|f(k_z)|^2}},\, 1\right)$. The first step for the study of optical properties of nanotubes is achievement of the electric dipole vector between an initial valence and a final conduction band.

$$\mathbf{D}(\mathbf{k'},\mathbf{k}) = \dfrac{-e}{m}\langle \psi_c(\mathbf{k'},\mathbf{r})|v|\psi_v(\mathbf{k},\mathbf{r})\rangle, \tag{24.14}$$

where e is the charge of an electron and m is the effective mass of an electron. We can use the velocity commutator from $v = \dfrac{i}{\hbar}[H,r]$ where H is the Hamiltonian operator. We can re-express the velocity matrix elements as

$$<\psi_c(\mathbf{k'},\mathbf{r})|v|\psi_v(\mathbf{k},\mathbf{r})> = \dfrac{i}{\hbar}<\psi_c|Hr|\psi_v> - \dfrac{i}{\hbar}<\psi_c|rH|\psi_v>. \tag{24.15}$$

We can use the unit operator that is given by $\displaystyle\sum_{m\in N}|\varphi_m><\varphi_m| = 1$. Furthermore, the atomic orbital wave function behaves as a delta function $<\varphi(\mathbf{r}-Rm)|r|\varphi(\mathbf{r}-\mathbf{R_n})> = \mathbf{R_n}\,\delta_{n,m}$ which can be used for changing Equation 24.15 as

$$<\psi_c(\mathbf{k'},\mathbf{r})|v|\psi_v(\mathbf{k},\mathbf{r})> = \dfrac{it}{\hbar}[(C_B^c)^*(C_A^V)\sum_{i=1}^{3} e^{iK.\mathbf{u}_i}\mathbf{u}_i - (C_A^c)^*(C_B^V)\sum_{i=1}^{3} e^{-iK.\mathbf{u}_i}\mathbf{u}_i], \tag{24.16}$$

where t and u_i are transfer integral and position vectors for the ith nearest neighbor, respectively. We can recast Equation 24.16 by using $[-(C_A^c)^*(C_B^V)]^* = (C_B^c)^* C_A^V$.

$$<\psi_c(\mathbf{k'},\mathbf{r})|v|\psi_v(\mathbf{k},\mathbf{r})> = -\dfrac{i2t}{\hbar}\Re\left[(C_A^c)^* C_B^V \sum_{i=1}^{3} e^{-ik.\mathbf{u}_i}\mathbf{u}_i\right]. \tag{24.17}$$

This equation results from selection rules for optical transition in zigzag nanotubes (Ahmadi et al. 2016) and it can be rewritten for curvature effect in nanotubes. Here, the nanotube axis direction is z, thus, for vertical transition (incident light polarized parallel) and considering the curvature effect we have

$$\mathbf{p} <\Psi_c(\mathbf{k},\mathbf{r})|v_x|\Psi_v(\mathbf{k},\mathbf{r}) = \dfrac{i}{\hbar} K.\Re\left[\dfrac{\tilde{f}^*(k_z)}{\sqrt{|f(k_z)|^2}}\sum_{i=1}^{3}\tilde{t}_i e^{ik.\mathbf{u}_i}\tilde{\mathbf{u}}_i\right], \tag{24.18}$$

where K is the unit vector of z and $\tilde{t}_i$ and $\tilde{u}_i$ are the ith modified transfer integral and position vector for the nearest neighbor under the curvature effect, respectively. $\tilde{f}(k_z)$ is given by $\tilde{f}(k_z) = t\exp\left(\dfrac{ik_z a}{\sqrt{3}}\right) + 2\tilde{t}_2\exp\left(\dfrac{-ik_z a}{2\sqrt{3}}\right)\cos(k_y a_y)$ where t and $\tilde{t}_2$ are transfer and modified transfer integrals, respectively. By substituting for $\tilde{u}_i$ and $\tilde{f}(k_z)$ and retaining the real term of Equation 24.18 one can obtain the matrix elements in the x component (Ahmadi et al. 2016).

$$\mathbf{P}.\langle\Psi_c(\mathbf{k},\mathbf{r})|v_x|\Psi_v(\mathbf{k},\mathbf{r})\rangle = \dfrac{i4\hbar v_f^2}{3\sqrt{3}a}\dfrac{1}{\sqrt{|\tilde{f}(k_z)|^2}}\left[1 - 2\left(\dfrac{\tilde{t}_2}{t}\right)^2\cos^2(k_y a_y) + \dfrac{\tilde{t}_2}{t}\cos\left(\dfrac{\sqrt{3}k_z a}{2}\right)\cos(k_y a_y)\right].$$

$$\tag{24.19}$$

Here vf is the Fermi velocity in Si NTs (Ahmadi et al. 2016). We could obtain a useful equation for optical absorption in low energy and considering curvature effect by using Equation 24.19.

The absorption probability per unit of time for an electron with the wave vector **k** can be obtained by time-dependent perturbation theory (Gruneis et al. 2003) as

$$W(\mathbf{k}) = \frac{4\hbar^4 I}{\tau \epsilon c^3 E_{laser}^2} \left| \mathbf{p.D} \; (\mathbf{k'},\mathbf{k}) \right|^2 \times \frac{\sin^2 \left[(E_c(\mathbf{k}) - E_v(\mathbf{k}) - E_{laser}) \frac{\tau}{2\hbar} \right]}{(E_c(\mathbf{k}) - E_v(\mathbf{k}) - E_{laser})^2}. \tag{24.20}$$

In Equation 24.20, E_{laser}, I, ϵ, and τ are incident laser energy, intensity of the pumping laser power, dielectric constant, and the time used for taking the average between an initial valence to final conduction band.

By substituting Equation 24.19 in Equation 24.20, a useful equation for optical absorption considering curvature effect can be obtained. For incident light polarized parallel the corresponding dipole matrix selection rule from $E_{l'}^v \rightarrow E_l^c$ is $l = l'$, while for a polarization perpendicular it is $l = l' \pm 1$.

Figure 24.9 shows the dipole matrix elements $\mathbf{D}(k_z, l)$ of zigzag Si h-NTs (6,0), (7,0), and (8,0) as a conductor and two semiconductor tubes for incident light polarized parallel to the nanotube axis with and without a geometry curvature effect. The dipole matrix elements have a maximum absolute value at the van-Hove singularities for each band. For light polarized parallel the value of $\mathbf{D}(k_z, l)$ disappears for the subband with the different values of subband index "l" in valence and conduction bands.

Moreover, the optical transition occurs between the subband with the same wave vector k_z. From Figure 24.9 it is obvious that the dipole elements have equal value for the degenerate bands in the first Brillouin zone when the curvature effect is neglected. Furthermore, for (6,0) Si h-NTs as a metallic nanotube there is no transition at the Γ point because the band numbers $\frac{2n}{3}$ and $\frac{4n}{3}$ cross the Fermi energy when neglecting the curvature effect and the transitions have small amplitude for other values of k_z. From Figure 24.9b we can see considering the curvature effect causes optical transition at the Γ point. On the other hand, for a semiconducting nanotube such as (7,0) and (8,0) there are optical transitions at the Γ point in the first Brillouin zone. These transitions have large amplitudes for other value of k_z unlike the metallic tubes. Figure 24.9d and f show using the curvature effect leads to the variations in the transition dipole matrix elements because the band degeneracy is broken under the curvature effect, and some new optical transitions occur. Figure 24.10 shows the optical absorption of zigzag Si h-NTs for incident photon polarized parallel to the tube axis in low energy from 1 to 4 eV with and without considering the curvature effect. The optical transitions occur between bands with the same subband index number l. Maximum values are obtained in the van-Hove singularity point around the Fermi level for each band. Panel "a" shows the absorption spectrum of (6,0) without considering the curvature effect. In this case, three peaks ($l = 3$, 5, *and* 6) are intense and other transition amplitudes are weak. By applying the curvature effect the band degeneracy is broken and some band-to-band transitions occur. Thus in panel (b) one can see six intense peaks ($l = 3, 4, 5, 6, 7$, *and* 8), especially between 2 and 4 eV. Within the tight-binding model, zigzag nanotubes with n even have a characteristic peak at $= \frac{n}{2}$. The absorption spectra of (6,0) and (8,0) include characteristic peaks about 1.9 eV and this is because the energy spectrum at $l = \frac{n}{2}$ is independent of k_z and causes an infinite density of state. Without the considering of curvature effect one can see four and five intense peaks at ($l = 4, 5, 6$, *and* 7) and ($l = 4, 5, 6, 7$, *and* 8) in the absorption spectra of (7,0) and (8,0) nanotubes, respectively. Splitting the band degeneracy under the curvature effect leads to increased optical transitions and we can see six and eight intense peaks for (7,0) and (8,0) at ($l = 4, 5, 6, 7, 8$, *and* 9) and ($l = 4, 5, 6, 7, 8, 9, 10$, *and* 11), respectively, and other peaks are weak.

24.3.2 OPTICAL PROPERTIES OF SILICON NANOTUBES: A SECOND-NEAREST NEIGHBOR TIGHT-BINDING APPROACH

Shokri and coworker (Shokri and Ahmadi 2015) studied the optical absorption of zigzag Si h-NTs using the tight-binding model up to the second-nearest neighbor. They could derive an easy-to-use equation for optical transition matrix elements for incident light polarized parallel to the tube axis.

They used Equation 24.14 and the same method, and obtained the optical transition matrix elements considering second-nearest neighbors as

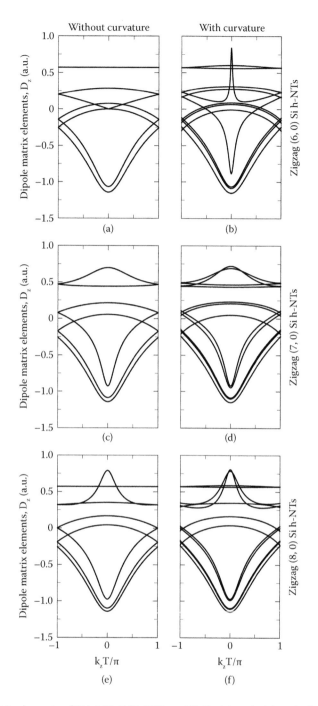

Figure 24.9 Dipole Matrix elements of Si h-NTs (6,0), (7,0), and (8,0) with and without including the curvature effect in the first Brillouin zone for light parallel polarization to the tube axis. (Ahmadi et al., 2016).

$$\mathbf{P.D}\ (\mathbf{k'},\mathbf{k}) = \mathbf{M}_{v \to c} = \frac{-ei}{m\hbar} \frac{1}{\omega(\mathbf{k})} \times \left(\frac{a\gamma_0}{\sqrt{3}} \left[1 - 2\cos^2\left(\frac{1}{2}k_y a\right) + \cos\left(\frac{\sqrt{3}}{2}k_z a\right)\cos\left(\frac{1}{2}k_y a\right) \right] \right.$$
$$\left. + \sqrt{3}a\gamma_1 \cos\left(\frac{1}{2}k_y a\right)\left(1 - 2\cos\left(\frac{1}{2}k_y a\right)\right)\left[\cos\left(\frac{\sqrt{3}}{6}k_z a\right) - \cos\left(\frac{5\sqrt{3}}{6}k_z a\right) \right] \right). \tag{24.21}$$

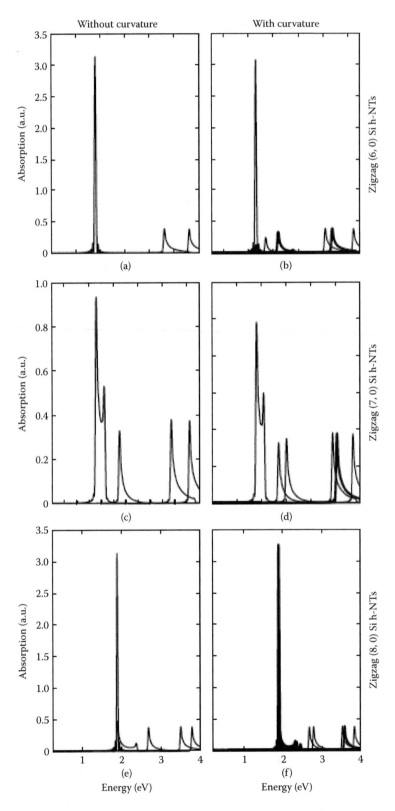

Figure 24.10 Variation in the optical absorption spectrum of Si h-NTs (6,0), (7,0), and (8,0) with considering the curvature effect. Left (right) panels show the results including without (with) curvature effect. (Ahmadi et al., 2016).

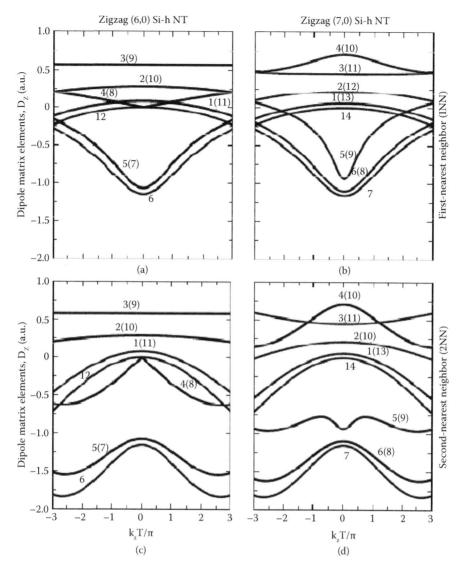

Figure 24.11 Dipole matrix elements of Si h-NTs (6,0) and (7,0) in the first Brillouin zone according to tight-binding model up to the second nearest-neighbor approximation (2NN). (From Shokri and Ahmadi, *Opt. Quant. Electron*, 47:2169–2179, 2015. With permission.)

Here $w(k) = |g_{12}(k)|$ and γ_0, γ_1 are the first- and the second-nearest neighbor transfer integrals, respectively (Table 24.1). This equation shows the selection rules for incident light polarized parallel to the tube axis for the 2NN approximation. Furthermore, by substituting Equation 24.21 in Equation 24.20 one can obtain a useful equation for optical absorption considering 2NN approximation. Figure 24.11 shows the dipole matrix elements of (6,0) and (7,0) as a metallic and semiconducting nanotube, respectively, considering first- and second-nearest neighbor approximation.

In the dipole matrix elements of (6,0) it is clear that at $l = 2$ and $l = 3$ there is not any difference between 1NN and 2NN approximation. But at $l = 1$ and $l = 12$ the slopes of the curves have been increased. The alterations in the slope of curve can be observed for $l > \dfrac{2n}{3}$. It is clear that at ($l = 4, 5,$ *and* 6) there is a maximum value at the Γ point.

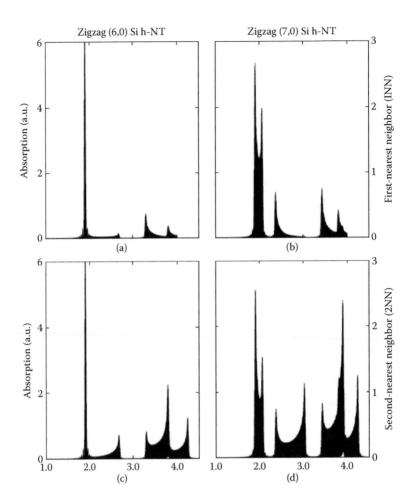

Figure 24.12 Variation of the optical absorption spectrum of Sih-NTs (6,0) and (7,0) on the incident light polarized parallel to the tube axis and subband index *l* with considering 1NN and 2NN approximation. (From Shokri and Ahmadi, *Opt. Quant. Electron*, 47:2169–2179, 2015. With permission.)

The behavior of the dipole matrix element of (7,0) as a semiconducting tube changes when considering the 2NN approximation and the slope of curves has been changed and maximum value at the Γ point for *l* = 6 *and* 7 is shown.

Figure 24.12 shows the optical absorption of (6,0) and (7,0) nanotubes in terms of incident photon energy and subband index for light polarized parallel in low energy from 1 to 4 eV. In this figure, the effect of first-nearest neighbor and second-nearest neighbor has been compared. It is clear that using the tight-binding model up to the second-nearest neighbor changes the shape of absorption spectrum especially between 2.5 and 4.5 eV. Furthermore, the number of peaks are increased in this region.

24.3.3 OPTICAL ABSORPTION OF NANOTUBES UNDER AXIAL MAGNETIC FIELD

Tight-binding calculations have also been used to predict the effect of an axial magnetic field on the optical properties of a nanotube. Ahmadi and coworker (Ahmadi and Shokri 2016) used the tight-binding model and obtained a useful equation for optical transition matrix elements under an axial magnetic field. This equation is given by

$$\mathbf{M}_{v \to c} = \mathbf{P.D}\left(k', k\right) = -i\frac{4e\hbar(v_f)^2}{3\sqrt{3}\ ma\gamma_0}\frac{1}{\sqrt{\left|f(k_z)^2\right|}}\left[1+\cos\left(\frac{\sqrt{3}ak_z}{2}\right)\cos\left(\frac{\pi(1+F)}{n}\right)-2\cos^2\left(\frac{\pi(1+F)}{n}\right)\right], \quad (24.22)$$

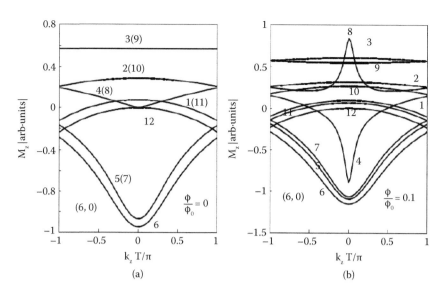

Figure 24.13 (a) Dipole matrix elements of Sih-NT (6,0) in the absence of magnetic field (b) in the presence of magnetic field, $\Phi = 0.1\Phi_0$ in the first Brillouin zone for light parallel polarization to the tube axis.

where $F = \dfrac{\phi}{\phi_0}$. Furthermore, according to previous descriptions $f(k_z)$ is given by $f(k_z) = \exp\left(\dfrac{iak_z}{\sqrt{3}}\right) +$ $2\exp\left(\dfrac{-iak_z}{2\sqrt{3}}\right)\cos\left(\dfrac{\pi k_y a}{2}\right)$, where $k_y = \dfrac{2\pi l}{na}$ and by applying the magnetic field l replace with $l + F$ as has been explained in Section 24.2.4.

By substituting Equation 24.22 in Equation 24.20 we can derive a useful equation for optical absorption of a nanotube under an axial magnetic field. Figure 24.13 shows the dipole matrix elements of (6,0) as a metallic Si h-NTs in the absence and presence of a magnetic field. As an example, we apply a magnetic field with value $\phi = 0.1\phi_0$. It is clear that the band degeneracy is broken and a metal-semiconducting phase transition occurs due to the band gap open at the Γ point by applying the magnetic field. In zigzag metallic nanotubes (n,0) the band number $\dfrac{2n}{3}$ and $\dfrac{4n}{3}$ cross the Fermi energy at the Γ point and the dipole matrix elements and optical transition are zero at this point. The magnetic field causes optical transitions to occur for $\dfrac{2n}{3}$ and $\dfrac{4n}{3}$ at the Γ point, thus this kind of nanotube behaves like a semiconducting tube.

Figure 24.14 shows the dipole matrix elements of (7,0) Si h-NTs as a semiconducting tube for different values of magnetic field. The behavior of semiconducting nanotubes is different from metallic. The band gap in semiconducting nanotubes decreases by increasing the magnetic field and becomes zero at $\phi = \dfrac{\phi_0}{3}$ and $\phi = \dfrac{2\phi_0}{3}$. According to Figure 24.14c, the optical transition between $9_c \rightarrow 9_v$ is forbidden at the Γ point and a semiconductor-metal phase transition has occurred. Furthermore, Figure 24.14b and c show by applying the magnetic field the band degeneracy is broken, and there is a proliferation of the number of bands.

In Figure 24.15 one can see the optical absorption of (6,0) zigzag Si h-NTs as a function of the incident photon energy polarized parallel to the tube axis, and the subband index in the absence and presence of a magnetic field in low energy. By applying the magnetic field, $\dfrac{\Phi}{\Phi_0} = 0.1$, as an example for the magnetic field value, the absorption spectrum has been changed and the number of peaks has been increased

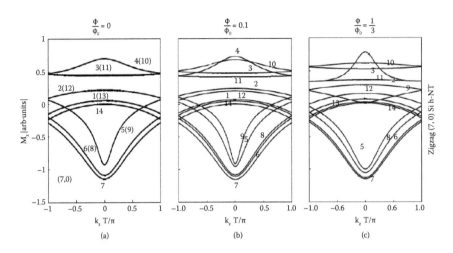

Figure 24.14 (a) Dipole matrix elements of Sih-NTs (7,0) in the absence of magnetic field (b) in the presence of magnetic field when $\Phi = 0.1\Phi_0$ (c) when $\Phi = \frac{1}{3}\Phi_0$ in the first Brillouin zone for light parallel polarization to the tube axis.

especially between 2 and 4 eV. It is obvious there is an intense peak for (6,0) at about 2 eV. This characteristic peak for zigzag nanotubes with even "n" within the tight-binding model is due to the band structure of zigzag nanotubes with "n" even independent of k_z at $1 = \frac{n}{2}$, and leads to an infinite density of state.

Figure 24.16 shows the optical absorption spectrum of (7,0) as a semiconductor nanotube in term of photon energy in the absence and presence of a magnetic field. Figure 24.16b shows the changes in the optical absorption spectrum in the presence of a magnetic field ($\phi = 0.1\phi_0$) compared with Figure 24.16a. The increasing of the peak number is obvious especially between 2 and 4 eV. By increasing the magnetic field at $\phi = \frac{1}{3}\phi_0$ from Figure 24.16c, one can see a sharp peak around 2 eV similar to a (6,0) as a conductor nanotube.

Proliferation of the number of peaks in the optical absorption spectrum is due to the presence of a magnetic field; the band degeneracy is broken and some band-to-band transition can be created.

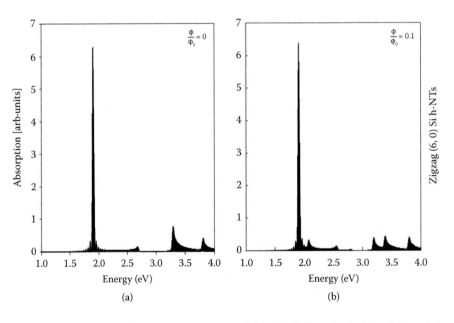

Figure 24.15 Variation of the optical absorption spectrum of Sih-NTs (6,0) on the incident light polarized parallel to the tube axis and subband index *l* with considering the magnetic field in the first Brillouin zone.

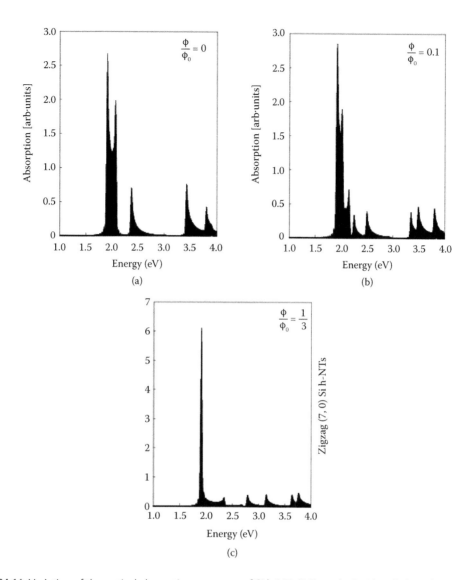

Figure 24.16 Variation of the optical absorption spectrum of Si h-NTs (7,0) on the incident light polarized parallel to the tube axis and subband index l (a) in the absence of magnetic field (b) with considering the magnetic field when $\Phi = 0.1\Phi_0$ (c) when $\Phi = \frac{1}{3}\Phi_0$ in the first Brillouin zone.

24.4 FABRICATION METHODS OF SILICON NANOTUBES AND EXPERIMENTAL EVIDENCES

The greatest challenge for using the SiNTs is the fabrication costs of SiNTs which is higher than for other silicon nanoparticles. Therefore, developing useful and easy methods to fabricate SiNTs is essential for industrial applications. The first fabrication method of SiNTs was CVD (chemical vapor deposition) which was reported by Sha and coworkers (Sha et al. 2002). In this method, Si is deposited into a nanoporous anodic aluminum oxide (AAO) template to obtain SiNTs. Furthermore AAO membranes, the sol-gel method and molecular beam epitaxy (Jeong et al. 2003) are used for fabrication of SiNTs. Moreover, AAO membrane, nanowires, and nanorods are used as a template to grow SiNTs (Park et al. 2009; Song et al. 2010) .

Yoo and coworkers (Yoo et al. 2012) reported a method for the synthesis of SiNTs. They used electrospun polyacrylonitrile nanowires as a template. According to this method the fabrication of SiNTs has four steps. Figure 24.17 shows these steps.

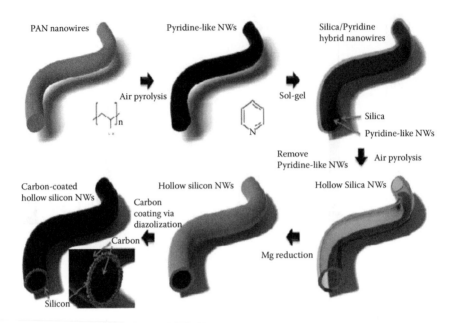

Figure 24.17 Schematic picture of the synthesis of Si nanotubes coated with carbon. (From Yoo et al., *Energy Storage. Adv. Mater,* 24:5452–5456, 2012. With permission.)

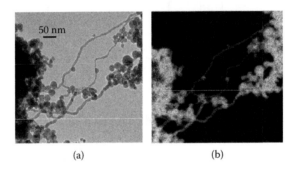

Figure 24.18 TEM image (650 nm × 650 nm) of nanoparticles and free-standing nanotubes dispersed through an isopropyl droplet on Au grid (mesh 1000). (b) Local distribution of silicon. (From Castrucci P et al., *Thin Solid Films* 508:226–230, 2006. With permission.)

First of all, electrospun polyacrylontrile nanowires were converted into pyridine nanowires through air pyrolysis. Second, tetraethyl orthosilicate was coated on the surface of nanowires to form tetraethyl orthosilicate/pyridine hybrid nanowire. Then pyridine nanowires were removed to produce silica (SiO_2) nanotubes. Finally, silica nanotubes were reduced by magnesium to obtain SiNTs. Recently, a new method for synthesis of SiNTs is developing. The name of this method is glancing angle deposition (Huang et al. 2009), and there is a great challenge to achieve large-scale production in this method. Wen and coworkers (Wen et al. 2013) reported a strong method to prepare gram-scale SiNTs by using a nanorod-like nickel-hydrazine complex as the template. Furthermore, SiNTs have been synthesized in gram quantities by the DC arc plasma method (Balasubramanian et al. 2004). Castrucci and coworkers (Castrucci et al. 2006) synthesized SiNTs in a DC arc plasma reactor, with direct arcing of graphite cathode and silicon anode. They reported transmission electron microscopy (TEM) analysis and Figure 24.18 shows spherical nanoclusters and tubular structure, approximately in the ratio of 10:1. In this case, the tubes had different diameters (≥ 5 *nm*) and their length amounted to hundreds of nanometers.

De Crescenzi and coworkers (De Crescenzi et al. 2005) reported the STM image (200 × 200 *nm²*) of two SiNTs with some silicon nanoparticles, Figure 24.19. These nanotubes have very different diameters and one of them is almost three times smaller than another one. It can mean the tubes experience a strong

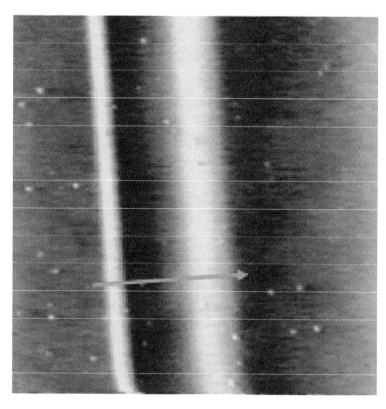

Figure 24.19 STM image (200 nm × 3200 nm) showing two silicon straight nanotubes, lying on the highly oriented pyrolic graphite surface together with some silicon nanoparticles. $I_{tunn} = 0.8$ nA; $V_{bias} = 0.7$ V. (From DeCrescenzi et al., *Appl. Phys. Lett.* 86: 231901–231903, 2005. With permission.)

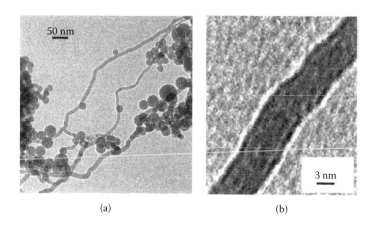

(a)   (b)

Figure 24.20 (a) TEM image reporting several silicon nanotubes and nanoparticles. (b) TEM image reporting a silicon nanotube with diameter of about 7 nm and a typical line profile. (From DeCrescenzi et al., *Appl. Phys. Lett.* 86: 231901–231903, 2005. With permission.)

radial compression. This compression is induced by the van der Waals interaction between the nanotubes and the substrate. Furthermore, they reported a TEM image for an SiNT and nanoparticles with a diameter about 7 nm, Figure 24.20. The length and outer (inner) diameter of SiNTs are the most important properties that determine the mechanical and electrochemical properties of SiNTs.

The SiNTs can also be synthesized by using AAO (Song et al. 2010) or ZnO rod, but there is a great challenge to achieve large-scale production using this method.

Gao and coworkers (Gao et al. 2011) reported a useful method for the fabrication of silica nanotubes on a large scale. This method motivated other groups to change silica nanotubes to SiNTs (Wen et al. 2013). They investigated a method to synthesize Si nanotubes through the thermal reduction of silica nanotubes with the assistance of magnesium powder.

24.5 APPLICATION OF SILICON NANOTUBES FOR LITHIUM-ION BATTERIES

Nowadays more than 85% of energy consumption is provided by fossil fuel. Using this fuel causes climate changes and health problems. Utilization of green energy such as solar and wind power are promising and appropriate for sustainable economic growth (Hoffert et al. 2002; Armand and Tarascon 2008; Thackeray et al. 2012; Etacheri et al. 2011; Goodenough 2013). Furthermore, driving electrical vehicles can relieve the environment issues (Thackeray et al. 2012; Goodenough 2013), although for using solar and wind power or electrical vehicles we need highly efficient energy storage devices. In this regard LIBs play an important role (Armand and Tarascon 2008; Goodenough 2013). LIBs have broad uses in portable electronic devices such as laptops, cell phones, cameras, and medical microelectronic devices, due to high energy density and long scale life, although there are still many challenges for using these batteries in electric vehicles and storing energy (Armand and Tarascon 2008; Thackeray et al. 2012; Choi et al. 2012; Armand and Tarascon 2001). These challenges include further increasing their energy and power densities, improving their safety, and lowering the cost (Armand and Tarascon 2008; Armand and Tarascon 2001). Presently the materials which fabricate negative electrodes (anode) and positive electrodes (cathode) in LIBs are graphite and lithium metal oxide (lithium cobalt oxide and lithium manganese oxide) or lithium iron phosphate, respectively. The theoretical capacities of these materials are 372 mAhg^{-1} and less than 200 mAhg^{-1}, respectively.

The discharge potential of Si is about 0.2 V, with respect to Li/Li$^+$ this is lower than most of other alloy-type and metal oxide anodes. On the other hand silicon is plentiful in the earth's crust. However, the most important challenges for using an Si anode are the huge volume variations during lithiation and delithiation processes, and unstable surface electrolyte interphase (SEI) films, low cycling efficiency, and permanent capacity losses. Furthermore, the electronic conductivity is quite low—about 10^{-3} Scm^{-1} which increases to $\cong$ 10^2 Scm^{-1} after lithiation (Kulova et al. 2007). Lithium diffusion in silicon material is very slow, with a diffusion coefficient between 10^{-14} and 10^{-13} cm^{2}s^{-1} (Kulova et al. 2007; Xie et al. 2010). These problems have hindered the rate of usage of silicon anodes (Aricò et al. 2005; Mukherjee et al. 2012).

In the past decade scientists have worked on these issues and they investigated the use of nanomaterials based on silicon for increasing cycle life and improving cycling rate performance (Magasinski et al. 2010; Song et al. 2010). Between various silicon nanostructures, SiNTs can improve electrochemical performance and potentially obtain reversible morphological change (Song et al. 2010). Using the SiNTs as anodes in LIBs, there are not as many of these silicon nanoparticles, nanowires, and nanorods because the synthesis of the nanotubes is difficult.

The first report of using SiNTs was given by Park and coworkers in 2009 (Park et al. 2009). After this report scientists paid attention to SiNTs for the anode in LIBs because using this kind of nanotube improved the specific capacity and cycling performance. Thus using the SiNTs as the LIBs' anode is more suitable than other kinds of silicon nanostructures, due to the existence of empty space inside the core which offers extra space for silicon expansion during cycling. This will prevent the cracking of silicon and the deformation of the SEI layer during charging and discharging. Furthermore, SiNTs have a higher active surface area and can produce a higher current density per unit area of electrode compared with other silicon nanostructures. In comparison with other kinds of nanotubes, SiNTs have lower mass density because they are highly porous. For this reason the volumetric capacity of the electrode and energy per unit volume of LIB cells will be reduced. In addition, using the SiNTs as an LIB anode causes impressive increases in the long cycling and rate capacity due to the unique tubular structure of these tubes which provides free space for volume variations. The hollow structure of these nanotubes can make shorter the lithium diffusion distance during lithium insertion (extraction) into (from) silicon material.

24.6 SUMMARY

In the present chapter the possible structures, optoelectronic properties, and application of SiNTs in LIBs have been systematically explained. Furthermore, the fabrication methods of SiNTs have been defined and the challenges of these ways have been demonstrated.

The dependence of electronic properties of SiNTs on their structure has been investigated. According to the tight-binding model, SiNTs follow Hamada's rule and these results are different to *ab initio* results.

Theoretical studies based on the tight-binding model show SiNTs have conductor and semiconductor behaviors according to Hamada's rule. We can see some changes in the electronic properties of SiNTs when considering the curvature effect or applying a magnetic field.

The changes in electronic properties cause changes in optical properties such as optical absorption spectrum of nanotubes.

Experimental evidences show that the fabrication of SiNTs is more difficult and more expensive than other kinds of silicon nanoparticles. Due to the application of these nanotubes as a good candidate for anodes in LIBs, nowadays scientists are working on the fabrication methods to solve the challenges in this field.

REFERENCES

Ahmadi N, Shokri AA (2016). Optical properties of silicon hexagonal nanotubes under axial magnetic field. DOI:10.1016/j.optcom.2016.07.049.

Ahmadi N, Shokri AA, Elahi SM (2016), Optical transition of zigzag silicon nanotubes under intrinsic curvature effect. *Silicon* 8: 217–224.

Arico A S, Bruce P, Scrosati B, Tarascon JM, Van Schalkwijk W (2005). Nanostructured materials for advanced energy conversion and storage devices. *Nat. Mater.* 4: 366–377

Armand M, Tarascon JM (2008). Building better batteries. *Nature.* 451: 652–657.

Bai J, Zeng X C, Tanaka H, Zeng JY (2003). Metallic single-walled silicon nanotubes. *Proc. Natl. Acad. Sci.* U S A. 101: 2664–2668.

Balasubramanian C, Godbole VP, Rohatgi VK, Das AK, Bhoraskar SV (2004). Synthesis of nanowires and nanoparticles of cubic aluminium nitride. *Nanotechnology.* 15: 370–373.

Castrucci P et al. (2006). Silicon nanotubes: Synthesis and characterization. *Thin Solid Films* 508: 226–230.

Castrucci P et al. (2012). Si nanotubes and nanospheres with two-dimensional polycrystalline walls. *Nanoscale.* 4: 5195–5201.

Chen YW, Tang YH, Pei LZ, Guo C (2005). Self-assembled silicon nanotubes grown from silicon monoxide. *Adv. Mater.* 17: 564–567.

Choi NS et al. (2012). Challenges facing lithium batteries and electrical double-layer capacitors. *Angew. Chem. Int. Ed.* 51: 9994–10024.

De Crescenzi M, et al. (2005). Experimental imaging of silicon nanotubes. *Appl. Phys. Lett.* 86: 231901–231903.

De Padova P, Quaresima C, Olivieri B, Perfetti P, Le Lay G (2011). sp2-like hybridization of silicon valence orbitals in silicene nanoribbons. *Appl. Phys. Lett.* 98: 081909-1–081909-3.

Dmitrii F, Perepichka, Federico Rosei (2006). Silicon nanotubes. *Small.* 2: 22–25. *Energy Environ. Sci.* 5: 7854–7463.

Etacheri V, Marom R, Elazari R, Salitra G, Aurbach D (2011). Challenges in the development of advanced Li-ion batteries: *A revie. Energ. Environ. Sci.* 4: 3243–2362.

Fagan S, Baierle R, Mota R, da Silva A J, Fazzio A (2000), Ab initio calculations for a hypothetical material: Silicon nanotubes. *Phys. Rev. B.* 61: 9994–9996.

Gao C, Lu Z, Yin Y (2011). Gram-scale synthesis of silica nanotubes with controlled aspect ratios by templating of nickel-hydrazine complex nanorods. *langmuir.* 27: 12201–12208.

Goldberger J, et al. (2003), Single-crystal gallium nitride nanotubes. *Nature.* 422: 599–602.

Goodenough JB (2013). Evolution of strategies for modern rechargeable batteries. *acc. Chem. Res.* 46: 1053–1061.

Grosso G, Piermarocchi C (1995). Tight-binding model and interactions scaling laws for silicon and germanium. *Phys. Rev. B.* 51: 16772–16776.

Gruneis A, et al. (2003). Inhomogeneous optical absorption around the K point in graphite and carbon nanotubes *Phys. Rev. B.* 67: 165402–165409.

Guzmán-Verri GG, LewYan Voon LC (2007). Electronic structure of silicon-based nanostructures. *Phys. Rev. B.* 76: 075131–075140.

Guzmán-Verri GG, Lew Yan Voon LC (2011). Band structure of hydrogenated Si nanosheets and nanotubes. *J. Phys. Condens. Matter.* 23: 145502–145507.

Hamada N, Sawada SI, Oshiyama A (1992). New one-dimensional conductors: Graphitic microtubules. *Phys. Rev. Lett.* 68: 1579–1581.

Harrison TG (1989). *Electronic Structure and the Properties of Solids.* Dover Press, New York.

Hatami H, Abedpour N, Qaiumzadeh A, Asgari R (2011). Conductance of bilayer graphene in the presence of a magnetic field: Effect of disorder. *Phys. Rev. B.* 83: 125433–125441.

Hoffert MI, et al. (2002.) Advanced technology paths to global climate stability: Energy for a greenhouse planet. Science. 298: 981–987.

Huang ZG, Harris KD, Brett MJ (2009). Morphology control of nanotube arrays. *Adv. Mater.* 21: 2983–2987.

Iijima S (1991). Helical microtubules of graphitic carbon. *Nature.* 354: 56–58.

Jeong SY et al. (2003). Synthesis of silicon nanotubes on porous alumina using molecular beam epitaxy. *Adv. Mater.* 15: 1172–1176.

Kara A, et al. (2009). Physics of silicene stripes. J. Supercond. Novel Magnetism. 22: 259–263.

Kulova TL, Skundin AM, Pleskov YV, Terukov EI, Konkov OI (2007). Lithium insertion into amorphous silicon thin-film electrodes. J. *Electroanal. Chem.* 600: 217–225.

Leobandung E, Guo L, Chou SY (1997). Observation of quantum effects and Coulomb blockade in silicon quantum-dot transistors at temperatures over 100 K. *Appl. Phys. Lett.* 67: 938–940.

Lu JP (1995). Novel magnetic properties of carbon nanotubes. *Phys. Rev. lett.* 74: 1123–1126.

Magasinski A, et al. (2010). High-performance lithium-ion anodes using a hierarchical bottom-up approach. *Nat. Mater.* 9: 353–358.

Mukherjee R, Krishnan R, Lu TM, Koratkar N (2012). Nanostructured electrodes for high-power lithium ion batteries. *Nano Energ* 1: 518–533.

Namastsu H, Takahashi Y, Nagase M, Murase K (1995). Fabrication of thickness-controlled silicon nanowires and their characteristics. J. *Vac. Sci. Technol.* B. 13: 2166–2169.

Ono T, Saitoh H, Esashi M (1997). Si nanowire growth with ultrahigh vacuum scanning tunneling microscopy. *Appl. Phys. Lett.* 70: 1852–1854.

Park MH et al. (2009). Silicon nanotube battery anodes. *Nano Lett.* 9: 3844–3847.

Pollak E, Salitra G, Baranchugov V, Aurbach D (2007). In Situ conductivity, impedance spectroscopy, and ex situ Raman Spectra of Amorphous Silicon during the Insertion/Extraction of Lithium. J. *Phys. Chem.* C. 111: 11437–11444.

Popov VN (2004). Curvature effects on the structural, electronic and optical properties of isolated single-walled carbon nanotubes within a symmetry-adapted non-orthogonal tight-binding model. *New J Phys.* 6: 17-1-17.

Saito R, Dresselhaus G, Dresselhaus MS (1998). *Physical Properties of Carbon Nanotubes.* Imperial College Press, London.

Schi SW, Zheng YF, Wang N, Lee CS, Lee ST (2001). A General synthetic route to III–V compound semiconductor nanowires. *adv. Mater.* 13: 591–594.

Seifert GK, Ohler Th, Urbassek HM, Hernandez E, Frauenheim Th (2001). Tubular structures of silicon Phys. *Rev. B.* 63: 193409-1-193409-4.

Sha J, et al. (2002). *Silicon nanotubes. Adv. Mater.* 14: 1219–1221.

Shokri AA, Ahmadi N (2015). Tight-binding description of optoelectronic properties of silicon nanotubes. Opt. Quant. Electron 47: 2169–2179.

Song T, et al. (2010). Arrays of sealed silicon nanotubes as anodes for lithium ion batteries. *Nano lett.* 10: 1710–1716.

Tang YH, Pei LZ, Chen YW, Guo C (2005). Self-assembled silicon nanotubes under supercritically hydrothermal conditions. *Phys. Rev. Lett.* 95: 116102–116105.

Tang YH, Zheng YF, Lee ST (2000). A simple route to annihilate defects in silicon nanowires. *Chem. Phys. Lett.* 328: 346–349.

Tarascon JM, Armand M (2001). Review article Issues and challenges facing rechargeable lithium batteries. *Nature.* 414: 359–367.

Thackeray MM, Wolverton C, Isaacs ED (2012). Electrical energy storage for transportation—Approaching the limits of, and going beyond, lithium-ion batteries. *Energ. Environ. Sci.* 5: 7854–7863

Vogl P, Hjalmarson H.P, Dow J D (1981) A semi-empirical tight-binding theory of the electronic structure of semiconductors? J. *Phys. Chem. Solids* 44: 365–378.

Vogt P, et al. (2012). Silicene: Compelling experimental evidence for graphenelike two-dimensional silicon. *Phys.Rev. Lett.* 108: 155501-5.

Wen Z, et al. (2013). Silicon nanotube anode for lithium-ion batteries. *Electrochem. Commun.* 29: 67–70.

Wu CX et al. (2001). Preparation and photoluminescence properties of amorphous silica nanowires. *Chem. Phys. Lett.* 336: 53–56.

Xi GC, Liu YK, Liu XY, Wang XQ, Qian YT (2006). Mg-catalyzed autoclave synthesis of aligned silicon carbide nanostructures. J. *Phys. Chem. B.* 110: 14172–14178.

Nanowires and nanotubes

Xie J. et al. (2010). Li-ion diffusion in amorphous Si films prepared by RF magnetron sputtering: A comparison of using liquid and polymer electrolytes. *Mater. Chem. Phys.* 120: 421–425.

Yang XB, Ni J (2005). Electronic properties of single-walled silicon nanotubes compared to carbon nanotubes. *Phys. Rev. B.* 72: 195426-1-195426-5.

Yoo JK, Kim J, Jung Y S, Kang K (2012). Scalable fabrication of silicon nanotubes and their application to energy storage. *Adv. Mater.* 24: 5452–5456.

Zhang RQ, Lee AT, Law CK, Li WK, Teo BK (2002). Silicon nanotubes: Why not? Chem. *Phys. Lett.* 364: 251–258.

Zhang XY et al. (2001). Synthesis of ordered single crystal silicon nanowire arrays. *Adv. Mater.* 13: 1238–1241.

Zhukovskii YF, Popov AI, Balasubramanian C, Bellucci S (2006). Structural and electronic properties of single-walled AlN nanotubes of different chiralities and sizes. J. *Phys. Condens. Matter.* 18: S2045–S2054.

25 Amorphous silicon nanotubes

Mirko Battaglia, Salvatore Piazza, Carmelo Sunseri, and Rosalinda Inguanta

Contents

25.1 INTRODUCTION

Usually, the birth of nanotechnology dates back to Feynman's lecture of 1959, reprinted in 1992 (Feynman 1992). The synthesis of fullerenes (Kroto et al. 1985) and carbon nanotubes (Iijima 1991) in 1985 and 1991, respectively, can be considered as the initial steps of the nanotechnology development. Over the successive years, research on nanomaterials was strongly boosted by possible innovative applications due to the peculiar features at nanoscale (Kuchibhatlaa et al. 2007).

Materials made with atomic accuracy show unique properties because of nanoscale-size confinement, predominance of interfacial phenomena, and quantum effects. Therefore, by reducing the dimensions of a structure to nanoscale, many inconceivable properties will appear which may lead to different applications, from nanoelectronics and nanophotonics to nanobiological systems and nanomedicine. Specific properties can be found at nanoscale which are absent at higher scale, such as reduction of the melting temperature of nanoparticles due to a surface energy effect, and lattice parameter change associated with surface stress effect. Consequently, for instance, fabrication of structural ceramics is strongly favored by reducing the grain size to several nanometers because the sintering temperature can be decreased and single-phase plasticity increased. In addition, multiphase materials formed by nanosized single-phase show specific properties, such as low magnetoresistance of multilayered magnetic thin films, with beneficial effects on magnetic recording read heads (Baibich et al. 1988; Hylton et al. 1993). Catalysis was the initial, almost exclusive, application field of nanomaterials because the surface area-to-volume ratio increases the chemical activity, determining significant cost advantages in catalyst fabrication. In view of current and future applications, fabrication methods of 2D and 3D architectures composed of 1D nanostructures were also investigated (Joshi and Schneider 2012).

In general, several studies were conducted for exploiting the physicochemical properties of specific materials at nanoscale, such as silica nanotubes which can be synthesized, and functionalizing by different methods for extending the application fields (Garcia-Calzon and Diaz-Garcia 2012). Over the years, great attention has been focused on the preparation of nanostructures as shown in the extensive review by Wu et al. (2010) who emphasized the advantage of the electrochemical techniques for synthesizing oriented and hierarchical quasi-1D semiconducting nanostructures.

In addition to the technological aspects, nanomaterials have also attracted great interest for basic scientific investigations. Surface stress and surface-free energy effects facilitating the sintering process of ceramic nanoparticles are both due to the reduced coordination of the surface atoms. The quantum confinement of electrons in a small geometry creates new energy states, with consequent modification of the optoelectronic properties of the semiconductors. In addition, the stability of the cluster formed by certain numbers of atoms was investigated (Knight et al. 1984), evidencing the increased stability of closed-shell electronic configurations for alkali metals or geometric effects in rare gas clusters. For instance, geometric effects are responsible for the stability of fullerene clusters. Voltage-dependent transmission characteristics of short silicon nanotubes (SiNTs) were investigated using first-principles methods to simulate their metallic behavior, in order to obtain their current–voltage characteristics (Yamacli 2014).

The rapid advancement in nanoscience has also been favored by the introduction of more and more sophisticated machines able to investigate the matter at atomic scale (Midgley and Weyland 2003). Instruments such as the atomic force microscope (AFM), scanning tunneling microscope (STM), 3D electron microscope, and so on, allow a more detailed and reliable characterization of the nanomaterials.

In the scenario of the nanomaterials, a relevant position is occupied by silicon. It can be considered as the innovative driving material of the high-tech industry and consequently strongly impacting our everyday lifestyle owing to its extensive applications such as microelectronics. Over the years, different forms of silicon were developed depending on the application. Starting from massive either n- or p-doped crystalline silicon (c-Si) for the first generation electronic components, amorphous silicon (a-Si) was successively studied in depth by Dong and Drabold (1998). Thin film of a-Si either hydrogenated or not was extensively investigated for solar cell technology in competition with monocrystalline and polycrystalline forms (Khanal et al. 2012; Mandal et al. 2016; Novák et al. 2016). The interest in amorphous and microcrystalline Si is evidenced by the studies conducted in the past on their properties, with attention also to the performance and cost advantages of a-Si over traditional c-Si, for fabricating photovoltaic modules (Iqbal and Vepiek 1982; Jansen et al. 2006). Coatings of a-Si on carbon nanostructures were also investigated for application in Li-ion batteries and photovoltaic cells (Kozinda et al. 2011; Nguyen et al. 2011; Zhou et al. 2012).

In the following, attention will be focused on the SiNT which is a highly desired form of silicon for its fundamental role in the miniaturization trend of electronic devices. After a description of the properties and applications of SiNTs and their fabrication methods, attention will be focused on chemical vapor deposition (CVD) template synthesis that is the most usual method for this material. Then, galvanic template synthesis will be described as a general method for the fabrication of different metal and oxide nanostructures, therefore the use of this technique for synthesizing SiNTs will be detailed. Characterization methods will be also described, confirming that template galvanic synthesis is a valuable tool for fabricating SiNTs because it has been found to be a very simple and cheap route in comparison with the other techniques. In addition, it is an easily scalable process that is the major advantage in light of industrial production.

25.2 PROPERTIES AND APPLICATIONS OF SiNTs

The advancements in the field of nanotechnology after the carbon nanotube discovery in 1991 have led to the synthesis and characterization of an assortment of quasi-one-dimensional (Q1D) structures, such as nanowires, nanoneedles, nanobelts, and nanotubes. These fascinating materials exhibit novel physical properties owing to their unique geometry with high aspect ratio (i.e., size confinement in two coordinates). They are the potential building blocks for a wide range of nanoscale electronics, optoelectronics, magnetoelectronics, and sensing devices (Lu et al. 2006).

In this context, the nanostructured morphology is the most recent desired form of silicon for electronics (Chen et al. 2007; Wingert et al. 2015), energy (Hu and Chen 2007), biomedical (Cheng et al. 2007), and environmental (Kunjie et al. 2011) applications. In the case of SiNTs, applications can be further extended by functionalizing their inner and outer surfaces with organic molecular layers. The functionalization induces selective modification with various functional groups which may lead to complex multifunctional nanostructures, while largely retaining their structural integrity (Ben-Ishai and Patolsky 2011).

The current great efforts in investigating nanoscale forms of silicon are due to the interest in further miniaturizing microelectronic devices and unveiling new properties that often arise at the nanoscale including the goal of integrating electronics and photonics on the same Si chip.

The key role that SiNTs can play in different areas of technological interest was well illustrated by Pei et al. (2010) who highlighted the principal potential applications from lithium-ion batteries to field-effect transistors, magnetic nanodevices, hydrogen storage, nanoscale electron, and field-emitting devices.

Grillet et al. (2012) showed the advantage of using amorphous nanostructured silicon in place of c-Si in waveguides combining high nonlinear figure of merit, high nonlinearity, and good stability at telecom wavelengths.

The adsorption capacity of hydrogen in SiNTs at 298 K and in the pressure range from 1 to 10 MPa was theoretically predicted with a multiscale method, showing that SiNT arrays exhibit much stronger attraction to hydrogen compared to the isodiameter carbon nanotubes (Jianhui et al. 2008).

An attractive core/shell nanostructured architecture consisting of c-Si/a-Si was investigated as a one-dimensional building block of single switch for nonvolatile memory and programmable nanoprocessors (Dong et al. 2008).

In addition, modification of the nanostructured morphology can induce significant properties. For instance, nanostructured silicon thermal conductivity can be modified by passing from nanowires to nanotubes so an Si-based thermal material can be fabricated which can provide electricity when subjected to a temperature gradient or provide cooling performance when electrical current is passed through it (Chen et al. 2010). A strict analysis of the thermal transport in c-Si and a-Si nanotubes showed that the first ones with shell thickness as thin as ~5 nm exhibit low thermal conductivity in contrast with the common opinion, according to which amorphous materials present the lower limit of thermal transport (Wingert et al. 2015). Optoacoustic effects of Si nanowires ignited in air and exhibited when exposed to a conventional photographic flash was attributed to changes of their microstructure with confinement of photoenergy from visible light (Wang et al. 2003).

Electrochemical storage and conversion of energy is another possible application of nanostructured Si. Exciting reviews on this application of SiNTs was made by several authors (Wua and Cuia 2012; Penga et al. 2013; Zamfir et al. 2013). In order to increase the energy density of Li-ion batteries, silicon is considered a promising alternative for advanced anode material owing to the theoretical capacity of 3950 mAhg^{-1} which is about 10 times greater than that of the currently used carbonaceous materials (Prosini et al. 2014; Ashuri et al. 2016). Extensive investigations were focused on overcoming the principal drawback that, at the moment, hinders the use of an Si-based anode in Li-ion battery. This is the limited life cycle of the battery owing to the pulverization of the anode due to the large volume expansion (about 300%) and contraction accompanying Li in de-intercalation (Hieu et al. 2014; Di Leo et al. 2015). The focus on the nanostructured morphology is driven by the possibility to better accommodate the fatigue stress accompanying the cycling (Song et al. 2010), thanks to surface effects, so SiNTs were proposed as a possible effective solution (Wu et al. 2012). In addition, nanostructured anodes present the advantage of shorter diffusion distances for Li species, leading to high charging and discharging rates, that implies significant improvement in the specific energy (Wh g^{-1}).

The principal advantages of Si one-dimensional nanostructured forms are summarized in Figure 25.1.

Synchrotron soft X-ray absorption spectroscopy and related techniques, such as X-ray emission spectroscopy and scanning transmission X-ray microscopy, were employed for revealing the nature of surface and electronic structures of the nanostructures. These play a key role in the development and improvement of nanosystem energy conversion and storage (Zhong et al. 2014). The impedance behavior of silicon nanowire has been investigated to understand the electrochemical process kinetics that influence the performance when used as an anode in an Li-ion battery (Ruffo et al. 2009).

Nanostructured Si is also one of the most investigated materials for application in solar cells. Morphology, electrocatalytic activity, and photovoltaic performances of dye-sensitized solar cell (DSSC) with silicon microstructures as a counter electrode were investigated, and the excellent findings were attributed to the good electrocatalytic activity of the microstructures, facile diffusion based on their well-connected framework, highly ordered microtexture, and large straight pores (Tao et al. 2013).

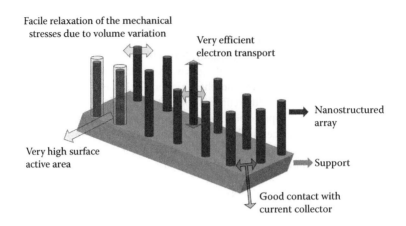

Facile relaxation of the mechanical stresses due to volume variation

Very efficient electron transport

Nanostructured array

Very high surface active area

Support

Good contact with current collector

Figure 25.1 Schematic representation of one-dimension nanostructure and its principal advantages.

A core/shell n–p junction of Si nanowires/amorphous SiNT arrays was fabricated using a solution-phase etching method for n-type Si core, and low-pressure chemical vapor deposition (LPCVD) with subsequent rapid thermal annealing (RTA) of p-type amorphous Si (Garnett and Yang 2008). The same authors successively showed that this architecture is able to increase the path length of incident solar radiation by up to a factor of 73 which was a light-trapping method far superior to any other (Garnett and Yang 2010).

25.3 FABRICATION METHODS

Barth et al. (2010) conducted a valuable review on the various methods for synthesizing the most popular inorganic semiconductors having high aspect ratio form. For each material, the most suitable synthetic methods for fabricating a specific morphology are shown.

The synthesis of SiNTs is strongly challenging owing to the tendency of silicon to make sp^3- rather than sp^2-like bonds, therefore more compact nanostructures such as nanowires are preferentially formed. Silicon hybridization (sp^2, sp^3, or sp^2–sp^3 mixed) is one of the most interesting topics addressed by a variety of theoretical studies exploring the stability, the structure, and the electronic properties of SiNTs. Perepichka and Rosei (2006) conducted an in-depth study on the SiNT structures, also showing the differences from nanostructures of carbon (fullerenes, nanotubes) which is an element of group IV, like Si.

A detailed description of the CVD process based on catalyzed vapor–liquid–solid mechanism for growing nanostructured Si forms was given in Zamfir et al. (2013). In the same paper, other fabrication techniques, among which electrodeposition in porous ion-track-etched polycarbonate membranes from ionic liquid containing $SiCl_4$, were presented.

Copper/silicon core/shell nanostructures were synthesized for application in Li-ion batteries, with a copper core acting as a current collector and an a-Si shell as a host material for Li intercalation. Amorphous silicon shell was prepared with plasma-enhanced chemical vapor deposition (PECVD) (Sun et al. 2014), while core/shell amorphous/crystalline Si nanostructures were fabricated starting from silicon nanowires synthesized by gold catalyzed thermal CVD, and successive crystallization of the shell under laser irradiation in an argon environment (Chang et al. 2011).

Si spherical nanoclusters and tubular structures, approximately in the ratio of 10:1, were fabricated in a DC arc plasma reactor, starting from silicon powder used as an anode (Castrucci et al. 2006). SiNT thin films for Li-ion battery applications were prepared using PECVD on ZnO nanowire thin films at temperatures around 400°C or lower and subsequent removal of the zinc oxide core (Carreon et al. 2015). The same technique was used for uniformly depositing amorphous Si on the CNT surface, at a low temperature (200°C) and a high rate (20 nm min⁻¹) (Liao et al. 2011).

The electrochemical deposition can be considered as one of the most attractive routes for fabricating either 1D or 2D silicon nanostructures because it is simple, easy to scale-up to large surface areas, and applicable to a wide variety of different shapes and surface geometries (Nishimura and Fukunaka 2007; Falola and Suni 2015). In particular, template synthesis is the most suitable technique for obtaining nanotubes.

25.4 TEMPLATE SYNTHESIS

The template synthesis is a powerful method for synthesizing nanostructures of different materials, including plasmonics, which are growing in scientific interest for valuable application in the high-tech microelectronic industry (Jones et al. 2011). Attention has also been addressed to the fabrication of the template, and, specifically, electrochemical techniques have been proposed as one of the most advantageous methods to fabricate highly ordered nanostructures to be used as templates for replicating other nanostructured materials and growing functionalized material arrays. An interesting review was proposed some years ago (Hernandez-Velez 2006).

One of the principal problems associated to the electrochemical synthesis in template is the mass transport, because it occurs in a confined ambient. The matter was investigated by Bograchev et al. (2013a), who proposed a simple and effective model which takes into account the kinetics of electrochemical reaction and the diffusion transfer of metal cations both in the pores and in the outer diffusion layer. The same author further scrutinized the matter in successive papers (Bograchev et al. 2013b, 2015, 2016). This technique has specifically attracted much attention for fabrication of tubular nanostructures because it offers a convenient way for producing structurally uniform nanostructures periodically aligned in template matrices (Lee et al. 2005).

In this context, template synthesis has been found as one of the most suitable techniques for fabricating SiNTs owing to the difficulties in fabricating them in comparison with other more compact nanostructured forms such as Si nanowires. The most popular processes for synthesizing SiNTs are the catalyzed CVD and molecular beam epitaxy (MBE). SiNTs can be obtained using different templates, such as nanopore Al_2O_3 template and removable templates (ZnS/Si nanowires and Ge/Si nanowires) (Pei et al. 2010).

Template synthesis consists of conformal covering or filling in a nanostructured mass having the desired features. Nanoparticles can act as a template, but more commonly hard nanostructured porous materials are used. A fundamental requirement of the template is its easy preferential removal, in order to expose the synthesized nanostructured morphology, which replicates the template one. Therefore this technique is highly preferred because it allows the tailoring of the nanostructured morphology by a suitable choice of the template. Another basic requirement for successful template synthesis is the chemical stability of the synthesized nanostructures during the chemical removal of the template.

Also either liquid droplets or gas bubbles can act as a template. In these cases, the growth mechanism together with the template removal is different. Imhof and Pine (1998) proposed the basic idea of using a highly uniform dispersion of liquid emulsion droplets of one fluid dispersed in a second immiscible fluid and stabilized by a surfactant as a template around which solid material is grown. CuS micrometer-sized hollow spheres were synthesized by a single-step procedure based on a template interface reaction between CuCl and sulfur liquid droplets in ethylene glycol solvent (Wan et al. 2004). A droplet template was also used for synthesizing Janus nanoparticles (Sun et al. 2016), whose physical properties and extensive applications in addition to all aspects from synthesis to self-assembly were presented by Walther and Müller (2013).

A physically patterned solid surface was also used as a template for assembling polystyrene beads and silica colloids (≥150 nm in diameter) into complex aggregates that include polygonal or polyhedral clusters, linear or zigzag chains, and circular rings (Yin et al. 2001). In general, the surface patterning by physical methods such as lithography leads to highly ordered arrays useful for applications in microelectronics industries.

Gas-bubble dynamic template deposition method for synthesizing $Ni/Ni(OH)_2$ composite with three-dimension porous hierarchal nanostructure was recently performed via an electrodeposition process followed by an etching step (Ashassi-Sorkhabi et al. 2016). In addition, hollow α-Fe_2O_3 (hematite) microspheres can be formed by the gas-bubble template method (Valladares et al. 2016), while bimetallic AuPt nanowire networks with tunable compositions were fabricated by using hydrogen bubble as a dynamic template (Liu et al. 2016).

The advantage of using electrochemical techniques in conjunction with CVD for fabricating nanostructure arrays was shown by Emilio Munoz-Sandoval et al. (2007). They reported the growth of nitrogen-doped multiwalled carbon nanotubes aligned on porous silicon surfaces, patterned by hydrogen bubbles covering the surface during electrochemical etching in HF-containing solutions. In practice, the gas bubbles act as a template, while the combination of the current and the way the electrolyte flows on the substrate resulted in the formation of SiOx-rich micropatterns with different shapes and size. A four-step fabrication method of nanostructured core/shell silicon array solar cell, including deep reactive ion etching and boron diffusion to form the radial p–n junction, was also proposed (Garnett and Yang 2010).

In order to overcome the hindrance due to sp³ hybridization, epitaxial template synthesis was proposed for fabricating SiNTs. According to lattice matching theory, SiNTs can be deposited on ZnS and vice versa because diamond-like cubic Si and ZnS have similar crystal structures and very close lattice constants, therefore ZnS nanowires were used as a one-dimensional template and crystalline SiNTs were obtained after chemical removal of the ZnS nanowire core (Hu et al. 2004).

Mbenkum et al. (2010) showed that using hydrogen instead of oxygen plasma during ordered gold nanoparticle array deposition on borosilicate glass and SiOx/Si substrates enabled Si 1D growth at temperatures as low as 320°C. In particular, SiNTs were obtained right up to 420°C on SiOx/Si, while a mixture of SiNTs and SiNWs was observed at 450°C and only SiNWs grew at 480°C.

A direct but rather complex and time-consuming template fabrication of SiNTs was proposed (Wen et al. 2013). It consists of synthesizing rod-like NiN_2H_4 to be used as a template for preparation of silica nanotubes that were converted to SiNTs by a thermal reduction, assisted with magnesium powder.

25.4.1 NANOPOROUS TEMPLATES

Hard nanoporous materials are the most fascinating templates, because they provide geometric confinement where either atoms or molecules aggregate so that bottom-up nanostructure growth, replicating the features of the host material, occurs.

Not all of the nanoporous materials are suitable for acting as templates, because some requirements must be satisfied; first of all, is the easy removal after the growth of the nanostructures without damaging them. Anodic alumina and polycarbonate membranes (Schönenberger et al. 1997; Bocchetta et al. 2002; Inguanta et al. 2009a, b; Battaglia et al. 2014 a, b) are the most popular templates for fabricating one-dimensional nanostructures. The comparison between them well evidences the role of the template, because they strongly differ in the morphology.

The template-based approach using porous anodic alumina (PAA) has been extensively investigated due to its highly ordered close-packed hexagonal structure, and the flexibility to modify it by adjusting the influencing factors during the anodization process such as the electrolyte nature, anodization voltage, process temperature, and duration. Especially, anodization voltage strongly affects the pore diameter and interpore distance (Wood and O'Sullivan 1970). It has been found that the most ordered structures with perfectly parallel cylindrical channels to be also used as a photonic crystal are formed after two-step anodization (Masuda and Fukuda 1995). The major advantage in using anodic alumina membrane as a template is the possibility of tuning the morphology features of the porous mass by adjusting the aluminum anodizing conditions (Butera et al. 2006; Inguanta et al. 2007a). By this way, different diameter channels can be formed by aluminum foil anodization in H_2SO_4, oxalic acid, and H_3PO_4, where the largest diameter up to about 200 nm can be obtained. Over the past years, Furneaux et al. (1989) conducted extensive investigations on the fabrication of anodic alumina membranes starting from the pioneering work. Figure 25.2 shows the typical morphology of PAA formed in 0.4 M H_3PO_4 at 180 V and room temperature (25.2a cross-sectional view, 25.2b top surface view). The section view (Figure 25.2a) evidences the parallelism of the channels, which are schematized in Figure 25.3 where the fabrication steps of nanostructures through electrochemical deposition are shown.

On the contrary, polycarbonate membranes are manufactured by track-etching of polycarbonate film through irradiation of heavy ions from nuclear fissions and have excellent chemical and thermal resistance with nominal diameters between 10 and 200 nm. Different from the alumina membrane, polycarbonate channels are highly interconnected, as shown in Figure 25.4 where typical section (Figure 25.3a) and top surface (Figure 25.3b) views of track-etched membranes from Whatman® (cyclopore type) are shown.

The interconnected morphology contributes to improve mechanical stability of the nanostructures after removal of the template (Silipigni et al. 2014), as evidenced in Figure 25.5 where the fabrication scheme of one-dimension nanostructures electrodeposited into the channels of a Whatman track-etched polycarbonate membrane is shown.

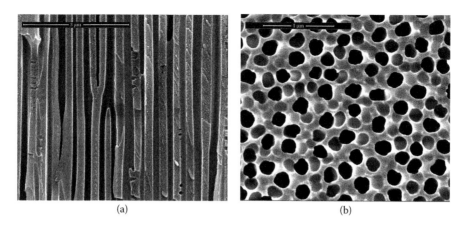

Figure 25.2 (a) Typical section and (b) top surface views of porous anodic membrane formed by aluminum anodization in 0.4 M H$_3$PO$_4$ at 180 V and room temperature.

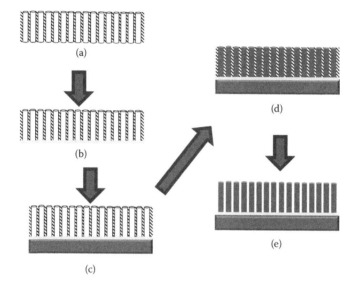

Figure 25.3 Schematics exemplifying nanostructure synthesis by electrodeposition into porous anodic alumina acting as a template. (a) Anodic alumina membrane. (b) Mt sputtering. (c) Current collector electrodeposition. (d) Nanostructures electrodeposition. (e) Nanostructured electrode.

The most significant features of PAA and polycarbonate membranes usually employed as templates are shown in Table 25.1.

Figures 25.2 through 25.5 clearly show that the electrochemical techniques, either direct electrodeposition or galvanic deposition (see below), for growing nanostructured material in template are valuable. Their peculiar advantage is the extreme versatility because different metals, alloys, and oxides can be synthesized by proper choice of the deposition conditions for a given precursor dissolved in the solution. Figure 25.6 shows materials for different applications which were synthesized in a nanoporous template by either electrodeposition or galvanic deposition.

Electrochemical deposition in PAA acting as a template was also investigated for fabricating mesoporous silica nanowires (Ren and Lun 2012).

Direct synthesis of Si nanowire arrays using gold nanoparticles as a catalyst and a PAA as a template is shown in Kim et al. (2012). Sha et al. (2002), who also proposed a model for explaining the formation of either SiNTs or SiNWs, investigated a similar procedure. Through MBE, Jeong et al. (2003) grew SiNTs on top of the protruding pore edges of PAA without using a gold catalyst.

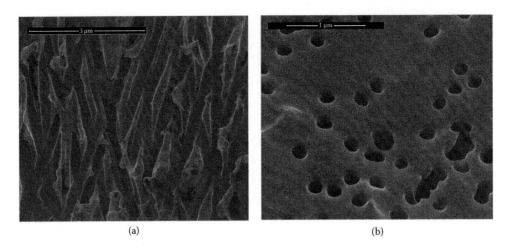

(a)   (b)

Figure 25.4 (a) Typical section and (b) top surface views of track-etched polycarbonate membrane from Whatman® (Cyclopore).

Table 25.1 Characteristic features of PAA and polycarbonate membranes used as templates

TEMPLATE	POLYCARBONATE	ALUMINA
Mean pore diameter (nm)	180	250
Pore density (pores/m^2)	$3.9*10^9$	$8*10^{12}$
Porosity (%)	15	30
Interpore distance (nm)	100–800	100–130
Thickness (μm)	15–20	50–60
Fabrication method	Track etched	Anodization

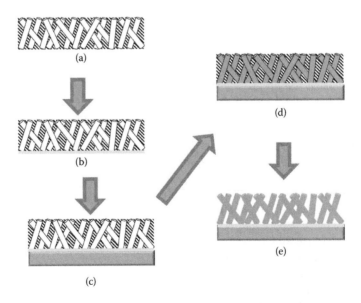

(a)

(b)

(c)

(d)

(e)

Figure 25.5 Steps of nanostructure synthesis by electrodeposition into track-etched polycarbonate membrane acting as a template. (a) Polycarbonate membrane. (b) Mt sputtering. (c) Current collector electrodeposition. (d) Nanostructures electrodeposition. (e) Nanostructured electrode.

Nanowires and nanotubes

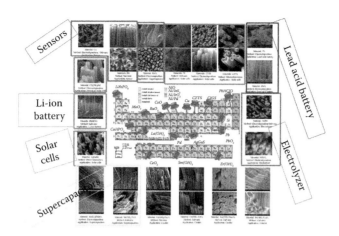

Figure 25.6 Pure and composite nanostructured materials for different applications synthesized through either electrochemical or galvanic deposition in PAA, or in polycarbonate membranes.

Huang et al. (2010) proposed a modification to the nanoporous template synthesis in the paper. They used PAA template to form arrays of metal nanodots that act as a hard mask for fabricating, by chemical wet etching uniform SiNWs with high density and highly controllable diameters through the fine control over the pore size of the PAA template used in the metal dot deposition.

25.4.2 GALVANIC TEMPLATE SYNTHESIS

Galvanic template synthesis is based on the electrical connection between two different materials which are shorted through a simultaneous ohmic and electrolytic connection. By this way, the two materials behave as electrodes, able to separately sustain oxidation and reduction reactions, whose difference in the electrochemical standard potentials gives the maximum driving force. Figure 25.7 shows two possible connections depending on dipping the materials in different solutions (Figure 25.7a) or not (Figure 25.7b).

The displacement reaction leading to the deposition of the desired material can be schematized as follows:

$$mM + nM'^{m+} \rightarrow mM^{n+} + nM' \tag{25.1}$$

where M' is the material of the interest.

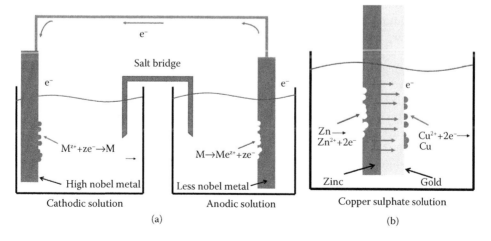

Figure 25.7 Schemes of possible galvanic connections: (a) anode (corroding) and cathode (thickening) dipped in different electrolytes, with salt bridge guaranteeing the electrolytic connection; (b) anode and cathode dipped in the same electrolytic solution.

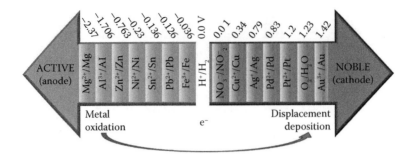

Figure 25.8 Electrochemical standard potential of some redox couples (Volt), and operation scheme of displacement reaction.

The electrochemical standard potential of the redox couples involved in the reaction gives the thermodynamic feasibility of a displacement reaction. In particular, reaction (25.1) occurs when the electrochemical standard potential of M^{n+}/M is less than one of M'^{m+}/M'. According to reaction (25.1), M behaves as a sacrificial anode and it is progressively consuming while M' is accumulating onto the cathodic area.

Figure 25.8 gives the electrochemical standard potentials of some redox couples, also showing a simple thermodynamic scheme of feasibility of a displacement reaction.

The principal parameters to be adjusted for controlling the galvanic process are i) the nature of the sacrificial anode, ii) the anode-to-cathode surface area ratio, iii) the temperature, and iv) time.

For a given displacement reaction, the choosing of the anode material is principally associated to the desired driving force. The most common materials used as sacrificial anodes are zinc, aluminum, and magnesium. Among these three materials, the thermodynamic driving force is the highest for magnesium, the lowest for zinc. In any case, care must be taken in avoiding the passivating conditions of the sacrificial anode. From this point of view, aluminum is the material at the highest risk of being covered by a nonconductive passivating film, even if it holds a great attraction due to its very low electrochemical standard potential and facile handling.

The anode-to-cathode area ratio determines the current density on each area. In turn, it controls the displacement reaction rate when the charge transfer is the controlling step. When the displacement reaction is occurring inside the nanopores of a template, mass transport can be the real factor controlling the rate of growth of the nanostructures owing to the frequent collisions of the diffusing species with the pore walls.

The temperature always plays a fundamental role, because both reaction rate and mass transport are enhanced as temperature is increased, therefore it could appear to be advantageous to operate at higher than room temperature. The principal drawback in increasing the operative temperature is the morphology of the deposit which can result in nonuniform height, but especially, it might result in being formed by heterogeneous aggregation of nanotubes and nanowires, when a nanoporous template is used. To avoid these problems, it is advisable to operate at room temperature even if deposition time is longer. Operation time also controls either nanostructure height, or film thickness and uniformity depending on whether 1D or 2D nanostructures are forming. Therefore, to adjust the deposition duration is of great importance for controlling characteristic features in view of the desired application.

Inguanta et al. (2007b) patented the use of the galvanic displacement reaction for depositing nanostructured material in a template. Initially, the attention was focused on the synthesis of copper (Inguanta et al. 2008a) and palladium (Inguanta et al. 2009a) nanowires starting from electrolytic solutions containing copper and palladium ions, respectively, and using anodic aluminum membrane as a template.

An interesting redox reaction-based method for depositing mono- and multisegment nanowires of various pure metals such as Au, Pt, Pd, Cu, Ni, Co, and their alloys with both linear and branched morphologies was proposed by Xu et al. (2009). The method is based on the simple aqueous solution infiltration of metal chloride salts into a native porous anodic aluminum oxide template with ring-shaped Al foil, so that electric power, organic surfactants, or modifying on the pore walls is not required.

The galvanic replacement reaction was also used for synthesizing bimetallic nanostructures with nonrandom metal atom distribution that are very important for various applications such as catalysis (Teng et al. 2008).

In addition to deposition of metallic nanostructures, galvanic template synthesis was found to be an effective method for growing oxy/hydroxide nanostructures into the pores of anodic alumina membranes (Inguanta et al. 2011a; Inguanta et al. 2012a, 2012b). In this case, the process consists of inducing a local pH increase at the electrode/solution interface (the well-known "electrogeneration of base") in order to precipitate more or less hydrated metal oxides directly into membrane pores, without external electric power. In the case of lanthanide oxy/hydroxides synthesis, the running reactions leading to their precipitation are (Inguanta et al. 2012a, 2012b)

$$NO_3^- + H_2O + 2e^- \rightarrow NO_2^- + OH^- \tag{25.2}$$

$$2H_2O + 2e^- \rightarrow H_2 + 2OH^- \tag{25.3}$$

$$Ln^{3+} + 3OH^- \rightarrow Ln(OH)_3 \downarrow \rightarrow Ln_yO_x \tag{25.4}$$

Structure and morphology of lanthanide oxy/hydroxide nanostructures in the form of both nanotubes and nanowires were investigated as a function of the deposition time and electrolytic bath temperature (Inguanta et al. 2012a, 2012b). It was found that independently of temperature, a crystalline structure was formed only in the case of CeO_2, while the other LnO oxy/hydroxides were amorphous or strongly disordered.

ZnO nanorods for optoelectronic applications were deposited by a galvanic cell-based approach on various conducting substrates at low temperature without the seed layer, which is usually required to improve the density and vertical alignment (Zheng et al. 2013).

In addition, 2D nanostructured deposits were fabricated by galvanic displacement reaction. Krishnamurthy et al. deposited nanoporous Si thin film onto 6061 Al alloy from dilute aqueous hydrofluoric acid solution at pH 2.5 containing SiF_6^{2-} ions, which were reduced to Si owing to simultaneous oxidation and dissolution of Al (Krishnamurthy et al. 2011). A similar procedure was proposed by Falola and Suni 2014 for depositing Ti onto Al 6061 alloy from an aqueous solution containing hydrofluoric acid and TiF_6^{2-} ions. The interest for this study is in showing the potentiality and versatility of the galvanic displacement because it evidences that it is possible to reduce titanium ions from an aqueous solution which is highly challenging owing to the very negative potential for reducing titanium ions. Water reduction leading to hydrogen production is the thermodynamic favorite reaction in comparison to the more negative one, and it becomes the prevailing, if not exclusive situation also at high current densities, when the usual electrochemical technique is applied.

In some cases, ionic liquid solutions were preferred to aqueous solutions where the desired process such as copper deposition onto aluminum could be easily conducted (Kang et al. 2014). In this case, the combination of ionic liquids with a suitable concentration of thiourea restrained the galvanic replacement process, resulting in obtaining a compact and uniform copper film.

PbTe and PbTe/Te nanostructured films for thermoelectric and optoelectronic applications were synthesized using a thin cobalt film as a sacrificial anode in acidic nitrate baths containing Pb^{2+} and $HTeO^{2+}$ ions, whose relative concentration strongly influences the composition of PbTe nanostructures (Chang et al. 2014).

Through a galvanic displacement, silver was deposited on both p- and n-silicon substrates from fluoride-free baths, with positive effects on the environment (Djokica and Cadien 2015). The overall displacement reaction proposed by the authors is

$$Si + 4OH^- + 2Ag(NH_3)_2^+ \rightarrow SiO_3^{2-} + 2Ag + H_2O + 4NH_3 + H_2 \tag{25.5}$$

Where silicon acts as an anode (E° = –1.69 V in alkaline solution), while silver amino complex ion is reduced to Ag simultaneously with OH^- to H_2 because the $2Ag(NH_3)_2^+/Ag$ couple is less noble than Ag^+/Ag (E° = 0.80 V).

Galvanic replacement reaction driving base electrogeneration was also proposed for depositing thin films of multielement compounds, such as brushite ($CaHPO_4 \cdot H_2O$)/ hydroxyapatite ($Ca_{10}(PO_4)_6(OH)_2$), onto 316LSS (Blanda et al. 2016). Zinc was used as a sacrificial anode, and the role of temperature and deposition time was investigated in depth for controlling coating morphology, structure, and composition in order to fabricate a composite material with valuable features for biomedical applications.

A last issue to be addressed deals with the difference between galvanic and electrochemical methods for fabricating nanostructures through deposition in a porous template. Apparently, the two methods appear identical because both are based on the ion reduction of the desired material. In practice, the electrochemical method requires an external power supply while the galvanic reduction is driven by a sacrificial anode which supplies the reducing electrons. Really, the action mechanisms are identical but the galvanic method gives a major advantage consisting of a finer control of the deposition parameters, so it is possible to attain nanostructures with distinctive features difficult or impossible to be obtained through conventional electrochemical methods.

The principal drawback of the galvanic method is the duration that is rather long, usually. However, if the mutual effects of the various parameters controlling a galvanic displacement reaction were scrutinized in depth, it can be concluded that the low rate of the process is the sign of the strength of the galvanic method.

25.5 PREPARATION OF AMORPHOUS Si NANOTUBES

For the template synthesis of SiNTs through galvanic displacement reaction, an appropriate experimental apparatus was made which was patented by Inguanta et al. (2013). As shown in Figure 25.9, the core of the apparatus is an aluminum tube for the double purpose of sacrificial anode and sustaining the template. The last one was a track-etched polycarbonate membrane, which was glued to the aluminum tube with a conductive paste in order to guarantee the ohmic continuity with the gold film sputtered on one side of the membrane. The principal advantage of this arrangement was the facility of depositing in succession a copper layer (Figure 25.9a) acting as a current collector and SiNTs (Figure 25.9b) after a simple rotation of the tube for dipping the membrane into the solution. Both copper layer and SiNTs were deposited by galvanic displacement reaction, using the same aluminum tube as a sacrificial anode. Figure 25.9c shows the scheme of the electrode at the end of the deposition process while Figure 25.9d shows it after template dissolution. As below, the presence or not of the copper current collector strongly influences the morphological quality of the SiNTs.

Initially, the tube was mechanically polished with a 1200 grade abrasive paper, then it was ultrasonically treated in pure water and acetone, and dried in air. The exposed aluminum surface was delimited by a lacquer. For deposition of SiNTs, a template surface of 2.5 cm^2 was exposed, while the sacrificial anode area was 50 cm^2. Really, the area ratio was less than 0.05, because 2.5 cm^2 was the geometric area of the template, while the real cathodic surface was less owing to the porosity of the template. A rough estimation of the real surface can be conducted assuming a porosity around 30%. In this case, the cathodic current density was 66.67-fold higher than the anodic one, being the true surface about 0.75 cm^2. This low cathodic-to-anodic area ratio guaranteed a very high cathodic current density, which is of advantage for reducing the silicon precursor, as will be discussed below.

The aqueous solution composition was 20 mM Na_2SiF_6, 0.2 M HF, and 80% (w/w) formic acid with an initial pH of 0.15. Na_2SiF_6 was the silicon source, while fluoride ions from hydrofluoric acid acted essentially as a stabilizer of silicon hexafluoride according to the equilibrium

$$SiF_6^{2-} \rightarrow SiF_4 + 2F^- \tag{25.6}$$

The beneficial stabilizing effect of fluoride ions, according to reaction (25.6), has been evidenced by conducting experiments at various concentrations of hydrofluoric acid, in otherwise identical conditions. It was found that fluoride concentration enhancement from 0.02 to 0.2 M favored compact texturing and thickening of the nanotube. The morphological characterization of the nanotubes was conducted by an FEG-ESEM

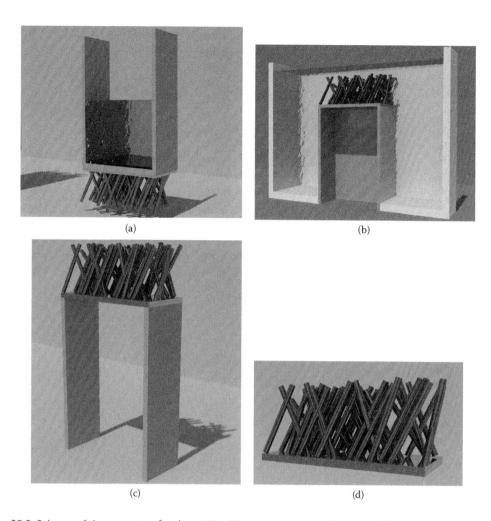

Figure 25.9 Scheme of the apparatus for depositing (a) copper current collector and (b) Si nanotubes, through galvanic displacement reaction. Scheme (c) shows the electrode at the end of deposition process while scheme (d) shows Si nanotubes after template dissolution.

200 QUANTA scanning electron microscope (SEM) equipped with energy dispersive spectrometry (EDS) for elemental characterization. Figure 25.10 shows the difference in morphologies of the SiNTs formed at different concentrations of hydrofluoric acid. At low concentration, the wall thickness was extremely thin so that the SiNTs were transparent. In addition, the mechanical stability of the array was scarce, and the nanostructures collapsed during template dissolution, so that the SiNTs were useless (Figure 25.10a). On the contrary, too high a concentration of hydrofluoric acid creates some problems in the managing the synthesis apparatus because special materials hardly resistant to fluoride ions are requested. In addition, the stability of the sacrificial anode is put at risk [see reaction (25.7), below]. At high F^- concentration (up to 1 M), reaction is favored toward products with large dissolution of aluminum, according with the law of mass action. Another negative effect, likely the most constraining, is due to the half-cathodic displacement reaction (25.8) (see below). In the presence of high F^- concentrations, equilibrium (25.8) is carried too far toward reagents, with consequent growing thermodynamic hindrance to silicon depositing, owing to the law of mass action, also in this case. After numerous experiments, hydrofluoric concentration was optimized at 0.2 M (Figure 25.10b).

The half-anodic displacement reaction, involving aluminum tube is

$$Al + 6F^- \rightarrow AlF_6^{3-} + 3e^- \quad E^\circ = -2.07 \text{ V (NHE)} \tag{25.7}$$

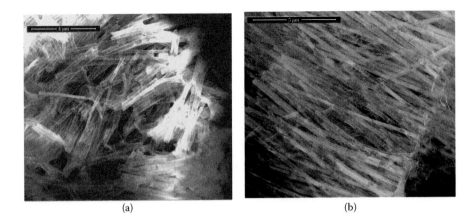

Figure 25.10 Influence of the hydrofluoric acid concentration on the morphology of the Si nanotubes: (a) 0.02 M and (b) 0.2 M.

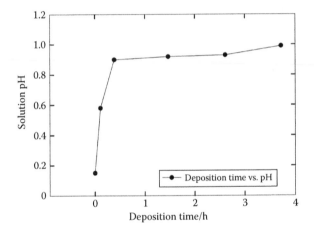

Figure 25.11 Bulk solution pH versus deposition time plot.

while the desired reaction leading to deposition of the SiNTs is

$$SiF_6^{2-} + 4e^- \rightarrow Si + 6F^- \quad E° = -1.37 \text{ V (NHE)} \tag{25.8}$$

Given the very negative standard potential of (25.8), the favorite reaction in acidic aqueous solution is

$$2H^+ + 2e^- \rightarrow H_2 \quad E° = 0.00–0.059pH \text{ V (NHE)} \tag{25.9}$$

In the practice, reaction (25.9) is by far the prevailing one, with abundant hydrogen evolution inside the template pores. Owing to this reaction, the solution pH at the interface progressively increases because the H^+ replenishment from the surrounding solution does not compensate its depletion, therefore the thermo-dynamic potential (E°) shifts toward more negative values. Figure 25.11, showing the bulk solution pH versus deposition time plot, clearly supports the occurrence of reaction (25.9). The figure evidences that solution pH out of the pores progressively increases with the advancement of the galvanic reaction. The increase is very fast within the first 33 minutes, then it slows down to about pH 1. This behavior can be explained considering such a vigorous hydrogen evolution to determine a local pH rise, quickly extending out of the pores up to the bulk of the solution. This means that H^+ consumption occurring at the interface of the growing nanostructures is not compensated by the supply from the surrounding solution, and, it is likely that a pH gradient is established through the pore length up to the bulk solution, where, according to

Figure 25.11, pH progressively rises with different rate from the initial 0.18 to 1 after about 3 hours 52 minutes. The high diffusion resistance of H+ ions through the pore length must also be considered as a further cause in establishing the pH gradient. It is likely that the hindrance to the diffusive transport is due to the high frequency of collisions with the pore walls, and bulk solution stirring is ineffective because the solution inside the pores is stagnant and is not involved in the external stirring owing to the small size of pore entrance. This interpretation was confirmed by proper experiments conducted at various solution stirring rates which showed the absence of any effect on the curve in Figure 25.11. If stirring had been effective in accelerating H+ diffusion, a different dependence of pH on time would be expected.

The pH increase is dangerous because it can trigger off the conversion of SiF_6^{2-} to SiO_2 according to

$$SiF_6^{2-} + 4OH^- \rightarrow SiO_2 + 2H_2O + 6F^- \tag{25.10}$$

In this scenario, formic acid plays a fundamental role. Being a weak acid, its dissociation

$$HCOOH \rightarrow H^+ + COOH^- \tag{25.11}$$

advances with the pH enhancement, slowing it down. From this point of view, formic acid behaves almost like a pH buffer of the interface. Despite the high concentration of formic acid, the pH of the bulk solution increases as shown in Figure 25.11. This behavior suggests that formic acid is effective in slowing down the pH rise in the initial stages of the galvanic reaction. As H+ ions are progressively consumed, fresh formic acid must diffuse from the next surrounding solution toward the interface. Since diffusive transport in the template pores is slow, the buffer action of formic acid progressively diminishes.

The influence of formic acid on deposition of SiNTs has been scrutinized in depth through experiments conducted at different concentrations. At 100%, the only possible reaction involving formic acid is its oxidation to CO_2 according to

$$HCOOH \rightarrow CO_2 + 2H^+ + 2e- \quad E° = -0.20 \text{ V(NHE)} \tag{25.12}$$

whose effect should be the generation of H+ undergoing the reduction of (25.9). Since both driving force and conductivity of the liquid phase are low, the possible H+ reduction can occur on surface cathodic sites close to the anodic ones, according to the model of local galvanic couple determining metallic material corrosion. Of course, deposition of silicon cannot be driven by reaction (25.12), because its standard potential is nobler than that one of reaction (25.8). About this, the driving force evaluation for a galvanic displacement reaction in nonaqueous solution is worth looking into. The thermodynamic feasibility of a redox reaction depends on the electrochemical standard potentials of both anodic and cathodic reactions. These potentials are evaluated in aqueous solution, therefore, strictly speaking, they hold only in this condition, whereas reaction (25.12) occurs in a nonaqueous solution. Nevertheless, going from aqueous to nonaqueous systems, the electrochemical redox scale is still valid, even if the numerical values are slightly different. A typical example is an Li-ion battery, where Li+/Li holds its negative redox value also in a nonaqueous electrolyte. On this basis, it is possible to infer a possible route leading to the observed deposition of silicon traces on the aluminum also in the presence of pure formic acid by invoking the local galvanic couple action on aluminum surface, where the overall reaction occurs:

$$4Al + 3Na_2SiF_6 \rightarrow 3Si + 6NaF + 4AlF_3 \tag{25.13}$$

This reaction can occur only locally where anodic and cathodic sites are strictly close each to other, because the liquid phase formed by 100% formic acid is scarcely conductive. Therefore, Si cannot be deposited inside the template pores, because of the high distance of pore cathodic area from the aluminum. Experimental findings have confirmed this interpretation.

Experiments in solutions containing 40% (w/w) formic acid were also conducted. In this condition, SiNTs were deposited into the pores, but their morphology was unsatisfactory. As shown in Figure 25.12a, they appeared as restricted with a weak texture in comparison to that attained at 80%, shown in Figure 25.12b. Consequently, the 80% (w/w) was considered as the best acid formic concentration for fabricating SiNTs.

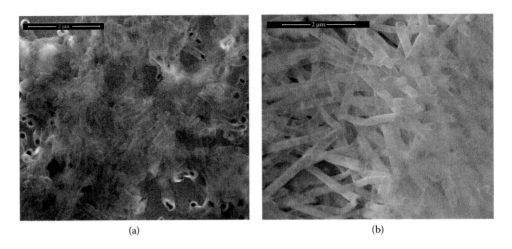

(a)   (b)

Figure 25.12 Influence of the formic acid concentration on the morphology of the Si nanotubes: (a) 40% (w/w), (b) 80% (w/w).

In addition, the deposition of a copper layer on the gold-sputtered film is fundamental for fabricating good quality SiNTs. In practice, it acts as both current collector and structural reinforcement of the nanotubes. It was observed that in the absence of the copper layer, the deposit easily broke as shown in Figure 25.13, where SEM images of SiNT array deposited into a template with and without an underlying copper layer is shown. Figure 25.13a clearly shows the structural role of the copper layer, because the deposit can be easily handled when copper layer is present, while it broke just handled for disassembling the cell when absent. Copper layer was deposited through a galvanic displacement reaction based on aluminum as a sacrificial anode and 0.2 M copper sulfate as a copper source.

About deposition of SiNTs, it must be observed that advancement of the reaction (25.9) is largely prevalent with respect to the reaction (25.8) because is thermodynamic favored and there is not any relevant kinetic hindrance. In these condition, deposition of SiNTs according to the reaction (25.8) can occur only in presence of a high cathodic current density which can be achieved by an adequately low cathodic-to-anodic area ratio. By this way, a large cathodic charge transfer overpotential is attained, according to the Tafel equation (Bockris and Reddy 1970). In the presence of concurrent cathodic reactions, when the overpotential curve of the favored reaction overcomes the thermodynamic value of the less noble reaction, the latter starts and advances faster if its overpotential as a function of the current density is less sloping than the other. This is the case of reactions

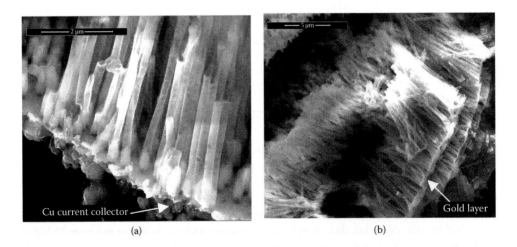

(a)   (b)

Figure 25.13 Role of the copper layer deposited on gold sputtered film (a) with and (b) without copper layer.

(25.8) and (25.9). At very high current density, reaction (25.8) occurs simultaneously with reaction (25.9), with current distribution depending on the overvoltage of each reaction. Therefore, it can occur that the partial current of the less favorite reaction is higher than that of the favorite one if the Tafel curve of the latter is more sloping. A simple scheme of this behavior is shown in Figure 25.14. This interpretation for deposition of SiNTs is fully supported by the voltage curve monitored during galvanic deposition and shown in Figure 25.15. Soon after the galvanic connection, the potential shifts from −0.1 to −0.8 V, remaining constant at this value for the whole deposition time. Since anode and cathode are electrically shorted, the potential of Figure 25.15 is a mixed potential, and its very negative value indicates that the galvanic process is under cathodic control—the Tafel line of the cathodic process is much more sloping than the anodic one. The high slope depends on charge transfer overvoltage that exponentially depends on the cathodic current density, confirming the need of establishing a very low cathodic-to-anodic area ratio for the advancement of reaction (25.8).

The hydrogen evolution reaction is fundamental for obtaining silicon in the form of nanotube, as nanowire is the favorite morphology owing to the preferred sp³ hybridization of silicon. The role of the hydrogen bubble in controlling the morphology of deposits into the nanostructured template was evidenced in the literature (Fukunaka et al. 2006; Inguanta et al. 2008b; Inguanta et al. 2011b; Battaglia et al. 2013; Battaglia et al. 2014). Since reaction (25.8) is the prevailing one, hydrogen bubbles accumulate inside the template channels, so that the silicon deposition is confined to the gap between the pore wall and the bubble, according to the model by Inguanta et al (2008b). Consequently, SiNT growth occurs through addition of single atoms, according to a typical bottom-up mechanism.

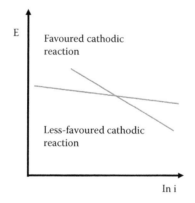

Figure 25.14 Schematic representation of the simultaneous advancement of galvanic cathodic reactions with different both standard potentials and Tafel line.

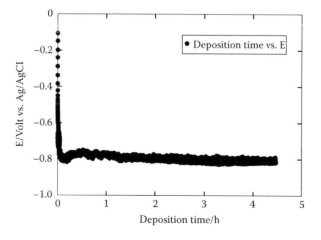

Figure 25.15 Mixed potential of the galvanic assembly for Si nanotube deposition as function of time.

25.6 CHARACTERIZATION OF AMORPHOUS Si NANOTUBES

Figure 25.16 shows the EDS spectra at the end of the deposition and after template dissolution. Silicon and gold peaks are well evident, where the first is due to the galvanic deposit, while the second ones are due to the sputtered film on one side of the template. The sample shows carbon and oxygen peaks due to the polycarbonate membrane, which partially remain also after its chemical dissolution.

The structure of the deposit was determined by X-ray diffraction (XRD) patterns which were recorded soon after electrodeposition by using a Philips generator (mod. PW 1130) and a PW goniometry (mod. 1050). Diffractograms were obtained in the 2θ range from 10° to 100° with a step of 0.02° and a measuring time of 0.5s per step, using the Cu Kα radiation (λ = 1.54A°). XRD patterns did not show any peak attributable to silicon, therefore it can be concluded that the nanostructured deposit was amorphous.

This conclusion was confirmed by Raman spectroscopy which was conducted by a Renishaw (inVia Raman microscope) spectrometer equipped with an He:Ne 532 nm laser. Analysis was carried out with 10% laser power and a spot of 2 μm performing three acquisitions for each analyzed point in the range between 1500 and 100 cm^{-1}. Prior to analysis, system was calibrated by means of the Raman peak of polycrystalline Si (520 cm^{-1}). All spectra were obtained at room temperature and the analysis was performed on several points to verify homogeneity of the samples. Raman bands were identified by comparison with the RRUFF™ database (Downs and Hall-Wallace 2003). Figure 25.17 shows the spectrum of the as-deposited sample (a) together with the spectra of the template (b) and crystalline silicon (c). The last was from instrument calibration, while the template spectrum was collected prior to depositing SiNTs. The formation of crystalline silicon can be excluded by the comparison of the as-deposited sample spectrum, where only peaks due to the template are present, with the crystalline silicon one which presents a strong peak at about 520 cm^{-1}.

The semiconducting nature of the SiNTS was investigated by photoelectrochemical spectroscopy carried out at room temperature in aerated 0.1 M Na$_2$SO$_4$ solution (pH = 5.6), using a cell having flat quartz windows for allowing sample illumination. A Pt net served as the counter electrode, and the reference was a mercurous sulfate electrode (MSE, E° = 0.65 V vs. NHE). Monochromatic irradiation was achieved using a UV-Vis Xenon lamp (Oriel) coupled to a UV-Vis monochromator (Bausch & Lomb), mounted in an optical line equipped with quartz optics. For improving photocurrent resolution, a two-phase lock-in amplifier (EG&G, mod. 5206) was used in connection with a mechanical chopper (frequency: 10 Hz). Data were acquired by a desk computer through an analogic interface using LABVIEW™ 7 software and processed according to home-written programs. Photocurrent

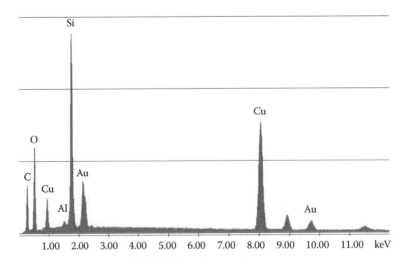

Figure 25.16 EDS spectra at the end of the Si nanotube deposition and after dissolution of the template.

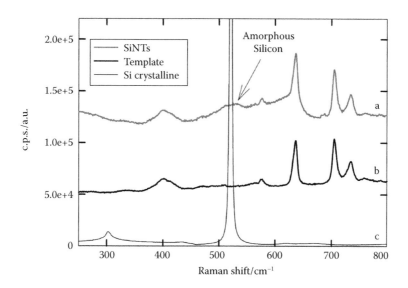

Figure 25.17 RAMAN spectra of (a) the as-deposited Si nanotubes (b) the template (c) and crystalline silicon.

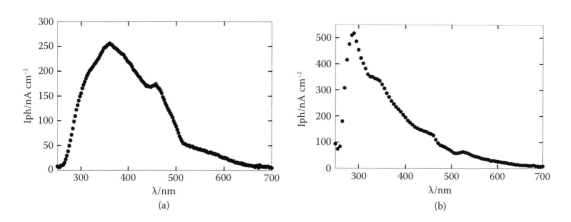

Figure 25.18 Photocurrent action spectrum of Si nanotube (a) before and (b) after correction for lamp emission.

spectra reported below were corrected for the photon emission at each wavelength of the lamp/mono-chromator system; the latter was detected using a calibrated thermopile (Newport). A typical photo-current spectrum collected in the wavelength ranging from 220 to 700 nm is shown in Figure 25.18a, while Figure 25.18b shows the same spectrum after correction for lamp emission. The optical gap of the deposit was evaluated by the extrapolation to zero photocurrent of the

$$(Iph*h*v)^{0.5} \text{ vs. } h*v$$

assuming nondirect optical transitions, as shown in Figure 25.19. A value of about 1.63 eV was obtained, which agrees well with the values in the literature (1.54–1.78 eV) (Fonash 2010). The semiconducting type was evaluated by current transients, generated under manually chopping the incident monochromatic light of different wavelengths. Figure 25.20 shows that the galvanic-deposited SiNTs behave as n-type semiconductors. The same figure also shows the absence of photocurrent spikes, indicating a low recombination rate at the surface of the electron–hole pairs generated by illumination.

Nanowires and nanotubes

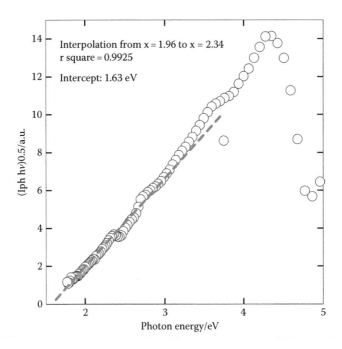

Figure 25.19 Determination of the optical gap of the amorphous Si nanotubes deposited through galvanic displacement reaction.

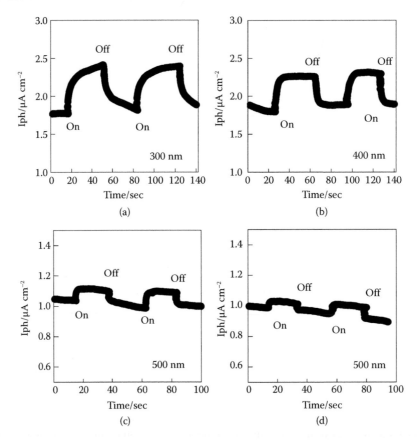

Figure 25.20 Transients of photocurrent at different wavelengths under intermittent monochromatic illumination of the amorphous nanotubes deposited through galvanic displacement reaction.

25.7 SUMMARY

SiNTs can be synthesized by a galvanic displacement reaction using aluminum as a sacrificial anode and a track-etched polycarbonate membrane as a template. This method is extremely advantageous because it is easy to conduct, cheap because it does not need neither sophisticated high-vacuum apparatus nor expensive reagents, and can be easily scaled at industrial level. The importance of scalability of SiNT production is a predominant feature, because this form of silicon is driving the innovation in several technological fields, such as microelectronics, optoelectronics, electrochemical energy accumulation, and nanomedicine.

The galvanic displacement reaction represents a very simple route for overcoming the principal difficulty in synthesizing SiNTs, consisting of the tendency of the silicon toward sp^3 hybridization. Nanowire is the preferred morphology of the silicon at nanoscale as shown by the easiness of its fabrication through various methods. In the case of the displacement reaction in hard template, the sp^2 hybridization is due to a steric effect, because the growth of SiNTs is confined to the gap between the pore wall and the hydrogen bubble. From this point of view, we could assign an indirect templating role to the hydrogen bubble. This finding confirms the utopic outlook by Feynman based on the role of the nanotechnology for fabricating challenging materials. The addition of single atoms of Si in a confined space allows bottom-up growth of a nanostructure with morphology very difficult to obtain through other methods.

A discriminant role for achieving the deposition of silicon is played by the proper choice of the cathodic-to-anodic area ratio, which must be as low as possible in order to kinetically drive the reduction of silicon, because it is less noble than the H^+/H_2 couple.

A strict control of the solution composition is also important for successful deposition of SiNTs. In particular, fluoride ions from hydrofluoric acid contributes to stabilizing SiF_6^{2-} which is the source of silicon, while formic acid slows down the pH increase due to hydrogen evolution, which can be dangerous because it favors the formation of undesired SiO_2.

A careful control of all these parameters allows the synthesis of amorphous n-type semiconducting SiNTs with an optical gap of 1.6 eV.

Also in this case, galvanic displacement reaction appeared as a valuable route for easily synthesizing materials usually requiring expensive and time-consuming procedures.

REFERENCES

Ashassi-Sorkhabi H, Badakhshan PL, Asghari E (2016). Electrodeposition of three dimensional-porous Ni/Ni(OH)2 hierarchical nano composite via etching the Ni/Zn/Ni(OH)2 precursor as a high performance pseudocapacitor. *Chemical Engineering Journal.* 299:282–291.

Ashuri M, Hea Q, Shaw LL (2016). Silicon as a potential anode material for Li-ion batteries: Where size, geometry and structure matter. *Nanoscale.* 8:74–103.

Baibich MN et al. (1988). Giant magnetoresistance of (001)Fe/(001)Cr magnetic superlattices. *Physical Review Letters.* 61:2472–2475.

Barth S, Hernandez-Ramirez F, Holmes JD, Romano-Rodriguez A (2010). Synthesis and applications of one-dimensional semiconductors. *Progress in Materials Science.* 55:563–627.

Battaglia M, Inguanta R, Piazza S, Sunseri C (2014a). Fabrication and characterization of nanostructured NiIrO2 electrodes for water electrolysis. *International Journal of Hydrogen Energy.* 39:16797–16805.

Battaglia M, Piazza S, Sunseri C, Inguanta R (2013). Amorphous silicon nanotubes via galvanic displacement deposition. *Electrochemistry Communication.* 34:134–137.

Battaglia M, Piazza S, Sunseri C, Inguanta R (2014b). CuZnSnSe nanotubes and nanowires by template electrosynthesis. *Advanced Science and Technology.* 93:241–246.

Ben-Ishai M, Patolsky F (2011). Wall-selective chemical alteration of silicon nanotube molecular carriers. *Journal of the American Chemical Society.* 133:1545–1552.

Blanda G et al. (2016). Galvanic deposition and characterization of brushite/hydroxyapatite coatings on 316L stainless steel. *Materials Science and Engineering C.* 64:93–101.

Bocchetta P et al. (2002) Asymmetric alumina membranes electrochemically formed in oxalic acid solution. *Journal of Applied Electrochemistry.* 32:977–985.

Bockris JO'M, Reddy AKN (1970). *Modern electrochemistry: An introduction to an interdisciplinary area.* New York: Plenum Press.

Bograchev DA, Volgin VM, Davydov AD (2013a). Simple model of mass transfer in template synthesis of metal ordered nanowire arrays. *Electrochimica Acta.* 96:1–7.

Bograchev DA, Volgin VM, Davydov AD (2013b). Simulation of inhomogeneous pores filling in template electrodeposition of ordered metal nanowire arrays. *Electrochimica Acta*. 112:279–286.

Bograchev DA, Volgin VM, Davydov AD (2015). Modeling of metal electrodeposition in the pores of anodic aluminum oxide. *Russian Journal of Electrochemistry*. 51:799–806.

Bograchev DA, Volgin VM, Davydov AD (2016). Mass transfer during metal electrodeposition into the pores of anodic aluminum oxide from a binary electrolyte under the potentiostatic and galvanostatic conditions. *Electrochimica Acta*. 207:247–256.

Butera M, Inguanta R, Sunseri C, Piazza S (2006). Nanoporous alumina membranes grown electrochemically: Fabrication and modification by metal deposition. In: *Proceedings of the 9th International Conference on Inorganic Membranes (ICIM9)*, SINTEF, Oslo, (Bredesen R, Raeder H, eds), pp. 78–81, ISBN 10: 82-14-04026-5.

Carreon ML, Thapa AK, Jasinski JB, Sunkara MK (2015). The capacity and durability of amorphous silicon nanotube thin film anode for lithium ion battery applications. *ECS Electrochemistry Letters*. 4:A124–A128.

Castrucci P et al. (2006). Silicon nanotubes: Synthesis and characterization. *Thin Solid Films*. 508:226–230.

Chang C-C et al. (2011). Tailoring the crystal structure of individual silicon nanowires by polarized laser annealing. *Nanotechnology*. 22:305709–305715.

Chang CH, Zhang M, Lim J, Choa Y, Park S, Myung NV (2014). Synthesis of PbTe and PbTe/Te nanostructures by galvanic displacement of cobalt thin films. *Electrochimica Acta*. 138:334–340.

Chen J, Zhang G, Li B (2010) Remarkable reduction of thermal conductivity in silicon nanotubes. *Nano Letters*. 10:3978–3983.

Chen T-M, Pan F-M, Hung J-Y, Chang L, Wu S-C, Chena C-F (2007). Amorphous carbon coated silicon nanotips fabricated by MPCVD using anodic aluminium oxide as the template. *Journal of the Electrochemical Society*. 154:D215–D219.

Cheng M, Qiang Z, Dongsheng X, Qiankun Z, Yuanhua S (2007). Silicon nanotube array/gold electrode for direct electrochemistry of cytochrome c. *Journal of Physical Chemistry B*. 111:1491–1495.

Di Leo CV, Rejovitzky E, Anand L (2015). Diffusion–deformation theory for amorphous silicon anodes: The role of plastic deformation on electrochemical performance. *International Journal of Solids and Structures*. 67–68:283–296.

Djokica SS, Cadien K (2015). Galvanic deposition of silver on silicon surfaces from fluoride-free aqueous solutions. *ECS Electrochemistry Letters*. 4:D11–D13.

Dong J, Drabold DA (1998). Atomistic structure of band-tail states in amorphous silicon. *Physical Review Letters*. 80:1928–1931.

Dong Y, Yu G, McAlpine MC, Lu W, Lieber C (2008). Si/a-Si core/shell nanowires as nonvolatile crossbar switches. *Nano Letters*. 8:386–391.

Downs RT, Hall-Wallace M (2003). The American Mineralogist crystal structure database. *American Mineralogist*. 88:247–250.

Falola BD, Suni II (2014). Galvanic deposition of Ti atop Al 6061 alloy. *Journal of the Electrochemical Society*. 161:D107–D110.

Falola BD, Suni II (2015). Low temperature electrochemical deposition of highly active elements. *Current Opinion in Solid State and Materials Science*. 19:77–84.

Feynman RP (1992). There's plenty of room at the bottom. *Journal of Microelectromechanical Systems*. 1:60–66.

Fonash SJ (2010) *Solar cell device physics*. Oxford, UK: Elsevier.

Fukunaka Y, Konishi Y, Ishii R (2006). Producing shape-controlled metal nanowires and nanotubes by an electrochemical method. *Electrochemical Solid State Letters*. 9:C62–C64.

Furneaux RC, Rigby WR, Davidson AP (1989). The formation of controlled-membranes from anodically oxidized aluminium. *Nature*. 337:147–149.

Garcia-Calzon JA, Diaz-Garcia ME (2012). Synthesis and analytical potential of silica nanotubes. *Trends in Analytical Chemistry*. 35:26–38.

Garnett EC, Yang P (2008). Silicon nanowire radial p-n junction solar cells. *Journal American Chemical Society*. 130:9224–9225.

Garnett EC, Yang P (2010). Light trapping in silicon nanowire solar cells. *Nano Letters*. 10:1082–1087.

Grillet C et al. (2012). Amorphous silicon nanowires combining high nonlinearity, FOM and optical stability. *Optics Express*. 20:22609–22615.

Hernandez-Velez M (2006). Nanowires and 1D arrays fabrication: An overview. *Thin Solid Films*. 495:51–63.

Hieu NT, Suk J, Kim DW, Park JS, Kang Y (2014). Electrospun nanofibers with a core-shell structure of silicon nanoparticles and carbon nanotubes for use as lithium-ion battery anodes. *Journal of Materials Chemistry A*. 2:15094–15101.

Hu J, Bando Y, Liu Z, Zhan J, Golberg D, Sekiguchi T (2004). Synthesis of crystalline silicon tubular nanostructures with ZnS nanowires as removable templates. *Angewandte Chemie International Edition*. 43:63–66.

Hu L, Chen G (2007). Analysis of optical absorption in silicon nanowire arrays for photovoltaic applications. *Nano Letters*. 7:3249–3252.

Huang J, Chiam SY, Tan HH, Wang S, Chim WK (2010). Fabrication of silicon nanowires with precise diameter control using metal nanodot arrays as a hard mask blocking material in chemical etching. *Chemistry of Materials.* 22:4111–4116.

Hylton TL, Coffey KR, Parker MA, Howard JK (1993). Giant magnetoresistance at low fields in discontinuous NiFe-Ag multilayer thin films. *Science.* 261:1021–1024.

Iijima S (1991). Helical microtubules of graphitic carbon. *Nature.* 354:56–58.

Imhof A, Pine DJ (1998). Uniform macroporous ceramics and plastics by emulsion templating. *Advanced Materials.*10:697–700.

Inguanta R, Battaglia M, Piazza S, Sunseri C (2013). *Italian Patent.* VI-2013-A000119.

Inguanta R, Butera M, Sunseri C, Piazza S (2007a). Fabrication of metal nano-structures using anodic alumina membranes grown in phosphoric acid solution: Tailoring template morphology. *Applied Surface Science.* 253:5447–5456.

Inguanta R, Ferrara G, Livreri P, Piazza S, Sunseri C (2011b). Ruthenium oxide nanotubes via template electrosynthesis. *Current Nanoscience.* 7:210–218.

Inguanta R, Ferrara G, Piazza S, Sunseri C (2009b). Nanostructures fabrication by template deposition into anodic alumina membranes. *Chemical Engineering Transactions.* 17:957–962.

Inguanta R, Ferrara G, Piazza S, Sunseri C (2011a). Fabrication and characterization of metal and metal oxide nanostructures grown by metal displacement deposition into anodic alumina membranes. *Chemical Engineering Transactions.* 24:199–204.

Inguanta R, Ferrara G, Piazza S, Sunseri C (2012b). A new route to grow oxide nanostructures based on metal displacement deposition. Lanthanides oxy/hydroxides growth. *Electrochimica Acta.* 76:77–87.

Inguanta R, Piazza S, Sunseri C (2007b). *Italian Patent.* VI-2007-A000275.

Inguanta R, Piazza S, Sunseri C (2008a). Novel procedure for the template synthesis of metal nanostructures. *Electrochemistry Communications.* 10:506–509.

Inguanta R, Piazza S, Sunseri C (2008b). Influence of electrodeposition techniques on Ni nanostructures. *Electrochimica Acta.* 53:5766–5773.

Inguanta R, Piazza S, Sunseri C (2009a). Synthesis of self-standing Pd nanowires via galvanic displacement deposition. *Electrochemistry Communications.* 11:1385–1388.

Inguanta R, Piazza S, Sunseri C (2012a). A route to grow oxide nanostructures based on metal displacement deposition: Lanthanides oxy/hydroxides characterization. *Journal of the Electrochemical Society.* 159:D493–D500.

Iqbal Z, Vepiek S (1982). Raman scattering from hydrogenated microcrystalline and amorphous silicon. *Journal of Physics C: Solid State Physics.* 15:377–392.

Jansen KW, Kadam SB, Groelinger JF (2006). The advantages of amorphous silicon photovoltaic modules in grid-tied systems. *2006 IEEE 4th World Conference on Photovoltaic Energy Conversion:* Waikoloa, Hawaii, May 7–12, 2006. pp. 2363–2366.

Jeong SY et al. (2003). Synthesis of silicon nanotubes on porous alumina using molecular beam epitaxy. *Advanced Materials.* 15:1172–1176.

Jianhui L, Daojian C, Dapeng C, Wenchuan W (2008). Silicon nanotube as a promising candidate for hydrogen storage: From the first principle calculations to grand canonical Monte Carlo Simulations. *Journal of Physical Chemistry C.* 112:5598–5604.

Jones MR, Osberg KD, Macfarlane RJ, Langille MR, Mirkin CA (2011). Templated techniques for the synthesis and assembly of plasmonic nanostructures. *Chemical Review.* 111:3736–3827.

Joshi KR, Schneider JJ (2012). Assembly of one dimensional inorganic nanostructures into functional 2D and 3D architectures. Synthesis, arrangement and functionality. *Chemical Society Reviews.* 41:5285–5312.

Kang R, Liang J, Liu B, Penga Z (2014). Copper galvanic replacement on aluminium from a choline chloride based ionic liquid: Effect of thiourea. *Journal of the Electrochemical Society.* 161:D534–D539.

Khanal RR et al. (2012). Single wall carbon nanotube electrodes for hydrogenated amorphous silicon solar cells. *Conference record of the IEEE Photovoltaic Specialists Conference.* pp. 57–61.

Kim KH, Lefeveure E, Chatelet M, Sorin Cojocaru C (2012). Porous alumina template based versatile and controllable direct synthesis of silicon nanowires. *MRS Online Proceedings Library Archive.* 1439:11–16.

Knight WD, Clemenger K, de Heer WA, Saunders WA, Chou MY, Cohen ML (1984). Electronic shell structure and abundances of sodium clusters. *Physical Review Letters.* 53:510.

Kozinda A, Jiang Y, Lin L (2011). Amorphous silicon-coated CNT forest for energy storage Applications. *2011 16th International Solid-State Sensors, Actuators and Microsystems Conference, TRANSDUCERS'11,* IEEE. Beijing, China, June 5–9, pp. 723–726.

Krishnamurthy A, Rasmussen DH, Suni II (2011). Galvanic deposition of nanoporous Si onto 6061 Al alloy from aqueous HF. *Journal of The Electrochemical Society.* 158:D68–D71.

Kroto HW, Heath JR, O'Brien SC, Curl RF, Smalley RE (1985). C60: Buckminsterfullerene. *Nature.* 318:162–163.

Kuchibhatlaa VNTS, Karakotia AS, Beraa D, Seala S (2007). One dimensional nanostructured materials. *Progress in Materials Science.* 52:699–913.

Kunjie L, Wenchuan W, Dapeng C (2011). Novel chemical sensor for CO and NO: Silicon nanotube. *The Journal of Physical Chemistry C.* 115:12015–12022.

Lee W, Scholz R, Nielsch K, Gösele U (2005). A Template-Based Electrochemical Method for the Synthesis of Multisegmented Metallic Nanotubes. *Angewandte Chemie International Edition.* 244:6050–6054.

Liao H, Karki K, Zhang Y, Cumings J, Wang Y (2011). Interfacial mechanics of carbon nanotube@amorphous-si coaxial nanostructures. *Advanced Materials.* 23:4318–4322.

Liu L, Chen L-X, Wang A-J, Yuan J, Shen L, Feng J-J (2016). Hydrogen bubbles template-directed synthesis of self-supported AuPt nanowire networks for improved ethanol oxidation and oxygen reduction reactions. *International Journal of Hydrogen Energy.* 11:8871–8880.

Lu JG, Chang P, Fan Z (2006). Quasi-one-dimensional metal oxide materials—Synthesis, properties and applications. *Materials Science and Engineering.* 52:49–91.

Mandal S, Dhar S, Das G, Mukhopadhyay S, Barua AK (2016). Development of optimized n-lc-Si:H/n-a-Si:H bilayer and its application for improving the performance of single junction a-Si solar cells. *Solar Energy.* 124:278–286.

Masuda H, Fukuda K (1995). Ordered metal nanohole arrays made by a two-step replication of honeycomb structures of anodic alumina. *Science.* 268:1466–1468.

Mbenkum BN et al. (2010). Low-temperature growth of silicon nanotubes and nanowires on amorphous substrates. *ACS Nano.* 4:1805–1812.

Midgley PA, Weyland M (2003). 3D electron microscopy in the physical sciences: The development of Z-contrast and EFTEM tomography. *Ultramicroscopy.* 96:413–431.

Munoz-Sandoval E et al. (2007). Architectures from aligned nanotubes using controlled micropatterning of silicon substrates and electrochemical methods. *Small.* 3:1157–1163.

Nguyen JJ, Evanoff K, Ready WJ (2011). Amorphous and nanocrystalline silicon growth on carbon nanotube substrates. *Thin Solid Films.* 519:4144–4147.

Nishimura Y, Fukunaka Y (2007). Electrochemical reduction of silicon chloride in a non-aqueous solvent. *Electrochimica Acta.* 53:111–116.

Novák P, Očenášek J, Prušáková L, Vavruňková V, Savková J, Rezek J (2016). Influence of heat generated by a Raman excitation laser on the structural analysis of thin amorphous silicon film. *Applied Surface Science.* 364:302–307.

Pei LZ, Wang SB, Fan CG (2010). Recent progress and patents in silicon nanotubes. *Recent Patents on Nanotechnology.* 4:10–19.

Penga K-Q, Wang X, Li L, Hu Y, Lee S-T (2013). Silicon nanowires for advanced energy conversion and storage. *Nano Today.* 8:75–97.

Perepichka DF, Rosei F (2006). Silicon nanotubes. *Small.* 2:22–25.

Prosini PP et al. (2014). Electrochemical characterization of silicon nanowires as an anode for lithium batteries. *Solid State Ionics.* 260:49–54.

Ren X, Lun Z (2012). Mesoporous silica nanowires synthesized by electrodeposition in AAO. *Materials Letters.* 68:228–229.

Ruffo R, Hong SS, Chan CK, Huggins RA, Cui Y (2009). Impedance analysis of silicon nanowire lithium ion battery anodes. *Journal of Physical Chemistry C.* 113:11390–11398.

Schönenberger C et al. (1997). Template synthesis of nanowires in porous polycarbonate membranes: Electrochemistry and morphology. *Journal of Physical Chemistry B.* 101:5497–5505.

Sha J et al. (2002). Silicon nanotubes. *Advanced Materials.* 14:1219–1221.

Silipigni L et al. (2014). Template electrochemical growth and properties of Mo oxide nanostructures. *Journal of Physical Chemistry C.* 118:22299–22308.

Song T et al. (2010). Arrays of sealed silicon nanotubes as anodes for lithium ion batteries. *Nano Letters.* 10:1710–1716.

Sun L, Wang X, Susantyoko RA, Zhang Q (2014). Copper–silicon core–shell nanotube arrays for free-standing lithium ion battery anodes. *Journal Materials Chemistry A.* 2:15294–15297.

Sun X-T, Yang C-G, Xu Z-R (2016). Controlled production of size-tunable Janus droplets for submicron particle synthesis using an electrospray microfluidic chip. *RSC Advances.* 6:12042–12047.

Tao B, Miao F, Chu J (2013). Structure and photoelectrochemical properties of silicon microstructures arrays. *Electrochimica Acta.* 108:248–252.

Teng X et al. (2008). Formation of Pd/Au nanostructures from Pd nanowires via galvanic replacement reaction. *Journal of the American Chemical Society.* 130:1093–1101.

Valladares LDLS et al. (2016). Preparation and crystallization of hollow α-Fe2O3 microspheres following the gas-bubble template method. *Materials Chemistry and Physics.* 169:21–27.

Walther A, Müller A (2013). Janus particles: Synthesis, self-assembly, physical properties, and applications. *Chemical Reviews.* 113:5194–5200.

Wan S et al. (2004). Single-step synthesis of copper sulfide hollow spheres by a template interface reaction route. *Journal of Materials Chemistry.* 14:2489–2491.

Wang N, Yao BD, Chan YF, Zhang XY (2003). Enhanced photothermal effect in Si nanowires. *Nano Letters.* 3:475–477.

Wen Z et al. (2013). Silicon nanotube anode for lithium-ion batteries. *Electrochemistry Communications*. 29:67–70.

Wingert MC, Kwon S, Hu M, Poulikakos D, Xiang J, Chen R (2015). Sub-amorphous thermal conductivity in ultra-thin crystalline silicon nanotubes. *Nano Letters*. 15:2605–2611.

Wood GC, O'Sullivan JP (1970). The anodizing of aluminium in sulphate solutions. *Electrochimica Acta*. 15:1865–1876.

Wu H et al. (2012). Stable cycling of double-walled silicon nanotube battery anodes through solid–electrolyte interphase control. *Nature Nanotechnology*. 35:1–6.

Wu X-J et al. (2010). Electrochemical synthesis and applications of oriented and hierarchically quasi-1D semiconducting nanostructures. *Coordination Chemistry Reviews*. 254:1135–1150.

Wua H, Cuia Y (2012). Designing nanostructured Si anodes for high energy lithium ion batteries. *Nano Today*. 7:414–429.

Xu Q et al. (2009). A generic approach to desired metallic nanowires inside native porous alumina template via redox reaction. *Chemistry of Materials*. 21:2397–2402.

Yamacli S (2014). Investigation of the voltage-dependent transport properties of metallic silicon nanotubes (SiNTs): A first-principles study. *Computational Materials Science*. 91:6–10.

Yin Y, Lu Y, Gates B, and Xia Y (2001). Template-assisted self-assembly: A practical route to complex aggregates of monodispersed colloids with well-defined sizes, shapes, and structures. *Journal American Chemical Society*. 123:8718–8729.

Zamfir MR, Nguyen HT, Moyen E, Lee YH, Pribat D (2013). Silicon nanowires for Li-based battery anodes: A review. *Journal of Material Chemistry A*. 1:9566–9586.

Zheng Z, Lim ZS, Peng Y, You L, Chen L, Wang J (2013). General route to ZnO nanorod arrays on conducting substrates via galvanic-cell-based approach. *Scientific Reports*. 3–2434:1–5.

Zhong J, Zhang H, Sun X, Lee S-T (2014). Synchrotron soft X-ray absorption spectroscopy study of carbon and silicon nanostructures for energy applications. *Advanced Materials*. 26:7786–7806.

Zhou H et al. (2012). Photovoltaic measurements in carbon nanotube—Amorphous silicon core/shell nanowire. *Conference record of the IEEE Photovoltaic Specialists Conference*. pp. 1944–1947.

26 Nanotubular-structured porous silicon

Chi-Pui Tang and Jie Cao

Contents

26.1 INTRODUCTION

In this chapter, we have calculated a novel allotrope of silicon that is characterized by nanotubular holes along the inplane axis using first-principle calculation. Such a porous structure of silicon (also named nt-p-Si) belongs to the I41/amd space group, and its band structure shows that it possesses a direct band gap in contrast to the indirect band gap of diamond cubic crystal silicon. The bulk modulus of nt-p-Si is 64.4 GPa and the density is 1.9 g/cm³. These parameters are lower than the diamond silicon and are presumably caused by the presence of nanotubular holes. With the direct band gap and specific nanotubular structure, the nt-p-Si may be widely used or replace the diamond silicon in many fields.

In 1956, Arthur Uhlir Jr. and Ingeborg Uhlir, from the famous Bell Labs, discovered porous silicon (p-Si) with nanoporous holes in its microstructure [1]. However, it did not cause widespread academic interest until the late 1980s. At 1990, Leigh Canham [2], from the Defence Research Agency in the UK, observed the photoluminescence of p-Si at room temperature fabrication in the electrochemical and chemical dissolution of wafers.

Although the crystal silicon (c-Si) with a diamond structure is a useful semiconductor, its indirect band gap restricts its application in electro-optical devices. Then p-Si aroused the attention of scientists with its stability, purity, and ease of processing properties [3].

For a long time, gallium arsenide (GaAs) was preferred over c-Si as it is relatively good at emitting light and has high efficiency as solar cells or detectors. Therefore, GaAs solar arrays have been used on the Spirit rover and the Opportunity rover for the exploration of Mars' surface. However, silicon has many advantages over GaAs. For example, silicon is cheaper to process and has a relatively stable structure. It can be grown into large diameter boules. Generally speaking, the wafers of silicon can be made to have a diameter of about 300 mm, while the GaAs wafers can only be made to at most 150 mm. With these advantages of silicon, some allotropes were designed and synthesized.

For example, amorphous silicon (a-Si) and nanocrystalline silicon (nc-Si) [3] have been used as solar cell materials as they are cheaper than c-Si. The guest-free silicon clathrate Si136 and open-framework Si24 were successfully synthesized by thermal the decomposition of alkali metal silicides under the high vacuum [4–6]. Both Si136 and Si24 have a wide band gap, for example, 1.9 eV, and 1.3 eV, respectively [4,6,7].

Some allotropes, such as allo-Si, Si-CFS, fourfold-coordinated clathrate Si, and other structures, were predated by computational calculations [7,8,9]. Almost all of these possesses an indirect band gap (except Si24 has a "quasidirect" band gap [6]) and such a band gap causes a poor carrier mobility and a low quantum efficiency. However, these unique structures would be of interest for gas and lithium storage or for molecular filters.

Besides the above allotropes, the silicene, the two-dimensional allotrope of silicon, may have had the most attention among the allotropes of silicon. It has a hexagonal honeycomb structure, which is similar to graphene. It has a much stronger spin-orbit coupling along with a sizable and tunable band gap. These properties of silicene are considered to be an advantage over graphene with a realization of quantum spin hall effect in the experimental temperature and a better tunable band gap. For a long time, silicene has been investigated regarding its stability and properties by computation. In 2010, researchers (Aufray et al.) observed self-assembled silicene nanoribbons and silicene sheets by using a scanning tunneling microscope. The structure of silicene is similar to graphene and represents interesting properties.

In our previous work [10], we predicted the nanotubular structure of the porous silicon (nt-p-Si) base on the density-functional theory (DFT). It is a new allotropic silicon crystal with a stable structure. It belongs to the I41/amd space group and has a direct band gap wider than the indirect gap of c-Si by 0.5 eV. The density of nt-p-Si is 1.9g/cm^3 and its bulk modulus is 64.4 GPa. These parameters are lower than those of c-Si. The structure of nt-p-Si is also remarkable as it has nanotubular holes along the two perpendicular directions at section area 28.8 Å^2 and penetrating the whole crystal. Such an unusual nanotubular structure of nt-p-Si crystal and its direct band gap will provide a strong prospect of applications in producing electro-optical devices, new-style solar cells, photocatalyst, molecular sieves, and aerospace materials.

26.2 CALCULATIONS

The structure and properties of nt-p-Si are calculated by DFT [9,11] using the Cambridge sequential total energy package (CASTEP) [12]. We used the ultrasoft pseudopotential method [13] with two approximation methods for comparison: the first one is local density approximation (LDA) that was developed by Ceperley and Alder [14] and parameterized by Perdew and Zunger [15] (CA-PZ); and the generalized gradient approximation (GGA) was developed by Perdew et al. (PBE) [16] with their kinetic energy cut-off taken at 400 eV. The $12 \times 12 \times 12$ k points sampling of the Monkhorst-Pack scheme is also employed in the selection of k points in the Brillouin zone (BZ) [17]. In order to reduce and compare the systematic underestimation of the energy gap in GGA and LDA calculations [18,19], the band structure of nt-p-Si is also calculated by the Becke, 3-parameter, Lee-Yang-Parr (B3LYP) hybrid functional [20] and screened exchange functional (sX-LDA) [21,22,23]. The Broyden–Fletcher–Goldfarb–Shanno (BFGS) minimizer [24] in CASTEP is then used for the geometry and the positions of atom optimization with the convergence tolerances set to 5.0×10^{-6} eV/atom for energy, 0.01 eV/Å for maximum force, 0.02 GPa for maximum stress, and 5.0×10^{-4} Å for maximum displacement. The ultrasoft pseudopotential is replaced by normconserving pseudopotential [25] in the calculations of B3LYP, sX-LDA functionals, and/or in the phonon calculation.

26.3 RESULTS AND DISCUSSION

26.3.1 CRYSTAL STRUCTURE AND ITS STABILITY

Figure 26.1a shows the structure of nt-p-Si as it has eight atoms in the primitive cell. It belongs to the body-centered crystal (BCC) and the space group I41/amd. Figure 26.1b also depicts the BZ and each silicon atom has four coordinates as shown in Figure 26.1c. After the geometric optimization, the lattice constants of the BCC cell are found to be a = b = 5.45 Å, c = 13.00 Å using the GGA calculation while a = b = 5.37 Å, c = 12.7 Å using the LDA calculation. In this structure, there are rows of octagons separated with rows of pentagons through the [110] direction as shown in Figure 26.1d. From another angle, we can also observed rows of nanotubes penetrating the whole crystal along the [110] direction as in Figure 26.1e. The octagonal section area of the nanotube is 28.8 Å^2. In the different planes, there are also rows of nanotubes showing smaller pentagonal sections. We can identify the same projection pattern along the [1$\underline{1}$0]

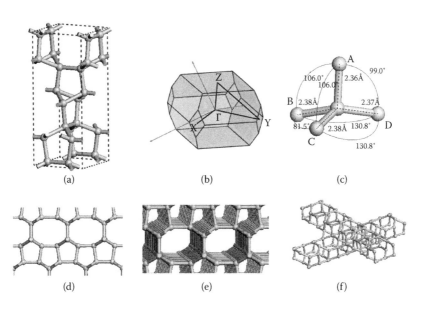

(a)   (b)   (c)

(d)   (e)   (f)

Figure 26.1 The crystalline structure of nt-p-Si. (a) The body-centered crystalline structure of nt-p-Si belonging to the I41/amd space group. The eight violet balls indicate the central part which forms a primitive cell. (b) The Brillouin zone. (c) The bond angles and bond lengths of an atom with its four coordinates. (d) Pattern viewed along the [110] direction. (e) A part of nt-p-Si crystal viewed along a direction slightly tilted from [110]. (f) Two criss-crossed octagonal nanotubes along the [110] and [1$\bar{1}$0] directions extracted from the structure.

Table 26.1 **The cell parameters and atomic positions (fractional coordinates) of primitive cell in nt-p-Si structure**

LENGTHS	A = 7.414Å	B = 7.414Å	C = 5.37Å
ANGLES	α = 68.76°	β = 68.76°	γ = 42.46°
POSITIONS	X	Y	Z
Si_1	0.0356	0.0356	0.2478
Si_2	0.0356	0.0356	0.6810
Si_3	0.2145	0.2145	0.0022
Si_4	0.2145	0.2145	0.5690
Si_5	0.5690	0.0022	0.0022
Si_6	0.0022	0.5690	0.7856
Si_7	0.6810	0.2478	0.0356
Si_8	0.2478	0.6810	0.0356

direction. As indicated in Figure 26.1f, the nanotubes along [110] and [1$\underline{1}$0] directions are perpendicularly criss-crossed with a shift along the c-axis. We have listed the cell parameters and the positions of all the atoms after geometric optimization in Table 26.1. Such a hollow structure is reminiscent of the pores in p-Si, however, nt-p-Si is wholly crystalline and has much longer nanotubes with smaller sections in comparison with p-Si.

The bond angles and bond lengths for four bonds of an atom in nt-p-Si are shown in Figure 26.1c. They are $\angle AOD$ = 99.0°, $\angle AOC = \angle AOB$ = 106.0°, $\angle COD = \angle DOB$ = 130.7°, $\angle COD$ = 81.5°, AO = 2.35Å, $CO = BO$ = 2.37Å, and DO = 2.36Å. The bonds are only slightly different from the regular tetrahedron structure in c-Si, with four bonds of the same length and the same bond angle ($\approx$ 109.5°). The obtained deviations of bond lengths in nt-p-Si from those of c-Si are lying within the range of 1 % but the deviations of bond angles may be rising up to 40 % based on LDA and GGA calculations. Therefore, it shows a difference in the short-range structure between nt-p-Si and c-Si as they come from the large deviations in bond angles, while the lengths of covalence bonds in Si are rigid.

After finding the nt-p-Si structure with the GGA method, we simulated the spectrum of the powder X-ray diffraction (XRD) of nt-p-Si with radiation from a copper X-ray source. The wavelengths are $\lambda_1 = 1.54056$Å and $\lambda_2 = 1.54439$Å and along with an intensity ratio $I_2/I_1 = 0.5$, where λ_1 and λ_2 are the intensities of components 1 and 2 in the sources. The XRD spectrum is shown in Figure 26.2 (2θ from 10.0° to 80.0°). It is possible to use this result to serve as the benchmark for the candidate samples in the future experiments.

Since nt-p-Si does not exist in nature, an estimation of the structural stability is necessary. We first calculated the cohesive energy per atom for both nt-p-Si and c-Si. The results are −5.24 eV for nt-p-Si and −5.46 eV for c-Si, using the GGA method. While using the LDA calculation, the results are −6.01 eV for nt-p-Si and −5.81 eV for c-Si. The difference between the two methods is so small—0.2 eV, and thus, c-Si is very stable. We can speculate that the stability of nt-p-Si is also high in the same environment. We then plotted the cohesive energy dependence for both nt-p-Si and c-Si in terms of volume per atom. As shown in Figure 26.3, both energy curves have minima occurring in a similar way. The only difference is that the minimum of nt-p-Si is slightly higher and has a larger volume per atom, suggesting that nt-p-Si is a metastable phase with rather high stability.

Furthermore, the phonon dispersion relation is calculated by the linear response method and the norm-conserving pseudopotential. The results are displayed in Figure 26.4. There is no complex frequency or any anomalies in the phonon bands; this strongly implies the mechanical stability of nt-p-Si.

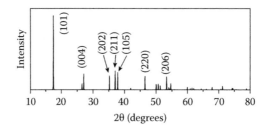

Figure 26.2 Simulated powder X-ray diffraction (XRD) under radiation from copper X-ray source of nt-p-Si structure obtained from the GGA method.

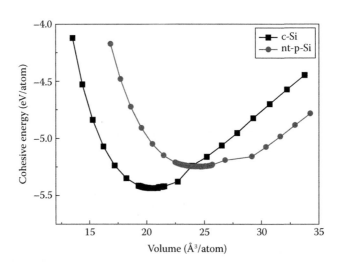

Figure 26.3 Cohesive energies per atom of ntc-Si and c-Si as functions of volume per atom under GGA calculation.

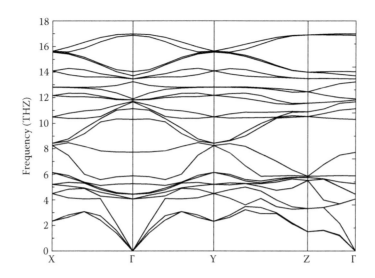

Figure 26.4 Phonon bands structure of ntc-Si calculated with the linear response method and normconserving pseudopotential.

Table 26.2 Properties of nt-p-Si: lattice constant l (Å), equilibrium density ρ (g/cm³), equilibrium cell volume V (Å³), volume per atom $V_{per\ atom}$ (Å³), atomic volume V_{atom} (Å³, hard-touching spheres model), bond length d (Å), atomic packing factor (APF), cohesive energy E_{coh} (eV/atom), band gap E_g (eV), bulk modulus B (GPa), and Young's modulus Y (GPa)

STRUCTURE	METHOD	l (A,B,C)	ρ	V	$V_{per\ atom}$	V_{atom}	d
nt-p-Si	GGA	5.45, 5.45, 12.99	1.93	385.8	24.11	7.0	2.36, 2.37, 2.38
nt-p-Si	LDA	5.37, 5.37, 12.74	2.03	367.3	22.96	6.6	2.31, 2.33, 2.33
nt-p-Si	B3LYP	5.45, 5.45, 12.99	-	-	-	-	-
c-Si	GGA	5.46, 5.46, 5.46	2.29	162.8	20.35	7.0	2.37
c-Si	LDA	5.38, 5.38, 5.38	2.40	155.7	19.46	6.6	2.33
c-Si	Exp[25,26]	5.43, 5.43, 5.43	2.33	160.1	20.01	6.8	2.35
c-Si	B3LYP	5.46, 5.46, 5.46	-	-	-	-	-
Structure	Method	APF	E_{coh}	E_g	B	Y	
nt-p-Si	GGA	0.29	-5.24	1.16	64.4	a = 74.8, b = 74.6, c =116.4	
nt-p-Si	LDA	0.29	-5.81	1.17	73.0	a = 68.4, b = 68.2, c =119.6	
nt-p-Si	B3LYP	-	-	1.25	-		
c-Si	GGA	0.34	-5.4	0.60	87.8	a = b = c = 120.1	
c-Si	LDA	0.34	-6.0	0.44	96.9	a = b = c = 126.7	
c-Si	Exp[25,26]	0.34	-	1.12	98.0	a = b = c = 130.0	
c-Si	B3LYP	-	-	0.71	-		

Next, we compared the physical properties of nt-p-Si and c-Si and then listed the results in Table 26.2. We found that the density of nt-p-Si is 1.93 g/cm³ and 2.03 g/cm³ from the GGA and LDA calculations, respectively. The difference is known to be due to the existence of hollow nanotubes. The volume per atom of nt-p-Si is 24.11 Å³ from GGA and 22.96 Å³ from LDA. It is slightly larger compared to c-Si, as it is 20.35 Å³ from GGA and 19.46 Å³ from LDA. For the same reason, the atomic packing factor (APF) of nt-p-Si is also less than that of a diamond structure, being 0.29 and 0.34, respectively. The bulk modulus of nt-p-Si is 64.4 GPa while the modulus of c-Si is 87.8 GPa. It is also worth noting that the Young's modulus in nt-p-Si is anisotropic, being 74.8 GPa, 74.6 Gpa, and 116.4 GPa in directions [100], [010], and [001], respectively, from the GGA calculation. The same calculation method gives the Young's modulus of c-Si as 120.1 GPa in all three directions. The results of modulus from the LDA calculation are also presented in Table 26.2. The anisotropy of nt-p-Si along with smaller values of the Young's modulus is expected due to the existence of the nanotubes along the [110] and [1$\underline{1}$0] directions.

26.3.2 ELECTRONIC AND OPTICAL PROPERTIES

Having a direct band gap is a remarkable property of nt-p-Si as it is absent in c-Si. We have calculated the band structure and partial densities of states (PDOS) using the GGA, LDA, B3LYP, and sX-LDA methods. Results from the B3LYP calculation are displayed in Figure 26.5. The band structure of nt-p-Si shows the presence of a direct band gap at 1.25 eV while the same calculation gives an indirect gap at 0.71 eV for c-Si. The direct band gap of nt-p-Si is also wider than the indirect one in c-Si by 0.5 eV. Furthermore, the sX-LDA calculation also produces some similar results. The GGA and LDA calculated band gap of nt-p-Si is 1.16 eV and 1.17 eV, while the band gap of c-Si is 0.60 eV and 0.44 eV.

Meanwhile, the DFT calculation is known to have a systematic underestimation of the band gap. We can, however, use the difference of the calculated band gap from the experimental values of c-Si as a corrector of the calculated band gap of nt-p-Si. The direct gap value of nt-p-Si is then similar to that of crystalline GaAs. The nt-p-Si may be able to be used as a new semiconductor with similar properties to GaAs, without toxic elements.

We then calculated the optical properties of nt-p-Si and c-Si using the B3LYP method. The absorption and the imaginary part of the dielectric function are shown in Figure 26.6. The imaginary part of the dielectric function of c-Si has a cut-off at 2.6 eV in the low-energy region, while the cut-off of the nt-p-Si function is extended to 1.5 eV. With the band gap of nt-p-Si larger than that of c-Si, we can conclude that the direct gap in nt-p-Si can tremendously improve the optical activity in the low-energy region. Such an improvement of the optical properties may extend the applications of silicon crystals in electro-optical devices.

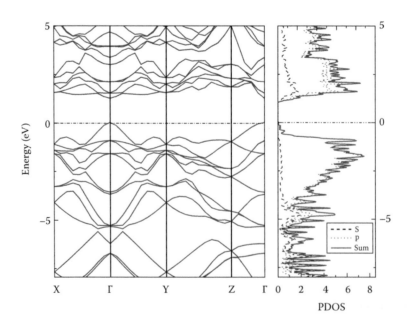

Figure 26.5 Electronic band structure and partial densities of states (PDOS) of ntc-Si calculated by using B3LYP functional and optimized structure. The direct band gap is 1.25 eV.

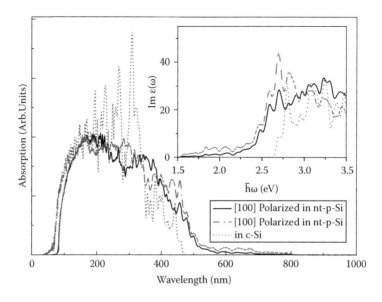

Figure 26.6 Optical absorption and the imaginary part of dielectric function of ntc-Si and c-Si calculated by using the B3LYP method.

26.3.3 DISCUSSION OF FAVORITE CONDITIONS FOR THE SYNTHESIS

Finally, we discuss the possible conditions for the synthesis of nt-p-Si. As discussed before, from the GGA and LDA calculations under ambient conditions, we found that the cohesive energy of c-Si is slightly lower than that of nt-p-Si by 0.2 eV. However, if we apply an anisotropic pressure in the [001] direction with the pressure greater than 16 GPa, the GGA calculation shows that the cohesive energy of nt-p-Si ends up lower than for c-Si (see Figure 26.7). This implies that the environment under anisotropic pressure in the [001] direction is favored for the synthesis of nt-p-Si.

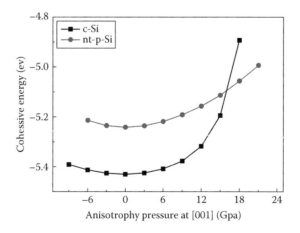

Figure 26.7 Cohesive energies per atom of c-Si and nt-p-Si as functions of anisotropy pressure in [001] direction. The curves intersect at 16 Gpa where the cohesive energy is 5.1eV.

26.4 SUMMARY

As a brief summary of this chapter, we have introduced a novel and stable crystal of silicon with hollow nanotubes penetrating through it. Its stability is testified with the calculations of the cohesive energy, geometry optimization, and phonon band structure. We have pointed out that the structure has remarkable properties, such as regular penetrating pores, direct band gap, wide frequency range optical activities, and anisotropy. The structure of nanotubes can store or transport small molecules or atoms, and the optical activities may be used in photo-chemical reactions. Therefore, nt-p-Si can serve as molecular sieves, electro-optical devices, new-style solar cells, photocatalysis, and aerospace materials.

REFERENCES

1. L. T. Canham, ed. *Properties of porous silicon*. The Institution of Electrical Engineers, London, UK, 1997.
2. L. T. Canham, *Appl. Phys. Lett.* 57 (1990): 1046.
3. W. Heywang, K. H. Zaininger, in *Silicon: Evolution and future of a technology*, eds. P. Siffert and E. F. Krimmel, Springer Verlag, Heidelberg, Berlin, 2004.
4. L. Bagolini, A. Mattoni, G. Fugallo, L. Colombo, E. Poliani, S. Sanguinetti, E. Grilli, *Phys. Rev. Lett.* 104 (2010): 176803.
5. J. Gryko, P. F. McMillan, R. F. Marzke, G. K. Ramachandran, D. Patton, S. K. Deb, O. F. Sankey, *Phys. Rev. B* 62 (2000): R7707.
6. A. Ammar, C. Cros, M. Pouchard, N. Jaussaud, J. M. Bassat, G. Villeneuve, M. Duttine, M. Menetrier, E. Reny, Solid state sciences 6(5) (2004): 393–400.
7. D. Y. Kim, S. Stefanoski, O. O. Kurakevych, T. A. Strobel, *Nat. Mater.* (2014).
8. J. C. Conesa, *J. Phys. Chem. B* 106(13) (2002): 3402–3409.
9. M. A. Zwijnenburg, K. E. Jelfs and S. T. Bromley, *Phys. Chem. Chem. Phys.* 12(30) (2010): 8505–8512.
10. C. P. Tang, C. Jie, and S. J. Xiong. *Physica. B: Condens. Matter* 466 (2015): 59–63.
11. P. Hohenberg, W. Kohn, *Phys. Rev. B* 136 (1964): 864.
12. M. Levy, *Proc. Natl. Acad. Sci. U. S. A.* 76 (1979): 6062.
13. M. D. Segall, P. L. D. Lindan, M. J. Probert, C. J. Pickard, P. J. Hasnip, S. J. Clark, M. C. Payne, *J. Phys.: Condens. Matter* 14 (2002): 2717.
14. D. Vanderbilt, *Phys. Rev. B* 41 (1990): 7892.
15. D. M. Ceperley, B. J. Alder, *Phys. Rev. Lett.* 45 (1980): 566.
16. J. P. Perdew, A. Zunger, *Phys. Rev. B* 23 (1981): 5048.
17. J. P. Perdew, K. Burke, M. Ernzerhof, *Phys. Rev. Lett.* 77 (1996): 3865.
18. J. D. Pack, H. J. Monkhorst, *Phys. Rev. B* 13 (1976): 5188.
19. J. D. Pack, H. J. Monkhorst, Phys. Rev. B 16 (1977) 1748.
20. G. Cicero, A. Catellani, G. Galli, *Phys. Rev. Lett.* 93 (2004): 016102.
21. A. D. Becke, *Phys. Rev. A* 38 (1988): 3098.
22. K. Kim, K. D. Jordan, *J. Phys. Chem.* 98 (1994): 10089.
23. P. J. Stephens, F. J. Devlin, C. F. Chabalowski, M. J. Frisch, *J. Phys. Chem.* 98 (1994): 11623.
24. A. Seidl, A. Gorling, P. Vogl, J. A. Majewski, M. Levy, *Phys. Rev. B* 53 (1996): 3764.
25. B. G. Pfrommer, M. Cote, S. G. Louie, M. L. Cohen, *J. Comput. Phys.* 131 (1997): 133.
26. A. M. Rappe, K. M. Rabe, E. Kaxiras, J. D. Joannopoulos, *Phys. Rev. B* 41 (1990): 1227.

27 Porous silicon nanotube arrays

Nguyen T. Le and Jeffery L. Coffer

Contents

27.1 INTRODUCTION

Silicon (Si) is the essential elemental semiconductor of modern microelectronic devices (Canham 2014). Since silicon itself does not conduct electricity as efficiently as conventional metallic conductors (e.g., copper), the doping of different elements in silicon, particularly arsenic and boron, to transform it into n-type and p-type, respectively, is a strategy to perturb the lattice structure. This enhances the conductivity of the material and permits operation most commonly as a transistor (Brus 1994; Canham 2014). Although three-dimensional bulk crystalline silicon has played a dominant role in the development of electronic technologies in the past few decades, an interest in nanoscale Si has received widespread investigation (Murphy and Coffer 2002; Elbersen 2015). When the Si feature size is in the 1–100 nm range, quantum confinement effects can emerge and significantly alter the properties and performance of the material (Pavesi and Lockwood 2004). Notably, in the 1990s, the discovery of efficient room temperature, visible light emission from nanostructured, two-dimensionally confined porous silicon (pSi) has intrigued scientists around the world. Since then, extensive research has been conducted to investigate new properties and applications of this material (Cullis and Canham 1991). Furthermore, studies of pSi have been extended to biorelevant applications including drug delivery, tissue engineering, and biosensing—a consequence of its biocompatibility and biodegradability (Coffer et al. 2005; Anglin et al. 2008; Salonen et al. 2008, Santos et al. 2014). The properties of pSi are well governed by its surface chemistry and porous morphology, thereby rendering an ability to modulate this material for a number of specific purposes (Canham 2014).

In terms of morphology, nanostructured porous silicon has shown a range of options—from dendtritic interconnected mesopores to large parallel macropores—the appearance of which is dictated by Si wafer resistivity and dopant identity, electrolyte composition, magnitude and duration of bias, and illumination (Smith and Collins 1992). Yet these morphologies are not without shortcomings; for example, the quantum sponge morphology of mesoporous materials often yields a range of crystalline Si thicknesses at the nanoscale, with a highly interconnected pore structure. At the other extreme, macroporous Si films do retain a straight columnar pore structure, but these pores are typically separated by relatively significant distances with an overall low porosity for the film (and associated low surface area and bioresorbability). An ideal scenario is one where the pores are arranged in uniform vertical columns of a uniform Si thickness and length at the nanoscale. The successful formation of one-dimensional (1D) hollow Si nanotubes with well-defined sidewalls and inner void space provides opportunities to investigate new properties and potential applications in a number of diverse fields, ranging from optoelectronics and energy to nanomedicine.

In this chapter, a comprehensive review of silicon nanotubes (SiNTs) is presented. We begin with preliminary computational studies predicting the properties of single-walled SiNTs (SWSiNTs), followed by an overview of the main fabrication routes of SiNTs currently in use. There are multiple detailed examples of their potentially useful properties in Li battery storage, templates for organometal perovskite formation, magnetically guided drug delivery, surface modification, and other potential uses.

27.2 EARLY COMPUTATIONAL STUDIES: SINGLE-WALLED SiNTs—INFLUENCE OF TUBE GEOMETRY ON NANOTUBE PROPERTIES

Given the historical dominance of interest in carbon nanotubes, both in single-walled and multiwalled architectures, it is logical to raise predictive comparisons with the heavier Group IV congener silicon. For the case of carbon nanotubes, the presence of strong π bonds and their intratube connectivity govern the stability and conduction of the entire tubular structure. In contrast to carbon's ability to easily undergo sp^2 hybridization, the lower energy difference between valence s and p orbitals in Si ($\Delta E_{3p} - \Delta E_{3s} = 5.66$ eV) suggests the latter element's tendency to form sp^3 hybrids. Additionally, due to the greater interatomic distance of Si relative to that of C, $p\pi$–$p\pi$ overlap is not as efficient for Si as in C, thereby explaining the inability of Si to form strong π bonds (Zhang et al. 2002; Perepichka 2006).

An assessment of the likelihood of SiNT formation was initially discussed using computational methods such as Hartree–Fock and molecular dynamics simulations (Perepichka et al. 2006). By constructing hypothetical carbon-like SiNTs, Fagan et al. have shown that from *ab initio* calculations based on density-functional theory, the electronic characteristics of SiNTs are largely dependent on the geometry of interatomic connectivity in a similar manner to CNTs (Fagan et al. 2000). In particular, SiNTs with an armchair structure appear to have metallic characteristics due to their zero band gap, whereas those with a zigzag structure behave more like semiconductors. However, in terms of nanotube formation, although the energy required to roll the sheets into nanotubes is the same for both Si and C, Si requires a higher energy than C to be able to form a graphite-like sheet as a precursor for nanotube formation. Hence, these preliminary studies confirm the challenges in trying to form Si into a stable nanotube structure.

Nevertheless, Seifert and coworkers have suggested that the existence of SiNTs is attainable if silicide (Si⁻) and SiH are utilized as the local geometric centers for the tubular network (Seifert et al. 2001). Taking into account the similarities with phosphorus nanotubes (PNTs) (which were demonstrated in earlier studies), this group points out that silicide NTs possess a lower strain energy (energy difference between a planar sheet and a curved surface) compared to that of PNTs, thereby implying a relatively stable structure for the SiNTs. Moreover, in terms of electronic characteristics, unlike the hypothetical SiNTs analyzed by Fagan, the sheets and nanotubes of silicide and SiH exhibit, regardless of chirality, semiconducting behavior with band gaps of 2.49 eV and 2.50 eV for Si⁻ and SiH sheets, respectively. Also, in another separate study, Zhang et al. suggest that under certain conditions, H-terminated SiNTs possess a local distortion away from planarity around a given Si center (a "puckered surface appearance") (Zhang et al. 2002).

A significant breakthrough in SiNT computational studies by Bai et al. proposed for the first time the existence of SWSNTs with Si possessing sp^3 hybridization and fourfold coordination (Bai 2004). Through the use of classical molecular dynamics simulation of the molten Si in a nanoporous environment, this group confirmed the possible formation of 1D SWSNTs in hexagonal, pentagonal, and square structures. In comparison to cubic diamond (bulk) silicon, these three structures of SWSNTs are considered to be metastable as a result of their higher values of energy per atom. According to these studies, while quantum confinement effects predict an increasing band gap for silicon nanowires (SiNWs) with decreasing nanowire diameter, such behavior is not observed in SWSNTs. Interestingly, considering nanotubes as finite stacks of pentagons, the computed HOMO-LUMO gap (using a B3LYP/6-31G(d) basis set) approaches zero with the increasing cluster size (smaller nanotubes). Further studies with hexagonal SiNTs show that the density of states associated with this morphology is nonzero at the Fermi level, suggesting that SWSNTs with this architecture have a zero band gap and therefore exhibit metallic characteristics.

Thus far, these initial computational studies predicting the properties of SWSiNTs have encouraged investigations of the possible experimental synthesis of this type of structure and also usefully assist in prediction of the possible physical and chemical properties of multiwalled materials of this element.

27.3 FABRICATION METHODS

Until now, multiple synthetic strategies of SiNTs have been attempted, and depending on the methodology, SiNTs of different morphologies are achieved. In this section, we will focus on three main routes that entail formation of SiNTs: (1) the use of alumina membrane templates, (2) gas phase condensation routes, and (3) the use of sacrificial ZnO templates with HCl etching.

27.3.1 ALUMINA MEMBRANE TEMPLATES

This synthetic route was inspired by the fabrication of SiNWs using chemical vapor deposition (CVD) (Sha et al. 2002). In this strategy, silane is deposited into the nanochannels of an Al_2O_3 (NCA) template, whose sidewall is coated with a gold catalyst in order to promote the growth of crystalline silicon. The CVD of silane is exploited for the formation of Si according to the reaction:

$$SiH_4 \rightarrow Si\,(g) + 2\,H_2\,(g) \tag{27.1}$$

A follow-up step involves etching with HCl that eventually yields crystalline hollow-needle Si nanostructures, with a distinct boundary between the core and shell. Further transmission electron microscope (TEM) analysis also suggests the absence of a hollow core, confirming Si is in the form of nanotube. A crystalline shell with small domains of amorphous Si can be detected, showing a lattice spacing of 0.315 nm, which is associated with the Si(111) lattice plane. Due to the relatively high chemical reactivity of Si, the oxidation of Si is inevitable resulting in native oxide layers of Si on the nanotube surface.

However, the main challenge in this method is the difficulty in controlling the deposition of gold catalyst in the NCA template in order to achieve a hollow Si structure. Although SiNTs are formed in this method possess well-defined nanotube structure, in some cases, when one end of the NCA is blocked, an overfill of the channel is unavoidable and eventually results in the formation of SiNWs along with the nanotubes. In this case, achieving high-uniform SiNT arrays demands a very careful preparation of the templates for the subsequent deposition of the silane vapor. Furthermore, depending on the application, the interfacial gold layer between Si and the alumina template may or may not be problematic to its use in an authentic platform.

27.3.2 GAS-PHASE CONDENSATION

Another synthetic strategy reported by Castrucci et al. is to perform gas-phase condensation without the addition of any catalyst to grow Si nanotubular structures (Castrucci 2006). In this method, the condensation of Si in the gas phase, followed by homogeneous nucleation of the deposited Si, yields SiNTs that have inner diameters greater than 5 nm and lengths of hundreds of nanometers. Encouragingly, unlike the nanotubes produced from the aluminum membrane template method as described above, TEM mapping in this case detects silicon as the main element in all the nanotubes, while the oxygen content is otherwise negligible. Significantly, those SiNTs appear to exhibit multiple properties similar to those reported in CNTs. In particular, the ability to image the edges of Si nanostructures based on focusing/defocusing conditions, and similarities to single-walled carbon nanotubes (SWCNTs) (only) implies the presence of SWSiNTs. Another critical observation is the structure of T-shaped SiNTs closely related to the Y and T-branched CNTs, thus implying that the obtained SiNTs are composed of sp^2-hybridized Si rather than sp^3-hybridized Si.

Nevertheless, the main drawback in this method is the ill-defined tubular structures. From scanning tunneling microcopy STM images, the tubes are a mix of bent, straight, and coiled forms. Furthermore, widespread contamination of the reaction products with numerous Si nanoparticles possessing high amounts of oxygen also poses critical concerns.

27.3.3 SACRIFICIAL ZnO TEMPLATES WITH HCL ETCHING

Compared to previously described methods in which the control of nanotube growth is still greatly limited, an alternative strategy of SiNT fabrication through the use of sacrificial preformed ZnO nanowire (NW) templates and CVD of silane successfully yields a broad range of tunable structural parameters (Hu et al. 2004).

This route involves three main processes as illustrated in Figure 27.1 (Huang et al. 2013). For ZnO removal, different methods have been extensively investigated. In one approach, ZnO removal can be achieved via thermal reduction in H_2 at high temperatures (600°C) (Equation 27.2) (Goldberger et al. 2003). However, operation at a relatively high temperature in the presence of H_2 gas for an extensively long period of time (24 h) poses another safety concern in carrying out the experiment. Instead, another viable protocol reported recently entails subsequent reaction of the Si/ZnO composite with an HCl/NH_3 etchant gas phase generated from precursor NH_4Cl at 450°C. This treatment can convert ZnO to a zinc amide species, which is readily removed through an evaporation process (Equations 27.3, 27.4) (Huang et al. 2013). If desired, an additional annealing step (in He) can be performed to further achieve highly crystalline SiNTs with nanocrystalline domains of (111) readily apparent.

$$ZnO + H_2 \xrightarrow{\Delta} Zn + H_2O \tag{27.2}$$

$$ZnO + 2HCl \leftrightarrow ZnCl_2 + H_2O \tag{27.3}$$

$$ZnCl_2 + NH_3 \leftrightarrow Zn(NH_2)Cl + HCl \tag{27.4}$$

In this synthetic procedure, details of the nanotube structure can be readily manipulated by controlling fabrication parameters (Huang et al. 2013). Specifically, by adjusting the ZnO NW growth conditions, a wide range of ZnO diameters (30–200 nm) and lengths (5 to 10 µm) have been achieved yielding length to diameter (L/D) ratios of ~10–25. On the other hand, the sensitivity of Si layer grown onto ZnO NWs under CVD conditions produces diverse options with regard to shell morphology. By selection of various deposition parameters, including time of silane exposure, local concentration of the Si, as well as the location of the sample inside the reactor, Si shell thicknesses as thin as 10 nm or as thick as 100 nm can be attained (Figure 27.2). Interestingly, under controlled conditions, the production of a wall thinner than 12 nm is often observed in conjunction with a uniform 5–10 nm mesoporous appearance (Figure 27.2d), which presumably results from an Ostwald-type coalescence of Si into islands grown over the ZnO template. On the other hand, if the shell becomes thicker than 25 nm, the porous feature of the nanotube surface disappears and the wall surface becomes smooth.

These initial studies involving use of a sacrificial ZnO/HCl template method indicate a viable route for obtaining uniform SiNT arrays that can provide opportunities of investigating novel emerging properties associated with these materials. Importantly, the notable discovery of porous SiNTs and a subsequent ability to infiltrate selected molecules of interest expands the range of potential versatile platforms for diverse applications. Some of these are described below in further detail.

Nanowires and nanotubes

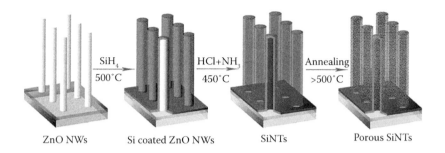

Figure 27.1 Silicon nanotube fabrication scheme—Sacrificial etching of a ZnO nanowire template using HCl.

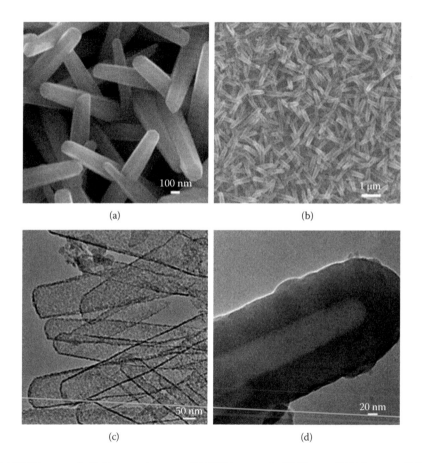

Figure 27.2 (a) SEM image of a ZnO nanowire array; (b) SEM image of a SiNT array; TEM images of SiNTs with various wall thicknesses: (c) 10 nm, (d) 70 nm.

27.3.4 STRUCTURAL DIVERSITY/OPPORTUNITIES

Using the ZnO NWs template/CVD method, manipulation of wall thickness and inner diameter attains flexible control of a wide range of overall nanotube widths (Huang et al. 2013). The tunable size of the hollow core through control of the diameter of the ZnO NW template can confine preformed loaded structures of various sizes as well as different molecular payloads of interest, such as superparamagnetic Fe_3O_4 nanoparticles (Section 27.5). Likewise, wall thickness is considered to be an important factor not only in determining the shell morphology, but also governing the associated properties that determine a particular behavior. For example, in the case of biomedically relevant applications, an understanding of the dissolution behavior of the SiNT variants in aqueous environments is critical for ultimate *in vivo* studies (Coffer 2016). In recent studies by Huang et al., the ability of the nanotubes to dissolve has been shown to be directly related to the wall thickness, such that the resorption of the NTs increases in conjunction with a thinner NT wall. Not surprisingly, the relative extent of Si crystallinity otherwise slows down the dissolution process as a consequence of the ordered structure of Si atoms. Incubation conditions such as higher temperature as well as the presence of salts in the medium facilitates the dissolution process. Using a molybdate-based spectroscopic assay for soluble Si, when porous SiNTs with a 10 nm shell thickness are exposed to buffered media at 37°C, more than 80% of the material is dissolved after 48 h, and complete resorption is observed after 72 h (Huang et al. 2013). Notably, the dissolution behavior of SiNTs as demonstrated by Huang et al. (2013) is comparable to that of anodized pSi nanoparticles, suggesting the possible utility of pSiNTs in future bio-related applications including drug delivery devices and tissue engineering.

27.4 USEFUL PROPERTIES 1: Li STORAGE/CYCLING–RELEVANT BATTERY SOURCES

Currently, rechargeable Li-ion batteries (LIBs) have become one of the most promising energy storage technologies as a consequence of their high electrochemical performance such as low self-discharge, and long life cycle (Linden and Reddy 2001; Nitta et al. 2015). Unlike traditional rechargeable lead acid batteries, which require lead metal and acid electrolytes, in LIBs intercalated Li is used instead, and the facilitated movement of ions between the electrodes by electrolyte assists charge/discharge cycles (Ahuja et al. 2011; Nitta et al. 2015). Nevertheless, the technology of LIBs in electric vehicles (EVs) is still limited as a consequence of low specific capacity offered by graphite (372 mA.h.g^{-1}) which is currently employed as the anode of commercial batteries (Goriparti et al. 2014; Nitta et al. 2015). In contrast, Si has been recognized as a promising substitute material for the anode. Theoretically, Si can be incorporated into the cell with a maximum of 4.4 Li atoms (Equation 27.4), thereby offering 10 times higher specific capacity (4200 mA.h.g^{-1}) which enhances the energy storage capacity of the battery significantly (Tesfaye et al. 2015):

$$Si + xLi^+ + xe^- Li_xSi \ (0 \leq x \leq 4.4) \tag{27.5}$$

However, the loss of electrical contact is often observed in the latter case after a few charge/discharge cycles as a result of large volume expansion (400%) and contraction of Si after lithiation (integration of Li), and delithiation (removal of Li), thus leading to inevitable pulverization of the Si structure (Canham 2014; Beattie et al. 2016). In order to overcome this issue, Si nanostructures have been proposed to serve as a substitute for the bulk counterpart (Rahman et al. 2016). In this regard, it has been pointed out that the presence of voids in the nanophase are capable of attenuating the stress caused by such increase in volume (Szczech and Jin 2011). Importantly and encouragingly, studies conducted so far with various nanostructured Si structures, such as 1D Si nanowires and 3D porous silicon have in fact shown enhanced efficiencies, suggesting promising opportunities of utilizing nanoscaled Si as a LIB anode (Chan et al. 2008; Rahman et al. 2016).

As part of this general approach, current developments using SiNTs of a well-defined hollow structure motivate investigations of their electrochemical properties in LIB designs. Recent efforts have demonstrated an impressive improved charge capacity with high Coulombic efficiency (CE) after testing at various cycle rates (Song et al. 2010; Wu et al. 2012). In contrast to the previously discussed Si nanomaterials (i.e., Si NWs and nanoporous Si), SiNTs with high surface areas theoretically enhance accessible contact to the electrolyte, thereby facilitating the incorporation of Li ions with Si while accommodating necessary volume expansion (Song et al. 2010). By using SiNT arrays sealed at one end as the anode, Song et al. have demonstrated that SiNT anodes exhibit CEs greater than 85% after the first cycle; importantly, such anodes are able to maintain CE values as high as 80% retention after 50 cycles in comparison to <25% observed in (solid) silicon nanospheres. The results of those studies amplify the advantages of the available empty space offered by a nanotube structure to compensate the volume increase that is detrimental to capacity and electrochemical cycling of the material, thereby improving such reversible morphological changes in Si (Song et al. 2010).

While nanoporous Si has demonstrated an improved performance in LIBs, the combination of porosity with a nanotube morphology can further support an enhanced electrochemical performance of the battery.

In this regard, recently published studies by Tesfaye et al. address the efficiency of LIBs using SiNTs with porous sidewalls, which are readily synthesized using the sacrificial ZnO template described in section III.C (Tesfaye 2015). Unlike the previously examined smooth-surface SiNTs whose shell thickness is approximately 30 nm, in this particular case, the crystalline SiNTs in this case possess ultrathin walls (~10 nm) with a porous structure. The mesopores (2–5 nm) of the nanotubes presumably provide extra free space to accommodate the necessary structural changes in the nanotubes. Studies of the electrochemical performance of these vertical arrays of porous SiNTs have demonstrated a much higher specific capacity after 30 cycles (1670 mAh g^{-1}) compared to the results reported by Wen (1158 mAh g^{-1} after 10 cycles), and Zhou groups (1398 mAh g^{-1} after 20 cycles). Such robust performance of the LIBs using these porous SiNT anodes is attributed to the enhanced surface area of the porous shell structure, which therefore facilitates strain relaxation and assists the storage of charge based on a pseudocapacitive effect.

While the use of bare SiNTs enables investigations of the fundamental role of the materials in the operation of LIBs (and therefore reveals key aspects for future optimization), further modification of the SiNTs into a nanocomposite can also address other underlying problems with such a system. In addition to the importance of high charge capacity of the anode in enhanced performance of LIBs, it has also been proposed that the unstable formation of a solid electrolyte interface (SEI) layer on the electrode surface is also responsible for the fading capacity and poor performance of the batteries (Wu et al. 2012; Guan et al. 2015). It has also been pointed out that the SEI layer produced from electrolyte decomposition serves as an electronic insulator that eventually impedes electrical conduction. The unstable interface between electrolyte and Si anode during the charge–discharge cycle promotes continuous SEI formation and subsequently results in degradation of the electrode material. In order to address those issues, a carbon coating layer on SiNTs can minimize the contact between silicon and electrolyte, and consequently improve the stability of SEI formation (Park et al. 2009). Recently reported carbon-coated SiNTs have demonstrated an impressive performance with high capacity retention. In a specific case, Park et al. has presented an LIB system that exhibits 89% retention capacity after up to 200 cycles (Park et al. 2009). Although such SiNT anodes have demonstrated efficient operation in terms of battery performance, it should be pointed out that for commercial requirements, further improvements still need to be made to achieve a Li-ion full cell that can maintain CEs of greater than 99.994% after 5000 cycles (Wu et al. 2012).

27.5 USEFUL PROPERTIES 2: A TEMPLATE FOR PEROVSKITE NANOSTRUCTURE FORMATION

Organic–inorganic halide perovskites with the generic formula RPbX$_3$ (where R = monovalent organic cations methylammonium CH$_3$NH$_3^+$ or formamidinium [HC(NH$_2$)$_2$]$^+$; X = halides) have been extensively investigated due to their unique properties and notable impact on high-performing solar cells, light-emitting diodes, and lasing platforms (Zhu et al. 2015; Chen et al. 2016). While CH$_3$NH$_3$PbI$_3$ is the most currently studied composition, investigations of various hybrid perovskites, such as CH$_3$NH$_3$PbI$_{3-x}$Cl$_x$ and CH$_3$NH$_3$PbI$_{3-x}$Br$_x$, are also underway (Chen et al. 2016; Xu et al. 2016). Compared to current solar cells based on monolithic Si designs, the use of cost-effective and abundant materials allows the inexpensive fabrication of perovskite-based solar cells (Fan et al. 2014).

Since the morphology and structure of perovskites critically influence photovoltaic performance, current research on the efficiency of perovskites-based solar cells have carefully focused on manipulating the growth conditions of the perovskites for an evaluation of the size-dependent properties of the materials (Heo et al. 2014; Ahn et al. 2015). In particular, the crystallization of the organic–inorganic halide hybrid into the perovskite phase is vital to the morphology of the film, thus dictating the performance of the devices (Heo et al. 2014; Jeon et al. 2016). In particular, careful selection of additives for the enhancement of perovskite crystallization, annealing temperature, etc. have demonstrated a striking impact on the morphological evolution of perovskite films (Dualeh et al. 2014; Chen et al. 2016). Although thin films are the dominant morphology of perovskites, other nanostructures such as 0D quantum dots, 1D nanowires, 2D nanoplates, as well as 3D crystals have also been recently recognized (Im et al. 2015; Zhang et al. 2015; Chen et al. 2016; Jaffe 2016).

One of the strategies to control growth and crystal domain size of perovskites is through a template approach, such that the growth of perovskite is allowed to proceed within a predesigned space

(Hörantner et al. 2015; Gonzalez-Rodriguez et al. 2016). In this regard, SiNTs with tunable hollow structures can serve as feasible reaction vessels to confine and modulate the growth of crystalline $CH_3NH_3PbI_3$ perovskites (Gonzalez-Rodriguez et al. 2016). By modifying the SiNT platform structures (inner diameter, length, and wall thickness), structural variations of the enclosed perovskite phase are therefore expected. It should also be pointed out that the formation of a perovskite nanostructure inside the tubes, along with the interaction between perovskite and the silicon platform, may also introduce new properties and influence performance. Recently, Gonzalez-Rodriguez et al. investigated the formation of methylammonium lead iodide ($CH_3NH_3PbI_3$) inside thin-shelled SiNTs (shell thickness ~10 nm) and an inner diameter of either 70 or 200 nm (note that the nanotubes were formed using sacrificial ZnO NW templates/HCl etching method) (Figure 27.3a and c). In their approach, the nanotubes are immersed in a mixture of CH_3NH_3I

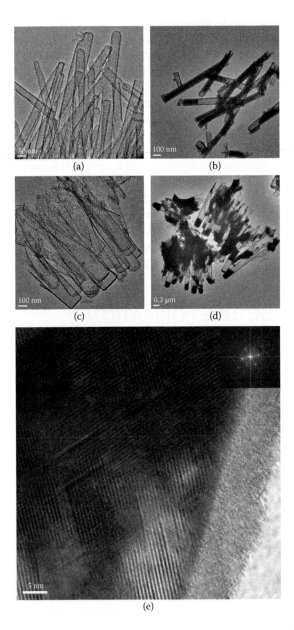

Figure 27.3 TEM images of: (a) empty 70 nm ID inner diameter (ID) porous SiNTs; (b) 70 nm ID porous SiNTs filled with $CH_3NH_3PbI_3$; (c) empty 200 nm ID porous SiNTs; (d) 200 nm ID porous SiNTs filled with $CH_3NH_3PbI_3$; (e) crystalline domains of perovskite confined within SiNTs. (Images courtesy of Roberto Gonzalez-Rodriguez.)

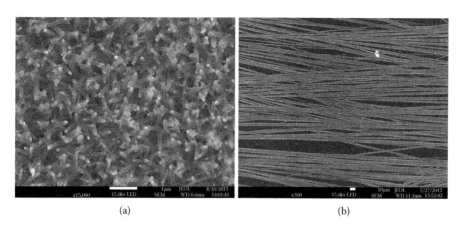

(a) (b)

Figure 27.4 FESEM images of (a) porous SiNTs filled with $CH_3NH_3PbI_3$. (b) $CH_3NH_3PbI_3$ perovskite microwires.

and PbI_2, so the porous nature of the nanotube (as described in the previous sections) facilitates infiltration of the reactants, permitting the formation of $CH_3NH_3PbI_3$ in the inner compartment of the nanotube. Subsequent spin coating and thermal annealing (95°C) produces sufficient crystalline perovskite nanorods confined at the tips of the nanotubes without formation of the compound on the exterior sidewall (Figures 27.3b,c, and 27.4a).

The photophysical properties of perovskite/SiNT nanocomposites can be assessed by a combination of photoluminescence (PL) and optical absorption techniques. From these measurements, the nanorod-like perovskites confined in 200 nm SiNTs behave in a manner similar to that of the pseudobulk perovskite microwires (Figure 27.4b) with a slight blue shift (5–10 meV) of the adsorption edge/PL maximum. However, an additional blue shift (of 10–15 meV) was observed in the case of perovskite-loaded 70 nm SiNTs. It has been suggested that the latter outcome might emerge from quantum confinement effects, but it is more likely a consequence of the large stress resulting from the relatively high surface area interface between silicon and perovskite. Interestingly, the phase transition behavior, from tetragonal to orthorhombic, of these nanoscale perovskites is very sensitive to the $CH_3NH_3PbI_3$ feature size (as detected by temperature-dependent PL measurements). For the case of bulk $CH_3NH_3PbI_3$ microwires and 200 nm wide perovskite nanostructures, the tetragonal to orthorhombic phase transition is cleanly detected by the appearance of an additional PL peak in the range of 1.68–1.7 eV when a temperature lower than 140K is reached. However, for smaller perovskite nanostructures of 70 nm and 30 nm width, this phase transition is strongly suppressed. Thus the structural confinement of these nanostructures of $CH_3NH_3PbI_3$ strongly influences the size-dependent phase transition behavior by a mechanism that likely involves strain at such curved interfaces.

Though additional studies and analysis are still required to rationalize such phenomena, preliminary results suggest novel opportunities of utilizing SiNTs as a tunable template that allows assessment of the unique properties of nanoscale perovskites. For future progress, further investigations of other organic–inorganic halide hybrid perovskites composed of halide elements other than I (e.g., Br, Cl) are also currently underway in order to evaluate the properties of the nanocomposites as a function of halide variants in the perovskites.

27.6 USEFUL PROPERTIES 3: INCORPORATION OF MAGNETIC NANOPARTICLES AND MATRIX DEPENDENCE EFFECTS

Smart biomaterials that are designed with a specific targeting ability have been extensively investigated in order to reach a given disease site while minimizing interference with other biological components (Canham 2014). In order to address this issue, functionalizing the material with a specific species for molecular recognition is a popular strategy to direct a therapeutic moiety toward a cancer cell, for example, onto which a complementary receptor is overly expressed (Castillo et al. 2013; Ma et al. 2015b; Yin et al. 2015). Magnetic field-assisted drug delivery is another delivery approach in which the movement of the drug-containing material can be readily controlled through the use of an external magnetic field (Kempe and Kempe 2010; Tewes et al. 2014). Among the available nanostructured candidates, superparamagnetic iron oxide (Fe_3O_4)nanoparticles (NPs) are of current widespread interest due to their ample advantages, including an absence of toxicity as well as an availability in nanoscale sizes appropriate for biological applications (Na et al. 2009; Weinstein et al. 2010). By incorporating Fe_3O_4 NPs in high surface area matrices, a high density of the magnetic nanoparticles can be packed inside a given carrier, which is potentially transformed into a targeting vehicle (Coffer 2016). It should be noted that in order to be qualified as an efficient magnetic-guided drug delivery system, a low blocking temperature, T_B, must be intrinsic to the structure (Granitzer et al. 2014; Rumpf et al. 2014). Since T_B is strongly dependent on the size of Fe_3O_4 as well as the relative intraparticle distance, the morphology of the carrier as well as the properties of the loaded NPs must be carefully considered in order to evaluate the role of each component in the behaviors of the overall nanocomposite structure (Granitzer et al. 2015b).

As pointed out above, pSi with high surface areas, well-established dissolution behavior, and biocompatibility has been widely studied and evaluated for possible drug delivery vehicles (Anglin et al. 2008; Salonen et al. 2008). In recent years, the loading of Fe_3O_4 into mesoporous Si has been demonstrated (Rumpf et al. 2014; Granitzer et al. 2015a). A rather low cytotoxicity of the nanocomposite unveiled using HEK293 cells suggests promising opportunities of using Fe_3O_4-loaded pSi structures in magnetically-guided drug delivery. These initial studies with pSi have prompted investigations of incorporating Fe_3O_4 NPs in SiNTs whose uniform arrays and clearly defined structures could provide possibilities for tuning the density of loaded Fe_3O_4 in a unit volume, in addition to investigating the fundamental magnetic features of the composite systems.

Recently, extensive efforts have been devoted to an evaluation of the influence of the morphology and structures of a given Si matrix toward the fundamental magnetic properties of the loaded superparamagnetic Fe_3O_4 NPs (Granitzer et al. 2014; Granitzer et al. 2015a). Different nanostructured silicon host materials, namely pSi and SiNTs, have been shown to have an impact on the magnetic properties of the NPs being confined. In terms of loading method, infiltration of Fe_3O_4 NPs into pSi is simply facilitated by the presence and proper location of an external magnetic field relative to the Si sample. In the case of SiNTs, two loading methods are available, depending on the absence or presence of a porous morphology in the SiNTs and the relative size of the Fe_3O_4 NPs. Particularly, the loading of small NPs into porous SiNTs can be readily achieved via simple diffusion, whereas a magnetic field is required to assist loading of large NPs into SiNTs that possess no porous sidewalls (Figure 27.5).

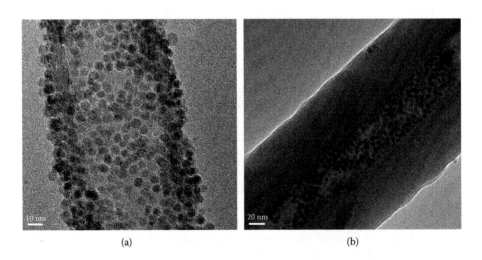

(a) (b)

Figure 27.5 TEM images of (a) porous SiNTs (10 nm shell) filled with 4 nm Fe_3O_4 NPs; (b) SiNTs (70 nm) filled with 4 nm Fe_3O_4 NPs.

Nanowires and nanotubes

For small Fe_3O_4 NPs (4 and 5 nm diameter), since the oleic acid coating layer (2 nm) can sufficiently inhibit the dipole coupling effect, T_B is insensitive to the morphology of the Si matrices; thus, T_B values are observed to be the same in both cases of Fe_3O_4 NPs loaded in pSi and SiNTs. In contrast, when larger Fe_3O_4 NPs (8 and 10 nm) are incorporated, the impact of oleic acid is no longer sufficient, resulting in a strong dependence of T_B on the morphology of the matrix. Interestingly, T_B is observed to be significantly lower in the Fe_3O_4 NP/SiNT system; for instance, when 70 nm inner diameter SiNTs are used to encapsulate 8 nm Fe_3O_4 NPs, the T_B value is 20K, compared to a value of 160K for the same nanocrystals loaded into mesoporous Si. It has been suggested that this significant increase in T_B in the case of pSi is due to an intrinsic dipole coupling, whereas this effect is absent in SiNTs. Since SiNTs offer a thick, relatively uniform separation barrier (140 nm), long-range coupling is presumably suppressed.

When considering the effect of Si wall thickness in the case of the nanotube platforms, comparisons of porous thin wall (~10 nm) with smooth thick wall (70 nm) SiNTs, measurements suggest the intrinsic magnetic properties of the isolated Fe_3O_4 NPs are still retained and no significant difference in T_B is observed. Hence, varying morphologies of the nanotubes do not critically influence this parameter of the confined NPs.

For both SiNT and pSi systems, T_B values have been shown to be significantly below room temperature, thereby satisfying a key requirement for biorelevant applications. Importantly, the tunable structures and the lower T_B values observed in SiNTs confirm the potential of utilizing this system in drug delivery studies. Although detailed investigations in the biodistribution and delivery of NP-loaded SiNTs also need to be addressed, recent reports of cytocompatibility of the Fe_3O_4/pSi nanocomposite certainly endorse further studies in this area. (Rumpf et al. 2014; Granitzer et al. 2015b).

27.7 USEFUL PROPERTIES 4: SURFACE FUNCTIONALIZATION AND POSSIBLE USE AS A BIOMATERIAL

27.7.1 SURFACE FUNCTIONALIZATION FOR BIOSENSING AND/OR THERAPEUTIC TARGETING

The well-established surface chemistry of silicon can clearly be exploited for utility as in theranostic platforms. With regard to nanostructured pSi, this versatility has been used in its coupling with numerous molecules of interest for serving multiple purposes: an enhancement of biocompatibility, biosensing/targeting, as well as drug delivery (Jane et al. 2009; Santos 2014). For the case of SiNTs, the native oxide surface specifically promotes the attachment of a uniform layer of multiple biomolecules possessing stable Si–O–Si surface linkages for various purposes, as extensively demonstrated earlier for the case of oxidized pSi (Sailor 2012).

The often-employed linker strategy is a logical approach for SiNT functionalization: one end of the linker is attached to the Si–O surface while the other end presents a moiety to the surroundings that either interacts with a biological target, or can be further chemically modified. Based on this approach, the Coffer group recently explored the coupling of fluorescent organic dyes to an oxidized SiNT surface through the widely used reagent 3-aminopropyltriethoxysilane (APTES) (Pasternack 2008; Sailor 2012). Hydrolysis of the three ethoxy groups on APTES allows the formation of Si–O bonds between the linker and the nanotube surface (Figure 27.6). On the other end of APTES, the reactive $-NH_2$ group can subsequently bond to a molecule of interest. To probe the uniformity of surface coverage of these APTES moieties on the nanotube surface, the amino group was allowed to react with fluorescent organic dyes such as the Alexa Fluor® series or fluorescein isothiocyanate (FITC) (Natte et al. 2012; Baumgärtel et al. 2013). A relatively uniform fluorescence emitted from the functionalized SiNTs (as monitored by confocal microscopy) is consistent with a high density of amino groups emanating from the surface, and also presents opportunities for tracking assays *in vitro* as well as *in vivo*. Other linkers are currently under consideration, including 3-aminopropyldimethylethoxysilane (APDMES), which provides a single Si–O linkage and better access to the interior spaces of the pores, and reacts with oxidized Si at those regions while avoiding cross-linking with other APDMES molecules (Sailor 2012). Other additional functional groups for producing more favorable biocompatibility and/or biodistribution are available for incorporation, including polyethylene glycol or carboxylate species (RCO_2^-) (Sailor 2012).

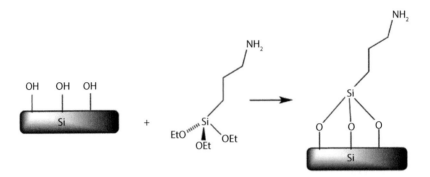

Figure 27.6 Functionalization of a Si oxide surface with APTES.

27.7.2 POSSIBLE APPLICATIONS IN GENE DELIVERY

Gene therapy is a novel method that involves the careful administration of genes to cells in order to modify and treat diseases at the genetic level (Naldini 2015). While only a few clinical successes have been recognized, multifaceted approaches involving various types of viral vectors (e.g., viruses, peptides) have been introduced (Heo 2002). With ongoing advances in nanoscience, different types of nanoparticles are under investigation as possible nonviral vectors in gene therapy. In the specific case of SiNTs, the large exterior surface area of the tubes can benefit a conjugation with targeted genes. Recent studies involving an evaluation of cytocompatibility of SiNTs with HEK 293 human cells suggests no deleterious effects are present, therefore encouraging the possible use of SiNTs in gene therapy (Coffer 2016). For designing SiNTs as vehicles for gene delivery, the primary amine moiety presented by APTES functionalized nanotubes, protonated at physiological pH, is ideal for electrostatic coupling with a desired polynucleotide species. In initial studies, SiNTs were projected to deliver plasmid DNA (pDNA) that encodes the gene for expression of enhanced green fluorescent protein (eGFP). After 72 h incubation of HEK 293 cells with SiNTs/pDNA, eGFP was expressed from HEK 293 cells as analyzed using confocal microscopy. Additional experiments involving modification of SiNT vectors are underway in order to shorten the transfection time and increase transfection efficiency.

27.7.3 LOADING KNOWN CHEMOTHERAPEUTIC DRUGS–CISPLATIN

In the context of drug delivery, the large interior surface of SiNTs can be exploited to enclose therapeutic molecules of interest. Selective modification of the nanotube surface can grant the nanotube targeting ability and subsequently deliver a desired payload to the site(s) of disease.

In terms of possible candidates, cisplatin is an FDA-approved platinum-based drug used to treat a variety of cancers (carcinoma, testicular cancer, lymphomas, etc.) (Tsang et al. 2009). Due to a lower concentration of chloride ions in the cancer cells compared to that in the extracellular fluid, chloride ligands on cisplatin are readily displaced by water-producing either cis-$[PtCl(NH_3)_2(H_2O)]^+$ or cis-$[Pt(NH_3)_2(H_2O)_2]^{2+}$ aquo complexes (Lau and Ensing 2010; Ma et al. 2015). Those active forms of cisplatin inhibit replication of DNA by cross-linking with DNA, particularly through binding at guanine, and eventually promote apoptosis (Ma et al. 2015a). Nevertheless, cisplatin still suffers several limitations that restrict its uses in clinical treatment. Since cancer cells have been reported to be capable of repairing DNA damage caused by cisplatin (Martin et al. 2008), the administration of high doses of cisplatin in order to offset this issue is required, but unfortunately introduces various adverse side effects including nausea, neurotoxicity, hair loss, etc. (Ma et al. 2015a; Chu et al. 2016). Moreover, a nonspecific targeting of healthy cells by cisplatin is also a major concern in this treatment (Tsang et al. 2009).

In order to enhance the effectiveness, the idea of clustering a concentrated number of cisplatin centers on an SiNTs carrier to the targeted disease sites is currently under investigation. SiNTs with porous sidewalls provide opportunities for facilitated infiltration and release of this chemotherapeutic. A significant loading of nanoscale palatinate species onto SiNTs is readily achievable; effective release and associated therapeutic activity remains to be demonstrated with this system, however.

Nanowires and nanotubes

27.8 CONCLUDING REMARKS AND FUTURE DIRECTIONS

To date, a number of key properties of SiNTs have been demonstrated, properties that have significant implications in applications ranging from battery science to biotechnology. Most functional properties investigated to date exploit both the controlled interior loading capacity of the nanotube, coupled with the fact that its composition remains semiconducting silicon. Additional opportunities remain, including those that take advantage of the use of the SiNT as a well-defined nanoscale reaction vessel. The ultimate challenge of forming SWSiNT structures of a targeted shape, and evaluating their associated physicochemical properties, also remains a long-term goal.

ACKNOWLEDGMENT

Financial support for fundamental research regarding the chemistry of silicon nanotubes by the Robert A. Welch Foundation (Grant P-1212 to JLC) is gratefully acknowledged.

REFERENCES

Ahn N, Kang SM, Lee J, Choi M, Park N. (2015). Thermodynamic regulation of $CH_3NH_3PbI_3$ crystal growth and its effect on photovoltaic performance of perovskite solar cells. *J Mater Chem A.* 3:19901–19906.

Ahuja R, Blomqvist A, Larsson P, Pyykko P, Zaleski-Ejgierd P. (2011). Relativity and the lead-acid battery. *Phys Rev Lett.* 106:018301.

Anglin EJ, Cheng L, Freeman WR, Sailor MJ. (2008). Porous silicon in drug delivery devices and materials. *Adv Drug Deliv Rev.* 60(11):1266–1277.

Bai J, Zeng XC, Tanaka H, Zeng JY. (2004). Metallic single-walled silicon nanotubes. *Proc Natl Acad Sci U S A.* 101(9):2664–2668.

Baumgärtel T, Borczyskowski C, Graaf H. (2013). Selective surface modification of lithographic silicon oxide nanostructure by organofunctional silanes. *Beilstein J Nanotechnol.* 4:218–226.

Beattie SD, Loveridge MJ, Lain MJ, Ferrari S, Polzin BJ, Bhagat R, Dashwood R. (2016). Understanding capacity fade in silicon based electrodes for lithium ion batteries using three electrode cells and upper cut-off voltage studies. *J Power Sources.* 302:426–430.

Brus L. (1994). Luminescence of silicon materials: Chains, sheets, nanocrystals, nanowires, microcrystals, and porous silicon. *J Phys Chem.* 98(14):3515–3581.

Canham LT. (2014). *Handbook of porous silicon.* Cham, Switzerland: Springer International Publishing AG.

Castillo JJ, Rindzevicius T, Novoa LV, Svendsen WE, Rozlosnik N, Boisen A, Escobar P, Martíneza F, Castillo-Leon J. (2013). Non-covalent conjugates of single-walled carbon nanotubes and folic acid for interaction with cells overexpressing folate receptors. *J Mater Chem B.* 1:1475–1481.

Castrucci P, Scarselli M, De Crescenzi M, Diociaiuti M, Chaudhari PS, Balasubramanian C, Bhave TM, Bhoraskar SV. (2006). Silicon nanotubes: Synthesis and characterization. *Thin Solid Films.* 508:226–230.

Chan CK, Peng H, Liu G, McIlwrath K, Zhang XF, Huggins RA, Cui Y. (2008). High-performance lithium battery anodes using silicon nanowires. *Nat Nanotechnol.* 3:31–35.

Chen Y, He M, Peng J, Sun Y, Liang Z. (2016). Structure and growth control of organic–inorganic halide perovskites for optoelectronics: From polycrystalline films to single crystals. *Adv Sci.* 3:1500392.

Chu Y, Sibrian-Vazquez M, Escobedo JO, Phillips AR, Dickey DT, Wang Q, Ralle M, Steyger PS, Strongin RM. (2016). Systemic delivery and biodistribution of cisplatin in vivo. *Mol Pharmaceutics.* 13:2677–2682.

Coffer JL. (2014). *Semiconducting silicon nanowires for biomedical applications.* Cambridge, UK: Woodhead.

Coffer JL. (2016). Mesoporous nanotubes as biomaterials. *Mesoporous Biomater.* 2(1):33–48.

Coffer JL, Whitehead MA, Nagesha DK, Mukherjee P, Akkaraju G, Totolici M, Saffie RS, Canham LT. (2005). Porous silicon-based scaffolds for tissue engineering and other biomedical applications. *Phys Status Solidi A.* 202:1451–1455.

Cullis AG, Canham LT. (1991). Visible light emission due to quantum size effects in highly porous crystalline silicon. *Nature.* 353:335–338.

Dualeh A, Tétreault N, Moehl T, Gao P, Nazeeruddin MK, Grätzel M. (2014). Effect of annealing temperature on film morphology of organic–inorganic hybrid pervoskite solid-state solar cells. *Adv Funct Mater.* 24:3250–3258.

Elbersen R, Vijselaar W, Tiggelaar RM, Gardeniers H, Huskens J. (2015). Fabrication and doping methods for silicon nano- and micropillar arrays for solar-cell applications: A review. *Adv Mater.* 27(43):6781–6796.

Fagan SB, Baierle RJ, Mota R, da Silva AJR, Fazzio A. (2000). Ab initio calculations for a hypothetical material: Silicon nanotubes. *Phys Rev. B* 61(15):9994–9996.

Fan J, Jia B, Gu M. (2014). Perovskite-based low-cost and high-efficiency hybrid halide solar cells. *Photon Res.* 2(5):111–120.

Goldberger J, He R, Zhang Y, Lee S, Yan H, Choi HJ, Yang P. (2003). Single-crystal gallium nitride nanotubes. *Nature.* 422:599–602.

Gonzalez-Rodriguez R, Arad-Vosk N, Rozenfeld N, Sa'ar A, Coffer JL. (2016). Control of CH$_3$NH$_3$ PbI$_3$ perovskite nanostructure formation through the use of silicon nanotube templates. *Small.* 12:4477–4480.

Goriparti S, Miele E, De Angelis F, Di Fabrizio E, Proietti Zaccaria R, Capiglia C. (2014) Review on recent progress of nanostructured anode materials for Li-ion batteries. *J Power Sour.* 257:421–443.

Granitzer P, Rumpf K, Gonzalez R, Coffer J, Reissner M. (2014). Magnetic properties of superparamagnetic nanoparticles loaded into silicon nanotubes. *Nanoscale Res. Lett.* 9:413.

Granitzer P, Rumpf K, Gonzalez-Rodriguez R, Coffer JL, Reissner M. (2015a). The effect of nanocrystalline silicon host on magnetic properties of encapsulated iron oxide nanoparticles. *Nanoscale.* 7:20220–20226.

Granitzer P, Rumpf K, Tian Y, Coffer J, Akkaraju G, Poelt P, Reissner M. (2015b). Assessment of cytocompatibility and magnetic properties of nanostructured silicon loaded with superparamagnetic iron oxide nanoparticles. *ECS Trans.* 64(47):1–7.

Guan P, Liu L, Lin X. (2015). Simulation and experiment on solid electrolyte interphase (SEI) morphology evolution and lithium-ion diffusion. *J Electrochem Soc.* 162(9):A1798–A1808.

Heo DS. (2002). Progress and limitations in cancer gene therapy. *Genet Med.* 4:52S–55S.

Heo JH, Song DH, Im SH. (2014). Planar CH$_3$NH$_3$PbBr$_3$ hybrid solar cells with 10.4% power conversion efficiency, fabricated by controlled crystallization in the spin-coating process. *Adv Mater.* 48:8179–8183.

Hörantner MT, Zhang W, Saliba M, Wojciechowski K, Snaith HJ. (2015). Templated microstructural growth of perovskite thin films via colloidal monolayer lithography. *Energ Environ Sci.* 8:2041–2047.

Hu J, Bando Y, Liu Z, Zhan J, Golberg D, Sekiguchi T. (2004). Synthesis of crystalline silicon tubular nanostructures with ZnS nanowires as removable templates. *Angew Chem Int Ed.* 116(1):65–68.

Huang X, Gonzalez-Rodriguez R, Rich R, Gryczynski Z, Coffer JL. (2013). Fabrication and size dependent properties of porous silicon nanotube arrays. *Chem Comm.* 49:5760–5762.

Im J, Luo J, Franckevičius M, Pellet N, Gao P, Moehl T, Zakeeruddin SM, Nazeeruddin MK, Grätzel M, and Park NG. (2015). Nanowire perovskite solar cell. *Nano Lett.* 15(3):2120–2126.

Jaffe A, Lin Y, Beavers CM, Voss J, Mao WL, Karunadasa HI. (2016). High-pressure single-crystal structures of 3D lead-halide hybrid perovskites and pressure effects on their electronic and optical properties. *ACS Cent Sci.* 2(4):201–209.

Jane A, Dronov R, Hodges A, Voelcker NH. (2009). Porous silicon biosensors on the advances. *Trends Biotechnol.* 27(4):230–239.

Jeon T, Jin HM, Lee SH, Lee JM, Park HI, Kim MK, Lee KJ, Shin B, Kim SO. (2016). Laser crystallization of organic-inorganic hybrid perovskite solar cells. *ACS Nano.* 10(8):7907–7914.

Kempe H, Kempe M. (2010). The use of magnetite nanoparticles for implant-assisted magnetic drug targeting thrombo-lytic therapy. *Biomaterials.* 31:9499–9510.

Lau JK, Ensing B. (2010). Hydrolysis of cisplatin—A first-principles metadynamics study. *Phys Chem Chem Phys.* 12:10348–10355.

Linden D, Reddy T. (2001). *Handbook of batteries.* New York: McGraw-Hill.

Ma P, Xiao H, Li C, Dai Y, Cheng Z, Hou Z, Lin J. (2015a). Inorganic nanocarriers for platinum drug delivery. *Mater Today.* 18(10):554–564.

Ma Y, Ding H, Xiong H. (2015b). Folic acid functionalized ZnO quantum dots for targeted cancer cell imaging. *Nanotechnology.* 26(30):305702.

Martin LP, Hamilton TC, Schilder RJ. (2008). Platinum resistance: The role of DNA repair pathways. *Clin Cancer Res.* 14(5):1291–1295.

Murphy CJ, Coffer J. (2002). Quantum dots: A primer. *Appl Spectrosc.* 56:16A–27A.

Na HB, Song IC, Hyeon T. (2009). Inorganic nanoparticles for MRI contrast agents. *Adv Mater.* 21:2133–2148.

Naldini L. (2015). Gene therapy returns to centre stage. *Nature.* 536:351–360.

Natte K, Behnke T, Orts-Gil G, Würth C, Friedrich JF, Österle W, Resch-Genger U. (2012). Synthesis and characterisa-tion of highly fluorescent core–shell nanoparticles based on alexa dyes. *J Nanopart Res.* 14:680.

Nitta N, Wu F, Lee JT, Yushin G. (2015). Li-ion battery materials: Present and future. *Mater Today.* 18(5):252–264.

Park M, Kim MG, Joo J, Kim K, Kim J, Ahn S, Cui Y, Cho J. (2009). Silicon nanotube battery anodes. *Nano Lett.* 9(11):3844–3847.

Pasternack RM, Amy SR, Chabal YJ. (2008). Attachment of 3-(Aminopropyl)triethoxysilane on silicon oxide surfaces: Dependence on solution temperature. *Langmuir.* 24:12963–12971.

Pavesi L, Lockwood DJ. (2004). *Silicon photonics*, Volume 1. Berlin, Germany: Springer Science & Business Media.

Perepichka DF, Rosei F. (2006). Silicon nanotubes. *Small.* 2(1):22–25.

Rahman MA, Song G, Bhatt AI, Wong YC, Wen C. (2016). Nanostructured silicon anodes for high-performance lithium-ion batteries. *Adv Funct Mater.* 26:647–678.

Rumpf K, Granitzer P, Tian Y, Coffer J, Akkaraju G, Poelt P, Michor H. (2014). Porous silicon with deposited iron oxide as vehicle for magnetically guided drug delivery. *ECS Trans.* 58(32):133–137.

Sailor MJ. (2012). *Porous silicon in practice: Preparation, characterization and applications*. Weinheim, Germany: Wiley-VCH.

Salonen J, Kaukonen AM, Hirvonen J, Lehto VP. (2008). Mesoporous silicon in drug delivery applications. *J Pharm Sci.* 97:632–652.

Santos HA. (2014). *Porous silicon for biomedical applications*. Cambridge, UK: Woodhead.

Santos HA, Mäkilä E, Airaksinen AJ, Bimbo LM, Hirvonen J. (2014). *Nanomedicine*. 9:535–554.

Seifert G, Köhler T, Urbassek HM, Hernández E, Frauenheim T. (2001). Tubular structures of silicon. *Phys Rev B.* 63(19):193409.

Sha J, Niu J, Ma X, Xu J, Zhang X, Yang Q, Yang D. (2002). Silicon nanotubes. *Adv Mater.* 14(17):1219–1221.

Smith RL, Collins SD. (1992). Porous silicon formation mechanisms. *J Appl Phys.* 71:R1–R22.

Song T, Xia J, Lee J, Lee DH, Kwon M, Choi J, Wu J, et al. (2010). Arrays of sealed silicon nanotubes as anodes for lithium ion batteries. *Nano Lett.* 10(5):1710–1716.

Szczech JR, Jin S. (2011). Nanostructured silicon for high capacity lithium battery anodes. *Energ Environ Sci.* 4:56–72.

Tesfaye A, Gonzalez R, Coffer J, Djenizian T. (2015). Porous silicon nanotube arrays as anode material for Li-ion batteries. *ACS Appl Mater Inter.* 7:20495–20498.

Tewes F, Ehrhardt C, Healy AM. (2014). Superparamagnetic iron oxide nanoparticles (SPIONs)-loaded trojan microparticles for targeted aerosol delivery to the lung. *Eur J Pharm Biopharm.* 86(1):98–104.

Tsang RY, Al-Fayea T, Au H. (2009). Cisplatin overdose: Toxicities and management. *Drug Saf.* 32(12):1109–1122.

Weinstein JS, Varallyay CG, Dosa E, Gahramanov S, Hamilton B, Rooney WD, Muldoon LL, Neuwelt EA. (2010). Superparamagnetic iron oxide nanoparticles: Diagnostic magnetic resonance imaging and potential therapeutic applications in neurooncology and central nervous system inflammatory pathologies, a review. *J Cereb Blood Flow Metab.* 30(1):15–35.

Wu H, Chan G, Choi JW, Ryu I, Yao Y, McDowell MT, Lee SW, et al. (2012). Stable cycling of double-walled silicon nanotube battery anodes through solid–electrolyte interphase control. *Nat Nanotechnol.* 7:310–315.

Xu W, Liu L, Yang L, Shen P, Sun B, McLeod JA. (2016). Dissociation of methylammonium cations in hybrid organic-inorganic perovskite solar cells. *Nano Lett.* 16(7):4720–4725.

Yin F, Zhang B, Zeng S, Lin G, Tian J, Yang C, Wang K, Xu G, Yong K. (2015). Folic acid-conjugated organically modified silica nanoparticles for enhanced targeted delivery in cancer cells and tumor in vivo. *J Mater Chem B.* 3:6081–6093.

Zhang F, Zhong H, Chen C, Wu XG, Hu X, Huang H, Han, J, Zou B, Dong Y. (2015). Brightly luminescent and color tunable colloidal $CH_3NH_3PbX_3$ (X = Br, I, Cl) quantum dots: Potential alternatives for display technologies. *ACS Nano.* 9(4):4533–4542.

Zhang RQ, Lee ST, Law C, Li W, Teo BK. (2002). Silicon nanotubes: Why not? *Chem Phys Lett.* 364:251–258.

Zhu H, Fu Y, Meng F, Wu X, Gong Z, Ding Q, Gustafsson M, Trinh M, Jin S, Zhu XY. (2015). Lead halide perovskite nanowire lasers with low lasing thresholds and high quality factors. *Nat Mater.* 14:636–642.

Index